Geophysical Monograph Series

Including

IUGG Volumes

Maurice Ewing Volumes
Mineral Physics Volumes

GEOPHYSICAL MONOGRAPH SERIES

Geophysical Monograph Volumes

1. Antarctica in the International Geophysical Year *A. P. Crary, L. M. Gould, E. O. Hulburt, Hugh Odishaw, and Waldo E. Smith (Eds.)*
2. Geophysics and the IGY *Hugh Odishaw and Stanley Ruttenberg (Eds.)*
3. Atmospheric Chemistry of Chlorine and Sulfur Compounds *James P. Lodge, Jr. (Ed.)*
4. Contemporary Geodesy *Charles A. Whitten and Kenneth H. Drummond (Eds.)*
5. Physics of Precipitation *Helmut Weickmann (Ed.)*
6. The Crust of the Pacific Basin *Gordon A. Macdonald and Hisashi Kuno (Eds.)*
7. Antarctic Research: The Matthew Fontaine Maury Memorial Symposium *H. Wexler, M. J. Rubin, and J. E. Caskey, Jr. (Eds.)*
8. Terrestrial Heat Flow *William H. K. Lee (Ed.)*
9. Gravity Anomalies: Unsurveyed Areas *Hyman Orlin (Ed.)*
10. The Earth Beneath the Continents: A Volume of Geophysical Studies in Honor of Merle A. Tuve *John S. Steinhart and T. Jefferson Smith (Eds.)*
11. Isotope Techniques in the Hydrologic Cycle *Glenn E. Stout (Ed.)*
12. The Crust and Upper Mantle of the Pacific Area *Leon Knopoff, Charles L. Drake, and Pembroke J. Hart (Eds.)*
13. The Earth's Crust and Upper Mantle *Pembroke J. Hart (Ed.)*
14. The Structure and Physical Properties of the Earth's Crust *John G. Heacock (Ed.)*
15. The Use of Artificial Satellites for Geodesy *Soren W. Henricksen, Armando Mancini, and Bernard H. Chovitz (Eds.)*
16. Flow and Fracture of Rocks *H. C. Heard, I. Y. Borg, N. L. Carter, and C. B. Raleigh (Eds.)*
17. Man-Made Lakes: Their Problems and Environmental Effects *William C. Ackermann, Gilbert F. White, and E. B. Worthington (Eds.)*
18. The Upper Atmosphere in Motion: A Selection of Papers With Annotation *C. O. Hines and Colleagues*
19. The Geophysics of the Pacific Ocean Basin and Its Margin: A Volume in Honor of George P. Woollard *George H. Sutton, Murli H. Manghnani, and Ralph Moberly (Eds.)*
20. The Earth's Crust: Its Nature and Physical Properties *John C. Heacock (Ed.)*
21. Quantitative Modeling of Magnetospheric Processes *W. P. Olson (Ed.)*
22. Derivation, Meaning, and Use of Geomagnetic Indices *P. N. Mayaud*
23. The Tectonic and Geologic Evolution of Southeast Asian Seas and Islands *Dennis E. Hayes (Ed.)*
24. Mechanical Behavior of Crustal Rocks: The Handin Volume *N. L. Carter, M. Friedman, J. M. Logan, and D. W. Stearns (Eds.)*
25. Physics of Auroral Arc Formation *S.-I. Akasofu and J. R. Kan (Eds.)*
26. Heterogeneous Atmospheric Chemistry *David R. Schryer (Ed.)*
27. The Tectonic and Geologic Evolution of Southeast Asian Seas and Islands: Part 2 *Dennis E. Hayes (Ed.)*
28. Magnetospheric Currents *Thomas A. Potemra (Ed.)*
29. Climate Processes and Climate Sensitivity (Maurice Ewing Volume 5) *James E. Hansen and Taro Takahashi (Eds.)*
30. Magnetic Reconnection in Space and Laboratory Plasmas *Edward W. Hones, Jr. (Ed.)*
31. Point Defects in Minerals (Mineral Physics Volume 1) *Robert N. Schock (Ed.)*
32. The Carbon Cycle and Atmospheric CO_2: Natural Variations Archean to Present *E. T. Sundquist and W. S. Broecker (Eds.)*
33. Greenland Ice Core: Geophysics, Geochemistry, and the Environment *C. C. Langway, Jr., H. Oeschger, and W. Dansgaard (Eds.)*
34. Collisionless Shocks in the Heliosphere: A Tutorial Review *Robert G. Stone and Bruce T. Tsurutani (Eds.)*
35. Collisionless Shocks in the Heliosphere: Reviews of Current Research *Bruce T. Tsurutani and Robert G. Stone (Eds.)*
36. Mineral and Rock Deformation: Laboratory Studies —The Paterson Volume *B. E. Hobbs and H. C. Heard (Eds.)*
37. Earthquake Source Mechanics (Maurice Ewing Volume 6) *Shamita Das, John Boatwright, and Christopher H. Scholz (Eds.)*
38. Ion Acceleration in the Magnetosphere and Ionosphere *Tom Chang (Ed.)*
39. High Pressure Research in Mineral Physics (Mineral Physics Volume 2) *Murli H. Manghnani and Yasuhiko Syono (Eds.)*
40. Gondwana Six: Structure, Tectonics, and Geophysics *Gary D. McKenzie (Ed.)*
41. Gondwana Six: Stratigraphy, Sedimentology, and Paleontology *Garry D. McKenzie (Ed.)*
42. Flow and Transport Through Unsaturated Fractured Rock *Daniel D. Evans and Thomas J. Nicholson (Eds.)*
43. Seamounts, Islands, and Atolls *Barbara H. Keating, Patricia Fryer, Rodey Batiza, and George W. Boehlert (Eds.)*

44 Modeling Magnetospheric Plasma   *T. E. Moore and J. H. Waite, Jr. (Eds.)*

45 Perovskite: A Structure of Great Interest to Geophysics and Materials Science   *Alexandra Navrotsky and Donald J. Weidner (Eds.)*

46 Structure and Dynamics of Earth's Deep Interior (IUGG Volume 1)   *D. E. Smylie and Raymond Hide (Eds.)*

47 Hydrological Regimes and Their Subsurface Thermal Effects (IUGG Volume 2)   *Alan E. Beck, Grant Garven, and Lajos Stegena (Eds.)*

48 Origin and Evolution of Sedimentary Basins and Their Energy and Mineral Resources (IUGG Volume 3)   *Raymond A. Price (Ed.)*

49 Slow Deformation and Transmission of Stress in the Earth (IUGG Volume 4)   *Steven C. Cohen and Petr Vaníček (Eds.)*

50 Deep Structure and Past Kinematics of Accreted Terranes (IUGG Volume 5)   *John W. Hillhouse (Ed.)*

51 Properties and Processes of Earth's Lower Crust (IUGG Volume 6)   *Robert F. Mereu, Stephan Mueller, and David M. Fountain (Eds.)*

52 Understanding Climate Change (IUGG Volume 7)   *Andre L. Berger, Robert E. Dickinson, and J. Kidson (Eds.)*

53 Plasma Waves and Instabilities at Comets and in Magnetospheres   *Bruce T. Tsurutani and Hiroshi Oya (Eds.)*

54 Solar System Plasma Physics   *J. H. Waite, Jr., J. L. Burch, and R. L. Moore (Eds.)*

55 Aspects of Climate Variability in the Pacific and Western Americas   *David H. Peterson (Ed.)*

56 The Brittle-Ductile Transition in Rocks   *A. G. Duba, W. B. Durham, J. W. Handin, and H. F. Wang (Eds.)*

57 Evolution of Mid Ocean Ridges (IUGG Volume 8)   *John M. Sinton (Ed.)*

58 Physics of Magnetic Flux Ropes   *C. T. Russell, E. R. Priest, and L. C. Lee (Eds.)*

59 Variations in Earth Rotation (IUGG Volume 9)   *Dennis D. McCarthy and Williams E. Carter (Eds.)*

60 Quo Vadimus Geophysics for the Next Generation (IUGG Volume 10)   *George D. Garland and John R. Apel (Eds.)*

61 Cometary Plasma Processes   *Alan D. Johnstone (Ed.)*

62 Modeling Magnetospheric Plasma Processes   *Gordon R. Wilson (Ed.)*

63 Marine Particles: Analysis and Characterization   *David C. Hurd and Derek W. Spencer (Eds.)*

64 Magnetospheric Substorms   *Joseph R. Kan, Thomas A. Potemra, Susumu Kokubun, and Takesi Iijima (Eds.)*

65 Explosion Source Phenomenology   *Steven R. Taylor, Howard J. Patton, and Paul G. Richards (Eds.)*

66 Venus and Mars: Atmospheres, Ionospheres, and Solar Wind Interactions   *Janet G. Luhmann, Mariella Tatrallyay, and Robert O. Pepin (Eds.)*

67 High-Pressure Research: Application to Earth and Planetary Sciences (Mineral Physics Volume 3)   *Yasuhiko Syono and Murli H. Manghnani (Eds.)*

68 Microwave Remote Sensing of Sea Ice   *Frank Carsey, Roger Barry, Josefino Comiso, D. Andrew Rothrock, Robert Shuchman, W. Terry Tucker, Wilford Weeks, and Dale Winebrenner*

69 Sea Level Changes: Determination and Effects (IUGG Volume 11)   *P. L. Woodworth, D. T. Pugh, J. G. DeRonde, R. G. Warrick, and J. Hannah*

70 Synthesis of Results from Scientific Drilling in the Indian Ocean   *Robert A. Duncan, David K. Rea, Robert B. Kidd, Ulrich von Rad, and Jeffrey K. Weissel (Eds.)*

71 Mantle Flow and Melt Generation at Mid-Ocean Ridges   *Jason Phipps Morgan, Donna K. Blackman, and John M. Sinton (Eds.)*

72 Dynamics of Earth's Deep Interior and Earth Rotation (IUGG Volume 12)   *Jean-Louis Le Mouël, D.E. Smylie, and Thomas Herring (Eds.)*

73 Environmental Effects on Spacecraft Positioning and Trajectories (IUGG Volume 13)   *A. Vallance Jones (Ed.)*

74 Evolution of the Earth and Planets (IUGG Volume 14)   *E. Takahashi, Raymond Jeanloz, and David Rubie (Eds.)*

75 Interactions Between Global Climate Subsystems: The Legacy of Hann (IUGG Volume 15)   *G. A. McBean and M. Hantel (Eds.)*

76 Relating Geophysical Structures and Processes: The Jeffreys Volume (IUGG Volume 16)   *K. Aki and R. Dmowska (Eds.)*

77 The Mesozoic Pacific: Geology, Tectonics, and Volcanism—A Volume in Memory of Sy Schlanger   *Malcolm S. Pringle, William W. Sager, William V. Sliter, and Seth Stein (Eds.)*

78 Climate Change in Continental Isotopic Records   *P. K. Swart, K. C. Lohmann, J. McKenzie, and S. Savin (Eds.)*

79 The Tornado: Its Structure, Dynamics, Prediction, and Hazards   *C. Church, D. Burgess, C. Doswell, R. Davies-Jones (Eds.)*

80 Auroral Plasma Dynamics   *R. L. Lysak (Ed.)*

81 Solar Wind Sources of Magnetospheric Ultra-Low Frequency Waves   *M. J. Engebretson, K. Takahashi, and M. Scholer (Eds.)*

82 Gravimetry and Space Techniques Applied to Geodynamics and Ocean Dynamics (IUGG Volume 17)   *Bob E. Schutz, Allen Anderson, Claude Froidevaux, and Michael Parke (Eds.)*

83 Nonlinear Dynamics and Predictability of Geophysical Phenomena (IUGG Volume 18)   *William I. Newman, Andrei Gabrielov, and Donald L. Turcotte (Eds.)*

84 Solar System Plasmas in Space and Time   *J. Burch, J. H. Waite, Jr. (Eds.)*

85 **The Polar Oceans and Their Role in Shaping the Global Environment** *O. M. Johannessen, R. D. Muench, and J. E. Overland (Eds.)*

86 **Space Plasmas: Coupling Between Small and Medium Scale Processes** *Maha Ashour-Abdalla, Tom Chang, and Paul Dusenbery (Eds.)*

87 **The Upper Mesosphere and Lower Thermosphere: A Review of Experiment and Theory** *R. M. Johnson and T. L. Killeen (Eds.)*

88 **Active Margins and Marginal Basins of the Western Pacific** *Brian Taylor and James Natland (Eds.)*

89 **Natural and Anthropogenic Influences in Fluvial Geomorphology** *John E. Costa, Andrew J. Miller, Kenneth W. Potter, and Peter R. Wilcock (Eds.)*

90 **Physics of the Magnetopause** *Paul Song, B.U.Ö. Sonnerup, and M.F. Thomsen (Eds.)*

91 **Seafloor Hydrothermal Systems: Physical, Chemical, Biological, and Geological Interactions** *Susan E. Humphris, Robert A. Zierenberg, Lauren S. Mullineaux, and Richard E. Thomson (Eds.)*

92 **Mauna Loa Revealed: Structure, Composition, History, and Hazards** *J. M. Rhodes and John P. Lockwood (Eds.)*

93 **Cross-Scale Coupling in Space Plasmas** *James L. Horwitz, Nagendra Singh, and James L. Burch (Eds.)*

94 **Double-Diffusive Convection** *Alan Brandt and H.J.S. Fernando (Eds.)*

95 **Earth Processes: Reading the Isotopic Code** *Asish Basu and Stan Hart (Eds.)*

Maurice Ewing Volumes

1 **Island Arcs, Deep Sea Trenches, and Back-Arc Basins** *Manik Talwani and Walter C. Pitman III (Eds.)*

2 **Deep Drilling Results in the Atlantic Ocean: Ocean Crust** *Manik Talwani, Christopher G. Harrison, and Dennis E. Hayes (Eds.)*

3 **Deep Drilling Results in the Atlantic Ocean: Continental Margins and Paleoenvironment** *Manik Talwani, William Hay, and William B. F. Ryan (Eds.)*

4 **Earthquake Prediction—An International Review** *David W. Simpson and Paul G. Richards (Eds.)*

5 **Climate Processes and Climate Sensitivity** *James E. Hansen and Taro Takahashi (Eds.)*

6 **Earthquake Source Mechanics** *Shamita Das, John Boatwright, and Christopher H. Scholz (Eds.)*

IUGG Volumes

1 **Structure and Dynamics of Earth's Deep Interior** *D. E. Smylie and Raymond Hide (Eds.)*

2 **Hydrological Regimes and Their Subsurface Thermal Effects** *Alan E. Beck, Grant Garven, and Lajos Stegena (Eds.)*

3 **Origin and Evolution of Sedimentary Basins and Their Energy and Mineral Resources** *Raymond A. Price (Ed.)*

4 **Slow Deformation and Transmission of Stress in the Earth** *Steven C. Cohen and Petr Vaníček (Eds.)*

5 **Deep Structure and Past Kinematics of Accreted Terranes** *John W. Hillhouse (Ed.)*

6 **Properties and Processes of Earth's Lower Crust** *Robert F. Mereu, Stephan Mueller, and David M. Fountain (Eds.)*

7 **Understanding Climate Change** *Andre L. Berger, Robert E. Dickinson, and J. Kidson (Eds.)*

8 **Evolution of Mid Ocean Ridges** *John M. Sinton (Ed.)*

9 **Variations in Earth Rotation** *Dennis D. McCarthy and William E. Carter (Eds.)*

10 **Quo Vadimus Geophysics for the Next Generation** *George D. Garland and John R. Apel (Eds.)*

11 **Sea Level Changes: Determinations and Effects** *Philip L. Woodworth, David T. Pugh, John G. DeRonde, Richard G. Warrick, and John Hannah (Eds.)*

12 **Dynamics of Earth's Deep Interior and Earth Rotation** *Jean-Louis Le Mouël, D.E. Smylie, and Thomas Herring (Eds.)*

13 **Environmental Effects on Spacecraft Positioning and Trajectories** *A. Vallance Jones (Ed.)*

14 **Evolution of the Earth and Planets** *E. Takahashi, Raymond Jeanloz, and David Rubie (Eds.)*

15 **Interactions Between Global Climate Subsystems: The Legacy of Hann** *G. A. McBean and M. Hantel (Eds.)*

16 **Relating Geophysical Structures and Processes: The Jeffreys Volume** *K. Aki and R. Dmowska (Eds.)*

17 **Gravimetry and Space Techniques Applied to Geodynamics and Ocean Dynamics** *Bob E. Schutz, Allen Anderson, Claude Froidevaux, and Michael Parke (Eds.)*

18 **Nonlinear Dynamics and Predictability of Geophysical Phenomena** *William I. Newman, Andrei Gabrielov, and Donald L. Turcotte (Eds.)*

Mineral Physics Volumes

1 **Point Defects in Minerals** *Robert N. Schock (Ed.)*

2 **High Pressure Research in Mineral Physics** *Murli H. Manghnani and Yasuhiko Syona (Eds.)*

3 **High Pressure Research: Application to Earth and Planetary Sciences** *Yasuhiko Syono and Murli H. Manghnani (Eds.)*

Geophysical Monograph 96

Subduction
Top to Bottom

Gray E. Bebout
David W. Scholl
Stephen H. Kirby
John P. Platt

Editors

American Geophysical Union

Published under the aegis of the AGU Books Board

Cover illustration based on the gravity map "Marine Gravity Anomaly From Satellite Altimetry" by David T. Sandwell and Walter H. F. Smith.

Library of Congress Cataloging-in-Publication Data
Subduction top to bottom / Gray E. Bebout ... [et al.], editors.
        p.   cm. -- (Geophysical monograph ; 96)
    Includes bibliographical references.
    ISBN 0-87590-078-X
    1. Subduction zones. I. Bebout, Gray E., 1958- . II. Series.
QE511.46.S83 1996
551.1'36--dc20   96-35932
                                                                           CIP

ISBN 0-87590-078-X
ISSN 0065-8448

Copyright 1996 by the American Geophysical Union
2000 Florida Avenue, N.W.
Washington, DC 20009

Figures, tables, and short excerpts may be reprinted in scientific books and journals if the source is properly cited.

Authorization to photocopy items for internal or personal use, or the internal or personal use of specific clients, is granted by the American Geophysical Union for libraries and other users registered with the Copyright Clearance Center (CCC) Transactional Reporting Service, provided that the base fee of $1.50 per copy plus $0.35 per page is paid directly to CCC, 222 Rosewood Dr., Danvers, MA 01923. 0065-8448/96/$01.50+0.35.

    This consent does not extend to other kinds of copying, such as copying for creating new collective works or for resale. The reproduction of multiple copies and the use of full articles or the use of extracts, including figures and tables, for commercial purposes requires permission from AGU.

Printed in the United States of America.

CONTENTS

Preface
Gray E. Bebout, David W. Scholl, Stephen H. Kirby, and John P. Platt xi

What Goes In

Thermo-mechanical Evolution of Oceanic Lithosphere: Implications for the Subduction Process and Deep Earthquakes (Overview)
Seth Stein and Carol A. Stein 1

Geochemical Fluxes During Seafloor Alteration of the Basaltic Upper Oceanic Crust: DSDP Sites 417 and 418 (Overview)
Hubert Staudigel, Terry Plank, Bill White, and Hans-Ulrich Schmincke 19

The First Squeeze

Accretionary Mechanics With Properties That Vary in Space and Time (Overview)
Dan M. Davis 39

Mountain Building in Taiwan and the Critical Wedge Model
Chi-Yuen Wang, Adam Ellwood, Francis Wu, Ruey-Juin Rau, and Horng-Yuan Yen 49

Sediment Pore-Fluid Overpressuring and Its Effect on Deformation at the Toe of the Cascadia Accretionary Prism From Seismic Velocities
Guy R. Cochrane, J. Casey Moore, and Homa J. Lee 57

Oblique Strike-Slip Faulting of the Cascadia Submarine Forearc: The Daisy Bank Fault Zone off Central Oregon
Chris Goldfinger, LaVerne D. Kulm, Robert S. Yeats, Cheryl Hummon, Gary J. Huftile, Alan R. Niem, and Lisa C. McNeill 65

Fabrics and Veins in the Forearc: A Record of Cyclic Fluid Flow at Depths of < 15 km (Overview)
Donald M. Fisher 75

Large Earthquakes in Subduction Zones: Segment Interaction and Recurrence Times (Overview)
Larry J. Ruff 91

What Controls the Seismogenic Plate Interface in Subduction Zones?
Larry J. Ruff and Bart W. Tichelaar 105

Displacement Partitioning and Arc-Parallel Extension: Example From the Southeastern Caribbean Plate Margin
Hans G. Avé Lallemant 113

The Big Squeeze: Back From the Pressure Cooker

Thermal and Petrologic Structure of Subduction Zones (Overview)
Simon M. Peacock 119

Contrasting P-T-t Histories for Blueschists From the Western Baja Terrane and the Aegean: Effects of Synsubduction Exhumation and Backarc Extension
Suzanne L. Baldwin 135

CONTENTS

Tectonic Uplift and Exhumation of Blueschist Belts Along Transpressional Strike-Slip Fault Zones
Paul Mann and Mark B. Gordon 143

Syn-Subduction Forearc Extension and Blueschist Exhumation in Baja California, México
Richard L. Sedlock 155

Slip-History of the Vincent Thrust: Role of Denudation During Shallow Subduction
Marty Grove and Oscar M. Lovera 163

A Thermotectonic Model for Preservation of Ultrahigh-Pressure Phases in Metamorphosed Continental Crust
W. G. Ernst and Simon M. Peacock 171

Volatile Transfer and Recycling at Convergent Margins: Mass-Balance and Insights From High-P/T Metamorphic Rocks (Overview)
Gray E. Bebout 179

The Big Squeeze: From Beneath the Arc

Intermediate-Depth Intraslab Earthquakes and Arc Volcanism as Physical Expressions of Crustal and Uppermost Mantle Metamorphism in Subducting Slabs (Overview)
Stephen Kirby, E. Robert Engdahl, and Roger Denlinger 195

Subducted Lithospheric Slab Velocity Structure: Observations and Mineralogical Inferences
George Helffrich 215

Plate Structure and the Origin of Double Seismic Zones
Geoffrey A. Abers 223

Phase Equilibria Constraints on Models of Subduction Zone Magmatism (Overview)
James D. Myers and A. Dana Johnston 229

Deciphering Mantle and Crustal Signatures in Subduction Zone Magmatism (Overview)
Jon P. Davidson 251

Describing Chemical Fluxes in Subduction Zones: Insights From "Depth-Profiling" Studies of Arc and Forearc Rocks
Jeff Ryan, Julie Morris, Gray Bebout, and Bill Leeman 263

Boron and Other Fluid-Mobile Elements in Volcanic Arc Lavas: Implications for Subduction Processes
William P. Leeman 269

Effect of Sediments on Aqueous Silica Transport in Subduction Zones
Craig E. Manning 277

Does Fracture Zone Subduction Increase Sediment Flux and Mantle Melting in Subduction Zones? Trace Element Evidence From Aleutian Arc Basalt
Bradley S. Singer, William P. Leeman, Matthew F. Thirlwall, and Nicholas W. Rogers 285

Experimental Melting of Pelagic Sediment, Constraints Relevant to Subduction
Geoffrey T. Nichols, Peter J. Wyllie, and Charles R. Stern 293

CONTENTS

**The Influence of Dehydration and Partial Melting Reactions on the Seismicity
and Deformation in Warm Subducting Crust**
Tracy Rushmer   299

**Contrasting Styles of Mantle Metasomatism Above Subduction Zones:
Constraints From Ultramafic Xenoliths in Kamchatka**
Pavel Kepezhinskas and Marc J. Defant   307

Suprasubduction Mineralization: Metallo-Tectonic Terranes of the Southernmost Andes (Overview)
Eric P. Nelson   315

Hazards and Climatic Impact of Subduction-Zone Volcanism: A Global and Historical Perspective
Robert I. Tilling   331

The Biggest Squeeze: Slab Structure and Deep-Focus Earthquakes

Eclogite Formation and the Rheology, Buoyancy, Seismicity, and H_2O Content of Oceanic Crust
Bradley R. Hacker   337

Double Seismic Zones, Compressional Deep Trench-Outer Rise Events, and Superplumes
Tetsuzo Seno and Yoshiko Yamanaka   347

Characteristics of Multiple Ruptures During Large Deep-Focus Earthquakes
Wang-Ping Chen, Li-Ru Wu, and Mary Ann Glennon   357

Imaging Cold Rock at the Base of the Mantle: The Sometimes Fate of Slabs? (Overview)
Michael E. Wysession   369

PREFACE

Perhaps no other plate tectonic setting has attracted as diverse multidisciplinary attention as convergent margins. This has in part been spurred by the extremely tangible hazards imposed by subduction, particularly in the form of earthquakes and tsunamis and arc volcanism. Concern regarding these hazards is heightened by the tendency of convergent margins to be heavily populated coastal regions. There has also been great interest in convergent margin settings for their potential (and demonstrated capability) of producing economically important oil and gas reservoirs and ore deposits. The cycling of materials (e.g., CO_2) at convergent margins has been recognized as potentially significantly effecting changes in our environment, in particular, impacting evolution of the hydrosphere and atmosphere. It is widely accepted that convergent margin accretion and arc magmatism have been largely responsible for continental crust formation over long periods of Earth's history.

It is critical that we be able to fathom the incredible global diversity in the geological and geophysical expressions of subduction zones, a diversity that is incompatible with generic models of plate boundary interactions involving simple thermal, structural, kinematic, or mineralogical approaches alone. In our opinion, further progress depends critically upon improved interdisciplinary studies of the subduction process. In this spirit, we organized a conference/workshop—SUBCON ("Subduction From Top to Bottom Conference")—to encourage Earth scientists from diverse backgrounds and interests to consider subduction as an interactive and evolutionary process that depends upon the prior history of lithosphere in the ocean basins and on the specific characteristics (properties and settings) of individual sectors (e.g., depth intervals) of convergent margins. Our goal was to bring scientific specialists to a forum where virtually all earth-science observations relevant to the progressive stages of subduction are considered, from the surface and shallow subsurface environment of the outer rise, trench and accretionary wedge to the mantle's transition zone and below where deep earthquakes occur and slab melting and assimilation occur.

Participants in SUBCON (held in Avalon, California, June 1994) were selected for the significance of their recent contributions to our understanding of convergent margin processes, but also based on their demonstrated abilities to consider diverse observations in novel multidisciplinary studies, that is, to listen to others approaching the same problem from a different perspective and using different techniques. The choice of the Santa Catalina Island venue was guided by its location within a complexly evolving convergent margin regime (i.e., largely reflecting Late Cretaceous accretion and Late Tertiary extensional processes). Our brief field trip to examine exposures of the Catalina Schist subduction-zone metamorphic complex not only served as a welcome break from the lengthy oral and poster sessions, but also focussed the group further on the complexity of the rock record and the need for multi-disciplinary interactions to cope with this complexity. The papers in this volume, written mainly by SUBCON participants, convey the multidisciplinary spirit of SUBCON and will key readers to the critical additional constraints that are lacking and the most exciting directions for future research. Authors have attempted to "put a different spin" on their discussions and presentation of data to reflect the multi-disciplinary flavor of SUBCON. However, these papers will also stand on their own as fundamental advancements in their respective subdisciplines. The papers in this volume are organized largely by the depths they consider in an idealized subduction zone, but the organization is also conceptual and reflects in part our biases regarding areas of potential research synergy. As an example, in the section "The Big Squeeze: From Beneath the Arc," papers range from those considering the geodynamics of the mantle wedge and slab-mantle interface [*Helffrich*] and the effect on arcs of varying thermal evolution [*Kirby et al.*], to those examining our state of knowledge regarding the petrology of melting in the slab and mantle wedge [*Myers and Johnston; Nichols et al.*] and the geochemistry of arcs reflecting processes in the subducting slab, the mantle wedge, and the overlying oceanic or continental crust [*Ryan et al.; Davidson; Leeman; Singer et al.; Kepezhinskas and Defant*]. We hope that the readers will examine all papers in these groupings rather than skipping directly to the paper in their specific discipline.

We have included papers that fall into two format categories. Some papers, labeled as "Overview" in the table of contents, contain a greater component of literature review and synthesis of recent investigation, and to a greater extent direct the reader to areas of remaining uncertainty and potentially fruitful future research. Other papers, which are on the average shorter, are more topical in content, providing "case studies" directed toward the understanding of key dynamic aspects of subduction. An attempt has been made by all authors to reduce the amount of scientific jargon in their papers and make the papers technically accessible to diverse readership. Thus, we have endeavored to find "common denominators" in our understanding and approaches with the hope of fueling future interdisciplinary collaboration.

Several dominant themes emerge from this collection of 35 papers (and in the SUBCON abstracts volume, which contains over 100 abstracts). As alluded to above, a particularly prominent theme that emerges from these and other recent papers considering subduction processes is that of comparative "subductology" (using the phrase of Seiya Uyeda) or the comparison among subduction zones of critical parameters affecting their dynamics and rock and earthquake manifestations at individual subduction zones [see, for example, *Stein and Stein; Peacock; Kirby et al.; Ryan et al.; Leeman; Bebout*]. Numerous papers in this volume showcase the spectacular recent advances in geophysical methods, in both the acquisition and the interpretation of the diverse data [*Kirby et al.; Helffrich; Abers; Chen et al.; Seno and Yamanaka; Wysession*]. Earthquakes are considered from the standpoint of societal hazard [*Tilling*] but also as records of deep slab rheology and mineralogical transitions [*Kirby et al.; Hacker; Abers*]. Also highlighted repeatedly are the critical constraints on convergent margin processes provided by the Deep Sea Drilling Program/Ocean Drilling Program [*Staudigel et al.; Bebout; Hacker*]. Unfortunately, several invited participants were unable to attend SUBCON because of their involvement in an ODP drilling leg in the Barbados accretionary prism.

Many of the papers in this volume consider the general theme of the evolving (and modern) structural/mechanical state of subduction zones, utilizing results ranging from the shallow and deep earthquake records [*Kirby et al.; Hacker; Ruff; Ruff and Tichelaar; Chen et al.*] to $^{40}Ar/^{39}Ar$ [*Grove and Lovera; Baldwin*], field structural [*Davis; Wang et al.; Ave Lallemant; Sedlock*] and petrologic constraints on the pressure-temperature-time and strain histories of subduction-related metamorphic rocks representing deep accretionary complex evolution. Included in this group are papers considering the makeup and evolution of the seismogenic interface and the depths and mechanical significance of seismic coupling in subduction zones [*Ruff; Ruff and Tichelaar; Helffrich*]. Several papers report on the record of subduction-zone dynamics which result from plate interactions deviating from the simplified SUBCON depth-profile of a ocean-continent subduction zone, in particular, in settings involving continental collision and oblique subduction [e.g., *Ernst and Peacock; Mann and Gordon; Ave Lallemant*]. The unique mechanics in these scenarios may be necessary to afford the rapid uplift and surficial exposure of some deeply subducted materials, including those containing coesite and diamond (i.e., reflecting metamorphism at depths of greater than 100 km [*Ernst and Peacock; Hacker*]).

Another theme which emerges in this volume concerns the "Top to Bottom" consideration of the fluxes (inputs and outputs) of energy and matter during subduction, first ("What Goes In" and "The First Squeeze") with attempts to constrain these fluxes through study of the geochemical, lithological, mineralogical, thermal, and mechanical state of the subducting slab and sediments outboard of trenches (i.e., before their subduction [*Stein and Stein; Staudigel et al.*]) and structural and fluid histories in very shallow parts of accretionary complexes [*Davis; Wang et al.; Fisher; Cochrane et al.; Goldfinger et al.*]. Included in the section "The Big Squeeze: Back from the Pressure Cooker" is an assessment of our state of knowledge, from a modeling approach, regarding the complex, varying thermal evolution (heat flux) in subduction zones [*Peacock*], and the use of high-P/T metamorphic suites to reconstruct the paleo-mechanics, thermal evolution, and geochemistry of convergent margins [*Baldwin; Grove and Lovera; Mann and Gordon; Sedlock; Ernst and Peacock; Bebout*]. Included in the section "The Big Squeeze: From Beneath the Arc" are papers focussing on the present constraints on the evolving slab and sediments during progressive subduction and the processes of mineral reactions, fluid release and transport, metasomatism, deformation, and melting, particularly those processes leading to convergent margin magmatic and metasomatic flux [*Kirby et al.; Helffrich; Manning; Nichols et al.; Leeman; Ryan et al.; Davidson; Singer et al.; Rushmer; Kepezhinskas and Defant*]. Also included in this section are a synthesis, using the Andes as an example, of the ore deposits produced in continental magmatic arcs [*Nelson*], and an assessment of the hazards and climatic impact (i.e., flux of volcanic gases) of subduction-zone volcanism [*Tilling*]. Later papers (in the section, "The Biggest Squeeze: Slab Structure and Deep-Focus Earthquakes") consider the structure of deeply subducing slabs [*Hacker; Seno and Yamanaka; Abers; Chen et al.*], and the ultimate fate of subducting slabs, many of which are likely stored in the deepest parts of the mantle [*Wysession*]. We would like to enthusiastically thank those organizations who provided the funding that enabled us to follow through with our ambitious project, in particular to attract the high-calibre, highly international crowd of participants and to hold the meeting on Santa Catalina Island. This funding was largely provided by the United States Geological Survey, specifically the Office of the Chief Geologist, the National Earthquake and Volcano Hazards Programs, the Deep Continental Studies Program, and the Office of Energy and Marine Geology, and JOI-USSAC, with some support also coming from NSF (Earth Sciences Division, Tectonics Program; EAR-9406056). We also extend special thanks to the Santa Catalina Island Conservancy, whose logistical support and patience afforded us access to the exposures of the Catalina Schist. A hearty thanks is due Susan Kalb of the USGS, without whose logistical framework, managing skills, and patience our idea would certainly have fizzled. Finally, we thank the many who reviewed manuscripts ultimately published in this volume for their efforts toward ensuring high-quality contributions in-line with the original SUBCON theme of multidisciplinarity.

Gray E. Bebout
David W. Scholl
Stephen H. Kirby
John P. Platt
Editors

Thermo-mechanical Evolution of Oceanic Lithosphere: Implications for the Subduction Process and Deep Earthquakes

Seth Stein

Department of Geological Sciences, Northwestern University, Evanston IL

Carol A. Stein

Department of Geological Sciences, University of Illinois at Chicago, Chicago, IL

Because subduction involves the return of cold oceanic lithosphere to the warmer mantle, much of our thinking about subduction reflects models of temperatures in subducting slabs. These models, in turn, rely on thermal models of the oceanic lithosphere before it subducts, developed using variations in ocean depth and heat flow with age. Such models predict that subducting slabs are colder, denser, and stronger than the surrounding mantle, in accord with evidence from seismic velocities and earthquake depths. Although the simple models describe the basic observable phenomena which reflect thermal structure, the variation in depth and heat flow between and among plates before they subduct, and in velocity structure and distribution of seismicity when they subduct, illustrates the need for improved models.

INTRODUCTION

Subduction zones are downgoing limbs of the mantle convection system, where slabs of cold oceanic lithosphere formed at midocean ridges return to the deep mantle. As discussed in this volume and reviewed elsewhere [e.g., *Kincaid*, 1995; *Kirby*, 1995; *Kirby et al.*, 1996a], many features of subduction reflect subducting slabs being much colder than surrounding mantle. Because slabs subduct rapidly compared to the time needed for heat conducted from the surrounding mantle to warm them up, they remain colder, denser, and mechanically stronger than the surrounding mantle. Consequently, slabs transmit seismic waves faster and with less attenuation than the surrounding mantle, making it possible to map slabs and to show that deep earthquakes occur within them. The negative thermal buoyancy of slabs should provide a major source of stress within downgoing slabs and appears to be the primary force driving plate motions. The cold slab should be mechanically stronger than its surroundings, and thus sustain higher stresses. This additional strength, and the fact that mineral reactions occur more slowly at lower temperatures, have been suggested as factors permitting earthquakes within slabs to occur to almost 700 km depth, far deeper than in the surrounding hotter mantle. Interaction between the slab and overlying mantle wedge gives rise to arc volcanism, apparently via metamorphism and dehydration of slab tops and partial melting and flow of the wedge.

As a result, considerable attention has been directed toward estimating temperature in subducting slabs. The temperature depends primarily on the temperature in plates when they enter the trench, and how the plates warm up as they descend. The temperatures, in turn, control the properties and behavior of slabs. Here, we review some basic observations and concepts that influence our thinking about these factors, and consider some of their implications for the subduction process.

THE OCEANIC LITHOSPHERE

Recognition of seafloor spreading led rapidly to the view of plates of strong lithosphere moving over softer asthenosphere [e.g., *Elsasser*, 1971]. This rheological stratification reflects the thermal evolution of oceanic lithosphere, which cools as it spreads away from midocean ridges and reheats upon subduction into the deep mantle. In this view, which relies heavily on a set of observations reviewed next, oceanic lithosphere is the relatively thin and cold upper boundary layer of the mantle convective system (Plate 1), the primary mode of heat transfer from the earth's interior [e.g., *Parsons and Richter*, 1981; *Jarvis and Peltier*, 1989; *Pollack et al.*, 1993]. The lithosphere cools such that when it reaches most subduction zones, it is thought to be about 100 km thick with a basal temperature exceeding 1000°C. Below the lithosphere, temperatures are thought to increase more slowly, rising to only about 1500°C in the mantle transition zone (400-700 km depth) [e.g., *Ito and Katsura*, 1989]. Hence in the lithosphere, where heat transfer occurs primarily by conduction, temperature gradients are much higher (about 10°C/km) than below it, where lower (about 0.3°C/km) adiabatic temperature gradients are expected. Because of heat loss at the sea floor, subducting slabs are much colder than the surrounding mantle.

This temperature structure has major consequences. Because rock strength decreases with temperature, the oceanic lithosphere is also a mechanical boundary layer, which is stronger, i.e. can sustain greater stress, than material below (Figure 1). It arises because strength increases with pressure at shallow depths, where rocks fail by fracture, and decreases with temperature at greater depths, where rocks deform by temperature-dependent creep. Hence once temperature reaches about 800°C, lithosphere should be too weak to support significant stress [*Kirby*, 1977, 1983; *Goetze and Evans*, 1979; *Brace and Kohlstedt*, 1980]. As we will see, this idea is consistent with observations of quantities reflecting strength at depth. The strong lithosphere over a weaker asthenosphere allows the lithosphere to act as a stress guide for horizontal forces [*Elsasser*, 1969] and to sustain vertical loads [e.g., *Turcotte*, 1979], giving rise to familiar aspects of plate tectonics and plate boundary processes.

In addition to forming a thermal and mechanical boundary layer, differentiation at spreading centers causes the crust and uppermost mantle to form a chemical boundary layer [*Oxburgh and Parmentier*, 1977]. As a result, different definitions of "the" lithosphere are used for different purposes. Although the term strictly refers to material strength, it is often applied to the different boundary layers, in part because strength as a function of depth is not directly measurable, and in part because strength is temperature controlled. Thus the "thickness" of "the" lithosphere depends on the property (temperature, strength, chemistry) under consideration, and thus on the criterion or set of observations used to infer its variation with depth. The mechanical thickness inferred from the response to applied loads can also be strain-rate- and time-dependent [e.g. *Turcotte*, 1979; *Kirby*, 1983; *Stein et al.*, 1989].

It is worth bearing in mind that although many tectonic discussions focus on the lithosphere, it is the upper boundary layer of the convective system. Hence although density and strength variations largely reflect contrasts between the lithosphere and the remainder of the convective system, the total variations of each quantity due to convection drive plate motions and control the style of plate tectonics [e.g., *Verhoogen*, 1980].

THERMO-MECHANICAL STRUCTURE OF OCEANIC LITHOSPHERE

Data

Temperatures in subducting slabs are inferred from thermal models of the oceanic lithosphere before it subducts. Because temperatures at depth are not directly measurable, simple models are used, which attempt to provide a general description of average thermal structure as a function of age. The primary surface observables constraining these models are variations in seafloor depth and heat flow with lithospheric age (Table 1). Subsidence relative to the ridge crest, and hence seafloor depth, depends on temperature integrated with depth in the lithosphere, whereas heat flow depends on the temperature gradient just below the seafloor. A third observable is the variation with lithospheric age of the geoid, an equipotential of the gravity field, which reflects a depth-weighted integral of the density distribution, and hence provides a third constraint on the geotherm.

The observation that depth and heat flow vary approximately with the square root of lithospheric age (Figure 2) led to the view that young lithosphere acts largely as the upper boundary layer of a cooling halfspace [*Turcotte and Oxburgh*, 1967]. However, for ages older than about 70 Myr, average depth and heat flow "flatten", varying more slowly with age than for a halfspace. It is thus often assumed that halfspace cooling stops for older ages because heat added from below balances heat lost at the seafloor, causing the geotherm to approach steady state and thus the depths and heat flow to flatten. The plate

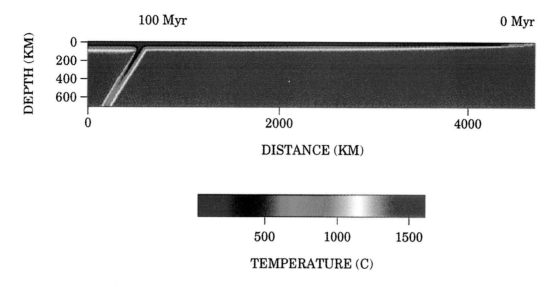

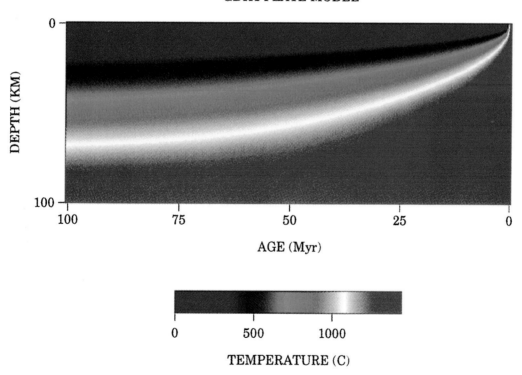

Plate 1. *Top:* Schematic illustration of the oceanic lithosphere as a thermal boundary layer, which cools as it moves away from midocean ridges and reheats as it subducts. Lithospheric temperatures are for the GDH1 thermal model [*Stein and Stein*, 1992] and a half-spreading rate of 4 cm/yr. Slab temperatures are from a finite difference calculation for a convergence rate of 8 cm/yr. Only lithospheric temperatures are calculated, so sublithospheric temperatures are shown as following an adiabatic gradient. *Bottom:* Thermal structure of the oceanic lithosphere for the GDH1 plate model.

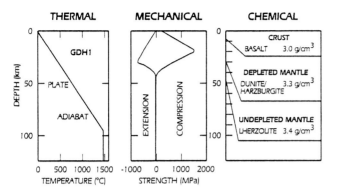

Fig. 1. The oceanic lithosphere forms a boundary layer, due to its thermal evolution. *Left:* The cooling plate forms a thermal boundary layer, illustrated by the asymptotic temperature structure for the GDH1 thermal model for lithosphere older than 70 Myr. The temperature structure controls the variations in depth, heat flow, gravity, seismic velocity and attenuation with age, and gives rise to the density variations causing plate driving forces. *Center:* The thermal boundary layer gives rise to a mechanical boundary layer, illustrated by a strength profile for old lithosphere, computed for a dry olivine flow law [*Brace and Kohlstedt*, 1980]. At shallow depth strength is controlled by brittle fracture, whereas at greater depth the ductile flow at high temperatures predicts rapid weakening. The strength profile controls flexure by vertical loads, horizontal stress transmission, plate boundary interactions, and maximum earthquake depths. *Right:* The crustal differentiation process at midocean ridges yields a chemical boundary layer, different in composition and density from the sublithospheric mantle [*Oxburgh and Parmentier*, 1977]. Because the three boundary layers differ in thickness, the "lithospheric thickness" depends on the property in question.

TABLE 1. Constraints on thermal models $T(z,t)$

OBSERVABLE	PROPORTIONAL TO	REFLECTS
Young Ocean Depth	$\int T(z,t)\,dz$	$k^{1/2}\alpha T_m$
Old Ocean Depth	$\int T(z,t)\,dz$	$\alpha T_m a$
Old Ocean Heat Flow	$\dfrac{\partial T(z,t)}{\partial z}\bigg|_{z=0}$	kT_m/a
Geoid Slope	$\dfrac{\partial}{\partial t}\int zT(z,t)\,dz$	$k\alpha T_m \exp(-kt/a^2)$

T	temperature		
z	depth	t	age
a	plate thickness	T_m	basal temperature
α	thermal expansion coefficient	k	thermal conductivity

model, a simple description of this perturbation, uses an isothermal base of the lithosphere to model its thermal equilibration (Plate 1, Figure 3) [*Langseth et al.*, 1966; *McKenzie*, 1967]. The plate model fits the data reasonably well, but does not directly describe how heat is added [e.g., *Crough*, 1977; *Parsons and McKenzie*, 1978; *Fleitout and Doin*, 1994]. Although plate and halfspace models are the same for young ages, they differ for ages old enough that the basal condition has an effect. In plate models the lithosphere tends to an equilibrium geotherm, whereas in halfspace models cooling continues for all ages.

Thermal Models

Thermal models are solutions to the inverse problem of finding the temperature T as a function of age t and depth z that best fits the variation in depth and heat flow with age. The data are used to estimate the primary parameters (plate thickness a, basal temperature T_m, thermal expansion coefficient α, and thermal conductivity k) characterizing halfspace and plate models. (A halfspace can be considered an infinitely thick plate.) Other parameters (densities, specific heat, and ridge depth) are generally specified a priori. For simplicity, conductivity and coefficient of thermal expansion are usually treated as depth-independent. An a priori value for conductivity is often used, because the improved fit from estimating it from the data is not meaningfully better [*Stein and Stein*, 1992].

As shown in Table 1, depth, heat flow, and geoid slope are nonlinear functions of the model parameters. The models have two parameters (plate thickness and basal temperature) reflecting thermal structure, and two parameters (conductivity and coefficient of thermal expansion) reflecting average physical properties of the lithosphere. The latter two are scale factors which map the thermal parameters into the primary observable features of the data.

It is useful to consider limiting features of the data which models seek to match. The first, the slope of the depths in young lithosphere versus square root of age, is proportional to $k^{1/2}\alpha T_m$. Because these depths can be equally well fit assuming a cooling halfspace, they are insensitive to plate thickness. In contrast, the predicted behavior at old ages depends on plate thickness. The asymptotic depth for old ocean is proportional to $\alpha T_m a$, the heat lost as the plate cools. Similarly, the asymptotic heat flow for old ocean, kT_m/a, is proportional to the asymptotic linear geotherm. Hence in a plate model depth and heat flow tend to asymptotic values depending on

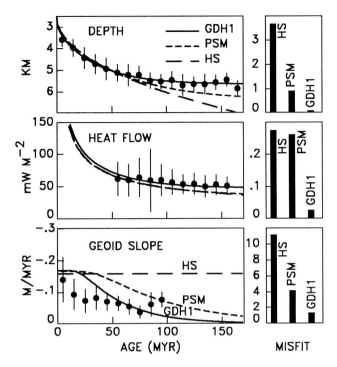

Fig. 2. Data used to constrain thermal models of the oceanic lithosphere. Comparison with the predictions of the three thermal models, which were not derived using these data, shows the general pattern that at older ages the data differ from the predictions of a cooling halfspace, and are better fit by plate models. Global depth data exclude hotspot swells [*Kido and Seno*, 1994]. Global heat flow are from *Stein and Stein* [1992], who used only North Pacific and Northwest Atlantic values to derive model GDH1. Geoid slopes across fracture zones are from *Richardson et al.* [1995]. The misfit values (*right panels*) show that the thin-plate GDH1 model fits better than either a halfspace (HS) or the thick-lithosphere PSM model.

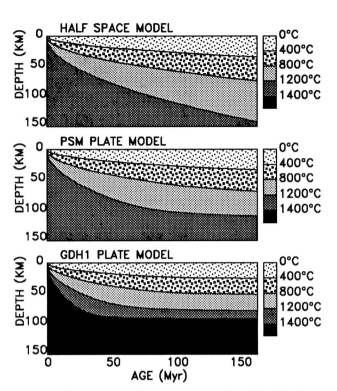

Fig. 3. Isotherms predicted by three thermal models. The lithosphere continues cooling for all ages for a halfspace model, reaches equilibrium for approximately 125 Myr old lithosphere in the thick-lithosphere (125 km) PSM model, and equilibrates for lithosphere about 70 Myr old in the thin (95 km) plate GDH1 model.

plate thickness and basal temperature, whereas in a halfspace model they continue to change with age. The derivative of the geoid with age (geoid slope) is also sensitive to the difference between models. It is constant for a halfspace model. For a plate model, the predicted slope is the same as for a halfspace at young ages, but "rolls off" at older ages at a rate depending inversely on plate thickness [*Cazenave*, 1984]. Like the "flattening" of depth and heat flow, this predicted deviation from halfspace behavior reflects the lithosphere approaching equilibrium thickness at older ages.

Figure 4 illustrates estimation of model parameters from data. The joint fit to depth and heat flow is shown as a function of assumed plate thickness and basal temperature. The best-fitting model, termed GDH1 (Plate 1, Table 2) [*Stein and Stein*, 1992], fits significantly better than either a halfspace (HS) model or a plate model with the parameters used by *Parsons and Sclater* [1977] (PSM). (The halfspace shown has parameters from *Carlson and Johnson* [1994] but the results would not differ significantly for other proposed parameters.) In particular, GDH1 reduces the systematic misfit to the depth and heat flow in older (>70 Myr) lithosphere (Figure 2), where PSM or a halfspace predict depths deeper and heat flow lower than generally observed.

The improved fit occurs because relative to PSM, GDH1 has thinner lithosphere with a higher basal temperature, and hence a steeper geotherm, higher heat flow, and shallower depths. An F-ratio test indicates the improved fit is significant at 99.9%. The improved fit going from PSM to GDH1 is comparable to that of PSM relative to a halfspace. Of the models, GDH1 fits geoid slope data best, though no model fits well.

The process of reestimating model parameters is conceptually the same as for global seismic velocity structure or relative plate motions. The goals are the same: to provide a better average description of the data and the pro-

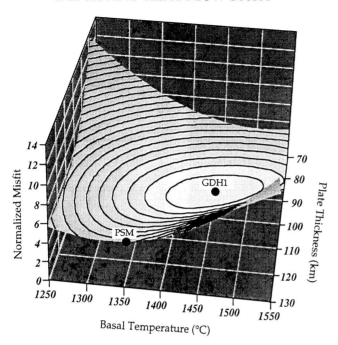

Fig. 4. Fitting process used for thermal model parameters. The misfit surface for the depth and heat flow data as a function of plate thermal thickness and basal temperature is shown, with the positions of the GDH1 and PSM models. Values are normalized to the GDH1 misfit, and the contour interval is 0.5. The misfit for PSM is five times that for GDH1. The surface is plotted for the GDH1 coefficient of thermal expansion, so PSM which has a different value plots slightly above the surface. The data and fitting function are discussed in *Stein and Stein* [1992].

TABLE 2. GDH1 model parameters

a	plate thickness	95 km
T_m	basal temperature	1450 °C
α	thermal expansion coefficient	3.1×10^{-5} °C^{-1}
k	thermal conductivity	3.138 W m^{-1} °C^{-1}
C_p	specific heat	1.171 kJ kg^{-1} °C^{-1}
ρ_m	mantle density	3330 kg m^{-3}
ρ_w	water density	1000 kg m^{-3}
d_r	ridge depth	2600 m

cess causing it, and facilitate study of regions still poorly fit by the new, better-fitting, model. The improvements can be significant; GDH1 fits about five times better than PSM, compared to the plate motion case where the recent NUVEL-1 model [*DeMets et al.*, 1990] gives a factor of three improvement over the earlier RM2 model [*Minster and Jordan*, 1978].

The resulting model should better describe the average thermal state of oceanic lithosphere. In addition, because the model better fits the data, it makes it easier to assess which regions are "anomalous", in that their depths and heat flow differ from most lithosphere of that age. A difficulty with using a halfspace or a thick plate as reference models is that because they systematically mispredict depth and heat flow for old lithosphere, almost any old lithosphere appears "anomalous" relative to these models, although it is normal for old lithosphere. It is thus harder to assess which areas differ from average

old lithosphere [*Stein and Stein*, 1993]. This situation is reminiscent of the mythical town of Lake Wobegon in the radio show *Prairie Home Companion,* "where all children are above average".*

Assessment of Models

Several points about such models are worth noting. Although the models fit data reasonably well, clear misfits remain, which presumably reflect both processes acting in addition to those incorporated in the simple thermal models, and variability between and within plates. For example, heat flow for ages younger than about 65 Myr is lower than predicted, presumably due to hydrothermal circulation [e.g., *Wolery and Sleep*, 1976; *Stein and Stein*, 1994a] (The extent to which this circulation affects temperatures is unknown, but may not be significant except very close to the ridge axis, [e.g., *Stein et al.*, 1995], so it is assumed to not affect depths significantly). The misfit to geoid slope data for young ages may reflect the geoid offset across fracture zones incorporating effects in addition to that purely of the thermal age contrast, such as flexure, thermal stresses, or local asthenospheric flow [*Sandwell*, 1984; *Parmentier and Haxby*, 1986; *Robinson et al.*, 1988; *Wessel and Haxby*, 1990]. Moreover, the thermal models describe only average temperature structure as a function of age, using a few depth- and age-independent parameters. Hence these models are simple representations of a complex thermal structure which do not address the variations in depth, heat flow, and geoid slope as functions of age between and within plates [e.g., *Calcagno and Cazenave*, 1994] which may reflect both variations in temperature and other perturbations such as those due to intraplate volcanism, crustal thickness, or asthenospheric flow.

Because these models are solutions to an inverse problem, we can assess how they fit data, but have no direct

---
* Similarly, S. Peacock pointed out to us in his review that 90% of motorists are said to consider themselves above-average drivers.

way of telling how well they describe temperature in the earth. Some insight can be derived by using models to predict data not used in deriving them. For example, GDH1 fits depth and heat flow data that were not inverted in deriving it (Figure 2) better than PSM or a halfspace model [*Johnson and Carlson*, 1992; *Stein and Stein*, 1993; *Shoberg et al.*, 1993; *Kido and Seno*, 1994]. Similarly GDH1, derived by inverting depth and heat flow data, predicts geoid data better than the other models. Nonetheless, GDH1 or any simple model does not fully describe the thermal structure, as illustrated by the misfit to the geoid data at young ages or the variations in depth, heat flow, and geoid slope as functions of age between and within plates.

Estimation of thermal structure of the lithosphere thus faces difficulties common to inverse problems. The models are oversimplifications of the real situation, and even for a given model, the parameters estimated depend on the choice of data and fitting function, and are nonunique [*Stein and Stein*, 1992, 1993]. The usual question arises when more complicated models better fit data, whether the improved fit exceeds that expected purely by chance from the model's having more free parameters [*Stein and Stein*, 1992, 1993, 1994b]. Similarly, there is the issue of how best to incorporate other information. For example, should parameters like the average coefficient of thermal expansion be determined from inversion, specified a priori from extrapolation of laboratory results, or estimated by combining these approaches? This question is illustrated by the observation that the GDH1 basal temperature is slightly (7%) higher than the approximately 1350°C often inferred for the temperature of midocean ridges from the thickness of oceanic crust [e.g., *Sleep and Windley*, 1982; *McKenzie and Bickle*, 1988]. However, the improved fit of GDH1 over a model with basal temperature fixed at 1350°C is significant as measured by F-ratio test [*Stein and Stein*, 1993]. If the ridge temperature is known to sufficient precision (a question beyond our scope here), the discrepancy could have several causes. For example, although the thermal model uses a single basal temperature for all ages, the estimate of this parameter (rather than its product with the coefficient of thermal expansion) depends largely on data for old ages. Hence the discrepancy may reflect heat addition to old lithosphere by a process analogous to mantle plumes, which are thought to be several hundred degrees hotter than ridges [*Sleep*, 1992].

Investigation of the thermal evolution of oceanic lithosphere remains an active research area. Even the basic question of whether old lithosphere approaches thermal equilibrium is still under discussion [e.g., *Carlson and Johnson*, 1994]. Although the most straightforward interpretation of the depth, heat flow, and geoid data is in terms of a plate model [*Richardson et al.*, 1995], other interpretations are possible. In particular, explanations other than thermal equilibration have been offered for flattening of the depth curve. In one, flattening is analogous to that associated with seafloor traces of mantle plumes [e.g., *Heestand and Crough*, 1981; *Davies and Pribac*, 1993], and so is due largely to dynamic pressure of plumes, with some heating of the lithosphere. Another possibility is that apparent flattening results from excess volcanism masking continued subsidence due to halfspace cooling. A third possibility is that depths are perturbed by pressure differences driving asthenospheric flow [*Schubert and Turcotte*, 1972; *Schubert et al.*, 1978; *Phipps Morgan and Smith*, 1992, 1994; *Stein and Stein*, 1994b]. Moreover, even if flattening is a thermal effect, the physical process of heat addition has yet to be satisfactorily explained. These questions remain unresolved because all proposed perturbations are at least qualitatively consistent with the observed flattening, because it is difficult to isolate the effects of possible different mechanisms, and because flattening differs enough in different locations [*Calcagno and Cazenave*, 1994] that multiple mechanisms may operate.

Implications for Subduction Zone Thermal Structure

Fortunately, for many subduction zone applications, the choice of temperature model is not crucial. For example, the model predictions in Figure 3 are similar, especially at shallow depths (consider the 400°C isotherm). The differences between models are, however, of possible significance for the common case of the subduction of old lithosphere. As discussed shortly, various subduction zone phenomena seem vary with the age of the subducting plate, and hence presumably its temperature structure. Thus where old lithosphere subducts, it matters somewhat whether we consider a plate model (in which all lithosphere older than about 70 Myr is similar) or a halfspace model (in which temperatures still vary with age for old lithosphere). The choice of model matters largely for the deepest portions of the lithosphere.

Mechanical Structure

The mechanical structure of the oceanic lithosphere before it subducts also has implications for the subduction process. The primary determinant of mechanical structure (Figure 1) is variation in strength with depth and age resulting from the temperature and pressure. A secondary factor is the presence of structural heterogeneities.

8 THERMO-MECHANICAL EVOLUTION OF LITHOSPHERE

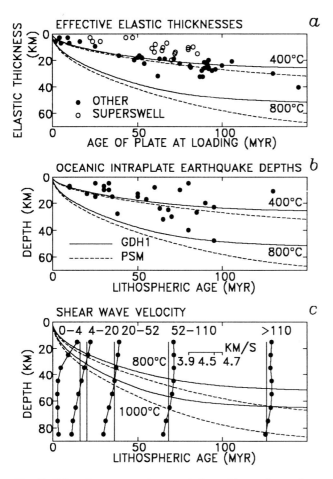

Fig. 5. Other data types whose variation with age is consistent with cooling of the lithosphere, as illustrated by isotherms for two thermal models. Except for the oldest lithosphere, the isotherms corresponding to the effective elastic thickness data [*Calmant et al.*, 1990] (a) and deepest intraplate seismicity [*Wiens and Stein*, 1983] (b) are similar for the two models. The difference between the temperatures for the low velocity zone (c) is greater. The vertical line for the velocity structure [*Nishimura and Forsyth*, 1989] in each age range (e.g., 0-4 Ma) corresponds to 4.5 km s^{-1}.

The predicted weakening of the lithosphere below a temperature-controlled depth is consistent with several observations (Figure 5). Effective elastic thickness inferred from loads on the lithosphere, a measure of the depth to which the lithosphere is strong enough to support significant stress, increases with age approximately as the 400°C isotherm [e.g., *Bodine et al.*, 1981; *Calmant et al.*, 1990]. The maximum depth of intraplate seismicity, which presumably reflects the depth to which the lithosphere is strong enough to support seismogenic stresses, increases with age approximately as the 700°C isotherm [*Wiens and Stein*, 1983; *Chen and Molnar*, 1983]. Similarly, the depth to the low velocity zone inferred from seismic surface wave dispersion increases with age [e.g., *Nishimura and Forsyth*, 1989]. Although these observations appear to reflect lithospheric cooling, using them to discriminate between thermal models is difficult, as it requires rheological models, assumptions about the strength required to support specific loads, and assumptions about the variation in seismic velocity with temperature. For subduction zone considerations, however, they jointly indicate that the upper portion of the slab should be strongest, and that the geometry of this strong region should be temperature- and pressure- controlled as the slab subducts.

In addition to the oceanic lithosphere being weak at depth, tectonic features may cause weak zones. Oceanic intraplate seismicity occurs preferentially on no-longer-active tectonic features, such as fossil spreading ridges and hotspot tracks, suggesting that these zones are weaker than normal lithosphere and move easily in response to intraplate stress [e.g., *Stein and Okal*, 1978; *Stein*, 1979; *Bergman and Solomon*, 1980; *Geller et al.*, 1983; *Wysession et al.*, 1991, 1995]. Similarly, unusual seismicity occurs where seafloor tectonic features enter trenches [e.g., *Vogt et al.*, 1976; *Chung and Kanamori*, 1978; *Stein et al.*, 1982]. In addition, earthquakes, some of which are large, occur in the subducting plate as it approaches the trench [e.g., *Kanamori*, 1971; *Chapple and Forsyth*, 1979], perhaps due to plate bending. Thus preexisting faults, also including those remaining from near-ridge processes, may survive as weak zones once the plate subducts and be the loci of intermediate [*Kirby et al.*, 1996b] and deep earthquakes [*Silver et al.*, 1995]. It thus seems that some of the variation in seismicity along subduction zones depends on weakness in the lithosphere before it subducts.

THERMAL STRUCTURE OF SUBDUCTING SLABS

Models

Predicting temperatures in subducting slabs is more challenging than in lithosphere before it subducts, because temperatures are not only unconstrained by direct observations, but less easily inferred indirectly. Hence caveats raised earlier about thermal models of the lithosphere apply even more strongly. Fortunately, the basic ideas about slab temperatures from simple models are relatively insensitive to the details of the model.

A simple analytical model, based on the time required for a slab to heat up by conduction as it subducts into a hotter isothermal mantle [McKenzie, 1969], illustrates several ideas. Isotherms in the slab extend downward, such that the maximum depth reached by a isotherm is proportional to the product of the vertical descent rate (trench-normal convergence rate times the sine of the dip) and the square of plate thickness. Hence for a halfspace thermal model, in which the thickness of the subducting lithosphere is proportional to the square root of its age at the trench, the depth reached by an isotherm is proportional to the product of the vertical descent rate and age, a quantity known as the thermal parameter.

Similar results emerge from numerical thermal models. Plate 2 shows thermal models from a program derived from one by N. H. Sleep, based on a finite difference algorithm [Toksöz et al., 1971] widely used in slab thermal modeling [e.g., Sleep, 1973; Hsui and Toksöz, 1979] The temperature structure of the lithosphere before it subducts and the thermal diffusivity are from GDH1. In such cooling plate models, temperatures approach steady state at about 70 Myr, such that all older lithosphere has about the same geotherm. Hence the geotherm in subducting lithosphere does not vary directly with thermal parameter, as would be true for a halfspace thermal model. For simplicity, lithosphere entering the trench is assumed to be of constant age. Models are computed by allowing subduction to go on long enough that a stable temperature structure results. The models shown are for a relatively younger and slower-subducting slab (thermal parameter about 2500 km), approximating the Aleutian arc, and an older and faster-subducting slab (thermal parameter approximately 17,000 km), approximating the Tonga arc. As expected, the slab with higher thermal parameter warms up more slowly, and is thus colder. These predicted temperature distributions are typical of slab thermal models, though the depth to individual isotherms vary, as shown by comparison of various models, including those listed by Helffrich et al. [1989] and a more recent study [Davies and Stevenson, 1992].

The thermal models also give insight into the rate at which slabs should equilibrate with the mantle. Figure 6 shows the predicted minimum temperature within a slab as a function of time since subduction. The coldest portion reaches half the mantle temperature in about 10 Myr, and 80% in about 40 Myr. Hence because most subduction zones have been active far longer than the time (about 10 Myr) required for the slab to first reach 670 km [e.g., Engebretson et al., 1992], the maximum depth of earthquakes in each subduction zone does not simply indicate the maximum depth that the slab has reached, a possibility suggested by Isacks et al. [1968] before an adequate magnetic anomaly record became available.

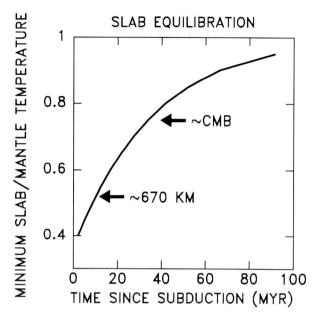

Fig. 6. Minimum temperature within a slab as a fraction of the mantle temperature, as a function of the time since subduction, computed using the analytic model of McKenzie [1969] for a slab with GDH1 model parameters. The coldest portion reaches half the mantle temperature in about 10 Myr, by which a typical slab is approximately at 670 km depth, and 80% in 40 Myr, by which a slab which continued descending at the same rate would reach the core-mantle boundary. Slabs can thus remain thermally distinct for long periods of time.

Moreover, because the time required for subducted material to reach 670 km depth is far less than that required for thermal equilibration with the surroundings, the restriction of seismicity to depths shallower than 670 km does not indicate that the slab is no longer a discrete thermal, and thus mechanical, entity. Hence from a thermo-rheological standpoint, there is no reason for slabs not to penetrate into the lower mantle, in accord with seismological observations and convection modeling, discussed shortly. In fact, if a slab descended through the lower mantle at the same rate, it would retain a significant thermal anomaly at the core-mantle boundary. As a result, cold slab remnants in the lower mantle are thought to give rise to thermal and thus density heterogeneity [e.g., Richards and Engebretson, 1992].

Such thermal models are used to predict approximate temperatures within slabs. Clearly the geometry assumed is simplified and the uniform thermal diffusivity adopted is an approximation. As the model shown only allows slab heating by conduction from the surrounding mantle and does not include shear or radiogenic heating or latent heat release, the predicted temperatures are lower bounds.

10 THERMO-MECHANICAL EVOLUTION OF LITHOSPHERE

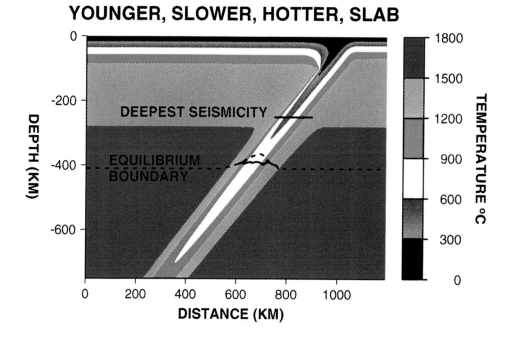

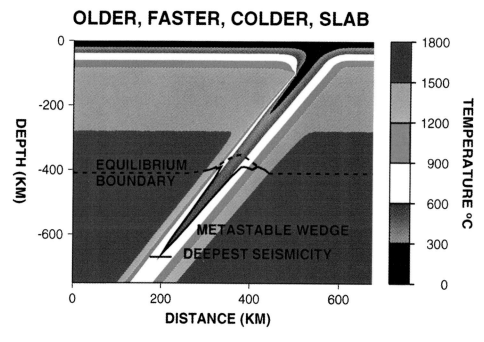

Plate 2. Comparison of thermal structure and predicted regions of metastability for a relatively younger and slower-subducting slab (thermal parameter about 2500 km), which approximates the Aleutian arc, and an older and faster-subducting slab (thermal parameter about 17,000 km) which approximates the Tonga arc. As expected, the slab with higher thermal parameter warms up more slowly, and is thus colder. A metastable wedge forms only for old and fast-subducting slabs which are sufficiently cold that kinetic hindrance prevents phase transformations from keeping pace with the descent rate. For the model parameters used, the metastable wedge is bounded on its sides and bottom by the 600°C isotherm, in those slabs cold enough for it to form. Deep earthquakes are presumed to occur by transformational faulting in the metastable wedges.

Hence these (or other) models' predicted temperatures are probably not accurate to better than about 200°C. The fact that the temperatures from such models predict seismic velocities similar to those inferred from observations (Plate 3) suggests that the models are at least reasonable approximations. As a result, it seems plausible to use such models to explore subduction zone processes.

Implications for Deep Earthquakes

As papers in this volume illustrate, thermal models of subducting slabs are used to study aspects of the subduction process [e.g. *Kirby et al.*, 1996b; *Peacock*, 1996]. The models are used in studies which characterize subduction zones by various parameters, such as the convergence rate and age of the subducting lithosphere, and investigate how processes vary among subduction zones [e.g., *Jarrard*, 1986; *Peacock*, 1992; *Davies and Stevenson*, 1992]. One striking example is the variation in the maximum depth of deep earthquakes (those below 325 km) as a function of thermal parameter (Figure 7). Although earthquakes are restricted to depths shallower than about 680 km, the maximum depth increases with thermal parameter. This observation argues for the maximum depth of earthquakes being controlled by a temperature-dependent mechanism.

No consensus, however, exists about what the thermal control mechanism may be. One possibility is that seismicity is limited by a thermally-controlled strength, such that at higher temperatures the slab is too weak to support seismic failure [*Molnar et al.*, 1979; *Wortel*, 1982; *Wortel and Vlaar*, 1988]. A difficulty, however, is that laboratory results predict that slabs should be strong well below the deepest earthquakes [*Brodholt and Stein*, 1988].

A second possibility is that faulting occurs by brittle fracture as for shallow earthquakes. Although high pressures would normally suppress fracture, it may occur once slabs become hot enough that water released by decomposition of hydrous minerals reduces effective stress on fossil faults formed before subduction [*Raleigh*, 1967; *Meade and Jeanloz*, 1991; *Silver et al.*, 1995]. Because shallow earthquakes in oceanic lithosphere only occur where the temperature is less than approximately 800°C [*Wiens and Stein*, 1983; *Chen and Molnar*, 1983], similar temperature control would be expected in the slab [*Stein*, 1995]. A possible problem with this dehydration embrittlement model is that hydrothermal circulation would not be expected to bring water to depths of more than a few km in oceanic plates [e.g., *Stein et al.*, 1995], whereas large earthquakes would require water in the cold interior of slabs. Moreover, it is unclear whether the hydrated minerals could survive to these depths [*Ulmer et al.*, 1994].

In the third hypothesis, deep earthquakes result from solid state phase changes, primarily that in which olivine transforms to a denser spinel structure [*Kirby*, 1987; *Green and Burnley*, 1989; *Kirby et al.*, 1991]. Thus deep seismicity occurs only in the depth range of the mantle transition zone, where phase changes should occur in downgoing slabs. Because the rate of the phase transformation depends exponentially on temperature, then in fast-subducting cold slabs the transformation cannot keep pace with the descent, and metastable olivine should persist below the equilibrium phase boundary [*Sung and Burns*, 1976ab; *Rubie and Ross*, 1994; *Kirby et al.*, 1996a] (Plate 2). Deep earthquakes are assumed to occur in the metastable material by a shear instability, known as transformational faulting, observed in the laboratory when metastable materials under stress undergo strongly exothermic reactions that isochemically transform one phase into a denser form. This mechanism resolves the objection traditionally raised to phase-change models for deep earthquakes, because the resulting motion would be slip on a fault rather than an implosion, in accord with seismological observations [*Kawakatsu*, 1991].

The metastability hypothesis makes several predictions generally consistent with various observations. The idea that deep earthquakes occur by a failure mechanism different from that for shallow and intermediate earthquakes is tempting, because seismicity as a function of depth has a minimum at about 350 km and then increases, suggesting deep earthquakes form a distinct population. The idea of deep earthquakes due to phase changes explains why these earthquakes coincide with the 400-700 depth range of the transition zone, where these phase changes are expected. In particular, it explains why deep earthquakes cease at the base of the transition zone, because phase changes associated with formation of the lower mantle mineral assemblage are endothermic and thus should not cause transformational faulting. This idea is significant for mantle dynamics, in that although the simplest explanation for the cessation of seismicity near 670 km is that slabs do not descend into the lower mantle, seismological observations are interpreted as indicating that some do [e.g., *Fischer et al.*, 1988; *Van der Hilst et al.*, 1991; *Fukao et al.*, 1992; *Van der Hilst*, 1995]. Such slab penetration is also predicted by mantle convection modeling [e.g., *Christensen and Yuen*, 1984].

Metastability may explain the observation (Figure 7) that deep seismicity occurs only for slabs with thermal parameter greater than 5000 km. Rapid deepening is hard

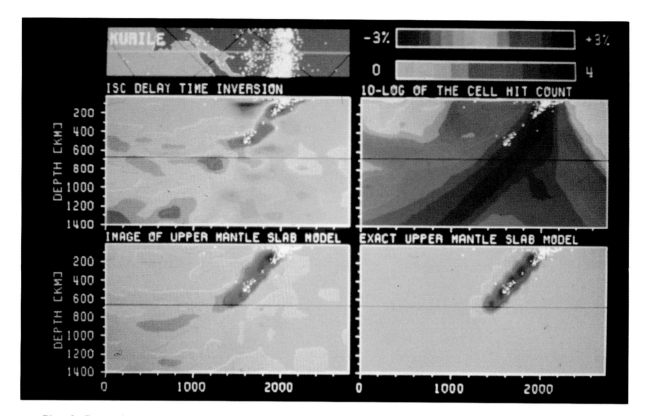

Plate 3. Comparison of seismic (P-wave) tomographic image of the subducting Kurile slab obtained from inversion of International Seismological Center delay time data (*center left*) to the image (*lower left*) predicted for a slab thermal model. The seismic velocity anomaly predicted by the thermal model (*lower right*) is imaged by a simulated tomographic study with the same seismic ray path sampling as used for the data. The ray path sampling is shown by the hit count (*center right*), the number of rays sampling each cell used in the inversion. White dots indicate earthquake hypocenters. As a result of ray geometry and noise added to the synthetic data, the exact model of the slab gives a somewhat distorted image (*lower left*), showing how the model would appear in such a tomographic study. The fact that the image of the slab model and the tomographic result (*center left*) are similar suggests that the slab model is a reasonable description of the major features of the actual slab. A similar conclusion emerges from the observation that the tomographic result also resembles parts of the model image which are resolution artifacts not present in the original model. These artifacts, generally of low amplitude, cause the slab to appear to broaden, shallow in dip, or "finger". [*Spakman et al.*, 1989].

to explain if seismicity is controlled directly by temperature, because temperatures vary smoothly as a function of thermal parameter. It is easier to explain as a consequence of metastability, as illustrated by the wedge-shaped region of predicted metastability in Plate 2. This region is delineated above by the equilibrium boundary for the olivine → spinel transition, which is elevated in the cold slab, and below by the 99% transformation contour where almost all metastable olivine has transformed. The younger, slower-subducting, and hence warmer slab is hot enough that phase transformation keeps pace with subduction, essentially no metastable wedge forms, transformational faulting should not occur, and deep earthquakes are not expected. In contrast, the older faster-subducting slab is cold enough that phase transformation cannot keep pace with subduction, a distinct metastable wedge forms, and deep earthquakes are expected. For the model parameters used, the wedge is bounded on its sides and bottom by the 600°C isotherm, so the occurrence of deep earthquakes is temperature-controlled in slabs cold enough for a metastable wedge to form.

Problem: a Fault too big?

Recent large deep earthquakes illustrate a potential difficulty with all three hypotheses for deep earthquakes. In these hypotheses, earthquakes should be restricted to the portion of the slab cooler than 600-800°C, which ther-

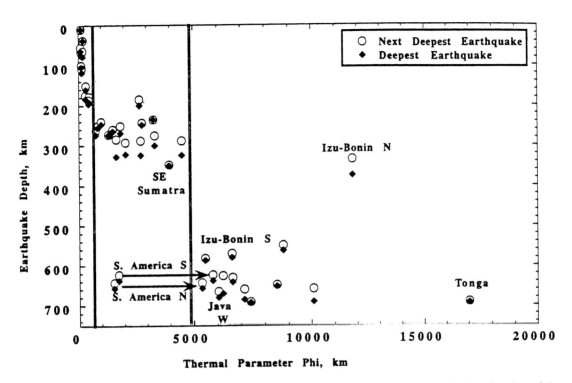

Fig. 7. Plot of earthquake depths for different subduction zones [Kirby et al., 1996a]. The plot is a function of thermal parameter, the product of vertical descent rate and lithospheric age. For simple thermal models, the maximum depth to an isotherm should vary with the thermal parameter. Hence if deep earthquakes were directly temperature limited, their maximum depth should be a smooth function of thermal parameter. Instead, the maximum depths seem divided into a group with thermal parameter less than about 5000 km, which do not have deep earthquakes, and a group with greater thermal parameters that do. This abrupt change is consistent with the predictions of the thermo-kinetic model in Plate 2 where deep earthquakes result from phase changes in metastable olivine, and so occur only in slabs where significant metastability is expected.

mal models predict should be a wedge narrowing to less than about 10 km at depths greater than 600 km. The fault areas of recent deep earthquakes, however, exceed the predicted wedge dimensions. A magnitude 7.6 earthquake beneath Tonga in 1994 had an usually large aftershock sequence, which defined a 50 x 65 km fault zone [Wiens et al., 1994]. Only a few months later, the largest deep earthquake instrumentally recorded occurred beneath Bolivia. Analyses of data for this earthquake indicate a near-horizontal fault area 30-50 km on a side [Kikuchi and Kanamori, 1994; Silver et al., 1995]. Hence although this interpretation is non-unique [Chen, 1995], large fault zones appear to cut across the predicted narrow wedge of material below 600-800°C.

The idea of a temperature-controlled process is hard to abandon, however, because deep earthquakes occur only in those slabs which are relatively colder. Thus the fault areas may indicate slab temperature structures more complicated than the simple models of essentially parallel isotherms in undeformed slabs. The high seismic energy release below about 600 km [e.g., Richter, 1979], earthquake mechanisms [Lundgren and Giardini, 1994], images of slabs from seismic tomography [e.g., Van der Hilst et al., 1991], and convection models [e.g., Kincaid and Olson, 1987; Tao and O'Connell, 1993] suggest that slabs deform due to interaction with the 670 km discontinuity, presumably because a major change in physical properties occurs at the base of the transition zone.

Kirby et al. [1995] hence suggest that the slab in the region of the Bolivian earthquake has a complex thermal structure because of variations in the age of the subducting plate over time and thickening due to slab deformation, causing a widened cold "pod". Large deep earthquakes could occur in this region, either due to metastability or another temperature-controlled process. The real geometry is presumably more complicated and varies within and among slabs. For the Tonga earthquake, characterized by many more aftershocks than usual for deep earthquakes, a different mechanism seems required, such as aftershocks by ductile faulting in the relatively cold spinel-rich region outside the wedge, perhaps triggered by a large transformational-faulting main shock in the wedge [*Kirby et al.*, 1996a].

Complex and variable deep slab thermal structure is plausible for several reasons. Although simple thermal models vary only slowly along strike for a given slab, deep seismicity is quite variable. Deep seismicity has distinct clusters and gaps where later large earthquakes can occur (as for the Bolivian earthquake) [*Kirby et al.*, 1995]. Tomographic images of deep slabs vary along strike and show more complexity [*Van der Hilst et al.*, 1991; *Fukao et al.*, 1992; *Engdahl et al.*, 1995; *Van der Hilst*, 1995] than simple thermal models predict [*Spakman et al.*, 1989]. As noted earlier, some of this complexity may result from deformation at the 670 km discontinuity. Moreover, in addition to mechanical perturbations to the slab, some of this variability may reflect metastability, because latent heat release would perturb thermal structure [*Daessler and Yuen*, 1993; *Kirby et al.*, 1996a]. These variations in both temperature and metastability would cause complex density variations, and thus affect slab stresses and driving forces [*Kirby et al.*, 1996a]. If wedges were large and continuous enough, their buoyancy might deflect the slab toward the horizontal [*Silver et al.*, 1995], as observed in some cases.

PROSPECTS

Ideas about the thermal structure of the oceanic lithosphere before it subducts, and the resulting thermal structure of slabs, seem poised for refinement. Both have been based on fairly simple models, as reviewed here. The models are surprisingly successful at describing the basic observed phenomena with a small number of parameters. This general success is striking, given that the models are two dimensional, include only the simplest thermal effects, and use temperature-, depth-, and pressure-independent physical properties.

The challenge is less how to pose more complex models, than to test and constrain them. Many features of the data illustrate the need for improved models. In particular, variations in ocean depth and heat flow about their mean values as functions of age, and the misfit to geoid data at fracture zones, illustrate the need for models in which depth and heat flow depend on more than age-dependent thermal structure. For this application, a best-fitting age-dependent thermal model can be used as a reference model to identify regions differing from the average at that age, and thus estimate the magnitude of the perturbing process [*Stein and Stein*, 1993]. Much needs to be done to characterize average lithosphere, and to investigate variations about the average. Our sense is that the primary deviations from halfspace cooling are thermal, and hence described on average by a plate model, whereas secondary regional deviations reflect temperature and pressure variations, perhaps due to both asthenospheric flow and local temperature variations. An important issue is the cause of variations in ridge crest depth [e.g., *Calcagno and Cazenave*, 1994] which provide differing initial conditions. We expect that models will continue to be posed and tested, hopefully by explicit numerical comparison to data [*Stein and Stein*, 1994b].

Similarly, the variation along subduction zones in phenomena including velocity structure and seismicity distribution illustrate the need for better thermal models. Because these variations can be attributed to effects not included in current models, such as three-dimensional slab geometry, slab mineralogy, and the variation of properties like thermal diffusivity with temperature and pressure, improved models are being suggested and it will be interesting to see which prove most successful.

Acknowledgements. We have benefited from discussions over the years with S. Kirby on aspects of the thermal evolution of the lithosphere. We thank him, S. Peacock, M. Wysession, and an anonymous reviewer for helpful comments. We benefited from the hospitality and support of the Laboratory for Terrestrial Physics, NASA Goddard Space Flight Center, where much of this research was done. Additional support came from NASA grant NAG 5-2003 and NSF grant EAR-9022476.

REFERENCES

Bergman, E. A., and S. C. Solomon, Oceanic intraplate earthquakes: Implications for local and regional intraplate stress, *J. Geophys. Res.*, 85, 5389–5410, 1980.

Bodine, J. H., M. S. Steckler, and A. B. Watts, Observations of flexure and the rheology of the oceanic lithosphere, *J. Geophys. Res.*, 86, 3695–3707, 1981.

Brace, W. F., and D. L. Kohlstedt, Limits on lithospheric stress imposed by laboratory experiments, *J. Geophys. Res.*, 85, 6248–6252, 1980.

Brodholt, J., and S. Stein, Rheological control of Wadati-Benioff zone seismicity, *Geophys. Res. Lett.*, 15, 1081–1084, 1988.

Calcagno, P., and A. Cazenave, Subsidence of the seafloor in the Atlantic and Pacific Oceans: Regional and large-scale variations, *Earth Planet. Sci. Lett.*, 126, 473–492, 1994.

Calmant, S., J. Francheteau, and A. Cazenave, Elastic layer thickening with age of the oceanic lithosphere, *Geophys. J. Int.*, *100*, 59–67, 1990.

Carlson, R. L., and H. P. Johnson, On modeling the thermal evolution of the oceanic upper mantle: An assessment of the cooling plate model, *J. Geophys. Res.*, *99*, 3201–3214, 1994.

Cazenave, A., Thermal cooling of the lithosphere: constraints from geoid data, *Earth Planet. Sci. Lett.*, *70*, 395–406, 1984.

Chapple, W. M., and D. W. Forsyth, Earthquakes and bending of plates at trenches, *J. Geophys. Res.*, *84*, 6729–6749, 1979.

Chen, W.-P., En echelon ruptures during the great Bolivian earthquake of 1994, *Geophys. Res. Lett.*, *22*, 2261–2264, 1995.

Chen, W.-P., and P. Molnar, Focal depths of intraplate earthquakes and implications for thermal and mechanical properties of the lithosphere, *J. Geophys. Res.*, *88*, 4183–4214, 1983.

Christensen, U. R., and D. A. Yuen, The interaction of a subducting lithospheric slab with a chemical or phase boundary, *J. Geophys. Res.*, *89*, 4389–4402, 1984.

Chung, W.-Y., and H. Kanamori, Subduction process of a fracture zone and aseismic ridges, *Geophys. J. R. astron. Soc.*, *54*, 221–240, 1978.

Crough, S. T., Approximate solutions for the formation of the lithosphere, *Phys. Earth Planet. Inter.*, *14*, 365–377, 1977.

Daessler, R., and D. A. Yuen, The effects of phase transition kinetics on subducting slabs, *Geophys. Res. Lett.*, *20*, 2603–2606, 1993.

Davies, G. F., and F. Pribac, Mesozoic seafloor subsidence and the Darwin Rise, past and present, in *The Mesozoic Pacific: Geology, Tectonics and Volcanism, American Geophysical Union Monograph 77 in memory of S. O. Schlanger*, edited by M. S. Pringle, W. W. Sager, W. V. Sliter and S. Stein, pp. 39–52, AGU, Washington, D.C., 1993.

Davies, J. H., and D. J. Stevenson, Physical model of the source regions of subduction zone volcanics, *J. Geophys. Res.*, *97*, 2037–2070, 1992.

DeMets, C., R. G. Gordon, D. F. Argus, and S. Stein, Current plate motions, *Geophys. J. Int.*, *101*, 425–478, 1990.

Elsasser, W. M., Convection and stress propagation in the upper mantle, in *The Application of Modern Physics to the Earth and Planetary Interiors*, edited by S. K. Runcorn, pp. 223–246, John Wiley, New York, 1969.

Elsasser, W. M., Seafloor spreading as thermal convection, *J. Geophys. Res.*, *76*, 1101–1112, 1971.

Engdahl, R., R. van der Hilst, and J. Berrocal, Tomographic imaging of subducted lithosphere beneath South America, *Geophys. Res. Lett.*, *22*, 2317–2320, 1995.

Engebretson, D., K. Kelly, H. Cashman, and M. Richards, 180 million years of subduction, *GSA Today*, *2*, 93–100, 1992.

Fischer, K. M., T. H. Jordan, and K. C. Creager, Seismic constraints on the morphology of deep slabs, *J. Geophys. Res.*, *93*, 4773–4784, 1988.

Fleitout, L., and M. P. Doin, Thermal evolution of the oceanic lithosphere: an alternative view (abstract), *Eos Trans. AGU, Fall meeting supplement*, 648, 1994.

Fukao, Y., M. Obayashi, and H. Inoue, Subducting slabs stagnant in the mantle transition zone, *J. Geophys. Res.*, *97*, 4809–4822, 1992.

Geller, C. A., J. K. Weissel, and R. N. Anderson, Heat transfer and intraplate deformation in the central Indian Ocean, *J. Geophys. Res.*, *88*, 1018–1032, 1983.

Goetze, C., and B. Evans, Stress and temperature in the bending lithosphere as constrained by experimental rock mechanics, *Geophys. J. R. astron. Soc.*, *59*, 463–478, 1979.

Green, H. W., II, and P. C. Burnley, Self-organizing mechanism for deep-focus earthquakes, *Nature*, *341*, 733–737, 1989.

Heestand, R. L., and S. T. Crough, The effect of hot spots on the oceanic age-depth relation, *J. Geophys. Res.*, *86*, 6107–6114, 1981.

Helffrich, G., S. Stein, and B. Wood, Subduction zone thermal structure and mineralogy and their relation to seismic wave reflections and conversions at the slab/mantle interface, *J. Geophys. Res.*, *94*, 753–763, 1989.

Hsui, A. T., and M. N. Toksöz, Evolution of thermal structure beneath a subduction zone, *Tectonophysics*, *60*, 43–60, 1979.

Isacks, B., J. Oliver, and L. R. Sykes, Seismology and the new global tectonics, *J. Geophys. Res.*, *73*, 5855–5899, 1968.

Ito, E., and T. Katsura, A temperature profile of the mantle transition zone, *Geophys. Res. Lett.*, *16*, 425–428, 1989.

Jarrard, R. D., Relations among subduction parameters, *Rev. Geophys.*, *24*, 217–284, 1986.

Jarvis, G. T., and W. R. Peltier, Convection models and geophysical observations, in *Mantle Convection*, edited by W. R. Peltier, pp. 479–594, Gordon and Breach, New York, 1989.

Johnson, H. P., and R. L. Carlson, The variation of sea floor depth with age: a test of existing models based on drilling results, *Geophys. Res. Lett.*, *19*, 1971–1974, 1992.

Kanamori, H., Seismological evidence for a lithospheric normal faulting, *Phys. Earth Planet. Inter.*, *4*, 289–300, 1971.

Kawakatsu, H., Insignificant isotropic component in the moment tensor of deep earthquakes, *Nature*, *351*, 50–53, 1991.

Kido, M., and T. Seno, Dynamic topography compared with residual depth anomalies in oceans and implications for age-depth curves, *Geophys. Res. Lett.*, *21*, 717–720, 1994.

Kikuchi, M., and H. Kanamori, The mechanism of the deep Bolivia earthquake of June 9, 1994, *Geophys. Res. Lett.*, *21*, 2341–2344, 1994.

Kincaid, C., Subduction dynamics, *Rev. Geophys., Supplement*, 401–412, 1995.

Kincaid, C., and P. Olson, Experimental study of subduction and slab migration, *J. Geophys. Res.*, *92*, 13,832–13,840, 1987.

Kirby, S. H., State of stress in the lithosphere: Inferences from the flow laws of olivine, *Pure and Applied Geophys.*, *115*, 245–258, 1977.

Kirby, S. H., Rheology of the lithosphere, *Rev. Geophys. Space Phys.*, *21*, 1458–1487, 1983.

Kirby, S. H., Localized polymorphic phase transitions in high-pressure faults and applications to the physical mechanism of deep earthquakes, *J. Geophys. Res.*, *92*, 13,789–13,800, 1987.

Kirby, S., Earthquakes and phase changes in subducting lithosphere, *Rev. Geophys., Supplement*, 287–297, 1995.

Kirby, S. H., W. B. Durham, and L. A. Stern, Mantle phase

changes and deep-earthquake faulting in subducting lithosphere, *Science, 252,* 216–225, 1991.

Kirby, S. H., E. A. Okal, and E. R. Engdahl, The great 9 June 1994 Bolivian deep earthquake, *Geophys. Res. Lett., 22,* 2233–2236, 1995.

Kirby, S. H., S. Stein, E. A. Okal, and D. Rubie, Deep earthquakes and metastable phase changes in subducting oceanic lithosphere, *Rev. Geop., 34,* 261–306, 1996a.

Kirby, S., E. R. Engdahl, and R. Denlinger, Intraslab earthquakes and arc volcanism, in *SUBCON: An Interdisciplinary Conference on the Subduction Process, Santa Catalina Island, CA, Geophys. Mono.* edited by G. Bebout, D. Scholl and S. Kirby, Am. Geophys. Un., Washington, D.C., 1996b.

Langseth, M. G., X. Le Pichon, and M. Ewing, Crustal structure of the mid-ocean ridges, 5, Heat flow through the Atlantic Ocean floor and convection currents, *J. Geophys. Res., 71,* 5321–5355, 1966.

Lundgren, P. R., and D. Giardini, Isolated deep earthquakes and the fate of subduction in the mantle, *J. Geophys. Res., 99,* 15,833–15,842, 1994.

McKenzie, D. P., Some remarks on heat flow and gravity anomalies, *J. Geophys. Res., 72,* 6261–6273, 1967.

McKenzie, D. P., Speculations on the consequences and causes of plate motions, *Geophys. J. R. astron. Soc., 18,* 1–32, 1969.

McKenzie, D. P., and M. J. Bickle, The volume and composition of melt generated by extension of the lithosphere, *J. Petrol., 29,* 625–679, 1988.

Meade, C., and R. Jeanloz, Deep-focus earthquakes and recycling of water into earth's mantle, *Science, 252,* 68–72, 1991.

Minear, J., and M. N. Toksöz, Thermal regime of a downgoing slab and new global tectonics, *J. Geophys. Res., 75,* 1379–1419, 1970.

Minster, J. B., and T. H. Jordan, Present-day plate motions, *J. Geophys. Res., 83,* 5331–5354, 1978.

Molnar, P., D. Freedman, and J. S. F. Shih, Lengths of intermediate and deep seismic zones and temperatures in downgoing slabs of lithosphere, *Geophys. J. R. astron. Soc., 56,* 41–54, 1979.

Nishimura, C., and D. Forsyth, Anisotropic structure of upper mantle in the Pacific, *Geophys Jour, 96,* 203–226, 1989.

Oxburgh, E. R., and E. M. Parmentier, Compositional and density stratification in oceanic lithosphere- causes and consequences, *J. Geol. Soc. Lond., 133,* 343–355, 1977.

Parmentier, E. M., and W. F. Haxby, Thermal stresses in the oceanic lithosphere: evidence from geoid anomalies at fracture zones, *J. Geophys. Res., 91,* 7193–7204, 1986.

Parsons, B., and D. P. McKenzie, Mantle convection and the thermal structure of the plates, *J. Geophys. Res., 83,* 4485–4496, 1978.

Parsons, B., and F. Richter, Mantle convection and oceanic lithosphere, in *Oceanic Lithosphere, (The Sea, vol. 7),* edited by C. Emiliani, pp. 73–117, Wiley-Interscience, New York, 1981.

Parsons, B., and J. Sclater, Variation of ocean floor bathymetry and heat flow with age, *J. Geophys. Res., 82,* 803–827, 1977.

Peacock, S. M., Blueschist-facies metamorphism, shear heating, and P-T-t paths in subduction shear zones, *J. Geophys. Res., 97,* 17,693–17,707, 1992.

Peacock, S. M., Thermal and petrologic structure of subduction zones, in *SUBCON: An Interdisciplinary Conference on the Subduction Process, Santa Catalina Island, CA, Geophys. Mono.* edited by G. Bebout, D. Scholl and S. Kirby, Am. Geophys. Un., Washington, D.C., 1996.

Phipps Morgan, J., and W. Smith, Flattening of the sea-floor depth age curve as a response to asthenospheric flow, *Nature, 359,* 524–527, 1992.

Phipps Morgan, J., and W. H. F. Smith, Correction to "Flattening of the sea-floor depth-age curve as a response to asthenospheric flow", *Nature, 371,* 83, 1994.

Pollack, H. N., S. J. Hurter, and J. R. Johnston, Heat loss from the earth's interior: analysis of the global data set, *Rev. Geophys., 31,* 267–280, 1993.

Raleigh, C. B., Tectonic implications of serpentinite weakening, *Geophys. J. R. astron. Soc., 14,* 113–118, 1967.

Richards, M. A., and D. C. Engebretson, Large-scale mantle convection and the history of subduction, *Nature, 355,* 437–440, 1992.

Richardson, W. P., S. Stein, C. A. Stein, and M. T. Zuber, Geoid data and thermal structure of the oceanic lithosphere, *Geophys. Res. Lett., 22,* 1913–1916, 1995.

Richter, F. M., Focal mechanisms and seismic energy release of deep and intermediate earthquakes in the Tonga-Kermadec region and their bearing on the depth extent of mantle flow, *J. Geophys. Res., 84,* 6783–6795, 1979.

Robinson, E. M., B. Parsons, and M. Driscoll, The effect of a shallow low-viscosity zone on the mantle flow, the geoid anomalies and geoid and depth-age relationships at fracture zones, *Geophys. J. R. astron. Soc., 93,* 25–43, 1988.

Rubie, D. C., and C. R. I. Ross, Kinetics of the olivine-spinel transformation in subducting lithosphere, *Phys. Earth Planet. Inter., 86,* 223–241, 1994.

Sandwell, D. T., Thermomechanical evolution of fracture zones, *J. Geophys. Res., 89,* 11,401–11,413, 1984.

Schubert, G., and D. L. Turcotte, One-dimensional model of shallow-mantle convection, *J. Geophys. Res., 77,* 945–951, 1972.

Schubert, G., D. A. Yuen, C. Froidevaux, L. Fleitout, and M. Souriau, Mantle circulation with partial shallow return flow, *J. Geophys. Res., 83,* 745–758, 1978.

Shoberg, T., C. Stein, and S. Stein, Constraints on lithospheric thermal structure for the Indian Ocean from depth and heat flow data, *Geophys. Res. Lett., 20,* 1095–1098, 1993.

Silver, P. G., S. L. Beck, T. C. Wallace, C. Meade, S. C. Meyers, D. E. James, and R. Kuehnel, Rupture characteristics of the deep Bolivian earthquake of 1994 and mechanism of deep-focus earthquakes, *Science, 268,* 69–73, 1995.

Sleep, N. H., Teleseismic P-wave transmission through slabs, *Bull. Seismol. Soc. Am., 63,* 1349–1373, 1973.

Sleep, N. H., Hotspots and mantle plumes, *Ann. Rev. Earth Planet. Sci., 20,* 19–43, 1992.

Sleep, N. H., and B. F. Windley, Archean plate tectonics: constraints and inferences, *J. Geol., 90,* 363–379, 1982.

Spakman, W., S. Stein, R. van der Hilst, and R. Wortel, Resolution experiments for NW Pacific subduction zone tomography, *Geophys. Res. Lett., 16,* 1097–1110, 1989.

Stein, C. A., and S. Stein, A model for the global variation in oceanic depth and heat flow with lithospheric age, *Nature, 359,* 123–129, 1992.

Stein, C. A., and S. Stein, Constraints on Pacific midplate swells from global depth-age and heat flow-age models, in *The Mesozoic Pacific, Geophys. Monogr. Ser. vol. 77,* edited by M. Pringle, W. W. Sager, W. Sliter and S. Stein, pp. 53–76, AGU, Washington, D. C., 1993.

Stein, C. A., and S. Stein, Constraints on hydrothermal heat flux through the oceanic lithosphere from global heat flow, *J. Geophys. Res.,* 3081–3095, 1994a.

Stein, C. A., and S. Stein, Comparison of plate and asthenospheric flow models for the thermal evolution of oceanic lithosphere, *Geophys. Res. Lett., 21,* 709–712, 1994b.

Stein, C. A., S. Stein, and A. M. Pelayo, Heat flow and hydrothermal circulation, in *Seafloor hydrothermal systems, Geophys. Mono. 91,* edited by S. Humphris, L. Mullineaux, R. Zierenberg and R. Thomson, pp. 425–445, 1995.

Stein, S., Intraplate seismicity on bathymetric features: The 1968 Emperor Trough earthquake, *J. Geophys. Res., 84,* 4763–4768, 1979.

Stein, S., Deep earthquakes: a fault too big?, *Science, 268,* 49–50, 1995.

Stein, S. and E. Okal, Seismicity and tectonics of the Ninetyeast Ridge area: evidence for internal deformation of the Indian plate, *J. Geophys. Res., 83,* 2233-2246, 1978.

Stein, S., J. F. Engeln, D. A. Wiens, K. Fujita, and R. C. Speed, Subduction seismicity and tectonics in the Lesser Antilles arc, *J. Geophys. Res., 87,* 8642–8664, 1982.

Stein, S., S. Cloetingh, N. Sleep and R. Wortel, Passive margin earthquakes, stresses, and rheology, in S. Gregerson and P. Basham (eds) *Earthquakes at North Atlantic Passive Margins,* 231-259, Kluwer, 1989.

Sung, C., and R. Burns, Kinetics of the olivine-spinel transition: Implications to deep-focus earthquake genesis, *Earth Planet. Sci. Lett., 32,* 165–170, 1976a.

Sung, C., and R. Burns, Kinetics of high-pressure phase transformations: implications for olivine-spinel transition in downgoing lithosphere, *Tectonophysics, 31,* 1–32, 1976b.

Tao, W. C., and R. J. O'Connell, Deformation of a weak subducted slab and variation of seismicity with depth, *Nature, 361,* 626–628, 1993.

Toksöz, M. N., J. W. Minear, and B. R. Julian, Temperature field and geophysical effects of a downgoing slab, *J. Geophys. Res., 76,* 1113–1138, 1971.

Turcotte, D. L., Flexure, *Advances in Geophysics, 21,* pp. 51–86, Academic Press, Inc., 1979.

Turcotte, D. L., and E. R. Oxburgh, Finite amplitude convective cells and continental drift, *J. Fluid Mech., 28,* 29–42, 1967.

Ulmer, P., V. Trommsdorf, and E. Ruesser, Experimental investigation into antigorite stability to 80 kilobars, *Mineralogical Mag., 58A,* 919–920, 1994.

van der Hilst, R., Complex morphology of the subducted lithosphere in the mantle beneath the Tonga trench, *Nature, 374,* 154–157, 1995.

van der Hilst, R., R. Engdahl, W. Spakman, and G. Nolet, Tomographic imaging of subducted lithosphere below northwest Pacific island arcs, *Nature, 353,* 37–43, 1991.

Verhoogen, J., *Energetics of the Earth,* National Academy of Sciences, Washington, D.C., 1980.

Vogt, P. R., A. Lowrie, D. R. Bracey, and R. N. Hey, *Subduction of aseismic oceanic ridges, Special Paper 172,* Geol. Soc. Amer., Boulder, CO, 1976.

Wessel, P., and W. Haxby, Thermal stress, differential subsidence, and flexure at oceanic fracture zones, *J. Geophys. Res., 95,* 375–391, 1990.

Wiens, D. A., and S. Stein, Age dependence of oceanic intraplate seismicity and implications for lithospheric evolution, *J. Geophys. Res., 88,* 6455–6468, 1983.

Wiens, D. A., J. J. McGuire, P. J. Shore, M. G. Bevis, K. Draunidalo, G. Prasad, and S. P. Helu, A deep earthquake aftershock sequence and implications for the rupture mechanism of deep earthquakes, *Nature, 372,* 540–543, 1994.

Wolery, T., and N. Sleep, Hydrothermal circulation and chemical flux at mid-ocean ridges, *J. Geol., 84,* 249–275, 1976.

Wortel, M. J. R., Seismicity and rheology of subducted slabs, *Nature, 296,* 553–556, 1982.

Wortel, M. J. R., and N. J. Vlaar, Subduction zone seismicity and the thermo-mechanical evolution of downgoing lithosphere, *PAGEOPH, 128,* 625–659, 1988.

Wysession, M. E., E. A. Okal, and K. L. Miller, Intraplate seismicity of the Pacific basin, *Pure Appl. Geophys., 135,* 261–359, 1991.

Wysession, M. E., J. Wilson, L. Bartko, and R. Sakata, Intraplate seismicity in the Atlantic Ocean basin: a teleseismic catalog, *Bull. Seismol. Soc. Am., 85,* 755-774, 1995.

C. A. Stein, Department of Geological Sciences (m/c 186), University of Illinois at Chicago, 845 W. Taylor St., Chicago, IL 60607-7059.

S. Stein, Department of Geological Sciences, Northwestern University, Evanston, IL 60208.

Geochemical Fluxes During Seafloor Alteration of the Basaltic Upper Oceanic Crust: DSDP Sites 417 and 418

Hubert Staudigel[1], Terry Plank[2], Bill White[3], Hans - Ulrich Schmincke[4]

Seafloor alteration of the basaltic upper oceanic crust provides one of the major geochemical pathways between the mantle, the ocean/atmosphere and subduction zone regimes. Yet, no reliable mass balances are available, in large part because of the extremely heterogeneous distribution of altered materials in the oceanic crust but also because of the limited availability of high recovery drill cores. In this paper, we document the feasibility of determining the bulk altered and fresh composition of the oceanic crust on a 10-500 m length scale, from a region in the western Atlantic Ocean (DSDP/ODP Sites 417-418). Unaltered compositions were obtained from glass and phenocryst data and altered compositions were determined through analysis of composite samples. Most of the alteration-related chemical inventory resides preferentially in the upper oceanic crust and in highly permeable volcaniclastics. Most major elements (Si, Al, Mg, Ca, and Na) and many trace elements (Sr, Ba, LREE's) experience substantial large scale redistribution, but fluxes are relatively low. Overall, 12 wt % are added to the crust, mostly H_2O, CO_2, and K, but the distribution varies widely. High field strength elements, Th, Ti and Fe remain essentially immobile during low temperature alteration, while most other elements are affected to some degree. While the total fluxes are relatively small, the re-distribution of alteration - sensitive elements in the ocean crust is much larger, even on length scales exceeding 100m. The bulk composition of the upper 500m at Sites 417/418 can be used to constrain the impact of ocean crust subduction on element recycling to volcanic arcs. Flux balances indicate that the altered domains within the upper basaltic crust may contribute a very large proportion of some element fluxes recycled to the arc (H_2O, CO_2, K, Rb, U), while other element fluxes require additional contributions from sediments and deeper oceanic crust.

1. INTRODUCTION

Generation, alteration and recycling of oceanic crust represents the largest chemical cycle in the dynamic solid earth

[1]Faculty of Earth Sciences, Free University Amsterdam, Amsterdam, Netherlands
[2]Department of Geology, University of Kansas, Lawrence, Kansas
[3]Dept. Geological Sciences, Cornell University, Ithaca, New York
[4]Abteilung Vulkanologie und Petrologie, Geomar Forschungszentrum, Kiel, Germany

Subduction: Top to Bottom
Geophysical Monograph 96
Copyright 1996 by the American Geophysical Union

whereby approximately 5×10^{16} grams of basalt are generated and recycled per year. This cycle provides a pathway for mantle components into the hydrosphere, and for sea water derived elements into subduction zones and the mantle. Chemical fluxes in these pathways are extremely poorly constrained, including the extent of high temperature alteration at ridges, as well as off-axis low temperature chemical exchange. This lack of data provides a major stumbling block in our understanding of earth chemical dynamics.

The alteration of the oceanic crust on the seafloor, and its subsequent metamorphism and chemical losses during subduction have a major impact on the loci and composition of arc magmatism. Several recent studies have pointed to the altered basaltic crust specifically as a major source of elements recycled to volcanic arcs during subduc-tion (e.g. H_2O [Peacock, 1990; Plank, 1994]; B [Ishikawa and Nakamura, 1993]; Pb [Miller et al, 1992; Peucker-Ehrenbrinck et al., 1995]). Despite the recognition of the altered oceanic crust as an important player in subduction zone recycling, the lack of detailed studies of the bulk composition of different basement sections has thwarted efforts to quantitatively mass balance fluxes through subduction zones. Thus, a major goal of the current study is to provide a comprehensive geochemcial data set for an upper oceanic crust section to use in subduction recycling studies. A few general implications of the compositional effects of seafloor alteration on the subducted sources of arc magmas is discussed in the final section of this paper.

The bulk effects of sea water-ocean crust exchange may be determined in two ways, through direct monitoring of hydrothermal activity on the ocean floor, or through a comparison of altered and fresh oceanic crust. In this study, we are following the latter approach and attempt to determine the bulk composition of the oceanic crust using composite samples that were combined to be representative on scale lengths ranging from tens to hundreds of meters. This scale length is manageable geochemically and it represents a lower size limit for distinct compositional domains that may impart a chemical imprint on partial melts or metasomatic fluids.

We carried out this investigation at Deep Sea Drilling Sites 417A, 417D and 418A, near Bermuda Rise in the West-Central Atlantic, because they are uniquely suited for this type of investigation:

- Up to 500m penetration into the basaltic basement, amongst three closely spaced sites provides a detailed and representative view of ocean crust in this region.
- The high crustal age (magnetic Anomaly MO, 118 Ma) suggests that all major hydrothermal and low temperature exchange processes have been completed.
- The abundance of fresh glass allows for an estimate of the unaltered rock composition
- High recovery rates (70%) allow quantitative assessment of the alteration inventory that is distributed very heterogeneously, and
- A representative spread in alteration behaviour can be found at these sites, including extremely oxidative alteration in an abyssal hill at Site 417A as well as more moderate alteration at Sites 417D and 418A. This allows for an assessment of diversity as well as consistency in the composition of altered oceanic crust.

In this study, we wish to demonstrate the feasibility of determining bulk composition of the oceanic crust, its large scale diversity, as well as fluxes between seawater and basalt. However, this is possible only through integration of structural and chemical techniques, and it requires deep ocean drill holes with high recovery rates. Given the fact that the required rotary drilling techniques have been available for decades, it is disappointing how few of such holes exist.

2. METHODS AND ANALYTICAL TECHNIQUES

Estimation of large - scale chemical and isotopic compositions requires a detailed geological analysis for representative sampling as well as an analytical laboratory effort, and both of them have their own characteristic uncertainties. We used composites in order to characterize the large-scale composition of the oceanic extrusives at DSDP/ODP Sites 417A, 417D and 418A. For this, we targeted a series of representative, evenly distributed core sections with particularly favourable recovery rates (80-100%) and the occurrence of all major lithological types. In these core sections, we selected typical samples from massive flows, pillow lavas, and volcaniclastic rocks, in a range of characteristic alteration conditions. For each sample we estimated its characteristic abundance in the core, using our own analysis and and published petrographic descriptions of the (entire) drill cores [Donnelly et al., 1979]. In these estimates, we balanced representative types of volcaniclastics, and to took into account the typical size variations of pillows and massive flows. The latter is particularly important for the volume fraction of the typically highly altered pillow margins which varies significantly with pillow tube diameter. Special care was also taken to include vein materials, inter- pillow hyaloclastites, pillow fragment breccias, massive flow materials, and pillow interiors and margins in their average proportions at these sites and in other submarine extrusive rocks. These estimates were used in determining the mixing proportions of individual samples for the volcaniclastic (VCL) and submarine extrusive (FLO) composites (Appendix 1). These "lithology" composites were prepared for three depth intervals including 20% of samples from Site 417A and 40 % each from Sites 417D and 418A. In addition to these lithology composites, we also prepared composites to represent the depth-related chemical variations. In these composites we kept the fractions of different lithologies constant (volcaniclastics, 6%, massive flows, 20% and pillows, 74%; following Robinson et al. [1979]), giving us an estimate of downhole chemical variation that is insensitive to lithology variations.

Sample powders were prepared for each individual sample using clean laboratory techniques. Surfaces were cleaned with SiC sandpaper, samples washed in distilled water and all subsequent sample contact was limited to materials that were cleaned prior to use. These materials include agate grinding vessels, polyethylene, and Teflon labware. However, it has to be pointed out that such cleaning procedures cannot remove pervasive contamination as it may be introduced, e.g., from drilling fluids into porous cores or through sorption to clays. Individual sample powders were weighed

Table 1 Composites Major Element Data

sample number	depth	SiO$_2$	TiO$_2$	Al$_2$O$_3$	Fe$_2$O$_3$	FeO	MnO	MgO	CaO	NaO$_2$	K$_2$O	P$_2$O$_5$	H$_2$O	CO$_2$	sum
417/418 super	-289.30	45.80	1.18	15.53	5.60	3.98	0.17	6.66	12.88	2.07	0.56	0.11	2.68	2.95	100.17
417A 24	118.33	45.20	1.49	16.21	10.96	1.13	0.48	3.47	7.67	1.42	4.38	0.22	4.93	3.41	100.97
417A 32	45.66	45.90	1.45	16.56	8.13	3.05	0.16	5.32	9.82	2.35	1.63	0.14	4.48	1.54	100.53
417A 44-46	-63.48	41.60	1.08	14.53	5.46	3.05	0.16	5.53	14.38	2.71	1.09	0.16	3.98	6.47	100.20
417D 22	-17.84	43.10	1.18	13.92	5.28	3.54	0.21	5.82	15.46	1.85	0.76	0.10	2.11	6.07	99.40
417D 39	-124.85	45.10	1.29	15.74	6.04	3.41	0.13	5.66	13.23	1.80	0.83	0.10	2.69	3.89	99.91
417D 59	-288.20	46.50	1.21	16.57	5.36	4.43	0.18	5.99	13.78	1.97	0.30	0.11	1.44	2.29	100.13
418A 15	-5.09	42.20	0.97	14.63	5.03	2.80	0.19	6.51	15.55	2.10	0.61	0.09	2.24	6.70	99.62
418A 40	-187.04	45.20	0.96	16.93	5.00	3.34	0.13	5.94	12.04	2.99	0.84	0.07	3.85	2.44	99.73
418A 73-75	-463.51	48.10	1.21	15.32	6.03	4.58	0.15	8.14	10.98	1.92	0.17	0.10	2.73	0.35	99.78
418A 86	-541.93	47.40	1.22	15.57	5.25	4.55	0.17	7.92	12.35	2.01	0.13	0.10	2.23	1.36	100.26
VCL top	-17.67	27.20	0.64	8.28	6.63	0.52	0.50	3.61	26.57	0.90	1.82	0.07	3.41	19.73	99.88
VCL 100	-190.52	43.80	1.00	13.99	7.60	2.26	0.10	6.39	10.43	2.38	2.06	0.08	4.78	4.84	99.71
VCL 300	-518.54	48.30	1.17	15.42	8.10	2.30	0.09	9.20	8.88	2.09	0.49	0.09	3.73	0.70	100.56
FLO top	-18.82	43.90	1.18	15.01	6.04	3.10	0.19	5.89	13.28	2.05	1.11	0.11	2.87	4.71	99.44
FLO 100	-189.92	43.10	1.08	15.29	5.07	3.40	0.15	5.45	14.41	2.65	0.78	0.12	3.32	5.03	99.85
FLO 300	-511.89	48.10	1.20	15.33	5.54	4.48	0.17	8.00	12.08	1.89	0.11	0.10	1.97	1.03	100.0

Chemical Analyses done by H. Niephaus at the Ruhr University Bochum using techniques by Flower et al., [1979]. Note that some of these data were previously reported with some typographic errors for CaO and CO$_2$ abundances by Staudigel et al [1989] and Spivack and Staudigel [1994] (note that the calculations and conclusions in these papers are not affected by these errors).

Table 2 Trace Elements of Composite Samples [in ppm]

Sample #	Y	Zr	Nb	Cs	Ba	La	Ce	Pr	Nd	Sm	Eu	Gd	Tb	Dy	Ho	Er	Yb	Lu	Hf	Ta	Th	U
SUPER	26.9	66.5	1.22	0.153	22.6	1.84	6.01	1.17	6.62	2.50	0.91	3.65	0.713	4.40	0.98	2.77	2.69	0.425	1.92	0.097	0.070	0.300
417A-24	37.5	90.1	1.54	0.788	109.3	3.69	8.55	1.78	9.71	3.44	1.22	4.97	0.957	5.94	1.28	3.61	3.45	0.540	2.47	0.125	0.084	0.259
417A-32	32.2	86.8	1.48	0.441	46.7	2.24	7.37	1.40	7.86	2.92	1.07	4.32	0.859	5.30	1.16	3.34	3.28	0.496	2.35	0.109	0.085	0.149
417A-44	26.9	63.3	1.08	0.345	14.1	2.30	6.12	1.28	7.11	2.56	0.94	3.78	0.739	4.56	0.99	2.81	2.74	0.423	1.67	0.078	0.099	0.102
417D-22	28.1	72.9	1.24	0.199	14.7	1.95	6.33	1.26	6.98	2.64	0.98	3.89	0.764	4.71	1.02	2.91	2.85	0.440	1.92	0.092	0.071	0.91
417D-39	27.1	75.9	1.25	0.340	10.2	1.86	6.02	1.15	6.50	2.45	0.87	3.55	0.711	4.40	0.97	2.77	2.80	0.444	2.10	0.108	0.068	0.462
417D-59	28.0	73.4	1.09	0.120	80.1	1.84	6.00	1.20	6.73	2.60	0.98	3.91	0.766	4.81	1.05	3.04	2.94	0.458	1.95	0.088	0.103	0.089
418A-15	22.2	53.2	1.18	0.059	20.6	1.51	4.98	0.97	5.46	2.04	0.84	3.15	0.623	3.76	0.82	2.33	2.31	0.359	1.50	0.091	0.058	0.611
418A-40	24.6	52.9	0.74	0.188	5.3	1.29	4.49	0.90	5.26	2.08	0.81	3.20	0.628	3.94	0.87	2.50	2.45	0.383	1.48	0.062	0.042	0.233
418A-73	25.5	66.8	1.32	0.069	16.4	1.84	6.36	1.22	6.86	2.51	0.95	3.75	0.745	4.58	1.00	2.81	2.73	0.433	1.91	0.105	0.071	0.182
418A-86	28.2	63.4	1.30	0.036	28.3	1.84	6.36	1.22	6.58	2.51	0.94	3.72	0.733	4.61	1.01	2.88	2.87	0.441	1.91	0.101	0.069	0.304
VCL-top	19.1	39.8	0.66	0.455	16.6	1.55	4.29	0.79	4.13	1.50	0.54	2.23	0.433	2.73	0.62	1.78	1.88	0.301	1.07	0.055	0.038	0.659
VCL-100	22.0	57.1	0.98	0.599	16.9	1.85	4.97	0.99	5.40	2.00	0.72	2.98	0.583	3.66	0.79	2.28	2.34	0.369	1.64	0.079	0.064	0.536
VCL-300	22.6	63.6	1.34	0.082	25.0	1.67	5.69	1.11	6.13	2.31	0.88	3.45	0.681	4.17	0.92	2.61	2.62	0.403	1.85	0.103	0.070	0.292
FLO-Top	28.0	70.3	1.32	0.219	31.5	2.05	6.46	1.26	7.05	2.61	0.98	3.87	0.767	4.73	1.04	2.95	2.87	0.444	1.98	0.097	0.079	0.635
FLO-100	27.5	64.4	1.01	0.247	12.1	1.90	5.71	1.16	6.52	2.45	0.90	3.66	0.735	4.49	0.99	2.84	2.77	0.433	1.76	0.075	0.084	0.197
FLO-300	27.6	67.6	1.30	0.037	25.7	1.81	6.28	1.19	6.82	2.51	0.93	3.73	0.730	4.59	0.99	2.81	2.77	0.440	1.95	0.107	0.070	0.204

Analyses by ICPMS by T. Plank at Cornell University

and mixed into composites using the proportions in Appendix 1. Each composite sample was homogenized in one last grinding step.

Composite samples were analyzed for their major element composition using a combination of X-ray fluorescence and rapid wet chemical techniques at the Ruhr University Bochum (Table 1 and 4; analyst H. Niephaus using techniques of Flower et al. [1979].). A subset of these data were previously reported [Spivack and Staudigel, 1994; Staudigel et al., 1995]. Trace elements were determined by ICP-MS at Cornell University using techniques of [Cheatham et al., 1992] (Table 2. and 3). These data include re-determination of earlier ICP-MS data [Staudigel et al., 1995] (REE, U, Th) and XRF data from Bochum (Rb, Sr). Agreement between the different generation ICP-MS data is very good, even though we prefer the newer data set because of better instrument performance as the result of an upgraded interface. Significant discrepancies were found only between ICP-MS and XRF Rb and Sr data (Table 3), resulting in significant differences in our calculation of initial $^{87}Sr/^{86}Sr$

Table 3 Isotopic Composition of Composite Samples

Composite	B [ppm]	$\delta^{11}B$	$\delta^{13}C$	$\delta^{18}O$	Rb [ppm]	Sr [ppm]	$^{87}Sr/^{86}Sr$	$^{143}Nd/^{144}Nd$
417/418 super	26.2	0.8	1.06	9.96	9.58	115	0.704575	0.513077
417A 24	104	1.6	0.07	18.05	57.51	96	0.707212	0.513006
417A 32	52	0.6	2.24	13.2	26.65	135	0.705879	0.513081
417A 44-46	69	2.4	1.81	13.65	19.06	141	0.705793	0.513002
417D 22	26	-1.7	1.32	11.21	16.28	104	0.704697	0.513061
417D 39	40	3.1	1.94	11.72	17.68	113	0.705058	0.513078
417D 59	25	-2.5	n.d.	7.87	7.61	104	0.704689	0.513005
418A 15	30	2.3	1.01	11.3	6.99	116	0.704359	0.513078
418A 40	23	3.8	0.66	10.69	10.38	176	0.705554	0.513082
418A 73-75	11.8	1.2	n.d.	8.8	3.25	111	0.703987	0.513101
418A 86	7.2	0.4	n.d.	8.27	2.21	102	0.703744	0.513087
VCL top	59	0.3	1.6	19.17	35.27	89	0.707437	0.513023
VCL 100	64	2.6	1.14	15.35	34.29	166	0.70681	0.513027
VCL 300	16	5.4	n.d.	11.6	5.30	118	0.70476	0.513083
FLO top	40	2	0.98	12	16.29	113	0.704873	0.513072
FLO 100	34	1.6	1.7	10.7	13.52	141	0.705336	0.513043
FLO 300	10.2	0.5	n.d.	7.72	2.20	108	0.703636	0.513069*

^a from Staudigel et al. [1995]; Smith et al. [1995]

Table 4 Major Elements of Volcaniclastics From DSDP Sites 417 and 418A [individual samples, wt %]

sample number	SiO_2	TiO_2	Al_2O_3	Fe_2O_3	FeO	MnO	MgO	CaO	NaO_2	K_2O	P_2O_5	H_2O	CO_2	sum
417A 24-2 52-54	42.10	0.98	13.21	9.67	0.43	1.66	3.07	10.29	1.24	3.64	0.03	5.03	9.39	100.74
417A 24-2 80-82	41.60	1.45	13.91	11.74	0.48	0.41	2.29	9.37	0.96	5.81	0.10	5.00	7.64	100.76
417A 32-4 114-116	49.30	1.45	18.07	11.67	0.57	0.05	4.09	2.11	1.47	4.26	0.01	7.30	0.37	100.72
417A 32-2 48-50	48.60	1.29	16.76	11.44	0.55	0.05	5.63	1.97	1.29	3.97	0.02	8.72	0.60	100.89
417A 46-2 15-17	38.40	1.09	14.18	7.25	1.83	0.18	4.12	13.87	3.31	1.66	0.16	5.75	8.85	100.65
417D 22-1 5-7	15.80	0.03	2.42	3.80	1.27	2.04	2.02	35.90	0.32	1.97	0.01	2.07	32.05	99.70
417D 26-1 9-12	40.30	0.82	10.07	7.39	3.18	0.20	9.00	13.80	1.77	1.07	0.13	3.23	8.51	99.47
417D 27-4 76-79	36.30	0.90	11.46	5.34	2.32	0.13	4.60	20.27	1.58	1.21	0.08	3.10	13.53	100.82
417D 39-1 30-40	47.80	1.31	15.86	8.01	2.47	0.08	6.61	8.22	1.87	1.47	0.05	4.80	2.05	100.60
417D 59-3 51-53	42.50	1.02	11.81	9.25	1.82	0.07	6.14	13.51	1.51	1.87	0.07	3.58	7.69	100.84
417D 60-5 97-100	41.10	1.05	13.10	7.11	1.87	0.12	6.74	16.16	1.61	0.53	0.07	3.38	8.03	100.87
418A 15-2 140-144 (A)	19.10	0.39	5.66	3.95	1.17	0.37	2.93	34.55	1.02	0.95	0.07	1.79	27.84	99.79
418A 15-2 140-144 (B)	19.00	0.39	5.62	3.87	1.19	0.36	2.94	32.93	1.02	0.94	0.07	1.68	29.16	99.17
418A 40-2 15-17	48.00	1.17	16.42	6.84	3.51	0.06	6.62	2.93	4.06	3.37	0.10	6.40	0.48	99.96
418A 40-2 52-56	47.10	1.05	15.54	5.71	3.44	0.12	5.12	9.93	3.14	2.50	0.11	3.21	2.25	99.22
418A 41-2 20-24	45.40	0.81	15.28	7.24	2.64	0.07	9.45	5.93	2.24	2.70	0.04	6.47	2.63	100.90
418A 41-2 92-96	48.30	0.89	16.29	6.18	3.57	0.12	6.93	9.67	2.78	1.52	0.07	3.57	0.89	100.78
418A 41-2 131-136	37.00	0.60	10.27	6.16	2.22	0.10	5.53	17.44	1.76	2.84	0.07	3.48	11.95	99.42
418A 42-1 50-54	46.60	0.75	13.26	9.33	2.36	0.08	7.48	7.10	2.95	2.66	0.06	4.88	3.20	100.71
418A 75-2 68-70	47.80	1.14	14.66	9.24	3.46	0.11	10.34	7.93	1.64	0.25	0.09	3.60	0.58	100.84
418A 75-5 42-45	47.70	1.19	14.49	8.40	3.87	0.12	9.38	7.98	1.52	0.46	0.07	4.87	0.57	100.62
418A 86-1 99-102	50.30	1.09	13.78	7.02	3.03	0.07	10.85	7.22	2.87	0.78	0.09	3.04	0.72	100.86
418A 86-1 43-45	50.00	1.10	13.68	6.98	3.05	0.07	10.84	7.18	2.85	0.78	0.09	3.38	0.72	100.72
418A 86-3 27-30	48.80	1.10	17.33	5.53	3.32	0.09	7.71	10.98	2.08	0.22	0.08	2.68	0.41	100.33

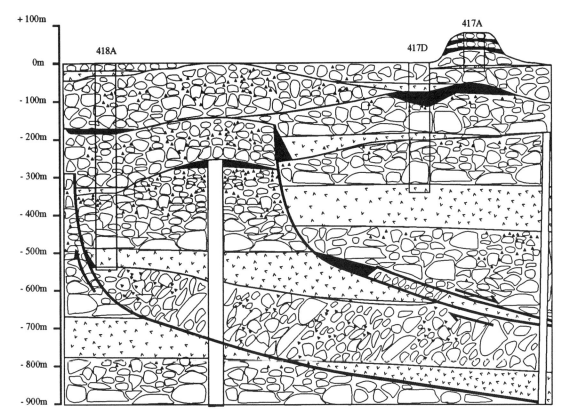

Fig. 1. Hypthetical lithological section through the oceanic crust at DSDP Sites 417A, 417D, and 418A to illustrate the extremely uneven distribution of lithologies with distinct alteration characteristics. Volcanic cycles typically begin with the deposition of massiv flows ("v" pattern), followed by pillows of decreasing diameter (pillow-shaped pattern). Towards the top of the pillow units, the number of brecciated particles increases, and the caps of pillow volcanoes are ofte entirely made of volcaniclastics (black pattern). Hyaloclastites often fill the pillow interstices (not shown). Amagmatic extension results in the formation of listric faults. Exposed fault scarps often also result in debris accumulations at their base. Note that highly altered volcaniclastics are distributed extremely irregular and they are slightly more common in abyssal hills.

ratios, and, thus, the mixing proportions between sea water and basalt Sr relative to Staudigel et al. [1995]. The newer ICP-MS analyses also provide data for Ba, Cs, Nb, Hf, Zr and Ta.

3. GEOCHEMICAL ALTERATION AS A FUNCTION OF DEPTH

The distribution of the sea water-derived chemical inventory in the oceanic crust is largely controlled by porosity and permeability as it is formed during volcanic and tectonic processes. For this reason, we have reviewed the lithological logs published for these sites [Donnelly et al., 1979a] and provided one possible structural scenario (Figure 1). Even though our structural interpretation in Figure 1 is not unique, it does serve to illustrate how the volcanic tectonic evolution of mid-ocean ridge volcanoes can influence the distribution of alteration. Oceanic crust is build in characteristic volcanic - tectonic cycles (e.g. [Schmincke and Bednarz, 1990; Robinson et al., 1979; Staudigel and Schmincke, 1985]). Each volcanic cycle begins with massive flows or very thick pillows (20%), continues with pillow lavas of decreasing diameter (74%), and ends with volcaniclastics (6%). This overall lithological variation is also characteristic for ophiolites and other deep water pillow lava sequences. After a relatively short-lived volcanic period, there may be an extended period of seafloor alteration of the volcanic surface and in zones of hydrothermal upwelling. Tectonic activity during amagma-tic spreading fragments and tilts the oceanic crust along listric faults [e.g. Karson, 1987; Varga, 1991], which may cause additional permeability along the fault itself and along tectonically brecciated zones. Furthermore, the surface expression of extensional tectonics causes fault scarlp that also are commonly associated with breccias. Subsequent to such a phase of amagmatic spreading, a new volcanic pulse may begin another cycle. Site 417D includes almost three complete cycles, while 418A includes four complete cycles and the top of a fifth (Figure 1). Overall, this discussion shows that permeability is very unevenly distributed and enhanced in breccia zones, fissures and possibly along listric fault systems. This causes a similarly complex, three-dimensional distribution of sea water-derived components in the upper oceanic crust and it greatly complicates large-scale chemical analysis and mass balance considerations. For these reasons, drill hole data from the oceanic crust cannot be simply extrapolated in a horizontally continuous geometry, not for distinct lithologies, nor for formation properties such as permeability.

The seafloor-alteration related chemical inventory in the oceanic crust is added in the form of vein and vug fillings and as replacement of igneous phases. The most common secondary mineral phases in the upper oceanic extrusives are clays, carbonates, and zeolites. Addition of these phases to mid-ocean ridge basalts is chemically correlated with enhanced abundances of H_2O and CO_2 and increases in $\delta^{18}O$ and $^{87}Sr/^{86}Sr_{init}$ isotopic ratios of bulk rock chemical data. In order to study the distribution of alteration effects in the oceanic crust, we have reviewed previously published major element bulk rock data from the Initial Reports of Deep Sea Drilling Legs 51-53 [Donnelly et al., 1979a; Donnelly et al., 1979b], and compared them to our data in a series of down hole plots (Figure 2 a-d).

The CO_2 and H_2O abundances of individual samples vary substantially, whereby the highest values are found generally in the upper 300m of the oceanic crust (Figure 2 a and b, up to 28 mole% H_2O^+ and 38 mole% CO_2 (some of the highest three values of Table 2 are not shown for illustrative purposes). The latter carbonate content indicates about 75% modal carbonate. Such mixtures of carbonates and basalt are commonly found in ophiolites (i.e. "ophicalcite"), even though they generally are not very abundant (<<5%). The total range of observed CO_2 and H_2O abundances in highly altered rocks decreases with depth. Due to the negligible CO_2 and H_2O inventory in fresh MORB, their abundances in ocean crustal rocks are essentially all derived from sea water and, thus, their abundance in rocks cumulatively indicates the extent of seafloor alteration. The maximum extent of alteration appears to decrease with depth, without any apparent relationship to the individual volcanic cycles.

Despite the very large scatter of CO_2 and H_2O in highly altered samples, there is a very large number of individual samples that form a relatively narrow band of samples. This group of minimally altered samples displays approximately 0-3 mole % CO_2 throughout the drill hole, whereby water decreases from about 2-7 mole% in the upper 400m to 1-5 mole% water in the lowermost 418A section. These samples provide a minimum baseline for the extent of pervasive alteration. However, it has to be noted that most of the analyses used in Figures 1 a-d were taken from petrogenetic studies where an attempt is made to analyze only the least altered samples. (e.g. [Flower et al., 1979]). Thus, a linear average of the lower range in CO_2 and H_2O probably gives a good estimate of the minimum degree of pervasive alteration, without the addition of obvious vein materials.

As it might be expected from the composite recipes, they show higher CO_2 and H_2O inventories than the minimally altered samples (Figure 2 a, b), confirming the composites as mixtures of less altered bulk rocks with the addition of vein materials. The CO_2 contents of the depth composites show a rather systematic decrease with depth, while the H_2O displays much more scatter. The uppermost 417D and 418A composites are actually lower in H_2O than samples at greater depth. The lowermost composite at Site 417D (-59) displays extremely low water contents while the intermediate 418A - 40 has rather high water contents. These systematics correlate with rather minor alteration in the deeper pillows and massive flows at the bottom of 417D, and breccia zones at 150 and 180 m depth at 417D and 418A, respectively. It is interesting to note that this lithological difference is not apparent in the down-hole variation of CO_2. Overall, H_2O in 417A extends the 417D and 418A downhole trend, displaying a very high degree of hydration, while the CO_2 at this highly altered site is relatively low when compared to the top of 417D and 418A. Thus, the highly oxidizing alteration environment at 417A is characterized by only minor carbonate precipitation.

The addition of vein minerals to the oceanic crust as reflected in the elevated CO_2 and H_2O abundances may also be seen in the SiO_2 and MgO abundances (Figure 2 c and d). The upper crust displays scatter towards lower Si contents which is consistent with a dilution from the addition of carbonates and hydrous secondary phases (Figure 2b). Overall, the MgO of bulk rocks does not show very large variation. In the upper portion of the oceanic crust, a large number of samples have very low MgO, and the diversity is relatively high. Composites also display low MgO near the top and show an increase with depth, they cluster to the left of the bulk of the individual rock data in the upper portion while they plot to the right in the lower crust. This possibly indicates addition of Mg-poor secondary phases in the upper portion, while adding Mg-bearing phases in the lower part of Site 418A.

Isotopic tracers of ocean floor alteration also show some systematic down-hole trends, in particular for $^{87}Sr/^{86}Sr_{init}$

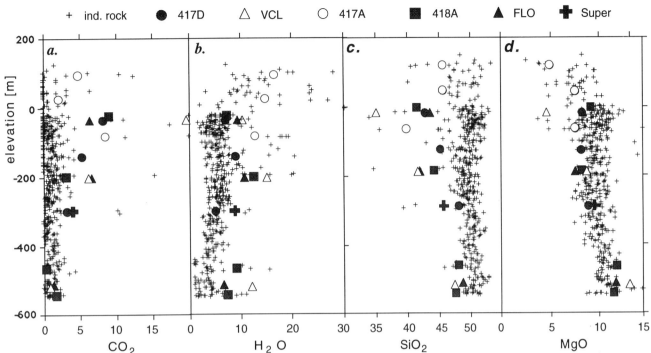

Fig. 2. CO_2 (a), H_2O (b), SiO_2 (c) and MgO (d), in mole%, versus elevation for composite samples and individual samples from the Initial Reports of Deep Sea Drilling [Donnelly et al 1979 a, b]. Symbols are explained above panels a and b; elevation is given in meters above or below the abyssal plain at 417D and 418A, analagous to the scale in Figure 1. Composite sample depths are calculated as weighted depths of all contributing samples. The tight cluster of individual rock samples at low CO_2 and H_2O contents indicates the pervasive minimum alteration, whereas composites indicate the bulk crustal composition including veins and vesicle fillings. SiO_2 and MgO of composites display lower abundances consistent with the addition of low Mg clays, and carbonates.

and $\delta^{18}O$ isotopic ratios (Figure 3 a-c). The $^{87}Sr/^{86}Sr_{init}$ directly reflects a mixing ratio between added sea water Sr ($^{87}Sr/^{86}Sr_{init}$ = 0.70735) and remaining basalt Sr (0.70295). Quite similar to the variation in H_2O, the $^{87}Sr/^{86}Sr_{init}$ of 417D and 418A depth composites displays relatively low values near the very top, increasing towards higher values near the major breccia zones in 417D and 418A, and then decreasing with depth to the lowermost values in the deepest portions of 418A. The overall similarity between 417D and 418A composites suggests that these trends are representative for this region. 417A and VCL composites display relatively high $^{87}Sr/^{86}Sr_{init}$ values, while the FLO composites are generally relatively little altered and the VCL's have substantially higher $^{87}Sr/^{86}Sr_{init}$. The difference between VCL's and FLO's decreases with depth. It is interesting to note that the water content of the top VCL's and FLO's is nearly identical, whereas the $^{87}Sr/^{86}Sr_{init}$ displays large differences. Both rock types may have interacted with similar quantities of water, but the water in the less permeable submarine FLO's must have already carried Sr with a significant basaltic component.

The $\delta^{18}O$ shows many similar patterns as the $^{87}Sr/^{86}Sr_{init}$, such as the consistency between 417D and 418A depth composites and the differences between VCL's and FLO's, but the dominant feature is a steady decrease in $\delta^{18}O$ with depth. This decrease was explained by decreasing intensity of alteration with depth as well as an increase in alteration - temperature [Staudigel et al., 1995]. The down-hole trend observed in these data is consistent with the trends observed in the Ibra Section of the Oman Ophiolite [Gregory and Taylor, 1981], even though this, more detailed sample set is slightly off-set towards lower values. The rather systematic variation of $\delta^{18}O$ shows that the estimate for the Super Composite is relatively robust, and it allows for extrapolation to greater depth.

The variation of $\delta^{11}B$ is much less systematic than other alteration parameters [Smith, et. al., in press]. The upper part of the crust displays scatter, volcaniclastics systematically increase in $\delta^{11}B$ with depth, while flows show a steady decrease with depth. Samples from 417D indicate ^{11}B depletion, while most other samples show enrichments. Overall, it appears that $\delta^{11}B$ displays a modest increase during alteration, although there is clearly no simple depth variation.

The distribution of H_2O, CO_2, $^{87}Sr/^{86}Sr_{init}$, $\delta^{18}O$ and $\delta^{11}B$ suggests that alteration in the uppermost 300m of the oceanic crust is enhanced and more heterogeneous, whereby

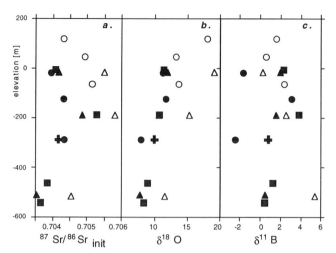

Fig. 3. Down-hole variation in initial $^{87}Sr/^{86}Sr_{init}$(a), $\delta^{18}O$ (b) $\delta^{11}B$ (c) of composite samples. The top 200 m of the oceanic crust is characterized by the largest extent of seafloor alteration and diversity in isotopic composition. Symbols and elevation scale as in Figure 2.

VCL's tend to be more altered than FLO's. This behavior was explained with an enhanced water transport, in particular in the shallow breccias and VCL's [Staudigel et al., 1995]. Despite the uneven distribution of the alteration-related ocean crust inventory, the Super composite defines an acceptable mean for the top 500m of the oceanic crust at these sites.

4. DISCUSSION

4.1 Processes of Bulk Seafloor Alteration

The most important seafloor alteration processes include the breakdown and dissolution of primary igneous phases and the deposition of secondary phases in voids and in fractures. Secondary phases are added to the crust but they derive much of their chemical inventory from the breakdown of basaltic material, in particular from glass but also olivine and plagioclase. Clinopyroxene is the only major phase in MORB that tends to be relatively stable during low temperature basalt alteration. The most common secondary phase at low temperatures are palagonite, clays, carbonates and zeolites, in order of typical abundances [Alt et al., 1992; Alt et al., 1986; Andrews, 1977; Humphris and Thompson, 1978].

The alteration of basaltic glass to palagonite involves uptake of H_2O and the alkalis K, Rb, Cs, with characteristically low K/Rb and Rb/Cs ratios [Staudigel and Hart, 1983]. Palagonite passively accumulates Ti and HFS elements during nearly complete loss of Na and Ca, and large scale removal of Si, Al, and Mg (e.g. [Hay and Iijima, 1968; Staudigel and Hart, 1983]). The precipitation of clays also involves uptake of H_2O, K, Rb and Cs in similar proportions as in [Staudigel et al., 1981a, 1981b;]. Precipitation of the calcium-carbonates is most obviously indicated by increases in CO_2, but not necessarily in CaO because much of their Ca is derived from basalt (e.g. [Staudigel et al., 1979]).

In Figure 4, we present correlation diagrams between the alteration sensitive isotopic tracers ($^{87}Sr/^{86}Sr$, $\delta^{18}O$, $\delta^{11}B$) and elemental abundances of H_2O, Rb and B. Good correlations between H_2O and Rb with $\delta^{18}O$ and $^{87}Sr/^{86}Sr$ in composite samples indicate that the extent of Sr exchange with sea water is probably dominated by palagonitization and the precipitation of clays (Figure 4). It is surprising that $\delta^{11}B$ shows only a very poor positive correlation with water, and none with Rb and B. B abundances correlate well with $\delta^{18}O$ within the VCL and within the rest of the samples as a group ([Smith et al., 1995], and Fig. 4) and less strongly with $^{87}Sr/^{86}Sr$. Thus, B uptake is crudely correlated with the extent of overall alteration, but the process of B isotopic fractionation is not simply controlled by addition of clays.

The dominating control of seafloor alteration by clays and palagonite is also supported by the generally preferred uptake of Cs over Rb and K, with characteristic K/Rb and Rb/Cs ratios of 510 and 65, respectively. However, even though the bulk of the crust has relatively consistently low K/Rb and Rb/Cs ratios, there are some compositional domains that display deviations from this behaviour, indicating that locally other controls may be active. This behavior is displayed in particular by the upper composite from 418A with K/Rb and Rb/Cs ratios of 725 and 120, respectively. Such elevated K/Rb and Rb/Cs ratios may indicate that some of the alkalis were added from seawater without fractionation (K/Rb$_{sw}$ = 3300; Rb/Cs$_{sw}$ = 410).

U is highly enriched in composites from the upper portion of the moderately altered 417D and 418A, while U uptake is minor at Site 417A ([Staudigel et al., 1995]; Table 2). This behavior was explained by the high solubility of U in the oxidizing conditions at 417A and its precipitation in the reducing chemical environments at 417D and 418A [Staudigel et al., 1995]. The dominant control on U abundances, however, appears to be calcium carbonate, as there is a good correlation between U and CO_2 (Figure 5). Such a correlation may be caused by preferred partitioning of U into carbonates, or by a second U-bearing phase that precipitated simultaneously with carbonate.

Surprisingly, Na is the only element that correlates well with Sr abundances (R value > 0.6; Figure 6). The Na-Sr

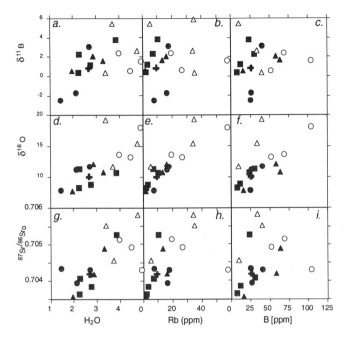

Fig. 4. Cross correlation diagram between the alteration sensitive isotope ratios of $^{87}Sr/^{86}Sr_{init}$, $\delta^{18}O$ and $\delta^{11}B$ and abundances of H_2O, Rb, and B. Best correlations are displayed between of $^{87}Sr/^{86}Sr_I$, $\delta^{18}O$ and and H_2O, Rb. This is consistent with the dominance of clay precipitation. Symbols as in Figure 2.

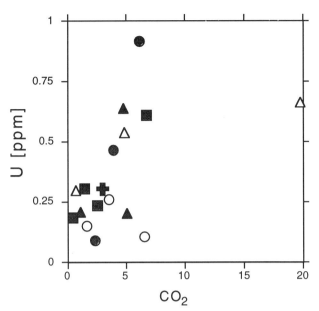

Fig. 5. Correlation of U content (in ppm) of composites and CO_2 (in wt. %), suggesting that the removal of U from seawater is linked to the precipitation of carbonates in the oceanic crust. Symbols as in Figure 2.

trend defined by the composites, however, does not include the (only) fresh glass analysis from these sites. Most data cluster at 1.7-2.2 wt. % Na_2O, less than the observed range of Na_2O in fresh glasses (2.2-2.4 wt. %; [Byerly and Sinton, 1979]). Thus, this trend cannot be explained by the addition or removal of a single component to and from fresh rock. Loss of Sr and Na is commonly observed during alteration of volcanic glass [Staudigel and Hart, 1983], while gain of Na is probably caused by the addition of Na-bearing secondary minerals, such as analcite and natrolite, common Na-bearing minerals at these sites (see analyses by [Pritchard, 1979]. Increases in Sr may be caused by a large number of processes, one of which is also the addition of secondary phases. Sr concentrations in analcite, however, are low [Staudigel et al., 1986]. Thus, although the correlation between Sr abundances and Na_2O is useful for characterization of the bulk composition of the crust, it is probably not controlled by a single processes.

Even though the REE and $^{143}Nd/^{144}Nd_0$ are often considered immobile during alteration, our composites display a good correlation between K_2O and La/Sm, and a weak inverse correlation between K_2O and $^{143}Nd/^{144}Nd$ ([Staudigel et al., 1995]. In Figure 7, we show that the Ce anomaly of the composites also correlates quite well with $^{143}Nd/^{144}Nd$.

Negative Ce anomalies are a characteristic of sea water, where Ce is depleted relative to the other REE due to the oxidation of Ce^{3+} to more insoluble Ce^{4+} [Elderfield and Greaves, 1982]. These REE systematics together can be taken as strong evidence for the uptake of sea water REE into the oceanic crust. However, due to the extremely low concentrations of REE in sea water, unreasonably high water rock ratios (i.e. $> 10^5$) would be necessary to explain the deviations in REE isotope and abundance characteristics by sea water addition [Staudigel et al., 1995]. It is more likely that REE's are added to the crust in the form of hydrothermal Fe-Mn oxide particles. Such particles are a natural fallout of hydrothermal effluent, may be strongly enriched in REE, and often show large negative Ce anomalies [Barrett and Jarvis, 1988]. Thus, our REE data probably do not provide very meaningful water - rock ratios, but they do indicate that a significant fraction of the sea water REE inventory is added to the oceanic crust.

The high field strength elements (HFSE: Nb, Ta, Hf, Zr) display linear correlations with each other as well as with TiO_2 and Th (not shown), and their abundance ratios remain nearly constant. The HFSE change in abundance only by igneous processes or by simple dilution/accumulattion due to the mobility of other elements, and, thus, are conserved during alteration of the oceanic crust.

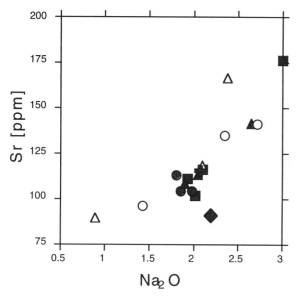

Fig. 6. Sr content (in ppm) of composites correlates well with Na_2O abundances. Symbols as in Figure 2, diamond indicates fresh glass composition. Note that the only fresh glass falls off this trend, suggesting that the correlation is not caused by a simple mixing relationship.

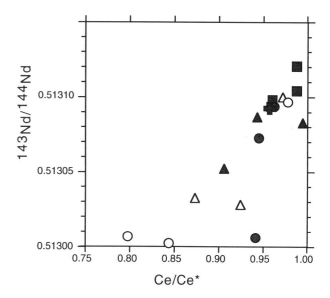

Fig. 7. The isotopic ratio of $^{143}Nd/^{144}Nd$ t in composites correlates well with the Ce anomaly (Ce/Ce*), suggesting the large fraction of hydrogenous REE in the highly altered composites (Ce* is the hypothetical Ce abundance, based on a linear extrapolation between La and Pr). Symbols are given in Figure 2

4.2. Chemical Fluxes between sea water and the oceanic crust.

Altered oceanic crust, as it is represented by the composites analyzed here, contains an integrated chemical record of all chemical exchange with sea water that occurred in its 118 Ma history. A comparison of composite data with their respective unaltered "MORB" composition can provide bulk chemical fluxes between sea water and basalt. For this, we have estimated fresh rock compositions for all our composites from compiled published glass and phenocryst microprobe analyses [Byerly and Sinton, 1979; Sinton and Byerly 1979] and phenocryst abundance data. Such a procedure generally results in bulk rock compositions that agree well with measured bulk rock compositions [Staudigel and Bryan, 1981].

In our procedure of estimating the unaltered composition of our composite samples, we first determined the fresh rock composition of each individual sample. Then, we combined these fresh rock estimates into "fresh composites" by using the "recipes" in Appendix 1. To calculate the unaltered composition for each individual sample, we combined glass and fresh phenocryst compositions in the proportions indicated by the samples. Glass compositions were estimated from the downhole variation of glass data published by Byerly and Sinton [1979], and phenocryst compositions were based on those analyzed by Sinton and Byerly [1979]. The abundances of phenocrysts were determined on the basis of our own and published thin section descriptions and point-count data [Donnelly et al. 1979 a; 1979b].

The errors in this procedure are largely controlled by uncertainties in phenocryst abundances. The chemistry of glasses and phenocrysts at these sites is rather systematic, and known with high confidence. For this reason, we made two estimates that reflect end members of the maximum range in the potential phenocryst population in a given section. For our fresh-rock estimate, we used the mean of these end members and the error gives their range (Table 5).

Before we can calculate fluxes from the difference in the fresh rock estimates (Table 5) and the altered compositions (Table 1), a common reference system has to be established. This is largely due to the fact that seafloor alteration occurs in an open chemical system, whereby basaltic material is dissolved and transferred to the oceans and lower crust, and sea water-derived components are added to the crust in void spaces. For this reason, one has to find monitors for the total quantity of basalt dissolution, as well as the amount of vesicle infill. One common approach is to use elements that are considered immobile, such as Fe or Ti, as these elements are passively accumulated during dissolution of basalt and diluted during the addition of vein minerals. In the following, we shall focus on Ti, because of the abun-

Table 5 Fresh Rock Composition Calculated From Fresh Glass Compositions and Phenocryst Abundances and Compositions

sample	SiO_2	TiO_2	Al_2O_3	Fe_2O_3	FeO	MnO	MgO	CaO	NaO_2	K_2O	P_2O_5	H_2O	CO_2
SUPER	50.14±0.18	1.32±0.07	15.47±0.37	0.79±0.03	9.41±0.30	0.19±0.01	7.66±0.19	12.29±0.10	2.17±0.05	0.09	0.12	0.2	0.05
417A 24	49.63±0.10	1.43±0.05	15.56±0.24	0.80±0.02	9.59±0.19	0.20±0.01	7.33±0.15	12.34±0.02	2.36±0.03	0.09	0.11	0.2	0.05
417A 32	49.30±0.16	1.30±0.06	14.99±0.04	0.80±0.01	9.62±0.16	0.18±0.01	9.24±0.48	12.12±0.04	2.14±0.06	0.09	0.10	0.2	0.05
417A 44	49.52±0.13	1.54±0.06	15.23±0.39	0.82±0.02	9.79±0.25	0.19±0.01	7.20±0.05	12.62±0.09	2.24±0.04	0.09	0.11	0.2	0.05
417D 22-27	49.93±0.22	1.32±0.10	16.54±0.67	0.76±0.04	9.12±0.46	0.18±0.01	7.16±0.15	12.19±0.12	2.22±0.05	0.08	0.12	0.2	0.05
417D 39	50.48±0.14	1.44±0.10	15.38±0.62	0.82±0.04	9.77±0.51	0.18±0.01	7.14±0.01	12.40±0.25	2.19±0.05	0.09	0.12	0.2	0.05
417D 59	49.88±0.28	1.43±0.12	14.98±0.60	0.83±0.04	9.96±0.51	0.18±0.01	8.21±0.24	12.01±0.27	2.08±0.10	0.07	0.12	0.2	0.05
418 A 15	50.91±0.11	1.20±0.02	15.02±0.17	0.77±0.01	9.23±0.11	0.20±0.00	7.59±0.12	12.47±0.02	2.18±0.02	0.10	0.10	0.2	0.05
418A 40	50.57±0.24	1.20±0.07	16.19±0.58	0.74±0.03	8.82±0.33	0.19±0.01	7.34±0.10	12.44±0.04	2.21±0.04	0.07	0.10	0.2	0.05
418A 73	50.49±0.19	1.38±0.05	15.40±0.36	0.81±0.02	9.69±0.26	0.19±0.01	7.29±0.13	12.28±0.12	2.18±0.06	0.10	0.12	0.2	0.05
418A 86	49.82±0.14	1.13±0.04	15.54±0.03	0.73±0.01	8.74±0.14	0.20±0.00	8.26±0.41	12.27±0.05	2.13±0.05	0.09	0.12	0.2	0.05
VCL-Top	50.41±0.13	1.28±0.05	15.37±0.26	0.78±0.02	9.32±0.19	0.20±0.01	7.54±0.17	12.38±0.01	2.22±0.03	0.09	0.11	0.2	0.05
VCL-100m	50.08±0.35	1.36±0.07	15.52±0.39	0.79±0.03	9.44±0.32	0.19±0.01	7.21±0.10	12.18±0.03	2.21±0.05	0.08	0.11	0.2	0.05
VCL-300m	50.11±0.09	1.18±0.04	15.74±0.28	0.73±0.02	8.76±0.23	0.19±0.01	7.69±0.06	12.59±0.12	2.12±0.04	0.09	0.12	0.2	0.05
FLO-top	50.21±0.15	1.28±0.06	15.72±0.38	0.77±0.02	9.25±0.27	0.19±0.01	7.50±0.15	12.31±0.05	2.22±0.03	0.09	0.11	0.2	0.05
FLO-100m	49.97±0.16	1.44±0.08	15.37±0.49	0.81±0.03	9.64±0.36	0.19±0.01	7.36±0.08	12.47±0.14	2.20±0.05	0.08	0.11	0.2	0.05
FLO-300m	49.99±0.15	1.21±0.04	15.36±0.08	0.76±0.01	9.10±0.14	0.19±0.00	8.13±0.36	12.24±0.01	2.13±0.05	0.09	0.12	0.2	0.05

dance of Fe bearing secondary phases (e.g. nontronitic clays) in veins suggests at least some mobility, while Ti-bearing secondary phases are much rarer. The arguments above for the coherency in the HFSE as a group also suggests that they (including TiO_2) are conservative. However, before we shall use this assumption, we wish to evaluate its validity.

Using the density estimates of fresh and altered basalt, as well as the density of secondary phases and their modal abundance data, we can make an independent estimate of the total gain or loss to a given reference volume. This predicted gain or loss can be compared to the apparent dilution or passive accumulation of Ti. Characteristic void-free densities for fresh and altered basalts are 2.95 and 2.914 g/cm^3, respectively [Hyndman and Drury, 1977] and 2.71 g/cm^3 and 2.35 g/ccm^3 for the major secondary phases, carbonate and smectite, respectively [Tröger, 1979]. The void-space at 418A includes 7.5 vol. % for smectite veins, 5 vol. % for carbonate veins and 6 vol. % open space [Johnson, 1979]. Assuming that the total porosity stayed constant, we can calculate the total weight of material in an initial reference volume before and after alteration. From this we calculate a net weight addition of 10.3 grams to each 100 g of original crust. If one includes the water that probably filled the void spaces, the total addition would be closer to 15.3 grams per 100 grams. The addition of 10.3 grams of sea water-derived materials to the crust would lead to a net dilution in TiO_2 from 1.32 wt% (fresh) to 1.19 wt% (altered), which is within error of the measured value of 1.18 wt. %. Based on this calculation, the assumption that Ti is conserved appears reasonable, and we have calculated accordingly all chemical fluxes with respect to a constant Ti (Table 6). Some of these data were used previously by Spivack and Staudigel [1994] to evaluate the impact on the alkalinity budget of the oceans.

To illustrate the effects of removal or addition of chemical components to different rock types we have plotted flux data from Table 6 in a series of down-hole plots (Figure 8 a-i). Due to the minimal inventories of K_2O, H_2O and CO_2 in fresh MORB, there is no significant error to the fluxes from errors in initial rock composition (Figure 8 a- c). CO_2 fluxes are by far the largest whereby average crust has acquired 3.26 wt%, followed by water (2.81 wt%). It may be pointed out here that this water flux is based only on H_2O^+, the structurally bound water (i.e. water that remains in a rock powder after it was heated to 110°C). Not included is loosely bound water and formation water that fills the open crack and void space in the crust (= 5 vol. %, Johnson, 1979). Using analytical data including H_2O^+ and H_2O^- as a guide ([Donnelly et al., 1979a]) and the open crack volumes of [Johnson, 1979], we estimate that the total water uptake in the oceanic crust can quite easily exceed 6 wt%. It may to be noted here that the low molecular weight of H_2O results in a much larger fraction on a molar basis. Water uptake broadly decreases with depth, even though there is substantial scatter. Although downward extrapolations appear poorly constrained, they allow for significant hydration with depth. Fluxes of K_2O and CO_2 in Depth Composites from 417D and 418A are quite consistent, decreasing with depth, whereby the top of the crust takes up about 0.9 g of K_2O per 100g and about 5 g of CO_2 (Fig. 8 a, c)

Fluxes of SiO_2, Al_2O_3, FeO, MgO, CaO and Na_2O are quite variable, including addition or removal at different depths in the core (Figure 8 d - i). While there is significant variations in the fluxes of different composite types and

Table 6 Fluxes of Oxides (in Grams Added per 100 Grams of Rock)

	SiO_2	Al_2O_3	FeO_{tot}	MnO	MgO	CaO	NaO_2	K_2O	H_2O	$CO2$	total gain/loss
SUPER	1.18	1.94	-0.01	0.00	-0.19	2.14	0.15	0.54	2.81	3.26	12.37
417A 24	-6.17	0.03	0.26	0.26	-4.00	-4.98	-1.00	4.13	4.56	3.24	-2.59
417A 32	-8.07	-0.12	-1.03	-0.04	-4.45	-3.30	-0.03	1.38	3.82	1.33	-9.83
417A 44	9.68	5.46	0.80	0.04	0.66	7.87	1.62	1.47	5.48	9.19	43.09
417D 22-27	-1.89	-1.02	-0.56	0.05	-0.67	5.05	-0.16	0.77	2.16	6.73	10.96
417D 39	-0.16	2.17	-0.63	-0.04	-0.82	2.36	-0.18	0.84	2.79	4.28	11.18
417D 59	5.07	4.60	0.22	0.03	-1.13	4.28	0.25	0.28	1.50	2.66	18.33
418 A 15	1.35	3.10	-0.85	0.03	0.47	6.79	0.42	0.66	2.57	8.25	23.34
418A 40	5.90	4.96	0.31	-0.03	0.08	2.59	1.53	0.98	4.60	2.99	24.45
418A 73	4.49	2.10	1.01	-0.02	2.01	0.27	0.01	0.10	2.91	0.35	13.84
418A 86	-6.01	-1.14	-0.83	-0.04	-0.94	-0.85	-0.28	0.03	1.88	1.22	-6.57
TopVCL	3.93	1.17	2.94	0.80	-0.32	40.73	-0.42	3.55	6.62	39.39	99.66
100mVCL	9.69	3.58	2.27	-0.06	1.51	2.05	1.04	2.75	6.35	6.59	36.73
300mvcl	-1.56	-0.24	0.22	-0.10	1.56	-3.68	-0.02	0.41	3.57	0.66	1.53
top FLO	-2.45	0.61	-0.66	0.01	-1.10	2.14	0.01	1.12	2.92	5.08	8.28
100mFLO	7.58	5.05	0.26	0.01	-0.08	6.77	1.34	0.96	4.24	6.67	33.45
300mFLO	-1.37	0.14	-0.21	-0.02	-0.04	-0.03	-0.22	0.02	1.80	1.00	1.52

Global flux, using an ocean crust production rate of 3.606×10^{15} g/year, in 10^{12} g/y

SUPER	42.59	69.9477	-0.28	-0.02	-6.99	77.27	5.394	19.54	0.23	101.2	117.54

depths, the total fluxes of these elements in the Super Composite are close to zero, except for Al and Ca which each show a net uptake of about 2 g/100g. Errors are particularly high for Mg, Fe and Al, because of uncertainties in modal abundances of olivine and plagioclase, respectively. Individual composites display rather high fluxes in SiO_2 (-8 to +9.6 wt%), MgO (-4..5 - +2 wt%), and CaO (5.0- + 40.7 wt%). This shows that while most elements display significant removal or addition in different portions of the oceanic crust, the net change is small.

Downhole variations in SiO_2, Al_2O_3, FeO, and Na_2O suggest relatively high dissolution rates near the top of the core, maximum addition at mid-levels, and increasing dissolution at the bottom of 418A. Thus, much of the alteration at 150-200m depths in 417D and 418A is based on material addition, whereas the top and the base of the section analyzed alters mostly by dissolution processes.

Ca and Mg fluxes between the oceanic crust and sea water are relatively small (with the exception of the Top VCL with 19.8 mole% addition of CaO), slightly negative for Mg and positive for Ca. Overall, there is a slight increase in the Mg flux down - section while Ca fluxes appear to decrease. Molar fluxes of CaO are generally less than the CO_2 fluxes (Table 6) indicating that the carbonate is not added as a simple $CaCO_3$ precipitation from sea water. This is consistent with the common assumption that the oceanic crust generally takes up Mg from the oceans and Ca is controlled by addition of $CaCO_3$ and removal from leaching. Thus, the Ca in the altered rock is probably a mixture of Ca from seawater and residual Ca from basalt. More Ca is exchanged between seawater and basalt than the the net fluxes indicate. However, in absence of an isotopic tracer for Ca, this exchange balance cannot uniquely be determined.

For Sr, however, the net fluxes in and out of a rock can be determined because basalt and seawater Sr have a distinct isotopic ratio. Unaltered basalt from Sites 417 and 418 has an $^{87}Sr/^{86}Sr$ of 0.70295 [Staudigel et al., 1979] and the $^{87}Sr/^{86}Sr$ of Cretaceous seawater is 0.70735 [Hess et al., 1986]. Intermediate Sr isotopic compositions provide a mixing ratio between these two end members. This mixing ratio, in combination with the total Sr abundance allows determination of the concentrations of basalt Sr and seawater -

STAUDIGEL ET AL. 31

Sr in any seafloor altered basalt. Differences in basalt- Sr concentration of fresh and altered basalt can directly be interpreted in terms of basaltic and sea water Sr-fluxes.

While the inventory of the altered rock can simply be measured, we have to estimate the Sr inventory of the fresh rock. For this, we used a similar approach as for the major elements. The fresh rock Sr abundances can be determined from the Sr inventory in fresh glass (89.5 ppm; [Staudigel and Hart, 1983]) and the phenocryst proportions used for the above fresh-rock estimates. From this, and published partition coefficients [Rollinson, 1993], we estimated original Sr compositions (e.g., 93 ppm for the Super composite, and 90-94 ppm for all remaining composites). Before we can calculate the final fluxes, however, we first have to "adjust" the altered Sr abundance to a constant TiO_2, similar to the calculations from the major element fluxes. After these calculations we can estimate for each composite the total amount of seawater Sr gained, and the amount of basalt Sr that has to be gained or lost to balance with the fresh rock composition. In Figure 9, we have plotted these values versus depth. All samples have lost basaltic Sr except sample FLO 300 that has gained only insignificantly (i.e. 1 ppm). Most losses are displayed by the uppermost volcaniclastics that show a very systematic relationship with depth. Depth and FLO composites do not lose as much Sr, but the losses also become less significant with depth. Sr uptake from seawater is also most significant near the top of the crust, with the enhancement in the volcaniclastic rich sections at about 150-200 m depth. Thus, the increases in $^{87}Sr/^{86}Sr$ at mid depth (Figure 3) is largely controlled by uptake of Sr from seawater. This gain is not accompanied by a contemporaneous loss of basaltic Sr. Overall, the downhole decreasing loss of basalt - Sr and gain of sea water Sr is consistent with a downhole increasingly exchanged hydrothermal fluid.

From the fluxes between fresh and altered basalts, we can calculate the "global" fluxes. Using an average ocean crust production rate of 3 km^2/year we estimate a volume production rate of 1.5×10^{15} cm^3/year (upper 500 m of crust only). An average density of fresh basalt of 2.95 g/cm^3 ([Hyndman and Drury, 1977]) and an open void volume of 18.5%, we determine the annual weight production rate of 3.606×10^{15} g/year. Using this rate and the fluxes for the Super composite, we estimate the annual fluxes of major elements (all in 10^{12} g/year): SiO_2 = 42.6; Al_2O_3 = 69.9. FeO_{tot}= 0.8, MnO =-0.242; MgO = -6.99, CaO= 77.3, Na_2O= 5.39 K_2O = 19.5, P_2O_5= 0.230, H_2O = 101; CO_2 = 118., where positive numbers indicate uptake and negative numbers suggest release from the crust. Amongst these, only the fluxes of Si and Al are significant when compared to hydrothermal and river fluxes (e.g. [van Damm et. al,

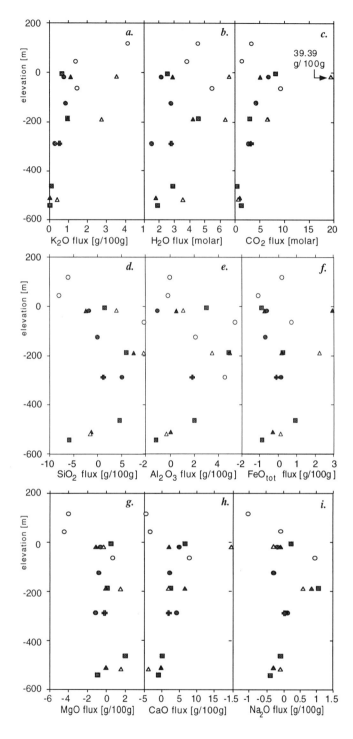

Fig. 8. Fluxes of K_2O (a), H_2O (b), CO_2 (c), SiO_2 (d) Al_2O_3 (e) FeO tot (f), MgO (g), CaO (h), Na2O (i), versus depth. These fluxes were calculated on the basis of our composite data (Table 1), fresh rock estimates (Table 5), assuming Ti to remain constant. Symbols as in Figure 2

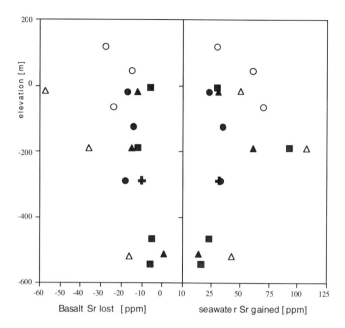

Fig. 9. Sr fluxes in composites versus depth. Most composites lost some basaltic Sr and gained seawater Sr. Losses and gains become less significant with depth. Symbols as in Figure 2

1988; Martin and Meybeck, 1979]). The fluxes of Al may be too high due to the uncertainties in our fresh rock estimate (Al is particularly sensitive to errors in plagioclase abundances).

Due to relatively low inventories in unaltered basalt, the uptake of Rb, Cs and U can be determined with relatively high confidence. As fresh rock composition we have chosen the Rb = 0.67 ppm and Cs =0.0096 ppm, from the freshest glass analyzed by [Staudigel and Hart, 1983]. For U, we chose 0.05 ppm, the lower intercept of the U-K correlation from [Staudigel et al., 1995]. From this, we obtain annual fluxes for Rb = 3.21×10^{10} g/y, for Cs x 5.16×10^8 g/y and for U 9.01×10^8 g/y. From a comparison of the fresh and unaltered Sr abundances and isotopic ratios and the isotopic ratio of Cretaceous Sea water, we estimate that 10 ppm of basaltic Sr are removed from the basalt and 32 ppm of sea water are added to the crust. This translates into a flux of 1.15×10^{11} g/y sea water Sr into the crust and 3.61×10^{10} g/y basaltic Sr into the deeper crust or the oceans. The flux of Sr into the crust accounts approximately for 10% of the global flux of Sr into the crust inferred by Raymo et al. [1988], slightly more than the volume fraction of the total ocean crust analyzed here (500m/6km). Our study suggests a minor imbalance of Sr fluxes, whereby the ocean crust is a sink for Sr, but less than 0.1% of the global Sr exchange between seaater and the crust [Raymo et al., 1988].

4.3 The Bulk Composition of the Upper Oceanic crust and Implications for Subduction Recycling to Volcanic arcs

Correlation diagrams and downhole plots in Figures 2-7 may also be used to estimate the robustness of our estimates of large-scale compositional domains. It is clear from the previous discussion that composites show an overall lower variance than individual samples and that the Super composite generally falls amongst the composites of moderately altered compositional domains. Most alteration-sensitive parameters (except $\delta^{11}B$) agree well between the similarly altered Sites 417D and 418A, suggesting that our estimates are at least internally consistent. Overall, we believe that these estimates are very close to an accurate representation of the upper oceanic crust at these sites, whereby the scatter between 417D and 418A composites reflects the uncertainty of this estimate. However, we caution that the deep drill sites studied here are the only ones with sufficient recovery to allow for such an estimate.

However, the crustal section studied was drilled in normal oceanic crust that may be subducted into the mantle. These compositions are reasonable bulk compositions of the oceanic crust, but it is uncertain at this time, how variable the crust might be. Similar flux estimates were also made at the Troodos ophiolite [Bednarz and Schmincke, 1989], where relatively large fluxes were explained by the rather high porosity and permeability of oceanic crust from a supra-subduction zone tectonic setting. In absence of other estimates on normal oceanic crust, our data should be considered the best approximation for upper ocean crust compositions for the purpose of subduction zone modeling and as a starting composition for experiments to simulate subduction zone processes.

The bulk altered ocean crustal composition is quite different from unaltered normal MORB. The biggest differences are in the rather high H_2O and CO_2 contents, in combination with locally elevated K_2O and Na_2O. In many compositional domains, these additions amount to more than 20 mole % of the rock, not including "free" water in void spaces (5% by volume; [Johnson, 1979]) and loosely bound water ("H_2O^-", i. e, water expelled at < 110°C). Addition of these elements to the ocean crust will make it an effective source of volatiles during dehydration events, and lower its melting point. The highly altered compositional domains will play an important role in the generation of metasomatic fluids and melts. It is probable that those fluids and melts may be slightly enriched in the alteration-related chemical inventory of the oceanic crust. However, since these melts and fluids will at least partially exchange with the ocean crust, it is unlikely that they remove the entire alteration related chemical inventory from the oceanic crust.

The bulk altered crust also has dramatically different trace element and isotopic compositions from unaltered MORB. The most noteworthy geochemical effects are the great enrichment in alkali elements (K, Rb and Cs) and U, generally more than ten times higher than original igneous abundances [Hart and Staudigel, 1989]. On the other hand, alkaline earth elements (Ba and Sr) are only a factor of two or so enriched over unaltered MORB, while the REE and Th are added to an even lesser degree. These distinctive geochemical signatures characterize the altered seafloor as it subducts into the mantle. The enrichment of Rb over Sr, and the decoupling of these elements during seafloor alteration mean that Sr isotopes are decoupled from Sr concentrations [Staudigel et al., 1995]. Sr isotopes follow Rb (parent) abundances (Fig. 4), which depend on the extent of palagonitization, clay mineral formation and hydration, while Sr abundances are mostly unaffected by these processes, and may vary subtly with Na_2O and Na-bearing phases such as zeolites (Fig. 6). Thus, characterizing the Sr isotopic budget in the subducted basaltic imput depends on the combination of processes that affect both Rb and Sr budgets. Another important geochemical effect of alteration is the strong decoupling of U and Th. U is enriched in carbonate-rich zones (Fig. 5) which largely dilute Th, thus leading to a virtually antithetic relationship. U is enriched more than ten times in the bulk crust, while Th is basically conserved. Thus seafloor alteration may increase U/Th ten fold or more, and lead to high $^{230}Th/^{232}Th$ in subducted inputs.

Strong decoupling also occurs between alkaline earths and alkali elements, such as Ba and Rb. The composite samples generally have Ba/Rb < 5 (Table 2 and 3), in stark contrast to MORB and ocean island basalts which have a virtually constant ratio of 11-13 [Hofmann and White, 1983]. Interestingly, this low Ba/Rb is not simply imparted to arc basalts, many of which have high ratios > 13 ([Plank and Langmuir, 1993]). This indicates that the fluids that are driven out of the oceanic crust in subduction zones may not simply transport the alteration chemistry of the oceanic crust to the arc, but may impose their own partitioning behaviour. Variable subducted sedimentary packages will also contribute Ba and Rb in differing ratios [Plank and Langmuir, 1993]). Marine sediments commonly show an opposite sense of enrichment to the oceanic crust: very high concentrations of Ba (in siliceous oozes) and Sr (in carbonates) relative to alkalis (in terrigenous clays) [Ben Othman et al., 1989; Plank and Langmuir, 1993]. Thus, the different sources of elements and transport processes in the subduction zone will combine to determine the composition of arc magmas.

Arc basalts are also notably enriched in Ba over La, generally with Ba/La higher than 20 [Morris and Hart, 1983]. Although some domains within the altered oceanic crust reach Ba/La of 20-30, most of the composites and have ratios < 20, and the Super composite is 12 (Table 2). Thus again, alteration processes alone cannot explain one of the unique geochemical features of arc basalts, and a combination of sediment sources and fractionation processes create the distinctive Ba/La in arc magmas.

We can also demonstrate that seafloor alteration fluxes are in most cases insufficient to balance arc output fluxes. Arc magmatic fluxes are ~ 30 $km^3/Ma/km$ arc length [Reymer and Schubert, 1984], which is of the same order as seafloor mass fluxes to the trench for the upper 500 m (0.5 km x an average convergence velocity of 70 km/Ma/km arc length). Thus, we can determine the flux balance simply from comparing alteration enrichments with arc enrichments. Alteration enrichments may be estimated by subtracting the pristine igneous composition at 417-418 [Hart and Staudigel, 1989] from the Super composite (Tables 1-3, 5). Arc enrichments may be estimated by subtracting an average N-MORB composition [Sun and McDonough, 1989] from an average arc basal composition; Plank and Langmuir [1993] provide such data for 8 arcs. In round numbers, the alteration flux into the upper 500m of the seafloor is ~ 70% of the arc enrichment flux for K, Rb and U. Thus the alteration fluxes are highly significant with respect to crustal recycling to the arc for these elements, and may even be sufficient to explain the total budget at some arcs. For the other elements, however, the alteration fluxes into the upper 500m are insufficient to account for much of the recycled budget: 35% of the Cs, ~ 5% of the Ba, Sr and La, and 1% of the Th. For these elements, additional inputs are required from 1) alteration components in the deeper oceanic crust, 2) MORB components in the oceanic crust, and/or 3) sedimentary components. This is no real surprise given the requirement of subducted sediments to explain ^{10}Be and Ba variations in some arcs [Plank and Langmuir, 1993; Tera et al., 1986], and the requirement of MORB components to explain Pb isotopes in others [Miller et al., 1994]. The point here is not that the alteration flux into the oceanic crust is the whole picture, but that it is possibly a large contributor to the total budget of some elements in arc magmas (alkalis, U). It may actually be the predominant source of recycled H_2O to arc volcanoes [Plank et al., 1994]. Seafloor alteration is thus a major part of subduction recycling, and there is a great need to explore the phenomenon in other locales in order to develop some predictive models for its distribution on the subducting seafloor.

Obtaining good estimates of crustal inputs is the necessary first step in attempting to model the subduction process geochemically. We have a fairly good handle on sediment inputs to several trenches about the globe [Plank and

Langmuir, 1993], but few estimate of the upper oceanic crust. Clearly, further drilling with good recovery in basement sections is critical to better modeling of subduction recycling.

5. CONCLUSIONS

One of the major results of this study is the successful demonstration that the bulk composition of in situ oceanic crust can be determined on scale lengths of 10-500 m, despite its extreme heterogeneity. This success is critically dependent the availability on drill cores with high recovery (> 70%) for a structural analysis and fresh glass fragments for a fresh rock estimates. Due to the recovery of fresh glass fragments from most sections of the drill core at Sites 417 and 418, we were able to reconstruct the fresh bulk composition also, allowing determination of low temperature alteration fluxes between the ocean crust and seawater.

The bulk major, trace element and isotope composition of the upper oceanic crust changes substantially during seafloor alteration. Most of the alteration-related chemical inventory can be found in the upper oceanic crust, and in highly permeable volcaniclastics. While high field strength elements, Th, Ti and Fe remain essentially immobile during low temperature alteration, most other elements show some sensitivity to alteration. Substantial redistribution occurs for most major elements (Si, Al, Mg, Ca, and Na) and many trace elements (Sr, Ba, LREE's). Significant uptake is observed for H_2O, CO_2, K, Rb, Cs, and U.

Observed chemical changes in the upper oceanic crust suggest that a very large component of many arc fluxes may be contributed from the upper oceanic crust (K, Rb, U). However, for most elements some sediment-derived and deeper ocean crustal sources must contribute as well.

Acknowledgements. This study was supported by the National Science Foundation, and samples were supplied by the DSDP/ODP. H. Staudigel acknowledges support from NSF, the Netherlands School for Sedimentary Geology and the Institute for Geophysics and Planetary Physics at Scripps Institution for Oceanography. HS also appreciated the support and patience of the ODP Leg 106 shipboard scientific party who acted as a sounding board for many of the ideas during preparation of composite samples. T. Plank gratefully acknowledges support from an NSF Postdoctoral Fellowship (EAR-9203151) and H.-U. Schmincke the support from the Deutsche Forschungsgemeinschaft. This paper benefitted from very thorough reviews by Patty Fryer and Alfred Hochstaedter.

REFERENCES

Alt, J.C., C. France-Lanord, P.A. Floyd, P. Castillo, and A. Galy, Low - temperature hydrothermal alteration of Jurassic ocean crust, Site 801, in *Proceedings of the Ocean Drilling Program Scientific Results Vol 129*, edited by R.L. Larson, Y. Lancelot, and et. al., pp. 415-427, Washington DC., 1992.

Alt, J.C., J. Honnorez, C. Laverne, and R. Emmermann, Hydrothermal Alteration of a 1 km section through the upper oceanic crust DSDP hole 504B: The mineralogy, chemistry and evolution of seawater - basalt interactions, *J. Geophys. Res.*, 91, 10309-10335, 1986.

Andrews, A.J., Low temperature fluid alteration of oceanic layer 2 basalts, DSDP Leg 37, *Can. J. Earth Sci.*, 14, 991-926, 1977.

Barrett, T.J., and I. Jarvis, Rare earth element geochemistry of metalliferous sediments from DSDP Leg 92: The East Pacific Rise., *Chem. Geol.*, 67: 243-259, 1988.

Bednarz, U., and H.-U Schmincke, Mass transfer during subseafloor alteration of the upper Troodos crust (Cyprus), *Contrib. Mineral. Petrol.*, 102, 93-101, 1989. .

Ben Othman, D., W.M. White, and J. Patchett, The geochemistry of marine sediments, island arc magma genesis and crust-mantle recycling, *Earth Planet. Sci. Lett.*, 94, 1-21, 1989.

Byerly, G.R., and J.M. Sinton, Compositional trends in natural basaltic glasses from Deep Sea Drilling Sites 417D and 418A, in *Initial Repts. Deep Sea Drilling Project Vol 51-53*, edited by T. Donnelly, J. Francheteau, W. Bryan, P.T. Robinson, M.F.J. Flower, M. Salisbury et.al., pp. 957-972, US Government Printing Office, Washington, 1979.

Cheatham, M.M., W.F. Sangrey, and W.M. White, Improved ICP-MS analytical precision using non-linar response drift corrections., in *Winter Conference on Plasma Spectrochemistry*, edited by R. Barnes, Univ. of Mass. Amherst MA, San Diego, CA, 1992.

Donnelly, T., A.J. Francheteau, W.B. Bryan, P.T. Robinson, M.F.J. Flower, M. Salisbury (Eds.), *Initial Reports of the Deep Sea Drilling Project*, 1613 pp., US Government Printing Office, Washington DC, 1979a.

Donnelly, T.J., G.Thompson, and M. Salisbury, The chemistry of altered basalts at Site 417, Deep Sea Drilling Project Leg 51., in *Initial Repts. Deep Sea Drilling Project Vol 51-53*, edited by T. Donnelly, J. Francheteau, W. Bryan, P.T. Robinson, M.F.J. Flower, M.Salisbury, et. al, pp. 1319-1330, US Government Printing Office, Washington, 1979b.

Elderfield H. and M.J. Greaves, The Rare Earth Elements in Seawater, *Nature*, 296, 214-219, 1982.

Flower, M.F.J., W. Ohnmacht, P.T. Robinson, G.Marriner, and H.-U. Schmincke, Lithologic and chemical stratigraphy at Deep Sea Drilling Project Sites 417 and 418, in *Initial Repts. Deep Sea Drilling Project Vol 51-53*, edited by T. Donnelly, J. Francheteau, W. Bryan, P.T. Robinson, M.F.J. Flower, M. Salisbury et. al., US Government Printing Office, Washington, 1979.

Gregory, R.T., and H.P. Taylor Jr., An oxygen isotope profile in a section of Cretaceous oceanic crust, Samail ophiolite, Oman: evidence for $\delta^{18}O$ buffering of the oceans by deep (>5 km) seawater-hydrothermal circulation at Mid-Ocean ridges, *J. Geophys. Res.*, 86, 2737-2755, 1981.

Hart, S.R., and H. Staudigel, Isotopic characterization and identification of recycled components., *NATO ASI Series, C: Math. and Phys. Sci.* 258. 15-28 pp., 1989.

Hay, R.L., and A. Iijima, Nature and origin of palagonite tuffs of Koko Craters, Oahu, Hawaii., *Geol. Soc. Amer. Mem.*, 116, 338-376, 1968.

Hess, J., B. M., and S. J.-G., Seawater $^{87}Sr/^{86}Sr$ evolution from Cretaceous to present, applications to paleoceanography, *Science*, 231, 979-984., 1986.

Humphris, S.E., and G. Thompson, Hydrothermal alteration of ocean basalts by seawater, *Geochim. Cosmochim. Acta*, 42, 107-125, 1978.

Hofmann, A.W. and White, W.M., Ba, Rb and Cs in the Earth's mantle. *Z. Naturforsch*, 38a, 256-266, 1983.

Hyndman, R.D., and M.J. Drury, Physical properties of basalts, gabbros, and ultramafic rocks from DSDP Leg 37, in *Initial Reports of the Deep Sea Drilling Project Vol 37*, edited by F. Aumento, W.G. Melson, and e. al., pp. 395-402, US Government Printing Office, Washington, DC, 1977.

Johnson, D.M., Crack Distribution in the upper oceanic crust and its effects upon seismic velocity, seismic structure, formation permeability, and fluid circulation., in *Initial Repts. Deep Sea Drilling Project v, 51, 52, 53 Part 2*, edited by T. Donnelly, J. Francheteau, W. Bryan, P.T. Robinson, M.F.J. Flower, M.Salisbury, and e. al., pp. 1473-1490, US Government Printing Office, Washington DC, 1979.

Karson, J.A., Factors controlling the orientation of dykes in ophiolites and oceanic crust., in *Mafic Dyke Swarms*, edited by H.C. Halls, and W.F. Fahrig, pp. 229-242, Geol. Assoc. Canada, 1987.

Miller, D.M., S.L. Goldstein, and C.H. Langmuir, Cerium/lead and lead isotope ratios in arc magmas and the enrichment of lead in the continents, *Nature*, 368, 514-520, 1994.

Martin, J.M. M. Meybeck, Elemental mass balance of Material carried by major world rivers, *Marine Geochem.*, 7, 173-191, 1979.

Morris, J.D., and S.R. Hart, Isotopic and incompatible element constraints on the genesis of island arc volcanics from Cold Bay and Amak island, Aleutians, and implications for mantle structure, *Geochim. Cosmochim. Acta*, 47, 2015-2030, 1983.

Plank, T., and C.H. Langmuir, Tracing trace elements from sediment input to volcanic output at subduction zones, *Nature*, 362, 799-742, 1993.

Plank, T., J.D. Morris, and G. Abers, Sediment water fluxes at subduction zones, *SUBCON Abstr.*, Catalina Island, CA, June, 1994

Pritchard, R.G., Alterations of basalts from Deep Sea Drilling Project Legs 51, 52, andd 53, Holes 417A and 418A, in *Initial Repts. Deep Sea Drilling Project v, 51, 52, 53 Part 2*, edited by T. Donnelly, J. Francheteau, W. Bryan, P.T. Robinson, M.F.J. Flower, M.Salisbury, and et. al.., pp. 1185-1199, US Government Printing Office, Washington, 1979.

Raymo, M.E., W.F. Ruddiman, P.N. Froehlich, 1988, Influence of late Cenozoic mountain building on ocean geochemical cycles, Geology, 16, 649-653

Reymer, A., and G. Schubert, Phanerozoic addition rates to the continental crust and crustal growth, *Tectonics*, 3, 63-77, 1984.

Robinson, P.T., M.F. Flower, H. Staudigel, and D.A. Swanson, Lithology and eruptive stratigraphy of Cretaceous oceanic crust, western Atlantic.., in *Init. Repts. Deep Sea Drilling Project v, 51, 52, 53 Part 2,*, edited by T. In Donnelly, and J. Francheteau, pp. 1535-1556, U.S. Govt. Print. Off, Washington, DC, 1979.

Rollinson, H., Using Geochemical Data: Evaluation, presentation, interpretation., *Longman Scientific and Technical* Essex, England, 352pp, 1993.

Schmincke, H.U and U. Bednarz, Pillow-, sheet flow-, and breccia volcano-tectonic-hydrothermal cycles in the Extrusive Series of the northwestern Troodos Ophiolite (Cyprus). in *Symposium Troodos 87 - Ophiolites and Oceanic Lithosphere, Proceedings*, edited by J. Malpas, E.M. Moores, A. Panayiotou, and C. Xenophontos, pp. 185 - 207, Geological Survey Department, Nicosia (Cyprus), 1990.

Sinton J.R and G.M. Byerly, Mineral Compositions and Crystallization trends in Deep Sea Drilling Project holes 417D and 418A., in *Initial Repts. Deep Sea Drilling Project t v, 51, 52, 53 Part 2*, edited by T. Donnelly, J. Francheteau, W. Bryan, P.T. Robinson, M.F.J. Flower, and e.a. M.Salisbury, pp. 1039-1053, US Government Printing Office, Washington, 1979.

Smith, H.J., A.J. Spivack, H. Staudigel, and S.R. Hart, The Boron Isotopic Composition of Altered Oceanic Crust, *Chemical Geol.* in press., 1995.

Spivack, A.J., and H. Staudigel, Low - temperature alteration of th upper oceanic crust and the alkalinity budget of seawater, *Chemical Geology*, 115, 239-247, 1994.

Staudigel, H., and H.-U. Schmincke, The Pliocene Seamount Series of La Palma/Canary Islands., *J. Geophys. Res.* 89, 11195 - 11215, 1984..

Staudigel, H., and W.B. Bryan, Phenocryst - Redistribution in pillow lavas from DSDP sites 417D and 418A, *Contrib. Mineral. Petrol*, 78, 255-262, 1981.

Staudigel, H., W.B. Bryan, and G. Thompson, Chemical variation in glass - whole rock pairs from individual cooling units in Holes 417D and 418A., in *Init. Repts. of Deep Sea Drilling Project v, 51, 52, 53 Part 2* , edited by T. Donnelly, and J.e.a. Francheteau, pp. 977-986, U.S. Govt. Print. Off., Washington, 1979.

Staudigel, H., G.R. Davies, S.R. Hart, K.M. Marchant, and B.M.;. Smith, Large scale isotopic Sr, Nd and O isotope anatomy of altered oceanic crust at DSDP/ODP sites 417/418., *Earth Planet. Sci. Lett.*, 130, 169-185, 1995.

Staudigel, H., K. Gillis, and R. Duncan, K/Ar and Rb/Sr ages of celadonites from the Troodos ophiolite, Cyprus, *Geology*, 14, 72-75, 1986.

Staudigel, H., and S.R. Hart, Alteration of basaltic glass: mechanisms and significance fro the oceanic crust-seawater budget, *Geochim. Cosmochim. Acta*, 47, 337-350, 1983.

Staudigel, H., S.R. Hart, and S. Richardson, Alteration of the oceanic crust: Processes and timing, *Earth and Planet. Sci. Lett.*, 52, 311-327, 1981a.

Staudigel, H., K. Muehlenbachs, S.H. Richardson, and S.R. Hart, Agents of low temperature ocean crust alteration, *Contrib. Mineral. Petrol.*, 77, 150-157, 1981b.

Sun, S.-S., and W.F. McDonough, Chemical and isotopic systematics of oceanic basalts: implications for mantle composition and process, in *Magmatism in the Ocean Basins.*, edited by A.D. Saunders, and M.J. Norry, pp. 313-345, 1989.

van Damm K.M., Systematics and postulated controls of submarine hydrothermal solution chemistry., *J. Geophys. Res.* 93, 4551-4461, 1988

Tera, F., L. Brown, J. Morris, I.S. Sacks, J. Klein, and R. Middleton, Sediment incorporation in island-arc magmas: Inferences from [10]Be, *Geochim. Cosmochim. Acta*, 50, 535-550, 1986.

Tröger, W.E., *Optische Bestimmungstabellen für Gesteinsbildende Mineralien*. Schweizerbart'sche Verlagsbuchhandlung Stuttgard, 188pp, 1979

Varga, R.J., Modes of extension at oceanic spreading centers: Evidence from the Solea Graben, Troodos Ophiolite, Cyprus, *J. Structur. Geol.*, 13, 517-537, 1991.

Hubert Staudigel[1], Faculty or Earth Sciences, Free University Amsterdam, De Boelelaan 1085, 1081HV Amsterdam, Netherlands; and Institute for Geophysics and Planetary Physics, Scripps Institution of Oceanography, La Jolla, CA 92093-0225, USA

Terry Plank, Department of Geology, 120 Lindley Hall, University of Kansas, Lawrence KS 66045

Bill White, Dept. Geological Sciences, Cornell University, Ithaca, NY 14853

H.-U. Schmincke, Abteilung Vulkanologie und Petrologie, Geomar Forschungszentrum, Wischhofstr. 1-3, D 2300 Kiel, Germany

Appendix 1 Composite Recipes.

Sample #		% in super	depth	Top FLO	100m FLO	300m FLO	Top VCL	100m VCL	300m VCL
417A									
1	24-4 104-108	0.51	24.97	3.55					
2	24-2 52-54	0.08	4.03				5.15		
3	24-1 135.137	0.43	21.02	3.55					
4	24-2 110-118	0.43	21.04	3.63					
5	24-2 25-28	0.52	25	1.07					
6	24-2 80-82	0.08	3.94				5.28		
1	32-2 84-86	0.61	29.9						
2	32-3 40+45	0.10	5.02	1.11					
3	32-4 114-116	0.05	2.5				5.15		
4	32-4 49-51	0.61	30.06	3.75					
5	32-5 48-50	0.05	2.56				5.94		
6	32-1 88-91	0.61	29.97	3.64					
1	44-3 109-111	2.05	27.03		11.51				
2	46-1 75-77	1.44	19.01		11.58				
3	46-1 99-101	1.36	17.99		11.57				
4	46-1 106-108	0.83	11.02		1.12				
5	46-2 15-17	0.45	5.98		11.47			9.18	
6	46-2 52+55	1.44	18.97						
417D									
1	22-1 5-7	0.16	2.01				7.86		
2	22-158-62	1.13	14.18	5.9					
3	22-5 17-19	0.35	4.42	1.52					
4	22-5 41-43	0.96	12.06	5.96					
5	26-1 9-12	0.16	2.04				7.86		
6	26-1 106-108	0.27	3.34	1.77					
7	26-1 112-115	0.27	3.34	1.6					
8	26-1 126-127	1.13	14.12	6.25					
9	27-1 79-82	1.13	14.16	5.89					
10	27-4 45-49	1.13	14.09	5.99					
11	27-4 76-79	0.16	1.99				7.84		
12	27-4 121-122	1.14	14.27	6.02					
1	39-1 30-40	0.83	8.33		0			15.44	
2	39-1 115-117	0.83	8.35		1.01				
3	39-1 121-125	0.83	8.35		1.06				
4	39-2 47-51	3.49	35.02		11.6				
5	39-4 22-27	3.99	39.95		11.62				
1	59-3 51-53	0.60	4.03		0			15.46	
2	59-4 40-42	13.76	92.01		11.55				
3	60-5 97-100	0.59	3.96		0			9.11	
418 A									
1	15-1 23-25	1.30	16.33	5.93					
2	15-1 123-125	0.27	3.39	1.61					
3	15-2 40-42	0.27	3.37	5.9					
4	15-2 58-60	0.27	3.36	1.59					

5	15-2 88-89	1.31	16.41	5.94		
6	15-2 140-144	0.64	8.01			
7	15-3 71-73	1.31	16.37	5.92		
8	15-4 8-10	1.31	16.36	5.99		
9	16-1 120-123	1.31	16.4	5.91		
1	40-2 15-17	0.10	1.31			9.01
2	40-2 52-56	0.10	1.35			9.09
3	41-2 113-118	1.90	25.37	1.15		
4	41-2 20-24	0.10	1.33			9.04
5A	41-2 92-96	1.89	25.32			
5B	41-2 92-96	0.00	0		5.48	
6	41-2 131-136	0.10	1.29		9	
7	42-1 17-21	0.43	5.78	1.13		
8	42-1 50-54	0.10	1.36			9.16
9	42-3 5-9	0.43	5.77	1.07		
10	42-3 57-60	0.43	5.76	1.1		
11	42-2 103	1.89	25.36	11.46		
1	73-4 38-42	0.42	2.08	2.4		
2	74-1 33-37	3.39	16.95	2.44		
3	75-1 50-53	3.40	17	2.45		
4	75-1 121-122	1.26	6.287	14.55		
5	75-2 68-70	1.19	5.97			15.04
6	75-3 75-78	3.38	16.9	14.9		
7	75-4 106-109	1.20	5.98			
8	75-5 42-45	0.40	2.02			15.09
9	77-5 100-105	5.37	26.88			
1	85-6 115-118	2.69	13.49	14.69		
2	85-7 42-46	2.70	13.52	14.57		
4	86-1 99-102	0.39	1.94		23.21	
5	86-1 43-45	0.40	2.01		23.22	
6	86-2 46-49	1.70	8.5	2.41		
7	86-2 141-146	4.99	25.01	14.43		
8	86-3 27 -30	0.41	2.05		23.44	
9	86-5 20-24	1.70	8.52	2.5		
10	86-5 45-48	4.98	24.96	14.65		

Accretionary Mechanics with Properties that Vary in Space and Time

Dan M. Davis

Department of Earth and Space Sciences, SUNY - Stony Brook, Stony Brook, NY

Sediments in accretionary wedges undergo a series of physical and chemical changes that eventually yield rocks whose mechanical behavior has changed greatly since the time of their accretion. These transitions in the shear localization, yielding, and seismic behavior of accreted sediments are all part of the evolution of a sedimentary packet from its initial accretion to its burial to the deeper parts of the accretionary forearc. The loss of fluid from pore spaces and the development of high fluid pressures appear to be the common factors linking several distinct aspects of the tectonics of forearcs. These factors appear to control the initial ability of strain to become concentrated on discrete faults. The transition to fully shear-localized, dilatant behavior is a necessary (but not sufficient) condition for the onset of seismic slip and is a prerequisite for the thrust-ramp geometry of the shortening in many accretionary wedges. Some other important factors, such as dewatering reactions and cementation are also intimately tied to the flow of fluids within the forearc. In addition, the variations of yield strength created by these same fluid-related processes of compaction, cementation, and failure mode evolution all exert important controls on the overall shape of and strain distribution within the forearc, as well as on the magnitudes, vergences, and loci of large-slip thrusts. Sharp lateral contrasts in strength, whatever their origin, lead to the formation of backstops different from those made by arc basement, which can dominate the structural geology of a forearc.

INTRODUCTION

Seismic reflection studies have discovered, and laboratory models have reproduced, a wide variety of tectonic styles at trenches. Margins have been found to be accretionary, with material added both frontally and at depth along the base of the overlying plate, and to undergo erosion of the upper plate both along the trench axis and at depth. A relatively small fraction of modern subduction zones can be described accurately as purely accretionary, and even their accretion and erosion histories may be neither simple nor easily determined. Nonetheless, because it can leave an enormous mass of sedimentary material (sometimes many hundreds of km^3 per km of trench) at and near the surface, accretion (and particularly frontal accretion) is well recorded in the geologic record. Simple mechanical models that balance forces above the plate-boundary décollement have been developed in order to understand the shape of the accretionary wedge and the general geometry of deformation within it [e.g., *Chapple*, 1978, *Davis et al.*, 1983; *Willett*, 1992]. Despite their simplicity, such models have been shown to be capable of explaining much about the growth of accretionary wedges. However, recent work emphasizes that it can be a mistake to view accretionary margins as simple in either space or time.

Even steady-state wedges that have had extremely simple growth histories have significant gradients in mechanically important properties, such as porosity, yield strength, failure mode, and fluid pressure [e.g., *Bangs et al.*, 1990; *Cochrane et al.*, 1994]. Furthermore, some processes related to compaction and lithification are neither steady-state nor simply linear in their effects. For example, subduction of basement topography, such as seamounts or seismic ridges, and climatic variation that affects the rates of erosion and sedimentation, can produce non-steady-state accretion. This can lead to properties across the forearc that vary in a complicated, stepwise manner. Even if the rate of accretion has been steady-state and the lithology of the accreted material has been constant through time, properties may vary discontinuously across the wedge. In this paper, I examine some ways in which even relatively simple accretion can lead to the existence of distinct structural and tectonic boundaries within an accretionary forearc.

A transition from non-localized to localized shear is required for the formation of discrete faults, and experimental evidence suggests that this transition occurs over a small range of properties, such as effective confining pressure, and porosity, that vary across a forearc. The same is probably true of the transition, generally much deeper, to seismic (as opposed to aseismic) frictional slip. An important control on all aspects of forearc mechanics is the distribution of elevated pore fluid pressures. Overpressuring is complicated by both the kinetics of dehydration reactions, which can produce an extremely uneven rate of production of new pore water across the forearc, and by the formation of discrete faults, which can cause highly anisotropic permeability and high overpressures near the toe of the accretionary wedge.

LIMITS OF SIMPLE MODELS

In light of the variety and complexity of natural accretionary wedges, it is remarkable that simple mechanical models tell us anything at all. Even under the least complicated of conditions, the porosity of an accretionary wedge will generally decrease with both depth and distance from the toe of the wedge [e.g., *Bray and Karig*, 1985; *Bangs et al.*, 1990; *Cochrane et al.*, 1994], as the accreted materials are compacted and begin to be lithified. In addition to having zones of sharp strength contrast along any lithologic boundaries or faults that influence hydraulic properties, the accreting mass will generally increase in frictional strength downward and landward of the deformation front because the sediments are better compacted and lithified there. Such 'steady-state' gradients in porosity, density, and yield strength can explain the usually concave cross-sectional shape of the frontal regions of most accretionary wedges [*Zhao et al.*, 1986; *Breen and Orange*, 1992]. Great progress has been made in using seismic techniques to estimate properties away from drill holes [e.g., *Bangs et al.*, 1990], but the absolute values of such parameters as yield strength and pore fluid pressure remain only roughly known across much of even the best studied forearcs.

Even when an accretionary wedge behaves in the most idealized of ways, its response to small perturbations in properties and boundary conditions may be disproportionate. *Dahlen* [1984] pointed out that for a given set of sediment strengths and degrees of overpressuring, there exists a stability field (in décollement dip-surface slope space) within which a stable wedge can exist. The extent of this field for typical modern accretionary wedges has been mapped out by *Lallemand et al.* [1994]. Wedges that have a surface slope angle smaller than that of the lower end of the stability field will shorten by thrust faulting when compressed horizontally, while those steeper than the upper end of the stable field will extend, generally by normal faulting. The size of the stability field separating those two types of deformation shrinks as fluid pressures increases. Thrusting in an accretionary wedge can be affected by stratigraphically or structurally induced perturbations in fluid pressures, or by changes in the geometry of the décollement due to underplating [e.g., *Platt*, 1986] or by the underthrusting of basement relief such as a ridge or a seamount. However, if the wedge is highly overpressured, the wedge may be weak enough and its stability field small enough that even small changes in the location of strength of the décollement may cause a contractional wedge to undergo extension by normal faulting. Such extension has been proposed for part of the Barbados Ridge wedge by *Brown et al.* [1990]. Equally important may be the ability of small perturbations in the boundary conditions of an overpressured wedge to cause the wedge to change temporarily from accreting to non-accreting, and to change the balance between frontally accreting and subducting sediments.

The size of the subduction channel [e.g., *Cloos and Schreve*, 1988] can have a huge impact upon the type of tectonics found in a forearc, altering the style of accretion or making the margin entirely non-accretionary. *Kukowski et al.* [1994] have demonstrated the importance of the relative sizes of the pile of sediments entering at the trench and the 'subduction gate' at depth, demonstrating experimentally how that sediment balance can affect accretion and erosion, which in some cases can occur simultaneously.

MAJOR DISCRETE MECHANICAL CONTRASTS (BACKSTOPS)

Both 'sandbox' and numerical modeling [*Malavielle*, 1984; *Byrne et al.*, 1988, 1993; *Willett*, 1992; *Wang and Davis*, 1996a,b] indicate that a sharp kinematic boundary at the front of the stable upper plate in a forearc dominated by friction can cause a pair of opposite-facing thrusts to form on either side of an outer-arc high above the kinematic boundary. Landward-vergent thrusts develop most strongly when the top of such a buttress or backstop is horizontal, and they fail to form if it has a large seaward dip. Similar results, along with extension near the surface of the high, are found in a viscous model [*Buck and Sokoutis*, 1994]. The material landward of such a kinematic discontinuity, which remains essentially fixed to the overlying plate instead of moving with the downgoing plate or with the accreting or underthrusting sediments, is called the backstop.

The horizontal strain rate, integrated across a path connecting a point seaward of the subduction zone on the downgoing plate and another point away from deformation on the overlying plate, must equal the plate convergence velocity. However, strain rate may be distributed across that path in any of an infinite number of possible ways. If the distribution is roughly uniform, with no sharp contrasts in strain rate, then there is nothing that can properly be called a backstop. However, if there is a transition over a relatively short distance between rapidly straining sediments and much slower deforming material just landward of it, then that less rapidly deforming material may be acting as a backstop (or buttress) for the sediments seaward of it. The imprecision inherent in any definition of a backstop (for example, what is a 'relatively short' distance?) makes it difficult to define what is and is not a backstop. Natural backstops appear in a number of forms, including continental or volcanic arc basement, accreted terranes, or even the relatively dewatered remains of an older prism after ridge subduction. There is a difference between a backstop as described here and a buttress as described by *von Huene and Scholl* [1991]. I use the term backstop to mean only the stronger material, at whatever depth, that supports a disproportionately large part of the compressional force in the forearc. However, any weaker material above the backstop that is protected from deformation by it might be described as part of the buttress.

Our modeling shows that the presence of any distinct landward increase in yield strength at depth in a forearc can produce such a contrast in strain rate and hence the development of a backstop. This then leads to a reversal in thrust vergence and the growth of a structurally and topographically elevated 'pop-up' zone. Depending upon the strength and geometry of the backstop, that pop-up ultimately develops into a structural high bounded by an asymmetric pair of contractional wedges (Figure 1). In nature, backstops are neither perfectly rigid nor perfectly horizontal, so the particular model shown here is an end-member case. A deforming backstop complicates the structures that form above it, and a seaward dip at the top of the backstop reduces the number of arcward-vergent

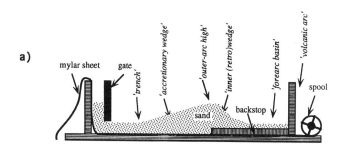

Fig. 1a. Schematic diagram illustrating how a sandbox model creates a model forearc. A Mylar sheet is pulled onto a spool at right, beneath a rigid, flat 'backstop'. A clear side wall allows us to see the model as it deforms.

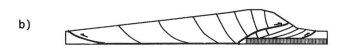

Fig. 1b. Line drawing showing the thrusts faults in the model shown above. The flattest active faults are indicated by thicker lines, and arrows indicate the thrust motion along them. Earlier faults are steepened and transported 'arcward' by subsequent deformation. *Wang and Davis* [1995].

Fig. 1c. A photograph of a sandbox model forearc. The growth of a broad accretionary wedge (to the left) and a narrower inner deformation belt (to the right of the 'outer-arc high') follow inevitably from the presence of a low-angle backstop that provides a good kinematic discontinuity.

structures produced there. An 'active buttress' as described by *von Huene and Scholl* [1991] (as opposed to their 'core buttress', or main backstop) is such as incomplete kinematic boundary.

There are two obvious way in which such an active buttress may form. It may result from a marginally stronger part of the wedge supporting much of the stress (a weak backstop). Conversely, it may be the result of extremely weak material, such as underplated sediments beneath the wedge having reduced plate boundary frictional coupling enough to have rendered the wedge above it locally supercritical (sufficiently flat in taper so that it does not undergo brittle deformation in compression). In either case, the affected part of the wedge could act as an active buttress for the rest of the wedge trenchward of it.

Modeling with different types of backstops indicates that above a sufficiently strong backstop the development of an outer-arc high and a relatively undeformed forearc basin behind it are inevitable [*Byrne et al.*, 1993; *Wang and Davis*, 1996a]. However, the appearance of these features may be altered or masked by either a dearth or an extreme excess of sediment, particularly if the source is terrigenous.

Some similar, but less dramatic, structural features may be produced in the wedge where a strength contrast develops. *Bangs et al.* [1990] imaged two relatively sharp seismic velocity gradients in the forearc of the Lesser Antilles. By far the more important of these is the upper and seaward surface of arc basement, which acts as the backstop beneath an outer-arc high (the Barbados Ridge) and the (Tobago) forearc basin. The other zone of velocity contrast is located within the accretionary wedge itself, beneath an inflection in the surface slope and just behind a major backthrust, both of which 'sandbox' and numerical modeling [e.g., *Byrne et al.*, 1993] predict to form over a minor seaward-dipping backstop. This minor zone of relatively low strain rate (a weaker backstop) may be a result of a non-uniform history of accretion. Because of the slowness of porosity loss and lithification, a hiatus, slowdown, or lithologic change in accretion at some time in the past can cause there to exist today a region of lower porosity, greater lithification, and higher strength that can act as a backstop.

Along with the chemical processes involved in lithification, porosity loss is one of the most important actors controlling the yield strength of sediments. Strength contrasts large enough to produce significant structural effects can result from porosity contrasts of only a few percent [e.g., *Hoshino et al.*, 1972]. Non-uniform porosity and overpressures can have very important effects upon the overall mechanics of an accretionary wedge. It is logical therefore to ask about how porosity loss and overpressuring affect the way in which that deformation occurs - by distributed strain or as discrete faults, as well as either seismically or aseismically.

INITIATION OF LOCALIZED SLIP

In order to understand what controls the ability of sediments to start to form discrete faults we must investigate the modes of failure of sediments under the range of conditions found at and near a trench. A transition between non-localized, non-dilatant, 'ductile' (strain hardening) deformation and shear-localized, dilatant 'brittle' (strain weakening) failure in porous sediments was mapped as a function of two parameters known to vary rapidly across accretionary wedges: porosity and effective pressure [*Zhang et al.*, 1993a,b]. To produce samples at a variety of initial porosities, we first compacted them isotropically. This was done slowly enough to allow water to drain from the shrinking pore spaces and to avoid significant overpressuring. We then removed the load from the samples, allowing them to attain the desired effective combination of confining pressure and initial porosity. We then deformed them under differential stress ($\sigma_1 > \sigma_2 = \sigma_3$) and noted whether their deformation was macroscopically 'brittle' or 'ductile'. When combined with data from other studies, this allowed us to map a 'failure mode' boundary between localized/dilatant and non-localized/non-dilatant behavior (e.g., Figure 2a). The formation of discrete faults occurs at relatively low values of porosity and effective pressure, and non-localized shear is found at higher porosities and effective pressures.

Similar conclusions have been drawn by *Hill and Marsters* [1990] using a different conceptual framework. They use critical state soil mechanics [*Schofield and Wroth*, 1968] to postulate the existence of a critical state line separating dilative and contractive behavior in void ratio/effective stress space. *Brandon* [1984] pointed out that the stress path of a sediment during deformation is a function of its density relative to its critical state as defined by *Schofield and Wroth* [1968]. Hill and Marsters suggest that small-scale faulting and scaly cleavage may represent opposite sides of the critical state line. In particular, they suggest that a sediment with a strong grain framework allows there to be only a small drop in porosity as pressure increases, with the result that at greater depth there is contractive, non-localized deformation, such as scaly cleavage. Presumably, the opposite could be true for a sediment with a weak grain framework for which porosity is a strong function of confining pressure.

Sediments from ODP leg 131 in the toe region of the Nankai accretionary wedge were found experimentally to undergo this transition at porosities of 40% to perhaps 20%, depending upon the effective confining pressure [*Zhang et al.*, 1993b]. At the high porosities typically found near the toe of an accretionary wedge [e.g., *Karig*, 1986; *Bangs et al.*, 1990; *Cochrane et al.*, 1994] the behavior is non-localizing, which means that the initial deformation should not be expected to be in the form of faults. Instead, the sediments are capable of strain in a less strain-localizing form as they compact. These results are

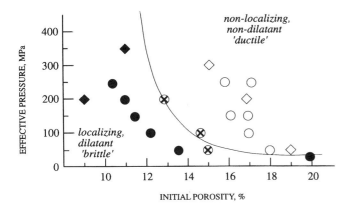

Fig. 2a. A 'brittle'-'ductile' transition map for Kayenta sandstone, separating the domain of localized faulting with dilatancy and that of non-localized, non-dilatant shear in terms of effective pressure and initial porosity [Zhang et al., 1993a]. Circles are the data of Zhang et al. [1993a] and diamonds are data from Jones [1980]. Solid shapes indicate samples that deformed in a macroscopically brittle manner, with shear localization. Open shapes indicate samples that showed non-localized shear, and the circles with crosses correspond to samples that deformed in an intermediate manner. The curved line is the surmised boundary between the two failure modes.

summarized in Figure 2b, which shows porosities of samples taken from depths greater than 200m (at shallower depths, drilling disrupted the poorly compacted samples enough to give inconsistent scattered results).

Each of the five large circles in Figure 2b represents the initial porosity and effective confining pressure for one of our Nankai samples. The solid circle indicates that deformation at 30% porosity and an effective confining pressure of 5 MPa produces a localized 'brittle' fault. The open circles represent the experiments at higher porosity and effective pressure at which the deformation was 'ductile' (non-localizing compactive shear). The two circles with crosses in them indicate intermediate behavior and therefore map out part of the failure mode transition. The small dots correspond to the measured porosities from the same Nankai drilling hole at Site 808 [Taira et al., 1992]. These data are plotted in terms of effective pressures calculated by integrating the overburden to the sample recovery depth, assuming that the bulk density of the sediment is 2 g/cm^3 and that pore pressures are everywhere hydrostatic. This latter oversimplified assumption will come into play later. The separation of the data in Figure 2a into distinct failure modes suggests that the way in which a sediment fails in compression at any given effective pressure depends largely upon its porosity.

Taken at face value, Figure 2b suggests that deformation in the Nankai toe should be essentially all non-localized.

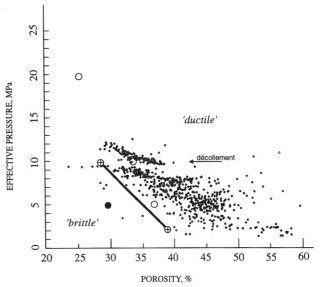

Fig. 2b. A failure mode transition map for ODP samples from the Nankai trough. The large solid circle indicates a sample that produced a localized 'brittle' fault. The large open circles represent experiments with non-localizing compactive shear. The two circles with crosses in them indicate intermediate behavior and therefore map out part of the failure mode transition.. The small dots indicate porosities measured on board the drilling ship from drilling at Site 808 [Taira et al., 1992]. Their effective pressures are plotted assuming that the bulk density is 2 g/cm^3 and that pore pressures are everywhere hydrostatic [Zhang et al., 1993b]. Data from depths less than 200m below sea floor are omitted because those samples are badly disrupted by drilling and show a great deal of scatter in their measured porosities. The depths of the décollement and the frontal thrust are indicated by horizontal arrows.

The primary exception is the basal décollement, which corresponds to the obvious offset in porosity plotted at an effective pressure of about 9.5 MPa in Figure 2b. Those points fall clearly on the shear-localizing 'brittle' side of the failure mode boundary. However, the décollement appears to be overpressured [Taira et al., 1992], so the data from the décollement should plot at lower effective stresses, deeper into the 'brittle' regime within which it must be in order to undergo discrete faulting. The same is probably true to some degree of the frontal thrust of the accretionary complex, data from which plot at about 3.6 MPa and show an offset similar to but smaller than that for the basal décollement.

These results suggest that localization of shear at the front of accretionary wedges may be a complicated process involving fluid loss and overpressuring, somewhat of a chicken-and-egg problem. Discrete faulting is aided by overpressures, but it is the faults themselves that are responsible for a large part of the overpressuring by increasing the overburden and by retarding fluid flow normal to the fault plane. Even if the sediments are too porous for the formation of discrete faults, then some sort of less localized compactive shear can still lead to kink bands and relatively broad shear bands that develop as protothrusts of limited net displacement. As compaction continues and these protothrusts develop and alter the hydrology of the front of the wedge by making the permeability tensor increasingly anisotropic, there will eventually occur transitory localized increases in fluid pressure sufficient to reduce the effective pressure enough to allow a transitional, partial localization of shear. That in turn will affect the permeability tensor sufficiently to allow the formation of localized shear zones and discrete faults.

Cochrane et al. [1994] show that seismic velocities increase rapidly landward in the protothrust zone of the Oregon wedge. At the seaward end of the protothrust zone, they show porosities ranging from about 40% near the top of the protothrust shear bands about 200m below the sea floor to about 20% near the bottom of the bands, just above the proto-décollement. The protothrusts are found landward to a point below the top of the frontal thrust, where *Cochrane et al.* [1994] estimate porosities to be as low as 18%. Small-scale faults are found in Nankai site 808 samples at all depths, but deformation bands are visible in drilling samples only to a depth of about 560m [*Maltman et al.*, 1992]. Drilling disturbances and porosity rebound are likely to bias sample porosities upward, so the in situ porosity at any given depth is most likely to be near the bottom of the range of the shipboard sample porosity measurements [*Taira et al.*, 1992] near that depth. Virtually all of the data points in Figure 2b from outside the fault zones fall well above and to the right of the failure mode boundary, into the macroscopically 'ductile' field, assuming hydrostatic fluid pressure. However, there is likely to be some amount of excess pore fluid pressures, and even modest overpressures (a modified Hubbert-Rubey normalized fluid pressure ratio λ [*Hubbert and Rubey*, 1959; *Davis et al.*, 1983] of $\lambda \approx 0.7$) will reduce the effective pressure by enough that many of the other sediments may actually fall on the 'brittle' side of our estimated failure mode boundary [*Zhang et al.*, 1993b].

The faults that ultimately form are true brittle/frictional faults for which the shear traction required for slip increases with increasing normal traction. In other words, they have a non-zero internal friction angle and are therefore likely to form with landward dips of substantially less than 45°, even if there is very little shear traction along the base of the accretionary wedge [*Hafner*, 1951; *Davis and von Huene*, 1987], as in the Nankai basal décollement which is apparently highly overpressured. In fact, *Taira et al.* [1992] give a value of 28° for the frontal thrust.

However, none of this is true of broad compactive shear zones seaward of the frontal thrust, because they are not truly brittle. If the Nankai basal décollement is truly weak, then seaward vergent (landward-dipping) shear zones would be expected to have mean dips near 45° (somewhat shallower if the anisotropy in sediment strength is large and somewhat steeper if there has been significant post-slip subhorizontal compactive shortening). Over a weak, nearly horizontal décollement, the ideally preferred shear zone dips in both directions are similar, and seaward-dipping zones are not heavily disfavored and thus may appear along with landward-dipping ones. Pairs of seaward- and landward-vergent thrusts have been described for the Kodiak margin by *Davis and von Huene* [1986] and for the Nankai margin by *Lallemand et al.* [1994]. Steeply-dipping, discontinuous shear zones are observed in seismic reflection lines in the protothrust zone just seaward of the frontal thrust in at least some accretionary wedges, including those at the Oregon [*Cochrané et al.*, 1994] and Nankai [*Moore et al.*, 1990] margins. Once a thrust fault has developed, it commonly thickens into a shear zone meters wide. However, it still remains distinct from a band of compactive shear in several ways. In particular, it becomes and remains weaker than the surrounding sediments, so it is the locus of ongoing strain concentration. In a thrust fault, that strain is accompanied by local dilatation. In addition, slip there occurs according to a frictional yielding criterion that increases in strength with confining pressure.

The fact that protothrusts are found where failure mode analysis suggests that non-brittle, largely non-localized, steeply-dipping shear zones would be expected to be located leads to a remarkable conclusion. In the region extending several kilometers either side of the deformation front in many accretionary wedges, there appears to be laid out before us a map to the failure modes of sediments at low-to-moderate porosities. The sediments from the seaward edge of the protothrust zone to well landward of the trench are all actively deforming. However, largely because of the landward gradient in porosity, we see a range of deformational styles starting with non-localized shear and ending in discrete faulting. A similar transition probably occurs in many basins, but in a vertical direction, and over a shorter distance, making its effects more difficult to image seismically.

Wang et al. [1994] calculated stresses in the footwall of a frontal thrust, at a variety of fluid pressures. They found that the width of the stressed zone (likely to be a protothrust zone) increases as the pore fluid pressure approached lithostatic and that preferred failure is on steeply dipping zones seaward of the frontal thrust scarp. *Shi and Wang* [1985] and *Wang et al.* [1990] assumed deformation in the wedge and calculated the pore fluid pressures that are likely to develop as a result, producing results that lend credence to the idea that elevated pore fluid pressures should develop near the toe of an accretionary wedge.

The creation of faults and the generation of overpressures are closely coupled phenomena in submarine accretionary wedges. To evaluate this coupling, we have developed a cyclically alternating undrained loading-diffusion model to simulate the mechanical processes of loading and fluid draining in natural forearcs. This cyclic loading-diffusion process approaches steady state for time steps that are small enough to produce negligible loading [*Wang and Davis*, 1996b]. Each undrained deformation step is then followed by a drained step.

A detailed review of the permeability of forearc materials was given by *Wang et al.* (1990). In producing Figure 3 we have used the same properties as in their study. We assume that the preferred orientation of a fault zone in the yielded element is described by Mohr-Coulomb criterion. Figure 3a shows the initial geometry of the model. Figure 3b is the result of a run in which there was diffusive, as opposed to fracture-dominated directed fluid flow. As *Wang et al.* [1990] did in their modeling, we assigned a hydraulic conductivity to the fault zone that is higher than that of the surrounding rocks or sediments (by a factor of 10^3) in creating Figure 3c. Our preliminary results for induced overpressuring in and in front of the deforming wedge bear considerable resemblance to those of *Wang et al.* [1990]. In our coupled models with frictionally-controlled wedge mechanics, tectonic stresses begin to have a significant effect on pore fluid pressures even at a very early stage in the accretion process (Figure 3b), particularly at and immediately in front of the toe of the wedge when fluid flow is dominated by faults. We find that from the earliest stages of deformation in numerical models that take into account the dominant effect of faults on fluid flow, overpressures start to concentrate near the toe of the wedge, instead of in front of the backstop. The generation of overpressures near and just in front of the frontal thrust is important to the continued development of the wedge, because that is precisely where the décollement is propagating farther seaward and undercompacted sediments are deformed by protothrusts. The influence of the tectonics upon pore fluids inevitably has corresponding effects upon both the generation of new thrust slices and upon the processes of diagenesis, strengthening, and the changes in failure mode that the sediments undergo during the process of accretion and wedge-building.

The sediments in the frontal part of an accretionary wedge are capable of producing thrust ramp-flat structures like those in subaerial fold-and-thrust belts, albeit with much more ongoing volume loss. However, over a very large distance from the deformation front, many accretionary wedges appear to be aseismic [*Byrne et al.*, 1988]. Within this region deformation is frictional and localized. However, that frictional slip apparently occurs either by frictional creep or by very small episodic slip events that do not release any significant elastic strain energy. At some depth, there must be another transition in failure mode to one that allows for stick-slip seismic failure. The location of that transition to seismically capable failure is called the seismic front.

SEISMIC FRONT

The main plate-boundary thrust at subduction zones has produced many of the largest and most damaging of earthquakes, including the 1960 Chile and 1964 Alaska events. Simple scaling laws suggest that the maximum moment for an earthquake at most margins is approximately proportional to the cube of the seismogenic zone width [e.g., *Byrne et al.*, 1988], so an appreciation of the factors controlling that width is important in understanding seismic risk.

The role of the high-temperature transition from seismic to aseismic slip at the deep end of the seismogenic zone at plate boundaries is well appreciated, but the nature of the shallow, trenchward limit to seismicity is less clear. The prevalence of aseismicity in sediment-rich accretionary wedges around the world suggests that the aseismic behavior of the frontal regions of forearcs is related to the physical state of the sediments common to that setting. The association of aseismic behavior with porous and relatively weak sediments is reinforced by the observation that backstops capable of producing pronounced outer-arc highs and forearc basins are usually seismic while the seismically slower and more porous sediments trenchward of them are invariably aseismic [*Byrne et al*, 1988].

Great plate-boundary thrust earthquakes cannot occur if the down-dip width of the seismogenic zone is small: in such a case there is simply not enough plate contact area to produce a large seismic moment. This would be the case in a subduction zone where the slab dips steeply and where there is little or no sediment to increase the plate contact area at shallow depths, such as the Marianas.

Even if the plate contact area is large, a great thrust event is impossible if most of that contact is incapable of storing elastic strain energy. Friction is not necessarily accompanied by stick-slip behavior: the stability or instability of slip depends upon both the frictional properties of the fault and the elastic properties of the

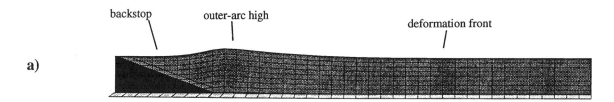

Fig. 3a. The finite element grid used in the numerical calculations. The dark triangular region at left is the relatively rigid backstop. The weak décollement along the bottom of the model is lightly shaded. There is a 2x vertical exaggeration.

Fig. 3b. Shading indicates relative excess pore fluid pressure in an early stage of deformation in a coupled mechanical/hydrological finite element model in which we assumes that fluid flow is diffusive, with isotropic permeability. Overpressures initially concentrate near the backstop [*Wang*, 1994].

Fig. 3c. The results of a finite element model that was identical to the one above, except that we assumed that faults would form in the ideal orientations for the state of stress, and that those faults would introduce highly anisotropic permeability, with fluid flow 1000 times easier along the faults that across them. Note that in this case, overpressures start to form near the toe of the wedge, facilitating the forward growth of the wedge.

surrounding rock that loads the fault elastically. Compilations of depth distributions of crustal earthquakes around the world [e.g., *Meissner and Strehlau*, 1982] are consistent with the idea that the seismogenic realm is generally bounded by the onset of crystalline plasticity at depth, but some other factor must be at work, keeping earthquakes from being nucleated within the top few kilometers. One factor could be how the elastic stiffness required for stick-slip behavior depends upon the normal stress at shallow depths, but earthquakes occur at extremely shallow depths in hard rock so there must be other issues involved [e.g., *Marone and Scholz*, 1988]. Instead, the seismicity or aseismicity of faults at shallow depths is likely to be strongly related to the degree of consolidation of the material in which the faulting occurs. Even when accreted sediments are sufficiently well consolidated to be able to form discrete faults, they are likely to be velocity-strengthening: velocity-weakening behavior is necessary for fault instability and seismic, stick-slip behavior. Therefore, even if the plate contact area is large, the seismogenic zone must be small if most of that boundary consists of relatively recently accreted, porous sediments. This appears to be the case at the Makran margin, where much of the broad accretionary wedge seaward of the coastline may be aseismic [*Byrne et al.*, 1992].

The most favorable condition for the generation of great plate-boundary earthquakes is that in which there is a huge body of accreted material that causes the plate contact area to be very large, but in which most of those accreted sediments have been accreted for a sufficiently long time as to have become dewatered and lithified. In such a case, the sediments can act as stick-slip, velocity-weakening seismically capable rock and a great plate-boundary thrust earthquake can occur [*Zhang et al.*, 1993a]. A prime example of this situation would be the Alaskan margin.

SUMMARY

The loss of fluid from pore spaces and the development of high fluid pressures appear to be the common factors linking several distinct aspects of the tectonics of forearcs. These factors appear to control the initial ability of strain to become concentrated on discrete faults. The transition to fully shear-localized, dilatant behavior explains the contrast in structural style between the protothrust zones observed seaward of some accretionary wedges and the dominance of faulting as a shortening mechanism in the wedge. Localized slip on faults is a necessary (but not sufficient) condition for the onset of seismic slip. Some other factors that may contribute to causing that transition, such as dewatering reactions and cementation are also intimately tied to the flow of fluids within the forearc. Finally, the variations of yield strength created by these same fluid-controlled processes of compaction and cementation as well as by lithologic contacts (e.g., the front of arc or continental basement) exert an important control on the overall shape of and strain distribution within the forearc and on the magnitudes and vergences of slip on thrusts.

Acknowledgments. This research was supported by NSF grant OCE9402008. Various parts of the work described here was done in collaboration with Wei-Hau Wang, Teng-fong Wong, and Jiaxiang Zhang. Nathan Bangs, Serge Lallemand, and Don Reed provided careful and very helpful reviews, all of which are appreciated.

REFERENCES

Bangs, N. L. B., G. K. Westbrook, J. W. Ladd, and P. Buhl: Seismic velocities from the Barbados Ridge: Indicators of high pore fluid pressures in an accretionary complex, *J. Geophys. Res., 95*, 8767-8782, 1990.

Brandon, M. T, Deformational processes affecting unlithified sediments at active margins: a field study and a structural model. (Ph.D. dissert.), Univ. of Washington, 1984.

Bray, C. J., and D. E. Karig, Porosity of sediments in accretionary prisms and some implications for dewatering processes, *J. Geophys. Res., 90*, 768-778, 1985.

Breen, N. A., and D. L. Orange, Effects of fluid escape on accretionary wedges, 1, Variable porosity and wedge convexity, *J. Geophys. Res. 97*, 9265-9276, 1992.

Brown, K. M., A. Mascle, and J. H. Behrmann, Mechanisms of accretion and subsequent thickening in the Barbados Ridge accretionary complex: Balanced cross sections across the wedge toe, in Moore, J. C., Mascle, A. et al., Proceedings of the Ocean Drilling Program, Scientific Results, Vol. 110, 209-226, 1990.

Buck, W. R., and D. Sokoutis, Analogoue model of gravitational collapse and surface extension during continental convergence, submitted to *Nature 369*, 737-740, 1994.

Byrne, D., D. M. Davis, and L. Sykes: Loci and maximum size of thrust earthquakes and the mechanics of the shallow region of subduction zones. *Tectonics, 7*, 833-857, 1988.

Byrne, D. E., L. R. Sykes, and D. M. Davis: Great thrust earthquakes and aseismic slip along the plate boundary of the Makran subduction zone, *J. Geophys. Res., 97* 449-478, 1992.

Byrne, D. E., W.-H. Wang, and D. M. Davis: Mechanical role of backstops in the growth of forearcs, *Tectonics, 12*, 123-144, 1993.

Chapple, W. M., Mechanics of thin-skinned fold-and-thrust belts, *Geol. Soc. Am. Bull., 89*, 1189-1198, 1978.

Cloos, M., and R. L. Schreve, Subduction-channel model of prism accretion, melange formation, sediment subduction, and subduction erosion at convergent plate margins: 2. Implications and discussion. *PAGEOPH, 128*, 501-545, 1988.

Cochrane , G. R., J. C. Moore, M. E. MacKay, and G. F. Moore, Velocity and inferred porosity model of the Oregon accretionary prism from multichannel seismic reflection data: Implications on sediment dewatering and overpressure, *J. Geophys. Res., 99*, 7033-7043, 1994.

Dahlen, F. A., Noncohesive critical Coulomb wedges: An exact solution, *J. Geophys. Res., 89*, 10125 10133, 1984.

Dahlen, F. A., J. Suppe, and D. Davis, Mechanics of fold-and-thrust belts and accretionary wedges: Cohesive Coulomb theory, *J. Geophys. Res., 89*, 10087-10101, 1984.

Davis, D. M., J. Suppe, and F. A. Dahlen, Mechanics of fold-and-thrust belts and accretionary wedges, *J. Geophys. Res., 88*, 1153-1172, 1983.

Davis, D. M., and R. von Huene, Inferences on sediment strength and fault friction from structures at the Aleutian Trench, *Geology, 15*, 517-522, 1987.

Hafner, W., Stress distributions and faulting, *Geol. Soc. Am. Bull., 62*, 373-398, 1951.

Hill, P. R., and J. C. Marsters, Controls on physical properties of Peru continental margin sediments and their relationship to deformation styles, in Suess, E., von Huene, R. et al., Proceedings of the Ocean Drilling Program, Scientific Results, Vol. 112, 623-632, 1990.

Hafner, W., Stress distributions and faulting, *Geol. Soc. Am. Bull., 62*, 373-398, 1951.

Hoshino, K. H. Koide, K. Inami, S. Iwamura, and S. Mitsui, Mechanical properties of Japanese Tertiary sedimentary rocks under high confining pressure, *Rep. 244*, 200 pp., Geol. Surv. Japan, Kawasaki, 1972

Hubbert, M. K., and W. W. Rubey, Role of fluid pressure in the mechanics of overthrust faulting, *Bull. Geol. Soc. Am., 70*, 115-166, 1959.

Jones, L. M., Cyclic loading of simulate fault gouge to large strains, *J. Geophys. Res., 85*, 1826-1832, 1980.

Karig, D. E., Physical properties and mechanical state of accreted sediments in the Nankai Trough, Southwest Japan Arc, in *Structural Fabrics in Deep-Sea Drilling Project Cores From Forearcs, Mem. 166*, edited by J. C. Moore, pp. 117-133, Geological Society of America, Boulder, Colo., 1986.

Kukowski, N., R. von Huene, J. Malavielle, and S. E. Lallemand, Sediment accretion against a buttress beneath the Peruvian continental margin as simulated with sandbox modeling, *Geologische Rundschau, 83,* 822-831, 1994.

Lallemand, S., P. Schnürle, and J. Malavielle, Coulomb theory applied to accretionary and nonaccretionary wedges: Possible causes for tectonic erosion and/or frontal accretion, *J. Geophys. Res., 99,* 12,033-12,055, 1994.

Malavieille, J.: Modelization experimentale des chevauchements imbriqués: Application aux chaines de mountagnes, *Bull. Soc. Geol. France, 7,* 129-138, 1984.

Maltman, A., T. Byrne, D. Karig, S. Lallemand, and Leg 131 Shipboard Party, Structural and geological evidence from ODP Leg 131 regarding fluid flow in the Nankai prism, Japan, *Earth. Planet. Sci. Letters, 109,* 463-468, 1992.

Marone, C. and C. H. Scholz, The depth of seismic faulting and the upper transition from stable to unstable slip regimes, *Geophys Res. Lett., 15,* 621-624, 1988.

Meissner, R., and J. Strehlau, Limits of stresses in continental crusts and their relation to the depth-frequency distribution of shallow earthquakes, *Tectonics, 1,* 73-90, 1982.

Moore, G. F., T. Shipley, P. Stoffa, D. Karig, A. Taira, S. Kuramoto, H. Tokuyama, and K. Suyehiro, Structure of the Nankai Trough accretionary zone from multichannel seismic reflection data, *J. Geophys. Res., 95,* 8753-8765, 1990.

Platt, J. P., Dynamics of orogenic wedges and the uplift of high-pressure metamorphic rocks, *Geol. Soc. Am. Bull., 97,* 1037-1053, 1986.

Schofield, A., and P. Wroth, *Critical State Soil Mechanics*: McGraw Hill, New York, 1968.

Shi, Y., C.-Y. Wang, High Pore Pressure Generation in Sediments in Front of the Barbados Ridge Complex, *Geophys. Res. Lett., 11,* 773-776, 1985.

Taira, A. and 18 others, Sediment deformation and hydrogeology of the Nankai Trough accretionary prism: Synthesis of shipboard results of ODP Leg 131, *Earth. Planet Sci. Letters, 109,* 431-450, 1992.

von Huene, R. and D. W. Scholl, Observations at convergent margins concerning sediment subduction, subduction erosion, and the growth of continental crust, *Reviews of Geophysics, 29,* 279-316, 1991.

Wang, C.-Y., Y. Shi, W.-T. Hwang, and H. Chen, Hydrogeologic Processes in the Oregon-Washington Accretionary Complex, *J. Geophys. Res., 95,* 9009-9023, 1990.

Wang, C.-y., W.-t. Hwang, and G. Cochrane, Tectonic dewatering and the mechanics of protothrust zones: Example from the Cascadia accretionary margin, *J. Geophys. Res., 99,* 20,043-20,050, 1994.

Wang, W.-H., Mechanical and hydrologic processes in the growth of forearcs, Ph.D. dissertation, 202 pp., State University of New York at Stony Brook, May 1994.

Wang, W.-H., and D. M. Davis. Sandbox model simulation of forearc evolution and non-critical wedges *J. Geophys. Res.,* 101, 11,329-11,339, 1996a.

Wang, W.-H., and D. M. Davis, Numerical Modeling of Coupled Fluid-Mechanical Controls on the Development of a Forearc, *submitted to J. Phys. Ear.,* 1996b.

Willett, S. D., Dynamic and kinematic growth and change of a Coulomb wedge, in *Thrust Tectonics,* edited by K. R. McClay, pp. 19-32, Chapman & Hall, New York, 1992.

Zhang, J., D. M. Davis, and T.-f. Wong: Brittle-ductile transition in porous sedimentary rocks: geological implications for accretionary wedge aseismicity, J. Structural Geol., 15, 7, 819-830, 1993a.

Zhang, J., D. M. Davis, and T.-f. Wong: Failure modes of tuff samples from Leg 131 in the Nankai accretionary wedge, in Hill, I. A., Taira, A., Firth, J. V. et al., Proceedings of the Ocean Drilling Program, Scientific Results, Vol. 131, 275-281, 1993b.

Zhao, W.-L., D. M. Davis, F. A. Dahlen, and J. Suppe, Origin of convex accretionary wedges: Evidence from Barbados, *J. Geophys. Res., 91,* 10,246-10,258, 1986.

D. M. Davis, Department of Earth and Space Sciences, SUNY - Stony Brook, Stony Brook, NY 11794-2100

Mountain-building in Taiwan and the Critical Wedge Model

Chi-Yuen Wang and Adam Ellwood

Department of Geology and Geophysics, University of California, Berkeley, CA 94720

Francis Wu and Ruey-Juin Rau

Department of Geological Sciences, State University of New York, Binghamton, NY 13902

Horng-Yuan Yen

Institute of Earth Sciences, Academia Sinica, Taipei, Taiwan, ROC

Recently accumulated geophysical data for Taiwan have provided an opportunity to examine the deep structure and the construction of an active fold-and-thrust mountain belt. Analyses of gravity anomalies and seismic tomography show the occurrence of high-density and high-velocity rocks beneath the Central Ranges extending from near the surface to depths of 40-70 km. Together with surface geology, fission-track data and radiometric data, the results show that large-scaled, autochthonous uplift of basement rocks from depths beneath the Central Ranges may have occurred since the last 1 Ma. These results are in contrast with the model of thin-skinned growth of a critical wedge, that was thought to occur across the entire Taiwan orogen.

INTRODUCTION

The critical wedge model has been widely taken as the paradigm for understanding the making of accretionary wedges and fold-thrust mountain belts. Specifically, the Taiwan orogen has been considered as a type-example for this model (e.g., Davies et al., 1983; Dahlen and Suppe, 1988). Lack of geophysical data for this orogen, however, has prevented an in-depth understanding of its deep structures and processes. Recently, a suite of geophysical data for the Taiwan Orogen have been accumulated; an opportunity thus presents itself for understanding the mountain-building processes. Because of its youth, the Taiwan Orogen preserves much geologic and geophysical signature in mountain-building, and is thus ideal for testing the hypotheses that have been advanced to explain these processes. Here we integrate the new results of gravity anomaly and seismic tomography in an effort to examine the processes and hypotheses.

GEOLOGIC MODELS

The Taiwan fold-and-thrust belt was formed by the collision between the Eurasian plate and the Luzon arc system on the Philippine Sea plate since 5 Ma (Figure 1a, Chi et al., 1981; Teng, 1990). East of Taiwan, the Philippine Sea plate subducts northwestward beneath the Ryukyu arc-trench system (Tsai, 1978; Teng, 1990). South of Taiwan, the plate boundary changes from active collisional to active subduction where the oceanic crust of the South China Sea subducts eastward beneath the Philippine Sea plate at the Luzon arc (Angelier, 1986).

The continental margin part of the island of Taiwan is made up by four NNE-SSW trending structural belts (Figure 1b): the Coastal Plain, Western Foothills, Hsuehshan Range (western Central Ranges), and the Backbone Range (eastern Central Ranges) (Ho, 1988). The Taiwan Strait is floored by a pre-Tertiary block-faulted basement and draped by flat-lying Cenozoic sedimentary sequences (Liou and Hsu, 1988; Tang, 1977; Teng 1992; Sun, 1982; Yuan et al., 1985). In eastern Taiwan, the convergent plate boundary is marked by the Taitong Longitudinal Valley, with the deformed Luzon arc system represented by the Coastal Range (Ho, 1988).

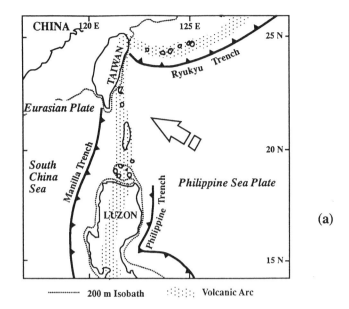
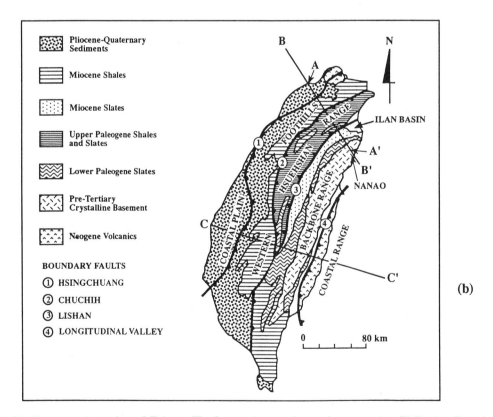

Figure 1. (a) Plate tectonic setting of Taiwan. The Luzon Arc on the northwest-moving Philippine Sea plate encounters the Asiatic continental margin along the Taiwan segment. Northeast of this collision boundary is the Ryukyu trench, with the Philippine plate subducting towards the north. Extending southward, this boundary changes from collision to active eastward subduction of the South China Sea plate beneath the Philippine Sea plate along the Manila Trench. (b) The Taiwan island with the major structural units of the Taiwan fold-and-thrust belt. The A-A' and C-C' lines show the geographic positions of Suppe's (1980, 1981) geologic cross-sections and the B-B' line shows the geographic position of Teng's (1992) geologic cross-section.

Two geologic cross-sectional models across the northern Taiwan are presented in Figure 2; their positions are marked in Figure 1a. These include Suppe's (1980) retrodeformable cross section along A-A', and Teng's (1992) cross section based on sedimentology and stratigraphy along B-B'. Suppe's (1980) model (Figure 2a) assumed initially flat-lying sedimentary units over a continuous basal decollement, uniaxial and continuous compression, and a limited amount of basement involvement; the retrodeformable constraint further required all autochthonous continental basement to exist below the basal decollement. On the other hand, Teng's (1992) model (Figure 2b) takes into account the pre-orogenic, rifted nature of the Chinese continental margin which consisted of block-faulted pre-Tertiary crystalline basement with grabens filled with thick Paleogene syn-rift deposits and covered by Neogene-Quaternary post-rift deposits (Liou and Hsu, 1988; Tang, 1977; Sun, 1982; Yuan et al., 1985; Teng, 1992). This model differs from Suppe's model in involving reactivated normal faults deeply rooted in the basement, thus allowing uplift of basement into the orogen. A regional tectonic cross-section (Figure 2c) along C-C' (Figure 1b) by Suppe (1981) is also included in the present gravity analysis; it is away from the tips of the Taiwan island and thus more suitable for two-dimensional modeling. It features a double-sided wedge underlain by two decollements, one westward dipping and the other eastward dipping.

Suppe's cross-sections (Figure 2a and 2c) were the basis for modeling the Taiwan orogen as a critical wedge (e.g., Davies et al., 1983), where the development of the orogen was assumed to resemble the thin-skinned growth of a wedge of soil pushed in front of a moving bulldozer over a basal detachment; the material in the wedge deforms until a critical taper is attained, after which it maintains this shape by continued internal adjustment (e.g., Davis et al., 1983; Dahlen and Suppe, 1988). Basic to this model is a minimal involvement of the basement.

GRAVITY EVIDENCE

A gravity survey of Taiwan was made by Yen (1992), with a total of 603 measurements, mainly at geodetic stations, at an average spacing of 7 km. Using Bouguer anomalies Yen (1992) modeled the configuration of the Moho. Because of the extreme topography of the island, the large amplitude of Bouguer anomalies tends to mask the lower-amplitude anomalies caused by density variations in the crust. The point of departure here is the use of the free-air anomaly, by which variations in the crust can be better delineated. A notable feature is a regional high gravity anomaly (Figure 2) which extends along the entire length of the Central Ranges (Yen, 1992).

A two-dimensional technique (Talwani et al., 1959), implemented in an interactive computer program (GRVMOD), was used in computing the gravity anomaly for the geologic cross-sections. The three geologic cross-sections (A-A', B-B' and C-C') were first digitized and input to the computer program. Average densities, based on rock types, were assigned to the major structural units (Figure 2). Since the emphasis of the present study is the variations in the crustal density, inhomogeneities in the deeper region may be neglected. The calculated anomalies are shown above their respective geologic cross-sections and compared with observed anomalies.

As shown in Figure 2a and c, the calculated gravity for cross-sections A-A' and C-C' deviates significantly from the observed gravity. Attempts to improve the "fit" by changing the assigned densities within reasonable ranges (as shown in Figure 2a and c by varying the density of the Miocene slates and/or the density of the Paleogene shales, or by varying the density of the accreted rocks in the southern model) failed to significantly improve the fit. On the other hand, the calculated gravity for Teng's (1992) model agrees much better with the observed gravity (Figure 2b). In making this calculation, we included in the eastern end of the cross section a block of accreted sediments corresponding to the accretionary wedge at the Ryukyu trench as imaged by seismic tomography of this region (Rau and Wu, 1995).

SEISMIC TOMOGRAPHY EVIDENCE

Detailed tomography was made possible by the recent expansion of a telemetered network on Taiwan and its neighboring islands (Figure 3). A dataset consisting of seismograms recorded for three and a half years was used for tomographically imaging the deep structure under the Taiwan area. Seismograms from 1197 events were chosen for this study. These events were recorded at more than 8 stations, with the largest azimuthal gap of less than 180° between stations (this criterion is relaxed for the deep events in the subduction zone immediately northeast of the island in order to increase the number of events). Tomographic inversion as formulated by Thurber (1983, 1993) was performed to determine the P-wave velocities at grid points in a three-dimensional space. Various grid orientations were used; more details can be found in Rau and Wu (1995).

Two velocity cross-sections and their corresponding spread functions (Menke, 1989; Toomey and Foulger, 1989) are shown in Figure 4. The hypocenters have been relocated during the iterative tomographic inversion. The most prominent feature in these sections (profiles A-A' through B-B') is the significant thickening of the crust

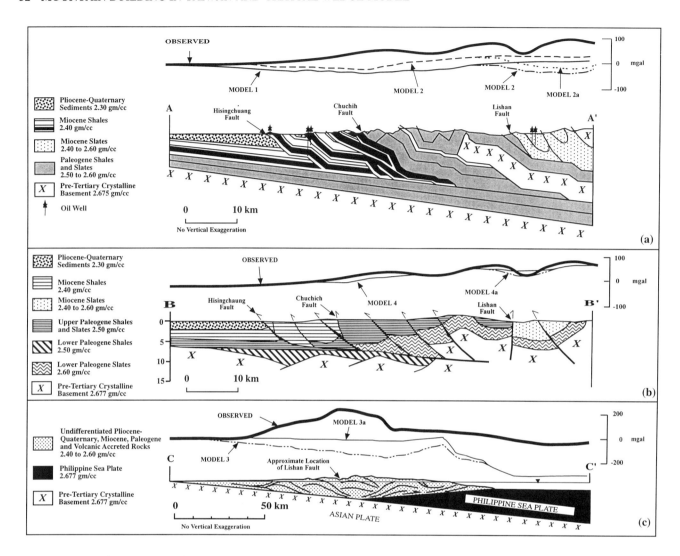

Figure 2. Three geologic cross sections and the corresponding gravity anomalies; the locations of the sections are shown in Figure 1b. (a) Suppe's (1980) geologic cross section along A-A'. Measured free-air gravity anomaly along this transect (thick solid line) is compared with the calculated anomaly for Suppe's section, in thin solid and dashed lines. A density of 2500 kg/m^3 has been assigned to the Paleogene rocks, while the assigned density of the Miocene Slates is varied from 2400 (thin dashed line) to 2600 (thin solid line) kg/m^3. (b) Teng's (1992) geologic cross section along the line B-B'. Measured free-air gravity anomaly is compared with the calculated anomaly for Teng's (1992) model, in thin solid and dashed line. The density of the Miocene Slates is varied from 2400 (thin dashed line) to 2600 (thin solid line) kg/m^3. (c) Suppe's (1981) regional cross-section along the line C-C'. Measured free-air gravity anomaly (thick solid line) is compared with the calculated anomaly; the assigned density of the accreted rocks is varied from 2400 (thin dashed line) to 2600 (thin solid line) kg/m^3.

under the higher elevations of the Central Ranges. This thickening is much more pronounced in the north where low velocity materials extend to a depth of about 75 km. This low velocity feature was also observed to some extent by Roecker et al. (1987), but due to the larger block size used in their work, the velocity contrasts are much attenuated. The high velocities below about 20 km in the eastern part of the A-A' and B-B' profiles are in sharp contrast to the low velocities in the west; the contact between them is quite steep. Also consistently seen in these profiles is the rise of the 5.5 km/s contours beneath the Central Ranges. On the other hand, low-velocity materials (<5 km/s) are imaged to a depth of 7-10 km beneath the Foothills and the Coastal Plain. The earthquakes shown in these sections occur mostly in the upper 40 km under the Western Foothills and in the top 40-70 km under the eastern Central Ranges, being deeper toward the north. The middle part of the Central Ranges is nearly aseismic.

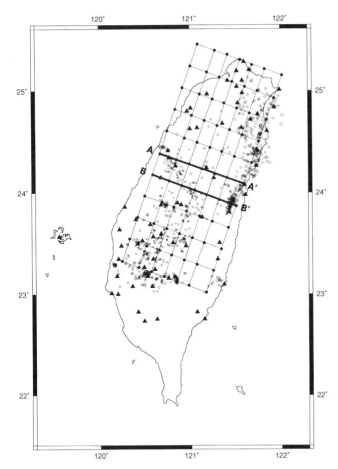

Figure 3. Map view of the grid configurations used in 3-D tomographic inversions. The grid system is setup with the horizontal orthogonal directions parallel (N20°E-S20°W) and perpendicular (S70°E-N70°W) to the strike of the island in the area of interest. Solid triangles are the CWBSN and TTSN seismic stations. The earthquake locations used in each system are shown as open circles. The nodes (the points where the grid lines crossed) are shown as solid dots. A-A' and B-B' indicate the positions of the tomographic profiles shown in Figure 5.

DISCUSSION

The distinct high density (Figure 2b) and high velocity (Figure 4) in the upper 10-15 km under the Central Ranges are important. Furthermore, the upper crustal high velocity under the Central Ranges is seen to be continuous with the lower crustal layer, which generally thickens to form a "root" under the high elevations of the Central Ranges. Thus the formation of the Central Ranges appears to be a result of uplift of the Pre-Tertiary basement, consistent with Teng's (1992) model (Figure 2b), as well as downwarping of the lower crust. Furthermore, the upper crustal high velocity in Figure 4a is seen to extend to the surface on the eastern end of the profile, where Pre-Tertiary high grade metamorphic rocks crop out. The high velocity (>7.5 km/s) under the eastern Central Ranges are somewhat intriguing. The fact that they start as shallow as 20 km and are next to the Philippine Sea plate make us suspect they are part of the oceanic lithosphere. The seismicity in this zone is also notably higher than in its neighboring area to the west. In profiles A-A' and B-B' (Figure 4) the crustal velocity root is quite deep, with the 7.5 km/s contour at around 55-65 km.

The crust beneath the middle part of the Central Ranges is mostly aseismic, which may be surprising in view of the active and rapid uplift in the area. The present study gives a natural explanation of this phenomenon. Given that faulting in the Central Ranges occurs on steep, reactivated normal faults, uplift of the basement can create locally high geothermal gradient. Thus temperatures on the faults beneath the Central Ranges can exceed the boundary between thermally induced stable-unstable sliding in rocks (Stesky et al., 1974) at relatively shallow depths. Seismicity would be totally absent if this depth becomes shallower than that for the upper, pressure-induced stable-unstable transition (Marone and Scholz, 1988). Thus the lack of seismicity beneath the Central Ranges may be a consequence of the steep uplift of the basement.

Also of interest is the occurrence of a strike-parallel belt of low-grade metamorphic rocks in the Central Ranges (Chen, 1979, 1981; Chen et al., 1983; Hsieh, 1990), which, in map view, is bound by rocks of higher metamorphic grades on both sides. Crespi (1991) and Crespi and Chen (1992) suggested that normal faulting on the Lishan Fault may have preserved this low-grade metamorphic belt. The present model provides an alternative explanation: If erosional rate is slope-dependent, the juxtaposition of low-grade and high-grade metamorphic rocks across major thrust faults may be expected as a natural consequence of discrete and sequential thrust faulting.

The present model is consistent with the studies of fission track ages of zircon and apatite (Liu, 1982) and Rb-Sr biotite ages (Lan et al., 1990) across the Central Ranges, which show that since ~1 Ma the hinterland in the Taiwan orogen has uplifted at an accelerated rate up to 10 mm/y, compared to ~1 mm/y during the period from the upper Miocene through Pliocene. Even through these rates may be affected by an assumed constant geothermal gradient, the basic conclusion that accelerated exhumation of basement in the Central Ranges has occurred in the last Ma remains sound.

In conclusion, recently available geophyscial data show strongly that mountain-building in Taiwan may have involved the autochthonous uplift of a great amount of basement rocks from depths beneath the Central Ranges.

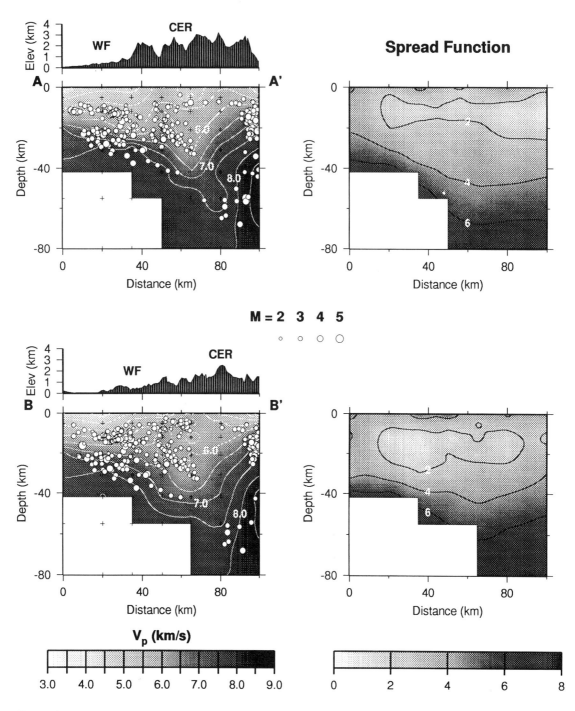

Figure 4. Two island-crossing tomographic profiles and their corresponding spread functions; the locations of the profiles are shown in Figure 4. The P-wave velocity distributions are shown on the left and the spread functions are shown on the right. A zero spread function implies a perfectly defined parameter, whereas large spread functions correspond to parameters having broad kernel shapes and small values of the resolving kernel; an acceptable spread function is 5. White areas mark unsampled regions. The velocity contour interval is 0.5 km/s. The position of the Moho corresponds approximately to the 7.5 km/s contour. The white circles are relocated hypocenters including events within 1 grid space of the profile. The crosses shown on the velocity profiles indicate the locations of the nodes. The topography corresponding to each profile is shown on top of the velocity section. CER = eastern Central Ranges; WF = Western Foothills.

Furthermore, the processes appear to be discontinuous in space and in time and accelerated in the last Ma. These results are in contrast with the model of thin-skinned growth of a critical wedge, that was applied across the Taiwan orogen.

Acknowledgements. This research was supported by NSF grant EAR92-18934. We thank Dan Davis, Louis S. Teng, Ben Page and J. G. Liou for helpful comments.

REFERENCES

Angelier, J., Preface in special issue of Geodynamics of the Eurasia-Philippine Sea Plate Boundary, *Tectonophysics, 125,* 1-2, 1986.

Chen, C.H., Geology of the East-West Cross-Island Highway in central Taiwan, *Mem Geol. Soc. China, 3,* 219-236, 1979.

Chen, C.H., Change of X-ray diffraction pattern of K-micas in the metapelites of Taiwan and its implication for metamorphic grade, *Mem. Geol. Soc. China, 4,* 475-490, 1981.

Chen, C.H., Chu, H.T. Liou, J.G. and ernst, W.G., Explanatory notes for the metamorphic facies map of Taiwan, *Special Publ. Central Geol. Survey, 2,* 32p. 1983.

Chen, C.H., Determination of lower greenschist facies boundary by K-mica--chlorite crystallinity in the Central Ranges, Taiwan, *Proc. Geol. Soc. China, 27,* 41-53, 1984.

Chi, W.R., J. Namson, and J. Suppe, Stratigraphic record of plate interactions in the Coastal Range of eastern Taiwan, *Mem. Geol. Soc. China, 4,* 491-529, 1981.

Crespi, J., Structural patterns in a belt of anomalously low-grade rocks in the hinterland of the Taiwan arc-continent collision, *EOS, 72,* 432-433, 1991.

Crespi, J., and C.H. Chen, Fabric development in a belt of anomalously low-grade rocks, central Taiwan, *EOS, 73,* 539, 1992.

Dahlen, F.A., and J. Suppe, Mechanics, growth and erosion of mountain belts, *Spec. Pap Geol. Soc. Am., 218,* 161-178, 1988.

Davis, D., J. Suppe, and F.A. Dahlen, Mechanics of fold-and-thrust belts and accretionary wedges, *J. Geophys. Res., 88,* 1153-1172, 1983.

Ho, C.S., An introduction to the geology of Taiwan:Explanatory text of the geologic map of Taiwan (2nd ed.), *Cent. Geol. Surv., MEOA, R.O.C.,* 192p., 1988.

Hsieh, S., *Fission-track Dating of Zircons from Several East-west Cross Sections of Taiwan Island,* M.S. thesis, National Taiwan, University, Taipei (in Chinese), 1990.

Lan, C.-Y., T. Lee, and C. W. Lee, The Rb-Sr isotopic record in Taiwan gneisses and its tectonic implication, *Tectonophysics, 183,* 129-143, 1990.

Liou, C.H. and C.Y. Hsu, Petroleum exploration on the southeastern plain of Taiwan, *Pet. Geol. Taiwan, 24,* 18-36, 1988.

Liu, T.K., Tectonic implication of fission track ages from the Central Ranges, Taiwan, *Proc. Geol. Soc. China, 25,* 22-37, 1982.

Menke, W., *Geophysical Data Analysis: Discrete Inverse Theory,* revised edition, Academic Press, 289 pp., 1989.

Marone, C., and C. Scholz, The depth of seismic faulting and the upper transition from stable to unstale regimes, *Geophys. Res. Lett., 15,* 621-624, 1988.

Rau, R.J., and F. T. Wu, Tomographic imaging of lithospheric structures under Taiwan, Earth and Planet. Sci. Letters, in press, 1995

Roecker, S.W., Y. H. Yeh and Y. B. Tsai, Three-dimensional P and S wave velocity structures beneath Taiwan: deep structure beneath an arc-continent collision, *J. Geophys. Res. 92,* 547-10,570, 1987.

Stesky, R., W. Brace, D. Riley, and P-Y Robin, Friction in faulted rock at high temperature and pressure, *Tectonophysics, 23,* 177-203, 1974.

Sun, S.C., The Tertiary basins of offshore Taiwan, *Proc. 2nd ASCOPE Conf. and Exibit.,* 125-135, 1982.

Suppe, J., A retrodeformable cross section of northern Taiwan, *Proc. Geol. Soc. China, 23,* 46- 55, 1980.

Suppe, J., Mechanics of mountain building and metamorphism in Taiwan, *Geol. Soc. China, Mem., 4,* 67-89, 1981.

Suppe, J., The active Taiwan mountain belt, in: *The Anatomy of Mountain Ranges,* J. P. Schaer and J. Rodgers, eds., pp. 277-293, Princeton Univ. Press, Princeton, 1987.

Talwani, M., J.L. Worzel, and M. Landisman, Rapid gravity computations for two-dimensional bodies with application to the Mendocino submarine fracture zone, *J. Geophys. Res., 64,* 49-59, 1959.

Tang, C.H., Late Miocene erosional unconformity of the subsurface Peikang High beneath the Chiayi-Yunlin coastal plain, Taiwan, *Mem. Geol. Soc. China, 2,* 155-168, 1977.

Teng, L.S., Geotectonic evolution of late Cenozoic arc-continent collision in Taiwan, *Tectonophysics, 183,* 57-76, 1990.

Teng, L.S., Geotectonic evolution of Tertiary continental margin basins of Taiwan, *Petrol. Geol. of Taiwan, 27,* 1-19, 1992.

Thurber, C.H., Earthquake locations and three-dimension crustal structure in the Coyote Lake area, central California, *J. Geophys. Res. 88,* 8226-8236, 1983.

Thurber, C.H. Local earthquake tomography: Velocities and Vp/Vs - Theory, in: *Seismic Tomography: Theory and Practice,* H. M. Iyer and K. Hirahara, eds., pp. 563-583, Chapman and Hall, London, 1993.

Toomey, D.R., and G. R. Foulger, Tomographic inversion of local earthquake data from the Hengill-Grensdalur central volcano complex, *J. Geophys. Res. 94,* 17,497-17,510, 1989.

Tsai, Y.B., Plate subduction and the Plio-Pleistocene orogeny in Taiwan, *Pet. Geol. Taiwan, 15,* 1-10, 1978.

Yen, H.-Y., *Gravity Anomaly in Taiwan and Its Geologic Implications,* Ph.D. thesis, National Taiwan University, 1992.

Yuan, J., Lin, S. J., S.T. Huang and C.L. Shaw, Stratigraphic study on the pre-Miocene under the Peikang area, Taiwan, *Pet. Geol. Taiwan, 21,* 115-128, 1985.

Chi-yuen Wang and Adam Ellwood, Department of Geology and Geophysics, University of California, Berkeley, CA 94720.

Francis Wu and Ruey-Juin Rau, Department of Geological Sciences, State University of New York, Binghamton, NY 13902.

Horng-Yuan Yen, Institute of Earth Sciences, Academia Sinica, Taipei, Taiwan, ROC

Sediment Pore-Fluid Overpressuring and Its Effect on Deformation at the Toe of the Cascadia Accretionary Prism From Seismic Velocities

Guy R. Cochrane

U.S. Geological Survey, Menlo Park, California

J. Casey Moore

University of California Santa Cruz, Santa Cruz, California

Homa J. Lee

U.S. Geological Survey, Menlo Park, California

Analysis of seismic velocity derived from a 1989 multichannel seismic (MCS) survey of the Oregon margin suggests that there is a combination of diffusive pore-fluid flow and fracture-controlled, fault-guided flow at the toe of the accretionary prism. In an area of incipient thrusting an increase in sediment velocity suggests that thrusts augment the diffusive flow responsible for consolidation of sediment further seaward. There is evidence, in the form of high pore-fluid pressure inferred from interval velocities, that the décollement forms in overpressured incoming sediments. Low seismic velocities in an area where a major strike-slip fault cuts the sedimentary section, and high seismic velocities adjacent to the fault, are interpreted to be the result of fracturing of consolidated sediment and escape of overpressured fluid through the fault zone.

1. INTRODUCTION

The importance of fluid pressure in the mechanics of overthrust faulting has been discussed in detail beginning with the paper by Hubbert and Rubey (1959). One of the theoretical requirements for reversal of vergence in a region of thrusting is a décollement with minimal shear strength, requiring fluid pressures near lithostatic pressure at the depth of the décollement (Seely, 1977). Several authors have noted reversals in the vergence of deformation from north to south along the Oregon margin (e.g.: Silver, 1972; Snavely et al., 1986). Moore et al. (1995b) have estimated fluid pressures greater than 90% of lithostatic below thrusts in the Barbados accretionary prism using density logs collected while drilling. Estimates of this type are lacking for the Cascadia accretionary prism. Sediment porosity anomalies can be indicative of overpressuring (Fertl, 1976). In this paper we estimate pore-fluid pressure from sediment porosity values (inferred from MCS P-wave velocity) and discuss the relationship between pore-fluid pressure and changes in the structural style of deformation along the toe of the Oregon accretionary prism (Figure 1).

In the study area (Figure 1) the prism toe is migrating seaward as thrusting rooted at the décollement steps out and incorporates new sediment into the prism (Silver, 1972). Prior to this brittle deformation, the sediment undergoes physical compaction and fluid loss induced by tectonic thickening (Carson, 1977; Bray and Karig, 1985; Cochrane et al., 1994a).

Numerous studies in recent years demonstrate the relationship between structural deformation and the compaction of sediments of the Cascadia accretionary system. Fluid venting occurs at numerous sites across the prism (Kulm and Suess, 1990). Near the toe of the prism, vents occur at the surface intersection of features such as depositional unconformities (Lewis and Cochrane, 1990), and bedding (Orange and Breen, 1992).

Subduction: Top to Bottom
Geophysical Monograph 96
Copyright 1996 by the American Geophysical Union

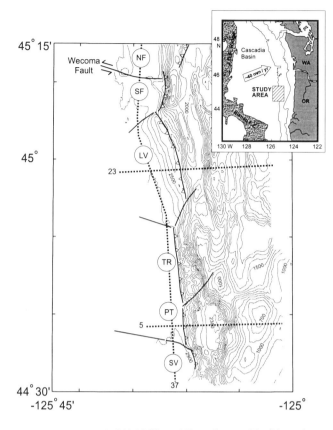

Figure 1. Map of 1989 MCS tracklines discussed in this study. Inset map show the location of the study area off Oregon. Seawardmost thrust faults and oblique-slip faults are from MacKay et al., (1992), and Tobin et al. (1993). NF= north of Wecoma fault, SF=south of Wecoma fault, LV=landward verging area, TR=area of transitional structure, PT=seaward verging area with protothrusts, SV=seaward verging area lacking protothrusts. SeaBeam bathymetry is contoured at 100 m intervals.

MacKay et al. (1992) identified deep-seated oblique strike-slip faults and blind incipient thrusts (protothrusts) seaward of the frontal thrust in Cascadia Basin using MCS data. Tobin et al. (1993) suggest that dewatering of the décollement in the area of one of the oblique strike-slip faults (the Wecoma fault) caused a local increase in décollement shear strength and a reversal of vergence. Carson et al. (1994) inferred that near-surface diagenetic carbonate deposits, associated with point-discharge of fluid are concentrated in the area of the strike slip faults and in the area of undeformed Cascadia Basin deposits which overlie the protothrusts. Cochrane et al. (1994b) have shown that seismic P-wave velocities increase landward, and infer that the protothrusts act as fluid conduits for pore fluid escape. Moore et al., (1995a) use synthetic-seismic models to show that reversed-polarity seismic reflections from the protothrusts originate from fault-zone overpressuring.

2. GEOLOGIC AND TECTONIC SETTING

The Cascadia subduction zone (Figure 1) is characterized by a slow convergence rate (42 mm/yr [DeMets et al., 1990]), young oceanic crust (8 Ma [Kulm et al., 1984]), and high sediment influx rates (1-2 km/my for unit 1 and 0.25 km/my for unit 2 [Moore et al., 1995a]). Consequently, water depths are in the range of 2-3 km at the deformation front; a bathymetric trench is absent nearly everywhere.

The total thickness of sediment in Cascadia Basin is approximately 3.5 km at the toe of the slope (Figure 2) (Cochrane et al., 1988). The sedimentary section is divided into two major lithologic units (Figure 2B) (Kulm et al., 1973). The upper half of the section (unit 1) is composed of sandy mudstones deposited as fan turbidites. The lower section (unit 2) is composed of more distal silty clays. In all areas the proto-décollement lies within the lower unit of silty clays (MacKay et al., 1992). Discontinuous reversed-polarity seismic reflections and a seismic interval-velocity reversal from the top of the protodécollement (Figure 2B) indicate that pore fluids may be overpressured (Cochrane et al., 1994b).

Figure 2 shows 1989 Cascadia MCS data and interpretations from areas of seaward and landward vergence. The location of the décollement (within unit 2) may be due to overpressuring beneath a cap of relatively more dewatered, low permeability clay as suggested by Bray and Karig (1988) for the Nankai Trough. In addition to the change from the seaward vergence to the landward vergence from south to north the depth of the décollement increases from near the unconformity between unit 1 and unit 2 to an interval of sediment just above oceanic crust (Figure 2B). We will show that changes in the depth of the décollement are associated with changes in the depth of maximum fluid overpressures inferred from seismic velocities.

3. MCS DATA VELOCITY ANALYSIS

The 1989 central Oregon MCS data were collected by Digicon aboard the M/V Geo Tide. The survey yielded approximately 2000 km of 144- channel seismic data. A 3800 m MCS streamer provided 72-fold common midpoint (CMP) data. Streamer feathering was limited to 10 degrees or less. Continuous satellite navigation (STARFIX) was used to maintain a shot spacing of 25 m. A tuned 75 liter air gun array produced a consistent spiky source wavelet.

The relatively shallow water depths in the Cascadia Basin combined with the long streamer yielded data with approximately 0.25 s of moveout at the base of the sedimentary section, sufficient moveout for the measurement

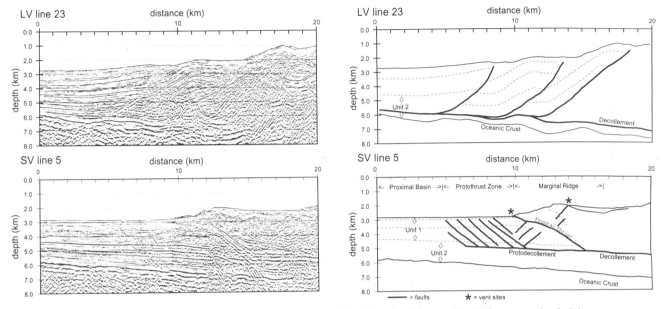

Fig. 2a. Depth sections of MCS line 5, in the PT area, and line 23 in the LV area. Vertical exaggeration is 1:1.

Fig. 2b. Line drawing interpretation of MCS data. Notice the difference in depth at which the protodécollement is propagating out into Cascadia Basin in the two areas.

of stacking velocity in the sedimentary section using the semblance method (Taner and Koehler, 1969).

Velocity analysis was performed along a strike line (line 37) that passes through several structural regions (Figure 1) including, from north to south: 1) areas north and south of the Wecoma fault, a northeast trending dip-slip fault that intersects the margin (NF and SF); 2) the landward verging area where thrusts dip west (LV, Figures 1 and 2); 3) an area of transition between landward vergence and seaward vergence (TR); 4) a seaward verging area where there are protothrusts in the basin (PT, Figures 1 and 2); 5) a seaward verging area (SV) where protothrusts are lacking.

4. DISCUSSION

In order to estimate pore pressure we converted stacking velocities to fluid pressures through a sequence of steps.

4.1 Regional Velocity Trends

There is an increase in stacking velocity from south to north along line 37 that exceeds the standard deviation from any one area within the study area (Figure 3). Similar changes in stacking velocity are seen in a west to east direction (perpendicular to the sediment transport direction) at the base of the slope by Cochrane et al. (1994b) and Yuan et al. (1994). We will similarly model changes in seismic velocity as changes in sediment physical properties.

The trend of northerly increasing velocity may be the result of changes in lithology of the sediment with proximity to the source of Astoria Fan sediments, or changes in the degree of sediment consolidation (decreasing porosity) related to tectonic stress and changes in structural style of deformation. Nelson (1976) has shown that the percentage of sand in unit 2 sediment does not vary appreciably over distances on the order of 100 km from the source. Because sand content is the main lithological factor controlling seismic velocity in sediment, we believe that varying tectonic stress caused the lateral variation in velocity in unit 2.

In order to show the changes in velocity with respect to the TR area, stacking velocity values were converted to interval velocity and normalized to the transitional area interval velocity profile (Figure 4). The trend of increasing velocity from south to north is still evident. Contrary to the trend, the velocity profile from the protothrust area exceeds the transitional profile in the upper section suggesting that the upper unit sediments in the protothrust area are more consolidated relative to sediments to the north and south. The higher degree of consolidation in the protothrust zone is probably due in part to dewatering along protothrusts. Although initial consolidation- and strength-heterogeneity may have initiated protothrusting at this location.

4.2 Porosity Derived From Velocity Data

In order to estimate porosity we used a velocity-porosity conversion derived by Hyndman et al. (1993) that predicted

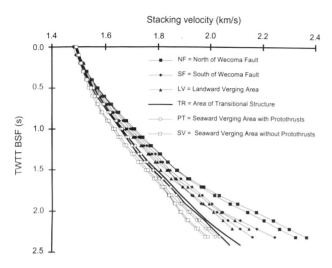

Figure 3. Graph showing stacking velocities for six areas of varying structure along line 37. Standard deviation for groups of 5 CMP's, is shown by a curve of the average velocity plus the deviation and a curve of the average minus the deviation for each area. Spacing between the CMP's within a group is 125 m. The maximum standard deviation for any group of picks is 0.4 km/s. Stacking velocities are time averaged values.

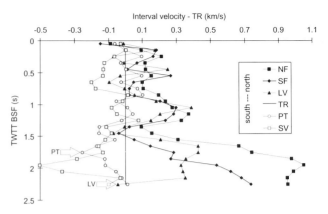

Figure 4. Interval velocity versus two-way traveltime curve normalized to the velocity profile in the transitional structural area. Derived from stacking velocities using Dix's (1955) equation. Since reflectors in Cascadia Basin are flat lying, no dip corrections of the resulting interval velocities were required. Arrows show where low velocity intervals in unit 2 correlate to the depth of the protodécollement determined from the MCS reflection profiles.

measured porosity for Nankai Trough sediments within a few percent (Figure 5). For velocities higher than 3.5 km/s we use the linear velocity-porosity relationship for shale of Vernik (1994). The Vernik (1994) curve, a velocity-porosity relationship for brine saturated shale at 40 MPa effective pressure, was used because the sediments of unit 2 are finer grained and more deeply buried than those of unit 1.

The contribution of protothrusts to consolidation of sediment in the upper section is shown in a porosity-depth plot (Figure 5). A regional decrease in porosity that corresponds to the velocity increase to the north is shown. Regardless of the cause of the regional trend, the porosity of the PT area sediment should be intermediate between sediment to the north and south in keeping with the trend. Figure 5 shows that in unit 1 the porosity is equal to or lower than the porosity in the adjacent areas north and south depending on the depth chosen for comparison. The porosity anomaly in the PT area is about 5% and decreases to zero at approximately 1400 m below the seafloor.

4.3 Pore Fluid Pressure Estimated From Porosity

Reversals in the porosity of unit 2 sediment (Figure 5) correspond to the depths of the protodécollement observed in the MCS data (Figure 2B) in the SV and LV areas. Because porosity anomalies can be indicative of overpressuring (Fertl, 1976) and because these anomalies correspond to structural detachments, we model the changes in porosity (derived from velocity) as fluid overpressures.

To determine excess fluid pressures we need to first establish an equilibrium porosity versus depth relationship that represents the porosity of sediment when fluids are at hydrostatic pressure. Figure 5 shows that the porosity change with depth curves roughly follow an exponential decrease. Hyndman et al.'s (1993) curve of porosity versus depth for the Nankai accretionary prism is approximately exponential but is not considered an equilibrium profile because rapid sediment deposition in the area of the Nankai Trough makes fluid pressures in excess of hydrostatic likely. Rubey and Hubbert's (1959) curve is also an exponential curve, and is considered to be an equilibrium porosity curve.

Pressures were calculated using Rubey and Hubbert's (1959) equation that converts porosity to pressure, given the equilibrium curve and assumptions about the density of the sediment grains and the pore water. The equation was derived from Athy's (1930) exponential relationship between Paleozoic shale and mudstone density and depth by substituting porosity terms for density and substituting stress for depth.

$$\lambda = 1 - \left[\frac{1}{c \log e} \times \frac{\overline{\rho_{bw}} - \rho_w}{\overline{\rho_{bw}}} \right] \times \frac{\log(f_0/f)}{z}$$

In this equation λ is the pore fluid pressure ratio, c is an exponential factor of dimension (length^{-1}) from the Athy

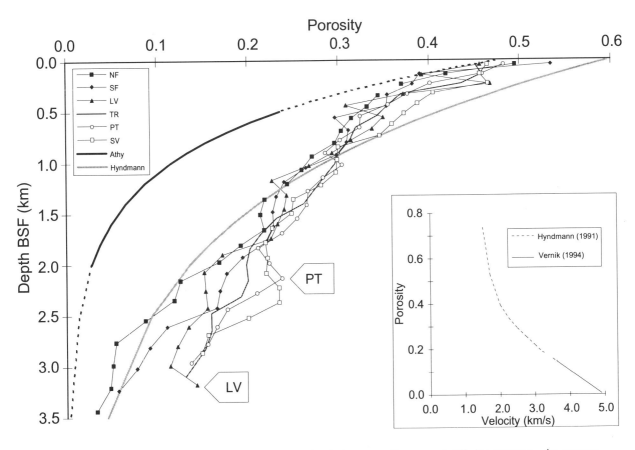

Fig. 5. Porosity versus depth curves derived from interval velocity curves for areas of differing structure. An average porosity-depth curve for the Nankai Trough by Hyndman (1993) is shown for comparison. The Athy curve is the equilibrium curve used in this study to convert porosity values to fluid pressure values in Cascadia Basin. The inset figure shows the velocity-porosity conversion curve using Hyndman's Nankai Trough relationship for unit 1 and Vernik's (1994) conversion for clays for unit 2.

(1930) empirical equation, e the base of Napierian logarithms. $\overline{\rho_{bw}}$ is the average density of the sediment above the depth z which we calculate at each depth using a grain density of 2700 kg/m^3 (Westbrook, et al., 1994) and a pore water density (ρ_w) of 1050 kg/m^3. f_0 is the equilibrium porosity at the sediment surface(0.48) and f is the porosity that varies with depth z. The fluid pressure P is derived from the fluid pressure ratio using $P = \lambda \overline{\rho}_{bw} g z$ where g is the gravitational acceleration.

This method of calculating the pore pressure assumes that all the stress on the sediment is derived from the load of the overlying sediment section and ignores tectonic stress. Because tectonic stresses act on the sediment along the Cascadia Margin, this method underestimates pore fluid pressures. The following examination of the magnitude of change of fluid pressure versus depth and from one structural setting to another is useful regardless of the underestimation of absolute fluid pressure. We are interested in the relationship between deformation and fluid pressure.

4.4 Interpretation of Pore Pressures: Protodécollement

Figure 6 shows the estimates of fluid pressure normalized to hydrostatic. Any values greater than hydrostatic are excess fluid pressures that are supporting some portion of the lithostatic load. Overpressuring of sediment from rapid sediment accumulation is expected (figure 6). Nonlinear fluid pressure versus depth is expected, even in areas of vertically homogeneous sediment. For instance, Barbados Ridge sediment fluid pressure increases suddenly at a critical depth where fluid flow can not keep pace with increasing overburden and tectonic pressure (Marlow et al., 1984).

In all the areas there is an elevation of the pore pressure at a depth of approximately 2.4 km (shaded area, figure 6). This

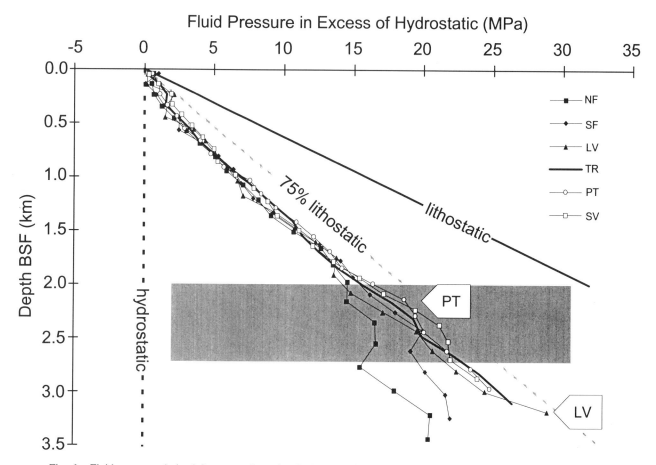

Fig. 6. Fluid pressure derived from porosity using Rubey and Hubbert's (1959) equation. Values are normalized to hydrostatic. Notice the reduction in fluid pressure from the protothrust area to the transitional area and the greater depth of the landward verging area fluid pressure maxima. Gray area represents the depth where fluid pressures in all areas are elevated suggesting a lithologic unit or other physical phenomena common to all areas.

suggests that a reduction in permeability either due to lithology or physical processes of consolidation is common to all the areas. We suggest that the change in lithology from unit 1 to unit 2 compounds the process described by Marlow et al. (1984). That is that the upper sediments of unit 2 are dewatering relative to the lower part of unit 2 because fluids can escape into the more permeable unit 1 sediment. Thus the upper part of unit 2 becomes more impermeable and forms a partial cap which traps water resulting in overpressuring of those sediments below the cap as loading increases with time (Cochrane et al., 1994b). The 2400 m depth is the depth of maximum fluid pressure in the SV, PT, and TR areas whereas the LV, SF, and NF areas have a fluid pressure maximum at the bottom of the sedimentary section. As discussed above it is not possible, without more extensive drilling data, to determine if these trends are the result of tectonic or lithologic variations along strike. Where the depth of the protodécollement is known it corresponds to the depth where inferred maximum fluid pressure occurs (PT and LV, figure 6). An estimate of the degree of overpressuring in the protodécollement is approximately 75% of lithostatic. If the effect of tectonic stress were added this value would increase.

In our conversion of stacking velocity to fluid pressure we use empirical relationships that were not supported by rigorous error analysis. The difference in fluid pressure between the LV and SV areas (Figure 6) is probably not statistically significant. This is consistent with landward vergence and seaward vergence being equally likely in areas where basal shear stress are minimal (Seely, 1977). In the area of the Wecoma Fault, however, low inferred fluid pressure in unit 2 (NF and SF, Figure 6) suggests that the fault may be responsible for increased fluid flow out of unit 2 and the local reversal from landward to seaward vergence.

Tobin et al. (1993) has suggested that the Wecoma strike-slip fault acts as a high-angle fluid conduit allowing escape of highly overpressured fluids from the protodécollement horizon producing an increase in wedge basal stress and causing the local change from landward vergence to seaward vergence (Figure 1), which is in agreement with the NF and SF pore pressure profiles (Figure 6). The high velocity values in unit 2 sediments adjacent to the fault (Figure 4) suggest that the fault is providing a path through the permeability cap for fluid to escape from unit 2 resulting in a decrease in porosity. Cementation accompanying fluid flow may also act to increase the rigidity of the sediment and thereby increase the seismic velocity.

A pore fluid pressure maximum of 70% of lithostatic, discussed above, occurs at a depth of approximately 2400 m below the seafloor in the NF and SF areas. If a décollement were to form in this setting it should form at 2400 m, where the pore fluid pressure is highest. Instead, the décollement in this area is at the base of the sedimentary section as in the LV area. Perhaps, in the area where our velocity measurements were taken, a pore pressure exceeding 70% of lithostatic (using our estimated values) was present at the time of formation of the décollement at the base of the sedimentary section. A subsequent pore-fluid pressure decrease accompanying fluid flow along the Wecoma fault would have the effect of strengthening the sediments and may explain why the deformation front has shifted seaward north of the Wecoma Fault (Figure 1).

5. SUMMARY

Analysis of seismic velocity values derived from a multichannel seismic survey of the Oregon margin implies that there is fracture-controlled, fault-guided flow at the toe of the Oregon accretionary prism. A lateral increase in sediment velocity in an area of protothrusting suggests that fluid flows along thrusts augmenting the diffusive flow assumed to be responsible for observed consolidation of sediment further seaward. There is evidence, in the form of high inferred pore fluid pressure premised on interval velocity variations, that the décollement is an overpressured section of the incoming sediment. High seismic velocities in an area where the Wecoma strike-slip fault cuts the sedimentary section probably results from reduced porosity due to flow of overpressured fluid from the décollement through the strike-slip fault zone.

The fluid pressure in unit 1 diverges from hydrostatic with depth below the seafloor. This suggests that during the rapid deposition of unit 1 the pore fluid flow could not maintain equilibrium consolidation. Within unit 2 there is an elevation of fluid pressure at approximately 2400 m below the sea floor. This may result from dewatering of the uppermost sediment of unit 2 augmented by the rapid loading and higher porosity of the overlying unit 1 sediment, forming a low permeability cap that traps the fluid below. The fluid pressure increases below this cap as the load increases with time and is approximately 75% of lithostatic at the protodécollement. More extensive drilling is required to strengthen our argument that fluid flow driven by tectonic stress is the primary cause of velocity variation in the sediment at the base of the slope rather than variation in several other properties of the sediment (e.g. lithology, cementation, and fracturing).

Acknowledgments. This work was supported by NSF grants OCE-8813907, and OCE-9116368. Some of the analysis was done using support from the U.S. Geological Survey Branch of Pacific Marine Geology. The manuscript benefited from reviews by Tom Shipley, Dan Orange, and an anonymous reviewer.

REFERENCES

Athy, L. F., Density, porosity, and compaction of sedimentary rocks, *Amer. Assoc. of Petroleum Geologists Bull.*, 14, 1-24, 1930.

Bray, C. J., and D. E. Karig, Porosity of sediments in accretionary prisms and some implications for dewatering processes, *J. Geophys. Res.*, 90, 768-778, 1985.

Bray, C. J., and D. E. Karig, Dewatering and extensional deformation of the Shikoku Basin hemipelagic sediments in the Nankai Trough, *Pure and Applied Geophysics*, 128, 725-747, 1988.

Carson, B., Tectonically induced deformation of deep-sea sediments off Washington and northern Oregon: mechanical consolidation, *Marine Geology*, 24, 289-307, 1977.

Carson, B., E. Seke, V. Paskevich, and M. L. Holmes, Fluid expulsion sites on the Cascadia accretionary prism: Mapping diagenetic deposits with processed GLORIA imagery, *J. of Geophys. Res.*, 99, 11,959-11,969, 1994.

Cochrane, G. R., B. T. R. Lewis, and K. J. McClain, Structure and subduction processes along the Oregon-Washington margin, *Pure and Applied Geophysics*, 128, 767-800, 1988.

Cochrane, G. R., M. E. MacKay, G. F. Moore, and J. C. Moore, Consolidation and deformation of sediments at the toe of the Central Oregon Accretionary Prism from multichannel seismic data, *Proceedings of the Ocean Drilling Program, Initial Reports*, 146, 421-426, 1994a.

Cochrane, G. R., J. C. Moore, M. E. MacKay, and G. F. Moore, Velocity and inferred porosity model of the Oregon accretionary prism from multichannel seismic reflection data: implications on sediment dewatering and overpressure, *J. of Geophys. Res.*, 99, 7033-7043, 1994b.

DeMets, C., R. G. Gordon, D. F. Argus, and S. Stein, Current plate motions, *Geophys. Jour. Int.*, 101, 425-478, 1990.

Dix, D. H., Seismic velocities from surface measurements, *Geophysics*, 20, 68-86, 1955.

Fertl, W. H., *Abnormal formation pressures*, 382 pp., Elsevier Scientific Publishing Co., New York, 1976.

Hubbert, M. K., and W. W. Rubey, Role of fluid pressure in mechanics of overthrust faulting, 1 Mechanics of fluid-filled solids and its application to overthrust faulting, *Geol. Soc. Amer. Bull.*, 70, 115-166, 1959.

Hyndman, R. D., G. F. Moore, and K. Moran, Velocity, porosity, and pore-fluid loss from the Nankai Subduction Zone accretionary prism., *Proceedings of the Ocean Drilling Program, Scientific Results*, 131, 211-219, 1993.

Kulm, L. D., R. A. Prince, and P. D., Snavely, Jr., Site survey of the Northern Oregon Continental Margin and Astoria Fan, *Init. Repts. Deep Sea Drilling Proj.*, 18, 979-987, 1973.

Kulm, L. D., et al., Western North American continental margin and adjacent ocean floor off Oregon and Washington, *Atlas 1 Ocean Margin Drilling Program*, Marine Sciences International, Woods Hole, Mass., 1984.

Kulm, L. D., and E. Suess, Relationship between carbonate deposits and fluid venting: Oregon accretionary prism, *J. of Geophys. Res.*, 95, 8899-8915, 1990.

Lewis, B. T. R., and G. R. Cochrane, Relationship between the location of chemosynthetic benthic communities and geologic structure on the Cascadia subduction zone, *J. of Geophys. Res.*, 95, 8787-8793, 1990.

MacKay, M. E., G. F. Moore, G. R. Cochrane, J. C. Moore, and L. D. Kulm, Landward vergence and oblique structural trends in the Oregon Margin Accretionary Prism: Implications and effect on fluid flow, *Earth and Planetary Sci. Let.*, 109, 477-491, 1992.

Marlow M. S., and A. W. Wright, Physical Properties of sediment from the Lesser Antilles Margin along the Barbados Ridge: results from Deep Sea Drilling Project Leg 78A, *Init. Repts. Deep Sea Drilling Proj.*, 78,549-558, 1984.

Moore, J. C., G. F. Moore, and G. R. Cochrane, Reversed-polarity seismic reflections along faults of the Oregon accretionary prism: Indicators of fault zone dilation, *J. of Geophys. Res.*, 100, 12,895-12,906, 1995a.

Moore, J. C., T. H. Shipley, et al., Abnormal fluid pressures and fault-zone dilation in the Barbados accretionary prism: Evidence from logging while drilling, *Geology*, 23, 605-608, 1995b.

Nelson, C. H., Late Pleistocene and Holocene depositional trends, processes, and history of Astoria deep-sea fan, northeast Pacific, *Marine Geology*, 20, 129-173, 1976.

Orange, D. L., and N. A. Breen, The effects of fluid escape on accretionary wedges, II, Seepage force, slope failure, headless submarine canyons, and vents, *J. of Geophys. Res.*, 97, 9277-9295, 1992.

Rubey, W. W., and M. K. Hubbert, Role of fluid pressure in mechanics of overthrust faulting, II, Overthrust Belt in geosynclinal area of Western Wyoming in light of fluid-pressure hypothesis, *Bull. of the Geological Soc. of Amer.*, 70, 167-206, 1959.

Seely, D. R., The significance of landward vergence and oblique structural trends on trench inner slopes, in *Island Arcs, Deep Sea Trenches, and Back-Arc Basins*, edited by M. Talwani, and W. C. I. Pitman, pp. 187-198, Amer. Geophys.. Union, Washington D.C., 1977.

Silver, E. A., Pleistocene tectonic accretion of the continental slope off Washington, *Marine Geology*, 13, 239- 249, 1972.

Snavely, P. D., Jr., R. von Huene, D. M. Mann, and J. Miller, The central Oregon continental margin, lines WO76-4 and WO76-5, in *Seismic Images of Modern Convergent Margin Tectonic Structures*, edited by R. von Huene, pp. 24-29, Amer. Assoc. Petroleum Geologists, Tulsa, Oklahoma, 1986.

Taner, M. T., and F. Koehler, Velocity spectra-digital computer derivation and applications of velocity functions, *Geophysics*, 34, 859-881, 1969.

Tobin, H. J., J. C. Moore, M. E. MacKay, D. L. Orange, and L. D. Kulm, Fluid flow along a strike-slip fault at the toe of the Oregon accretionary prism: implications for the geometry of frontal accretion, *Geol. Soc. of Amer. Bull.*, 105, 569-582, 1993.

Vernik, L., Predicting lithology and transport properties from acoustic velocities based on petrophysical classification of siliciclastics, *Geophysics*, 59, 420-427, 1994.

Westbrook, G., B. Carson, R. J. Musgrave, et al., *Proc. ODP, Init. Repts. of the Ocean Drilling Project*, 611 pp., Ocean Drilling Program, College Station, TX, 146 (Pt. 1), 1994.

Yuan, T., G. D. Spence, and R. D. Hyndman, Seismic velocities and inferred porosities in the accretionary wedge sediments at the Cascadia margin, *J. of Geophys. Res.*, 99, 4413-4427, 1994.

G. R. Cochrane, and H. J. Lee, U.S. Geological Survey, MS-999, 345 Middlefield Rd., Menlo Park, CA 94025. (email: guy@octopus.wr.usgs.gov)

J. C. Moore, Earth Sciences, University of California, Santa Cruz, CA 95064 (email: casey@earthsci.ucsc.edu)

Oblique Strike-Slip Faulting of the Cascadia Submarine Forearc: The Daisy Bank Fault Zone off Central Oregon

Chris Goldfinger[1], LaVerne D. Kulm[1], Robert S. Yeats[2], Cheryl Hummon[1], Gary J. Huftile[2], Alan R. Niem[2], and Lisa C. McNeill[2]

The Cascadia submarine forearc off Oregon and Washington is deformed by numerous active WNW-trending, left-lateral strike-slip faults. The kinematics of this set of sub-parallel left-lateral faults suggests clockwise block rotation of the forearc driven by oblique subduction. One major left-lateral strike-slip fault, the 94 km-long Daisy Bank fault, located off central Oregon, was studied in detail using high-resolution AMS 150 kHz and SeaMARC-1A sidescan sonar, swath bathymetry, multichannel seismic reflection profiles and a submersible. The Daisy Bank fault zone cuts the sediments and basaltic basement of the subducting Juan de Fuca plate, and the overriding North American plate, extending from the abyssal plain to the upper slope-outer shelf region. The Daisy Bank fault, a near-vertical left-lateral fault striking 292°, is a wide structural zone with multiple scarps observed in high-resolution sidescan images. From a submersible, we observe that these scarps offset late Pleistocene gray clay and overlying olive green Holocene mud, dating fault activity as post-12 ka on the upper slope. Vertical separation along individual fault scarps ranges from a few centimeters to 130 meters. Using a retrodeformation technique with multichannel reflection records, we calculate a net slip of 2.2 ± 0.5 km. Fault movement commenced at about 380 ± 50 ka near the western fault tip, based upon an analysis of growth strata and correlation with deep-sea drill hole biostratigraphy. We calculate a slip rate of 5.7 ± 2.0 mm/yr. for the Daisy Bank fault at its western end on the Juan de Fuca plate. The motion of the set of oblique faults, including the Daisy Bank fault, may accommodate a significant portion of the oblique component of plate motion along the central Cascadia margin. We propose a block rotation model by which the seawardmost part of the forearc rotates clockwise and translates northward.

INTRODUCTION

The Cascadia subduction zone has received considerable attention in recent years due in part to the enigmatic aseismic behavior of this convergent margin. The quantity and quality of geologic and geophysical data available for Cascadia make possible an improved level of understanding of subduction processes, particularly the response of the forearc to oblique subduction.

Using high-resolution AMS 150 and SeaMARC-1A sidescan sonar, swath bathymetry, multichannel seismic reflection profiles, and submersibles, we have surveyed in detail nine suspected zones of oblique left-lateral strike-slip faulting on the abyssal plain, continental slope and shelf off Oregon and Washington (four are shown in Figure 1). Five of these faults cut both the Juan de Fuca (JDF) and North American (NOAM) plates, extending from the abyssal plain to the upper slope-outer shelf region [*Appelgate et al.*, 1992; *Goldfinger et al.*, 1992; *Goldfinger et al.*, 1996]. The faults strike 298° to 283°, with obliquity to the margin generally increasing to the south [*Goldfinger*, 1994; *Goldfinger et al.*, 1996, 1996a]. The faults are commonly expressed in swath bathymetry as irregular ridges composed of en echelon folds and sigmoidal bends and or offsets of throughgoing accretionary wedge folds. SeaMARC-IA sidescan records of these structures reveal steep scarps cutting accretionary wedge folds, and commonly show straight traces across topography and reversals of vertical separation, characteristic of strike-slip faulting. Where topographic features are offset, the separations are left-lateral. Subsequent analysis of bathymetric and sidescan data has identified four additional structures with left offsets of the accretionary wedge and continental slope channels off northern Washington and northern California.

[1]College of Oceanic and Atmospheric Sciences, Oregon State University, Corvallis, Oregon 97331
[2]Department of Geosciences, Oregon State University, Corvallis, Oregon 97331

Subduction: Top to Bottom
Geophysical Monograph 96
Copyright 1996 by the American Geophysical Union

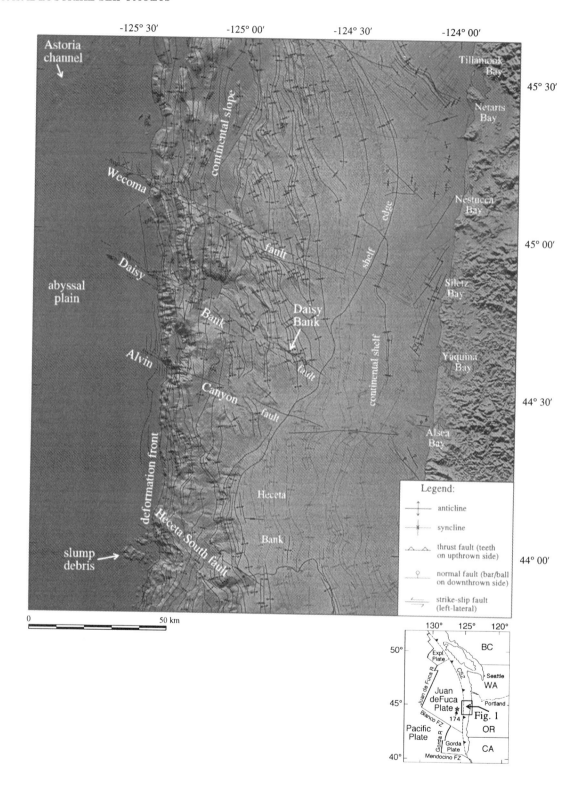

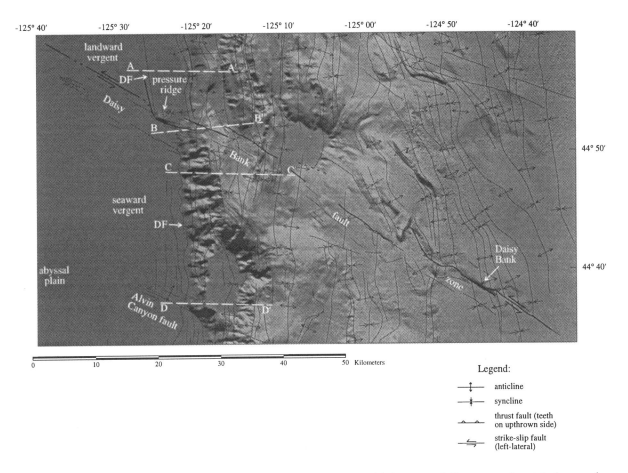

Fig. 2. Shaded relief bathymetry of the Daisy Bank fault and vicinity, central Oregon continental slope and abyssal plain. Rough-textured areas in eastern part of image result from noisy data. Active structures of the accretionary wedge are shown. DF = deformation front. A-A', B-B', C-C', and D-D' indicate locations of multichannel seismic (MCS) profiles shown in Figure 6. See text for discussion of named features.

In this paper we describe in detail one of these active strike-slip faults, the Daisy Bank fault (DBF), off central Oregon. We map and characterize this active structure, determine its sense of motion and rate of slip, and speculate on its origins and the role of such faults in the deformation of the Cascadia forearc.

DAISY BANK FAULT

The Daisy Bank fault [fault B of *Goldfinger et al.*, 1992; Figure 2] was surveyed along its entire length with SeaMARC-1A 30 kHz sidescan sonar at a 5 km swath width from the abyssal plain to the continental shelf. On the upper slope, we imaged the fault zone with both SeaMARC-1A 5 km and AMS 150 kHz 1 km sidescan swaths (2.5 m and 0.5 m resolution respectively). The fault zone and surrounding structures were also imaged by NOAA using SeaBeam swath bathymetry on the abyssal plain and continental slope. We conducted a series of dives using the DELTA submersible in 1992-1993 to ground truth the sonar imagery and investigate the DBF at outcrop scale. The DBF extends 94 km from the abyssal plain (Juan de Fuca plate), across the deformation front and continues across the continental slope landward to the continental shelf (North American plate). It probably terminates against the western edge of the deeply buried basement block of lower Eocene Siletz River Volcanics oceanic basalt [*Snavely*, 1987; *Tréhu et al.*, 1994] which underlies the forearc basins of the continental shelf (Figure 1). The landward terrane, called Siletzia, forms a high-strength backstop for the accretionary wedge. This fault is more prominently expressed on the outer shelf-upper slope (accretionary wedge-forearc basin) than any of the other strike-slip faults in the Cascadia subduction zone.

Upper Continental Slope and Shelf

Daisy Bank is one of several uplifted Neogene structural highs located on the upper slope to outer shelf off Oregon [*Kulm and Fowler*, 1974]. The DBF bounds the southern

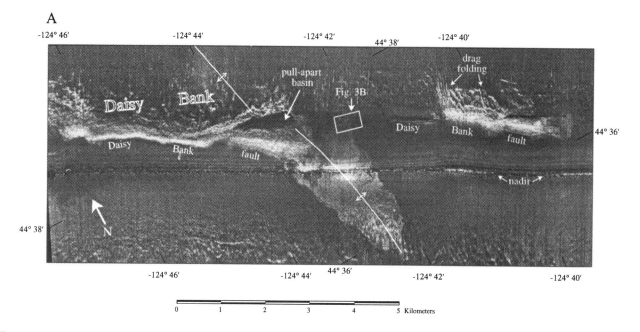

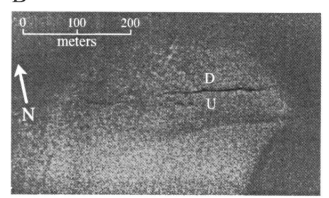

Fig. 3. Sidescan images of the Daisy Bank fault (DBF), upper continental slope. High backscatter areas are lighter tones in both images. A - SeaMARC 1A image (5 km swath) shows Daisy Bank in upper left of image. Total relief is 130 m on the western scarp and 47 m on the eastern scarp of the DBF. Left-lateral offset of anticline by DBF is shown. A small pull-apart basin formed across a left step on the DBF. Left-lateral drag folding of bedding is also visible along the eastern scarp. B - High-resolution AMS 150 kHz image shows details of secondary Holocene scarp shown in Figure 4.

flank of Daisy Bank; a second less prominent strand of the fault bounds the northern flank. SeaMARC-1A sidescan imagery, multichannel (MCS) and single channel (SCS) seismic reflection data show that the Daisy Bank fault is a wide structural zone, within which Daisy Bank is uplifted as a horst (Figure 2). The main fault zone is 5-6 km wide northwest of Daisy Bank, widening around the oblong bank, then narrowing to a single strand to the southeast. Multiple scarps are evident, with variable vertical separation southeast of Daisy Bank (Figure 3). Northwest of the bank, two main strands are evident, both up to the north. The traces of the fault strands are straight, implying a near vertical fault. Sidescan imagery shows reversals of vertical separation along strike, a characteristic common only to strike-slip faults. Probable drag folds of exposed strata, with a left-lateral sense of motion, are visible in sidescan imagery southeast of the bank (Figure 3).

Mapping from seismic reflection profiles indicates 2-3 km left lateral offsets of NNW-trending accretionary wedge fold axes at Daisy Bank [*Goldfinger et al.*, 1992a; 1996a]. Scarp heights measured from the submersible range from tens of centimeters to 47 m. The net uplift of the southern flank of Daisy Bank by both folding and faulting is about 130 m. The most prominent scarp is a steep (25°-50°) debris-covered slope, the debris typically consisting of a chaotic arrangement of angular gray mudstone blocks on the lower slopes and angular to tabular carbonate-cemented mudstone slabs on the upper slopes.

From the submersible DELTA, we observed a fresh scarp striking 290° across the unconsolidated Holocene mud (Figure 4), which is also visible in the AMS 150 kHz sidescan images (Figure 3B). At the outcrop scale, we observed both right-and left-stepping en echelon fault traces, although left-stepping was dominant. This fresh

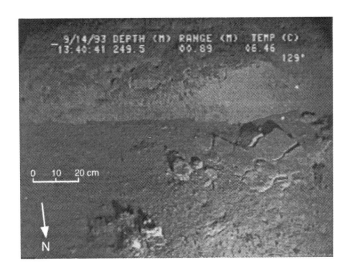

Fig. 4. Video image taken from DELTA submersible. Holocene scarp on a subsidiary fault to the Daisy Bank fault. The scarp is up to the south, with a maximum height of ~1 m. A thin layer of Holocene olive-gray clay is cut by the fault. The underlying late Pleistocene gray clay is exposed in the fault scarp. This constrains the most recent fault motion to post-12 ka. The lack of bioturbation of the scarp face, compared with the highly burrowed surrounding seafloor, suggests movement on the fault within the last few hundred years.

fault scarp ranges from a few centimeters to 1.0 m in height, dips steeply south with the south side up, and offsets cohesive gray late Pleistocene clay and olive-green Holocene mud (Figure 4). This sharp change in color occurs at about 12 ka in sediments on the upper slope off Oregon-Washington [*Barnard and McManus*, 1973], indicating post 12 ka motion on this segment of the Daisy Bank fault. Abrupt vertical changes in oxidation color of the late Pleistocene clay, and corresponding abrupt upward increases in bioturbation by benthic animals inhabiting the scarp face suggest that the scarp may represent multiple Holocene tectonic events.

We observed and mapped spectacular carbonate chimneys, doughnuts, and slabs within 100-150 m of the traces of the DBF. Their occurrence decreases rapidly with distance from the fault. Several investigators have observed a close association between fluid venting, methane-derived carbonate deposition, and active faulting along the DBF and other active faults in the Cascadia forearc [*Kulm and Suess*, 1990; *Sample et al.*, 1993]. The main fault scarp southeast of Daisy Bank localizes the largest concentration of carbonate deposits yet found on the Oregon margin. Tabular carbonate blocks cover much of the flat top of Daisy Bank. Near the top of the scarp slope, carbonate-bearing boulders up to 6 m in the longest dimension are common. Near fault zones, we observed tabular bodies 10-30 cm in thickness are a common mode of carbonate occurrence over wide areas of otherwise unconsolidated Holocene sediment. These slabs were broken and disrupted in a pattern similar to a parking lot excavated by a bulldozer. We presume that this pattern of disruption is tectonic in nature. We observed minor breakage of carbonate bodies by bottom-fishing activity, but conclude that this breakage is clearly distinguishable, and could not be responsible for the widespread and pervasive disruption we observed near the DBF.

Lower Slope

Using SeaMARC sidescan imagery, we traced the Daisy Bank fault zone across the lower continental slope (Figures 2, 5). The fault morphology is subdued on the lower slope relative to the upper slope or the abyssal plain. The DBF is characterized by discontinuous fault traces that disrupt thrust anticlines, and to a lesser degree, the intervening basins. One 3-4 km long strand terminates at the foot of a thrust ridge, producing gullies and a prominent slump scarp. Farther seaward, several splays of the DBF truncate the frontal thrust anticline of the accretionary wedge (Figure 5) with several tens of meters of relief evident along the main splay. The DBF crosses the plate boundary in a 1 km-wide fault zone that appears to have localized slumping of the seaward limb of the frontal thrust (Figure 5). The intersection of the DBF and the accretionary wedge marks a transition in structural domains between seaward-vergent thrusts (seaward directed thrusting) to the south and landward-vergent thrusts (landward directed thrusting) to the north [*MacKay et al.*, 1992; *MacKay*, 1995; *Goldfinger et al.* 1992]. The initial thrust ridge in this region has apparently undergone a progressive vergence reversal from south to north along the margin (Figure 6). The vergence transition is complex, and occurs over a 15 km length of the initial thrust ridge. Fifteen kilometers south of the DBF, the vergence direction is entirely seaward. The transition from seaward to landward is manifested as progressive undercutting of the originally landward vergent ridge by a seaward vergent thrust that is dies northward. The transition to landward vergence is complete at the intersection of the leading ridge with the DBF, where the undercutting seaward-vergent thrust terminates.

Abyssal Plain

In sidescan images, the DBF crosses the deformation front without interruption (Figure 5), extending 21 km seaward onto the abyssal plain where surface and sub-surface expression die out (Figure 2). The main trace intersects a 150 m-high ridge along the boundary between the landward-vergent thrust ramp and the fault. MCS lines 37 (Figure 5) and 19 (not shown) show this ridge to be a southwest-vergent thrust ridge bounded by the DBF on its southern flank. The main strand of the DBF steps to the right at the western end of this anticlinal ridge and continues to the northwest (Figure 5 cutaway). We interpret the ridge as a pressure ridge developed between the two overlapping fault strands.

Fig. 5. Perspective view of the Daisy Bank fault zone - accretionary wedge intersection, viewed from the west-southwest. The seafloor is represented by a 5 km-wide SeaMARC sidescan sonar image draped over SeaBeam swath bathymetry. High backscatter areas are light tones in the sidescan image. Multiple fault traces on the first ridge are shown by small white arrows. In the cutaway section, seismic reflection line MCS-37 shows the internal structure of the pressure-ridge anticline. Depth in two-way time (seconds) is shown at left margin. Thrust vergence in the accretionary wedge is landward to the north of the fault zone, and seaward to the south. Kilometer scale is approximate for foreground only.

A similar pressure ridge occurs at a right step of the Wecoma fault 34 km to the north (Figure 1; *Goldfinger et al.*, 1992, 1996a). Basement reflectors show two offsets, one up-to-the-north and one up-to-the-south, that define a "popup" of the basement across the DBF on MCS line 37. We observe the same vertical structure across the Wecoma and Alvin Canyon strike-slip faults, also on line 37 [Goldfinger et al., 1996]. The DBF is up to the north in both seismic data and sidescan images, the same vertical separation as observed for the other four strike-slip faults found on the Juan de Fuca plate.

While there are no surficial piercing points from which to determine horizontal offset, we reconstruct the fault's displacement history from the abyssal plain sedimentary section. To estimate the overall net slip and slip-rate for the Daisy Bank fault, we applied a geometric technique for restoration of strike-slip fault motion described by *Goldfinger et al.* [1992, 1996a]. This method utilizes the geometry of trenchward thickening abyssal plain sediment wedges in at least one trench-parallel and at least two trench-normal seismic reflection profiles. This configuration is the minimum required to establish 3D piercing points based on the geometry of abyssal plain sedimentary units. Isopachs in this area trend uniformly north-south in pre-faulting sediments, based on the 1989 MCS survey. We determined the position in the sedimentary section at which the fault-related growth strata first appears in MCS line 37. We converted the sediment thickness in two-way time to depth using an average velocity of 1680 m/s for Astoria Fan sediments [*Goldfinger*, 1994]. To determine the net-slip on the fault, we measured the horizontal separation of isopachs in units deposited prior to faulting. This method yields a best fit net slip of 2.2 ± 0.5 km for the Daisy Bank fault on the Juan de Fuca plate. The uncertainty represents the maximum and minimum allowable fault slip that does not produce a severe mismatch

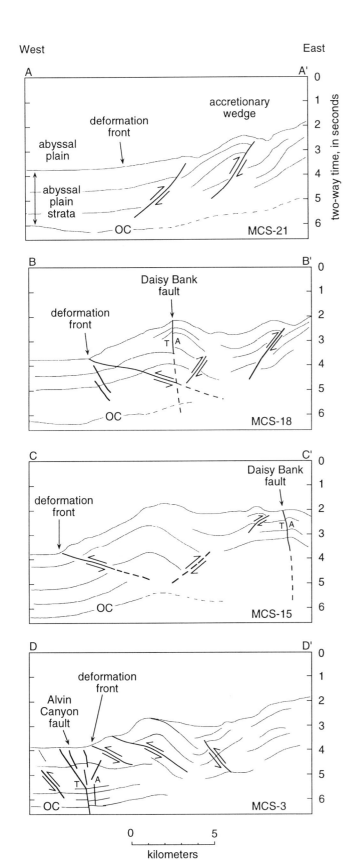

of thickness units across the fault. The direction of slip is left-lateral, the same as that shown by the drag folds on the upper slope (Figure 3A). We estimate that faulting began at 380 ± 50 ka, based upon the initiation of fault-related growth sedimentation in seismic records, and correlation with dated strata in DSDP drill hole 174 [see Figure 1 inset for location; *Goldfinger*, 1994]. This age yields an average slip rate of 5.7 ± 2 mm/yr for the DBF near the deformation front. We assume that the initiation of horizontal and vertical slip were coincident. If this assumption is incorrect, the calculated age is a minimum age. We used the same technique to calculate slip rates for the other strike-slip faults where warranted by data availability [*Goldfinger et al.*, 1996; *McCaffrey and Goldfinger*, 1995].

DISCUSSION

Tectonic Model

The DBF and similar oblique strike-slip faults off Oregon and Washington appear to play an important role in the deformation of the marine portion of the Cascadia convergent margin [*McCaffrey and Goldfinger*, 1995; *Goldfinger et al.*, 1992, 1996]. The set of nine sub-parallel faults with the same sense of slip implies clockwise rotation of forearc blocks about vertical axes of the style first proposed by *Freund* [1974]. This style of deformation has been proposed for the Oregon and Washington onshore forearc [*Wells and Coe*, 1985], and for the Aleutian forearc [*Geist et al.*, 1988; *Ryan and Scholl*, 1993] albeit on a much larger scale. We postulate that these WNW-trending strike-slip faults are R' Riedel shears within an overall dextral shear couple driven by oblique subduction [Figure 7; *Goldfinger et al.*, 1992]. Several of these transverse strike-slip faults and associated folds cross the plate boundary and accretionary prism, extending to the continental shelf [*Goldfinger et al.*, 1992, 1996]. Paleomagnetically-determined clockwise rotations in coastal basalts in Oregon and Washington suggest that similar processes have operated throughout the Tertiary, with Miocene Columbia River Basalts (12-15 Ma) in western Oregon and Washington rotated 10-30° clockwise [*England and Wells*, 1991], and Eocene Siletz River Volcanics rotated up to 90° [*Wells*, 1990], although some of this may be due to microplate rotation and opening of the Basin and Range [*Magill et al.*, 1982]. The location of

---

Fig. 6. Interpretation of migrated 144-channel seismic reflection profiles, near the intersection of the Daisy Bank fault (DBF) and the accretionary wedge. Seismic lines are arranged from north (A-A') to south (D-D'). Vertical scale is in two-way time (seconds). Location of seismic lines is shown on Figure 2. A-A' is in the landward-vergent province north of the DBF. B-B' and C-C' show transition from landward to seaward vergence. D-D' is entirely seaward-vergent. The DBF is shown in B-B' and C-C', but is out of view in A-A' and D-D'. OC = oceanic crust, A = away, T = toward.

the axes of rotation (pivot points) of the offshore blocks is presumably at or near their eastern ends, although this is not required. If the pivots are fixed to the North American plate, compression between the blocks is required. We observe evidence of compression across the three central Oregon faults, including the Daisy Bank horst, but not along the Washington faults. Without compression along the block edges, the pivots must translate northward relative to the North American plate. The model shown in Figure 7 suggests areas of compression and extension at the eastern ends of the rotated blocks. Currently we are using new seismic reflection data to test the model by evaluating the late Quaternary depocenters on the Oregon shelf. However, the complexity of this deformation superimposed on the accretionary wedge may preclude a definitive test of the model. This type of model, originally envisioned in a simple shear strike-slip environment, requires an overall dextral shear couple. However, we find little evidence for arc-parallel dextral faulting in Cascadia. In this case, the dextral shear couple is apparently being provided by an obliquely subducting slab rather than a pair of parallel faults. The mechanics of such a system are poorly known [*England and Wells*, 1991].

Fault Origin

What is the origin of the transverse strike-slip faults? Certainly their presence within an active accretionary wedge indicates partitioning of strain into arc-parallel (strike-slip) and arc-normal (thrust) components of oblique convergence. We suggest two classes of origins for the faults: (1) Intraplate stresses within the Juan de Fuca plate, transmitted across the plate boundary to the North American plate, and (2) Deformation driven by interplate coupling of the subducting and overriding plates. The high-resolution seismic data available for the three central Oregon strike-slip faults shows strong evidence of offset of the basaltic slab for all three of these faults [*Appelgate et al.*, 1992; *Goldfinger et al.*, 1992, 1996; *MacKay*, 1995]. Evidence of slab rupture includes: vertical separation of basement reflectors, some of which do not underlie the surficial structure, and thus could not be velocity pull-up; pop-ups of basement reflectors with fault splays branching upward into the sedimentary section; and abrupt changes in the character of basement reflectors across the Wecoma fault. The style of deformation of two of the Washington faults is virtually identical to the three central Oregon faults, and we suspect that they are analogous to the Oregon faults. The remaining four Cascadia faults show no evidence of surface rupture on the abyssal plain, and thus may either be limited to the upper plate.

Goldfinger et al. [1996] estimate that the interplate coupling stress in the Cascadia accretionary wedge is probably insufficient to rupture the subducting slab. *Wang et al.* [1995], using geodetic data, also infer very low coupling

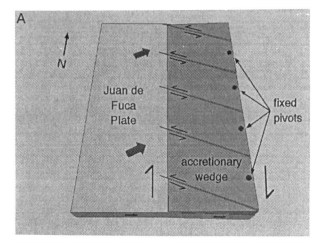

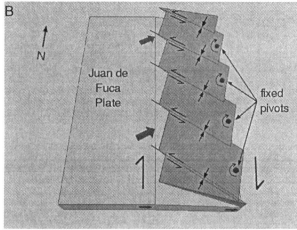

Fig. 7. A - Block rotation model for the central Cascadia forearc. The arc-parallel component of oblique subduction creates a right-lateral shear couple, which is accommodated by WNW-trending left-lateral strike-slip faults. B - Slip on nine mapped WNW-trending left-lateral faults results in clockwise block rotation and northward transport of the forearc. With fixed pivots, compression occurs along the block edges. We infer that Daisy Bank (off central Oregon) has been uplifted by this mechanism.

stress on the interplate thrust. We therefore favor an origin involving deformation of the subducting slab that is then partially transmitted into the weaker North American plate. For a fuller discussion of these issues, see Goldfinger et al. [1996]. We note a longitudinal pattern of deformation along the DBF and most of the other transverse faults: strong expression on the plain, poor expression on the lower slope, strong expression on the upper slope. This decrease of deformation from the plain to lower slope is also consistent with a lower plate origin. We speculate that reduced coupling across the overpressured and poorly coupled décollement beneath the lower slope results in weak expression of the overlying upper plate fault. Rejuvenation of the faults on the upper slope is consistent with progressive dewatering of the wedge, and stronger interplate coupling as would be expected for the rearward part of the wedge. Landward widening of the fault zones (Figure 1) is also consistent with fault slip transmitted upward through the eastward thickening accretionary wedge.

The observation that the frontal thrust has undergone a progressive vergence reversal, terminating at the DBF, suggests that passage of the DBF beneath the wedge (if the DBF moves NE with the Juan de Fuca plate) may have initiated the vergence reversal. *Tobin et al.* [1993] suggested that a similar vergence reversal at the intersection of the Wecoma fault and the accretionary wedge is due to fluid pressure loss through the vertical fault. They conclude that a local reduction in pore fluid pressure on the basal décollement results in an increase in basal shear stress, promoting the switch to seaward vergence. While the vergence reversal is probably the result of changes in sediment supply and basal shear stress along the margin [*MacKay*, 1995], the DBF clearly localizes the vergence change.

Plate Interaction

The observation that five of the nine mapped faults cross the plate boundary is difficult to reconcile with northeasterly subduction at 40 mm/yr [*DeMets et al.*, 1990], since the JDF plate should have traveled 10-24 km to the northeast during the 0.2 to 0.6 Ma elapsed since the initiation of faulting [from 5 faults with known ages; *Goldfinger*, 1996]. The arc-parallel offset due to northeasterly plate motion during this time would be 4-11 km. However, the deformation front also advanced rapidly westward during this period. Assuming the DBF moved about 15 km northeastward since 380 ka, this is resolved as 13.2 km normal convergence, and 7.0 km trench parallel motion. We estimate that the westward advance of the deformation front during this period was 8.5-12.7 km, based on microfossil ages of uplifted strata from DSDP site 175 [*Kulm, von Huene et al.*, 1973]. Since the deformation front is constantly renewed by rapid westward advance, the age of its intersection with the DBF is always young, minimizing the expected offset at the deformation front.

These factors act to reduce the expected fault offset at the deformation front. However, we conclude that the apparent lack of *any* measurable offset and the relatively straight trends of the fault zones strongly imply that the lower slope is moving with the subducting plate to some extent. We suggest that the forearc is deforming as a wide shear zone by distributed deformation and translating northward driven by oblique subduction [*Pezzopane and Weldon*, 1991; *McCaffrey and Goldfinger*, 1995; *England and Wells*, 1991]. The exponential arcward die out of clockwise rotations observed in Columbia River Basalt onshore strongly supports a model of distributed deformation driven by the dextral component of oblique subduction [*England and Wells*, 1991]. The net arc-parallel rate of deformation estimated from slip rates on the offshore strike-slip faults is sufficient to absorb 50-100% of the oblique component of plate convergence [*Goldfinger*, 1994; *McCaffrey and Goldfinger*, 1995]. Similarly, *England and Wells*, [1991] inferred that the onshore forearc may be absorbing most of the tangential component of oblique convergence, based on the rotation rate of the Columbia River Basalt. Although the mechanics of rotation for Siletzia and for the accretionary wedge are probably different in detail, the general agreement of the two independent datasets supports a model of distributed deformation by dextral shear. We suggest that this process may be occurring in both plates, as indicated by the rupture of the slab by three and possibly five of the transverse faults, and by the presence of faults that are probably limited to the upper plate. Further study is needed to test these hypotheses and shed more light on the complex mechanics of oblique subduction.

Acknowledgments. We thank the crews of the research vessels *Thomas Thompson* (University of Washington), and support vessels *Cavalier* and *Jolly Roger*, pilots of the submersible DELTA, Hiroyuki Tsutsumi, Craig Schneider, Margaret Mumford and members of the Scientific Party on cruises from 1992-1993 during which most of the data were collected. Thanks to Guy Cohrane (UCSC), Mary MacKay and Greg Moore (SOEST) for processing the 1989 MCS data. Multibeam bathymetry data was collected by NOAA and processed by the NOAA Pacific Marine and Environmental Laboratory, Newport OR. Thanks to Chris Fox, and Steve Mutula of NOAA for their assistance with the multibeam data. Thanks to Greg Moore, Ray Wells, and Holly Ryan for helpful reviews. This research was supported by National Science Foundation grants OCE-8812731 and OCE-9216880; U.S.G.S. National Earthquake Hazards Reduction Program awards 14-08-0001-G1800, 1434-93-G-2319, and 1434-93-G-2489, and the NOAA Undersea Research Program at the West Coast National Undersea Research, University of Alaska grants UAF-92-0061 and UAF-93-0035.

REFERENCES

Appelgate, B., C. Goldfinger, L. D. Kulm, M. MacKay, C. G. Fox, R. W. Embley, and P. J. Meis, A left lateral strike slip fault seaward of the central Oregon convergent margin, *Tectonics*, 11, 465-477, 1992.

Barnard, W. D., and D. A. McManus, Planktonic foraminiferan-Radiolarian stratigraphy and the Pleistocene-Holocene boundary in the northeast Pacific, *Geol. Soc. Am. Bull.*, 84, 2097-2100, 1973.

DeMets, C., R. G. Gordon, D. F. Argus, and S. Stein, Current plate motions, *Geophys. J. Int.*, 101, 425-478, 1990.

England, P., and R. E. Wells, Neogene rotations and quasicontinuous deformation of the Pacific Northwest continental margin, *Geology*, 19, 978-981, 1991.

Freund, R., Kinematics of transform and transcurrent faults, *Tectonophys*, 21, 93-134, 1974.

Geist, E. L., J. R. Childs, and D. W. Scholl, The origin of summit basins of the Aleutian Ridge: Implications for block rotation of the arc massif, *Tectonics*, 7, 327-341, 1988.

Goldfinger, C., Active deformation of the Cascadia forearc: Implications for great earthquake potential in Oregon and

Washington, PhD thesis, 202 pp., Oregon State Univ., Corvallis, 1994.

Goldfinger, C., L. D. Kulm, and R. S. Yeats, Neotectonic map of the Oregon continental margin and adjacent abyssal plain, scale 1:500,000, *Or. Dept. of Geol. Min. Ind.*, Open-File Report O-92-4, 1992a.

Goldfinger, C. L. D. Kulm, R. S. Yeats, L. McNeill, and C. Hummon, Oblique Strike-Slip Faulting of the Central Cascadia Submarine Forearc, *J. Geophys. Res.* in press, 1996.

Goldfinger, C., L. D. Kulm, and R. S. Yeats, B. Appelgate, M. E. MacKay, and G. R. Cochrane, Active strike-slip faulting and folding of the Cascadia plate boundary and forearc in central and northern Oregon, in *Assessing and Reducing Earthquake Hazards in the Pacific Northwest*, edited by A. M. Rogers, W. J. Kockelman, G. Priest, and T. J. Walsh, U.S.G.S. Professional Paper 1560, in press, 1996a.

Goldfinger, C., L. D. Kulm, R. S. Yeats, B. Appelgate, M. MacKay, and G. F. Moore, Transverse structural trends along the Oregon convergent margin: Implications for Cascadia earthquake potential, *Geology, 20*, 141-144, 1992.

Kulm, L. D., and G. A. Fowler, Oregon continental margin structure and stratigraphy: A test of the imbricate thrust model, in *The Geology of Continental Margins,* edited by C. A. Burke, C. A., and C. L. Drake, p. 261-284, Springer-Verlag, New York, 1974.

Kulm, L. D., and E. Suess, Relation of carbonate deposits and fluid venting: Oregon accretionary prism, *J. Geophys. Res., 95*, 8899-8915, 1990.

Kulm, L. D., R. von Heune et al., Site 174, in *Initial Reports of the Deep Sea Drilling Project,* edited by L. D. Kulm, and R. von Heune, p. 97-167, U.S. Gov. Printing Office, Washington, 1973.

MacKay, M. E., Structural variation and landward vergence at the toe of the Oregon accretionary prism, *Tectonics*, 14, 1309-1320, 1995.

MacKay, M. E., G. F. Moore, G. R. Cochrane, J. C. Moore, and L. D. Kulm, Landward vergence and oblique structural trends in the Oregon margin accretionary prism: Implications and effect on fluid flow, *EPSL, 109*, 477-491, 1992.

Magill, J. R., R. E. Wells, R. W. Simpson, and A. V. Cox, Post 12 M.Y. rotation of southwest Washington, *J. Geophys. Res., 87*, 3761-3777, 1982.

McCaffrey, R., and C. Goldfinger, Forearc deformation and great earthquakes: Implications for Cascadia earthquake potential, *Science, 267*, 856-860, 1995.

Pezzopane, S. K., and R. J. Weldon II, Tectonic role of Holocene fault activity in Oregon, *Tectonics, 12*, 1140-1169, 1991.

Ryan, H. F., and D. W. Scholl, Geologic implications of great interplate earthquakes along the Aleutian arc, *J. Geophys. Res., 98*, 22,135-22,146, 1993.

Sample, J. C., M. R. Reid, H. J. Tobin, and J. C. Moore, Carbonate cements indicate channeled fluid flow along a zone of vertical faults at the deformation front of the Cascadia accretionary wedge (northwest U. S. coast), *Geology, 21*, 507-510, 1993.

Snavely, P. D., Jr., Tertiary geologic framework, neotectonics, and petroleum potential of the Oregon-Washington continental margin, in *Geology and Resource Potential of the Continental Margin of Western North America and Adjacent Ocean Basins-Beaufort Sea to Baja California,* edited by D. W. Scholl, A. Grantz, and J. G. Vedder, p. 305-335, Circum-Pacific Council for Energy and Mineral Resources, Houston, TX., 1987.

Tobin, H. J., J. C. Moore, M. E. MacKay, D. L. Orange, and L. D. Kulm, Fluid flow along a strike-slip fault at the toe of the Oregon accretionary prism: Implications for the geometry of frontal accretion, *Geol. Soc. Am. Bull., 105*, 569-582, 1993.

Tréhu, A., I. Asudeh, T. M. Brocher, J. H. Luetgert, W. D. Mooney, J. L. Nabelek, and Y. Nakamura, Crustal architecture of the Cascadia forearc, *J. Geophys.. Res., 265*, 237-143, 1994.

Wang, K., T. Mulder, G. C. Rogers, and R. D. Hyndman, Case for very low coupling stress on the Cascadia subduction fault, *J. Geophys. Res., 100*, 12, 907-12,918, 1995.

Wells, R. E., Paleomagnetic rotations and regional tectonics of the Cascade arc, Washington, Oregon, and California, *J. Geophys. Res., 95*, 19,409-19,418, 1990.

Wells, R. E., and R. S. Coe, Paleomagnetism and geology of Eocene volcanic rocks of southwest Washington: Implications for mechanisms of tectonic rotation, *J. Geophys. Res., 90*, 1925-1947, 1985.

C. Goldfinger, L. D. Kulm, C. Hummon, College of Oceanic and Atmospheric Sciences, Oregon State University, 104 Ocean Admin Bldg, Corvallis, OR 97331-5503

R. S. Yeats, G. J. Huftile, A. R. Niem, and L. C. McNeill, Department of Geosciences, Oregon State University, 104 Wilkinson Hall, Corvallis, OR 97331-5506

Fabrics and Veins in the Forearc: a Record of Cyclic Fluid Flow at Depths of <15 km

Donald M. Fisher

Department of Geosciences, Pennsylvania State University, University Park, Pennsylvania

Fluid flow through much of the forearc is channeled along faults and fractures, so the distribution and textural history of veins and deformation fabrics can be used to evaluate forearc plumbing. From shallow to deeper levels, fabric elements depict a network of fluid conduits that fluctuate between dilatancy and collapse. In general, there is a transition from particulate flow in soft sediments deformed on the upper slope or near the toe of the prism to grain scale diffusive mass transfer associated with metamorphism of low porosity rocks in the interior of the forearc. With increasing depth of burial, veins vary from zones of distributed grain boundary failure ("mud-filled veins") to cracks that follow grain boundaries and fill with dirty carbonate to cracks that break across grains and fill with clean calcite or quartz. Deformation fabrics vary with depth from semipervasive mud-filled veins, kink bands, shear bands, microfaults, and scaly foliation to slaty cleavage. Fault zones display a scaly fabric that records dissolution or local collapse of a more open grain network. Dilatancy within these zones is inferred based on observations of carbonate and quartz veins and the requirement that the scaly fabrics act as fluid conduits capable of maintaining observed geochemical and thermal anomalies. Much of the fluid expelled from the downgoing slab may migrate upward within the scaly fabric at the top of the underthrust sediment pile to be ultimately vented near the toe of the prism. Observations of vein distributions within the more metamorphosed wedge interior, however, indicate diffuse movement of fluid along systems of hydrofractures that develop in regional zones of low permeability. Differences in hydrogeology between some active convergent margins reflect the range of incoming sediment packages as well as the episodic nature of fluid flow. Fluid flow may be cyclic, in which case the frequency of events may depend on the transient distribution of open fractures, the rate at which excess fluid pressures develop within fracture arrays, the rate at which fractures seal, the strain rate, and the nature of the relationship between the seismic events and fluid flow events.

INTRODUCTION

Most of the volatiles in a subducting plate are either returned to the mantle or expelled within the forearc through a plumbing system that is influenced by the rapid evolution of sediment physical properties [*Ito et al.*, 1983; *Moore and Vrolijk*, 1992]. At the toe of the accretionary wedge, diffuse fluid flow and porosity reduction occurs within the protothrust zone (e.g., Nankai and Oregon [*Bray and Karig*, 1985; *Taira et al.*, 1992; *Morgan et al.*, 1994; *Cochrane et al.*, 1994a]) and the underthrust sediment pile (e.g., Middle America [*Shipley et al.*, 1990]). As sediments become progressively more consolidated and less permeable, fluid migration changes from distributed flow through a grain-scale permeability to channeled flow along networks of interconnected faults and fractures.

Near the toe of active convergent margins, faults are sites of low Cl⁻ anomalies (e.g., Barbados and Nankai [*Moore et al.*, 1988; *Geiskes et al.*, 1990; *Vrolijk et al.*, 1991; *Kastner et al.*, 1991], thermal anomalies [*Fisher and Hounslow*, 1990], negative polarity seismic reflections (e.g., Nankai, Barbados, and Oregon [*Moore G. F. et al.*, 1990; *Bangs and Westbrook*, 1991; *Shipley et al.*, 1994, *Moore, J. C. et al.*, 1995], mineralized fractures (e.g., Barbados and Oregon-Washington [*Brown and Behrmann*, 1990; *Vrolijk and Sheppard*, 1991; *Tobin et al.*, 1993]), and upward deflection of the gas hydrate reflector (e.g., Oregon, [*Cochrane et al.*, 1994b]). These observations indicate movement of fluid along faults, which requires a higher permeability within fault zones, at least temporarily, than in the adjacent wall rock [*Moore et al.*, 1991). Lithostatic fluid pressures and geochemical anomalies along the Barbados décollement require a three orders of magnitude difference in permeability between the prism and the décollement zone [*Screaton et al.*, 1990]. In the absence of fractures, shearing of wet sediments results in a permeability reduction both perpendicular and parallel to the shear plane [*Brown et al.*, 1994]. Thus the reduction in tortuosity caused by grain alignment (which enhances permeability parallel to the fabric [*Arch and Maltman*, 1990]) is more than offset by the collapse of the pore network; active faults must experience episodes of dilatancy where fluid flow occurs along networks of fractures [*Moore*, 1989; *Brown et al.*, 1994].

The history of particulate flow, fracturing, and precipitation recorded by fabrics and veins can be used to evaluate some of the characteristics of forearc fluid flow regimes. For example, the orientation and distribution of veins can be used to depict the orientation of principal stresses and the geometry of fluid conduits (i.e. spacing, tortuosity, and interconnectivity) during episodes of dilatancy. Fluid inclusions trapped by veins place constraints on the trapping conditions and the composition and source of fluids. Vein textures can be used to estimate fracture apertures and to reconstruct complicated histories of cracking and crack closure (sealing or collapse).

Along active submarine accretionary prisms, the tectonic and stratigraphic setting of vein and fabric development is known, but the three-dimensional geometry of these structures is poorly constrained. In contrast, exposed accretionary complexes can be used to reconstruct the geometry of the vein network at depths not yet accessed by drilling along active convergent margins, but the tectonic setting of vein development must be inferred based on fluid inclusion analyses and timing of veining with respect to a structural and metamorphic history. Analyses of veins and fabrics from active and ancient accretionary prisms have led to the consideration of numerous questions about fluid flow in the forearc such as: how does fluid flow vary as a function of depth and sediment properties? Is flow continuous or episodic? If episodic, what is the frequency and/or duration of fluid migration events? Finally, what is the relationship between veins and faults, and what role do fluids play in the evolution of fault systems? At present, the relationship between fluid flow near the toe and fluid flow deeper in the subduction zone is largely unknown.

In the following forearc environments (Figure 1), veins and fabrics are described in terms of the burial conditions, deformation mechanisms, and the geometry of the fabric network. Observations from active margins are used to establish the nature of forearc fluid flow in offscraped or underthrust sediments at the toe of the accretionary wedge. Inferences about the underthrust sediment pile and interior of the wedge are largely based on numerous studies of underplated sequences within the Kodiak accretionary complex in southwest Alaska (Figure 2).

THE ACCRETIONARY WEDGE

Shallow Structural Levels- Development of Scaly Fabric, Deformation Bands, and Carbonate Veins

Sediments that are offscraped at the toe of the accretionary wedge undergo rapid changes in material properties due to diffuse dewatering. This dewatering is in some cases related to the development of semipervasive deformation fabrics as incoming trench sediments enter the prism through the deformation front (defined by a break in slope and the onset of tectonic deformation). The Nankai protothrust zone (i.e., between the deformation front and the frontal thrust) displays distributed conjugate deformation bands or kink bands that accommodate layer parallel shortening [*Karig and Lundberg*, 1990]. The material within bands is similar in grain size and composition but 5% lower in porosity than material outside bands. Arcward of the frontal thrust, the Nankai prism displays a gradation of fabrics related to layer parallel shortening including kink bands, shear bands, and small faults. When crosscutting relationships are observed, the faults are the latest fabrics to develop. The general progression through time (or arcward) from semipervasive ductile to brittle deformation involves a decrease in porosity; thus, there is a close relationship between the diffuse dewatering of the prism and the evolution of fabrics.

An imbricate fan typically develops at the toe of accretionary wedges that consists of emergent thrust faults that connect at some depth with a basal décollement. Along the Barbados margin, the prism is also dissected by low-angle, out-of-sequence faults [*Brown et al.*, 1990]. Fault zones that cut the prism have the potential to channel the upward flow of deeply derived fluids. The dominant fabric within fault zones near the toe of the prism is a scaly foliation that consists of an anastomosing web-like array of polished, striated surfaces. Individual scaly folia are marked by a local collapse of a more open grain network [*Moore et al.*, 1986].

Along the Barbados margin, scaly foliations are associated with development of carbonate veins; there is an arcward increase in the intensity of scaly fabric that corresponds with an increase in the abundance of calcite and rhodochrosite veins [*Brown and Behrmann*, 1990; *Vrolijk and Sheppard*, 1991]. Veins in some cases contain chunks of undeformed mud, and the carbonate within veins is "dirty" or cloudy due to dispersed traces of a silicate mineral [*Vrolijk and Sheppard*, 1991]. The

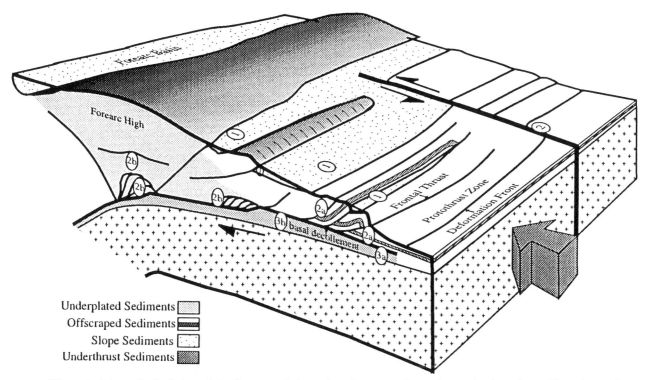

Figure 1: Schematic depiction of the forearc and the various forearc settings where veins have been either observed or inferred based on observations of exposed accretionary complexes. Numbers refer to: 1) upper and lower slope sediments, 2a) imbricate thrust faults and strike slip faults at the toe of the wedge, 2b) underplated rocks in the wedge interior, 3a) shear zone at the top of the underthrust sediment pile-shallow levels, and (3b) shear zone at the top of the underthrust sediment pile- deeper levels. This paper focusses on fabric and vein development within accreted rocks (i.e. forearc settings 2 and 3).

veins typically lie along scaly folia and are in some cases intensely folded or disrupted by faulting [*Vrolijk and Sheppard*, 1991]. Calcite veins are also observed along an out-of-sequence thrust from the Oregon Margin [*Westbrook et al.*, 1994], although veins were absent within the frontal thrust zone along the Nankai margin [*Maltman et al.*, 1992]. Observations of veins, scaly fabrics, and geochemical anomalies along faults suggest that thrust faults near the toe of the accretionary prism can act as fluid conduits. The absence of geochemical anomalies within fault zones along the Nankai margin [*Taira et al.*, 1991] indicates either that fluid flow from depth in this case is diffuse and is not channeled upward along faults (e.g., *Maltman et al.* [1992]) or that fluid flow is episodic and that geochemical anomalies along faults dissipate rapidly between flow episodes (e.g., *Yamano et al.* [1992]).

Regular systems of carbonate-filled veins have also been observed in gullies where a basement-involved left-lateral fault intersects the toe of the Oregon accretionary prism [*Tobin et al.*, 1993]. The veins are associated with carbonate crusts and biological communities, suggesting that the fault trace is a site of fluid venting. The frontal thrust near the strike slip fault displays anomalous seaward vergence, an observation that is attributed to increased shear stress on the décollement due to draining of fluid upward along the strike slip zone [*Tobin et al.*, 1993]. Thus, the strike slip fault at the toe of the Oregon accretionary prism has allowed upward flow of fluid from the décollement or deeper. Faults act as fluid conduits when shear failure is accompanied by hydrofracturing [*Sibson*, 1983; *Behrmann*, 1991]. This condition occurs in thrust fault systems only when the fluid pressure exceeds lithostatic. In strike slip or normal fault systems, simultaneous shear and tensile failure will occur at significantly lower excess fluid pressures [*Sibson*, 1983; *Behrmann*, 1991]. Thus, a strike slip fault in a given area is likely to be favored as a conduit over a nearby thrust fault. In general, forearcs that have numerous strike slip faults and normal faults at high angles to the margin may be better drained than forearcs that are dominated by thrusts. This may be the case along thinly sedimented margins, where the roughness on the incoming seafloor can indent and locally uplift the overriding forearc [*Von Huene and Scholl*, 1991]. In this setting, the spacing of fluid conduits that drain the forearc (i.e. strike slip or normal faults) may reflect the roughness on the downgoing plate.

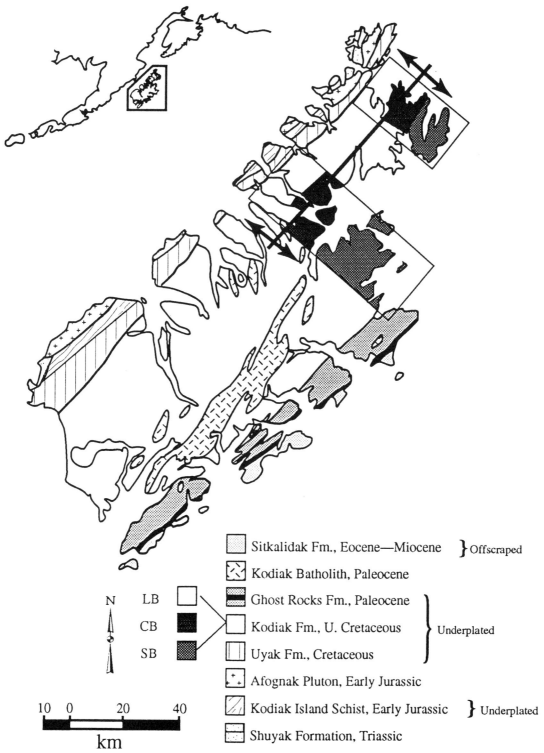

Figure 2: Geologic map of the Kodiak accretionary complex in southwest Alaska. Boxes show two mapped transects where the Kodiak Formation has been subdivided into the landward belt (LB), central belt (CB), and seaward belt (SB). The underplated sequences are differentiated from offscraped rocks of the Sitkalidak Formation based on higher metamorphic grades and the absence of overlying slope basins. In addition, the Sitkalidak Fm. experienced an early history of layer-parallel shortening that included development of landward verging folds [*Moore and Allwardt*, 1980], whereas underplated units record an early history of layer-parallel shear and extension during underthrusting followed by imbrication and seaward vergent thrusting under metamorphic conditions [*Byrne and Fisher*, 1987]. The melange zone in the Ghost Rocks Formation is shown in black.

Observations near the toe of a several convergent margins have shown that, in this zone of diffuse dewatering where porous sediments become consolidated and experience rapid changes in material properties, there are a number of possible micro- and meso-scale mechanisms of subhorizontal shortening, and these different responses are critical for defining the plumbing network that allows for the escape of pore fluids. This variability may in part be related to the large differences in permeability between mud-dominated (e.g., northern Barbados) and sandy (e.g., Nankai) margins. In both cases, the escape of pore fluids near the toe leads to a reduction in permeability, and the forearc plumbing network quickly evolves arcward from grain scale fluid flow to channeling along tectonic fabrics.

Deeper Structural Levels- Development of Slaty Cleavage and Quartz Veins

Sediments that are underthrust beneath the offscraped sediments in the toe of the accretionary wedge are either subducted or accreted beneath the forearc. The arcward portion of large accretionary prisms (e.g., the Kodiak Formation, Figure 2) is composed of underplated rocks which are differentiated from offscraped sequences on the basis of metamorphic grade, structural style (seaward vergence only) and the absence of overlying slope basins or unconformities. Duplex accretion is the underplating model that is most consistent with both seismic reflection profiles of modern margins [*Silver et al.*, 1984; *Brown and Westbrook*, 1988; *Brown et al.*, 1990] and structural observations and inferences from ancient margins [*Sample and Fisher*, 1986; *Platt*, 1986; *Byrne and Fisher*, 1990].

The premise of this model is that the basal décollement consists of flats where the fault follows a given horizon for long horizontal distances (e.g. Barbados, *Westbrook et al.*, [1982]) and ramps where it cuts across the underthrusting sediment pile. Along these ramps, the flatlying sediments on the downgoing plate are imbricated and incorporated into the overriding wedge. This process results in the seaward growth of duplexes, which consist of thrust slices bounded below by an active décollement and above by a deactivated or fossil décollement. Because duplex accretion can occur at a considerable depth and distance arcward of the deformation front, our understanding of the structural and hydrologic processes that operate during duplex accretion are largely based on observations of underplated rocks exposed on land.

In the Kodiak accretionary complex, the Uyak Complex and Kodiak Formation were imbricated after the stratal disruption and veining that was associated with scaly fabric development and underthrusting [*Sample and Moore*, 1987; *Byrne and Fisher*, 1990]. In the Kodiak Formation, imbrication and hence accretion occurred at depths of 8-12 km based on analyses of vitrinite reflectance, illite crystallinity, fluid inclusions, and transitional graphite. The Kodiak Formation has been subdivided into three belts based on structural distinctions (Figure 2) [*Sample and Moore*, 1987]. Mesoscale duplexes are numerous within the central belt, or the deepest structural level of the Kodiak Formation where measured strain magnitudes are larger, strain is noncoaxial, folds are overturned or recumbent, and both faults and cleavage are subhorizontal [*Fisher and Byrne*, 1992]. Larger duplexes are inferred in the landward belt based on the geometry of fault-related folds and cutoffs associated with steep thrust faults. Duplexing was broadly contemporaneous with slaty cleavage formation; cleavage is axial planar to fault-related folds.

There are two types of quartz veins that provide evidence for fracture-channeled fluid flow during duplex formation: (1) continuous laminated veins along fault surfaces and (2) crack seal veins (e.g., *Ramsay* [1980]; *Etheridge et al.* [1984]; *Cox and Etheridge* [1989]) at an angle to fault surfaces. The laminated veins exhibit slickenlines and mineralized steps at their margins as well as internal laminations that are defined by layers of deformed quartz and in some cases calcite that are separated by dark insoluble residues. Crack seal veins around thrust faults are typically rotated during cleavage development in response to noncoaxial strain consistent with trench-directed overthrusting. Although internally deformed, these veins retain chlorite inclusions embedded within the vein quartz that grew off wall rock seed crystals and record cracking parallel to the vein-wall rock interface followed by sealing of cracks with quartz (Figure 3).

The spatial relationship between laminated veins and crack seal veins suggests that fluid flow along fractures plays an important role in the development of mesoscale thrust systems. For example, Figure 4 shows a mesoscale duplex within a sequence of siltstones. The basal layer in each horse, or the layer overlying the active floor thrust, is a fine grained shale that contains dispersed bedding-parallel laminated quartz veins and elliptical vugs. These veins show evidence for both dilation (precipitation) and collapse (dissolution) during episodes of cracking along the active floor thrust. In units beneath the duplex, steeper crack-seal veins are observed. No veins are observed in a similar bed overlying the roof thrust.

The spatial distribution of veins and faults is consistent with a model whereby the veins in strata beneath the duplex are hydrofractures that provided conduits for upward migrating fluid lenses during crack-seal events (e.g., *Fisher and Brantley* [1992]). As a consequence, the low permeability basal layer experiences a cyclic history in which the fluid pressure in pores and vugs builds until the system of fractures

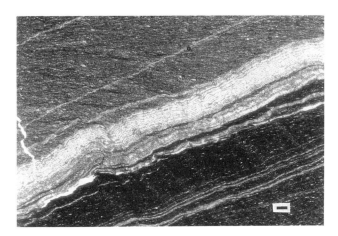

Figure 3: A quartz vein with crack seal bands composed of chlorite inclusions (scale bar=100μm). Vein is contemporaneous with imbrication and has been rotated in response to noncoaxial strain during trenchward overthrusting.

Figure 4: A mesoscale duplex from the central belt of the Kodiak Formation.

becomes interconnected, fault slip occurs, and the fluid bleeds off. Fractures subsequently seal or collapse, and the fault locks. This sequence of events may occur periodically, with fluid pressure fluctuating up and down at near-lithostatic values. Thus, lithologically controlled permeability variations may control the distribution of hydrofractures and the development of thrust systems. Development of slaty cleavage involves silica-producing reactions [Knipe, 1981] and dissolution of silica grains, processes which can provide a local source of silica that is transported to fractures by grain boundary diffusion [Fisher et al., 1995] or grain scale fluid advection [Etheridge et al, 1984]. Under these conditions, the permeability may be regulated by the sealing rate.

A large percentage of the quartz veins in the Kodiak Formation postdate duplex formation as well as most of the deformation associated with cleavage development [Fisher and Byrne, 1990]. These veins are regionally pervasive in the Central belt over an exposed area of 15 km across strike and over 100 km along strike [Fisher et al., 1995]. Thus, a significant portion of the veining in the Kodiak Formation has occurred in the interior of the accretionary wedge after the juxtaposition of accreted packages. The central belt is an ideal place to consider the fluid flow network in the low porosity, metamorphosed interior of the wedge. In particular, what are the attributes of the central belt that lead to such pervasive veining in the late stages of cleavage development?

There are two types of textures observed in these veins that indicate periodic fluid flow [Fisher and Byrne, 1990; Fisher and Brantley, 1992; Fisher et al., 1995] : (1) crack-seal bands spaced 10 μm apart and (Figure 5) (2) jagged collapse selvages spaced 20 μm to mm's apart (Figure 6). The difference in texture reflects the closure of cracks: crack-seal microstructures record the chemical sealing of the crack after each fracture event, so the spacing of bands gives a minimum estimate of the crack aperture prior to sealing. The collapse features record longer fluid-filled periods followed by more rapid draining of fractures. High concentrations of immobile elements in collapse selvages (e.g., the presence of rutile, monazite, and apatite) indicate that these fractures closed by collapse and dissolution, with penetration of quartz crystals into wall rock [Fisher et al., 1995].

Thick crack-seal veins typically display asymmetric growth textures; one side of the vein is straight and has fine-grained quartz, but the grain size of quartz increases across the vein, and the other side of the vein consists of euhedral crystal terminations [Fisher and Brantley, 1992]. In these cases, the straight side of the vein was sealed throughout vein development, and the euhedral side was open, with unidirectional growth toward the open side. The wall rock adjacent to the open side of the vein is depleted in silica with respect to the side which was sealed and can account for ~80% of the quartz in the vein [Fisher et al., 1995]. These silica depletion zones indicate that silica in veins was derived locally, probably by grain boundary diffusion from matrix to crack. Thus, local diffusion of silica from the matrix to cracks helps to regulate the flow of locally or externally derived fluid along the cracks.

The geometry of the vein system is consistent throughout the central belt, with vertical closely spaced (~5 mm) thin veins in the wall rock between more widely spaced (~ 500 mm) en echelon arrays of thicker veins [Fisher et al., 1995]. The en echelon arrays typically occur in southeast-dipping bands that indicate a top-to-the trench sense of shear. The vein textures also

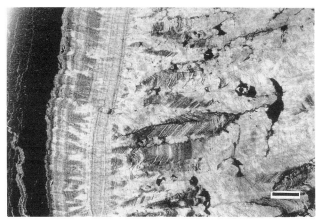

Figure 5: Crack seal vein within the Kodiak Formation with bands of chlorite inclusions that grew off wall rock seed crystals during cracking events (scale bar=100µm).

Figure 6: A quartz vein which evolved texturally (left to right) from continuous crack-seal to discontinuous crack seal to euhedral growth with collapse selvages (scale bar=1 mm).

vary systematically with respect to the vein network and record a range of crack behavior including complete sealing of fractures after cracking at the vein-wall rock interface, partial sealing of fractures after cracking, and crystal growth into open voids (Figure 7). In addition, thick veins within en echelon arrays record a progressive evolution from periodic sealing of fractures to the maintenance of open fractures with periodic collapse.

The thickness distribution for these veins is defined by a power law [*Fisher et al.*, 1995]. Power law distributions can reflect a runaway process whereby the growth rate of individual veins is proportional to vein thickness; thus, larger veins grow at faster rates [*Clark et al.*, 1995]. The largest veins in the Kodiak system are preferentially located within en echelon sets, and textures indicate that these veins experience runaway increases in crack aperture as they evolve from continuous crack seal veins to euhedral growth veins. Since the cracks which serve as sinks for fluid are also sites of silica precipitation, these veins experience runaway growth relative to fractures that seal completely early in the history. The coalescence of fluid within en echelon arrays of hydrofractures may be an important precursor to faulting within the forearc at depths of 10 to 14 km [*Clark et al.*, 1995].

This vein network is restricted to the central belt of the Kodiak Formation where the subhorizontal fabric and abundance of massive shale beds may have restricted upward fluid flow and provided a regional subhorizontal zone of low permeability beneath the forearc high. Local development of excess fluid pressures within this region led to distributed hydrofracturing followed by organization into en echelon arrays which provided reservoirs for fluid. Crack-seal events within these arrays are punctuated by less frequent events where the system

links up over a greater distance, fractured reservoirs become interconnected, and the fluid within reservoirs is drained upward or laterally [*Fisher et al.*, 1995]. Periodic inflation and deflation of en echelon arrays may reflect periodic rupture of the material that separates open fractures. Thus, the fracture network in the central belt valved fluid through a regionally extensive zone of low permeability. This area may be an exposed analog for deeply buried subhorizontal reflections observed in profiles of large accretionary prisms (e.g., *Fisher et al.*, [1989]). Bedding and cleavage are more typically steeply dipping within the arcward of the toe of the accretionary wedge, so regionally extensive subhorizontal zones of low permeability may limit the rate of upward fluid flow within the bulk of the forearc.

UNDERTHRUST SEDIMENT PILE

Development of Vein Structure, Scaly Fabrics, and Carbonate Veins Beneath the Toe of the Wedge

The décollement along convergent margins is a sharp boundary that decouples the imbricated sediments of the accretionary wedge from the relatively undeformed sediments of the underthrust sediment pile. There have been two active margins where drill sites penetrated the basal décollement into the underthrust sediment pile: Barbados [*Moore et al.*, 1988] and Nankai [*Byrne et al.*, 1993; *Moore and Shipley*, 1993]. Along both margins, the sediments just beneath the décollement are characterized by a pervasive scaly fabric defined by anastomosing arrays of polished and striated surfaces. The complex network of microfaults arises because individual scaly folia are abandoned after very small displacements, with slip transferred to new folia that

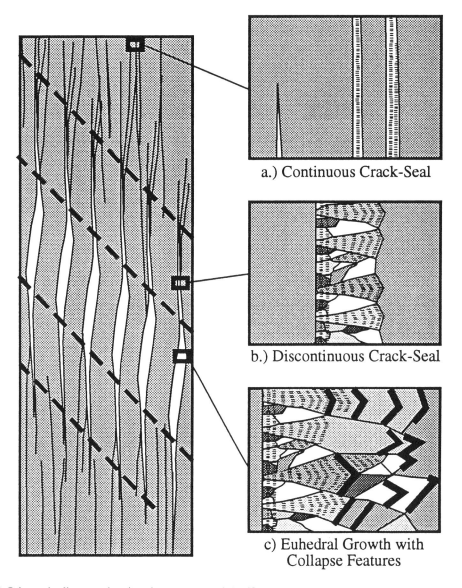

Figure 7: a) Schematic diagram showing the geometry of the Kodiak veins and the distribution of different textures (after *Clark et al.* [1995]). b) Continuous crack-seal veins. c) Discontinuous crack-seal veins. d) Euhedral-growth veins. Thin lines represent bands of phyllosilicate inclusions in the quartz grains, and the thick, black lines represent collapse features composed primarily of insoluble residue.

form in the less deformed material between folia [*Moore et al.*, 1986]. The transfer from one folia to the next reflects strain hardening of individual folia and may be a consequence of [*Moore and Byrne*, 1987]: 1) strengthening of folia due to collapse of the grain network, 2) reduction of fluid pressure within folia, and/or 3) reorientation of folia during slip on adjacent curviplanar microfaults. These processes cumulatively lead to a strengthening of the shear zone.

Across the toe of the Barbados accretionary prism, the décollement zone (as defined by the zone of pervasive scaly fabric) widens from a few meters at the deformation front to about 60 m where the displacement on the fault is 2200 m [*Brown and Behrmann*, 1990]. The shear zone at the top of the underthrusting sediment pile is not a site of extensive veining within a few kilometers of the deformation front, although dispersed veinlets of rhodochrosite were observed beneath the Barbados décollement (sites 675, 676, *Vrolijk and Sheppard*, 1991) and mud-filled veins are observed within the décollement horizon seaward of the deformation front [*Brown and Behrmann*, 1990]. The top of the underthrusting pile may be overpressured as indicated by logging while drilling results along the Barbados décollement [*Moore et al.*, 1995] and an observed 10% increase in porosity beneath the Nankai décollement [*Taira et al.*, 1992]. Thus, the décollement is an important hydrologic boundary; fluid that drains off the underthrusting plate may be channeled along the subhorizontal zone of scaly foliation below the décollement.

Development of scaly fabrics and carbonate/quartz veins at deeper levels (8-15 km)

In many exposed accretionary complexes, there are vein sets that record fluid flow within the underthrust sediment pile. There are several observations that support this interpretation: 1) the veins are concentrated within zones of mud matrix melange that can extend 100's of km along strike [*Connelly*, 1978; *Fisher and Byrne*, 1987; *Vrolijk et al.*, 1988] (Figure 2), 2) the veins are associated with a pervasive scaly fabric [*Byrne*, 1984], 3) the melanges are associated with regional faults that place older rocks over younger [*Fisher and Byrne*, 1987], and 4) the melanges in some cases display a ghost stratigraphy that matches the stratigraphy of an underthrusting sediment pile (i.e., greenstones, pelagic sediments, and turbidites) [*Byrne and Fisher*, 1987; *Agar*, 1990].

The thickness of these melange zones (several kilometers) is typically greater than the thickness of the Barbados décollement zone near the toe of the Barbados margin (meters to tens of meters). This observation can in part be a consequence of strain hardening-- greater displacement across ancient shear zones results in widening of fault zones beyond the thickness observed near the toe of active margins. However, much of the thickening is the result of subsequent imbrication and folding as indicated by repetition of a ghost stratigraphy [*Byrne and Fisher*, 1990]. The thickness of melange in individual thrust sheets from the Uyak Complex (100's of meters to 2 kilometers) is consistent with approximately two orders of magnitude more displacement than is observed along the Barbados décollement [*Brown and Behrmann*, 1990].

In the Kodiak archipelago, melanges are characterized by a block-in-matrix fabric composed of sandstone, chert, and/or greenstone blocks in a mud matrix with scaly fabric (Figure 8a). The melanges are stratally disrupted but structurally systematic [*Moore and Wheeler*, 1978; *Fisher and Byrne*, 1987]. For example, the melange inclusions display a shape fabric with the long axis of inclusions parallel to either a downdip stretching lineation (the Uyak Complex) or an along-strike lineation defined by intersecting cataclastic shear zones (the Ghost Rocks Formation, i.e. webstructure [*Byrne*, 1984]). The fabric is regionally consistent, with

Figure 8: a) Outcrop photo of Ghost Rocks melange. b) Outcrop photo of scaly fabric in the Ghost Rocks Melange. c) Photomicrograph of a scaly fabric network (scale bar=100 μm).

asymmetric melange inclusions, asymmetric folds, and shear bands that indicate layer parallel extension in a zone of layer-parallel shear [*Fisher and Byrne*, 1987]. The textural history reflects reduction in pore space during progressive lithification [e.g., *Orange et al.*, 1993], with a conversion from distributed grain scale particulate flow in the muds to localized slip on scaly microfaults [*Fisher and Byrne*, 1987]. The present orientation of the melange fabric is typically steep, but the scaly fabric and veins predate the steepening of fabrics caused by imbrication, folding, and cleavage development within the accretionary wedge. Based on these considerations, it is likely that the melange fabrics (the inclusion shape fabric and the scaly fabric) were subhorizontal (i.e. parallel to a basal décollement) at the time they developed.

The veins in melange zones are typically composed of carbonate or quartz. H_2O-rich fluids trapped within veins are low in Cl^- (e.g., *Geddes*, [1993])) as has been observed for fluids from the décollement along modern margins [*Kastner et al.*, 1991]. An important source of H_2O-rich fluids with lower salinity than seawater is dehydration reactions involving hydrous minerals [*Kastner et al.*, 1991]. In quartz veins from melange zones on Kodiak Island (the Ghost Rocks, Kodiak and Uyak Formations), H_2O-rich and CH_4-rich fluid inclusions are present as a consequence of simultaneous trapping of two immiscible fluids [*Vrolijk*, 1987; *Vrolijk et al.*, 1988]. Under these circumstances, the homogenization temperature of H_2O-rich inclusions is the trapping temperature [*Vrolijk et al.*, 1988]. Variation in the density of CH_4 inclusions is attributed to cyclic fluctuations in fluid pressure during quartz precipitation [*Vrolijk*, 1987]. The fluid inclusion analyses from melanges on Kodiak Island indicate temperatures of 215°-290° and depths of 10 to 14 km during vein formation [*Vrolijk et al.*, 1988]. These conditions are warmer than would normally be expected in a forearc, suggesting that warm fluids have percolated through melange zones [*Vrolijk et al.*, 1988].

Quartz and carbonate veins are observed in both melange blocks and in the matrix but are thicker and more conspicuous within sand beds or blocks. Within the sands, veins are typically perpendicular to either the long or intermediate axis of melange blocks and are typically truncated at sand-shale boundaries. The vein margins are irregular at the grain scale and follow sand grain boundaries without breaking grains within the wall rock [*Byrne*, 1984; *Fisher and Byrne*, 1987]. As in the case of veins from active margins, the veins from ancient melange zones display a "dirty" or cloudy appearance, with floating sand grains or chunks of wall rock [*Orange et al.*, 1993; *Byrne*, 1984; *Fisher and Byrne*, 1987]. Blocky crystals within the veins reflect growth of crystals into open voids. Mud from the adjacent matrix intrudes into the ends of carbonate veins as a result of diagenetic reactions in the muds that give off CO_2 and dissolve the carbonate along the vein-mud interface [*Byrne*, 1984; *Fisher and Byrne*, 1987; *Byrne*, 1994). In total, these observations indicate that neither the mud nor the sand was strongly cemented during veining.

Scaly fabrics in the melanges are defined by anastomosing arrays of polished surfaces (Figure 8b and c), and there are in places thin veins that lie along scaly folia. These scaly folia show evidence for both dissolution and precipitation (see also the Okitsu melange in the Shimanto belt, *Agar*, [1990]). Dissolution and sealing may be additional mechanisms for strengthening scaly folia within the décollement zone relative to undeformed material. The network of scaly folia may behave as an anastomosing fault-parallel network of fluid conduits during periods of hydrofracturing, fluid flow and quartz precipitation. Fault-parallel fluid migration may be absent during periods of collapse and dissolution or when cracks are sealed. Under these circumstances, fluid lenses move episodically along the web-like network as dilational waves (e.g., *Moore* [1989]).

DISCUSSION: FLUID FLOW IN THE FOREARC OF DEPTHS OF <15 KM

From shallow to deeper levels, fabric elements in the forearc depict a fracture permeability that alternates locally between dilatancy and collapse. Within a given rock volume, the fault and fracture system can behave as either a conduit or a barrier to fluid migration. This paradox is due to the feedbacks between pore fluid pressure, hydraulic fracture, and fluid flow. Low grain scale permeability in forearcs, coupled with rapid tectonic loading and devolitization, leads to a buildup of fluid pressure. Thrust faults will not experience simultaneous shear failure and tensile failure unless fluid pressure exceeds lithostatic pressure [*Sibson*, 1983; *Behrmann*, 1991], so fluid pressure can rise locally until it equals σ_3 + T (where T is the tensile strength) at which point hydraulic fracturing results in rupture of the material that separates fluid-filled fractures. Crack closure as a consequence of collapse or sealing during draining of fluid subsequently reduces the permeability and fluid pressure may again begin to rise. In this way, the fluctuations in fluid pressure in the forearc may be buffered at near-lithostatic values [*Platt*, 1990]. In areas of strike slip or normal faulting, the fluid pressure may fluctuate around a lower excess fluid pressure.

Dilatancy and collapse events within the fluid flow network occur in a variety of ways, depending on the sediment properties and the physical conditions associated with different locations within the forearc (Table 1). In general, there is a transition from

TABLE 1: Characteristics of forearc plumbing depicted by observed fabrics. Observations of mud-filled veins within slope sediments are from Cowan, 1982. Lundberg and Moore, 1986, Knipe, 1986, Kemp et al., 1990, Lindsley-Griffin et al., 1990, and Pickering et al., 1991.

Forearc Setting	Fabrics	Composition of Veins	Depth of Vein Formation	Geometry of Network	Dilation Events	Collapse Events
Upper and Lower Slope	"mud-filled" veins	in situ sediment, disseminated pyrite, and authigenic carbonate	<10 m to 420 m	fluid flow within low permeability layers restricted to bedding-parallel en echelon arrays of 0.5- 2mm-thick, steep zones spaced 1mm-3 cm apart	disaggregation of weakly cemented sediment within narrow zones,	loss of porosity and alignment of grains
Offscraped Sediments	"mud-filled" veins, scaly fabrics, kink bands, shear bands, microfaults, and blocky to fibrous carbonate veins	in situ sediment, "dirty" to "clean" calcite and rhodochrosite	< 5 km	-diffuse loss of pore fluid through pore network and semipervasive arrays of conjugate shear zones and small faults -fault-parallel fluid flow along web-like anastomosing arrays of microfaults	hydrofracturing along scaly folia in muds and cracking around grains in sands	loss of porosity and alignment of grains
Underplated Rocks	slaty cleavage, laminated veins, and continuous crack-seal, discontinuous crack seal and euhedral growth veinsscaly fabrics and blocky veins	clean calcite and quartz with chlorite inclusions	5-15 km	-fluid flow along faults and shallow to steep dipping planar cracks. Regional subhorizontal shear zones act as seals and develop an extensive fracture network that consists of steep veins~0.5 cm apart that are periodically sealed and en echelon arrays of thicker cracks spaced 50 cm apart that remain open	hydrofracturing at the margins of crack-seal veins and opening of en echelon gashes, with precipitation of quartz in open voids	dissolution of wall rock adjacent to impinging quartz crystal terminations
Underthrust Sequence- Shallow levels	"mud-filled" veins, scaly fabrics and blocky to fibrous carbonate veins	"dirty" calcite and rhodochrosite	1-5 km	fault-parallel fluid flow along web-like anastomosing arrays of microfaults that are pervasive at the scale of a thin section	hydrofracturing along scaly folia in muds and cracking around grains in sands	loss of porosity and alignment of grains
Underthrust Sequence- Deeper levels	scaly fabrics and blocky to fibrous veins	"dirty" to "clean" calcite and quartz	5-15 km	fault-parallel fluid flow along web-like anastomosing arrays of microfaults that are pervasive at the scale of a thin section	hydrofracturing along scaly folia in muds with precipitation of calcite or quartz	dissolution along scaly folia and at the margins of thin veins

particulate flow in soft sediments deformed on the upper slope or near the toe to diffusive mass transfer or pressure solution within low porosity rocks in the interior of the forearc. With increasing depth of burial, veins vary from zones of distributed grain boundary failure to cracks that follow grain boundaries and fill with dirty carbonate to cracks that break across grains and fill with clean calcite or quartz (Table 1).

Near the toe, sediments that enter the accretionary wedge undergo rapid changes in material properties due to diffuse dewatering. There is an evolution arcward from semipervasive ductile deformation to brittle deformation as sediments are progressively consolidated. Observations of veins, scaly fabrics, and geochemical anomalies along faults near the toe of the wedge suggest that thrust faults can act as fluid conduits. Observed differences between active convergent margins (e.g., Barbados vs. Nankai) could be due to episodic fluid flow as well as variations in plumbing between mud- and sand-dominated prisms.

Further arcward, the wedge interior consists of strongly cemented, low grade metamorphic rocks with very little porosity [*Sample*, 1990]. Consequently, much of the fluid expelled from the downgoing slab may migrate upward and laterally along the active décollement fracture network [*Bebout*, 1991] to be ultimately vented at the toe of the prism. Locally, the volume of fluid that passes through the wedge material may be low, but local variation in fluid pressure associated with local and regional permeability variations influences the distribution of faults. Duplex accretion and wedge deformation is accompanied by local development of crack seal veins and laminated veins that show evidence for precipitation during episodes of dilatancy. Closure of cracks occurs by both sealing of gaps with quartz and collapse of fracture space.

Deformation within the underthrust pile beneath the toe of the wedge is largely accommodated by localized particulate flow and development of scaly fabric. Scaly fabrics show evidence for collapse of the more open grain network. Dilatancy and abnormal fluid pressures in these zones is inferred along the Barbados margin based on logging-while-drilling results [*Moore et al.*, 1995], the rare observation of dirty carbonate veins [*Vrolijk and Sheppard*, 1991; *Brown and Behrmann*, 1990], and the requirement that the scaly fault zone behave as a fluid conduit capable of maintaining observed geochemical and thermal anomalies [*Brown et al.*, 1984]. Moreover, thrust faults near the toe of many accretionary prisms have seismic reflections with reversed polarity that cannot be explained solely by density inversion during thrusting and must be related to fault-parallel dilatant zones [*Moore et al.*, 1995]. These reflections are discontinuous and can vary laterally into normal polarity reflections [*Shipley et al.*, 1994). Episodic fluid flow within and between fault-parallel zones of dilatancy may occur parallel to the anastomosing fault-parallel network of scaly folia.

At greater depths, collapse of scaly folia along the décollement coincides with dissolution and development of selvages. The pervasive veining and scaly fabric development observed in melanges exposed on land may record a long-lived cyclical history of dilatancy and collapse. Under these circumstances, the onset of veining within melanges reflects in part a change in the mechanism of crack collapse due to the onset of pressure solution within the underthrusting sediment pile.

Each of the forearc fluid flow regimes described in this paper differ in terms of the orientation, and distribution of cracks as well as the mechanisms of crack closure (collapse of the grain network vs. sealing and dissolution), but in most cases, there is evidence that fluid flow is episodic or in some examples, cyclic (Table 1). There are two types of cyclic deformation and fluid flow [*Knipe et al.*, 1991]: (1) externally imposed cyclicity where a deformation front migrates into the volume of material under consideration and (2) internally generated cyclicity where deformation and fluid flow is driven by linkage of arrays as a consequence of local increase in fluid pressures or differential stress. Fluid flow may be characterized by externally imposed cyclicity at the deformation front where an underthrust turbidite section provides a source of fluid and the seaward propagation of the frontal thrust causes episodic migration of fluid sources [*Wang et al.*, 1990; *Knipe et al.*, 1991]. Duplex accretion may result in similar pulses of fluid flow. In both cases the frequency of fluid expulsion events is determined by the frequency of thrust events that produce seaward migration of the deformation front or the leading branch line of the duplex. Pulses of fluid could also be produced as a consequence of the narrow range of P-T conditions associated with some dehydration reactions [*Moore and Vrolijk*, 1992]

Alternatively, the cycles of cracking and crack closure recorded by fabrics from forearcs could be a consequence of periodic linkage of fluid open fractures when local increases in fluid pressure induce hydraulic fracture across impermeable material. Linkage may occur over a variety of different scales, so the frequency of fluid flow cycles may be controlled by a number of variables that are presently enigmatic: the transient distribution of open fractures, the rate at which excess fluid pressures develop within fracture arrays, the rate at which a seal is reestablished through precipitation of silica and carbonate, and perhaps most importantly, the nature of the relationship between the seismic events and fluid flow events. Are the asperities that govern the seismic cycle also barriers or valves that restrict fluid flow? Some of these variables could be addressed from long term monitoring of fluid flow, seismicity, and the distribution of fault-parallel dilatant zones along active convergent margins; others may be elucidated through analysis of fracture systems in exposed ancient forearcs.

Acknowledgments. This work was funded by NSF grant EAR-93-05101. I would also like to thank T. Byrne, J. Casey Moore, and P. Vrolijk for helpful reviews and S. Brantley and T. Engelder for useful discussions. Field work on Kodiak was aided by the U. S. Coast Guard and the Randall's of Seal Bay.

REFERENCES

Agar, S. M., The interaction of fluid processes and progressive deformation during shallow level accretion: examples from the Shimanto belt of SW Japan, *J. Geophys. Res., 95,* 9133-9147, 1990.

Arch, J. and A. J. Maltman, Anisotropic permeability, and tortuosity in deformed wet sediments, *J. Geophys. Res., 95,* 9035-9046, 1990.

Bangs, N. L. B., and G. K. Westbrook, Seismic modeling of the décollement zone at the base of the Barbados Ridge accretionary complex, *J. Geophys. Res., 96,* 3853-3866, 1991.

Bebout, G., Geometry and mechanisms of fluid flow at 15 to 45 km depths, *Geophys. Res. Lett., 18,* 923-926, 1991.

Behrmann, J. H., Conditions for hydrofracture and the fluid permeability of accretionary wedges, *Earth Plan. Sci. Lett., 107,* 550-558, 1991.

Bray, C. J. and D. E. Karig, Porosity of sediments in accretionary prisms and some implications for dewatering processes, *J. Geophys. Res., 90,* 768-778, 1985.

Brown, K. M. and G. Westbrook, Mud diapirism and subcretion in the Barbados Ridge complex, *Tectonics, 7,* 613-640, 1988.

Brown, K. M. and Behrmann, J., Genesis and evolution of small-scale structures in the toe of the Barbados Ridge accretionary wedge, *Proc. Ocean Drill. Prog., Sci. Results, 110,* Ocean Drilling Program, College Station, TX, 229-243, 1990.

Brown, K. M., Mascle, A., and J. H. Behrmann, Mechanics of accretion and subsequent thickening in the Barbados Ridge accretionary complex: balanced cross sections across the wedge toe, *Proc. Ocean Drill. Prog., Sci. Results, 110,* Ocean Drilling Program, College Station, TX, 209-228, 1990.

Brown, K. M., Bekins, B. Clennel, B., Dewhurst, D., and G. Westbrook, Heterogeneous hydrofracture development and accretionary fault dynamics, *Geology, 22,* 259-262, 1994.

Byerlee, J., Model for episodic flow of high-pressure water in fault zones before earthquakes, *Geology, 21,* 303-306, 1993.

Byrne, T. Early deformation in melange terranes of the Ghost Rocks Formation, Kodiak Islands, Alaska, in *Melanges: there nature origin, and significance,* edited by L. A. Raymond, *Spec. Pap. Geol. Soc. Am., 198,* 21-52, 1984.

Byrne, T. and D. Fisher, Episodic growth of the Kodiak convergent margin, *Nature, 325,* 338-341, 1987.

Byrne, T. and D. Fisher, Evidence for a weak and overpressured décollement beneath sediment-dominated accretionary prisms, *J. Geophys. Res., 95,* 9081-9097, 1990.

Byrne, T., Maltman, A., Stephenson, E., and R. Knipe, Deformation structures and fluid flow in the toe region of the Nankai accretionary prism, *Proc. Ocean Drill. Prog., Sci. Results, 131,* Ocean Drilling Program, College Station, TX, 83-92, 1993.

Byrne, T., Sediment deformation, dewatering and diagenesis: illustrations from selected melange zones, in *The Geological Deformation of Sediments,* edited by A. Maltman, pp. 239-260, Chapman and Hall, London, 1994.

Clark, M. B., Brantley, S., and D. Fisher, Power law vein thickness distributions and runaway vein growth, *Geology, 23,* 975-978, 1995.

Cochrane, G. R., Mackay, M. E., Moore, G. F., and J. C. Moore, Consolidation and deformation of sediment at the toe of the central Oregon accretionary prism from multichannel seismic data, *Proc. Ocean Drill. Prog., Sci. Results, 146,* Ocean Drilling Program, College Station, TX, 421-426, 1994a.

Cochrane, G. R., Moore, J. C., Mackay, M. E., and G. F. Moore, Velocity and inferred porosity model of the Oregon accretionary prism from multichannel reflection data: implications on sediment dewatering and overpressure, *J. Geophys. Res., 99,* 7033-7043, 1994b.

Connelly, W., Uyak Complex, Kodiak Island, Alaska: a Cretaceous subduction complex, *Geol. Soc. Am. Bull., 89,* 765-769, 1978.

Cowan, D. S., The origin of vein structure in slope sediments on the inner slope of the Middle America Trench off Guatemala, edited by J. Aubouin et al., *Initial Rep. Deep Sea Drill. Proj. 67,* 645-650, 1982.

Cox, S. F. and M. A. Etheridge, Coupled grain-scale dilatancy and mass transfer during deformation at high fluid pressures: Examples from Mount Lyell, Tasmania, *J. Struct. Geol., 11,* 147-162, 1989.

Etheridge, M. A., V. J. Wall, and S. F. Cox, High fluid pressures during regional metamorphism: Implications for mass transport and deformation mechanisms, *J. Geophys. Res., 89,* 4344-43 58, 1984.

Fisher A. and M. Hounslow, Transient fluid flow through the toe of the Barbados accretionary complex: constraints from Ocean Drilling Program Leg 110 heat flow studies and simple models, *J. Geophys. Res., 95,* 8845-8858, 1990.

Fisher, D. and T. Byrne, Structural evolution of underthrusted sediments, *Tectonics, 6,* 775-793, 1987.

Fisher, D. and T. Byrne, The character and distribution of mineralized fractures in the Kodiak Formation, Alaska: implications for fluid flow in an underthrust sequence, *J. Geophys. Res., 95,* 9069-9080, 1990.

Fisher, D. M. and Brantley, S. L., Models of quartz overgrowth and vein formation: deformation and episodic fluid flow in an ancient subduction zone, *J. Geophys. Res. 97,* 20,043-20,061, 1992.

Fisher, D. and T. Byrne, Strain variations in an ancient accretionary wedge: implications for forearc evolution, *Tectonics, 11,* 330-347, 1992.

Fisher, D., Brantley, S., Everett, M., and J. Dzvonik, Cyclic fluid flow through a regionally extensive fracture network within the Kodiak accretionary prism, *J. Geophys. Res., 100,* 12,881-12,894, 1995.

Fisher, M. A., Brocher, T. M., Nokleberg, W. J., Plafker, G. and G. L. Smith, Seismic reflection images of the crust of the northern part of the Chugach Terrane, Alaska: results of a survey for the Trans-Alaska Crustal Transect (TACT), *J. Geophys. Res.*, *94*, 3813-4708, 1989.

Geddes, D., Carbonate cement in sandstone as an indicator of accretionary complex hydrogeology, M.S. thesis, Univ. California Santa Cruz, 1993.

Gieskes, J. M., Vrolijk, P., and G. Blanc, Hydrogeochemistry of the northern Barbados accretionary complex transect: ocean drilling project leg 110, *J. Geophys. Res.*, *95*, 8809-8818, 1990.

Ito, E., Harris, D. M., and A. T. Anderson, Jr., Alteration of oceanic crust and geologic cycling of chlorine and water, *Geochimica et Cosmochimica Acta*, *47*, 1613-1624, 1983.

Kastner, M. Elderfield, H., and J. B. Martin, Fluids in convergent margins: what do we know about their composition, origin, role in diagenesis and importance for oceanic chemical fluxes?, *Phil. Trans. R. Soc. Lond. A*, *335*, 261-273, 1991.

Kemp, A. E. S., Fluid flow in "vein structures" in Peru forearc basins: evidence from back-scattered electron microscope studies, *Proc. Ocean Drill. Prog., Sci. Results, 112*, Ocean Drilling Program, College Station, TX, 33-41, 1990.

Knipe, R. J., The interaction of deformation and metamorphism in slates, *Tectonophysics*, *78*, 249-272, 1981.

Knipe, R. J., Microstructural evolution of vein arrays preserved in Deep Sea Drilling Project cores from the Japan Trench, Leg 57, *Geol. Soc. Am. Mem. 166*, 13-44, 1986.

Knipe, R. J., Agar, S. M., and D. J. Prior, The microstructural evolution of fluid flow paths in semi-lithified sediments from subduction complexes, *Phil. Trans. R. Soc. Lond. A*, *335*, 261-273, 1991.

Lindsley-Griffin, N., Kemp, A., and J. F. Swartz, Vein structures of the Peru margin, leg 112, *Proc. Ocean Drill. Prog., Sci. Results, 112*, Ocean Drilling Program, College Station, TX, 3-11, 1990.

Lundberg, N. and Moore, J. C., Macroscopic structural features in Deep Sea Drilling Project cores from forearc regions, *Geol. Soc. Am. Mem. 166*, 13-44, 1986.

Maltman, A., Byrne, T., Karig, D., Lallemant, S., and Leg 131 shipboard party, Structural geological evidence from ODP Leg 131 regarding fluid flow in the Nankai prism, Japan, *Earth Planet. Sci. Lett.*, *109*, 463-468, 1992.

Maltman, A., Byrne, T., Karig, D. E., and S. Lallemant, Deformation at the toe of an active accretionary prism: synopsis of results from ODP Let 131, Nankai, SW Japan, *J. Struct. Geol.*, *15*, 949-964, 1993.

Moore, G. F., Shipley, T. H., Stoffa, P. L., Karig, D. E., Taira, A., Kuramoto, S., Tokuyama, H., and K. Suyehiro, Structure of the Nankai trough accretionary zone from multichannel seismic reflection data, *J. Geophys. Res.*, *95*, 8753-8765, 1990.

Moore, G. F., and T. H. Shipley, Character of the décollement in the leg 131 area, Nankai trough, *Proc. Ocean Drill. Prog., Sci. Results, 131*, 73-82, 1993.

Moore, J. C. and R. Wheeler, Structural fabric of a melange, Kodiak Island, Alaska, *Am. J. Sci.*, *278*, 739-765, 1978.

Moore, J. C. and A. Allwardt, Progressive deformation of a Tertiary Trench slope, Kodiak Islands, Alaska, *J. Geophys. Res.*, *85*, 4741-4756, 1980.

Moore, J. C., Roeske S., Lundberg, N., Schoonmaker, J., Cowan, D. S., Gonzales, E., and Lucas, S. E., Scaly fabrics from Deep Sea Drilling Project cores from forearcs, *Geol. Soc. Am. Mem. 166*, 55-73, 1986.

Moore, J. C. and T. Byrne, Thickening of fault zones: a mechanism of melange formation in accreting sediments, *Geology*, *15*, 1040-1043, 1987.

Moore, J. C. et al., Tectonics and hydrogeology of the northern Barbados Ridge: results from Ocean Drilling Program Leg 110, *Geol. Soc. Am. Bull.*, *100*, 1578-1593, 1988.

Moore, J. C., Tectonics and hydrogeology of accretionary prisms; role of the décollement zone, *J. Struct. Geol.*, *11*, 95-106, 1989.

Moore, J. C., Brown, K. M., Horath, F., Cochrane, G., MacKay, M. and G. Moore, Plumbing accretionary prisms: effects of permeability variations, *Phil. Trans. R. Soc. Lond. A*, *335*, 275-288, 1991.

Moore, J. C. and P. Vrolijk, Fluids in accretionary prisms, *Reviews of Geophysics*, *30*, 113-135, 1992.

Moore, J. C., Moore, G. F., Cochrane, G. R., and H. J. Tobin, Negative-polarity seismic reflections along faults of the Oregon accretionary prism: indicators of overpressuring, *J. Geophys. Res.*, *100*, 12,895-12,906, 1995.

Moore, J. C., et al., Abnormal fluid pressures and fault-zone dilation in the Barbados accretionary prism: Evidence from logging while drilling, *Geology*, 605-608, 1995.

Morgan, J. K., Karig, D. E., and A. Maniatty, The estimation of diffuse strains in the toe of the western Nankai accretionary prism: a kinematic solution, *J. Geophys. Res.*, *99*, 7019-7032, 1994.

Orange, D., Geddes, D., and Moore, J. C., Structural and fluid evolution of a young accretionary complex: the Hoh rock assemblage of the western Olympic Peninsula, Washington, *Geol. Soc. Am. Bull.*, *105*, 1053-1075, 1993.

Pickering, K. T., Agar, S. M., and Prior, D. J., Vein structure and the role of pore fluids in early wet sediment deformation, *Phil. Trans. R. Soc. Lond. A*, *335*, 417-430, 1991.

Platt, J. P., Dynamics of orogenic wedges and the uplift of high-pressure metamorphic rocks, *Geol. Soc. Am. Bull.*, *97*, 1037-1053, 1986.

Platt, J. P., Thrust mechanics in highly overpressured accretionary wedges, *J. Geophys. Res.*, *95*, 9025-9034, 1990.

Ramsay, J. G., The crack seal mechanism of rock deformation, *Nature*, *284*, 135-139, 1980.

Sample, J. C., The effect of carbonate cementation of underthrust sediments on deformation styles during underplating, *J. Geophys. Res.*, *95*, 9111-9121, 1990.

Sample, J. and D. Fisher, Duplex accretion and underplating in an ancient accretionary complex, Kodiak Islands, Alaska, *Geology*, *14*, 160-163, 1986.

Sample, J. and J. C. Moore, Structural style and kinematics of an underplated slate belt, Kodiak and adjacent islands, Alaska, *Geol. Soc. Am. Bull.*, *99*, 7-20, 1987.

Screaton, E. J., D. R. Wuthrich, and S. J. Dreiss, Permeabilities, fluid pressures, and flow rates in the Barbados Ridge Complex, *J. Geophys. Res.*, *95*, 8997-9007, 1990.

Shipley, T. H., Stoffa, P. L., and D. F. Dean, Underthrust sediments, fluid migration paths, and mud volcanoes associated with the accretionary wedge off Costa Rica: Middle America Trench, *J. Geophys. Res.*, *95*, 8743-8752, 1990.

Shipley, T H., Moore, G. F., Bangs, N. L., Moore, J. C., and P. L. Stoffa, Seismically inferred dilatancy distribution, northern Barbados Ridge décollement: implications for fluid migration and fault strength, *Geology, 22,* 411-414, 1994.

Sibson, R. H., Controls on low-stress hydro-fracture dilatancy in thrust, wrench and normal fault terrains, *Nature, 289,* 665-667, 1983.

Silver, E. A., Jordan, M., Breen, N. and T. Shipley, Growth of accretionary prisms, *Geology, 13,* 6-9.

Taira, A., Hill, I., Firth, J., et al., Sediment deformation and hydrogeology at the Nankai accretionary prism: syntheses of ODP Leg 131 shipboard results, *Proc. Ocean Drill. Prog., Sci. Results, 131,* Ocean Drilling Program, College Station, TX, 273-285, 1991.

Tobin, H. J., Moore, J. C., MacKay, M. E., Orange, D. L., and L. D. Kulm, Fluid flow along a strike-slip fault at the toe of the Oregon accretionary prism: implications for the geometry of frontal accretion, *Geol. Soc. Am. Bull., 105,* 569-582, 1993.

Von Huene, R. and D. W., Scholl, Observations at convergent margins concerning sediment subduction, subduction erosion, and the growth of continental crust, *Reviews of Geophysics, 29,* 279-316, 1991.

Vrolijk, P. J., Tectonically driven fluid flow in the Kodiak accretionary complex, *Geology, 15,* 466-469, 1987.

Vrolijk P., G. Myers, and J. C. Moore, Warm fluid migration along tectonic melanges in the Kodiak accretionary complex, *J. Geophys. Res., 93,* 10,313-10,324, 1988.

Vrolijk, P., Fisher, A., and Gieskes, J., Geochemical and geothermal evidence for fluid migration in the Barbados accretionary prism (ODP Leg 110), *Geophys. Res. Lett., 18,* 947-950, 1991.

Vrolijk, P. and S. M. F. Sheppard, Syntectonic carbonate veins from the Barbados accretionary prism (ODP Leg 110)-- record of paleohydrology, *Sedimentology, 38,* 671-690, 1991.

Wang, C.-Y., Y. Shi, W.-T. Hwang, and H. Chen, Hydrogeologic processes in the Oregon- Washington Accretionary Complex, *J. Geophys. Res., 95,* 9009-9023.

Westbrook, G. K., M. J. Smith, J. H. Peacock, and M. J. Poulter, Extensive underthrusting of undeformed sediment beneath the accretionary complex of the Lesser Antilles subduction zone, *Nature, 300,* 625-628, 1982.

Yamano, M. Foucher, J.-P., Kinoshita, M., Fisher, A., Hyndman, R. D., and ODP Leg 131 shipboard scientific party, Heat flow and fluid flow regime in the western Nankai accretionary prism, *Earth Planet. Sci. Lett., 109,* 451-462.

D. M. Fisher, Department of Geosciences, Pennsylvania State University, University Park, PA 16802

Large Earthquakes in Subduction Zones: Segment Interaction and Recurrence Times

Larry J. Ruff

Department of Geological Sciences, University of Michigan, Ann Arbor, Michigan

Subduction zones generate most of the world's seismicity, and all of the largest earthquakes. This overview of large earthquakes in subduction zones consists of two parts: a review of the occurrence of large events in different tectonic regimes of subduction zones, and the timing of large interplate underthrust events. Our global review shows that large earthquakes have occurred in all intra-plate environments from the outer-rise down to 650 km depth, *except* for the fore-arc region of the upper plate in mature subduction zones. It seems that the seismogenic plate interface is an efficient concentrator of seismicity, though large earthquakes do occur just trenchward and also downdip of the interplate coupled zone. We focus on two aspects of the temporal occurrence of interplate events: a brief analysis of the composite global occurrence of great events, and then a brief review and analysis of the methodology of long-term earthquake forecasting, followed by a suggestion to improve the methodology. Occurrence times of the greatest interplate events in the 20th century are clustered more than expected from random occurrence. However, a "waiting time" analysis of the 40 great interplate events ($M \geq 8$) in the 20th century shows that their origin times are consistent with a model of independent random occurrence. The key to earthquake forecasting methodology is the accurate determination of recurrence time for each plate boundary segment for the current earthquake cycle. Observations of large earthquake occurrence show great variability in rupture mode and recurrence times. Mechanical models that include interaction between adjacent plate boundary segments produce synthetic event catalogs with variable rupture modes and recurrence times, similar to observed earthquake sequences. One robust "rule" extracted from these simulations is that if the rupture mode changes from one great event to several smaller events, then the first smaller event will occur in the epicentral segment of the great event with a recurrence time that is shorter than the average time for that segment. This "rule" appears to explain four examples of sooner-than-expected large earthquakes: the 1942 event in Ecuador; the 1986 event in central Aleutians; the 1994 event in Sanriku, Japan; and the 1995 event in Kuriles Islands.

1. INTRODUCTION

Most large earthquakes occur in subduction zones. A global view of seismicity shows these large earthquakes are distributed in depth from the surface down to seven hundred kilometers, the geographic range includes most of the world's subduction zones, and focal mechanisms range over all types. To classify all these large earthquakes, the key discriminant is to separate the events into either intra-plate or inter-plate events. Most large earthquakes, and indeed all the very largest ones, are interplate events. On the other hand, much of the geographic range and diversity of seismicity is removed if we just focus on the plate boundary earthquakes. In particular, the depths of plate interface events are less than 50 km, and their focal mechanisms are underthrusting on a fault with shallow dip angle. Subduction zones still show great diversity as measured by the size of their largest interplate earthquakes. The "great" interplate events are restricted to a subset of the world's subduction zones, while "large" ($M_w \geq 7.5$) interplate events occur more widely. At the other extreme, a few subduction zones are devoid of even "large" events. There is also considerable diversity in the temporal record of large interplate earthquakes. Indeed, the temporal occurrence of large interplate earthquakes is one of the most important unsolved puzzles in seismology.

In this overview, we first consider the diversity of tectonic environments of the large earthquakes in subduction zones, but then quickly move to the restricted environment of shallow plate interface events. We then review the temporal and spatial occurrence of these events. Next, we look at some of the basic observed features of sequences of large subduction events. We then briefly summarize the "seismic gap hypothesis" and the various ideas advanced to improve estimates of large earthquake recurrence time. We end this overview by emphasizing along-strike segment interaction, and showing how it may help explain some of the variability in large earthquake recurrence times.

2. OVERVIEW OF LARGE EARTHQUAKES IN SUBDUCTION ZONES

There is no clear consensus on the definition of what is a "large" earthquake. For the sake of internal consistency, I shall refer to earthquakes with magnitude of 7.5 or more as "large"; and I shall reserve the term "great" for those earthquakes with magnitude of 8 or more. In detail, one should also define which magnitude scale we use. Ideally, we would like to quote the moment magnitude, M_w, for all earthquakes, but we must use the surface-wave magnitude, M_s, for many events in the first part of the 20th century. In some cases, M_w has been estimated from the tsunami waves, or aftershock area, or felt area. Of course, the societal impact or "significance" of an event may be greater than its quantitative size. An unfortunate recent example of this is the Kobe, Japan earthquake of Jan. 16, 1995, with a $M_s = 7$, is more "significant" than the much larger event that occurred off Sanriku, Japan on Dec. 28, 1994, with $M_s = 7.8$. Nonetheless, events considered in our overview are strictly based on earthquake size, not significance.

Figure 1 shows an idealized cross-section of the composite subduction zone, with the various "modes of occurrence" of large earthquakes in subduction zones. It attempts to collapse the great diversity of subduction zones onto a single cross-section. Thus, while Figure 1 is all-inclusive, it is important to emphasize that no single subduction zone has all the large earthquake occurrences depicted there. Figure 2 shows the geographic distribution of the large and great earthquakes in the various tectonic categories. Recall that the primary classification of events is interplate versus intraplate; there is only one place in Figure 1 where interplate events occur. We briefly discuss intraplate events in the down-going oceanic lithosphere,

shallow to deep, and then discuss intraplate events in the overlying plate, and conclude with interplate earthquakes.

2.1. *Intraplate Seismicity*

2.1.1. Outer-rise. Outer-rise earthquakes occur within the oceanic plate, with most epicenters in the vicinity of the trench axis. A global survey of outer-rise events shows that the focal mechanisms of most events are normal faulting with the tension axis perpendicular to the trench, and the focal depths are quite shallow; these aspects are basically consistent with the bending lithosphere interpretation [*Chapple and Forsythe*, 1979]. However, an alternative interpretation is that compressional outer-rise events in "coupled" subduction zones may indicate high stress levels on the plate interface just prior to a large interplate event [*Christensen and Ruff*, 1988]. The largest outer-rise events have tensional focal mechanisms, and tend to occur in uncoupled subduction zones, e.g. the 2 Mar 1933 Sanriku (M_w 8.4) and 19 Aug 1977 Sumbawa (M_w 8.3) events. There is still some disagreement over the depth extent of the largest tensional and compressional outer-rise events [*Lynnes and Lay*, 1988; *Tichelaar et al.*, 1992].

2.1.2. Beneath the coupled plate interface. In "strongly coupled" subduction zones (i.e., those that produce great interplate events), the down-going oceanic lithosphere tends to be aseismic beneath the interplate contact zone. Even detailed microearthquake surveys show a paucity of events in this region. Thus, the recent occurrence of the great earthquake of Oct. 4, 1994 in the Kuriles (M_w 8.3) is quite significant and puzzling. Although the waveform and geodetic data show conclusively that this event is *not* an interplate event, the focal mechanism of the Oct. 4 event has a large component of dip-slip thrusting. Hence, as discussed in *Kikuchi and Kanamori* [1995] and *Tanioka et al.* [1995b], the Oct. 4 event forces us to the consider the possibility that some of the underthrust interplate events in the 20th century catalog might actually be "Oct. 4 type" intraplate events.

2.1.3. Wadati-Benioff seismicity Wadati-Benioff zone seismicity presumably occurs within the subducting oceanic lithosphere, though its explanation is still a controversial matter. From the earliest catalogs of *Gutenberg and Richter* [1952], there was a clear minimum in seismicity at a depth of about 300 km. Some subduction zones have an even deeper Wadati-Benioff zone that extends to nearly 700 km. As deduced in *Isacks et al.* [1968], the Wadati-Benioff seismicity from the deepest edge of the coupled interface to 300 km depth is dominated by down-dip tension. One more recent discovery is the existence of a double Benioff zone in some subduction zones in the depth

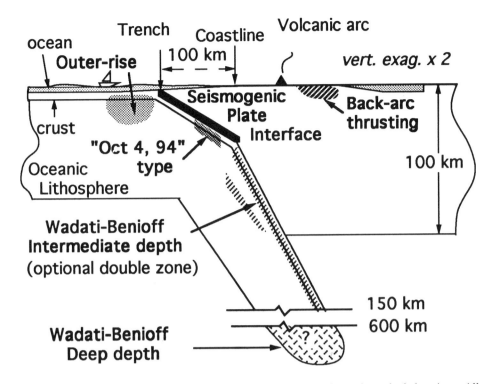

Fig. 1. Subduction zone cross-section with seismicity zones plotted as the various shaded regions. All zones are intra-plate seismicity, except for the seismogenic plate interface (bold bar). All the zones plotted have had at least one large ($M \geq 7.5$) earthquake somewhere in the world. On the other hand, no individual subduction zone has had large earthquakes in all the possible zones.

range of 50 to 200 km. While the faulting depth range for some of the largest intermediate-depth events is still a contentious issue, they seem to be in the range of 50 to 100 km. The largest documented intermediate-depth event is the 4 Nov 1963 Banda Sea event with M_W of 8.3 [*Welc and Lay*, 1987].

In several subduction zones, the seismicity increases from the minimum at a depth of 300 km to a peak at about 600 km, and then rapidly declines. A variety of explanations have been offered for this behavior, from the sinking slab and variable viscosity [*Isacks et al.*, 1968, *Vassiliou and Hager*, 1988] to delayed phase changes (see papers in this volume). The largest deep earthquakes tend to occur at the same depth of the global seismicity peak, about 570 to 650 km. Focal mechanisms of these large deep events could be characterized as vertical dip-slip (with many exceptions), though the interpretation is not always clear. In particular, detailed studies of the great 9 Jun 1994 Bolivia deep event (M_W 8.2) show that the horizontal nodal plane is the fault plane (see papers in this volume).

2.1.4. *Overlying Plate.* The general level of seismicity is less in the overlying plate as compared to the Wadati-Benioff zone, and hence there are fewer large events as well.

The tectonic environment is split into the fore-arc (between trench and volcanic arc, where present) and back-arc (behind the volcanic arc). The volcanic arc itself is characterized by numerous small earthquakes associated with the volcanoes, and arc-parallel strike-slip faults in a few cases (e.g., Sumatra).

In general, active accretionary prisms are devoid of seismicity [see *Byrne et al.*, 1988], thus the potential for earthquakes starts behind the "backstop" of the accretionary prism. Given the presumed high stress levels, it is somewhat surprising to realize that there are few examples of large intraplate events in the fore-arc region. Thus, it seems that interplate events efficiently concentrate the seismic strain release onto the plate interface. A possible exception to this rule is the North Island, New Zealand subduction zone, where several large earthquakes with unknown focal mechanism have occurred in the forearc [*Smith et al.*, 1989].

Some subduction zones show evidence of active compression in the back-arc area. In a few places, large thrust earthquakes have occurred in the back-arc (for example, the back-arcs of Honshu, Mindanao, and Indonesia). Back-arc thrust events tend to occur just at the

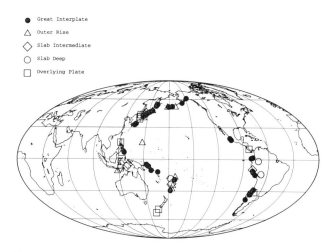

Fig. 2. Epicenter map of global occurrence of large and great earthquakes associated with subduction zones. Closed circles are the 40 great ($M \geq 8$) interplate events from 1900 to 1995. Intraplate seismicity in various subduction zone settings are plotted with different open symbols.

back-arc coastline, thus they can be quite hazardous due to shaking and tsunami generation. This tectonic setting becomes difficult to characterize if we allow for the possibility of subduction initiation along these boundaries. This allowance is particularly important for the Sea of Japan margin of Honshu, where a nearly continuous rupturing of this margin has now occurred in the 20th century, and thus appears to be a newly forming subduction zone [e.g., *Seno*, 1985; *Tanioka et al.*, 1995a]. Thus, our last category of intraplate seismicity crosses the boundary into interplate seismicity.

2.2. Interplate Seismicity

Interplate seismicity is rather simple as compared to the tectonic diversity encountered with intraplate seismicity. Large interplate events occur on the seismogenic portion of the plate interface, which has a shallow dip of between 10° to 30°, and the down-dip edge is at a depth of about 40 km for most subduction zones; it may go as deep as 55 km in a few places, and be as shallow as 20 km in places such as Mexico or Cascadia [*Hyndman and Wang*, 1993]. *Tichelaar and Ruff* [1993] present a global review of the seismogenic depth (also see a related paper by *Ruff and Tichelaar* in this volume). Coseismic slip may extend up to the trench axis during great events, but subsidiary faults in the shallow fore-arc may accommodate some of the slip. The size of the greatest interplate events is more related to the along-strike rupture length rather than to the down-dip rupture width [*Ruff*, 1989]. The main focus of this paper is on the temporal occurrence of large interplate events. We first discuss temporal occurrence and then briefly look at the global spatial occurrence, before turning to some details of large earthquake recurrence models.

3. GLOBAL TEMPORAL OCCURRENCE OF LARGE INTERPLATE EARTHQUAKES

The very largest earthquakes in the 20th century have been identified [e.g., *Kanamori*, 1983; *Nishenko*, 1991; and *Pacheco and Sykes*, 1992]. Specialized catalogs for particular tectonic regimes can be found in, for example, *Dmowska and Lovison* [1988] and *Christensen and Ruff* [1988]. These lists include all events above some magnitude threshold, mostly the moment, surface-wave, or tsunami magnitude (M_W, M_S, and M_t, respectively). *Ruff* [1989] compiled a catalog of the nineteen largest underthrust subduction events, with $M_W \geq 8.2$, from the above lists. The catalog of *Pacheco and Sykes* [1992] attempts to be globally complete down to a magnitude of low 7's, and they try to identify focal mechanism type; unfortunately, the mechanisms of most of the large events in the first half of the century are "unknown". As a compromise, I have tried to extend the list of great interplate subduction events down to a magnitude of 8.0 by simply assuming an underthrust mechanism for those events with a suitable location and tectonic environment (Table 1). As discussed in much more detail in *Nishenko* [1991], there are many uncertainties in such identifications for some regions, such as in the Samoa and Tonga region. Thus, I may have excluded many events with M_S of 8 in the early 20th century that may be underthrust events. Nonetheless, in lowering the magnitude threshold from 8.2 to 8.0, the number of events increases from 19 to 40. Table 1 should be complete from the 1950's to present.

3.1. Global Temporal Variation

One well-known feature of 20th century seismicity is the temporal cluster of largest events in the 1950's and 1960's. Most of the Alaska-Aleutian through Kamchatka to Kuriles subduction zones ruptured in the time interval from 1952 to 1969. These observations are commonly cited as evidence for earthquake clustering or triggering. For example, *Kanamori* [1983] comments: "It is clear that the global seismic activity is very non-uniform in time, at least at a time scale of 100 years or so". Since a random process can produce clustering, we apply two simple quantitative tests to the observed earthquake sequence. One key difference between the data in *Kanamori* [1983] and that in Table 1 is the demotion of the 1957 Aleutian earthquake

TABLE 1. The "Great" (M≥8) Interplate Thrust Earthquakes in the 20th Century (through 1995).

Date(1900+) Yr Mo Dy	Lat	Location Lon	Region	depth*	Mag^ (M_w)
04 06 25	52.	159.	Kamchatka	sh	8.0 s
06 01 31	1.0	-81.3	Colombia	sh	8.8
06 08 17	51.	179.	Aleutian	sh	8.2 s
06 08 17	-33.0	-72.0	Chile	sh	8.2
15 05 01	47.	155.	Kuriles	sh	8.0 s
18 09 07	45.5	151.5	Kuriles	sh	8.2 s
19 04 30	-19.	-172.5	Tonga	sh	8.2 s
22 11 11	-28.5	-70.0	Chile	sh	8.5
23 02 03	54.0	161.0	Kamchatka	sh	8.5
24 04 14	6.5	126.5	Mindanao	sh	8.3 s
32 06 03	19.5	-104.3	Mexico	sh	8.1 s
34 07 18	-11.8	166.5	Santa Cruz Is	sh	8.1 s
38 11 10	55.5	-158.	Alaska	sh	8.2
39 01 30	-6.5	155.5	Solomon Is	sh	8.0 s
39 04 30	-10.5	158.5	Solomon Is	sh	8.2 s
40 05 24	-11.2	-77.8	Peru	sh	8.2
42 08 24	-14.5	-74.8	Peru	sh	8.2
43 04 06	-31.0	-71.3	Chile	sh	8.2
44 07 12	33.8	136.	Honshu	sh	8.1
46 12 20	32.5	134.5	Honshu	sh	8.1
52 03 04	42.5	143.	Hokkaido	sh	8.1
52 11 04	52.8	160.	Kamchatka	sh	9.0
57 03 09	51.6	-175.	Aleutians	30	8.6
58 11 06	44.4	148.6	Kuriles	sh	8.3
59 05 04	53.2	159.8	Kamchatka	sh	8.2
60 05 22	-38.2	-73.5	Chile	40	9.5
63 10 13	44.9	149.6	Kuriles	35	8.5
64 03 28	61.1	-148.	Alaska	35	9.2
65 02 04	51.3	178.6	Aleutians	35	8.8
66 10 17	-10.7	-78.6	Peru	sh	8.1
68 05 16	40.9	143.4	Honshu	35	8.2
69 08 11	43.6	147.2	Kuriles	35	8.2
71 07 14	-5.5	153.9	Solomon Is	sh	8.0
71 07 26	-4.9	153.2	Solomon Is	sh	8.1
74 10 03	-12.2	-77.6	Peru	30	8.1
76 01 14	-29.5	-178.	Kermadec	30	8.1 s
79 12 12	1.6	-79.4	Colombia	sh	8.2
85 03 03	-33.1	-71.9	Chile	30	8.0
85 09 19	18.1	-103.	Mexico	30	8.1 s
86 05 07	51.3	-175.	Aleutian	30	8.0

General references are: *Kanamori*, 1983; *Pacheco and Sykes*, 1992; and *Nishenko*, 1991.
* "sh" means shallow depth, probably less than 70 km, likely to be less than 40 km.
^ Magnitude is M_w, unless followed by s, then M_s.

from M_w of 9.1 to 8.6 [*Johnson et al.*, 1994]. Even so, the interval from 1952 to 1964 contains the three largest earthquakes of the 20th century. What are the odds of this happening for a random Poisson process? If one supposes that three events occur randomly over the 95 year interval from 1900 to 1995, then you would expect them to be clustered within a 12 year interval in about 5% of the random trials. Twentieth century interplate seismicity also contains three intervals of 15 years or more with no great earthquakes (M≥8.2): August 1906 to November 1922; February 1923 to November 1938; and from December 1979 to present (note that the Oct. 4, 1994 event has a M_w of 8.2, but it is not an interplate event). What are the odds of this happening for a random process? To answer this question, we must analyze the expected distribution of "waiting times" between events. It is well-known that a random Poisson point process produces waiting times between events that follow the exponential distribution. Given the occurrence of 19 great (M≥8.2) interplate events over a 95 year time span, the cumulative exponential distribution predicts that the longest "waiting time" would be between 14 and 15 years, on average. But of course, any single trial can violate this average due to the small number of events. If we perform many random simulations, we find that the three longest "waiting times" can all be excess of 15 years in about 2% of the trials. In other words, if the global great earthquake catalog could be extended back for 5,000 years and about twenty events occur randomly in each century, then we would expect to see the clustering observed in the 20th century for one or two of the 50 centuries. Thus, the largest interplate events of the 20th century are "unusually" clustered in the sense that only a few percent of random trials would produce the 12-year-cluster of the three largest events and the three long quiet periods. On the other hand, one can still argue that the 20th century happens to be one of those rare centuries.

The number of interplate events is doubled by decreasing the magnitude threshold from 8.2 to 8.0. The three 15-year-long blank intervals discussed above are subdivided by the addition of these smaller events. There are three intervals with duration more than 8 years, including 1986 to 1995. However, simulations show that three intervals of such duration are "expected" for a random process. Thus, lowering the magnitude threshold quickly obscures the improbably long quiet periods seen for the very largest earthquakes.

3.1.1. *"Wait times" of the forty great events.* To further analyze the temporal occurrence of events in Table 1, we consider the full spectrum of "wait times". The times between successive events (the "wait" times) are ordered from the shortest to longest time. I then "integrate" this distribution from the longest time down to zero time to obtain the cumulative number of wait times versus wait time. We can see that the observed cumulative wait times conform rather closely to the theoretical curve for random occurrence (Figure 3). The three wait times of 8 years or more are consistent with the theoretical expectations, as discussed above. Figure 3 illustrates yet another example of a random Poisson process that produces apparent clustering of events. While the average wait time is 2.3 years, about half of the wait times are less than 1.5 years, and ten of the wait times are about 6 months or less. In conclusion, while the clustering of the greatest earthquakes in the 1950's and 1960's is unlikely to occur by random occurrence, the catalog of large subduction events of magnitude 8.0 or larger (Table 1) does conform to a model of random occurrence.

3.2. Spatial Clustering

Spatial clustering of seismicity is readily proven by examination of global seismicity maps (see Fig. 2). Large interplate events are not evenly distributed amongst the world's subduction zones [*Uyeda and Kanamori*, 1979]; they are concentrated in the subduction zones of South America, Alaska-Aleutian, and Kamchatka-Kuriles-Japan. In contrast, the numerous subduction zones in the southwestern Pacific lack great interplate earthquakes. *Ruff and Kanamori* [1980] offered a tectonic basis for these observations, and this subject has received considerable work and speculation since then. We can summarize these results by stating that larger great earthquakes are correlated with the fast subduction of young oceanic lithosphere, but a clear physical explanation still eludes us [see discussion in *Ruff*, 1989; and *Scholz*, 1990].

4. RECURRENCE TIMES

While consideration of global "wait times" is of some intellectual interest, the question of great practical importance is: When will the next large earthquake occur *here*? This final topic of our overview brings us into the realm of recurrence times, the seismic gap hypothesis, and long-term earthquake forecasting. This topic must be approached with care and skepticism, yet it is always one of the most important aspects of any scientific discussion of subduction zone seismicity. Here, I shall offer one perspective on the estimation of large earthquake recurrence times, with the suggestion that along-strike segment interaction explains some of the variability in observed recurrence times.

4.1. Overview of Long-term Earthquake Forecasting

Plate tectonics provides the underlying kinematic framework for the succession of ideas from the seismic gap

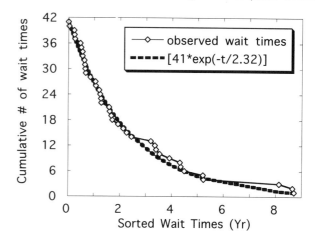

Fig. 3. Cumulative distribution of "wait times" between the 40 great interplate events in Table 1. Observed wait times are sorted and the numbers are accumulated from the largest inter-event time down to zero time. Independent random occurrence of events produces wait times with an exponential distribution, (dashed line) a relation that fits the data well.

hypothesis [*Fedotov*, 1965; *Mogi*, 1968; *Kelleher et al.*, 1973] to the most recent versions of long-term earthquake forecasting [*Nishenko*, 1991]. The basic idea is that large earthquakes occur again and again along the same plate boundary segment with approximately the same co-seismic slip and recurrence time. Steady plate motions provide a constant rate of stress increase on a locked plate interface segment, and the assumption of constant plate interface properties and segmentation then results in regular earthquake behavior. From an observational perspective, one merely needs to find the plate boundary segments that rupture as large events and the average recurrence time for each segment, then one can forecast the time and size of the next large earthquake. If you cannot determine the average recurrence time for that segment, then use an average recurrence time from somewhere else, or possibly calculate the recurrence time based on the time-predictable model (described in section 4.2). This methodology is remarkably simple, and it was remarkably successful with a string of fulfilled forecasts in the 1970's decade [see *McCann et al.*, 1979]. Detailed studies showed that there were many violations of the above assumptions, as discussed below. Nonetheless, it was difficult to argue with the "short-term" successes of long-term forecasting in the 1970's.

McCann et al. [1979] published a map that summarized their long-term forecasts for most of the world's subduction zones. In their color scheme, red regions were those segments where great earthquakes are expected soon, green regions should be safe for many decades. *Kagan and Jackson* [1991] have caused considerable debate and controversy with their statistical test of the forecast performance of the *McCann et al.* map over the 1980's decade. They conclude that the *McCann et al.* forecasts have performed rather poorly: large events occurred in green zones, and did not occur in red zones. The main criticism of the *Kagan and Jackson* test is that there were only two or three "characteristically large" earthquakes in the decade of the 1980's, thus any statistical test is either premature, or if a statistical test uses the smaller events in subduction zones, then it is not a proper test of long-term forecasting [see *Nishenko and Sykes*, 1993]. At this point, I shall not consider this issue any further, but note that everyone involved agrees that the societal implications of long-term earthquake forecasting require close scrutiny and testing of any methodology, and that we should strive to improve the methodology.

This overview is focused on just the largest earthquakes, hence I shall consider only the six largest subduction earthquakes that occurred since 1979. The 1980 Santa Cruz Is. earthquake represents the worse type of failure of long-term forecasting: the characteristic large event occurring in a "safe" region (i.e. a green zone in the *McCann et al.* map). The 1985 Chile earthquake represents a success for long-term forecasting [*Nishenko*, 1985], as does the 1985 Michoacan earthquake. The 1986 Aleutian earthquake represents another failure for long-term forecasting in that it occurred in a green zone. On the other hand, once could argue that the 1986 event is not a "characteristically large" earthquake in this region since the preceding event was the great 1957 Aleutian earthquake. Most recently, the 3 Dec 1995 Kuriles event re-ruptured the epicentral asperity of the 1963 great Kuriles earthquake, a green zone in the *McCann et al.* map. To conclude, three of the six largest subduction events since 1979 occurred in segments that had great earthquakes just 15 to 32 years before the most recent events. Can we understand this extreme variation in great earthquake recurrence times?

4.2. *Time/Slip Predictable Model*

The time-predictable and slip-predictable models [*Shimazaki and Nakata*, 1980] were devised to provide a deterministic, yet ad hoc, explanation for variability in recurrence times. These two models are the end-member cases of a class of models that relate variations in recurrence times to variations in seismic slip. For reference, Figure 4a shows how a hypothetical "uniform" seismic segment might rupture through time: the earthquake occurs when a constant failure stress is achieved,

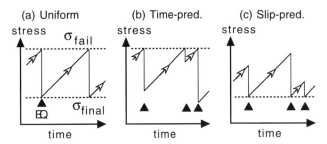

Fig. 4. Idealized time histories of earthquake sequences for a single fault segment. Diagrams show stress as a function of time, occurrence of earthquakes is denoted by solid triangles. It is assumed that stress increases at a steady rate for all three models. For the uniform model (a), both the failure and final stress levels are constant (dashed lines). This model produces a sequence of identical earthquakes with constant recurrence times. The time-predictable (b) and slip-predictable (c) models both produce the same earthquake sequences in terms of recurrence times. The time-predictable model states that the failure stress is constant, but the final stress varies. The slip-predictable model states that the final stress is constant, but the failure stress varies.

and the stress drop is constant so that the final stress is uniform. With a constant stress accumulation rate, the "ideal" earthquake behavior would be a sequence of identical earthquakes with a constant recurrence time. But, earthquakes do not behave this way. Figure 4(b,c) shows a hypothetical sequence of three earthquakes, the two recurrence times are different. The time/slip-predictable model offers two end-member explanations for variable recurrence time: (i) the failure stress is always the same, but the final stress can be different (due to unknown factors), (ii) the failure stress can vary, but the final stress is always the same. It is possible to test which end-member model is preferred if we know the stress-drop associated with each earthquake. Due to several technical problems in obtaining the stress-drop, it is common to use co-seismic slip as a proxy for the stress-drop of a segment. Even with this simplification, it is difficult to obtain slip estimates for older earthquakes. Several workers have tried to analyze sequences of earthquakes and slip estimates to decide whether the slip-predictable or time-predictable cases are preferred. Perhaps the best case is the sequence of three earthquakes that occurred in a segment of the Nankai trough region, *Shimazaki and Nakata* [1980] used measurements of coastline uplift as a proxy for the displacements on the fault plane and found a slight preference for the time-predictable model. Overall, there is not enough statistical evidence to globally prefer either end-member model; earthquake sequences are too irregular [see *Sykes and Quittmeyer*, 1981; *Nishenko*, 1985].

98 SEGMENT INTERACTION AND RECURRENCE TIMES

4.2.1. Beyond the Time/Slip predictable model. At this point, one can either probe the plate interface properties to follow a deterministic approach to the scatter in recurrence times, or follow a statistical approach. In a key paper, *Nishenko and Buland* [1987] argue that recurrence times around the world all follow a single "generic" log-normal distribution. In particular, the global scatter in recurrence times is about 20% of the average recurrence time. This statistical approach allows *Nishenko* [1991] to estimate when the next large earthquake will occur, plus and minus a certain number of years, for fault segments where there is only *one* observed recurrence time. Where there is *no* observed recurrence time, *Nishenko* [1991] assumes the time-predictable model to get a recurrence time, with the uncertainty again provided by the generic distribution. In this manner, *Nishenko* [1991] has provided recurrence time estimates for a total of 96 segments around the world. How can we do better with a deterministic approach? I show in a later section that a simple model of segment interaction can explain irregular earthquake sequences. But first, the next section reviews some of the above-discussed observed features of great earthquake occurrence.

4.3. *Some Additional Characteristics of Large Interplate Earthquake Occurrence*

Great earthquakes rupture the full width of the seismogenic zone, the along-strike rupture length and average slip are the key determinants of overall seismic moment. In short, greater earthquakes result from greater along-strike rupture length. Hence, much of the effort in characterizing great earthquake occurrence has been focused on the two parameters: (i) along-strike rupture length, and (ii) recurrence time between great earthquakes. Aftershock areas can be used to estimate rupture length for most of the great twentieth-century interplate earthquakes. The along-strike rupture length can be estimated for older earthquakes by examination of intensity maps. The recurrence of great earthquakes over time spans of a few hundred years has been established in several subduction zones through the study of historical documents. Perhaps the best example of well-documented great earthquake occurrence over several hundred years is for the Nankai subduction zone along the coast of southern Honshu, Japan [e.g. *Utsu*, 1974; *Ando*, 1975; *Ishibashi*, 1981]; several investigators found that the Nankai subduction zone could be divided into four or five seismic segments, though it can also be characterized as just two primary segments (i.e., the rupture zones of the great 1944 and 1946 earthquakes). In detail, these two primary segments tend to rupture as "doublets" [*Nomanbhoy and Ruff*, 1996]; in addition to the 1944-1946 doublet, the previous earthquake cycle consisted two earthquakes separated by only 32 hours in 1854. Two important facts emerge from studies of earthquake occurrence along the Nankai and other subduction zones: (1) the recurrence time for any one seismic segment is variable, and (2) the rupture lengths of large earthquakes is variable, i.e. adjacent seismic segments will sometimes rupture as individual earthquakes, other times a larger "multiple-event" earthquake will rupture two or more adjacent segments [see e.g., *Thatcher*, 1990, for more discussion]. Much research effort has been spent trying to understand the cause of observations (1) and (2). Related to these two observations, seismic slip can also vary between successive earthquakes in a subduction zone segment; stress drop may or may not vary in accordance with slip variations. Thus, any detailed consideration of observed earthquake sequences shows many violations of the simple assumptions listed in section 4.1 that form the basis for the simple methods of long-term forecasting. Indeed, some observers conclude that variability may be more typical than regularity in earthquake sequences [*Thatcher*, 1990]. To improve our earthquake forecasting methodology, we must move beyond the simplest models of earthquake occurrence.

5. MODELS FOR LARGE EARTHQUAKE RECURRENCE

The simplest physical model for large earthquake recurrence employs one frictional slider block with a simple frictional failure criterion. The sketch in Figure 5 implies a geometry for shallow underthrusting events in subduction zones. This simple model assumes that: plate motion is steady, earthquakes occur when the stress reaches σ_{fail}, and stress drops to σ_{final}, and that these stress levels are nearly constant over several earthquake cycles. The prediction of the model for isolated frictional sliders is obvious: uniform earthquakes occurring with constant recurrence times.

If we now extend the model to accommodate adjacent segments along the plate boundary, we see that it is still the same idea (Fig. 5b). By showing two independent sliders side-by-side, we simply allow for the fact that each plate boundary segment might have a different recurrence interval, though they are still constant within each segment. This model assumes that major plate boundary segments remain constant, at least over a few earthquake cycles. The model in Fig. 5b may seem absurdly simple, but it is the underlying mechanical model for two decades of work in long-term earthquake forecasting. What is wrong with the mechanical model of Figure 5b? From the

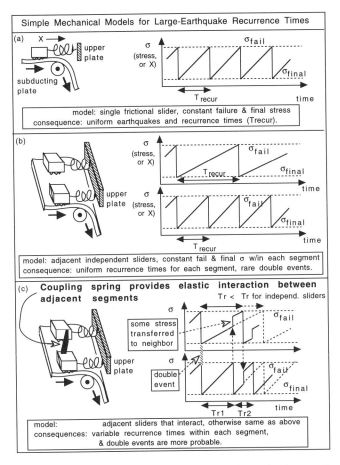

Fig. 5. Simple mechanical models used to simulate and forecast earthquake sequences. Each plate boundary segment can rupture as a large earthquake, and is viewed as a slider block in frictional contact with a subducting plate "conveyer belt". The sliders remain stuck to the moving "conveyer belt" until the increasing elastic strain causes the plate interface stress to reach σ_{fail}. An earthquake then occurs, with the stress falling to σ_{final}. T_{recur} is the recurrence time, and X is the position of the slider with respect to the upper plate. (a) An isolated segment. (b) Two adjacent segments, but without interaction. (c) Adjacent plate boundary segments that interact via the coupling spring. The relative stiffness of the coupling spring depends on the geometry. In (c), the occurrence of a large earthquake on one segment transfers some of its stress drop to adjacent segments, thereby affecting their recurrence times. For example, the two recurrence times for the lower segment, Tr1 and Tr2, are different. A double event is shown by the hachured bar.

observational viewpoint, it does not satisfy the observations of variable rupture mode and recurrence time. From the mechanical viewpoint, this simplest model ignores the fact that slip in adjacent fault segments will increase the stress level above that from tectonic loading.

5.1. *Simplest Model with Segment Interaction*

The simplest mechanical model with segment interaction consists of two adjacent frictional slider blocks connected by springs to the upper plate; the blocks are also connected to each other by a "leaf" spring (Fig. 5c). The basic effect of this addition is that slip in one segment increases the averaged stress level in the adjacent segment. If the stress increase places the adjacent segment above its failure stress, then we would obtain a "double event", where the rupture of one segment triggers the rupture of an adjacent segment on a time scale of seconds to a minute. Thus, this direct coupling dramatically enhances the likelihood of variable "rupture mode", i.e. the occurrence of double events in some earthquake cycles. Even if a double event does not occur, there are still important effects due to the coupling. Figure 5c also shows an idealized earthquake sequence. Just as for the previous models, plate motion accumulates linearly, and there is a constant σ_{fail} and σ_{final}. However, now there is an extra stress increase due to slip in the neighboring segment. The figure shows a certain percentage of the stress drop transferred to the adjacent segment, and this will cause that segment to fail sooner that it would without the interaction (see the dashed lines in the figure). Of course, allowance of this interaction tends to reduce the average recurrence time for both segments. Nonetheless, failure in *adjacent* plate boundary segments does influence the recurrence time. The size of this influence depends on the strength of segment-to-segment coupling compared to segment-to-upper plate coupling. *Ruff* [1992] investigated the simple two-block system of Fig. 5c, where the blocks represented large asperities that define the large-scale segmentation of the plate boundary. It is possible to determine many of the system parameters for many subduction segments around the world. As discussed in *Ruff* [1992], this segment-to-segment coupling can cause 10% to 80% of the static stress drop in one segment to be transferred to its neighbors. With such strong coupling, the net effects cannot be viewed as a minor perturbation to the independent sliders. It is necessary to run many numerical simulations to fully characterize the behavior. We have done that, and we show Figure 6 as one example that displays many of the general features. This figure shows the displacement of segments 1 and 2 as a function of time. When there is a double event, a solid line is drawn through both upper and lower parts, with the arrow indicating the direction of rupture triggering. Numbers give the recurrence time between events (e.g., you could multiply all these numbers by 100 to think of them as years). Finally, the horizontal dashed lines should go through the top/bottom corners of the displacement

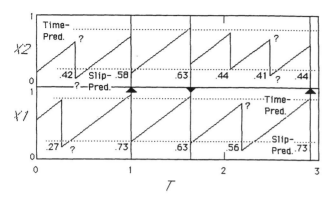

Fig. 6. A synthetic earthquake sequence produced for the model of Figure 5c, with realistic system parameters. *X1* and *X2* track the displacement for each segment, as a function of time, *T*. *X1* and *X2* are normalized by maximum displacement, *T* is normalized by maximum recurrence time. The numbers at each event are the recurrence times between events, double events are shown by the through-going vertical line, arrow points toward the triggered segment. Dashed lines labeled "Time-pred ?" and "Slip-pred" show that this synthetic earthquake sequence is too irregular to be adequately explained by either the slip- or time-predictable models.

histories if the time/slip-predictable model were correct. We see that this synthetic earthquake sequence is too irregular to be explained by the time- or slip-predictable models, just as observed earthquake sequences are too irregular. Although the model of Fig. 5c is the simplest mechanical model with segment interaction, it produces synthetic earthquake sequences that resemble observed sequences.

The discrete element model of Figure 5c cannot properly model the detailed three-dimensional elastic interactions of a real subduction zone boundary [see, e.g., *Rundle and Kanamori*, 1987; *Stuart*, 1988]. However, one advantage of the simple discrete element model is that we can produce earthquake sequences for all regions of parameter space. One key modeling result is that the assumption about multiple-event failure mode is crucial. Two extreme failure modes are the zero final stress assumption, and the constant co-seismic slip assumption [see *Ruff*, 1992, for details]. The zero final stress assumption resets the stress levels to zero for all segments that fail. This simplifying assumption produces the simplest possible earthquake sequences, and thus allows the overall system behavior to be easily analyzed [e.g., as in *Lomnitz-Adler and Perez-Pascual*, 1989]. The alternative failure mode is that a segment "heals" during a multi-segment rupture [e.g., *Huang and Turcotte*, 1992]. Since a segment is "stuck again" while slip occurs elsewhere along the plate boundary, then previously ruptured segments have a non-zero stress at the conclusion of the multi-segment rupture.

This *rapid healing* rupture mode results in earthquake sequences that show more variability than for the *zero final stress* rupture mode. Our simulations adopt the rapid healing rupture mode for multi-segment failures. The results discussed below are fairly robust with respect to all other parameter choices.

We do not believe that any particular synthetic earthquake sequence (as that in Fig. 6) can be interpreted literally as particular earthquake sequences for subduction zones. Instead, we seek to find some general rules that apply to all such synthetic earthquake sequences. While *Ruff* [1992] focused on quantitative statistical summaries based on thousands of synthetic earthquake sequences, here we extract some robust qualitative behaviors for successive earthquakes. Here are some of the general behaviors:

(1a) If the previous cycle was a multiple-segment rupture, then the next earthquake will occur in the segment that contains the epicenter (location of rupture initiation) of the previous multi-segment rupture.

(1b) The recurrence time to the next earthquake in the above scenario will be shorter than the average recurrence time.

(2) During an earthquake cycle in which adjacent segments break in separate events:

(2a) If the larger segment breaks first, then the smaller segment tends to break sooner than its average recurrence time.

(2b) If the smaller segment breaks first, then the recurrence time for the adjacent larger segment may or may not be shorter than its average recurrence time.

Some of these general rules can be seen in Fig. 6, and rules (1a) and (1b) are graphically summarized in Figure 7. However, it is more interesting and important to compare these results with observed earthquake sequences.

6. SOME OBSERVATIONS OF EARTHQUAKE RECURRENCE WITH SIGNIFICANT SEGMENT INTERACTION

We present several interesting examples of earthquake sequences that switch rupture modes. While these cases are problematic for anyone trying to estimate future earthquake occurrence, we shall see that the above general rules allow us to make some sense of these complicated sequences.

6.1. *Ecuador-Colombia*

The subduction zone off the coast of Colombia-Ecuador ruptured in one great earthquake in 1906, and has subsequently re-ruptured as three separate events in 1942, 1958, and 1979 (see Fig. 8). The prior earthquake history

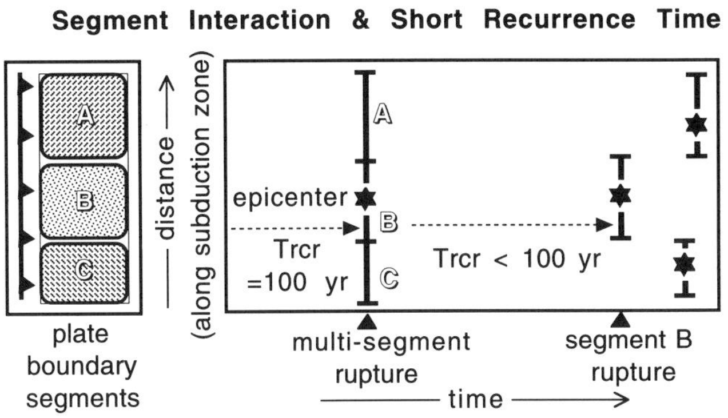

Segment Interaction & Short Recurrence Time

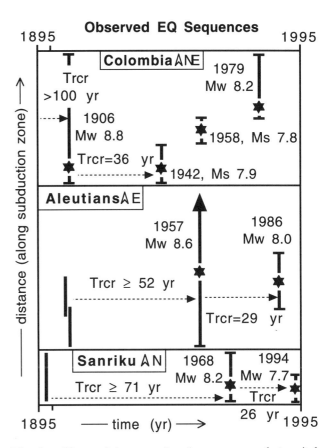

Fig. 7. Graphical summary of "rules" (1a) and (1b) obtained from synthetic seismicity produced by simple two-segment mechanical models with segment interaction. These models assume rapid healing such that the plate boundary segment that initiated the great earthquake, segment B in the above example, heals after the rupture front propagates on to segments A and C. Then, slip in adjacent segments reloads segment B to a high stress level at the conclusion of the great earthquake. If the rupture mode switches from a great multi-segment rupture to individual segment ruptures, then: ("rule" 1a) segment B will rupture before segments A and C; and ("rule" 1b) the recurrence time (Trcr) will be shorter than the average time.

is not well-known for this region; the previous great earthquake possibly occurred in 1797. The recurrence time before the 1906 event is more than one hundred years, while the recurrence time for the 1942 event was just 36 years. Of course, the 1942 event was much smaller than the preceding 1906 event, so one could argue that it was not the "characteristic" great earthquake. But the following two events clearly show that the rupture mode changed for the recent earthquake cycle. The 1906 epicenter [*Kanamori and McNally*, 1982] is located within the 1942 rupture zone. The 1906 rupture started in the southern-segment (the 1942 segment), and then continued to the north. If the southern-segment healed before the 1906 rupture was finished, then the southern-segment would be reloaded by slip in adjacent segments. Thus, the observed Colombia earthquake sequence displays the general behaviors (1a) and (1b) of interacting segment rupture: the next event occurred in the epicentral region of the multi-segment rupture; and the recurrence time for this next event is shorter than the average recurrence time.

6.2. Aleutian Islands

One of the notable problems with the seismic potential map of *McCann et al.* [1979] was the occurrence of the 1986 Aleutian event in the middle of the 1957 Great Aleutian Earthquake rupture zone. Given the size of the

Fig. 8. Observed large earthquake sequences that switch rupture modes from great multi-segment events to large individual-segment events. The space-time plot shows event occurrence time along the horizontal axis, with the earthquake rupture lengths plotted along the vertical direction. For reference, the great 1906 Ecuador-Colombia earthquake has a rupture length of 500 km. The 1957 Aleutian earthquake has a rupture length of more than a thousand kilometers, thus we show only part of its length. Stars are epicenters, and subduction zone strikes (arrows) are identified, e.g. "NE" is northeast. Recurrence times between events (Trcr) show that after a great earthquake, the next large event re-ruptures the epicentral segment with a short recurrence time.

1957 event, it was thought that the recurrence time for the next great event should be on the order of one hundred years or more. However, it was noted that the 1957 zone experienced several smaller earthquakes in the early 1900's [see *Nishenko*, 1991]. Thus, we can say that the first problem in the Aleutians is that the rupture mode appears to change between earthquake cycles. In detail, the segment that ruptured in the 1986 event includes the epicentral area of the preceding great 1957 event [*Johnson et al.*, 1994]. Again, rules (1a) and (1b) would seem to explain the location and sooner-than-expected occurrence time of the 1986 event.

6.3. Santa Cruz Islands

As mentioned above, another problem with the seismic potential map of *McCann et al.* [1979] was the occurrence of the July 17, 1980 Santa Cruz Is. event (M_S 7.7) so soon after the Dec. 31, 1966 event (M_S 7.9). A detailed study of this sequence by *Tajima et al.* [1990] shows that there is indeed considerable overlap between the aftershock areas of these two events, but that the 1966 and 1980 events ruptured different distinct asperities within this region. This case is different from the above two examples where the "unexpected" event clearly occurred within the rupture zone of the preceding great earthquake. Instead, this case may be better characterized as confusion in the segment identification. Referring to our above "rules", the 1966 and 1980 event sequence would fall under behavior (2), and possibly behavior (2a) since the 1966 event was larger than the 1980 event. While the 14 year recurrence time between 1966 and 1980 is certainly consistent with behavior (2a), we need to know more about the preceding great event in 1934 before we can be confident of our understanding of segment interaction in this subduction zone.

6.4. Sanriku

This subduction zone segment is off the eastern coast of northern Honshu, and was ruptured by the great Tokachi-Oki earthquake of 1968 (M_W 8.2). The previous large earthquake occurred in the 19th century. For the 1968 event, the main rupture started near the southern end of the aftershock zone, and then ruptured to the north [see, e.g., *Schwartz and Ruff*, 1985]. Many seismologists were surprised by the occurrence of the Dec. 28, 1994 large earthquake (M_W 7.7) off the Sanriku coast. Preliminary results show that the 1994 event re-ruptured the southern part of the 1968 rupture zone; a short recurrence time of only 26 years! But once again, we see a change in rupture mode, and the following smaller event (the 1994 event) occurred in the epicentral region of the preceding event (the 1968 event), with a shortened recurrence time. Thus, the Sanriku example is similar to the above first two examples, and it is explained by rules (1a) and (1b) from the simulations of interacting segments.

6.5. Kuriles Islands

The southern part of the Kuriles Islands subduction zone was ruptured by a sequence of large earthquakes from 1958 to 1973. The largest of these events was the 13 Oct 1963 event (M_W 8.5). The northern part of the 1963 rupture zone was previously ruptured in 1918, thus the minimum recurrence time is 45 years. Due to the large size of the 1963 great event, most workers assigned low probability to large event recurrence within the 1963 rupture zone. But on 3 Dec 1995, a large (M_W 7.9) event re-ruptured the 1963 epicentral asperity region. Since the 1995 event is much smaller than the 1963 event, one could argue that it is not the "characteristic" event for this segment and thus not worthy of consideration. Alternatively, the interacting segment model again offers an explanation for this sooner-than-expected large event in the epicentral segment of the previous great earthquake.

6.6. Summary of Segment Interaction and Recurrence Times

Several examples of shortened recurrence times can be understood by consideration of plate boundary segment interaction. If this model can be applied to most subduction zones, then we have the exciting potential of finding a deterministic explanation for the variability in large earthquake recurrence times. To utilize this potential, we must know whether a particular earthquake cycle shall occur as one great multi-segment rupture or as a sequence of single-segment ruptures. Some features of the interacting segments model can be tested with detailed specific studies; for example, studies of the co-seismic slip distribution and the rupture process can help define the effective failure mode. In particular, we want to know if the epicentral segments "heal" while other segments are still slipping. Also, the model can be tested by re-evaluating historical records for the occurrence of large events in the epicentral region of preceding great earthquakes. Of course, the ultimate test would be to use the interacting segment model to forecast future earthquakes. This model has the potential to switch rupture modes between earthquake cycles and thus forecast rupture length in addition to recurrence times. There are a few subduction zones that offer suitable experimental conditions, but we shall not make any specific earthquake forecasts here.

7. CONCLUSIONS

Large earthquakes have occurred in many different tectonic environments in and around subduction zones. Nearly every intraplate environment has generated at least one great ($M \geq 8$) earthquake. Perhaps the most notable exception is the fore-arc (trench to volcano) of the overlying plate in mature subduction zones; there are no verified great earthquakes in this region. Until the 4 Oct 1994 great earthquake, we also thought that no large earthquakes occurred within the subducting slab beneath the coupled plate interface. Unless we have misidentified many other large events, it is still a good generalization to conclude that few intraplate events occur above or below the seismogenic portion of the plate interface.

The tectonic regime that produces the most earthquakes and the largest earthquakes is the seismogenic plate interface. The seismogenic depth range extends down to about 40 km for most subduction zones. Underthrust events with magnitude larger than 7.7 typically rupture the full downdip width of the seismogenic plate interface. Thus, the size of the great underthrust events is determined by the along-strike rupture length and the average coseismic slip. In this overview, we present and discuss a catalog of 40 great subduction events with magnitude of 8 or larger from 1900 to 1995. These great events are mostly concentrated in a few subduction zones. Occurrence times of the very largest earthquakes are clustered more than would be expected from random occurrence, but the catalog of 40 great interplate events has a "wait time" distribution consistent with a Poisson process, i.e. independent random occurrence.

Long-term earthquake forecasting attempts to determine the plate boundary segments and recurrence times for each segment. The primary information source for long-term earthquake forecasting is the previous earthquake history. Several observers of global seismicity conclude that large earthquake occurrence is quite irregular in at least two important ways: (1) rupture mode can vary between successive earthquake cycles, i.e. different combinations of segments rupture; (2) the recurrence time for each segment can vary.

The simplest mechanical models that connect plate tectonic motions to recurrence times do not offer an explanation for variable rupture mode or recurrence times. These mechanical models do not incorporate along-strike elastic interactions between segments. If we add adjacent segment interaction to the mechanical models, then variable rupture mode and recurrence times are readily produced. Thousands of synthetic event simulations indicate a few robust conclusions can be reached: (1a) the first event that follows a multi-segment rupture occurs in the epicentral segment of the preceding multi-segment event, and (1b) its recurrence time will be shorter than the average recurrence time. Observed earthquake sequences in four subduction zones -- Colombia, Aleutians, Sanriku, and Kuriles Islands -- have one unusually short recurrence time that follows the above rules. Thus, our overview shows that great earthquake occurrence in subduction zones is remarkably complex and diverse, but segment elastic interaction offers the exciting potential to understand and predict some of the variability in recurrence times.

Acknowledgments. Thanks to C. Frohlich and S. Kirby for their comments and suggestions. Earthquake research is supported at the University of Michigan by the National Science Foundation (EAR94-05533).

REFERENCES

Ando, M., Source mechanisms and tectonic significance of historical earthquakes along the Nankai Trough, *Tectonophysics*, 27, 119-140, 1975.

Byrne, D. E., D. M. Davis, and L. R. Sykes, Loci and maximum size of thrust earthquakes and the mechanics of the shallow region of subduction zones, *Tectonics*, 7, 833-857, 1988.

Chapple, W.M., and D.W. Forsyth, Earthquakes and bending of plates at trenches, *J. Geophys. Res.*, 84, 6729-6749, 1979.

Christensen, D.H., and L.J. Ruff, Seismic coupling and outer rise earthquakes, *J. Geophys. Res.*, 93, 13421-13444, 1988.

Dmowska, R., and L. Lovison, Intermediate-term seismic precursors for some coupled subduction zones, *PAGEOPH*, 126, 643-664, 1988.

Fedotov, S., Regularities of the distribution of strong earthquakes in Kamchatka, Kurile Islands, and northeast Japan, *Trudy Inst. Fiz. Acad. Nauk SSSR*, 36, 66-93, 1965.

Gutenberg, B., and C.F. Richter, *Seismicity of the earth and associated phenomena*, Princeton University Press, Princeton, NJ, 1952.

Huang, J., and D. Turcotte, Chaotic seismic faulting with a mass-spring model and velocity-weakening friction, *PAGEOPH*, 138, 569-589, 1992.

Hyndman, R.D., and K. Wang, Thermal constraints on the zone of major thrust earthquake failure: The Cascadia subduction zone, *J. Geophys. Res.*, 98, 2039-2060, 1993.

Isacks, B., J. Oliver, and L. Sykes, Seismology and the new global tectonics, 73, *J. Geophys. Res.*, 5855-5899, 1968.

Ishibashi, K., Speculation of a soon-to-occur seismic faulting in the Tokai district, central Japan, based on seismotectonics, in *Earthquake prediction - an international review*, edited by D. Simpson and P. Richards, 297-332 pp., AGU, Washington D.C., 1981.

Johnson, J., Y. Tanioka, L. Ruff, K. Satake, H. Kanamori, and L. Sykes, The 1957 Great Aleutian earthquake, *PAGEOPH*, 142, 3-28, 1994.

Kagan, Y.Y., and D.D. Jackson, Seismic gap hypothesis: ten years after, *J. Geophys. Res.*, 96, 21,419-21,431, 1991.

Kanamori, H., Global seismicity, in *Earthquakes: Observation, Theory, and Interpretation*, edited by H. Kanamori and E.Boschi, pp. 596-608, North-Holland, Amsterdam, 1983.

Kanamori, H. and K. C. McNally, Variable rupture mode of the subduction zone along the Ecuador-Colombia coast, *Bull. Seism. Soc. Am.*, 72, 1241-1253, 1982.

Kasahara, H., and T. Sasatani, Source characteristics of the Kunashiri strait earthquake of Dec. 6, 1978 as deduced from strain seismograms, *Phys. Earth Planet. Int.*, *37*, 124-134, 1985.

Kelleher, J., L. Sykes, and J. Oliver, Possible criteria for predicting earthquake locations and their application to major plate boundaries of the Pacific and Caribbean, *J. Geophys. Res.*, *78*, 2547-2585, 1973.

Kikuchi, M., and H. Kanamori, The Shikotan earthquake of October 4, 1994: A lithosphere earthquake, *Geophys. Res. Lett.*, *22*, 1025-1028, 1995.

Lomnitz-Adler, J., and R. Perez Pascual, Exactly solvable two-fault model with seismic radiation, *Geophys. J. Int.*, *98*, 131-141, 1989.

Lynnes, C.S., and T. Lay, Source process of the great 1977 Sumba earthquake, *J. Geophys. Res.*, *93*, 13407-13420, 1988.

McCann, W., R. Nishenko, L. Sykes, and J. Kraus, Seismic gaps and plate tectonics: Seismic potential for major plate boundaries, *PAGEOPH*, *117*, 1087-1147, 1979.

Mogi, K., Some features of recent seismic activity in and near Japan, *Bull. Earthquake Res. Inst., Tokyo*, *46*, 1225-1236, 1968.

Nishenko, S. P., Seismic potential for large and great interplate earthquakes along the Chilean and southern Peruvian margins of South America: A quantitative reappraisal, *J. Geophys. Res.*, *90*, 3589-3615, 1985.

Nishenko, S. P, Circum-Pacific seismic potential: 1989-1999, *PAGEOPH*, *135*, 169-259, 1991.

Nishenko, S., and R. Buland, A generic recurrence interval distribution for earthquake forecasting, *Bull. Seism. Soc. Am.*, *77*, 1382-1399, 1987.

Nishenko, S., and L. Sykes, Comment on: "Seismic gap hypothesis: ten years after" by Y. Kagan and D. Jackson, *J. Geophys. Res.*, *98*, 9909-9916, 1993.

Nomanbhoy, N., and L. J. Ruff, A simple discrete element model of large multiplet earthquakes, *J. Geophys. Res.*, *101*, 5707-5724, 1996.

Pacheco, J.F., and L.R. Sykes, Seismic moment catalog of large shallow earthquakes, 1900 to 1989, *Bull. Seismol. Soc. Am.*, *82*, 1306-1349, 1992.

Ruff, L. J., Do trench sediments affect great earthquake occurrence in subduction zones?, in *Subduction Zones Part II*, edited by L. Ruff & H. Kanamori, *PAGEOPH*, *129*, 263-282, 1989.

Ruff, L. J., Asperity distributions and large earthquake occurrence in subduction zones, *Tectonophysics*, *211*, 61-83, 1992.

Ruff, L., and H. Kanamori, Seismicity and the subduction process, *Phys. Earth Planet. Int.*, *23*, 240-252, 1980.

Rundle, J., and H. Kanamori, Application of an inhomogeneous stress (patch) model to complex subduction zone earthquakes: A discrete interaction matrix approach, *J. Geophys. Res.*, *92*, 2606-2626, 1987.

Scholz, C. H., *The mechanics of earthquakes and faulting*, 439 pp., Cambridge Univ. Press, Cambridge, 1990.

Schwartz, S., and L. Ruff, The 1968 Tokachi-Oki and 1969 Kurile Is. earthquakes: Variability in the rupture process, *J. Geophys. Res.*, *90*, 8613-8626, 1985.

Seno, T., Is northern Honshu a microplate? *Tectonophysics*, *115*, 177-196, 1985.

Shimazaki, K., and T. Nakata, Time-predictable recurrence model for large earthquakes, *Geophys. Res. Lett.*, *7*, 279-282, 1980.

Smith, E.G., T. Stern, and M. Reyners, Subduction and back-arc activity at the Hikurangi convergent margin, New Zealand, *PAGEOPH*, *129*, 203-231, 1989.

Stuart, W.D., Forecast model for great earthquakes at the Nankai Trough subduction zone, *PAGEOPH*, *126*, 619-642, 1988.

Sykes, L. R., and R. Quittmeyer, Repeat times of great earthquakes along simple plate boundaries, in *Earthquake prediction - an international review*, edited by D. Simpson and P. Richards, 217-247 pp., AGU, Washington D.C., 1981.

Tajima, F., L. Ruff, H. Kanamori, Z. Zhang, and K. Mogi, Earthquake source processes and subduction regime in the Santa Cruz Islands region, *Phys. Earth Planet. Int.*, *61*, 269-290, 1990.

Tanioka, Y., K. Satake, and L. Ruff, Total analysis of the 1993 Hokkaido Nansei-oki earthquake using seismic wave, tsunami, and geodetic data, *Geophys. Res. Lett.*, *22*, 9-12, 1995a.

Tanioka, Y., L. Ruff, and K. Satake, The great Kurile earthquake of October 4, 1994 tore the slab, *Geophys. Res. Lett.*, *22*, 1661-1664, 1995b.

Thatcher, W., Order and diversity in the modes of circum-Pacific earthquake recurrence, *J. Geophys. Res.*, *95*, 2609-2623, 1990.

Tichelaar, B.W., D.H. Christensen, and L.J. Ruff, Depth extent of rupture of the 1981 Chilean outer-rise earthquake as inferred from long-period body waves, *Bull. Seismol. Soc. Am.*, *82*, 1236-1252, 1992.

Tichelaar, B.W., and L.J. Ruff, Depth of seismic coupling along subduction zones, *J. Geophys. Res.*, *98*, 2017-2037, 1993.

Utsu, T., Space-time pattern of large earthquakes occurring off the Pacific coast of, the Japanese islands, *J. Phys. Earth*, *22*, 325-342, 1974.

Uyeda, S. and H. Kanamori, Back-arc opening and the mode of subduction, *J. Geophys. Res.*, *84*, 1049-1061, 1979.

Vassiliou, M., and B.H. Hager, Subduction zone earthquakes and stress in slabs, *PAGEOPH*, *128*, 547-624, 1988.

Welc, J.L., and T. Lay, The source rupture of the great Banda Sea earthquake of Nov. 4, 1963, *Phys. Earth Planet. Int.*, *45*, 242-254, 1987.

L. J. Ruff, Dept. of Geological Sciences, University of Michigan, Ann Arbor, MI, 48109-1063.

What Controls The Seismogenic Plate Interface In Subduction Zones?

Larry J. Ruff and Bart W. Tichelaar[1]

Department of Geological Sciences, University of Michigan, Ann Arbor, Michigan

The greatest earthquakes occur on the seismogenic plate interface of subduction zones. We need to understand what controls the updip and downdip edges of this seismogenic zone. For the circum-Pacific subduction zones that generate great earthquakes, the downdip edge is at a depth of about 40 km, but with significant variations. Several mechanisms might control this transition from seismogenic to aseismic slip. *Tichelaar and Ruff* [1993] argue that one or two critical temperatures can explain the global observations. The model with two critical temperatures invokes two different upper plate rock types, crust and mantle rocks. In this paper, we investigate the correlation between the location of the downdip edge with the location of the coastline. We find a statistically significant correlation between these two variables for the major circum-Pacific subduction zones. Is this correlation a coincidence, or is it indicative of deeper fundamental processes? We offer a simple unifying explanation: the intersection of the overlying plate's Moho with the top of the subducting slab determines both the downdip edge of the seismogenic zone, and the coastline above. This explanation implies that rocks in the subduction zone mantle wedge are aseismic.

1. INTRODUCTION

Most of the world's great earthquakes are interplate underthrusting events in subduction zones. Generations of seismologists and geologists have speculated on what controls the occurrence of these great earthquakes. One key aspect is the fact that just a small fraction of the plate interface is seismogenic. It is important to understand the physical mechanisms that control the updip and downdip extent of the seismogenic plate interface, in addition to the along-strike segmentation of the seismogenic zone. There are two parts of this paper: a short review of the characteristics of the seismogenic plate interface; and a "suggestion" that a rather obvious feature is correlated with the downdip edge of the seismogenic plate interface.

[1] Now at Shell Research, PO Box 60, 2280 AB, Rijswijk, Netherlands

2. LARGE UNDERTHRUSTING EARTHQUAKES

Plate tectonics provides the kinematic explanation for great underthrusting earthquakes; they represent the subduction of oceanic lithosphere beneath the overlying plate. An important discovery was that deep Wadati-Benioff zone earthquakes were intraplate events within the subducted plate [*Isacks and Molnar*, 1971], and that only the shallow earthquakes occurred on the plate interface. On the other hand, these shallow underthrusting earthquakes dominate global seismicity [*Kelleher et al.*, 1973; *Kanamori*, 1977]. Numerous detailed studies of subduction zone seismicity and focal mechanisms show that the dip angle of the subducting plate is typically 20° or so for the shallow seismogenic plate interface, then steepens at greater depths. Another important realization was that subduction zones are quite diverse, and we can exploit the variation in characteristics to better understand the subduction process [e.g., *Uyeda and Kanamori*, 1979]. Application of this "comparative subductology" approach has strengthened the notion that such diverse features as subducting lithosphere age, convergence rate, seismicity, sediment subduction, volcanism, and back arc spreading, may be related to each other [see review by *Jarrard*, 1986].

To focus on underthrusting earthquakes, an important task has been to "map" the seismogenic plate interface; both its along-strike and along-dip extent and characteristics. To explain many features of seismicity, seismologists have invoked various heterogeneities within the seismogenic plate interface [see review in *Scholz*, 1990]. The simplest characterization of heterogeneity is a binary classification of strong and weak sub-regions within the seismogenic plate interface. Dependent on the particular role that they play in the earthquake cycle, the strong areas have been called barriers, asperities, and patches. There are many suggestions that the along-strike distribution of asperities may control several features of seismicity [see discussions in *Kanamori*, 1986; *Thatcher*, 1990; *Ruff*, 1992]. Other papers in this volume will further discuss asperities and the along-strike segmentation of the plate interface.

3. DOWN-DIP EDGE OF SEISMOGENIC INTERFACE

The downdip width of the seismogenic plate interface also plays a key role in subduction seismicity. In general, the plate interface is seismogenic between two depths, referred to here as the "updip edge" and "downdip edge" of the seismogenic plate interface. In some subduction zones, the updip edge extends to the surface at the trench axis. Dependent on the nature of trench sediments, the updip edge may be several kilometers deep [*Byrne et al.*, 1988; *Pacheco et al.*, 1993]. The updip edge is particularly important for tsunami generation.

Recent specialized studies of the downdip edge now offer a sharper view of the global variations in depth to the downdip edge. For example, *Tichelaar and Ruff* [1993] found that the downdip edge in most of the subduction zones is at a depth of 40 ± 5 km (Figure 1). They found a few places that have a slightly deeper downdip edge, northern Chile [also see *Comte and Suarez*, 1995] and the Honshu-Hokkaido corner, but perhaps the most significant variation is the shallow depth of about 25 km for the downdip edge in Mexico. It is important to understand what controls the plate interface transition from seismogenic to aseismic. Inspired by the accepted explanation of a critical temperature that controls the seismicity cut-off depth in strike-slip environments [*Chen and Molnar*, 1983; *Bergman and Solomon*, 1988; *Wiens and Stein*, 1983], recent studies have focused on subduction zone thermal models to control the downdip edge depth.

Tichelaar and Ruff [1993] explored the hypothesis that a critical temperature could explain both the average depth and the depth variations in the down-dip edge of the coupled interface. They used observations of the down-dip edge from the "coupled" subduction zones of Figure 1 (i.e.,

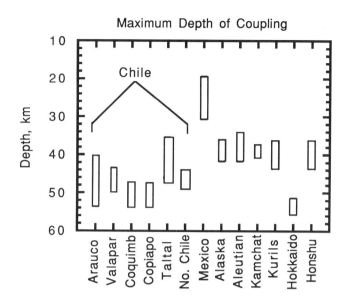

Fig. 1. Summary of seismological determinations of the depth to seismogenic interface downdip edge. See *Tichelaar and Ruff* [1993] for details of the definitions and determinations.

zones with great interplate earthquakes). To test this hypothesis, the temperature at the plate interface must be calculated down to the depth of the down-dip edge. Thermal calculations in subduction zones must include many processes and parameters, though the key unknown is frictional heating along the plate interface [see *Molnar and England*, 1990; *Kao and Chen*, 1991; *Peacock et al.*, 1994; *Molnar and England*, 1995; and *Peacock*, this volume]. Another important parameter is the radiogenic heat production in the overlying wedge [*Furukawa and Uyeda*, 1989]. *Tichelaar and Ruff* [1993] use the heat flow observations to invert for the frictional shear stress and, in turn, obtain the temperature at the down-dip edge of the seismogenic zone. For the subduction zones of Figure 1, they find the global average frictional shear stress to be:

(i) 14 ± 5 MPa, assuming stress is constant throughout the seismogenic zone;

(ii) 30 ± 5 MPa, assuming a constant friction coefficient, with stress increasing linearly through the seismogenic zone.

We now focus on the global variation of temperatures at the down-dip edge. *Tichelaar and Ruff* [1993] found two different distributions for the two different assumed stress functions (see Figure 2). For a constant stress, they found a distribution of downdip edge temperatures with a single peak at 250 °C; while for the constant coefficient of friction, they found a bimodal distribution of down-dip edge temperatures with peak values at 400 °C and 550 °C.

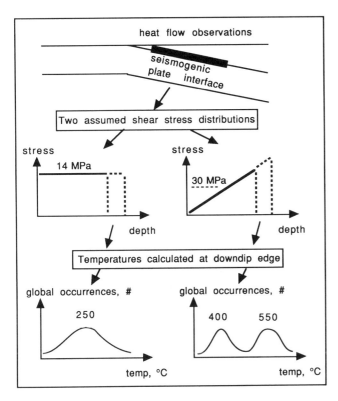

Fig. 2. Schematic representation of the modeling results of *Tichelaar and Ruff* [1993]. The global average level of frictional shear stress is determined for two different stress distributions: constant stress, and linear increase with depth. The stylized histograms show the subsequent results for model temperatures at the downdip edge for the circum-Pacific subduction zones.

Tichelaar and Ruff [1993] preferred the latter case, and explained the two different critical temperatures with two different rock types.

Control of the seismogenic downdip edge by one or more critical temperatures is consistent with observations. However, the scientific procedure that we must follow is to test all possible explanations, and try to eliminate all explanations except one. At this point, we have merely tested just one of the possible explanations. To adopt a critical view of the critical temperature hypothesis, we should not accept this hypothesis until it survives a competition against other hypotheses [see *Tichelaar and Ruff*, 1993, for discussion of other mechanisms and statistics of temperature distributions]. While the notion of a critical temperature has some intuitive appeal, we must continue to test other possibilities. We shall pursue a different explanation that appears whimsical at first glance, but will lead us back to the potential role of rock composition.

4. DOES THE COASTLINE CONTROL THE DOWNDIP EDGE OF SUBDUCTION EARTHQUAKES?

Maps of the downdip edge of the seismogenic plate interface show a rather curious coincidence. Projection of the downdip edge onto the surficial features show that the coastline and downdip edge are in close proximity for many subduction zones. In particular, the coastline and downdip edge plot close together for much of the Chile, Colombia-Ecuador, Mexico, and northern Honshu subduction zones [see detailed maps in *Tichelaar and Ruff*, 1993]. In some subduction zones, the coastline is complex and hence difficult to define (e.g., Alaska and southern Chile). In island arc subduction zones, the downdip edge plots close to the edge of the island arc platform rather than at the coastline of the volcanic islands. Is this observation indicative of some causal connection, or simply another demonstration of a coincidence with no significance? First, we shall test whether this apparent coincidence can be explained by random chance. Of course we expect these two properties of subduction zones - a coastline and the downdip edge - to closely parallel each other. At a global scale, the curves for the downdip edge and coastline overlap, while at the local scale the curves might be offset by many kilometers. How close should they be? We have used the trench-to-volcanic arc distance as the "measuring stick" to assess how close together are the coastline and downdip edge (see Figure 3). This choice is based on the fact that both the coastline and downdip edge must be located somewhere between the trench and volcanos. Hence, we normalize the trench-to-downdip edge (*Edge*) and trench-to-coastline (*Coast*) distances by the trench-to-volcano distance (*Volcano*). Table 1 lists the *Edge*, *Coast*, and *Volcano* distances for the well-determined subduction zone segments in *Tichelaar and Ruff* [1993]. For quantitative comparison, we use the ratios *Edge/Volcano* and *Coast/Volcano*, which must lie between 0.0 and 1. We find considerable variation in these ratios from about 0.3 to 0.9; and there is a significant correlation between the coastline and downdip edge ratios (Figure 4). If the downdip edge was always located directly beneath the coastline, then all points in Figure 4 would fall on the diagonal line. A least-squares line fit to all observations has a correlation coefficient of 0.77, but the slope differs from one. To probe the deviations, the individual subduction zones are plotted with different symbols in Figure 4. The wide bars for Aleutian-Alaska reflect the ambiguity over the coastline: do we choose the beach of Kodiak Island or Alaska Peninsula? One way to resolve this ambiguity is to replace the coastlines with the continental shelf breaks - which may have a deeper physical significance as well - but

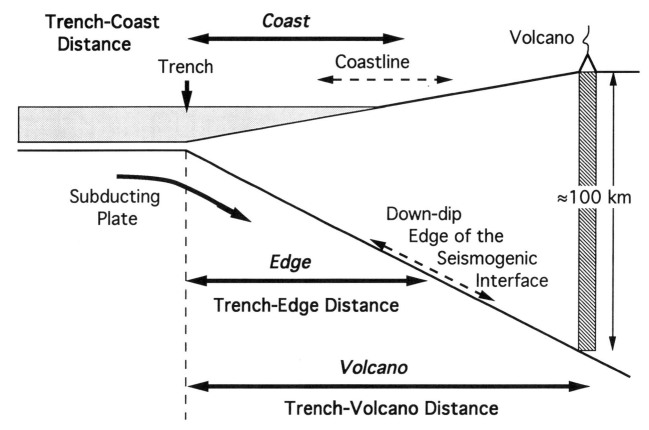

Figure 3. Sketch showing how we measure the coastline and the downdip edge. We use the Trench-Volcano distance (*Volcano*) as our measuring stick. Both the coastline and downdip edge are between the trench and volcanic arc, as shown in the above cross-section. We use ratios of the Trench-Coast (*Coast*) and Trench-Edge (*Edge*) distances to the Trench-Volcano distance.

we leave this refinement for future work. Overall, note the Chilean subduction zone plots above the reference diagonal (downdip edge is landward of coastline), while Kuriles-Kamchatka plots below the line (downdip edge is oceanward of coastline). Thus, there is a systematic tectonic component to the scatter from the reference line.

Since this statistical evidence for a correlation between the coastline and downdip edge is based on very few examples, one may wish to dismiss it as a coincidence. Even so, it is interesting to ponder possible physical connections between the locations of the coastline and the downdip edge. We offer the suggestion that crustal thickness provides a simple connection between the coastline and the downdip edge of the seismogenic interface. This is illustrated in Figure 5, where we show a cross-section that emphasizes the density structure of the ocean to continent transition. Recall that in a passive margin setting, the coastline is located where the crust has thickened to its typical continental value of about 40 km. Let us simplify the density structure of a subduction zone to only three materials: crust, mantle, and water. Then, we would expect to see the coastline approximately above the point of intersection of the continental Moho with the subducting oceanic Moho, as depicted in Figure 5. While the trench, sediments, and deeper mass excesses all affect the details of subduction zone structure and compensation, the simple picture of Figure 5 is still approximately correct. Thus, we can understand why the coastline should be above the intersection of the continental Moho and the subducting slab, but why would this intersection point also correspond to the downdip edge of the seismogenic plate interface? To carry this idea to its logical conclusion, we would have to contend that the seismogenic plate interface requires the overlying plate to be predominately crustal rocks, and that the encounter with mantle rocks causes the interface to become aseismic. Is this conclusion reasonable? One fact is that crustal rocks are certainly capable of seismogenic behavior, e.g., all the shallow continental earthquakes, in addition to subduction zone shallow seismicity. Mantle rocks might be "stronger" than

TABLE 1. Subduction Zone Profiles and Distances from Trench to:
Down-dip Edge of Coupled Zone; Coastline; and Volcanic Line.

Region	Location of profile (key Lat or Lon)	Trench-to-Edge (km)	Trench-to-Coast (km)	Trench-to-Volcano (km)
No. Honshu	37°N	204.00	222.00	295.00
No. Honshu	38.5°N	178.00	220.00	290.00
No. Honshu	39°N	198.00	198.00	290.00
No. Honshu	40.3°N	195.00	220.00	300.00
Hokkaido	43°N	131.00	158.00	210.00
Kuriles Is.	45°N	106.00	191.00	212.00
Kuriles Is.	47°N	118.00	177.00	200.00
Kamchatka	51°N	170.00	200.00	220.00
Aleutian Is.	175°E	280.00	310.00	320.00
Aleutian Is.	178°W	208.00	208.00	258.00
Aleutian Is.	175°W	170.00	170.00	200.00
Alaska	153°W	244.00	204.00	325.00
Mexico	103°W	85.00	85.00	265.00
Colombia	2°N	100.00	100.00	250.00
Chile	22°S	140.00	132.00	310.00
Chile	25°S	135.00	120.00	300.00
Chile	33°S	130.00	105.00	290.00
Chile	39°S	200.00	150.00	280.00

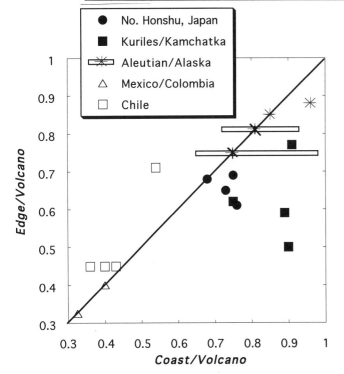

Fig. 4. Observations of the *Edge/Volcano* ratios plotted versus the *Coast/Volcano* ratio. The through-going diagonal line is for the coastline exactly above the downdip edge. Data points are identified by subduction zones, the bars for Aleutian-Alaska is due to uncertainty in coastline choice. The correlation between *Edge* and *Coast* ratios is apparent, and the deviations are consistent within the different subduction zones.

crustal rocks, but perhaps they fail by creep events rather than seismic stick-slip events. On the other hand, transform fault seismicity indicates that oceanic mantle rocks can be seismogenic. Thus, we must argue that rocks in the subduction zone mantle wedge behave different than those beneath fracture zones. It should be emphasized that this contention is pure speculation, but it is weakly supported by the fact that seismicity is not found within the mantle wedge rocks. The notion that mantle wedge rocks could be both strong and aseismic is somewhat counter-intuitive, but is physically possible [see *Scholz*, 1990]. Also, mantle wedge rocks could be altered and thus behave quite different from typical oceanic upper-mantle rocks [*Hyndman et al.*, 1996]. The most direct experiment to test this contention is to make a global map of the intersection of the down-going slab with the overlying plate's Moho, then see if this corresponds to the downdip edge of the seismogenic plate interface.

5. CONCLUSIONS

The greatest earthquakes occur on the seismogenic plate interface of subduction zones, which is a narrow ribbon that wraps around the Pacific basin. We need to understand what controls the updip and downdip edges of this seismogenic zone, in addition to the along-strike segmentation. *Tichelaar and Ruff* [1993] found that the

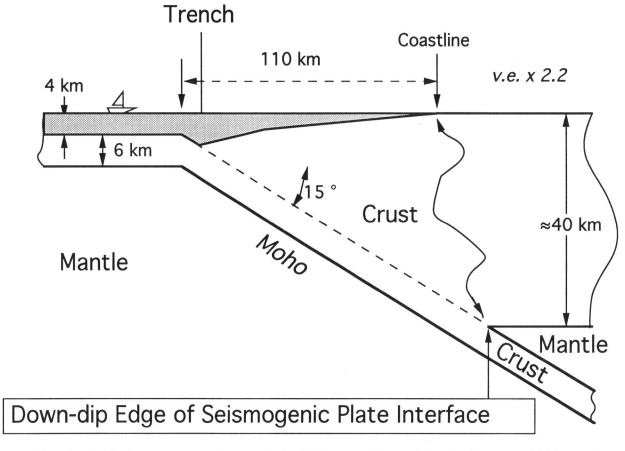

Figure 5. Subduction cross-section down to a depth of 50 km or so. The overlying plate has a crustal thickness of 40 km, a typical continental value. Moho is also indicated in the down-going oceanic slab. Water is shaded. Coastline and downdip edge are marked. Coastline is fixed to the point where crustal rocks achieve 40 km thickness. See text for discussion on the downdip edge of seismogenic zone.

downdip edge is at a depth of about 40 km, but with significant variations. What mechanisms control the transition from seismogenic to aseismic slip? Several mechanisms that might control this transition are: critical temperature, rock composition of the lower plate, the upper plate, or the interface itself. Global observations are satisfied by the hypothesis that one or two critical temperatures control the seismogenic to aseismic transition. The model with two critical temperatures invokes two different upper plate rock types, crust and mantle rocks. The new research result in this paper is a test of the correlation between the location of the downdip edge with the location of the coastline. We find a statistically significant correlation between these two locations for the major circum-Pacific subduction zones. Is this correlation a coincidence, or is it indicative of deeper fundamental processes? We offer one suggestion for a simple unifying explanation: the intersection of the overlying plate's Moho with the top of the subducting slab determines both the coastline and the downdip edge of the seismogenic zone. This particular suggestion of a rock-composition control of the seismogenic zone may be wrong, but we must test and eliminate all competing hypotheses before we can conclude that a critical temperature controls the downdip edge of the seismogenic plate interface.

Acknowledgments. Thanks to S. Kirby, W. Thatcher, D. Comte, and an anonymous reviewer for useful comments, and to R. Hyndman for advance copies of their work. Thanks to the organizers and participants of SUBCON for a wonderful subduction retreat. Earthquake studies at the University of Michigan supported by National Science Foundation (EAR94-0553).

REFERENCES

Bergman, E., and S. Solomon, Transform fault earthquakes in the North Atlantic: source mechanisms and depth of faulting, *J. Geophys. Res., 93,* 9027-9057, 1988.

Byrne, D., D. Davies, and L. Sykes, Loci and maximum size of thrust earthquakes and the mechanics of the shallow region of subduction zones, *Tectonics*, *7*, 833-857, 1988.

Chen, W.P., and P. Molnar, Focal depths of intracontinental and intraplate earthquakes and their implications for the thermal and mechanical properties of the lithosphere, *J. Geophys. Res.*, *88*, 4183-4214, 1983.

Comte, D., and G. Suarez, Stress distribution and geometry of the subducting Nazca plate in northern Chile using teleseismically recorded earthquakes, *Geophys. J. Int.*, *122*, 419-440, 1995.

Furukawa, Y., and S. Uyeda, Thermal structure under the Tohoko arc with consideration of crustal heat generation, *Tectonophysics*, *164*, 175-187, 1989.

Hyndman, R., M. Yamano, and D. Oleskevich, The seismogenic zone of subduction thrust faults, *Island Arc*, in press, 1996.

Isacks, B., and P. Molnar, Distribution of stresses in the descending lithosphere from a global survey of focal mechanism solutions of mantle earthquakes, *Rev. Geophys. Space Phys.*, *9*, 103-174, 1971.

Jarrard, R., Relation among subduction parameters, *Rev. Geophys.*, *24*, 217-284, 1986.

Kao, H., and W.P. Chen, Earthquakes along the Ryukyu-Kyushu arc: strain segmentation, lateral compression, and the thermo-mechanical state of the plate interface, *J. Geophys. Res.*, *96*, 21443-21485, 1991.

Kanamori, H., The energy release in great earthquakes, *J. Geophys. Res.*, *82*, 2981-2987, 1977.

Kanamori, H., Rupture process of subduction zones earthquakes, *Annu. Rev. Earth Planet. Sci.*, *14*, 293-322, 1986.

Kelleher, J., L. Sykes, and J. Oliver, Possible criteria for predicting earthquake locations and their application to major plate boundaries of the Pacific and Caribbean, *J. Geophys. Res.*, *78*, 2547-2585, 1973.

Molnar, P., and P. England, Temperatures, heat flux, and frictional stress near major thrust faults, *J. Geophys. Res.*, *95*, 4833-4856, 1990.

Molnar, P., and P. England, Temperatures in zones of steady-state underthrusting of young oceanic lithosphere, *Earth Planet. Sci. Lett.*, *131*, 57-70, 1995.

Pacheco, J., L. Sykes, and C.H. Scholz, Nature of seismic coupling along simple plate boundaries of the subduction type, *J. Geophys. Res.*, *98*, 14,133-14,159, 1993.

Peacock, S.M., T. Rushmer, and A.B. Thompson, Partial melting of subducting oceanic crust, *Earth Planet. Sci. Lett.*, *121*, 227-244, 1994.

Ruff, L.J., Asperity distributions and large earthquake occurrence in subduction zones. *Tectonophysics*, *211*, 61-83, 1992.

Thatcher, W., Order and diversity in the modes of circum-Pacific earthquake recurrence. *J. Geophys. Res.*, *95*:, 2609- 2623, 1990.

Tichelaar, B., and L. Ruff, Depth of seismic coupling along subduction zones, *J. Geophys. Res.*, *98*, 2017-2037, 1993.

Scholz, C., *The mechanics of earthquakes and faulting*, 439 pp., Cambridge University Press, New York, 1990.

Uyeda, S., and H. Kanamori, Back-arc opening and the mode of subduction, *J. Geophys. Res.*, *84*, 1049-1061, 1979.

Wiens, D.A., and S. Stein, Age dependence of oceanic intraplate seismicity and implications for lithosphere evolution, *J. Geophys. Res.*, *88*, 6455-6468, 1983.

L. J. Ruff, B. W. Tichelaar, Dept. of Geological Sciences, University of Michigan, Ann Arbor, MI, 48109.

Displacement Partitioning and Arc-Parallel Extension: Example From the Southeastern Caribbean Plate Margin

Hans G. Avé Lallemant

Department of Geology and Geophysics, Rice University, Houston, Texas 77005-1892

Focal-mechanism studies of earthquakes in the forearc of the southern Lesser Antilles and in northeastern Venezuela [*Russo et al.*, 1992, 1993] and kinematic studies of deformation structures in Late Cretaceous and Tertiary allochthonous terranes along the Caribbean/South American plate boundary in Venezuela suggest that at any site in arcuate, obliquely convergent plate boundary zones displacement partitioning results in three modes of deformation which may operate simultaneously: (1) plate-boundary- or arc-(sub)normal contraction, (2) plate-boundary- (arc-) parallel shear, and (3) plate-boundary- (arc-) parallel stretching. The latter mode of deformation may contribute to the ascent, decompression, and exhumation of high-pressure/low-temperature metamorphic assemblages typical of subduction zones (blueschists and eclogites).

1. INTRODUCTION

It is well known that in obliquely convergent plate boundary zones the relative convergence vector is generally partitioned into a plate-boundary- or arc-(sub)normal thrust component and an arc-parallel strike-slip component [e.g. *Ekström and Engdahl*, 1989; *McCaffrey*, 1991; *Kelsey et al.*, 1995]. Typically, the arc-normal component is expressed in the forearc and subduction zone by thrust faults and folds trending (sub)parallel to the arc, whereas the arc-parallel component results in one or more arc-parallel strike-slip faults. Thus, the two modes of deformation are operative simultaneously, but, generally, in separate areas in the subduction zone/forearc complex. It has also been shown [e.g. *Walcott*, 1978; *Kelsey et al.*, 1995] that an arc-parallel belt in which arc-normal contraction has occurred, subsequently may become a belt in which only arc-parallel shear occurs, and vice versa.

Where, in addition, the plate boundary is arcuate and convex toward the subducting plate, arc-parallel stretching (extension) may take place resulting in normal faults perpendicular to the plate boundary, as has been demonstrated in the Kuril arc [*Kimura*, 1986], the Ryukyu arc [*Kuramoto and Konishi*, 1989], and the Aleutian arc [*Geist et al.*, 1988]. It has been proposed [*Avé Lallemant and Guth*, 1990; *Avé Lallemant*, 1991] that this extension is the result of the increase of the arc-parallel strike-slip component of the convergence vector and causes thinning of the forearc terrane, and consequently decompression and uplift of deep-seated rocks such as eclogites and blueschists. In the rare instances where the arc is concave toward the subducting plate, arc-parallel contraction may occur (e.g. where the Kuril arc turns into Japan [*Kimura*, 1986]).

It can be demonstrated in the Lesser Antilles and in northeastern Venezuela that displacement partitioning and arc-parallel extension occur at present as suggested by earthquake focal-mechanism studies [*Russo et al.*, 1992, 1993] and have occurred as well in Cretaceous and Tertiary time [*Avé Lallemant*, 1991; *Avé Lallemant and Sisson*, 1993].

2. TECTONIC SETTING

The Caribbean/South American plate boundary zone in northern Venezuela is about 300 km wide. It consists of the following three east-west-trending lithotectonic belts: (1) the Leeward volcanic arc in the north, (2) the Caribbean Mountains system, and (3) the Serranía del Interior foreland fold and thrust belt in the south (Figure 1). The Caribbean Mountains system has been divided into several sub-belts of which only the Cordillera de la Costa and Araya-Paria belts will be discussed here.

The Leeward volcanic arc is the southern extension of the Lesser Antilles arc and the remnant Aves Ridge arc.

Magmatic activity in the Leeward Antilles started in mid-Cretaceous time [e.g. *Pindell*, 1993]. Magmatism ceased gradually from west to east: at about 85 Ma on the Dutch Leeward Antilles [*Priem et al.*, 1979] and at about 45 Ma on the Los Testigos Islands (Figure 1) [*Santamaría and Schubert*, 1974], but is, of course, still active in the Lesser Antilles.

The Cordillera de la Costa belt consists mainly of high to intermediate grade metamorphic rocks (epidote amphibolite and greenschist facies) which include eclogite and blueschist knockers [e.g. *Maresch*, 1975; *Guth and Avé Lallemant*, 1991]. The age of the epidote amphibolite metamorphism is about 85 Ma (amphibole and white mica $^{40}Ar/^{39}Ar$ ages [*Avé Lallemant*, unpublished data]).

The Araya-Paria belt consists mainly of low-grade (lower greenschist facies) metamorphic rocks. The age of metamorphism is about Oligocene [*Avé Lallemant*, unpublished $^{40}Ar/^{39}Ar$ data] to Early Miocene [*Foland et al.*, 1992].

The Serranía del Interior fold and thrust belt consists of Cretaceous and Paleocene passive margin sedimentary rocks. They were folded and thrust south- southeastward since Early Miocene time [*Rossi et al.*, 1987; *Bally et al.*, 1995].

The boundaries between all belts were originally east- to northeast-trending, south-to southeast-vergent thrust faults. Many of these faults were reactivated as dextral strike-slip faults.

In all recent plate tectonic models, the Caribbean has formed in the Pacific realm as part of the Farallon plate [e.g. *Pindell*, 1993]. In mid-Cretaceous time the plate started moving toward the northeast to southeast in between the North and South American plates [*Pindell*, 1993]. The direction of the present relative plate motion is not well constrained.

2.1 *Structural Geology*

Deformational structures in the Caribbean Mountains system formed during two tectonic events. The first (D_1) is synmetamorphic and is related to subduction of the Proto-Caribbean lithosphere beneath the Caribbean plate and the second (D_2) is postmetamorphic and related to obduction of the Caribbean Mountains system onto the South American craton.

The D_1 structures in the Cordillera de la Costa belt originated during the upward (retrograde) migration of the unit from great (eclogite facies) to intermediate (blueschist facies) depth, and subsequently to relatively shallow (epidote amphibolite to greenschist facies) depth [e.g. *Avé Lallemant and Sisson*, 1993]. As eclogites and blueschists occur only as knockers, no kinematic interpretation of their internal structure is possible. The structures in the epidote amphibolite- and greenschist-grade rocks are penetrative and occur in the entire Cordillera de la Costa belt from Puerto Cabello to the Araya Peninsula (Figure 1). They are metamorphic cleavages, isoclinal folds, and lineations (stretching, mineral, and intersection lineations), all trending east to northeast parallel to the plate boundary and presumably formed by plate-boundary-normal contraction (Figure 2). Asymmetric stretched particles (Figure 2B), pressure shadows (Figure 2C), and rotated boudins (Figure 2A) indicate a component of dextral shear roughly parallel to the plate boundary. Rotation may have occurred simultaneously with contraction [e.g. *Tikoff and Fossen*, 1993], but, in some cases, it may have postdated the formation of the cleavage (Figure 2A). Plate-boundary-parallel extension may also have occurred during cleavage formation (Figure 2D), but often is clearly later (Figure 2E).

The structures in the low-grade Araya-Paria belt are virtually identical to those in the Cordillera de la Costa belt. However, they seem to have formed in the Oligocene whereas in the Cordillera de la Costa belt these structures formed in mid-Cretaceous time.

D_2 structures in the metamorphic belt are correlated with the structures in the Serranía del Interior fold and thrust belt. They are characterized by south- to southeast-vergent thrust faults, east-trending dextral strike-slip faults, and southeast-trending normal faults. They formed during the Eocene in the west and Miocene in the east [*Pindell*, 1993], but seismic activity indicates that these structures are still active. Focal-mechanism studies [e.g. *Russo et al.*, 1992, 1993] indicate that the thrust faults, strike-slip faults, and normal faults are all active, but operate at different localities (Figure 1).

However on 2, 4, and 6 October, 1957, three strong earthquakes (M = 5.5, 6.7, and 5.1 respectively) occurred in the southern Lesser Antilles (Figure 1) [*Russo et al.*, 1992]. The first might have formed by oblique displacement along an arc-parallel dextral strike-slip fault; the second by oblique slip along an arc-parallel thrust fault; and the third by slip along an arc-normal normal fault. The epicenters of the three faults are located within a circle with a radius of 8 km; the focal depths are 10, 6, and 10 km, respectively [*Russo et al.*, 1992]. Thus, the three modes of deformation (arc-normal contraction, arc-parallel shear, and arc-parallel extension) operated simultaneously and virtually in the same area, but on different faults.

3. MODEL FOR DISPLACEMENT PARTITIONING

3.1 *Oblique Plate Convergence*

A model for displacement partitioning in a right-oblique subduction zone/forearc terrane/volcanic arc system, as ap-

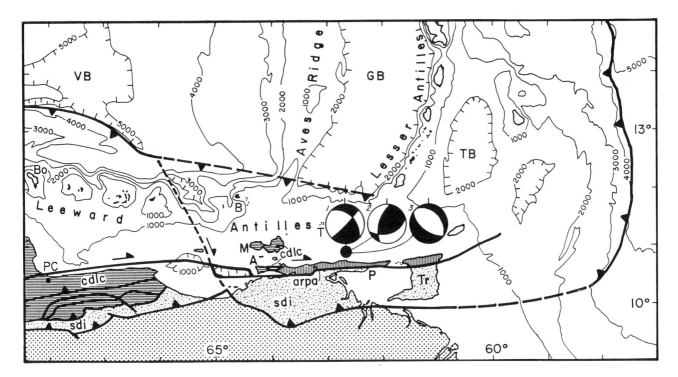

Fig. 1. Map of the southeastern Caribbean [after *Case and Holcombe*, 1980]. Horizontal ruling: Caribbean Mountains system with cdlc = Cordillera de la Costa belt; vertical ruling: Araya-Paria belt (arpa); fine stipple pattern: Serranía del Interior foreland fold and thrust (sdi); coarse stipple pattern = foreland basins. Geographic abbreviations: A = Araya Peninsula; B = La Blanquilla Island; Bo = Bonaire (Dutch Leeward Antilles); GB = Grenada Basin; M = Margarita Island; P = Paria Peninsula; PC = Puerto Cabello; T = Los Testigos Islands; TB = Tobago Basin; Tr = Trinidad; VB = Venezuelan Basin. Bathymetric contours in meters. Focal-mechanism solutions *1*, *2*, and *3* (equal-area, lower-hemisphere projections) for three earthquakes which occurred on 2 (*1*), 4 (*2*), and 6 (*3*) October 1957 in small area, north of the Paria Peninsula, and at depths between 6 and 10 km [*Russo et al.*, 1992]; black fields = compressional quadrants.

plicable to the southeastern Caribbean plate, is shown in Figure 3. An oceanic plate converges on a volcanic arc at a rate of **V** cm/year. The subduction zone (SS), forearc terrane, and volcanic arc have a rectilinear map trace between A and B and the obliquity α is 40°. The segment BC is arcuate and the obliquity α increases from 40° to 85°. The trace of segment CD is rectilinear and the obliquity remains at 85° (the plate boundary along CD is almost a transform fault).

The relative plate-convergence vector **V** is partitioned into a subnormal component ($\mathbf{V_n}$) and a tangential one ($\mathbf{V_t}$). The normal component results in arc-subparallel thrust faults in the forearc region and earthquake focal mechanisms are as in *1* in Figure 3. The arc-parallel component causes right-lateral strike-slip displacement along the arc-parallel fault ABCD (earthquake mechanisms as in *2* in Figure 3). The forearc terrane may move at a rate $\mathbf{V_t}$ toward the SW. In the arcuate segment BC, $\mathbf{V_t}$ increases and the forearc terrane migrates at increasing rate toward the SW and W, causing arc-parallel extension by arc-perpendicular normal faulting (*3* in Figure 3). In this segment strike-slip deformation is increasing and arc-normal contraction becomes less important. Along segment CD, $\mathbf{V_n}$ is very small, the arc may become extinct, and the forearc terrane may migrate westward at the total convergence rate **V**.

In several arcuate volcanic island arcs (e.g. Aleutian arc [*Geist et al.*, 1988]) arc-parallel strike-slip faulting occurs within the arc or in the backarc (PQ in Figure 3). Here the arc complex is extended parallel to itself and deep basins form between the islands.

Generally, at shallow depths in the brittle regime, the three modes of deformation are simultaneously active, but at very different sites in the forearc: thrusting occurs typically in the frontal part of the forearc; strike-slip displacements occur generally more toward the active arc; and extension occurs only in the arcuate portion of an arc over-printing the

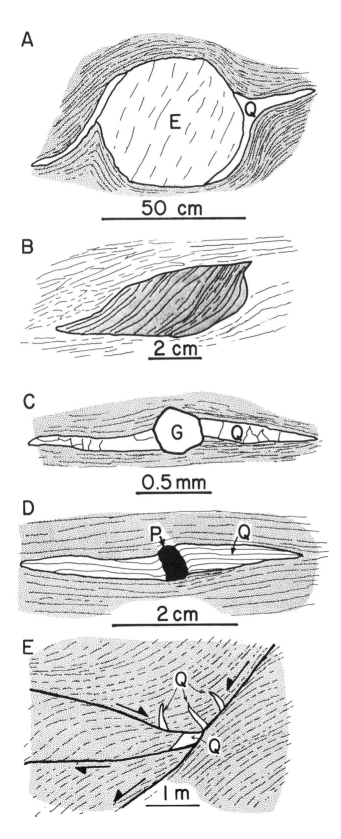

previous two modes of deformation. In contrast, the earthquakes of 1957 in the southern Lesser Antilles (Figure 1) occurred virtually simultaneously at approximately the same spot, suggesting that the three faults are related and merging into a single décollement surface [e.g. *Oldow et al.*, 1991].

At greater depth, in the ductile regime, displacement partitioning may not occur along an obliquely convergent boundary. A particle may be displaced parallel to the convergence vector, but due to the boundary conditions (a high viscosity contrast between the two convergent plates), this particle may undergo strain partitioning: boundary-normal shortening, boundary-parallel shear [*Tikoff and Fossen*, 1993], and boundary-parallel extension.

3.2 Exhumation of Eclogites and Blueschists

Many models for the uplift, decompression, and exhumation of high-pressure metamorphic rocks formed in subduction and collision zones have been proposed. The model, as discussed above, adds another mechanism. Plate-boundary-parallel stretching in the Venezuelan eclogite- and blueschist-bearing Cordillera de la Costa belt occurred simultaneously with decompression [*Avé Lallemant and Sisson*, 1993] suggesting that the stretching has caused thinning of the forearc terrane and, thus, decompression; simultaneous uplift may have been the result of isostatic rebound. This mechanism may be applicable to many forearc terranes as most at least modern volcanic arcs are arcuate and convex toward the subducting plate.

3.3 Tectonic Evolution of the Southeastern Margin of the Caribbean Plate

According to most plate models [e.g. *Pindell*, 1993] the Caribbean plate was part of the Farallon plate in mid-Cretaceous and earlier time. Eastward subduction occurred along the entire west coast of North, Central, and South America. At about 120 to 100 Ma the subduction polarity along the Central American portion of the plate boundary reversed and the Caribbean/Farallon plate with its fringing

Fig. 2. Sketches after photographs (A, B, D, and E) and one photomicrograph (C) of D_1 structures from the Cordillera de la Costa belt, Venezuela; all figures are oriented with West to the left and East to the right. (A) rotated boudin of eclogite (E) with quartz-filled pressure shadows (Q) in mica schist; (B) boudin of graphite schist in mica schist; (C) quartz (Q) pressure shadows around garnet (G) in mica schist; (D) quartz (Q) pressure shadows around pyrite (P) grain in phyllite; (E) east- and west-dipping semibrittle normal faults in phyllite with quartz veins (Q).

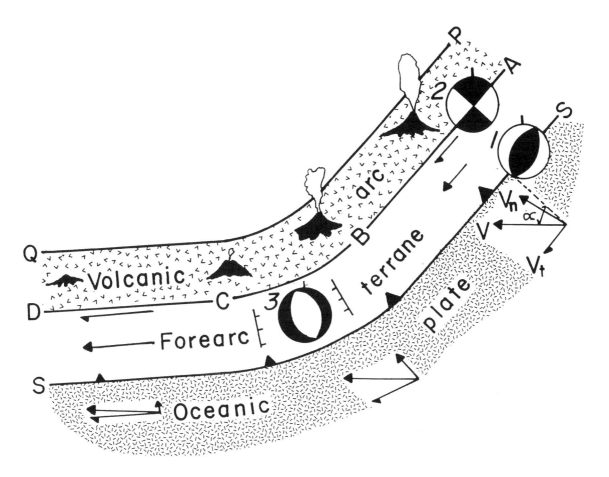

Fig. 3. Hypothetical map of volcanic island arc/forearc/subduction zone complex and model for displacement partitioning; hypothetical focal mechanisms as in Figure 1 (see text for explanation).

volcanic arc migrated northeastward in between the North and South American plates. In the southern part of the subduction zone plate convergence was highly right-oblique and the obliquity increased continuously. The D_1 structures in the Cordillera de la Costa belt formed here while the eclogites and blueschists slowly ascended and were decompressed. In Eocene time the Caribbean plate collided with the Bahamas and started moving east-southeastward. The Cordillera de la Costa terrane was carried passively to the east-southeast until parts of it were thrust (D_2) southward onto the South American continent; parts of the belt continued moving east-southeastward with the Caribbean plate along east-trending dextral strike-slip faults.

Meanwhile minor subduction along the southern part of the Lesser Antilles arc continued; the Araya-Paria terrane was subducted to only shallow depths and ascended during Oligocene time; deformation structures indicate that it was deformed in a similar tectonic environment as the Cordillera de la Costa belt. In the Miocene the Araya-Paria belt and a portion of the older Cordillera de la Costa belt were thrust southward onto South America.

The east-southeast migration of the Caribbean plate continues today. D_1 structures may still form at depth. Earthquake focal-mechanism studies clearly indicate that displacement partitioning takes place in the arc, but also in the thrust belt.

The structural evolution of the Cretaceous Cordillera de la Costa belt, the Tertiary evolution of the Araya-Paria belt, and the recent activity in the southern Lesser Antilles are an example of extreme diachronism of deformation resulting from highly oblique plate convergence.

Acknowledgments. This study was supported by National Science Foundation grants EAR-8517383, EAR-9019243, and EAR-9304377. This paper benefited enormously from critical reviews by Susan Cashman, Chris Goldfinger, and Richard Sedlock.

REFERENCES

Avé Lallemant, H.G., The Caribbean-South American plate boundary, Araya Peninsula, eastern Venezuela, *Trans. 12th Caribbean Geol. Conf.*, edited by D.K. Larue and G. Draper, pp. 461-471, Miami Geol. Soc., 1991.

Avé Lallemant, H.G., and L.R. Guth, Role of extensional tectonics in exhumation of eclogites and blueschists in an oblique subduction setting: Northwestern Venezuela, *Geology, 18*, 950-953, 1990.

Avé Lallemant, H.G., and V.B. Sisson, Caribbean-South American plate interactions: Constraints from the Cordillera de la Costa belt, Venezuela, in *Mesozoic and Early Cenozoic development of the Gulf of Mexico and the Caribbean region, Proc. 13th Ann. Res. Conf.*, edited by J.L. Pindell and B.F. Perkins, pp. 211-219, Gulf Coast Section S.E.P.M., 1993.

Bally, A.W., J. Di Croce, R.A. Ysaccis, and E. Hung, The structural evolution of the East Venezuela transpressional orogen and its sedimentary basins, *Geol. Soc. Am. Abstr. with Progr., 27*, A-154, 1995.

Case, J.E., and T.L. Holcombe, Geologic-Tectonic Map of the Caribbean Region, scale 1:2,500,000, *Misc. Invest. Ser. Map I-1100*, U.S. Geol. Surv., Reston, Va., 1980.

Ekström, G., and E.R. Engdahl, Earthquake source parameters and stress distribution in the Adak Island region of the central Aleutian Islands, Alaska, *J. Geophys. Res., 94*, 15,499-15,519, 1989.

Foland, K.A., R. Speed, and J. Weber, Geochronologic studies of the hinterland of the Caribbean Mountains of Venezuela and Trinidad, *Geol. Soc. Am. Abstr. with Progr., 24*, A-148, 1992.

Geist, E.L., J.R. Childs, and D.W. Scholl, The origin of summit basins of the Aleutian Ridge: Implications for block rotation of an arc massif, *Tectonics, 7*, 327-341, 1988.

Guth, L.R., and H.G. Avé Lallemant, A kinematic history for eastern Margarita Island, Venezuela, *Trans. 12th Caribbean Geol. Conf.*, edited by D.K. Larue and G. Draper, pp. 472-480, Miami Geol. Soc., 1991.

Kelsey, H.M., S.M. Casman, S. Beanland, and K.R. Berryman, Structural evolution along the inner forearc of the obliquely convergent Hikurangi margin, New Zealand, *Tectonics, 14*, 1-18, 1995.

Kimura, G., Oblique subduction and collision: Forearc tectonics of the Kuril arc, *Geology, 14*, 404-407, 1986.

Kuramoto, S., and K. Konishi, The southwest Ryukyu arc is a migrating microplate (forearc sliver), *Tectonophysics, 163*, 75-91, 1989.

Maresch, W.V., The geology of northeastern Margarita Island, Venezuela: a contribution to the study of Caribbean plate margins, *Geol. Rundschau, 64*, 846-883, 1975.

McCaffrey, R., Slip vectors and stretching of the Sumatran forearc, *Geology, 19*, 881-884, 1991.

Oldow, J.S., A.W. Bally, and H.G. Avé Lallemant, Transpression, orogenic float, and lithospheric balance, *Geology, 18*, 991-994, 1991.

Pindell, J.L., Regional synopsis of Gulf of Mexico and Caribbean evolution, in Mesozoic and Early Cenozoic Development of the Gulf of Mexico and Caribbean Region, *Proc. 13th Ann. Res. Conf.*, edited by J.L. Pindell and B.F. Perkins, pp. 251-274, Gulf Coast Section S.E.P.M., 1993.

Priem, H.N.A., P.A.M. Andriessen, D.J. Beets, N.A.I.M. Boelrijk, E.H. Hebeda, E.A.T. Verdurmen, and R.H. Verschure, K-Ar and Rb-Sr dating in the Cretaceous island-arc succession of Bonaire, Netherlands Antilles, *Geologie en Mijnbouw, 58*, 367-373, 1979.

Rossi, T., J.-F. Stephan, R. Blanchet, and G. Hernandez, Etude géologique de la Serranía del Interior Oriental (Venezuela) sur le transect Cariaco-Maturín, *Revue Inst. franc. pétrole, 42*, 3-30, 1987.

Russo, R.M., E.A. Okal, and K.C. Rowley, Historical seismicity of the southeastern Caribbean and tectonic implications, *Pure Appl. Geophys., 139*, 67-120, 1992.

Russo, R.M., R.C. Speed, E.A. Okal, J.B. Shepherd, and K.C. Rowley, Seismicity and tectonics of the southeastern Caribbean, *J. Geophys. Res., 98*, 14,299-14,319, 1993.

Santamaría, F., and C. Schubert, Geochemistry and geochronology of the southern Caribbean - northern Venezuela plate boundary, *Geol. Soc. Am. Bull., 85*, 1085-1098, 1974.

Tikoff, B., and H. Fossen, Simultaneous pure and simple shear: the unifying deformation matrix, *Tectonophysics, 217*, 267-283, 1993.

Walcott, R.I., Geodetic strains and large earthquakes in the axial tectonic belt of North Island, New Zealand, *J. Geophys. Res., 83*, 4419-4429, 1978.

Hans G. Avé Lallemant, Department of Geology and Geophysics, Rice University, Houston, Texas 77005-1892.

Thermal and Petrologic Structure of Subduction Zones

Simon M. Peacock

Department of Geology, Arizona State University, Tempe, Arizona

The subduction of oceanic lithosphere depresses isotherms on a regional scale resulting in large thermal anomalies in the upper mantle. The thermal structure of a subduction zone depends on many parameters including the thermal structure (age) of the incoming lithosphere, convergence rate, geometry of subduction, radioactive heating, induced convection in the overlying mantle wedge, and rate of shear heating along the subduction shear zone. Numerical calculations suggest that subduction shear-zone temperatures beneath the volcanic front generally lie between 500 and 700 °C, but considerable uncertainty in the thermal structure of subduction zones results from uncertainties in the rate of shear heating and mantle-wedge convection. Low forearc heat flow (25-50 mW/m^2), seismic coupling to depths of ~40 km, and high-pressure, low-temperature metamorphic rocks all indicate that subduction zones are cool and constrain subduction zone shear stresses in the brittle (frictional) regime to 10-30 MPa or 1-5% of lithostatic pressure. In most subduction zones, the subducting oceanic crust passes through the blueschist → eclogite metamorphic facies transition where continuous dehydration reactions may release large amounts of H$_2$O. Integrated over time, aqueous fluids released from the subducting slab cause extensive hydration of the overlying mantle wedge and trigger partial melting in the core of the convecting mantle wedge. Partial melting of subducting oceanic crust occurs only under rare circumstances such as near the subduction of a spreading ridge. Uncertainties in the petrologic structure of subduction zones result from uncertainties in the thermal structure, the distribution of hydrous minerals in the oceanic lithosphere, and possible kinetic barriers to metamorphic reactions. Detailed seismological investigations that illuminate the velocity and attenuation structure of the subducting slab and mantle wedge have the potential to better define the thermal and petrologic structure of subduction zones.

INTRODUCTION

Much of Earth's geologic activity occurs in subduction zones where oceanic lithosphere descends into the mantle. Spectacular arc volcanism is associated with most subduction zones, despite the pronounced cooling of the mantle by subducting lithosphere. The largest historic earthquakes (1960 Chile, 1964 Alaska) have occurred along subduction zones and in many western Pacific subduction zones seismicity extends from the surface to nearly 700 km depth. Arc magmatism, subduction-zone earthquakes, and other subduction-zone processes such as crust-mantle recycling are intimately linked to the thermal and petrologic evolution of subduction zones. In this contribution, I discuss (1) the thermal structure of subduction zones based on two-dimensional heat-transfer models; (2) constraints on thermal models provided by geological and geophysical observations; and (3) the petrologic structure of subducting oceanic crust and the overlying mantle wedge.

THERMAL STRUCTURE OF SUBDUCTION ZONES

Analytical vs. Numerical Solutions

The complex nature of subduction zone processes, as illustrated in Figure 1, rule out the possibility of deriving exact analytical solutions that describe the thermal structure of a subduction zone. However, accurate analytical approximations exist to calculate temperatures at shallow

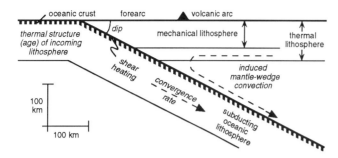

Figure 1. Schematic cross section showing important features and processes that determine the thermal structure of a subduction zone.

levels in a subduction zone and numerical solutions can yield accurate results at deeper levels.

At depths less than 50 km, subduction-zone temperatures may be readily calculated using analytical expressions derived by *Molnar and England* [1990]. Steady-state temperatures along the subduction shear zone (i.e., the slab-mantle interface) are closely approximated by [*Molnar and England*, 1990, eqns. 16 and 23]:

$$T = \frac{(Q_0 + Q_{sh}) z_f / k}{S} \qquad (1)$$

where T = temperature (K), Q_0 = basal heat flux (W/m^2), Q_{sh} = rate of shear heating (W/m^2), z_f = depth to the fault (m), k = thermal conductivity (W/(m-K)), and S = a divisor that accounts for advection given by:

$$S = 1 + b \sqrt{(V z_f \sin\delta) / \kappa} \qquad (2)$$

where b = a constant ($\approx$1 based on numerical experiments), V = convergence rate (m/s), δ = angle of subduction, and κ = thermal diffusivity (m^2/s). Equations (1) and (2) show that higher convergence rates and steeper subduction angles lead to cooler subduction shear-zone temperatures. Conversely, warmer incoming slabs (higher Q_0) and higher rates of shear heating lead to warmer subduction shear-zone temperatures. Thus, we would expect the Mariana subduction zone to be relatively cool compared to the Cascadia subduction zone. Additional analytical expressions presented by *Molnar and England* [1990] and *Peacock* [1992, 1993] allow temperatures to be calculated within the subducting slab and the overriding plate.

At depths greater than 50 km, subduction-zone temperatures may be more accurately predicted by numerical models [e.g., *Peacock et al.*, 1994] that simulate induced flow in the mantle wedge and the transition from brittle to plastic rheology in the subduction shear zone. A range of numerical modeling techniques, including finite-difference, finite-element, and spectral methods, can be used to solve the time-dependent heat transfer equation:

$$\frac{\partial T}{\partial t} = \kappa \nabla^2 T - \vec{v} \cdot \nabla T + \frac{A}{\rho C} \qquad (3)$$

where T = temperature (K), t = time (s), κ = thermal diffusivity (m^2/s), v = velocity field (m/s), A = volumetric heat production (W/m^3), ρ = density (kg/m^3), and C = heat capacity (J/(kg-K)). The three terms on the right side of equation (3) represent heat conduction, heat advection and heat sources/sinks, respectively.

The numerical model used in this paper is described in detail in *Peacock* [1991] and *Peacock et al.* [1994]. Briefly, equation (3) is solved using finite-difference techniques within a 400-km by 600-km grid oriented parallel to the subduction shear zone. The angle of subduction is fixed at 26.6° which approximates the average dip of Wadati-Benioff zones from 0 to 100 km depth [*Jarrard*, 1986]. The initial thermal structure of the subducting lithosphere is defined by *Stein and Stein*'s [1992] global depth and heat flow model (GDH1); for most simulations the age of the subducting lithosphere was fixed at 50 Ma. Below the base of the lithosphere we use an adiabatic gradient of 0.3°C/km. A steady-state 200 Ma geotherm is used for the initial thermal structure of the hanging wall. Flow in the mantle wedge is simulated using *Batchelor*'s [1967] analytical solution for two-dimensional incompressible, constant-viscosity flow in a corner driven by constant slip equal to the slab velocity at the top of the subducting oceanic lithosphere. A no-slip boundary condition is used at the base of the mechanical (rigid) lithosphere defined as the 1000 °C geotherm (equal to a depth of ~65 km). Each numerical simulation was run until all temperatures within 10 km of the subduction shear zone changed < 1% over a 1 m.y. interval (steady state).

Important Variables

The subduction of cool oceanic lithosphere exerts a primary control on the thermal structure of subduction zones by depressing isotherms in the region of the subducting slab (Fig. 2). On a regional scale, isotherms are subparallel to flow lines in the subducting slab and the overlying mantle wedge. In steady state, inverted isotherms (hotter rocks lying above cooler rocks) occur only within certain parts of the convecting system, i.e., near the top of the subducting slab and the base of the convecting mantle wedge. Within the "rigid" part of the overriding plate (mechanical lithosphere), inverted isotherms may occur during the early (transient) stages of subduction, but thermal steady state is characterized by temperatures that increase monotonically downward.

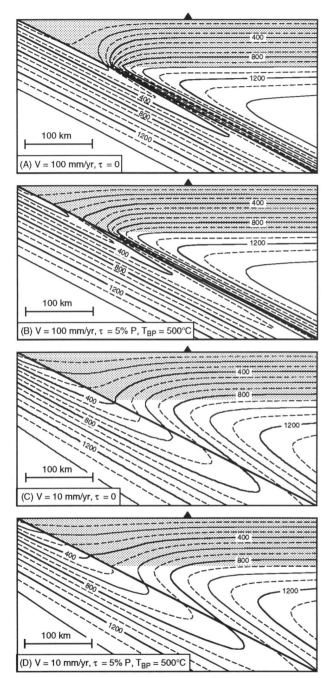

Figure 2. Steady-state subduction-zone thermal structures calculated using a two-dimensional numerical model for four different sets of parameters. Age of incoming lithosphere = 50 Ma. Solid and dashed lines are isotherms labeled in °C. Shaded area represents rigid, 65-km-thick, mechanical lithosphere. Solid triangles represents location of a typical arc volcano located 125 km above the top of the subducting slab. (A) $V = 100$ mm/yr, $\tau = 0$ (no shear heating). (B) $V = 100$ mm/yr, $\tau = 5\%\ P$, $T_{BP} = 500$ °C. (C) $V = 10$ mm/yr, $\tau = 0$ (no shear heating). (D) $V = 10$ mm/yr, $\tau = 5\%\ P$, $T_{BP} = 500$ °C.

Subduction zone models presented over the past 25 years demonstrate that the thermal structure of a subduction zone depends on numerous parameters including: the thermal structure (age) of the incoming lithosphere, the convergence rate, the geometry of subduction, the distribution of radiogenic heat-producing elements, the rate of shear heating along the subduction shear zone, and the geometry and vigor of induced convection in the overlying mantle wedge (Fig. 1) [e.g., *Oxburgh and Turcotte*, 1970; *Hasebe et al.*, 1970; *Toksöz et al.*, 1971; *Turcotte and Schubert*, 1973; *Anderson et al.*, 1978; *Honda and Uyeda*, 1983; *Hsui et al.*, 1983; *Wang and Shi*, 1984; *Cloos*, 1985; *Honda*, 1985; *van den Beukel and Wortel*, 1988; *Peacock*, 1990a, b; 1991; *Dumitru*, 1991; *Davies and Stevenson*, 1992; *Staudigel and King*, 1992; *Furukawa*, 1993; *Peacock et al.*, 1994]. The results of selected numerical models are presented in Table 1. These examples illustrate the most important factors that determine the thermal structure of a subduction zone are the rate of shear heating along the subduction thrust and induced convection in the mantle wedge. In the absence of shear heating, estimates of the temperature in the subduction shear zone at 100 km depth range from 300 to 750 °C (Table 1, shear stress = 0). The lowest calculated temperatures result from experiments in which the mantle wedge is assumed to be rigid. Shear heating, caused by shear stresses of the order of 100 MPa, dramatically increase calculated temperatures in the subduction shear zone to values in excess of 1000 °C at 100 km depth [e.g., *Toksöz et al.*, 1971; *Turcotte and Schubert*, 1973].

The thermal effects of convergence rate, slab age, and induced mantle convection are depicted in Figure 3 based on the numerical model described above. In the absence of shear heating, faster convergence rates result in cooler subduction shear-zone temperatures (Fig. 3B). In contrast, for high subduction-zone shear stresses, faster convergence rates result in higher rates of shear heating and warmer subduction shear-zone temperatures (Fig. 3B). The thermal structure (age) of the oceanic lithosphere prior to subduction also influences the thermal structure of the subduction zone; younger subducting slabs result in warmer subduction shear-zone temperatures (Fig. 3C).

The subduction of oceanic lithosphere induces convection in the overlying mantle wedge [*McKenzie*, 1969]. From a thermal point of view, the base of the mantle wedge becomes part of, and subducts with, the cool slab (Fig. 2). Induced mantle convection, which brings warmer mantle material into close proximity to the top of the subducting slab, increases temperatures in the subduction shear zone by 200-350 °C (Fig. 3D). Analytical and numerical solutions predict similar subduction shear-zone temperatures at shallow depths (Fig. 3D), but at deeper levels the effects of

TABLE 1. Comparison of selected subduction-zone thermal models.

Reference	Computational method, spatial resolution	Mantle flow model	Slab dip (°)	Convergence rate (mm/yr)	Shear stress (MPa)	Temp. at 100 km[a] (°C)	Comments
Toksöz et al. [1971]	Finite difference 10 km	No induced mantle-wedge flow	45 45 45 45 45	10 10 80 80 80	0 504 0 63 252	350 1250 300 520 900	Experiments with shear heating also include adiabatic compression, phase changes, and radioactivity.
Honda [1985]	Finite difference 5 km	T- and P-dependent viscosity	27 27	100 100	0 100	750 1050	Rigid mantle wedge corner.
Davies and Stevenson [1992]	Finite element 3 to 40 km	Constant viscosity	30 60	72 72	0 0	600 400	No frictional heating or viscous dissipation.
Furukawa [1993]	Finite difference 5 km	T- and P-dependent viscosity	30 60 30 60	40 40 80 80	0 0 0 0	650 620 620 600	No frictional heating along the slab interface.
Peacock et al. [1994]	Finite difference 1 to 5 km	Analytical corner flow solution	27 27 27 27 27 27	10 10 30 30 100 100	0 100 0 100 0 100	545 710 485 800 475 990	No viscous dissipation within mantle wedge. Additional calculations evaluate slab age, transient regimes, and different shear heating models.

[a] Temperature at the slab-mantle interface at 100 km depth. The interpolation of temperature values from contour diagrams in the original references results in uncertainties of approximately ±50 °C.

induced mantle convection, which are currently simulated only in numerical models, must be included.

Shear Heating

Much of the variation in subduction zone thermal structures presented in the literature results from different rates of shear heating (Table 1). Equations (1) and (2) explicitly show that temperatures in the subduction shear zone reflect a trade-off between shear heating and advective cooling, which increase and decrease temperatures, respectively. The rate of shear heating (Q_{sh}) is given by:

$$Q_{sh} = \tau V \quad (4)$$

where τ = shear stress (Pa) and V = convergence rate (m/s) [*Turcotte and Schubert*, 1973]. Convergence rates for modern subduction zones are well constrained, but the magnitude of subduction-zone shear stresses is poorly known. Recent estimates of shear stresses in subduction shear zones, based on surface heat flow measurements and petrologic arguments, range from ~100 MPa [*Scholz*, 1990; *Molnar and England*, 1990] to several tens of MPa [*van den Beukel and Wortel*, 1988; *Peacock*, 1992; *Tichelaar and Ruff*, 1993] to approximately zero [*Hyndman and Wang*, 1993]. To illustrate the potential importance of shear heating, consider that for V = 50 mm/yr and τ = 40 MPa, Q_{sh} = 60 mW/m^2, which is comparable to the average conductive heat flow out of oceanic crust. For plate tectonic rates of 10-100 mm/yr, advective cooling and shear heating are approximately balanced for average shear stresses of ~20 MPa. If average shear stresses along the slab-mantle interface are < 20 MPa, then faster convergence rates lead to cooler subduction shear-zone temperatures; if average shear stresses are > 20 MPa, then faster convergence rates lead to warmer shear zone temperatures.

In many thermal models of subduction zones, one of two different shear stress formulations are employed: (A)

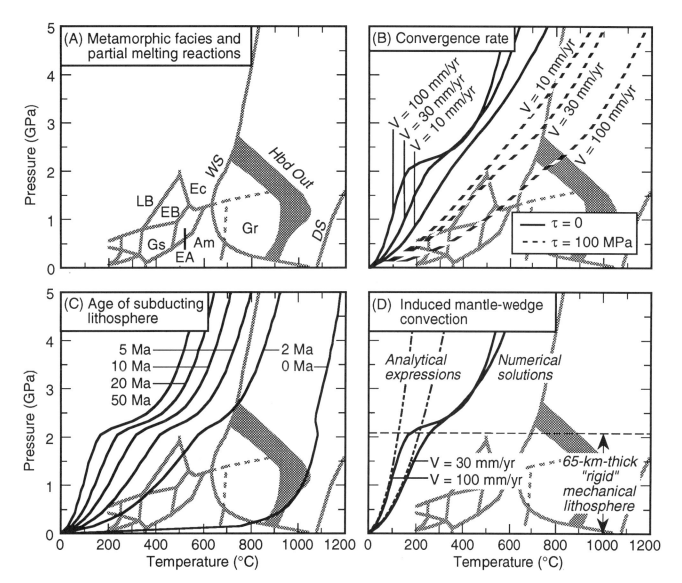

Figure 3. Steady-state pressure-temperature (*P-T*) conditions along the subduction shear zone calculated using a two-dimensional numerical model [*Peacock et al.*, 1994] overlain on basaltic phase equilibria diagram. (A) Metamorphic facies and partial melting reactions for basaltic compositions taken from sources described in *Peacock* [1993] and *Peacock et al.* [1994]: Am, amphibolite facies; DS, dry basaltic solidus; EA, epidote-amphibolite facies; EB, epidote-blueschist facies; Ec, eclogite facies; Gr, granulite facies; Gs, greenschist facies; Hbd out, fluid-absent partial melting region associated with breakdown of hornblende; LB, lawsonite-blueschist facies; WS, wet basaltic solidus. (B) Subduction shear zone *P-T* conditions for three different convergence rates (10, 30, and 100 mm/yr) and two different shear stresses (0, 100 MPa). Rapid increase in temperature at $P \sim 2.3$ GPa for the $\tau = 0$ curves results from induced convection in the mantle wedge. Age of incoming lithosphere = 50 Ma. (C) Subduction shear zone *P-T* conditions as a function of age of the subducting lithosphere. V = 100 mm/yr, $\tau = 0$. (D) Calculated steady-state *P-T* conditions along a subduction shear zone based on analytical expressions [*Molnar and England*, 1990] and numerical calculations [*Peacock et al.*, 1994]. Analytical and numerical solutions agree well at depths less than 65 km ($P < 2.1$ GPa) where the subducting lithosphere is in contact with the rigid mechanical lithosphere of the overriding plate. At depths greater than 65 km induced convection in the mantle wedge, which is simulated only in the numerical model, raises temperatures in the subduction shear zone 250-350 °C.

constant shear stress (τ = constant); or (B) shear stresses increasing linearly with lithostatic pressure (P) or depth ($\tau = \gamma P$ where γ is a proportionality constant). These simple formulations permit analytical approximations to be derived. The constant shear stress formulation (A) provides a feel for average shear stresses in a subduction zone, but conflicts with our knowledge that stresses vary with depth. Shear stress formulation (B) approximates brittle (frictional) deformation. Calculated steady-state subduction shear-zone temperatures for $V = 100$ mm/yr are shown for the two different shear stress formulations in Figs. 4A and 4B. Higher shear stresses result in higher rates of shear heating and warmer subduction shear-zone temperatures.

Neither shear stress formulation (A) nor (B) accurately simulates the transition from brittle (frictional) to plastic (creep) behavior as temperature increases in the subduction shear zone with increasing depth. In the plastic regime, rock strength and the maximum shear stress that can be supported drops dramatically with increasing temperature [e.g., *Kirby*, 1983]. Because subduction shear zones may occur in metasedimentary, metabasaltic, or hydrous ultramafic rocks, the choice of an appropriate plastic flow law is uncertain. Rather than specifying the material in the subduction shear zone, I choose to define the temperature of the brittle-plastic transition, T_{BP}, above which shear stresses decrease exponentially with increasing temperature according to the relation:

$$\tau = \tau_{BP} \exp[(T - T_{BP})/L] \quad (5)$$

where τ_{BP} = shear stress (Pa) at the brittle-plastic transition, T_{BP} = temperature (K) of the brittle-plastic transition, and L = characteristic 1/e relaxation scale (fixed at 75 K). This simple expression closely approximates the power-law rheology of plastic (ductile) materials [*Peacock et al.*, 1994]. Calculated subduction shear-zone temperatures for three different values of T_{BP}: 300°C (wet quartzite), 500°C (dry diabase, marble, and dry quartzite), and 800°C (dry dunite), are depicted in Fig. 4C. The increase in shear heating that occurs in the brittle regime drops off rapidly at $T > T_{BP}$. Partial melting temperatures are achieved in the shear zone only if the brittle-plastic transition lies close to the melting temperature of the rocks.

The range in predicted subduction shear-zone temperatures depicted in Fig. 4 illustrates the importance of constraining the rate of shear heating in subduction zones. Later in this paper, I describe different types of data that constrain the thermal structure of subduction zones. Most data suggest shear stresses in subduction zones lie in the range of 10 to 30 MPa or a few percent of lithostatic pressure. Given shear stresses in this range, I present a range in probable subduction shear-zone temperatures in Fig. 4D for $V = 10$, 30, and 100 mm/yr. For each convergence rate, the range in subduction shear-zone temperatures is bracketed by two shear-heating scenarios. Complete two-dimensional thermal structures for $V = 10$ and 100 mm/yr and the two bracketing shear stress end-members are depicted in Fig. 2. Predicted temperatures in the subduction shear zone at depths of 100-125 km range between 500 and 700 °C suggesting that the top of the subducting slab lies at subsolidus conditions beneath the volcanic front. Calculated temperatures within the upper part of the subducting slab below the subduction shear zone are even cooler (Fig. 2).

Induced Flow in the Mantle Wedge

Critical to our understanding of arc magma genesis is the thermal structure of the mantle wedge which is controlled primarily by the vigor and geometry of induced flow in the mantle wedge above the subducting slab. Flow in the mantle wedge is driven by viscous coupling to the subducting slab, thermal buoyancy, and petrologic buoyancy and has been investigated analytically and numerically by *McKenzie* [1969], *Andrews and Sleep* [1974], *Bodri and Bodri* [1978], *Marsh* [1979], *Anderson et al.* [1980], *Hsui et al.* [1983], *Honda* [1985], *Davies and Stevenson* [1992], and *Furukawa* [1993] among others. All of these studies predict qualitatively similar flow fields, but significant differences in the calculated thermal structure arise from different boundary conditions and viscosity formulations. Qualitatively, induced convection in the mantle wedge brings warm mantle material into close proximity to the subducting slab; induced convection warms the subducting slab at the expense of cooling the adjacent mantle wedge. Most models suggest that induced mantle-wedge convection heats the top of the slab by several hundred degrees; for example, the two-dimensional model described above suggests induced convection heats the top of the slab from ~250 °C to ~600 °C at a depth of 120 km (Fig. 3D). A cool boundary layer forms in the mantle wedge adjacent to the slab, the thickness of which depends primarily on the viscosity of the mantle wedge, the subduction velocity, and thermal parameters. Dynamical calculations by *Kincaid* [1995] which solve for the mantle-wedge flow field using a temperature-dependent rheology show that the cool boundary layer (viscous blanket) keeps the subducting slab at subsolidus conditions.

Within a subduction zone, viscosity varies over many orders of magnitude making dynamical calculations computationally difficult, particularly at the corner of the mantle wedge and at the boundary between the high-viscosity "rigid" subducting plate and the low-viscosity convecting mantle wedge. In contrast to most mantle-

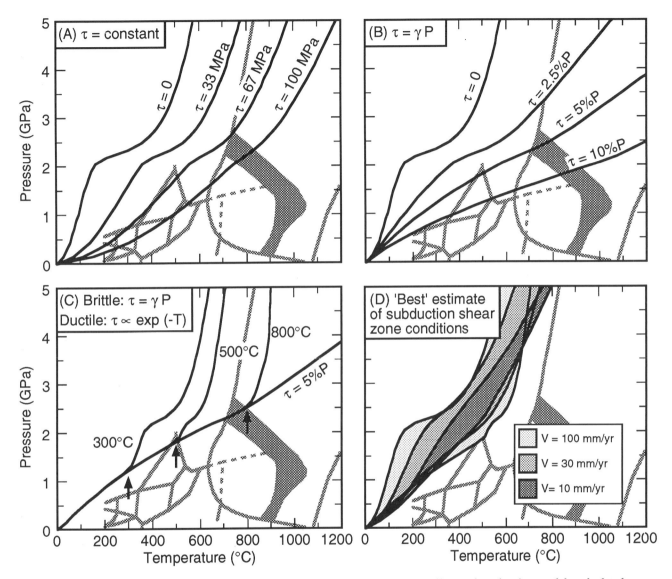

Figure 4. Steady-state (P-T) paths along the subduction shear zone for different shear heating models calculated using a two-dimensional numerical model [*Peacock et al.*, 1994]. Age of incoming lithosphere = 50 Ma. See Figure 3A for description of metamorphic facies and partial melting reactions. (A) Subduction shear zone P-T conditions calculated for different values of constant shear stress along the subduction shear zone. $V = 100$ mm/yr. (B) Subduction shear zone P-T conditions calculated for different rates of shear heating where shear stresses increase linearly with pressure. $V = 100$ mm/yr. (C) Subduction shear zone P-T conditions calculated for different rates of shear heating where shear stresses increase linearly with pressure in the brittle regime and decrease exponentially with increasing temperature in the plastic (ductile) regime. P-T paths depart from the $\tau = 0.05\ P$ curve at the temperature of the brittle-plastic transition (300, 500, or 800 °C) marked by arrows. $V = 100$ mm/yr. (D) Estimated range in steady-state P-T conditions along subduction shear zones for $V = 10$, 30, and 100 mm/yr. For each convergence rate, the shaded region is bounded by two numerical experiments: (1) no shear heating and (2) relatively high shear heating where $\tau = 5\%\ P$ for $0 < T < 500$ °C (brittle-plastic transition) and shear stresses decrease exponentially for $T > 500$ °C [*Peacock et al.*, 1994]. See text for discussion.

wedge calculations, *Marsh* [1979] argues that induced convection can effectively bring hot rocks into contact with the subducting slab, raising the temperature of the top of the slab to ~1400 °C at 120 km depth. This conflicting result suggests that the thermal effects of induced flow in the mantle wedge are still uncertain. Future work will also need to evaluate the effects of hydration and partial melting on the rheology of the mantle wedge.

Additional Uncertainties in the Thermal Models

Additional variables that lead to significant uncertainties in calculating the thermal structure of subduction zones include hydrothermal circulation in the oceanic lithosphere, the variation of thermal parameters with T and P, the amount of heat consumed and released by metamorphic reactions, and the effects of fluid flow at shallow depths.

In most thermal models of subduction zones, the thermal structure of the subducting lithosphere is defined by a purely conductive plate model. Hydrothermal circulation near spreading ridges removes substantial amounts of heat from the upper parts of the oceanic lithosphere [e.g., *Stein and Stein*, 1994], but this effect is not generally taken into account in subduction zone models. For example, the very warm P-T paths depicted in Figure 3C for young subducting slabs (curves labeled 0, 2, and 5 Ma) assume conductive geotherms for the incoming oceanic lithosphere. Because hydrothermal circulation cools the incoming lithosphere, actual P-T paths for young slabs may be substantially cooler than the paths depicted in Figure 3C.

In most thermal models thermal diffusivity (or thermal conductivity) is assumed to be constant throughout the model. Experimental data show that the thermal diffusivity of geologic materials decreases significantly with increasing temperature [*Clauser and Huenges*, 1995]. The decrease in thermal diffusivity with increasing temperature may be largely balanced by the increase in radiative heat transport. Because the effective length scale of radiative heat transport in the mantle may be limited by scattering at grain boundaries and absorption by Fe-bearing minerals, radiative heat transfer may be effectively modeled as a diffusive heat transport mechanism. I am not aware of any thermal modeling study that systematically explores the uncertainties introduced by variations in thermal parameters.

Metamorphic dehydration reactions in the subducting slab may consume significant amounts of heat [*Anderson et al.*, 1976, 1978; *Delany and Helgeson*, 1978]. Similarly, hydration reactions in the overlying mantle wedge may release significant amounts of heat [*Peacock*, 1987]. Together, these two effects will act to retard the thermal evolution of a subduction zone, i.e., endothermic dehydration reactions retard warming of the subducting slab and exothermic hydration reactions retard cooling of the mantle wedge [*Peacock*, 1987]. The magnitude of this effect depends critically on the poorly constrained amount and distribution of hydrous minerals in the oceanic lithosphere prior to subduction (see petrology section below). Thermal calculations by *Anderson et al.* [1978] that assumed a subducting crustal mineralogy containing 6 wt % bound H_2O suggested that the heat consumed by metamorphic reactions could depress isotherms several tens of kilometers. Thermal calculations by *Peacock* [1990b] that assumed 2 wt % bound H_2O in the subducting oceanic crust indicate that metamorphic reactions have only a minor effect on the calculated thermal structure (<5 °C). *Anderson et al.* [1978] assumed that the removal of heat by advecting fluids would effectively double the metamorphic heat sink. Thermal models that explicitly incorporate the flow of metamorphic fluids as an advective heat transfer mechanism do not support this assumption [*Peacock*, 1987].

At shallow depths (<10-20 km) the thermal structure of subduction zones may be perturbed by hydrothermal circulation. Fluid flow in accretionary prisms can strongly perturb the thermal structure because large volumes of pore fluids are expelled during sediment compaction [e.g., *Reck*, 1987; *Wang et al.*, 1993]. Metamorphic reactions also contribute to the fluid flux, particularly at shallow depths where clay minerals dehydrate to form micas. At depths > 50 km, advection of heat by the flow of metamorphic fluids is limited by the small volume of released fluid and does not affect the deep thermal structure significantly [*Peacock*, 1987; 1990b]. Within the arc crust, the thermal structure will be altered by the vigorous hydrothermal circulation associated with the emplacement of magmas which themselves obviously perturb the thermal structure.

CONSTRAINTS ON THE THERMAL STRUCTURE OF SUBDUCTION ZONES

1. Heat Flow Measurements

Surface heat flow measurements in the forearcs of subduction zones typically range from 25 to 50 mW/m^2 reflecting relatively cool temperatures at depth. The low heat flow values indicate that the subduction zone thermal structure is dominated by the advection of cool oceanic lithosphere and not by shear heating. However, forearc heat flow measurements are commonly greater than predicted by simple conduction-advection models of subducting lithosphere requiring a contribution to surface heat flow from shear heating, radiogenic heat production, fluid advection, or the thermal effects of forearc deformation.

Using heat-flow measurements to constrain shear heating in subduction zones suffers from three problems. First, heat flow data typically exhibit large scatter; single values must be interpreted cautiously. Second, heat flow measurements only reflect the local conductive geotherm which may be perturbed by advective processes such as fluid flow, deformation, and erosion/deposition. Third, heat flow data are acquired either in sediments or in relatively shallow boreholes. In order to constrain the rate of shear heating, these data must be extrapolated tens of kilometers downward to the subduction shear zone.

Perhaps the best heat flow data set comes from the northern Cascadia subduction zone, where surface heat flow measurements decrease systematically from ~120 mW/m² in the Cascadia basin to ~80 mW/m² on the slope of the accretionary prism to <40 mW/m² on Vancouver Island (250 km east of the trench) indicating that isotherms dip eastward in the same direction as the subducting slab [e.g., *Hyndman and Wang*, 1993]. Thermal modeling of the Cascadia subduction zone by *Hyndman and Wang* [1993] successfully matches the heat flow constraints without requiring any frictional heating.

Tichelaar and Ruff [1993] inverted heat flow measurements from nine subduction zones characterized by large thrust earthquakes and found that τ = 14 to 27 MPa (constant shear stress) or τ = 0.06 to 0.10 P (shear stress proportional to depth) for different radiogenic heat production models. For the case in which upper-plate radiogenic heating decreases exponentially with depth, *Tichelaar and Ruff* [1993] estimated subduction zone shear stresses at τ = 14 MPa (constant shear stress) and τ = 0.059 P (shear stress proportional to depth).

Using selected heat flow measurements, *Molnar and England* [1990] calculated relatively high shear stresses of 84 ± 18 and 73 ± 17 MPa for the Japan subduction zone and 55 ± 24 MPa for the Peru subduction zone. As recognized by *Molnar and England* [1990], their calculations do not include radioactive heat production which would reduce the calculated shear stresses. *Tichelaar and Ruff* [1993] included radioactive heat production in their inversion of heat flow data and calculated substantially lower shear stresses of 16-31 MPa for the Japan (Honshu) subduction zone. For the Japan subduction zone, *Molnar and England's* [1990] calculation implicitly assumes that the surface heat flow data can be reliably extrapolated downward to depths of 60-70 km.

2. Maximum Depth of Seismic Coupling

Most of the Earth's seismic energy is released in subduction zones where seismic coupling (frictional, stick-slip behavior) extends to greater depths than in other tectonic environments. In subduction zones characterized by large earthquakes, *Tichelaar and Ruff* [1993] determined the maximum depth of the seismic coupling to be 40 ± 5 km as compared to the 15 km depth observed in continental strike-slip fault zones [*Sibson*, 1984]. In continental fault zones, the maximum depth of the seismogenic zone correlates well with the ~300 °C isotherm presumably because the rheologic behavior of quartz changes at T > 300 °C [*Sibson*, 1984]. The maximum depth of the seismic coupling in subduction zones may also be controlled by a critical temperature [*Tichelaar and Ruff*, 1993; *Ruff and Tichelaar*, this volume].

In many subduction zones, weak quartz- and calcite-bearing sediments cap the subducting plate and may control the seismic behavior of the subduction shear zone. The brittle-plastic transition in wet quartzite occurs at ~300 °C [*Kirby*, 1983] for strain rates on the order of 10^{-13}/s. Assuming that the 300 °C isotherm controls the maximum depth of the seismic coupling in subduction zones, then the calculated shear stresses range from τ = 11 to 20 MPa (constant shear stress) or τ = 1.1% to 3.5% P (shear stress proportional to depth) using data for the ten subduction zones considered by *Tichelaar and Ruff* [1993]. Radiogenic heating was not included in these calculations, so these shear stresses should be considered maximum upper bounds.

In a subduction zone, as in other shear zones, deformation will be concentrated in the weakest rocks [e.g., *Yuen et al.*, 1978]. Deformation in most subduction zones probably will be concentrated in the sedimentary layer at the top of the subducting plate. In those subduction zones where all incoming sediments are offscraped or underplated at shallow depths, the shear zone may develop in the mafic subducting oceanic crust. Basaltic rocks are stronger than sediments [e.g., *Kirby*, 1983] and the brittle-plastic transition occurs at higher temperatures. If the critical temperature is substantially greater than 300 °C, then shear stresses calculated assuming T_c = 300 °C will be too low.

As discussed earlier, *Tichelaar and Ruff* [1993] used heat flow data to estimate subduction zone shear stresses of τ = 14 MPa or τ = 0.059 P using a model in which upper plate radiogenic heating decreases exponentially with depth. Using these shear stress values, they calculated the critical temperature (T_c) at the base of the seismogenic zone in different subduction zones, finding $T_c \approx$ 250 °C for a constant shear stress and $T_c \approx$ 400 (thick crust) or $\approx$ 550 °C (thin crust) for shear stresses that increase with depth [*Tichelaar and Ruff*, 1993].

3. High-Pressure, Low-Temperature Metamorphic Rocks

The low temperatures recorded by blueschist-facies metamorphic rocks place an upper bound on the magnitude of shear stresses in subduction zones at depths of 15-50 km [*Peacock*, 1992]. Calculated steady-state *P-T* paths intersect the blueschist facies, broadly defined, for τ = 10 to 60 MPa for V = 100 mm/yr and τ = 0 to 100 MPa for V = 30 mm/yr [*Peacock*, 1992]. If shear stresses increase linearly with depth, blueschist-facies P-T paths require τ = 0.01 to 0.09 P and τ = 0 to 0.14 P for V = 100 and 30 mm/yr, respectively [*Peacock*, 1992]. Inclusion of radiogenic heat production reduces the estimated shear stresses. The common occurrence of blueschist-facies metamorphic rocks in ancient subduction zones suggests that temperatures are relatively cool beneath forearcs, but uncertainties in the convergence rate during blueschist formation translate to considerable uncertainties in calculated shear stresses.

Recently, *Maekawa et al.* [1993] described clasts of blueschist-facies metabasalts that were brought to the surface by a serpentinite diapir in the forearc of the active Mariana subduction zone. Mineral assemblages and mineral compositions suggest peak metamorphic conditions of T = 150-250 °C and P = 0.5-0.6 MPa for the blueschist clasts [*Maekawa et al.*, 1993]. Assuming the blueschists formed in the present-day subduction zone (V = 80 to 100 mm/yr; dip = 12°), the *P-T* conditions require shear stresses of 18±8 MPa in the Mariana subduction zone. This value agrees well with *Bird's* [1978] estimate of τ = 16.5±7.5 MPa for the Mariana subduction zone based on a force-balance calculation. Alternatively, if shear stresses are proportional to depth, then the blueschist *P-T* conditions require τ = 2.4 to 4.9% P in the Mariana subduction zone.

Low-temperature mafic eclogites occur in many paleo-subduction zones and record peak metamorphic conditions of 500-600 °C at 2 GPa based on mineral assemblages and geothermobarometry [e.g., *Carswell*, 1990]. Coesite-bearing eclogites, mostly derived from continental crust, are exposed in the Alpine, Qinling-Dabie-Sulu (China), Caledonian, and Ural orogenic belts and record peak metamorphic temperatures of 550-900 °C at pressures > 2.5 GPa (depth > 90 km). The low temperatures recorded by coesite-bearing eclogites strongly suggest formation in a subduction zone where the downward advection of cool material depresses isotherms on a regional scale [e.g., *Ernst and Peacock*, this volume]. Although we do not know the rate at which the eclogite protoliths were subducted, they confirm that the subducting slab is several hundreds of degrees cooler than the surrounding mantle.

PETROLOGIC STRUCTURE OF SUBDUCTION ZONES

Petrologic models of subduction zones may be constructed by combining calculated thermal structures for the subducting oceanic crust and overlying mantle wedge with phase equilibria for mafic and ultramafic systems [e.g., *Wyllie*, 1979, 1988; *Peacock*, 1993]. The location of phase boundaries and the resultant petrologic structure are sensitive to the thermal structure; in other words, uncertainties in the thermal structure translate to uncertainties in the petrologic structure. In general terms if the subducting oceanic crust is relatively warm, then the subducting oceanic crust may undergo partial melting [*Wyllie*, 1988]. Alternatively, if the subducting oceanic crust is relatively cool, then the subducting oceanic crust will undergo subsolidus metamorphic dehydration reactions rather than partial melting [*Wyllie*, 1988].

Peacock [1990a, 1991, 1993] emphasized the importance of constructing petrologic models based on pressure-temperature (*P-T*) *paths* because rocks move through the subduction system and the thermal structure changes with time during the early stages of subduction. In a mature subduction zone that has reached thermal steady state, the *P-T* path followed by the top of the subducting oceanic crust corresponds to the *P-T* conditions along the subduction shear zone. It takes ~5-20 million years to reach steady-state conditions in a subduction zone. During this early, transient period subduction shear-zone temperatures are slightly warmer than depicted in Figures 3 and 4 and *P-T* paths deviate from the instantaneous *P-T* conditions along the shear zone [*Peacock*, 1992; *Peacock et al.*, 1994].

In steady-state subduction zones characterized by low to moderate rates of shear heating, subducting oceanic crust passes through the blueschist → eclogite metamorphic facies transition (Fig. 4D), consistent with the widespread occurrence of blueschists and eclogites in former convergent plate margins [e.g., *Ernst*, 1973; *Evans and Brown*, 1986]. Calculated *P-T* paths for subducting oceanic crust do not pass through the higher temperature greenschist, amphibolite, or granulite metamorphic facies as suggested by previous workers [e.g., *Wyllie*, 1988]. At the blueschist → eclogite transition hydrous minerals stable in the basaltic oceanic crust, such as lawsonite, sodic amphibole, and chlorite, breakdown via continuous reactions to form a largely anhydrous mineral assemblage of garnet + omphacite (Na-Ca clinopyroxene) (Fig. 5A) [*Peacock*, 1993]. The location of the blueschist → eclogite transition is poorly constrained at P > 2 GPa and no phase equilibria experiments have been conducted at the P > 2 GPa and T < 600 °C. Thermodynamic calculations by *Evans* [1990] suggest that the blueschist → eclogite transition is approximately isothermal at ~500 °C at P > 2 GPa. Recent experimental work [e.g., *Pawley and Holloway*, 1993] demonstrate that several hydrous phases, including lawsonite, mica, and chlorotoid may be stable

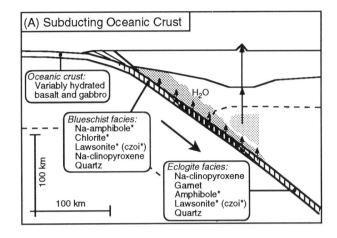

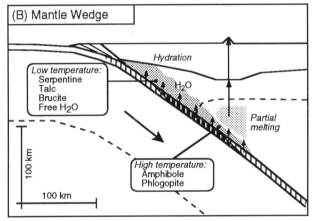

Figure 5. Petrologic model of a mature subduction zone. (A) Mineralogic changes in subducting oceanic crust. Large amounts of H_2O are released by continuous dehydration reactions that occur in subducting oceanic crust during blueschist → eclogite facies metamorphism. Proposed mineralogy of the subducting slab is shown in boxes; hydrous minerals are marked by asterisks. (B) Mineralogic changes in mantle wedge. Integrated over time, H_2O released from the subducting oceanic crust causes extensive hydration of the mantle wedge at shallow depths and adjacent to the subducting slab. Possible hydrous minerals stable at different depths are shown in boxes. Water-rich fluids that infiltrate the core of the convecting mantle wedge may trigger partial melting.

within the eclogite facies (P = 2-3 GPa) at 650 °C. Further experiments should help delineate the specific reactions that occur in subducting oceanic crust and the depth to which H_2O can be subducted.

Several thermal models of subduction zones presented in the early 1970's assumed *a priori* that arc magmas were derived by direct melting of the subducting slab [e.g., *Oxburgh and Turcotte*, 1970; *Turcotte and Schubert*, 1973]. In order to achieve the high temperatures necessary for slab melting, these early thermal models incorporated high rates of shear heating with τ = 100-500 MPa. *Defant and Drummond* [1990] suggest that the trace element and REE patterns of recent andesitic to dacitic volcanics (adakites) present in several modern arcs may represent partial melts derived from the subducting oceanic crust. Partial melting of subducting oceanic crust only occurs in numerical experiments that incorporate high rates of shear heating, which require shear stresses ≥ 100 MPa to be maintained by rocks close to their melting temperature (Fig. 4), or that simulate subduction of very young (<5 Ma) oceanic lithosphere (Fig. 3C) [*Peacock et al.*, 1994]. Other potential source regions for adakites exist besides the subducting slab; if it can be conclusively demonstrated that a specific arc magma was derived from the subducting slab, then we will have a valuable constraint on the thermal structure of the subduction zone.

Over time, large amounts of H_2O-rich fluids, released by blueschist → eclogite dehydration reactions, may cause extensive hydration of the overlying ultramafic mantle wedge (Fig. 5B). *Fyfe and McBirney* [1975] suggested that regions of forearc uplift may be related to the volume increases associated with hydration reactions in the mantle wedge. Serpentinite diapirs in the Mariana and Izu-Bonin forearcs [*Fryer et al.*, 1985] provide direct evidence of mantle-wedge hydration. The petrologic structure of the mantle wedge will depend on the time-integrated P-T conditions, hydration history, and chemistry of the hydrating fluids. At shallow depths the mantle wedge may contain large amounts of serpentine, talc, and chlorite. At deeper levels induced mantle convection results in large temperature gradients in the mantle wedge perpendicular to the subducting slab. Subsolidus conditions occur in the mantle wedge adjacent to the subducting slab where amphibole, chlorite, and possibly phlogopite are important stable hydrous phases capable of incorporating fluids driven out of the underlying slab. Because the mantle wedge is convecting the hydrated base of the mantle wedge will move downward with the slab [e.g., *Tatsumi*, 1989].

Geochemical, isotopic, and petrologic data suggest that most arc magmas are generated by partial melting of the mantle wedge induced by the infiltration of H_2O derived from the subducting slab [e.g., *Gill*, 1981; *Bebout*, 1991; *Davidson*, 1992; *Hawkesworth et al.*, 1993], although a minority view holds that partial melting of the subducting oceanic crust is an important process [e.g., *Marsh*, 1979; *Myers and Johnston*, this volume]. In the core of the convecting mantle wedge beneath the volcanic front, temperatures lie well above the wet peridotite solidus and approach 1100-1200 °C at ~90 km depth in subduction zones with rapid convergence rates (Fig. 2). Certain elements and isotopes that are enriched in arc lavas, such as B and ^{10}Be, are clearly derived from the subducting oceanic crust and sediments [*Morris et al.*, 1990]. How is the distinct geochemical signature of the subducting slab transferred into the core of the mantle wedge where arc magmas appear to form? The most likely transporting agent is an aqueous fluid derived from dehydration reactions in the subducting oceanic crust and sediments. The slab "component" may be transported laterally across the dynamic mantle wedge to the region of arc magma genesis via successive dehydration and hydration reactions [*Davies and Stevenson*, 1992; *Davies*, 1994] or partial melting reactions [*Wyllie*, 1988; *Tatsumi*, 1989].

Uncertainties in the Petrologic Structure

Aside from the uncertainty in the thermal structure, perhaps the biggest uncertainty in predicting the phase changes in the subducting slab and the sites of volatile release is our lack of knowledge regarding the amount and distribution of hydrous phases in the oceanic crust and uppermost mantle prior to subduction. Hydrothermal circulation and submarine weathering add CO_2 and H_2O to the oceanic crust through the formation of carbonates and hydrous minerals such as chlorite, amphibole, and serpentine. Carbonates and hydrous minerals are not distributed homogeneously throughout the oceanic crust [e.g., *Alt et al.*, 1986], but rather occur preferentially in, and adjacent to, fractures and permeable zones. Based primarily on dredge haul samples, *Anderson et al.* [1976] estimated that the basaltic layer of the oceanic crust contains 1.7 to 4.9 wt % H_2O (best estimate = 3.5 wt % H_2O) and that the gabbroic layer contains 1.6 to 5.8 wt % H_2O (best estimate = 2.5 wt % H_2O). Based on analyses of DSDP/ODP drill cores, dredge hauls, and ophiolite samples, *Peacock* [1990a] estimated that the basaltic layer contains an average of ~2 wt % H_2O and the gabbroic layer an average of ~1 wt % H_2O.

The amount and distribution of serpentinized ultramafic rocks in the oceanic crust and uppermost mantle is poorly constrained. Serpentinites have been recovered from oceanic fracture zones and rift valleys [e.g., *Cannat*, 1992], but we have no direct samples of the oceanic mantle. The basal ultramafic section of ophiolites is commonly serpentinized, but much of the serpentinization may have occurred during emplacement of the ophiolite. Heat flow data constrains the amount of hydrothermal circulation through the oceanic lithosphere, but does not constrain the depth of hydrothermal penetration [*Stein and Stein*, 1994]. Because serpentine minerals contain ~13 wt % H_2O bound in the crystal structure, even minor amounts of serpentinite in the oceanic crust and uppermost mantle would significantly increase, and conceivably dominate, the H_2O budget of the subducting slab. Recent experiments by *Ulmer and Trommsdorff* [1995] have shown that serpentine is stable to 620 °C at 5 GPa. Serpentinite, if present, will occur in a relatively cool part of the slab (5-20 km below the subduction shear zone) and may remain stable to depths of 100 km to >200 km based on the thermal structures depicted in Fig. 2.

A second major uncertainty lies in the common assumption that metamorphic reactions in subduction zones occur at the equilibrium phase boundaries in *P-T* space. Because temperatures in the subducting slab are quite cold, the possibility exists that reactions may be kinetically hindered. Laboratory experiments have demonstrated that reactions involving only solid phases, such as the important eclogite-forming reaction albite → jadeite + quartz, require substantial overstepping in *P-T* space in order to proceed [*Hacker et al.*, 1992a, b]. In contrast, reactions proceed rapidly in the laboratory in the presence of a free fluid phase. Deep-focus earthquakes may be caused by transformational faulting caused by metastable overstepping of the olivine → spinel reaction within the cold core of the subducting slab [*Green and Burnley*, 1989; *Kirby et al.*, 1991]. Intermediate-focus earthquakes that tend to occur at 90-170 km depth may reflect metastable overstepping of the gabbro → eclogite reaction within the subducting oceanic crust [*Kirby*, 1995]. This hypothesis is supported by the observation of a low seismic-velocity waveguide (interpreted as untransformed oceanic crust) that persists to depths of up to 75 to 150 km in the northeast Japan and other subduction zones [see review by *Kirby et al.*, this volume]. An important unresolved question is the extent to which fluids liberated by dehydration reactions in the subducting oceanic crust may trigger solid-solid reactions in adjacent rocks [*Kirby et al.*, this volume].

In the discussion above it was implicitly assumed that altered oceanic crust will transform to a blueschist-facies mineralogy during subduction. While this transformation may be likely for the uppermost oceanic crust, which contains pore fluids and very hydrous minerals such as smectite, kinetics may hinder transformation of deeper parts of the oceanic crust. Because of the lack of free water, altered oceanic gabbro (amphibolite) in the subducting slab may transform directly to eclogite without first transforming to blueschist at shallower depths.

Perhaps the best hope of constraining the thermal and petrologic structure of subduction zones lies in detailed seismological investigations that illuminate the velocity and attenuation structure of the subducting slab and overlying mantle wedge [e.g., *Helffrich et al.*, 1989; *Zhao* et al., 1994]. By investigating different seismic parameters in the same region we may be able to separate the effects of temperature, bulk composition, mineralogy, and seismic anisotropy. For example, shallow-level V_P/V_S and attenuation anomalies observed in the southern Alaska forearc mantle may reflect large-scale serpentinization [*Ponko et al.*, 1995]. Ideally, seismic studies should be interpreted in light of thermal models specifically tailored for individual subduction zones as has recently been done for Cascadia [*Lewis et al.*, 1988; *Hyndman et al.*, 1993], southwest Japan [*Wang et al.*, 1995], and southern Alaska [*Ponko and Peacock*, 1995]. Through detailed seismological studies we may well find that different subduction zones have substantially different thermal and petrologic structures.

CONCLUSIONS

The thermal structure of subduction zones is dominated by the downward advection of cool oceanic lithosphere. For a wide range of parameters, thermal models predict temperatures of 500 to 700 °C for the subducting oceanic crust at depths of 125 km beneath volcanic arcs; such temperatures are hundreds of degrees cooler than the surrounding mantle. Considerable uncertainty surrounds the importance of shear heating in subduction zones. Low forearc heat flow (25 to 50 mW/m^2), seismic coupling to ~40 km depth, and high P, low T metamorphic rocks constrain shear stresses to 1-5% of lithostatic pressure in the brittle (frictional) regime. The rate of shear heating drops off rapidly in the plastic (ductile) regime and the temperature of the brittle-plastic transition may be as low as 300 °C in sediment-dominated subduction zones. Additional uncertainties in the thermal structure of subduction zones arise from induced convection in the mantle wedge, hydrothermal circulation in the oceanic lithosphere, the variation of thermal parameters with T and P, and the effects of fluid flow at shallow levels.

For a wide range in parameter space, calculated P-T paths suggest that subducting oceanic crust does not melt, but rather passes through the subsolidus blueschist → eclogite facies transition. Only under rare circumstances, such as near the subduction of a spreading ridge, does subducting oceanic crust undergo partial melting. The dehydration of minerals such as Na-amphibole, lawsonite, and chlorite in the subducting oceanic crust releases large amounts of H_2O-rich fluids that may cause extensive hydration of the overlying mantle and trigger partial melting in the core of the convecting mantle wedge. Uncertainties in the thermal structure of subduction zones translate to uncertainties in the petrologic structure. Additional uncertainties in constructing an accurate petrologic model of subduction zones arise from our lack of knowledge regarding the depth and extent to which the oceanic crust and uppermost mantle is hydrated as a result of hydrothermal convection near spreading ridges, and the extent to which metamorphic reactions may be kinetically hindered.

Acknowledgments. I thank Dave Scholl, Gray Bebout, and Steve Kirby for organizing the highly-stimulating 1994 Interdisciplinary Conference on the Subduction Process (SUBCON). The original manuscript was greatly improved through constructive reviews by Steve Kirby and several anonymous reviewers. This research was supported by the National Science Foundation through grants EAR 91-05741 and EAR 93-03945.

REFERENCES

Alt, J. C., J. Honnorez, C. Laverne, and R. Emmermann, Hydrothermal alteration of a 1 km section through the upper oceanic crust, Deep Sea Drilling Project Hole 504B: Mineralogy, chemistry, and evolution of seawater-basalt interactions, *J. Geophys. Res.*, *91*, 10,309-10,335, 1986.

Anderson, R. N., S. Uyeda, and A. Miyashiro, Geophysical and geochemical constraints at converging plate boundaries — Part I: Dehydration in the downgoing slab, *Geophys. J. R. Astr. Soc.*, *44*, 333-357, 1976.

Anderson, R. N., S. E. DeLong, and W. M. Schwarz, Thermal model for subduction with dehydration in the downgoing slab, *J. Geol.*, *86*, 731-739, 1978.

Anderson, R. N., S. E. DeLong, and W. M. Schwarz, Dehydration, asthenospheric convection and seismicity in subduction zones, *J. Geol.*, *88*, 445-451, 1980.

Andrews, D. J., and N. H. Sleep, Numerical modelling of tectonic flow behind island arcs, *Geophys. J. R. Astr. Soc.*, *38*, 237-251, 1974.

Batchelor, G. K., *An Introduction to Fluid Dynamics*, 615 pp., Cambridge Univ. Press, Cambridge, 1967.

Bebout, G.E., Field-based evidence for devolatilization in subduction zones: Implications for arc magmatism, *Science*, *251*, 413-416, 1991.

Bird, P., Stress and temperature in subduction shear zones: Tonga and Mariana, *Geophys. J. R. Astr. Soc.*, *55*, 411-434, 1978.

Bodri, L., and B. Bodri, Numerical investigation of tectonic flow in island-arc areas, *Tectonophys.*, *50*, 163-175, 1978.

Cannat, M., D. Bideau, and H. Bougault, Serpentinized peridotites and gabbros in the mid-Atlantic ridge axial valley at 15°37'N and 16°52'N, *Earth Planet. Sci. Lett.*, *109*, 87-106, 1992.

Carswell, D. A. (Ed.), *Eclogite Facies Rocks*, 396 pp., Blackie, London, 1990.

Clauser, C., and E. Huenges, Thermal conductivity of rocks and minerals, in *Rock Physics and Phase Relations*, edited by

T.J. Ahrens, pp. 105-126, Am. Geophys. Union, Washington, D.C., 1995.

Cloos, M., Thermal evolution of convergent plate margins: Thermal modeling and reevaluation of isotopic Ar-ages for blueschists in the Franciscan complex of California, *Tectonics*, *4*, 421-433, 1985.

Davidson, J. P., Continental and island arcs, in *Encyclopedia of Earth System Science*, pp. 615-626, Academic Press, Inc., 1992.

Davies, J. H., Lateral water transport across a dynamic mantle wedge: A model for subduction zone magmatism, in *Magmatic Systems*, edited by M.P. Ryan, pp. 197-221, Academic Press, San Diego, CA, 1994.

Davies, J. H., and D. J. Stevenson, Physical model of source region of subduction zone volcanics, *J. Geophys. Res.*, *97*, 2037-2070, 1992.

Defant, M. J., and M. S. Drummond, Derivation of some modern arc magmas by melting of young subducted lithosphere, *Nature*, *347*, 662-665, 1990.

Delany, J. M., and H. C. Helgeson, Calculation of the thermodynamic consequences of dehydration in subducting oceanic crust to 100 kb and > 800°C, *Amer. J. Sci.*, *278*, 638-686, 1978.

Dumitru, T. A., Effects of subduction parameters on geothermal gradients in forearcs, with an application to Franciscan subduction in Calif., *J. Geophys. Res.*, *96*, 621-641, 1991.

Ernst, W. G., Blueschist metamorphism and P-T regimes in active subduction zones, *Tectonophys.*, *17*, 255-272, 1973.

Ernst, W. G., and S. M. Peacock, A thermotectonic model for preservation of ultrahigh-pressure mineralogic relics in metamorphosed continental crust, in *Dynamics of Subduction*, this volume, edited by G.E. Bebout, D. Scholl, S. Kirby, and J.P. Platt, Am. Geophys. Union, Washington, D.C., in press, 1996.

Evans, B. W., Phase relations of epidote blueschists, *Lithos*, *25*, 3-23, 1990.

Evans, B. W., and E. H. Brown (Eds.), *Blueschists and Eclogites*, 423 pp., Geological Society of America Memoir 164, Boulder, Colorado, 1986.

Fryer, P., E. L. Ambos, and D. M. Hussong, Origin and emplacement of Mariana forearc seamounts, *Geology*, *13*, 774-777, 1985.

Furukawa, Y., Depth of the decoupling plate interface and thermal structure under arcs, *J. Geophys. Res.*, *98*, 20,005-20,013, 1993.

Fyfe, W.S., and A.R. McBirney, Subduction and the structure of andesitic volcanic belts, *Am.J.Sci.*, *275-A*, 285-297, 1975.

Gill, J., *Orogenic Andesites and Plate Tectonics*, 390 pp., Springer-Verlag, New York, 1981.

Green, H. W. I., and P. C. Burnley, A new self-organizing mechanism for deep-focus earthquakes, *Nature*, *341*, 733-737, 1989.

Hacker, B. R., S. H. Kirby, and S. R. Bohlen, Time and metamorphic petrology: calcite to aragonite experiments, *Science*, *258*, 110-112, 1992a.

Hacker, B. R., S. R. Bohlen, and S. H. Kirby, Mechanisms and kinetics of the albite → jadeite + quartz transformation in rock(abstr.), *EOS, Trans.Am.Geophys.Un.*,*74*, 611, 1992b.

Hasebe, K., N. Fujii, and S. Uyeda, Thermal processes under island arcs, *Tectonophys.*, *10*, 335-355, 1970.

Hawkesworth, C. J., K. Gallagher, J. M. Hergt, and F. McDermott, Mantle and slab contributions in arc magmas, *Annu. Rev. Earth Planet. Sci.*, *21*, 175-204, 1993.

Helffrich, G. R., S. Stein, and B. J. Wood, Subduction zone thermal structure and mineralogy and their relationship to seismic wave reflections and conversions at the slab/mantle interface, *J. Geophys. Res.*, *94*, 753-763, 1989.

Honda, S., Thermal structure beneath Tohoku, northeast Japan—A case study for understanding the detailed thermal structure of the subduction zone, *Tectonophys.*, *112*, 69-102, 1985.

Honda, S., and S. Uyeda, Thermal process in subduction zones—A review and preliminary approach on the origin of arc volcanism, in *Arc Volcanism: Physics and Tectonics*, edited by D. Shimozuru, and I. Yokoyama, pp. 117-140, TERRAPUB, Tokyo, 1983.

Hsui, A. T., B. D. Marsh, and M. N. Toksöz, On melting of the subducted oceanic crust: effects of subduction induced mantle flow, *Tectonophys.*, *99*, 207-220, 1983.

Hyndman, R. D., and K. Wang, Thermal constraints on the zone of major thrust earthquake failure: The Cascadia subduction zone, *J. Geophys. Res.*, *98*, 2039-2060, 1993.

Jarrard, R. D., Relations among subduction parameters, *Rev. Geophys.*, *24*, 217-284, 1986.

Kincaid, C., Subduction dynamics: From the trench to the core-mantle boundary, *Rev. Geophys. Spec. Suppl., U.S. Natl. Report to the I.U.G.G. 1991-1994*, 401-412, 1995.

Kirby, S. H., Rheology of the lithosphere, *Rev. Geophys. Sp. Phys.*, *21*, 1458-1487, 1983.

Kirby, S. H., Intraslab earthquakes and phase changes in subducting lithosphere, *Rev. Geophys.Spec. Suppl., U.S. Natl. Report to the I.U.G.G. 1991-1994*, 287-297, 1995.

Kirby, S. H., W. B. Durham, and L. A. Stern, Mantle phase changes and deep-earthquake faulting in subducting lithosphere, *Science*, *252*, 216-225, 1991.

Kirby, S. H., E. R. Engdahl, and R. Denlinger, Intraslab earthquakes and arc volcanism: dual physical expressions of crustal and uppermost mantle metamorphism in subducting slabs, in *Dynamics of Subduction*, this volume, edited by G.E. Bebout, D. Scholl, S. Kirby, and J.P. Platt, Am. Geophys. Union, Washington, D.C., in press, 1996.

Lewis, T. J., W. H. Bentkowski, E. E. Davis, R. D. Hyndman, J. G. Souther, and J. A. Wright, Subduction of the Juan de Fuca plate: Thermal consequences, *J. Geophys. Res.*, *93*, 15,207-15,225, 1988.

Maekawa, H., M. Shozui, T. Ishii, P. Fryer, and J. A. Pearce, Blueschist metamorphism in an active subduction zone, *Nature*, *364*, 520-523, 1993.

Marsh, B. D., Island-arc volcanism: Volcanic arcs ringing the Pacific owe their origin to magma generated in a mysterious fashion when tectonic plates are consumed by the inner earth, *Amer. Scientist*, *67*, 161-172, 1979.

McKenzie, D. P., Speculations on the consequences and causes of plate motions, *Geophys. J. R. Astr. Soc.,18*, 1-32, 1969.

Molnar, P., and P. C. England, Temperatures, heat flux, and frictional stress near major thrust faults, *J. Geophys. Res., 95*, 4833-3856, 1990.

Morris, J. D., W. P. Leeman, and F. Tera, The subducted component in island arc lavas: constraints from Be isotopes and B-Be systematics, *Nature, 344*, 31-36, 1990.

Myers, and Johnston, Phase equilibria constraints on models of subduction zone magmatism, in *Dynamics of Subduction*, this volume, edited by G.E. Bebout, D. Scholl, S. Kirby, and J.P. Platt, Am. Geophys. Union, Washington, D.C., in press, 1996.

Oxburgh, E. R., and D. L. Turcotte, Thermal structure of island arcs, *Geol. Soc. Amer. Bull., 81*, 1665-1688, 1970.

Pawley, A. R., and J. R. Holloway, Water sources for subduction zone volcanism: new experimental constraints, *Science, 260*, 664-667, 1993.

Peacock, S. M., Thermal effects of metamorphic fluids in subduction zones, *Geology, 15*, 1057-1060, 1987.

Peacock, S. M., Fluid processes in subduction zones, *Science, 248*, 329-337, 1990a.

Peacock, S. M., Numerical simulation of metamorphic pressure-temperature-time paths and fluid production in subducting slabs, *Tectonics, 9*, 1197-1211, 1990b.

Peacock, S. M., Numerical simulation of subduction zone pressure-temperature-time paths: constraints on fluid production and arc magmatism, *Phil. Trans. R. Soc. Lond. A, 335*, 341-353, 1991.

Peacock, S. M., Blueschist-facies metamorphism, shear heating, and P-T-t paths in subduction shear zones, *J. Geophys. Res., 97*, 17,693-17,707, 1992.

Peacock, S. M., The importance of the blueschist → eclogite dehydration reactions in subducting oceanic crust, *Geol. Soc. Am. Bull., 105*, 684-694, 1993.

Peacock, S. M., T. Rushmer, and A. B. Thompson, Partial melting of subducting oceanic crust, *Earth Planet. Sci. Lett., 121*, 227-244, 1994.

Ponko, S. C., and S. M. Peacock, Thermal modeling of the Alaska subduction zone: Insight into the petrology of the subducting slab and overlying mantle wedge, *J. Geophys. Res., 100*, 2117-, 1995.

Ponko, S. C., S. M. Peacock, and C. O. Sanders, Thermal structure and hydration conditions in the southern Alaska subduction zone from thermal modeling and seismic tomography (abstract), *Abstracts to the IUGG XXII General Assembly*, A469, 1995.

Reck, B. H., Implications of measured thermal gradients for water movement through the northeast Japan accretionary prism, *J. Geophys. Res., 92*, 3683-3690, 1987.

Ruff, L., and B. Tichelaar, What controls the seismogenic plate interface in subduction zones?, in *Dynamics of Subduction*, this volume, edited by G.E. Bebout, D. Scholl, S. Kirby, and J.P. Platt, Am. Geophys.Union, Wash. D.C., in press, 1996.

Scholz, C. H., *The Mechanics of Earthquakes and Faulting*, 439 pp., Cambridge Univ. Press, Cambridge, 1990.

Sibson, R. H., Roughness at the base of the seismogenic zone: contributing factors, *J. Geophys. Res., 89*, 5791-5799, 1984.

Staudigel, H., and S. D. King, Ultrafast subduction: the key to slab recycling efficiency and mantle differentiation?, *Earth Planet. Sci. Lett., 109*, 517-530, 1992.

Stein, C. A., and S. Stein, A model for the global variation in oceanic depth and heat flow with lithospheric age, *Nature, 359*, 123-129, 1992.

Stein, C. A., and S. Stein, Constraints on hydrothermal heat flux through the oceanic lithosphere from global heat flow, *J. Geophys. Res., 99*, 3081-3095, 1994.

Tatsumi, Y., Migration of fluid phases and genesis of basalt magmas in subduction zones, *J. Geophys. Res., 94*, 4697-4707, 1989.

Tichelaar, B. W., and L. J. Ruff, Depth of seismic coupling along subduction zones, *J. Geophys. Res., 98*, 2017-2037, 1993.

Toksöz, M. N., J. W. Minear, and B. R. Julian, Temperature field and geophysical effects of a downgoing slab, *J. Geophys. Res., 76*, 1113-1138, 1971.

Turcotte, D. L., and G. Schubert, Frictional heating of the descending lithosphere, *J. Geophys. Res., 78*, 5876-5886, 1973.

Ulmer, P., and V. Trommsdorff, Serpentine stability to mantle depths and subduction-related magmatism, *Science, 268*, 858-861, 1995.

van den Beukel, J., and R. Wortel, Thermo-mechanical modeling of arc-trench regions, *Tectonophys., 154*, 177-193, 1988.

Wang, C.-Y., and Y.-L. Shi, On the thermal structure of subduction complexes: A preliminary study, *J. Geophys. Res., 89*, 7709-7718, 1984.

Wang, K., R. D. Hyndman, and E. E. Davis, Thermal effects of sediment thickening and fluid expulsion in accretionary prisms: model and parameter analysis, *J. Geophys. Res., 98*, 9975-9984, 1993.

Wang, K., R. D. Hyndman, and M. Yamano, Thermal regime of the Southwest Japan subduction zone: effects of age history of the subducting plate, *Tectonophys., 248*, 53-70, 1995.

Wyllie, P. J., Magmas and volatile components, *Amer. Mineral., 64*, 469-500, 1979.

Wyllie, P. J., Magma genesis, plate tectonics, and chemical differentiation of the earth, *Rev. Geophys., 26*, 370-404, 1988.

Yuen, D. A., L. Fleitout, G. Schubert, and C. Froidevaux, Shear deformation zones along major transform faults and subducting slabs, *Geophys. J. R. Astr. Soc., 54*, 93-119, 1978.

Zhao, D., A. Hasegawa, and H. Kanamori, Deep structure of Japan subduction zone as derived from local, regional, and teleseismic events, *J. Geophys. Res., 99*, 22,313-22,329, 1994.

S. M. Peacock Department of Geology, Arizona State University, Box 871404, Tempe, Arizona 85287-1404. (e-mail: peacock@asu.edu)

Contrasting P-T-t Histories for Blueschists From the Western Baja Terrane and the Aegean: Effects of Synsubduction Exhumation and Backarc Extension

Suzanne L. Baldwin

Department of Geosciences, University of Arizona, Tucson, Arizona

Thermochronologic studies of high P, low T metamorphic rocks provide constraints on protolith ages, subduction burial rates, the timing of subduction related metamorphism, and the timing of subsequent exhumation. Temperature-time data is essential for reconstruction of accurate P-T-t paths, unambiguous interpretation of the tectonic significance of P-T paths, and to test thermal models which predict the time scales required for the preservation of high P, low T assemblages. The P-T-t histories of high P metamorphic rocks from the western Baja terrane and the Aegean are used as examples to illustrate how thermochronologic data can be used to gain insight into forearc processes. Differences in exhumation rates and the degree to which high P, low T metamorphic rocks are overprinted reflect the variability in processes affecting convergent margins. Steady-state subduction characterized by low geothermal gradients in western Baja enables slow blueschist exhumation to occur without overprinting of high P, low T mineral assemblages. In contrast, subduction followed by back arc extension in the Aegean requires rapid exhumation of high P metamorphic rocks to prevent overprinting by higher temperature mineral assemblages during subsequent higher T metamorphic events. P-T-t and structural data in other areas (e.g., Papua New Guinea, New Caledonia) indicate extension tectonics has played a key role in the exhumation of high P metamorphic rocks.

1. INTRODUCTION

It is well known that blueschists and related high pressure metamorphic rocks form at convergent margins. However the mechanisms by which high P, low T mineral assemblages escape overprinting by lower P, higher T assemblages during exhumation remains a topic of considerable debate [see *Platt*, 1993 for a review]. Rocks metamorphosed in subduction zones commonly record significant variations in metamorphic conditions. Thermal models predict the time scales required for high-P, low-T assemblages to avoid partial or complete conversion to higher temperature phases upon decompression [*England and Richardson*, 1977; *Draper and Bone*, 1981; *England and Thompson*, 1984; *Peacock*, 1992]. However, the details of the timing of subduction metamorphism and subsequent exhumation of rocks in most ancient convergent margins are not well known.

Thermochronology [e.g., *McDougall and Harrison*, 1988] can be used to determine protolith ages, age(s) of metamorphism, and subsequent cooling histories of subduction complexes. Exhumation rates can be determined using calculated cooling rates together with an assumed average geothermal gradient. This approach is valid for geologic histories in which geothermal gradients have remained constant, and/or are known, during the time period investigated. In this paper thermochronologic constraints on the high-P metamorphic rocks from the western Baja terrane and the Aegean are discussed. Contrasts in their corresponding P-T-t histories reflect the effects of synsubduction exhumation and backarc extension, respectively.

2. SLOW SYNSUBDUCTION EXHUMATION, WESTERN BAJA TERRANE

The western Baja terrane [*Sedlock*, 1988a,b; this volume] contains high P, low T metamorphic rocks which were slowly exhumed during steady-state subduction [*Baldwin and Harrison*, 1989]. This Franciscan-type [*Ernst*, 1988] subduction complex was produced as a result of oblique convergence of paleo-Pacific plates with the North American plate during Jurassic-Cretaceous time. The lowest structural units (i.e., lower plate) exposed in west central Baja California [*Sedlock*, 1988b] consist of coherent metabasites and metasediments regionally metamorphosed under blueschist facies conditions [P=5-10+ kbar and 150-300°C; *Sedlock*, 1988a] in Early Cretaceous time [*Baldwin and Harrison*, 1989]. On Cedros Island the upper plate consists of the Choyal terrane, a 165-175 Ma arc-ophiolite complex overlain by late Jurassic volcanic and volcaniclastic rocks [*Boles and Landis*, 1984; *Kimbrough*, 1985; *Busby-Spera*, 1987]. Fault zones containing serpentinite-matrix melange separate the lower and upper plates. Blocks within serpentinite-matrix melange were derived from at least two sources within the subduction complex; (1) ~115-95 Ma blueschist facies rocks and (2) ~170-160 Ma epidote-amphibolite and amphibolite facies rocks [*Baldwin and Harrison*, 1992]. Some epidote-amphibolite facies blocks are partially overprinted by blueschist facies mineral assemblages and record both metamorphic events.

The western Baja terrane and serpentinite melange record initiation of subduction, terrane accretion, and synsubduction exhumation of high P, low T metamorphic rocks (Figure 1). Initiation of subduction resulted in growth of the Choyal oceanic island arc and metamorphism of ocean floor basalts and overlying sediments in Mid-Jurassic time. Tonalite and granodiorite plutons were emplaced at shallow crustal levels and cooled rapidly as indicated by nearly concordant U-Pb zircon [*Kimbrough*, 1985] and $^{40}Ar/^{39}Ar$ feldspar ages [166-160 Ma; *Baldwin*, 1988].

Relatively small volumes of sediment were supplied to the trench as subduction continued during Late Jurassic time. Some epidote-amphibolite facies "blocks" moved to shallow, cooler levels (< 100°C) of the accretionary wedge in serpentinite diapirs as indicated by $^{40}Ar/^{39}Ar$ and apatite fission track ages [*Baldwin and Harrison*, 1992]. The Choyal arc collided with the North American continent and arc magmatism ceased by latest Jurassic time [*Boles and Landis*, 1984]. Sediments that eventually formed the western Baja terrane were deposited on the ocean floor at some distance from the trench.

From 150-120 Ma oceanic crust and overlying sediments were subducted and high P, low T metamorphic rocks formed at depth. The accretionary wedge widened as sediment supply to the trench increased. Underplating of coherent sequences of pillow basalts, cherts, and thin-bedded turbidites at depth was accompanied by extension in the upper portion of the wedge so that stability was maintained [*Platt*, 1993]. Epidote-amphibolite facies blocks

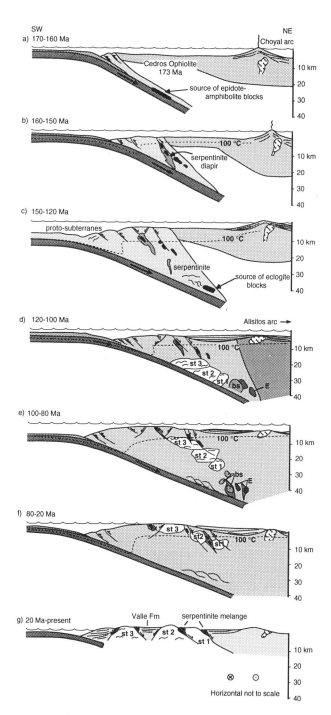

Fig. 1. Schematic evolution of the subduction complex in west central Baja California since 170 Ma. St 1-3 denote subterranes [after *Sedlock*, 1988b] of the western Baja terrane. Bs and E are blueschist and eclogite facies rocks, respectively. The 100°C isotherm is indicated. See text for discussion.

continued to move from depth to shallower, cooler portions of the wedge as suggested by apatite fission track ages.

From 120-100 Ma some eclogite facies rocks moved to shallower levels of the subduction complex and were subsequently overprinted by blueschist facies mineral assemblages. Erosion of the Choyal terrane contributed a distinctive detrital fraction to the Valle Formation turbidites [Gastil et al., 1981]. Arc magmatism continued on the mainland and formed the Alisitos arc of Baja California [Gastil et al., 1981].

Subduction and arc magmatism continued from 100-20 Ma. Coherent blueschists (i.e., subterranes 1-3), as well as blueschist and eclogite blocks moved to shallower, cooler portions of the accretionary wedge, the later entrained in serpentinite diapirs as indicated by apatite fission track ages and $^{40}Ar/^{39}Ar$ white mica and feldspar ages [Baldwin and Harrison, 1989]. Exhumation was accommodated by continued erosion and normal faulting [Sedlock, this volume]. Subduction continued until ~20 Ma when a transform plate boundary zone formed in the region [Atwater, 1970; Atwater and Molnar, 1973]. By mid-late Pliocene time the subduction complex was undergoing subaerial erosion [Kilmer, 1984].

Integrated P-T-t data provide quantitative constraints on burial and exhumation rates for some of the coherent blueschists. Maximum burial rate for subterrane 2 was ~0.7 mm/yr based on protolith ages, the timing of peak metamorphism, and peak pressure estimates (Figures 1 and 2). The presence of aragonite places quantitative limits on geothermal gradients and P-T-t paths followed by these rocks during exhumation [Carlson and Rosenfeld, 1981]. An exhumation rate of 0.1 mm/yr is obtained assuming an average geothermal gradient of 9°C/Ma [Baldwin and Harrison, 1989]. In Miocene time an increase in the exhumation rate (to ~1.0 mm/yr) is coincident with the change from a convergent to a transform plate boundary in this region [e.g., Atwater and Molnar, 1973].

3. SUBDUCTION FOLLOWED BY BACKARC EXTENSION, THE AEGEAN

The Cycladic blueschist belt of the central Aegean (Figure 3) is an Alpine-type blueschist belt [Ernst, 1988]. Eocene collision was followed by Miocene extension and since ~13 Ma the region has been dominated by backarc extension likely driven by southward retreat of the subducted slab or collapse of overthickened crust [Lister et al., 1984]. At least four metamorphic events are recognized in basement rocks of the Cyclades. These include Paleozoic and Mesozoic pre-Alpine gneissic basement (M_0), Eocene high P, low T metamorphic rocks (M_1), Oligocene-Miocene greenschists and amphibolites (M_2), and contact metamorphic rocks (M_3) associated with Mid-Late Miocene

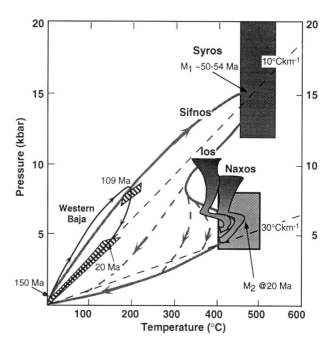

Fig. 2. Compilation of P-T-t paths for subterrane 2 of the western Baja terrane [after Baldwin, 1989], and the Aegean region [after van der Maar and Jansen, 1983; Schliestedt and Matthews, 1987; Okrusch and Brocker, 1990; Maluski et al., 1987; Wijbrans and McDougall, 1988; Wijbrans et al., 1990]. Note the difference in timescales indicated for slow synsubduction Cretaceous-Tertiary exhumation of the western Baja terrane (~105 Ma) compared to that indicated for Tertiary metamorphism and exhumation of Aegean (~50 Ma) high P metamorphic rocks.

granitoid intrusions. K-Ar, $^{40}Ar/^{39}Ar$, and Rb-Sr analyses have been used to establish the age of high P, low T metamorphism at ~50 Ma, subsequent overprinting by medium pressure Barrovian metamorphism at ~20-30 Ma [van der Maar and Jansen, 1983; Okrusch and Brocker, 1990; Wijbrans et al., 1990, and references therein] and intrusion of granites at ~16-10 Ma [Altherr et al., 1982; Wijbrans and McDougall, 1988; Baldwin and Lister, 1994]. The higher temperature post-Eocene metamorphic and plutonic events obscure much of the record of Eocene (M_1) high-P metamorphism.

$^{40}Ar/^{39}Ar$ data on white micas from M_1 rocks on Syros, and blueschists partially overprinted by the M_2 event from Ios and Syros illustrate typical apparent age variations for these polymetamorphosed rocks (Figures 3 and 4; Tables 1 and 2[*]). Blueschist and eclogite facies rocks from Syros that were not overprinted by retrograde M_2 assemblages yielded undisturbed $^{40}Ar/^{39}Ar$ white mica age spectra with apparent ages of ~50-54 Ma interpreted to date subduction

---
[*] Table 2 available from author on request.

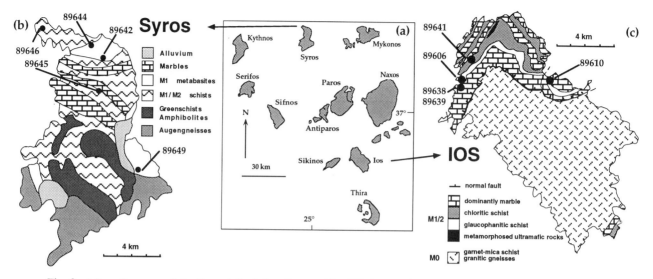

Fig. 3. a) Location map of the Central Cyclades, Greece. Simplified geologic maps of b) Syros and c) Ios showing sample localities [after *van der Maar and Jansen*, 1983; *Okrusch and Brocker*, 1990]. See *Vandenberg and Lister* [1995] for Ios cross sections.

zone metamorphism. In contrast, samples from partially overprinted blueschist facies rocks from Ios and Syros yielded $^{40}Ar/^{39}Ar$ apparent age gradients ranging from ~25 to 49 Ma, similar to $^{40}Ar/^{39}Ar$ white mica apparent ages from Naxos [*Wijbrans and McDougall*, 1988] and Sifnos [*Wijbrans et al.*, 1990].

Preservation of M_1 ages and mineral assemblages indicates that some blueschists moved to shallow crustal levels prior to M_{2-3} metamorphic and intrusive events (i.e., the argon systematics have not been reset). These rocks likely followed isothermal decompression paths and/or cooled during decompression (Figure 2). However, the majority of white micas yielded disturbed $^{40}Ar/^{39}Ar$ age spectra interpreted to reflect variable partial outgassing and/or recrystallization of M_1 white micas during subsequent M_2 and possibly M_3 metamorphism. $^{40}Ar/^{39}Ar$ apparent ages associated with the low T steps either approximate the time of post-M_1 mineral growth and/or loss of argon via volume diffusion due to a thermal pulse [c.f., *Wijbrans and McDougall*, 1988]. These age gradients however, are not completely meaningless as the apparent ages fall between ages corresponding to M_1 and M_2 metamorphic events.

4. DISCUSSION AND CONCLUSIONS

In the western Baja terrane thermochronologic data for coherent blueschists and blocks in melange indicate mid-Jurassic initiation of subduction was followed by a progressive decrease in geothermal gradients during early cooling of the subduction zone. Mid-Jurassic amphibolite and epidote-amphibolite facies blocks were metamorphosed during early stages of subduction prior to establishment of cooler steady-state conditions [c.f., *Platt*, 1975; *Peacock*, 1987]. By Cretaceous time steady-state subduction was established in the region and continued until Early Miocene time.

Within the western Baja terrane some rocks were buried (~0.7 mm/yr) and exhumed (~0.1-0.8 mm/yr) at similar rates. Relatively slow exhumation (0.1 mm/yr) following mid-Cretaceous peak metamorphism and extending into the Miocene, as well as the pristine preservation of blueschist facies assemblages, indicate that these coherent blueschists were transported through an accretionary wedge characterized by low (cold) geothermal gradients during steady-state subduction. Slow exhumation may be a result of erosional processes, structural data [*Smith and Busby*, 1993; *Sedlock*, this volume] indicate that synsubduction extensional mechanisms also contributed to blueschist exhumation.

In contrast, Aegean high P metamorphic rocks record significantly different P-T-t histories resulting from an apparent increase in geothermal gradients subsequent to Eocene metamorphism. Petrologic relationships and preserved mineral assemblages associated with M_{1-3} metamorphism indicate geothermal gradients increased from $\leq 10°C/km$ to $\geq 30°C/km$. The steepening of geothermal gradients is likely related to rollback of the subducting slab and backarc extension [e.g., *Lister et al.*, 1984]. The preservation of older (i.e., Eocene) apparent ages and

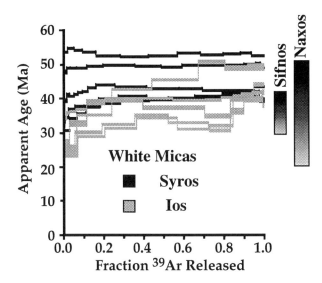

Fig. 4. Composite $^{40}Ar/^{39}Ar$ age spectra for M_1 and M_1/M_2 white micas from Syros [c.f., *Maluski et al.*, 1987] and Ios. The range in $^{40}Ar/^{39}Ar$ white mica apparent ages obtained from M_1 and M_1/M_2 rocks on Naxos [*Wijbrans and McDougall*, 1988] and Sifnos [*Wijbrans et al.*, 1990] is shown for comparison.

blueschist and eclogite facies assemblages for rocks from Syros and Naxos indicates that at least a portion of the subduction complex had been exhumed to shallow, cool crustal levels prior to M_{2-3} metamorphism and intrusion of Miocene granitoids. However, some of the high P rocks were affected by Miocene thermal events as indicated by partially reset $^{40}Ar/^{39}Ar$ apparent ages and overprinting by higher temperature mineral assemblages (Sifnos, Naxos, and Ios; Figure 4).

Argon systematics in white micas from high P, low T metamorphic rocks are not well understood. Several mechanisms may produce $^{40}Ar/^{39}Ar$ apparent age gradients including loss due to volume diffusion and/or recrystallization caused by a thermal pulse [e.g., *Wijbrans and McDougall*, 1988], as well as slow cooling resulting in argon loss from the least retentive sites [e.g., *Baldwin and Harrison*, 1992]. In addition depressurization during exhumation can lead to reopening of minerals to volume diffusion and result in argon loss [*Lister and Baldwin*, 1995] and may be significant in cases involving exhumation of high P metamorphic rocks. The compositional dependence on argon diffusivities in phengites [*Scaillet et al.*, 1992; *Grove*, 1993] may also lead to variations in $^{40}Ar/^{39}Ar$ apparent ages. All of these mechanisms can potentially result in $^{40}Ar/^{39}Ar$ age gradients such that petrologic, structural, and geologic constraints are essential for interpretation of argon data.

Thermochronologic data is essential for reconstruction of accurate P-T-t paths, unambiguous interpretation of the tectonic significance of P-T paths, and to test thermal models which predict the time scales required for the preservation of high P, low T assemblages during exhumation. Exhumation rates and the degree to which high P, low T metamorphic rocks are overprinted reflect the variable nature of subduction zones. P-T-t data can

TABLE 1. Summary of $^{40}Ar/^{39}Ar$ analyses for white micas from Ios and Syros.

Sample #	Lithology	$^{40}Ar/^{39}Ar$ apparent ages	Comments/interpretations
Ios Samples			
88606	qtz phengite schist	30.5 to 41.9 Ma[a]; 39.0 ± 0.2 Ma*	M_1/M_2
88610	glauc schist	25.6 to 39.8 Ma[a]; 29.9 ± 0.4 Ma*	M_1/M_2
89638	glauc schist	25.0 to 44.1 Ma[a]; 33.6 ± 0.2 Ma*	M_1/M_2
89639	glauc schist	23.5 to 42.1 Ma[a]; 31.7 ± 0.2 Ma*	M_1/M_2
89641	gt-glauc schist	25.3 to 49.3 Ma[a]; 42.2 ± 0.5 Ma*	M_1/M_2
Syros Samples			
89642	retrograde eclogite	49.2 ± 0.2 Ma[b]	phengite; flat spectra
89644	glauc-marble schist	52.4 to 55.0 Ma[a]; 53.1 ± 0.2 Ma*	phengite
89645	retrograde blueschist	34.8 to 42.4 Ma[a]; 39.6 ± 0.1 Ma*	M_1/M_2
89646	quartzite	31.0 to 41.2 Ma[a]; 39.6 ± 0.1 Ma*	M_1/M_2
89649	retrograde blueschist	40.0 to 44.2 Ma[a]; 43.05 ± 0.12 Ma*	M_1/M_2

Analytical procedures followed those described by *McDougall* [1985] and $^{40}Ar/^{39}Ar$ analytical data are in Table 2 (available from the author on request). Notes: white mica (wm); garnet (gt); quartz (qtz); glaucophane (glauc). [a]Gradients in $^{40}Ar/^{39}Ar$ apparent ages; [b]weighted mean; * indicates $^{40}Ar/^{39}Ar$ total fusion age. M_1/M_2 indicates partial loss profile/recrystallization of M_1/M_2 white micas.

provide insight into changes in subduction zone geothermal gradients with time. Long-lived subduction (e.g., western Baja terrane) may allow slow blueschist exhumation without overprinting of high P, low T mineral assemblages to occur. In subduction zones characterized by non steady-state geothermal gradients (e.g., the Aegean), calculation of exhumation rates using thermochronologic data is not straightforward. Rollback of the subducting slab resulting in back arc extension requires relatively rapid exhumation in order to escape overprinting by higher temperature mineral assemblages. In other areas (e.g., the D'Entrecasteaux Islands and New Caledonia) extension tectonics has played a key role in the rapid exhumation of high-P metamorphic rocks [*Hill et al.*, 1992; *Hill and Baldwin*, 1993; *Baldwin et al.*, 1993; *Rawling et al.*, 1995]. Multidisciplinary studies of subduction complexes which integrate thermochronologic data will continue to provide further insight into the timing and nature of subduction processes throughout Earth history.

Acknowledgements. Thanks to R.L. Sedlock and G.S. Lister for numerous stimulating discussions regarding the western Baja terrane and the Aegean, T. Rawling for assisting with figures, and P.G. Fitzgerald, M. Grove, H.G. Avé Lallemant, and an anonymous reviewer for their comments which helped to significantly improve the manuscript. Technical assistance from R. Myer, J. Mya, H. Kokkonen, and J. Overs and support from the Australian National University, University of Arizona, NSF grants EAR9316418 and OPP9316720 is gratefully acknowledged.

REFERENCES

Altherr, R., H. Kreuzer, I. Wendt, H. Lenz, G. A Wagner, J. Keller, W. Harre, A. Hohndorf, A Late Oligocene/Early Miocene high temperature belt in the Attic-Cycladic crystalline complex (SE Pelagonian, Greece), *Geol. Jb.*, *E23*, 97-164, 1982.

Atwater, T., Implications of plate tectonics for the Cenozoic evolution of western North America, *GSA Bull.*, *81*, 3513-3536, 1970.

Atwater, T., and P. Molnar, Relative motion of the Pacific and North American plates deduced from sea-floor spreading in the Atlantic, Indian, and South Pacific oceans, *Stanford Univ. Publ. Geol. Soc.*, *13*, 136-148, 1973.

Baldwin, S. L., Thermochronology of a subduction complex in western Baja California, Ph.D. thesis, 247 pp., State Univ. of N. Y., Albany, 1988.

Baldwin, S. L., and T. M. Harrison, Geochronology of blueschists from west-central Baja California and the timing of uplift in subduction complexes, *J. Geol.*, *9*, 149-163, 1989.

Baldwin, S. L., and T. M. Harrison, The P-T-t history of blocks in serpentinite-matrix mélange, west-central Baja California, *GSA Bull.*, *104*, 18-31, 1992.

Baldwin, S. L., G. S. Lister, E. J. Hill, D. A. Foster, and I. McDougall, Thermochronologic constraints on the tectonic evolution of an active metamorphic core complex, D'Entrecasteaux Islands, Papua New Guinea, *Tectonics*, *12*, 611-628, 1993.

Baldwin, S. L. and G. S. Lister, P-T-t paths of Aegean metamorphic core complexes: Ios, Paros, and Syros, *U.S.Geol. Surv. Circ.*,*1107*, 19, 1994.

Boles, J. R., and C. A. Landis, Jurassic sedimentary melange and associated facies, Baja California, Mexico, *GSA Bull.*, *95*, 513-521, 1984.

Busby-Spera, C. J., Lithofacies of deep marine basalts emplaced on a Jurassic backarc apron, Baja California (Mexico), *J. Geol.*, *95*, 671-686, 1987.

Carlson, W. D., and J. L. Rosenfeld, Optical determination of topotactic aragonite-calcite growth kinetics: metamorphic implications, *J. Geol.*, *89*, 615-638, 1981.

Draper, G., and R. Bone, Denudation rates, thermal evolution, and preservation of blueschist terrains, *J. Geol*, *89*, 601-613, 1981.

England, P. C., and S. W. Richardson, The influence of erosion upon the mineral facies of rocks from different metamorphic environments, *J. Geol. Soc. Lond.*, *134*, 201-213, 1977.

England, P. C. and A. B. Thompson, Pressure-temperature-time paths of regional metamorphism I. heat transfer during the evolution of regions of thickened continental crust, *J. Petrol.*, *25*, 894-928, 1984.

Ernst, W. G., Tectonic history of subduction zones inferred from retrograde blueschist P-T paths, *Geology*, *16*, 1081-1084, 1988.

Gastil, G., G. J. Morgan, and D. Krummenacher, The tectonic history of peninsular California and adjacent Mexico, in *The geotectonic development of California*, edited by W. G. Ernst, pp. 284-305, Prentice-Hall, New Jersey, 1981.

Grove, M., Thermal histories of southern California basement terranes, Ph.D. thesis, 419 pp., Univ. of California, Los Angeles, 1993.

Hill, E. J. and S. L. Baldwin, Exhumation of high pressure metamorphic rocks during crustal extension in the D'Entrecasteaux region, Papua New Guinea, *J. Met. Geol.*, *11*, 261-277, 1993.

Hill, E. J., S. L. Baldwin, and G. S. Lister, Unroofing of active metamorphic core complexes in the D'Entrecasteaux Islands, Papua New Guinea, *Geology*, *20*, 907-910, 1992.

Kilmer, F. H., *Geology of Cedros Island, Baja California, Mexico*, 69 pp., Humboldt State Univ., Arcata, California, 1984.

Kimbrough, D. L., Tectonostratigraphic terranes of the Vizcaino Peninsula and Cedros and San Benitio Islands, Baja California, Mexico, in *Tectonostratigraphic terranes of the circum Pacific region: Circum-Pacific Council for Energy and Earth Science Series*, no.1, edited by D.G. Howell, pp. 285-298, Houston, TX, 1985.

Lister, G. S., G. Banga, and A. Feenstra, Metamorphic core complexes of the Cordilleran type in the Cyclades, Aegean Sea, Greece, *Geology*, *12*, 221-225, 1984.

Lister, G. S., and S. L., Baldwin, Modelling the effect of arbitrary P-T-t histories on argon diffusion in minerals using the MacArgon program for the Apple Macintosh, *Tectonophysics*, 1995, in press.

Maluski, H., M. Bonneau, J. R. Kienast, Dating the metamorphic events in the Cycladic area: $^{39}Ar/^{40}Ar$ data from metamorphic rocks of Syros (Greece), *Bull. Soc. Geol. France, 8*, 833-842, 1987.

McDougall, I., K-Ar and ^{40}Ar/^{39}Ar dating of hominid-bearing Pliocene-Pleistocene sequence at Koobi Fora, Lake Turkana, northern Kenya, *GSA Bull., 96*, 159-175, 1985.

McDougall, I., and T. M. Harrison, *Geochronology and thermochronology by the ^{40}Ar/^{39}Ar method*, 212 pp., Oxford Univ. Press, 1988.

Okrusch, M., and M. Bröcker, Eclogites associated with high-grade blueschists in the Cyclades archipelago, Greece: a review, *Eur. J. Mineral., 2*, 451-478, 1990.

Peacock, S. M., Creation and preservation of subduction-related inverted metamorphic gradient, *J. Geophys. Res., 92*, 12,763-12,791, 1987.

Peacock, S. M., Blueschist-facies metamoprhism, shear heating, and P-T-t paths in subduction shear zones, *J. Geophys. Res., 97*, 17,693-17,707, 1992.

Platt, J. P., Exhumation of high-pressure rocks: a review of concepts and processes, *Terra Nova, 5*, 119-133, 1993.

Rawling, T.J., L. A. Verts, and S. L. Baldwin, Constrasts in P-T-t paths within the Tertiary high pressure metamorphic belt, New Caledonia: Implications for exhumation of coherent crustal blocks (abstract), *EOS Trans. AGU*, Fall Meeting Suppl., 1995.

Scaillet, S., G. Féraud, M. Ballèvre, and M. Amouric, Mg/Fe and [(Mg,Fe)Si-Al$_2$] compositional control on argon behaviour in high-pressure white micas: A ^{40}Ar/^{39}Ar continuous laser-probe study from the Dora-Maira nappe of the internal western Alps, Italy, *Geochim. Cosmochim. Acta, 56*, 2851-2872, 1992.

Schliestedt, M., R. Altherr, and A. Matthews, Evolution of the Cycladic crystalline complex: petrology, isotope geochemistry, and geochronology in *Chemical transport in metasomatic processes*, edited by H.C. Helgeson, pp. 389-428, Reidel, Dordrecht, 1987.

Sedlock, R. L., Metamorphic petrology of a high pressure, low temperature subduction complex in west-central Baja California, Mexico, *J. Met. Geol., 6*, 205-233, 1988a.

Sedlock, R. L., Tectonic setting of blueschist and island arc terranes of west-central Baja California, Mexico, *Geology, 16*, 623-626, 1988b.

Sedlock, R. L., Field evidence for mechanisms of blueschist exhumation, this volume, 1996.

Smith, D. P., and C. J. Busby, Mid-Cretaceous crustal extension recorded in deep-marine half-graben fill, Cedros Island, Mexico, *GSA Bull, 105*, 547-562, 1993.

Vandenberg, L. C., and G. S. Lister, Structural analysis of basement tectonites from the Aegean metamorphic core complex of Ios, Cyclades, Greece, *J. Struct. Geol.*, in press, 1995.

van der Maar, P. A., and J. B. H. Jansen, The geology of the polymetamorphic complex of Ios, Cyclades, Greece and its significance for the Cycladic Massif, *Geologische Rundschau, 72*, 283-299, 1983.

Wijbrans, J. R., and I. McDougall, 1988, Metamorphic evolution of the Attic Cycladic metamoprhic belt on Naxos (Cyclades, Greece) utilizing ^{40}Ar/^{39}Ar age spectrum measurements, *J. Met. Geol., 6*, 571-594, 1988.

Wijbrans, J. R., M.Schliestedt, and D. York, 1990, Single grain argon laser probe dating of phengites from the blueschist to greenschist transition on Sifnos (Cyclades, Greece), *Contrib. Mineral. Petrol. 104*, 582-593, 1990.

S. L. Baldwin, Department of Geosciences, University of Arizona, Tucson, AZ 85721

Tectonic Uplift And Exhumation Of Blueschist Belts Along Transpressional Strike-Slip Fault Zones

Paul Mann

Institute for Geophysics, The University of Texas at Austin, Austin, Texas 78759-8397

Mark B. Gordon

Department of Geology and Geophysics, Rice University, Houston, Texas 77251-1892

Transpressional strike-slip fault zones closely control the distribution, shape, tectonic uplift and exhumation of blueschist belts in the northern Caribbean and circum-Pacific. Tectonic uplift and exhumation of blueschist belts are proposed to occur in all examples by reverse or thrust faults within the upthrown, commonly convex sides of presently active, transpressional strike-slip fault systems that accompany or post-date the phase of subduction responsible for blueschist formation. We identify three fault geometries of transpressional strike-slip faults responsible for the crustal thrust and reverse faulting and consequent uplift and exhumation of blueschist and other deeper crustal rocks: (1) relatively straight strike-slip faults oblique to the direction of relative plate motion; these faults produce elongate, topographic uplifts and exposures of thin belts of deeper crustal rocks; (2) gentle restraining bends with angular curvatures ranging from 10-50° that produce domal areas of topographic uplift and exposure of deeper crustal rocks; the fault bend is much more oblique to the direction of relative plate motion than the adjacent straight fault segments; and (3) sharp restraining bends, or push-up blocks at distinct fault stepovers that produce rectangular areas of topographic uplift and exposure of deeper crustal rocks. In the circum-Caribbean, convergence related to late Cretaceous-Paleogene arc-continent collision events appears to play an important role in uplifting the blueschist rocks to crustal levels affected by Neogene transpressional strike-slip faults.

INTRODUCTION

Previous workers have proposed a variety of tectonic models to explain the tectonic uplift and exhumation of blueschist belts from depths of formation 30-55 km beneath forearc basins and accretionary wedges. Models of blue-schist uplift and exhumation have focussed

[1] By "tectonic uplift" we mean tectonically-induced displacement of the Earth's surface with respect to sea level and by "exhumation" we mean displacement of rocks relative to sea level (England and Molnar, 1990).

Subduction: Top to Bottom
Geophysical Monograph 96
Copyright 1996 by the American Geophysical Union

largely on syn-subduction and subcrustal processes that involve: (1) buoyant rise of subducted material along the approximate plane of subduction [*Ernst*, 1971; *Cloos* and *Shreve*, 1988]; (2) regional uplift driven by underplating of material at the base of the accretionary wedge [*Platt*, 1975]; (3) normal or reverse faulting related to the maintenance of a critically tapered accretionary wedge [*Platt*, 1986]; (4) normal faulting related to subduction shallowing [*Krueger* and *Jones*, 1989]; (5) normal faulting related to post-orogenic collapse [*Dewey*, 1988]; (6) normal faulting related to oblique subduction [*Avé Lallemant* and *Guth*, 1990]; and (7) isostatic rebound of the lower subducted plate [*Michard et al.*, 1994].

Previous tectonically-oriented studies of blueschists have traditionally focussed on outcrop to microscopic study of structural and petrologic features associated with blueschist mineralogies. Few studies have explored scale brittle faults on the crustal uplift of blueschist belts because many previous workers have assumed that brittle faults can account for only the relatively small crustal uplift of blueschists already elevated into the crust by other mechanisms. Karig (1979) and Roeske (1989) have pointed out the possible role of active and ancient strike-slip faults on the crustal uplift of blueschists in forearc and volcanic arc settings based on examples in Sumatra and Alaska. In this paper, we examine the close control that active transpressional strike-slip faults have on the distribution and shapes of blueschist belts in the northern Caribbean and circum-Pacific.

SIGNIFICANCE OF STRIKE-SLIP-RELATED UPLIFT AND EXHUMATION OF BLUESCHIST BELTS

Three basic observations suggest to us that strike-slip faulting provides an important mechanism to uplift blueschists and other deep crustal rocks:

(1) There is a common spatial association of blueschist outcrops with the traces of active or recently active strike-slip faults. This spatial association has been recognized by field workers for many years and suggests that strike-slip faults play a significant but poorly understood role in their crustal uplift and exhumation. For example, Ernst (1975) stated: "Perhaps it is not fortuitous that almost all blueschist terranes are bounded by - or at least closely associated with - post-metamorphic strike-slip faults of considerable movement."

(2) Studies of reworked blueschist clasts have shown that some blueschists are exhumed within a phase of strike-slip faulting that accompanies or post-dates subduction. There are commonly significant time gaps between the blueschist-forming subduction phase and the later strike-slip-related exhumation phases. These time gaps suggest that subduction-related uplift within accretionary wedges may be relatively slower and less effective in exhuming blueschists than syn- or post-subduction strike-slip-related uplift. Examples of lengthy time gaps separating blueschist crystallization and exhumation include: central California [*Cloos*, 1986: Late Jurassic-Late Cretaceous crystallization, mainly early to late Tertiary syn- to post-subduction exposure and erosion]; southern California borderlands [*Bebout et al.*, 1994: Early Cretaceous crystallization, syn- to post- subduction early Middle Miocene exposure and erosion]; Baja California [*Baldwin* and *Harrison*, 1989: mid-Jurassic crystallization, late Early Cretaceous crystallization, syn- to post subduction Neogene exposure], the Sanbagawa belt of Japan [*Yokoyama* and *Itaya*, 1990: Late Cretaceous crystallization, syn-subduction (Eocene) uplift and erosion]; the western Alps of France and Switzerland [*Mange-Rajetsky* and *Oberhänsli*, 1982: Late Cretaceous crystallization, syn- to post-subduction Oligocene exposure]; and the Kodiak Islands of Alaska [*Roeske*, 1989; *Clendenen* and *Byrne*, 1989: Early Jurassic crystallization, Oligocene uplift and exposure]. Available apatite fission track ages on blueschists in Baja California [*Baldwin* and *Harrison*, 1989] and California [*Dumitru*, 1989] indicate that uplift and exhumation accompanied Neogene strike-slip faulting in the region and post-dated Cretaceous subduction. Post-subduction exhumation of blueschist belts along strike-slip faults in California is consistent with the small amount of blueschist clasts found in arc-related basins of Cretaceous and Paleogene age [*Cloos*, 1986]. The significant time lag between blueschist formation and exhumation and their periods of exposure with periods of strike-slip faulting accompanying or post-dating subduction suggests strike-slip faulting may play a significant role in their uplift at crustal levels.

(3) Recent geophysical studies of strike-slip faults in a variety of settings in continental crust (see Beaudoin, 1994, for review) are showing that in many cases these faults are sub-vertical surfaces that offset the base of the crust and therefore are potential upward paths for materials from as deep as 20-30 km with accretionary prisms.

The field examples of strike-slip-related blueschist uplifts reveal three common strike-slip fault geometries that we infer to be responsible for crustal uplift of the blueschist belts described (Fig. 1). Terminology for these strike-slip fault geometries is modified from Crowell (1974). We subdivide our discussion of examples into two categories: blueschist belts in post-subduction tectonic settings and blueschist belts in oblique syn-subduction tectonic settings of the type previously discussed by Karig (1979).

SOME EXAMPLES OF BLUESCHISTS IN ACTIVE AND ANCIENT TRANSPRESSIONAL SETTINGS

Hispaniola, Northern Caribbean

The Samana Peninsula and Cordillera Septentrional of northern Hispaniola in the Dominican Republic are on the northern, upthrown side of the Septentrional fault zone, the main, active strike-slip fault separating the North America and Caribbean plates (Fig. 2A). The fault zone forms a prominent arcuate, topographic limit separating older Cretaceous-Early Pliocene igneous, metamorphic, and sedimentary rocks of the Cordillera Septentrional and Samana Peninsula [*de Zoeten* and *Mann*, 1991] from a Miocene-late Quaternary basinal section in the Cibao Valley (Fig. 2B). The direction of Caribbean-North America relative plate motion is roughly east-west at rates

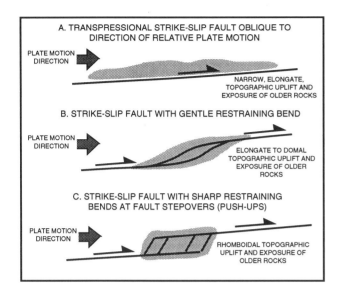

Fig. 1. Three mechanisms for exhumation of deeper crustal rocks along transpressional strike-slip faults. A. Tranpressional strike-slip fault oblique to the direction of relative plate motion. B. Strike-slip fault with gentle restraining bend. C. Strike-slip fault with sharp restraining bends (push-ups). Bend terminology is modified from Crowell (1974). Note the close control of the bend type on the areal shape of the topographic uplift and exposure of older rocks.

ranging from about 10-20 mm/yr [*DeMets et al.*, 1990]. Mann et al. (1990) and de Zoeten and Mann (1991) inferred that the central Cordillera Septentrional constitutes a gentle restraining bend because topographic relief is greatest in the central and most curved part of the fault (Fig. 2). The correspondence of structural and topographic doming suggests that the restraining bend structure is active. Structural doming does not match exactly topographic doming in the Cordillera Septentrional because of the complicating effects of older, now inactive fault systems.

The 10° maximum curvature of the fault in the central Cordillera Septentrional decreases to the east where exposure of a blueschist and eclogite assemblage on the Samana Peninsula are not related to a bend but occur along a relatively straight, transpressional fault segment that is slightly oblique to the east-west direction of relative plate motion (Fig. 1A). Remnants of an extensive Upper Miocene-Lower Pliocene shallow-water carbonate platform sequence suggest that tectonic uplift and exhumation at the gentle bend occurred over the last 5 m.y. in both the Cordillera Septentrional and Samana Peninsula [*de Zoeten and Mann*, 1991].

Blueschists on the Samana Peninsula consist of coherent schist and marble units characterized by the presence of lawsonite and/or pumpellyite [*Joyce*, 1991] (Fig. 2B).

Joyce (1991) estimates that prograde recrystallization of blueschist and eclogite assemblages took place at 400-500° C and 10 kbar. Sm-Nd and K-Ar whole-rock ages from the eclogite blocks suggest that high-pressure mineral assemblages equilibrated in late Cretaceous time but the cooling required for argon retention was not reached until late Eocene. Eclogites occur as boudins in calcite marble and as inclusions within metasedimentary rocks [*Giaramita and Sorensen*, 1994].

Blueschists in the Rio San Juan area of the Cordillera Septentrional occur as two varieties of incoherent blueschist-eclogite melange in a serpentinite matrix that protrude one variety of coherent fine-grained blueschist-greenschist rocks [*Draper* and *Nagle*, 1991]. $^{40}Ar/^{39}Ar$ cooling ages of hornblende yield ages of 90 ± 5 Ma and are thought to record crystallization during late Cretaceous subduction [*Draper* and *Nagle*, 1991]. Serpentinite fragments in Eocene sedimentary rocks overlying the incoherent blueschist-greenschist rocks are interpreted as a brief period of subaerial exposure of the complex related to the collision of the arc with the southeastern edge of the Bahama Platform. This collision event terminated arc activity along this margin and was followed by an Eocene to present phase of left-lateral strike-slip faulting [*Joyce*, 1991; *Draper* and *Nagle*, 1991]. During this prolonged strike-slip phase, blueschist rocks were shallowly buried beneath a veneer of Eocene to Miocene sedimentary rocks and subsequently tectonically uplifted and exhumed by restraining bend tectonics in Plio-Pleistocene time.

Jamaica, Northern Caribbean

The Blue Mountains of eastern Jamaica are within the convex-northward, upthrown side of the Enriquillo-Plantain Garden and Yallahs fault zones, the main, active left-lateral strike-slip faults separating the Caribbean plate from the Gonave microplate to the north [*Mann et al.*, 1985] (Fig. 3A). The Gonave microplate is an elonagate microplate formed in Late Neogene time between the much larger North America and Caribbean plates.

This complex fault zone forms a prominent topographic limit separating older, exhumed Cretaceous-Paleocene igneous, metamorphic, and sedimentary rocks of the Blue Mountains from Eocene-late Quaternary clastic and carbonate units fringing the domal uplift (Fig. 3B). The direction of Caribbean-North America relative plate motion is roughly east-west at rates ranging from about 4 mm/yr averaged over the last 10 my [*Mann et al.*, 1990]. Mann et al. (1985) inferred that the Blue Mountains constituted a gentle restraining bend (Fig. 1B) because topographic relief was greatest near the most curved segments of the Yallahs

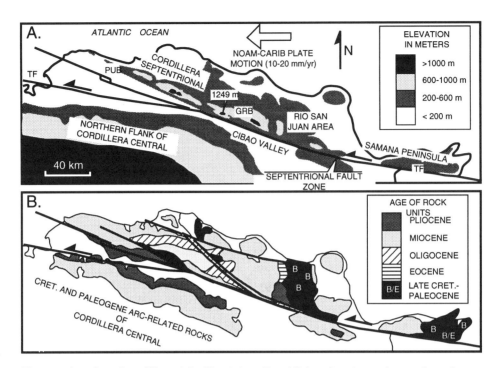

Fig. 2 A. Topography of northern Hispaniola (Dominican Republic) and active and recently active traces of the left-lateral strike-slip fault (Septentrional fault) separating the North America and Caribbean plates. Note that the direction of relative plate motion of North America-Caribbean relative motion is east-west and that the area of highest topography is adjacent to the segment of the fault most oblique to this direction. GRB = gentle restraining bend; PUB = push-up block; TF = transpressional fault zone (compare to Figure 1). B. Map showing the ages of exposed rock units in the bend area. Note the presence of blueschists (B) along the upthrown eastern edge of the gentle restraining bend (GRB) in the central Cordillera Septentrional and the presence of blueschists (B) and eclogites (E) along the transpressional fault (TF) bounding the southern edge of the Samana Peninsula.

and Blue Mountains fault zones and because the strikes of folds and thrust faults mapped in Tertiary sedimentary rocks are parallel to the bend. Structural doming seen in the outcrop pattern of older rocks matches well the topographic doming of the Blue Mountains and suggests the present activity of the bend. Structural and topographic doming in Jamaica is more pronounced than Hispaniola because there is much greater fault curvature of faults through an angle of about 50° (Fig. 3B). Structural doming and clast compositions of Neogene sedimentary rocks indicate exhumation of deeper crustal rocks in the restraining bend area initiated in Late Miocene time and remains active to the present [Mann et al., 1985].

Blueschists occur in several discontinuous outcrops about 0-10 km north of the major strike-slip fault zones (Fig. 3B). Blueschists consist of metamorphosed basalts, gabbros and peridotites thought to have been metamorphosed in an Early Cretaceous (pre-Campanian) subduction zone [Draper, 1986]. With one minor exception, these are the only outcrops of metamorphic rocks on Jamaica. The close spatial association of these rocks with the restraining bend faults suggests that the metamorphic rocks were uplifted as a result of transpression at the Neogene restraining bend, although Draper (1986) notes that blueschist clasts appear in post-subduction clastic sediments of Paleocene and early Eocene age. The ending of subduction in Jamaica may be related to either post-Campanian readjustments in an intraoceanic, southward-facing arc [Draper, 1986] or collision of an intraoceanic, northward-facing arc with continental crust in Central America [Mann and Burke, 1990].

The strong foliation and metamorphic minerals of the blueschist belt were formed in an early Cretaceous accretionary prism at depths greater than 20 km but early fabrics defined by oriented minerals were subsequently folded in Neogene time [Draper, 1986]. We infer that this Neogene folding is related to restraining bend transpression and uplift because folds in the schists have northwest-trending axes sub-parallel to the curvature of the bend and are more common adjacent to the Yallahs and Blue Mountain fault zones than north of the more east-west striking Enriquillo-Plantain Garden fault zone. Outcrops of

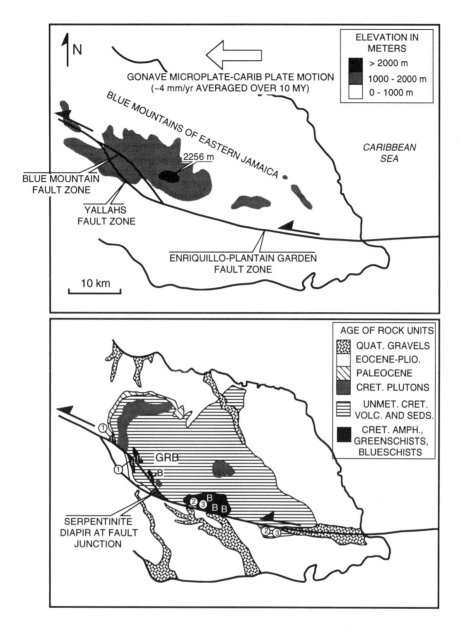

Fig. 3. A. Topography of eastern Jamaica and its relation to active traces of the left-lateral strike-slip faults separating the Gonave microplate from the Caribbean plate. Note that the direction of relative plate motion of North America-Caribbean relative motion is east-west and that the area of highest topography is adjacent to the segment of the fault most oblique to this direction. B. Map showing the ages of exposed rock units in the bend area. Numbers indicate points that have been left-laterally offset by 3.5 to 12.8 km. Note the presence of Cretaceous amphibolites, greenschists, and blueschists along the upthrown side of the gentle restraining bend (GRB).

blueschist rocks are also more elongate adjacent to the Blue Mountain and Yallahs fault zones, reflecting a dense network of predominately northwest-striking faults bounding the outcrops. The continuity of foliation and lineation measurements across the extent of the blueschist outcrops suggest that they represent coherent blocks at the scale of individual blocks. A serpentinite diapir occurs at the eastern junction of the Yallahs and Blue Mountain fault zone and probably represents the filling of a divergent gap formed by divergent fault motion at the junction (Fig. 3B).

Guatemala, Northern Central America

Jadeitite, a rock composed principally of the high pressure mineral jadeite, is found as blocks within a 1-4 km wide, 15 km long belt of serpentinite melange along the active, left-lateral Motagua fault zone separating the North America and Caribbean plates in northern Central America [*Harlow*, 1994] (Fig. 4A). The belt of serpentinite melange is located at the foot of the highest peak (Cerro Raxón at 3015 m) in the elongate and extremely steep-sided Sierra de las Minas. Mann et al. (1990) interpreted the formation of the Sierra de Minas and its exposures of serpentinite and deeper crustal rocks as the result of tectonic uplift in a sharp restraining bend or push-up block between the Motagua and Polochic fault zones. The outcrop pattern and trends of measured foliation planes in Mesozoic metamorphic rocks [*Kesler*, 1971] outlines the 300 X 100 km area (Fig. 4B) that best corresponds to the bend type shown in Figure 1C. Curvature of the Motagua fault may add to the bend effects in this area by the gentle restraining bend mechanism (Fig. 1B).

Harlow (1994) reports that jadeitite can only be loosely constrained with blueschist conditions (100° < 400°C; 5 < P < 11 kb) and interprets jadeitites as metasomatic alterations of a felsic knocker protolith affected by fluid flow in a high P/T setting. V. B. Sisson (pers. comm., 1995) proposes that jadeitites, while not typical blueschist rocks, do have peak PT conditions that do overlap with the blueschist facies field. The occurrence of jadeitite is rare worldwide but is commonly associated with ultramafic belts, ophiolites, and blueschists [*Harlow*, 1994]. The age of crystallization of the Motagua jadeitite is thought to be late Cretaceous and uplift is assumed to be from late Cretaceous to late Cenozoic [*Donnelly et al.*, 1990; *Fourcade et al.*, 1994]. This age of uplift of based on $^{40}Ar^{39}Ar$ and $^{40}K^{36}Ar$ age dates of mafic and metamorphic rocks in the Motagua fault zone that suggesting that blocking temperatures of these minerals were reached at 58.5 ± 3.7 m.y. ($^{40}Ar^{36}Ar$ and $^{40}K^{36}Ar$ dating by Bertrand et al., 1978) and between 78.0 and 63.7 m.y. ($^{40}Ar^{39}Ar$ dating by Sutter, 1979). Sutter (1979) notes that the oldest $^{40}Ar^{39}Ar$ ages of 78 m.y. (Campanian) are from the highest tectonic units and are probably are the closest estimate of the time of collision between continental crust in northern Central America and a north-facing arc. Nearby continental deposits assumed to be of Miocene age contain numerous cobbles of serpentinite.

Southern California Borderlands and Los Angeles Basin, USA

The California borderlands physiographic province extends from the Transverse Ranges of south-central California to the central part of the Baja Peninsula of Mexico and is characterized by elongate, sinuous, northwest-southeast trending ridges separated by flat-floored basins (Fig. 5A). Tectonically, the borderlands province forms part of a diffuse, right-lateral shear zone between the Pacific and North America plates that is centered on the San Andreas fault (Fig. 5B). The predicted N35°W direction of Pacific plate motion relative to North America and total rate of 56 mm/yr [*DeMets et al.*, 1990] is manifested across a broad zone of sub-parallel active, northwest-striking right-lateral faults [*Feigl et al.*, 1993]. The northwest direction of motion causes fault blocks bounded by right-lateral faults in the borderlands and in southern California to intersect the more westerly striking, reverse faults bounding the Santa Monica Mountains and Transverse Ranges (Fig. 5B). Active seismicity of the borderlands province and southern California indicates right-lateral displacement along N30°W-striking faults and thrusting and oblique-slip movement along more westerly striking thrusts at the base of the Santa Monica Mountains and Transverse Ranges [*Hutton et al.*, 1991].

Offshore faults in the borderlands province of southern California mapped by mostly non-systematic seismic reflection profiling have been compiled by Jennings (1975) and are reproduced on Figure 5A. Faults considered by Jennings to be Quaternary or younger than 2 million years include only about 15% of the faults shown on Figure 5A but are present in segments of each of the main fault zones summarized on Figure 5B. Major strike-slip faults include: the Newport-Inglewood-Rosewood and Malibu Coast-Santa Monica fault zones [*Hutton et al.*, 1991]; the Palos Verde Hills-Coronado Bank fault zone [*Legg*, 1985; *Ward* and *Valensise*, 1994]; the San Diego Trough fault zone [*Legg*, 1985]; and the San Clemente fault zone [*Legg et al.*, 1989]. Based on the regional alignments of fault segments, we propose that the San Diego Trough fault zone of Legg (1985) steps left through an angle of about 15° to form a major gentle restraining bend (Fig. 1B) with its topographic and structural culmination marked by Catalina Island. An unnamed zone of faults to the northwest of Santa Catalina Island is proposed to represent the continuation of the San Diego Trough fault zone (Fig. 5B).

Hauksson and Jones (1988) proposed that the 1986 Oceanside earthquake (M_L = 5.3) dips to the north and strikes parallel to the San Diego Trough fault zone (Fig. 5B). The main shock of this event showed reverse faulting on an east-southeast striking plane. These authors concluded that the Oceanside event ruptured a small left-stepping restraining bend just south of Crespi Knolls on the southeastward continuation of the Santa Catalina bathymetric ridge (Fig. 5A).

A gentle restraining bend is also present on the Palos Verdes Hills -Coronado Bank fault zone at the Palos Verdes

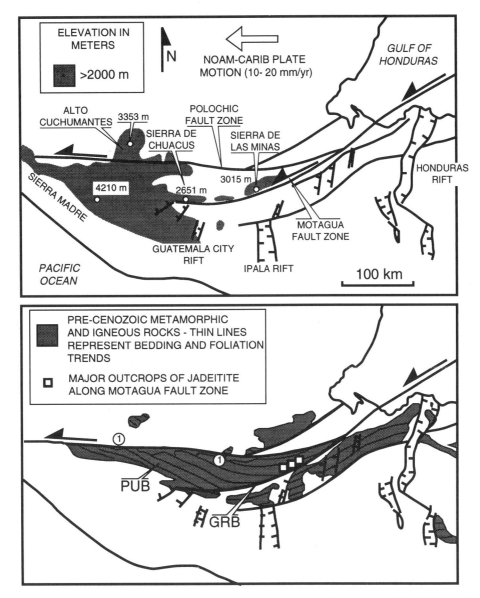

Fig. 4. A. Topography of northern Central America and its relation to active traces of the left-lateral strike-slip faults separating the North America and Caribbean plates. Note that the direction of relative plate motion of North America relative to the Caribbean is approximately east-west and that the area of highest topography is adjacent to the segment of the fault most oblique to this direction. B. Map showing the ages of exposed rock units in the bend area. Numbers 1-1 indicate points that have been left-laterally offset by 130 km along the Polochic fault [*Burkart*, 1983]. We propose that the topographic highlands and exposures of deeper crustal rocks in the Sierra de las Minas, Sierra de Chuacus, and the eastern part of the Sierra Madre may be related to the combined effects of a sharp stepover or push-up block (PUB) between the Motagua and Polochic fault zones and a gentle restraining bend (GRB) along the Motagua fault zone. Jadeitite outcrops are restricted to the area of the three points shown in the Sierra de las Minas.

Hills. Buried reverse faults at this gentle restraining bend uplift a staircase of marine terraces of Quaternary age at rates of about 0.2 mm/yr and exhume a core of blueschist rocks of the Catalina Schist [*Ward* and *Valensise*, 1994]. We propose that the Lasuen Knolls along the southeastern extension of the Palos Verdes-Coronado Bank fault zone is a similar, though less topographically prominent, gentle restraining bend. The Newport-Inglewood-Rose Canyon and San Clemente fault zones appear to lack gentle restraining bends (Fig. 5B). These faults appear to be

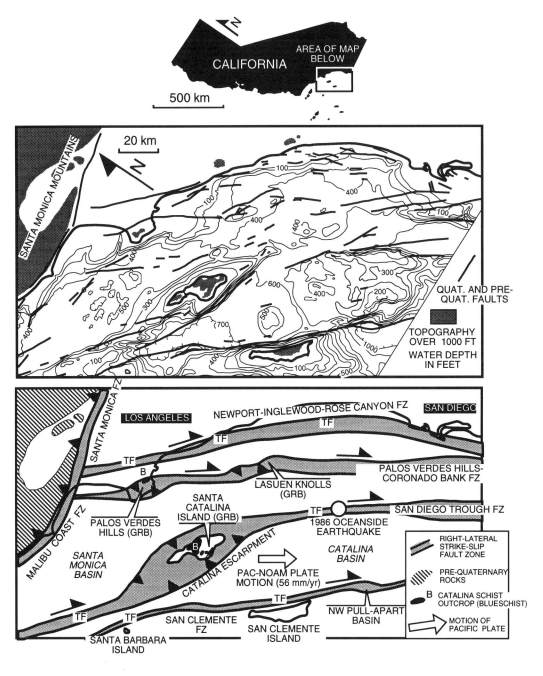

Fig. 5. A. Topography of the coastal area of southern California, the southern California borderlands and and the relation of topographic and bathymetric relief to active traces of the right-lateral strike-slip faults separating the North America and Pacific plates. Note that the direction of relative plate motion of the Pacific plate relative to North America is approximately southeast and that the areas of highest topography and shallowest water depths are adjacent to the east-southeast-trending segments of faults that are more oblique to this direction. B. Map showing the locations of pre-Quaternary rocks and Cretaceous blueschist rocks. We propose that the topographic highlands and exposures of blueschist rocks on the Palos Verdes Peninsula and Santa Catalina Island are related to gentle restraining bends (GRB) on the faults shown. The Lasuen Knolls is a shallowly submerged bank that may also be a gentle restraining bend. Transpressional faults (TF) slightly oblique to the direction of plate motion but lacking perceptible gentle bends include the Newport-Inglewood-Rose Canyon fault zone with elevations along the California coastline and the San Clemente fault zone with elevations on Santa Barbara and San Clemente Islands.

relatively straight transpressional faults oblique to the direction of relative plate motion (Fig. 1A). This fault geometry may account for the relatively straight and topographically elevated coastline between Los Angeles and San Diego. The upthrown side of the transpressional fault corresponds to the northeastern, higher block of the Newport-Inglewook-Rose Canyon fault zone. On the San Clemente fault zone, the upthrown side of the transpressional fault corresponds to the relatively straight and narrow uplift exposing Santa Barbara and San Clemente Islands. The lack of gentle bends or push-up blocks on the Newport-Inglewood-Rose Canyon and San Clemente fault zones may explain why neither of these faults exposes blueschist rocks of the Santa Catalina Schist. The simple pattern of exhumed deeper crustal rocks and topographic highs along transpressional strike-slip faults does not support the model by Crouch and Suppe (1993) for regional, low angle normal faults extending in a northeast-southwest direction and exhuming deeper crustal rocks in their footwall uplifts.

Rocks collectively assigned to the Catalina Schist and locally exposed at the Palos Verdes Hills and on Santa Catalina Island include blueschist-facies metagraywacke and metavolcanic rocks and also glaucophanic greenschist and amphibolite-facies rocks [*Crouch* and *Suppe*, 1993; *Bebout et al.*, 1994]. The pre-Late Cretaceous age of the Catalina Schist, the rock types present, and the style of deformation and metamorphism of these rocks indicate that they can be correlated with the Franciscan subduction complex of central California. The conditions of formation for the Catalina Schist included regional metamorphism at pressures of about 7-14 kb that correpond to paleodepths of 15-45 km and temperatures ranging from 300-600° C [*Bebout et al.*, 1994]. The Catalina Schist is overlain by the middle Miocene San Onofre breccia which contains angular clasts of the schist.

Japan

Blueschist rocks in Japan are concentrated along major strike-slip zones or "tectonic lines" in northern and southern Japan that are associated with active strike-slip faults that accommodate oblique plate convergence (Fig. 6). The central part of Japan which is undergoing convergence in an orthogonal direction lacks significant active strike-slip faults and blueschist exposures.

In southern Japan, blueschists of the Chicibu-Sanbagawa belts are located between the Median and Butsuzo tectonic lines. The Median tectonic line exhibits evidence for several tens of kilometers of left-lateral displacement during Cretaceous and Paleogene time with right-lateral motion

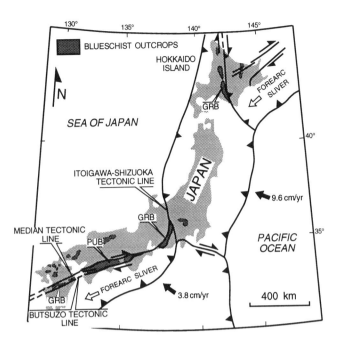

Figure 6. Map of major blueschist belts in Japan (from Banno, 1986; Banno and Nakajima, 1992) and their relation to active or recently active strike-slip faults (from Taira et al., 1983). Filled arrows indicate plate convergence directions with rates in cms/yr from DeMets et al. (1990). Open arrows indicate motions of forearc slivers driven by oblique plate convergence. We propose that the majority of blueschist outcrops occurs along the following two types of transpressional strike-slip faults: GRB = gentle restraining bend; PUB = push-up block. Note that few blueschists are exposed in central Japan where the Pacific Ocean is subducting orthogonally and there are few Neogene strike-slip faults.

beginning in Neogene time [*Taira et al.*, 1983; *Banno*, 1986] and culminating in about 10 km of right-lateral displacement in the Quaternary [*Itoh* and *Takemura*, 1993]. High P/T metamorphism occurred during early Cretaceous oblique subduction [*Hara et al.*, 1983; *Faure et al.*, 1986]. Yokoyama and Itaya (1990) identified blueschist clasts with isotopic ages of 120-85 Ma in middle Eocene conglomerate that unconformably overlies the Sanbagawa belt. They interpreted the conglomerate as a record of strike-slip-related exhumation of the belt during an Eocene phase of left-lateral strike-slip movement along the Median fault zone.

We propose that blueschist belts along the two tectonic lines occur as either gentle restraining bends (Fig. 1B) on the Butsuzo tectonic line or as a push-up blocks in a right-stepping push-up block between the Butsuzo and Median tectonic lines (Fig. 6). We interpret the blueschist belt at the northern end of the Median tectonic line near its intersection with the Itoigawa-Shizuoka tectonic line as a left-stepping gentle restraining bend.

Blueschists in northern Japan (Hokkaido Island) are concentrated along the central Hokkaido where a forearc sliver driven by oblique subduction along the Kuril trench is impinging the Japanese arc [*Taira et al.*, 1983; *Kimura*, 1986] (Fig. 6). The largest area of blueschist exposures is concentrated along a north-striking Miocene to Recent right-lateral fault that extends 2000 km northward from Hokkaido through Sakhalin Island [*Fournier et al.*, 1994]. The largest blueschist belt in northern Japan appears to be located on a gentle restraining bend of the active right-lateral fault zone that extends northward through Sakhalin Island.

Sumatra

Karig (1979) proposed the idea that strike-slip faults could serve as sub-vertical conduits for the uplift and exhumation of blueschist rocks formed deep within accretionary prisms. He pointed out that such conduits were necessary for deep-seated rocks to penetrate the commonly thick slope section blanketing the underlying accretionary prism. His observational basis for this conceptual model were exposures of high pressure amphibolites in blocks emplaced along the Neogene left-lateral Batee strike-slip fault zones that cuts accretionary prism sedimentary rocks on the island of Nias in the Sumatran forearc. In this region of the Sumatra forearc the Indian plate is obliquely converging in a northerly direction at a rate of about 10 cm/yr [*DeMets, et al.*, 1990]. The Batee fault has a strong, right-stepping curvature indicative of a gentle restraining bend at Nias Island (see Figure 3 of Karig, 1979). Rock types adjacent to the Batee fault include amphibolite, gabbros, serpentinized harzburgite, and metagraywacke. No true blueschists were identified although its possible that the amphibolite was retrogressively metamorphosed from eclogite. Like the Japanese example, the Nias Island example illustrates the possibility of transpressional strike-slip faults forming in oblique subduction settings and acting as conduits along which accretionary prism material can be exhumed.

PROPOSED SYN- TO POST-SUBDUCTION MODEL FOR THE UPLIFT OF BLUESCHISTS

An observation-based, conceptual model for the uplift and exhumation of blueschists modified from Platt (1986) using the examples discussed here is shown in Figure 7A-D. The first stage represents early to intermediate stages of prism formation when underplating of subducted material leads to its blueschist metamorphism under conditions > 5.5 kb pressure that corresponds to depths > 20 km (Fig. 7A).

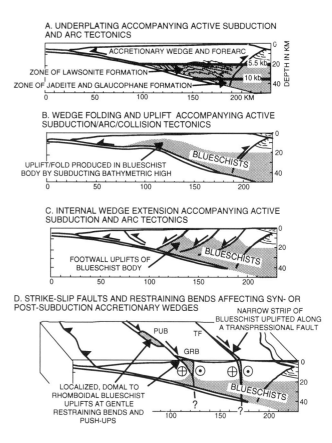

Fig.7. Proposed model for tectonic uplift and exhumation of blueschist rocks that involves a continuum between syn-subduction shortening (B) or lengthening (C) of the accretionary wedge [*Platt*, 1986] followed by a syn- or post-subduction phase of tectonic uplift related to activity of syn- to post-strike-slip faults cutting the accretionary wedge. In D, strike-slip faults are hypothesized as sub-vertical planes penetrating the entire crust and subduction is not necessarily active.

Subduction proceeds in a steady-state mode and there is no differential uplift of the blueschist body. The second and third stages shown in Figures 7B and C represent tectonic events perturbing the steady state subduction system and causing the accretionary wedge to either shorten (Fig. 7B) or lengthen (Fig. 7C). Such differential movement results in the upward displacement of the blueschist body and could occur by reverse faulting in response to the subduction of bathymetric highs as shown in Figure 7B or by normal faulting during lengthening of the wedge towards critical taper Figure 7C [*Platt*, 1986]. Upward movement could be either as an en bloc uplift within a transpressional strike-slip zone or diapiric rise of blueschist blocks entrained in a serpentinite matrix deformed by the transpressional strike-slip fault zone.

Our three dimensional modification of the Platt's (1986) two dimensional model for differential vertical movements of blueschists along strike-slip faults is shown in Figure 7D. We show local exposures of blueschist rocks resulting from tectonic uplift at either gentle bends (Fig. 1A), push-ups (Fig. 1C), or relatively straight transpressional faults (Fig. 1A). Blueschist uplift is proposed to occur in all three types of structures by crustal reverse or thrust faults within the upthrown, commonly convex sides of presently active, transpressional strike-slip fault systems. Note that the transpressional strike-slip faults could serve to accommodate either the upward flow of incoherent blueschist bodies within serpentinite/shaly matrices or the upward brittle thrusting of coherent blueschist bodies. The subduction phase responsible for producing the blueschists as seen in Figure 7A may or may not be active during the strike-slip phase represented in Figure 7D.

Acknowledgments. We thank G. Harlow for kindly providing preprints and S. Roeske and an anonymous reviewer for providing comments. UTIG contribution no. 1187.

REFERENCES

Ave Lallemant, H., and Guth, L., 1990, Role of extensional tectonics in exhumbation of eclogites and blueschists in an oblique subduction setting: Geology, v. 18, p. 950-953.

Baldwin, S., and Harrison, T., 1989, Geochronology of blueschists from west-central Baja California and timing of uplift in subduction complexes: Journal of Geology, v. 97, p. 149-163.

Banno, S., 1986, The high-pressure metamorphic belts of Japan: A review, in Evans, B. W., and Brown, E. H., eds., Blueschists and Eclogites, Geological Society of America Memoir, v. 164, p. 365-374.

Banno, S., and Nakajima, T., 1992, Metamorphic belts of Japanese islands: Annual Reviews of Earth and Planetary Sciences, v. 20, p. 159-179.

Beaudoin, B., 1994, Lower-crustal deformation during terrane dispersion along strike-slip faults: Tectonophysics, v. 232, p. 257-266.

Bebout, G. E., Grove, M., Sorensen, S. S., and Platt, J. P., 1994, Field trip guidebook: The Catalina Schist - A 15 to 45 Kilometer "Slice" of an Early Cretaceous Accretionary Complex, SUBCON: Subduction Top to Bottom Meeting, Avalon, California, June 12 to 17, 1994, 56 p.

Bertrand, J., Delaloye, M., Fontingnie, D., and Vaugnat, M., 1978, Ages (K-Ar) sur diverses ophiolites et roches associées de la Cordillére centrale du Guatémala: Bulletin Suisse de Mineralogie et Petrographie, v. 58, p. 405-412.

Burkart, B., 1983, Neogene North America-Caribbean plate boundary across northern Central America: Offset along the Polochic fault: Tectonophysics, v. 99, p. 251-270.

Clendenen, W., and Byrne, T., 1989, Late Cretaceous increase in uplift and cooling rates of the Kodiak accretionary prism and evidence for a change from an intraoceanic to a continental margin setting: Geological Society of America Abstracts with Programs, v. 21, p. A314.

Cloos, M., 1986, Blueschists in the Franciscan complex of California: Petrotectonic constraints on uplift mechanisms, in Evans, B., and Brown, E., eds., Blueschists and Eclogites, Geological Society of America Memoir 164, p. 77-93.

Cloos, M., and Shreve, R., 1988, Subduction-channel model of prism accretion, melange formation, sediment subduction, and tectonic erosion at convergent plate margins: 1. Background and description: Pure and Applied Geophysics, v. 128, p. 455-500.

Crouch, J. K., and Suppe, J., 1993, Late Cenozoic tectonic evolution of the Los Angeles basin and inner California borderland: A model for core complex-like crustal extension: Geological Society of America Bulletin, v. 105, p. 1415-1434.

Crowell, J., 1974, Origin of late Cenozoic basins in southern California, in Dickinson, W. R., ed., Tectonics and Sedimentation: Society of Economic Paleontologists and Mineralogists Special Publication 22, p. 190-204.

DeMets, C., Gordon, R., Argus, D., and Stein, S., 1990, Current plate motions: Geophysical Journal International, v. 101, p. 425-478.

Dewey, J. F., 1988, Extensional collapse of orogens: Tectonics, v. 7, p. 1123-1139.

de Zoeten, R., and Mann, P., 1991, Structural geology and Cenozoic tectonic history of the central Cordillera Septentrional, Dominican Republic, in Mann, P., Draper, G., and Lewis, J. F., eds., Geologic and Tectonic Development of the North America-Caribbean Plate Boundary in Hispaniola, Geological Society of America Special Paper 262, p. 265-279.

Donnelly, T., Horne, G., Finch, R., and López-Ramos, E., 1990, Northern Central America; the Maya and Chortis blocks, in Dengo, G., and Case, J. E., eds., The Caribbean region: Boulder, Colorado, Geological Society of America, The Geology of North America, v. H., p. 37-76.

Draper, G., 1986, Blueschists and associated rocks in eastern Jamaica and their significance for Cretaceous plate-margin development in the northern Caribbean: Geological Society of America Bulletin, v. 97, p. 48-60.

Draper, G., and Nagle, F., 1991, Geology, structure, and tectonic development of the Rio San Juan complex, northern Dominican Republic, in Mann, P., Draper, G., and Lewis, J. F., eds., Geologic and Tectonic Development of the North America-Caribbean Plate Boundary in Hispaniola, Geological Society of America Special Paper 262, p. 77-95.

Dumitru, T., 1989, Constraints on uplift in the Franciscan subduction complex from apatite fission track analysis: Tectonics, v. 8., p. 197-220.

England, P., and Molnar, P., 1990, Surface uplift, uplift of rocks, and exhumation of rocks: Geology, v. 18, p. 1173-1177.

Ernst, W., 1971, Metamorphic zonations on presumably subducted plates from Japan, California, and the Alps: Contributions to Mineralogy and Petrology, v. 34, p. 43-59.

Ernst, W., 1975, Introduction, in Ernst, W., ed., Subduction Zone Metamorphism: Stroudsburg, Pennsylvania, Dowden, Hutchinson, and Ross, p. 1-14.

Faure, M., Caridroit, M., and Charvet, J., 1986, The Late Jurassic oblique collisional zone of SW Japan: New structural data and synthesis: Tectonics, v. 5., p. 1089-1114.

Feigl, K., and 14 others, 1993, Space geodetic measurement of crustal deformation in central and southern California, 1984-1992: Journal of Geophysical Research, v. 98, p. 21,677-21,712.

Fournier, M., Jolivet, L., Huchon, P., Sergeyev, K., and Oscorbin, L., 1994, Neogene strike-slip faulting in Sakhalin and the Japan Sea opening: Journal of Geophysical Research, v. 99, p. 2701-2725.

Fourcade, E., Mendez, J., Azéma, J., Cros, P., De Wever, P., Duthou, J. L., Romero, J., and Michaud, F., 1994, Age présantonien-campanien de l'obduction des ophiolites du Guatemala: Comptes Rendus de la Académie des Sciences Paris, v. 318, series II, p. 527-533.

Giaramita, M. J., and Sorensen, S. S., 1994, Primary fluids in low-temperature eclogites: Evidence from two subduction complexes (Dominican Republic, and California, USA): Contributions Mineralogy and Petrology, v. 117, p. 279-292.

Hara, I., Shyoji, K., Sakurai, Y., Yokoyama, W., and Kide, K., 1980, Origin of the Median tectonic line and its original shape: Memoirs of the Geological Society of Japan, no. 18, p. 27-49.

Harlow, G. E., 1994, Jadeitites, albitites, and related rocks from the Motagua fault zone, Guatemala: Journal of Metamorphic Geology, v. 12, p. 49-68.

Hauksson, E., and Jones, L., 1988, The July 1986 Oceanside (M_L = 5.3) earthquake sequence in the continental borderland, southern California: Bulletin of the Seismological Society of America, v. 78, p. 1885-1906.

Hutton, L. K., Jones, L. M., Hauksson, E., and Given, D. D., 1991, Seismotectonics of southern California: in Slemmons, D. B., Engdahl, E. R., Zoback, M. D., and Blackwell, D. D., eds., Neotectonics of North America: Boulder, Colorado, Geological Society of America, Decade Map Volume 1, p. 133-152.

Itoh, Y., and Takemura, K., 1993, Quaternary geomorphic trends within southwest Japan: Extensive wrench deformation related to transcurrent motions of the Median tectonic line: Tectonophysics, v. 227, p. 95-104.

Jennings, C. W., 1975, Fault map of California with locations of volcanoes, thermal springs, and thermal wells, California Geologic Data Map Series, Map no. 1, scale 1:750,000 (fourth printing, 1988).

Joyce, J. J., 1991, Blueschist metamorphism and deformation of the Samana Peninsula - A record of subduction and collision in the Greater Antilles, in Mann, P., Draper, G., and Lewis, J. F., eds., Geologic and Tectonic Development of the North America-Caribbean Plate Boundary in Hispaniola, Geological Society of America Special Paper 262, p. 47-76.

Karig, D., 1979, Material transport within accretionary prisms and the "knocker" problem: Journal of Geology, v. 88, p. 27-39.

Kesler, S. E., 1971, Nature of ancestral orogenic zone in nuclear Central America: American Association of Petroleum Geologists Bulletin, v. 55, p. 2116-2129.

Kimura, G., 1986, Oblique subduction and collision: Forearc tectonics of the Kuril arc: Geology, v. 14, p. 404-407.

Krueger, S., and Jones, D, 1989, Extensional fault uplift of regional Franciscan blueschists due to subduction shallowing during the Laramide orogeny: Geology, v. 17, p. 1157-1159.

Legg, M. R., 1985, Geologic structure and tectonics of the Inner Continental Borderland of southern California and northern Baja California, Mexico, Ph. D. thesis, University of California at San Diego.

Legg, M. R., Luyendyk, B. P., Mammerickx, J., de Moustier, C., and Tyce, R. C., 1989, Sea Beam survey of an active strike-slip fault: The San Clemente fault in the southern California continental borderland: Journal of Geophysical Research, v. 94, p. 1727-1744.

Mange-Rajetsky, M., and Oberhänsli, R., 1982, Detrital lawsonite and blue sodic amphibole in the molasse of Savoy, France, and their significance in assessing Alpine evolution: Schweizer Mineralogische und Petrographiches Mittleidungen, v. 62, p. 415-436.

Mann, P., Draper, G., and Burke, K., 1985, Neotectonics of a strike-slip restraining bend system, Jamaica: in Biddle, K. T., and Christie-Blick, N., eds., Strike-slip Deformation, Basin Formation, and Sedimentation, SEPM Special Publication, No. 37, p. 211-226.

Mann, P., Schubert, C., and Burke, K., 1990, Review of Caribbean neotectonics, in Dengo, C., and Case, J. E., eds., The Caribbean region: Boulder, Colorado, Geological Society of America, The Geology of North America, v. H., p. 307-338.

Mann, P., and Burke, K., 1990, Transverse intra-arc rifting: Palaeogene Wagwater belt, Jamaica: Marine and Petroleum Geology, v. 7, p. 410-427.

Michard, A., Goffé, B., Saddiqi, O., Oberhänsli, R., and Wendt, A., 1994, Late Cretaceous exhumation of the Oman blueschists and eclogites: A two-stage extensional mechanism: Terra Research, v. 21, p. 404-413

Platt, J., 1975, Metamorphic and deformational processes in the Franciscan complex, California: Some insights from the Catalina schist terrane: Geological Society of America Bulletin, v. 86, p. 1337-1347.

Platt, J., 1986, Dynamics of orogenic wedges and the uplift of high-pressure metamorphic rocks: Geological Society of America Bulletin, v. 97, p. 1037-1053.

Roeske, S. 1989, Presence of high P/low T metamorphic terranes adjacent to island arc basement: Evidence for uplift during oblique subduction?: Geological Society of America Abstracts with Programs, v. 21, p. A215.

Sutter, J., 1979, Late Cretaceous collisional tectonics along the Motagua fault zone, Guatemala: Geological Society of America Abstracts with Programs, v. 11, p. 525-526.

Taira, A., Saito, Y., and Hashimoto, M., 1983, The role of oblique subduction ans strike-slip tectonics in the evolution of Japan: in Hilde, T., and Uyeda, S., editors, Geodynamics of the Western Pacific-Indonesia region, Geodyaamics Series, v. 11, American Geophysical Union, Washington, D. C., p. 303-316.

Ward, S. N., and Valensise, G., 1994, The Palos Verdes terraces, California: Bathtub rings from a buried reverse fault: Journal of Geophysical Research, v. 99, p. 4485-4494.

Yokoyama, K., and Itaya, T., 1990, Clasts of high-grade Sanbagawa schist in Middle Eocene conglomerates from the Kuma Group, central Shikoku, south-west Japan: Journal of Metamorphic Geology, v. 8, p. 467-474.

Mark Gordon, Department of Geology and Geophysics, P. O. Box 1892, Rice University, Houston, Texas 77251-1892

Paul Mann, Institute for Geophysics, The University of Texas at Austin, 8701 Mopac Boulevard, Austin, Texas 78759-8397.

Syn-Subduction Forearc Extension and Blueschist Exhumation in Baja California, México

Richard L. Sedlock

Department of Geology, San José State University, San José, California

Regionally metamorphosed, structurally coherent Cretaceous blueschists in western Baja California, México form the footwalls of major shallowly-dipping normal fault systems. The hanging walls of these major normal faults consist of Mesozoic arc, ophiolite, and forearc basin rocks that formed part of the Cretaceous forearc above an active subduction zone. All Mesozoic rocks underwent extensional strain of probable Late Cretaceous to Paleogene age. Exhumation of the blueschists from depths of 15-30+ km is interpreted to have occurred during syn-subduction extension of the North American forearc between >95 Ma and about 40-30 Ma, and thus is unrelated to Basin and Range extension or the opening of the Gulf of California. The structural style of extension is similar in many ways to that of metamorphic core complexes. This study is the first to document extension and exhumation of this age in Baja.

1. INTRODUCTION

Most surface exposures of blueschists are inferred to represent exhumed parts of ancient accretionary prisms because appropriate P-T conditions (roughly 0.4-1.2 GPa, 170-350°C) are attained only in subduction zones. Such blueschists must have been subducted to blueschist-facies depths while attached to the subducting plate, transferred at depth to the overriding plate, and then exhumed to shallow crustal levels. Blueschist preservation implies one or both of the following: (1) subducted rocks were exhumed so rapidly that higher-temperature overprints (e.g., greenschist-facies assemblages) had insufficient time to develop; (2) subducted rocks were exhumed during steady-state subduction and continual depression of forearc isotherms, i.e., constant refrigeration of the forearc.

Blueschists occur in Baja, as in orogenic belts throughout the world, in two forms. Isolated *tectonic blocks* are suspended in a matrix of serpentinite or argillite, and typically have long dimensions of meters to hundreds of meters. In contrast, *regionally metamorphosed coherent tracts* may have long dimensions of kilometers to hundreds of kilometers, and consist of structurally coherent sequences of basalt, ribbon chert, and terrigenous turbidites.

Most block-sized blueschists may have risen through the crust within diapirs of serpentinite or argillite, or within strike-slip fault zones [e.g., *Karig*, 1980]. However, exhumation of large tracts of regionally metamorphosed blueschists poses difficult kinematic problems. Explanations based on the buoyant rise of subducted material and erosion of the overburden [*Ernst*, 1971] may not satisfactorily explain the high structural level of dense units, lack of a higher-temperature overprint, and the timing of uplift and erosion. An alternative explanation, first discussed by *Lister et al.* [1984] and *Platt* [1986], is that underplating and overthickening of the forearc lead to gravitational collapse by extension, and to exhumation of blueschists in the footwalls of normal faults. This interpretation has gained support in the Franciscan Complex [*Jayko et al.*, 1987; *Harms et al.*, 1992], the Alps [*Selverstone*, 1988], Australia [*Little et al.*, 1993], and Japan [*Kimura*, 1994].

Disrupted Mesozoic blueschist, ophiolite, and arc terranes in western Baja California, México are similar in many respects to Franciscan Complex, Coast Range

Ophiolite, and other rock units in California, but are better exposed and little affected by later tectonism. The goals of this paper are to argue that Baja blueschists were exhumed by syn-subduction extension, and to highlight the effects of extension on rocks that structurally overlie the blueschists.

2. MESOZOIC ROCK UNITS OF WESTERN BAJA

Mesozoic rocks that crop out on San Benito, Cedros, Magdalena, and Santa Margarita Islands and the Vizcaíno Peninsula in western Baja (Fig. 1) are divided into three structural units [Sedlock 1988a, 1993]. The structurally highest unit, or upper plate, consists of several distinct terranes that crop out in all areas except the San Benito Islands (e.g., Cedros Island, Fig. 2). Upper-plate terranes contain basement of Late Triassic to Middle Jurassic island arcs, ophiolites, or both, and overlying Jurassic volcanogenic strata [Kimbrough, 1984; Moore, 1985]. Upper-plate terranes evolved in the Pacific oceanic realm until their attachment to North America in the latest Jurassic or earliest Cretaceous. Magmatism ceased in these terranes after attachment, but was active to the east in the Cretaceous arc of mainland Baja. The arc and ophiolite terranes in western Baja formed part of the forearc lid above an east-dipping subduction zone that fed this Cretaceous arc, and the terranes were overlapped by forearc basin turbidites that range in age from lower Aptian to latest Cretaceous [Smith and Busby, 1993; D. Smith, unpublished data; R. Sedlock, unpublished data).

The structurally lowest unit, or lower plate, consists of regionally-metamorphosed assemblages of blueschist-facies metabasite and metasedimentary rocks that crop out on Cedros, San Benito, and Santa Margarita Islands. These rocks are not melange, but rather a faulted, originally continuous stratigraphic sequence of ocean floor or intraplate basaltic flows, pillows, and breccias, Late Triassic to Early Cretaceous red ribbon radiolarian chert, and terrigenous turbidites. The entire oceanic package was subducted, deformed, and metamorphosed under blueschist facies conditions (P = 0.5-1.0+ GPa, T = 150-300°C) [Sedlock, 1988b] at 115-105 Ma ($^{40}Ar/^{39}Ar$) [Baldwin and Harrison, 1989].

All contacts between the lower plate and upper plate are fault zones up to 500 m thick that are occupied by serpentinite-matrix melange (not visible at scale of Fig. 2). The melange consists of sheared serpentinite containing tectonic blocks that include wall rocks and a diverse suite of Mesozoic metamorphic rocks [Baldwin and Harrison, 1992; Bonini and Baldwin, 1994].

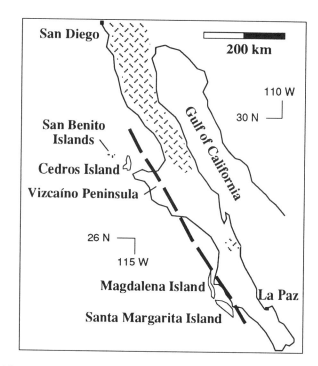

Fig. 1. Location map of Baja California. Mesozoic oceanic rocks on San Benito, Cedros, Magdalena, and Santa Margarita Islands and the Vizcaíno Peninsula are separated from the granitic basement (stipple pattern) of mainland Baja by an inferred major fault (heavy dashed line) that has been buried by Cenozoic sedimentary rocks.

3. SYN-SUBDUCTION EXTENSION

3.1 Evidence for Extension

Structural features in most upper-plate terranes indicate a single extensional deformation event; in some terranes on Magdalena and Santa Margarita Islands, extension is superposed on earlier, probably syn-accretion, contractional strains. Normal faults dip 0-70°, normal separations range from a few mm to at least 2 km (Fig. 3), and crustal thinning across several faults ranges from 1 to 3 km. Individual beds show up to 65% extension along sets of subparallel planar normal faults. Widespread veins are up to 1 m thick and record up to 15% bulk extension. In the lower-plate blueschists, extensional strain (normal faults, vein systems) is the youngest deformation event, overprinting contractional strain interpreted as syn-accretion. Normal separations on these faults range from a few mm to at least 10 m.

Earlier workers interpreted the faults between the upper and lower plates as thrusts [Rangin, 1978] or possible strike-slip faults [Kilmer, 1979], but I interpret them as normal faults for the following reasons.

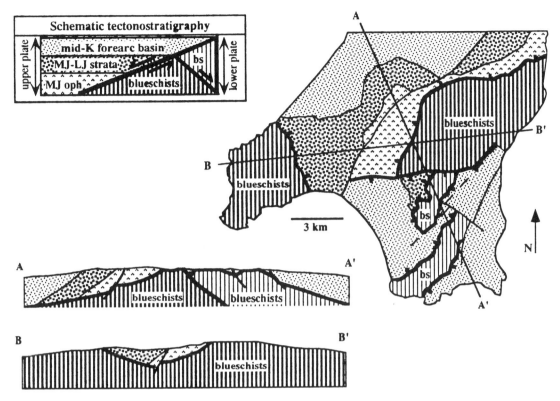

Fig. 2. Simplified geologic map and sections of southern Cedros Island. Patterns on map and sections correspond to units identified in schematic tectonostratigraphy (inset). Blueschists with heavier pattern were exhumed from greater depths (20-30+ km) than those with lighter pattern (15 km). Blueschists occupy footwalls of major shallowly-dipping fault zones (heavy lines) interpreted as normal faults (see text). Tick marks on fault zones point

1. Fault striae rake 45-90°, indicating dip slip or oblique slip.
2. Upper-plate normal faults and vein systems merge with the faults that separate upper and lower plates, suggesting a synkinematic origin.
3. Geobarometric estimates indicate differential maximum pressures of 0.1-0.6 GPa across the major faults, with lower-pressure rocks everywhere in the upper plate; this relation indicates crustal thinning of 3-20 km and net normal displacements of 5-40 km (Fig. 4).

3.2 Geometry and Kinematics of Extension

The geometry and kinematics of faulting are complex and have not yet been systematically studied. Faults between upper and lower plates, and most major upper-plate faults, are grouped into three sets: (1) dip <20°; undulatory, possibly domal (Cedros and Santa Margarita Islands, Vizcaíno Peninsula); (2) NE strike, dip 20-45°, undulatory (Cedros Island); (3) E-W strike, dip 30-55°, planar (Cedros and Magdalena Islands). Paleostresses cannot be inferred from fault strike because striae on most faults indicate oblique slip. Preliminary kinematic analysis of faults and striations [R. Sedlock, unpublished data] indicate that, whereas σ_1 is everywhere within 15° of vertical, σ_3 spans a range of N55E-S55W ± 50°. Systematic structural studies are in progress to determine whether the range of σ_3 directions represents heterogeneous faulting during a protracted episode of extension, superposed fault systems that developed under different stress regimes, or post-faulting rotations. The data also indicate that extension was not simply margin-normal.

3.3 Age of Extension and Exhumation

Geologic and geochronologic data indicate Late Cretaceous to Cenozoic extension and exhumation of lower-plate blueschists. Extension probably started at least as early as the Cenomanian, as indicated by active fault scarps within the Cenomanian forearc [Smith and Busby, 1993], and continued into the Cenozoic, as indicated by forearc basin strata as young as Maastrichtian that are cut by normal faults. Lamprophyre dikes on

Fig. 3. Photographs of normal faults within upper plate. A. Shallowly dipping normal fault separates mid-Cretaceous turbidites in hanging wall (bedding dips to left) from subhorizontal Jurassic shale in footwall; hanging wall is cut by numerous synthetic normal faults (planar, dip to right). Hill is about 100 m high. Cedros Island. B. Normal fault zone within Cretaceous turbidites. Sledgehammer in center of photo is 40 cm. Santa Margarita Island. C. Planar normal fault dips 45° to right above Jurassic arc basement of footwall (darker rocks left of main valley); bedded rocks in hanging wall are Aptian turbidites. Magdalena island.

Fig. 4. Normal fault zone marked by 8-m-thick band of light-colored serpentinite-matrix melange juxtaposes very shallow-level (maximum depth about 3-4 km) Jurassic volcaniclastic rocks of the upper plate and deep-level blueschists of the lower plate (exhumed from 20-25 km). Fault dips about 10° into hill, cuts subhorizontal bedding in dark upper-plate rocks. Southwestern Cedros Island.

Santa Margarita Island that have yielded K-Ar and ^{40}Ar/^{39}Ar ages of 30 Ma [*Forman et al.*, 1971; *Bonini and Baldwin*, 1994] intrude and are not cut by major upper-plate normal faults. This strongly suggests that widespread extension and exhumation ceased by 30 Ma.

Several lines of evidence suggest post-30 Ma extension and exhumation in western Baja, but none effectively constrains the magnitude or timing.

1. Normal faults with unknown magnitudes of net slip currently are active on or offshore Cedros, Magdalena, and Santa Margarita Islands. Slip on these faults probably has contributed to the modern topographic relief (up to 1 km).

2. Pliocene conglomerates containing blueschist clasts crop out on Cedros [*Kilmer* 1979], indicating surface exposure of blueschists by about 4 Ma. However, this is a minimum age because clast-bearing strata of Eocene-Miocene age are absent from western Baja.

3. *Baldwin and Harrison* [1989] (see also *Baldwin*, this volume) interpreted ^{40}Ar/^{39}Ar data from a lower-plate blueschist, which had experienced peak metamorphism of 170-220°C and 0.7-0.8 GPa (20-25 km) about 110 Ma, to indicate cooling below 145°C about 20 Ma. Based on this interpretation and on exhumation of these aragonite-bearing rocks into the calcite stability field at a temperature of 125-175°C [*Carlson and Rosenfeld*, 1981], they inferred average exhumation rates of 0.1 mm/yr from about 110 Ma to 20 Ma and >0.8 mm/yr from 20 Ma to 5 Ma. The rocks thus would have been exhumed from depths of 12-15 km during the Neogene. However, a different interpretation of the same observations is that, following peak metamorphism about 110 Ma, the rocks were exhumed at about 0.3 mm/yr from about 110 Ma to 30 Ma and at <0.1 mm/yr since 30 Ma. The rocks thus would have been within a few km of the surface by 30 Ma, allowing little subsequent exhumation. These two very different but equally permissible interpretations suggest that these data do not effectively constrain the age of extension and exhumation.

In summary, I infer that syn-subduction extension and blueschist exhumation began by about 95 Ma and effectively ceased in the mid-Tertiary. Crosscutting relationships and geochronologic data suggest that most extension and exhumation had occurred by 30 Ma; plate tectonic arguments (see below) may indicate completion by 40 Ma. In any case, extension and exhumation were not associated with the Basin & Range province or the opening of the Gulf of California. It is unclear whether extension and exhumation were continuous or episodic.

4. DISCUSSION

4.1 Exhumation of blueschists

The preservation of blueschists in western Baja California probably reflects protracted exhumation in the continuously refrigerated forearc of a steady-state subduction zone. The 2-dimensional sections in Figure 5 schematically portray only the margin-normal component of extension; as noted above, extension in Baja probably was not simply margin-normal. Geophysical and geologic evidence show that oceanic lithosphere (Farallon, Kula, or both) was subducted beneath North America from at least 100 Ma until about 30 Ma. Continuous

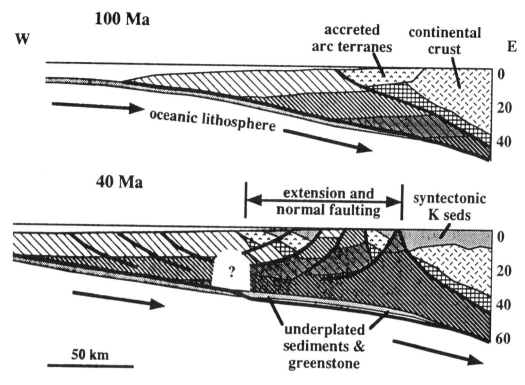

Fig. 5. Generalized, unbalanced cross-sections of western margin of Baja at 100 Ma (top) and about 40 Ma (bottom). Cross-hatched unit is peridotite. Accretionary prism consists of four diagonally-lined units divided at intervals of 0.4 GPa, 0.8 GPa, and 1.2 GPa. Underplating led to extension and normal faulting during Late Cretaceous and Early Tertiary (see text). Queried blank area is transition between extensional and contractional regimes.

subduction of cold lithosphere would have maintained low geothermal gradients and prevented blueschists from warming during exhumation. However, by about 40 Ma, young Farallon lithosphere entering the subduction zone may have caused a slowdown in the rate of subduction [*Ward*, 1991] and probably caused heating of the forearc and overprinting of blueschists not already at near-surface depths.

4.2 Similarity to Core Complexes

In many respects, the geometry of extended Mesozoic rocks in western Baja resembles that of metamorphic core complexes in the North American Cordillera (Fig. 6). Both consist of shallowly-dipping zones of great strain that separate brittlely extended, shallow-level rocks from deeper-level metamorphic or plutonic rocks. Extension in both settings may have resulted from vertical σ_1 caused by underplating (magmas or upper mantle added to crust in core complexes; subducted material added to hanging wall in subduction zones). Obviously, there also are notable differences between core complexes and western Baja, such as the nature of lower-plate rocks and of the intervening high-strain zones. However, there are no essential differences in the structural style of upper-plate rocks in the two settings.

4.3 Applications to Other Areas

Results from Baja California suggest that any rock unit at shallow to mid-crustal levels of a similar forearc may undergo large extensional strains. The geologic evolution of individual convergent margins determines what rock types may undergo this strain: in Baja these happened to be accreted arc and ophiolite terranes, but at other margins they might include flysch sequences, miogeoclinal strata, or granitoids.

Many workers have noted that ophiolites display incomplete sections that have been tectonically thinned by extension. Extensional strains in ophiolites may have been acquired in either or both of two environments: (1) shortly after formation, near an oceanic spreading center; (2) after obduction/accretion, within an extending

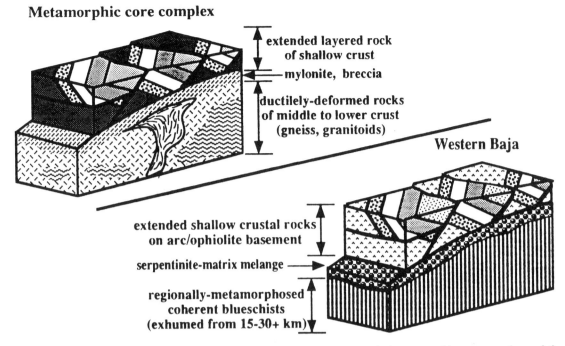

Fig. 6. Schematic block diagrams showing geometric similarity between typical metamorphic core complex and the Mesozoic rocks of western Baja California. Symbols in western Baja diagram correspond to those in Figure 2.

forearc. This study has found clear evidence of post-accretion extension.

In many ways the geology of western Baja is similar to that of Mesozoic rocks in western California (e.g., Franciscan Complex, Coast Range Ophiolite). Thanks to excellent exposure and the lack of tectonic overprint, it has been possible to document syn-subduction extension that is inferred but not easily demonstrable in related units in California.

Acknowledgments. Co-workers in Baja have included Suzanne Baldwin, Jennifer Bonini, Jon Hagstrum, Yukio Isozaki, and Dave Larue. My understanding of western Baja geology owes much to discussions with Gordon Gastil, Dave Kimbrough, Tom Moore, and Doug Smith. Work has been funded by NSF grants EAR85-18871 (Larue) and EAR91-04771.

REFERENCES

Baldwin, S. L., and T. M. Harrison, Geochronology of blueschists from west-central Baja California and the timing of uplift of subduction complexes, *J. Geol., 97,* 149-163, 1989.

Baldwin, S. L., and T. M. Harrison, The P-T-t history of blocks in serpentinite-matrix melange, west-central Baja California, *Geol. Soc. America Bulletin, 104,* 18-31, 1992.

Bonini, J. A., and S. L. Baldwin, $^{40}Ar/^{39}Ar$ geochronology of accreted terranes from southwestern Baja Califorrna Sur, Mexico, *U.S. Geol. Survey Circular 1107,* p. 34, 1994.

Carlson, W. D., and Rosenfeld, J. L., Optical determination of topotactic aragonite-calcite growth kinetics: Metamorphic implications, *J. Geol., 89,* 615-638, 1981.

Ernst, W. G., Metamorphic zonations on presumably subducted lithospheric plates from Japan, California, and the Alps, *Contrib. Mineralogy Petrology, 34,* 43-59, 1971.

Forman, J. A., Burke, W. H., Jr., Minch, J. A., and Yeats, R. S., Age of the basement rocks at Magdalena Bay, Baja California, México, *Geol. Soc. Amer. Abstr. Prog., 3,* 120, 1971.

Harms, T. A., A. S. Jayko, and M. C. Blake, Jr., Kinematic evidence for extensional unroofing of the Franciscan Complex along the Coast Range fault, northern Diablo Range, California, *Tectonics, 11,* 228-241, 1992.

Jayko, A. S., M. C. Blake, Jr., and T. Harms, Attenuation of the Coast Range Ophiolite by extensional faulting, and nature of the Coast Range "Thrust", California, *Tectonics, 6,* 475-488, 1987.

Kilmer, F. H., A geological sketch of Cedros Island, Baja California, México in *Baja California Geology,* edited by P. L. Abbott and R. G. Gastil, pp. 11-28, Dept. Geol. Sci., San Diego State University, 1979.

Kimbrough, D. L, Paleogeographic significance of the Middle Jurassic Gran Cañon Formation, Cedros Island, Baja California Sur, in *Geology of the Baja California Peninsula,* edited by V. A. Frizzell, Jr., pp. 107-117, Pacific Section, Society of Economic Paleontologists and Mineralogists, Book 39, 1984.

Kimura, G., The latest Cretaceous-early Paleogene rapid growth of accretionary complex and exhumation of high pressure series metamorphic rocks in northwestern Pacific margin, *J. Geophys. Research*, 99, 22,147-22,164, 1994.

Lister, G. S., G. Banga, and A. Feenstra, Metamorphic core complexes of Cordilleran type in the Cyclades, Aegean Sea, Greece, *Geology*, 12, 221-225, 1984.

Little, T. A., R. J. Holcombe, and R. Sliwa, Structural evidence for extensional exhumation of blueschist-bearing serpentinite matrix melange, New England orogen, southeast Queensland, Australia, *Tectonics*, 12, 536-549, 1993.

Moore, T. E., Stratigraphy and tectonic significance of the Mesozoic tectonostratigraphic terranes of the Vizcaino Peninsula, Baja California Sur, México, in *Tectonostratigraphic terranes of the Circum-Pacific region*, edited by D. G. Howell, pp. 315-329, Circum-Pacific Council for Energy and Mineral Resources, Earth Science Series, Number 1, Houston Texas, 1985.

Platt, J. P., Dynamics of orogenic wedges and the uplift of high-pressure metamorphic rocks, *Geol. Soc. America Bulletin*, 97, 1037-1053.

Rangin, C., Speculative model of Mesozoic geodynamics, central Baja California to northeastern Sonora (México), in *Mesozoic Paleogeography of the Western United States*, edited by D. G. Howell and K. A. McDougall, pp. 85-106, Pacific Section, Soc. Econ. Paleontolologists and Mineralologists, 1978.

Sedlock, R. L., Tectonic setting of blueschist and island-arc terranes of west-central Baja California, Mexico, *Geology*, 16, 623-626, 1988a.

Sedlock, R. L., Metamorphic petrology of a high-pressure, low-temperature subduction complex in west-central Baja California, Mexico, *J. Metam. Geol.*, 5, 205-233, 1988b.

Sedlock, R. L., Mesozoic geology and tectonics of blueschist and associated oceanic terranes in the Cedros-Vizcaíno-San Benito and Magdalena-Santa Margarita regions, Baja California, México, in *Mesozoic paleogeography of the western United States-II*, edited by G. C. Dunne and K. McDougall, pp. 113-125, Pacific Section, Society of Economic Paleontologists and Mineralogists, Book 71, 1993.

Selverstone, J., Evidence for east-west crustal extension in the Eastern Alps: Implications for the unroofing history of the Tauern Window, *Tectonics*, 7, 87-105, 1988.

Smith, D. P., and C. J. Busby, Mid-Cretaceous crustal extension recorded in deep-marine half-graben fill, Cedros Island, Mexico, *Geol. Soc. America Bulletin*, 105, 547-562, 1993.

Ward, P. L., On plate tectonics and the geologic evolution of southwestern North America, *J. Geophys. Res.*, 96, 12,479-12,496, 1991.

Richard L. Sedlock, Department of Geology, San José State University, San José, CA 95192-0102, sedlock@geosun1.sjsu.edu.

Slip-History of the Vincent Thrust: Role of Denudation During Shallow Subduction

Marty Grove and Oscar M. Lovera

Department of Earth & Space Sciences, University of California, Los Angeles

$^{40}Ar/^{39}Ar$ age and ^{39}Ar kinetic studies performed with K-feldspars sampled from above the Vincent Thrust (VT) allow reconstruction of its slip history during Late Cretaceous-Early Tertiary shallow subduction. Variational methods were applied to the multiple diffusion domain model to produce best fit thermal histories. The K-feldspar T-t results indicate that a temperature difference of ~150°C was maintained from >60 Ma to <55 Ma between positions that are presently at nearly the same elevation and are located 5 and 15 km west of the VT. Using a geothermal gradient and dip angle estimated from geologic constraints, simple numerical heat-flow models were used to determine the slip velocity and relative vertical separation of the samples during thrusting by requiring calculated T-t results to fit the K-feldspar thermal histories. For models in which cooling was due solely to subduction of colder rocks (hanging wall stationary), solutions most compatible with the K-feldspar results were yielded by underthrusting rates of ~1.4 cm/yr. and a vertical separation during the Late Cretaceous/Early Tertiary of 8.5 km. Allowing denudation of the hanging wall during thrusting (footwall stationary) provides somewhat more satisfactory fits to the data. For these models, ~0.2-0.4 cm/yr. displacement occurs along the VT from 65 Ma to 50 Ma along a fault plane inclined at 15°. Because only net vertical displacement can be constrained by K-feldspar data, our calculated slip rate is only a relative value that is inversely proportional to the dip angle at the time of thrusting.

1. INTRODUCTION

Knowledge of the slip history of subduction-related thrust faults is fundamental to understanding convergent margin evolution. Since tectonic movements modify the distribution of thermal energy within the crust, combined thermochronology and numerical heat-flow analysis offer a powerful means of evaluating tectonic models. Within subduction complexes, attention has generally been focused upon constraining thermal histories of accreted materials. In this contribution, we employ thermochronologic results from the Vincent Thrust (VT) of southern California together with simple numerical heat-flow calculations to demonstrate that knowledge of the temperature-time evolution above subduction complexes can also provide important insights. Our methods [Lovera et al., 1989] feature thermochronology based upon $^{40}Ar/^{39}Ar$ age and ^{39}Ar kinetic experiments performed with K-feldspar that are capable of constraining temperature-time histories of processes active within the middle crust (~400-150°C).

2. GEOLOGIC OVERVIEW

The VT within the eastern San Gabriel Mountains (Fig. 1) juxtaposed a Cretaceous magmatic arc and older crystalline rocks of the San Gabriel terrane over a subduction assemblage (the Pelona Schist) during the Late Cretaceous-Early Tertiary. The Pelona Schist and its regional equivalents (Fig. 1) consist of metamorphosed flysch, mafic, and minor ultramafic assemblages recrystallized under amphibolite, epidote amphibolite, greenschist, or rarely epidote blueschist facies conditions [Graham and Powell, 1984; Jacobson and Dawson, 1995]. These rocks are widely believed to have been accreted beneath the VT and related faults during Late Cretaceous-Early Tertiary shallow subduction [Coney and Reynolds, 1977; Graham and England, 1976; Burchfiel and Davis, 1981; Dickinson, 1981]. While Late Tertiary reactivation of this suture was widespread [Jacobson and Dawson, 1995 and references therein], structural, petrologic, and thermochronologic evidence all imply that an unmodified

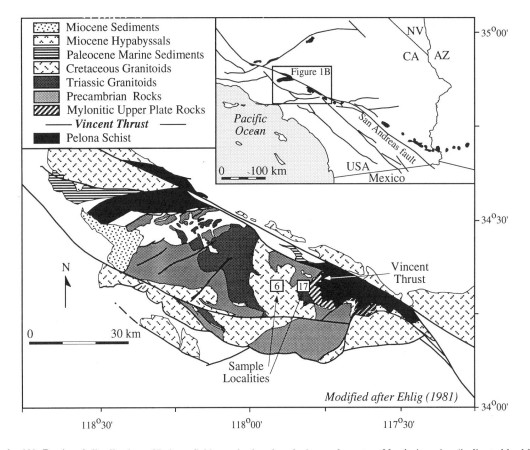

Fig. 1: (A) Regional distribution of Pelona Schist and related rocks in southwestern North America (indicated by black filled regions). Rocks exposed in the San Gabriel mountains (inset) have been displaced ~240 km along the San Andreas fault from their equivalents in SE California. (B) Geologic sketch map of the San Gabriel mountains illustrating the position of the Vincent thrust and samples discussed in the text.

thrust fault contact is preserved across the VT in the northeastern San Gabriel Mountains. Significant reactivation of the VT is seemingly precluded by the pattern of inverted metamorphism beneath the fault and development of isofacial mineral assemblages and concordant fabrics in the underlying schist and mylonitic upper plate rocks at the contact [Ehlig, 1981, Jacobson, 1983, 1995]. Similarity of mica K-Ar cooling ages above and beneath the VT also implies shared histories for the upper and lower plates during the early Tertiary [Jacobson, 1990 and references therein].

3. TESTING TECTONIC MODELS

Previous numerical modeling studies that have sought to explain the inverted metamorphic zonation beneath the VT have offered significantly different estimates of its slip history. Using relatively high estimates for shear heating (125-250MPa @ 3cm/yr = 0.12-0.24 W/m^2), Graham and England [1977] produced models that required displacement of ~90 km accompanied by 0.2 mm/yr erosion to avoid temperature increases unsupported by geologic observations. Peacock [1987] alternatively found that underflow of many hundreds of kilometers of oceanic crust was required to create and preserve the inverted metamorphic gradient across the VT if frictional heating was negligible. Similarly, Dumitru [1991] has invoked refrigeration of the Cretaceous magmatic arc during rapid, shallow-angle Laramide subduction beneath western North America. Note that although denudation of the VT hanging wall was potentially significant [Mahaffie and Dokka, 1986; Dillon, 1986; May and Walker, 1989], no numerical studies of the VT have explicitly considered it.

In the present paper, we discuss results from two K-feldspar samples supplied by A.P. Barth. The specimens were obtained at positions 5 and 15 km west of, and above, the presently exposed trace of the VT (Fig. 1). Both sample localities occur approximately at the same elevation. The specimens examined consist of homogeneous orthoclase (93-NG-17) and microcline with minor perthite (93-NG-6). Both were obtained from intrusions that are texturally similar to the youngest dated plutons (~80-75 Ma) emplaced into the VT hanging wall prior to thrusting [May and Walker, 1989; Barth et al., 1995]. A sample adjacent to 93-NG-17 has yielded a U-Pb sphene age of 77±2 Ma [Joe Wooden, unpublished data] and a hornblende inverse isochron age of 71±2 Ma [sample 93-NG-13; ^{40}Ar/^{39}Ar data tables and plots for this and other samples discussed in the text are available at the following WWW site: http://oro.ess.ucla.edu/argon.html].

4. THERMAL HISTORY RECONSTRUCTION

K-feldspar ^{40}Ar/^{39}Ar age and ^{39}Ar kinetic data allow reconstruction of thermal histories of mid-crustal rocks (~150-400°C) through use of the multiple diffusion domain (MDD) model [Lovera et al., 1989]. In this model, differing intra-sample argon retentivities are assumed to result from a discrete distribution of non-interacting diffusion domains that vary in dimension (r). The form of the Arrhenius plot ($\log(D/r^2)$ vs. 1/T; see Fig. 2b) or its associated $\log(r/r_o)$ plot [Fig. 2c; see Richter et al., 1991] is a function of the parameters that characterize the individual diffusion domains (activation energy (E), frequency factor (D_o), domain size (ρ), and volume fraction (ϕ)). Although the age spectrum (Fig. 2a) also depends upon these parameters, its shape is further modulated by the thermal history.

We have recently incorporated new automated routines [Lovera et al., 1995] that allow thorough analysis of the K-feldspar results. Levenberg-Marquardt variational methods (a generalization of least squares routines to non-linear cases) are used to find the maximum likelihood estimate of the model parameters [Press et al., 1988]. From the ^{39}Ar data, values for E and $\log(D_o/r_o^2)$ are found from a linear, weighted, least-squares fit to the initial, low-temperature $\log(D/r^2)$ values (Fig. 2b). The maximum likelihood estimate of the distribution parameters (ρ, ϕ) is then obtained by applying the variational method. With these values, the Levenberg-Marquardt method is again applied to model the measured ^{40}Ar/^{39}Ar age spectrum by varying the coefficients of Chebyshev polynomials used to approximate the thermal history until an acceptable solution is returned. In the present study, we have restricted our solutions to monotonic cooling by constraining the first derivative of the thermal history to be negative. Since uncertainties in age and non-linearity of the diffusion process give rise to multiple solutions (Fig.2d), we use contour plots to indicate the probability distribution of the thermal history.

5. THRUST FAULT MODELS

Simple, two-dimensional, numerical heat-flow models with thrust fault geometry and kinematics have been investigated. Calculations involved finite-difference methods using direct and alternating-direction implicit techniques. Thermal diffusivity was set at an average crustal value $\kappa = 10^{-6}$ m^2/sec. Constant temperature was maintained both at the surface (25°C) and base (1025°C) of the grid while a zero heat flux condition was imposed over the lateral boundaries. Frictional and internal heating (due to radioactive decay and metamorphic reactions) were not considered. Although frictional heating can significantly influence temperature distributions adjacent to the fault [Molnar and England, 1990], our calculations using shear stress values we consider reasonable (~25 MPa; cf. Peacock, 1992) indicate that samples situated more than ~3 km from the fault are only mildly affected by the frictional heating. The effect of frictional heating along the fault can essentially be compensated by an equivalent decrease of the slip rate (i.e., neglecting internal heating could lead to a slight underestimation of the slip rate).

6. DISCUSSION

Combined K-feldspar thermochronology and numerical heat-flow calculations allow us to estimate the displacement history of the VT. Given a geothermal gradient and dip angle, we are able to use the numerical heat-flow models to constrain the slip velocity and relative vertical separation of the samples by requiring the T-t results to agree with the K-feldspar thermal histories. A 25°C/km geothermal gradient and 15° dip angle have been used in all of the numerical heat-flow models. The geothermal gradient can be estimated from available petrologic and thermochronologic data. Assuming ~20 km emplacement depths [Barth, 1990; Barth et al., 1995], closure systematics of Pb in sphene and Ar in hornblende imply both a ~25°C/km geothermal gradient and slow cooling of 93-NG-17 through 600-450°C from >77 Ma to <70 Ma. The original dip of the VT is more uncertain since Neogene doming of the VT due to transpressional deformation resulting from San Andreas-related, strike-slip faulting precludes precise determination of this parameter (Fig. 1). Although the sinuous map pattern of the VT (Fig. 1) indicates that it dips gently at the surface, geologic relationships suggest that the fault rotates to a steeper orientation in the west beneath our samples. To the extent that the VT originally dipped at a shallower angle, this relationship implies that the WNW-trending domal uplift shown in Fig. 1 has exposed progressively deeper rocks moving eastwards toward the VT.

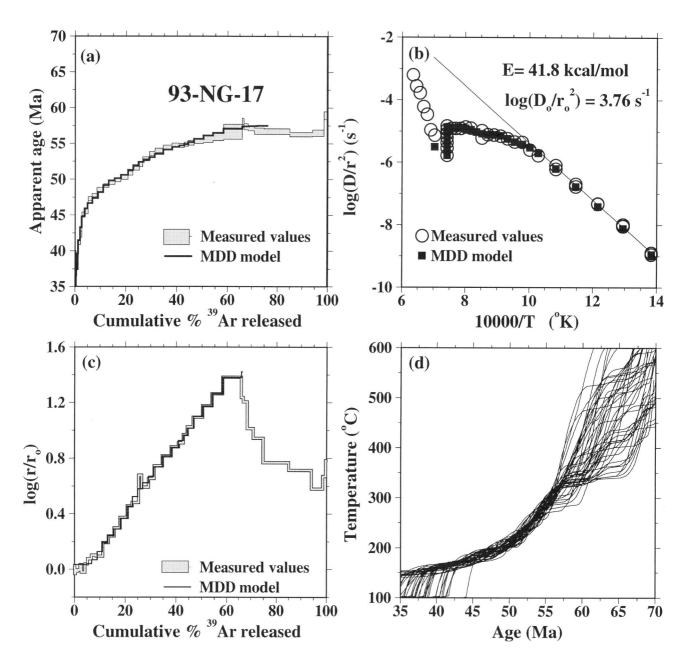

Fig. 2: Thermochronologic constraints from 93-NG-17 K-feldspar. (a) Measured and calculated age spectra; (b) Arrhenius plot calculated from ^{39}Ar loss (circles) and from the MDD model fit (squares). Isothermal duplicates have been performed at low temperatures (<700°C); (c) log(r/r_o) plot showing values calculated from laboratory data and MDD model fit. Results are modeled only to 1100°C since melting has occurred at higher temperatures. A cross correlation value of 0.99 was obtained between the log(r/r_o) plot and age spectra (to 1100°C); (d) Calculated thermal histories which fit the measured age spectrum in (a) above. The solid curve in (a) is a representative solution obtained from one of these monotonic cooling curves. The cooling thermal histories are constrained to within ±10°C between 350-150°C. A contour plot of the calculated thermal histories is shown in Fig.4a.

Thermochronologic constraints provided by our samples and by previous studies are most consistent with motion along the VT initiating sometime between ~63-69 Ma and continuing to ~50 Ma. As mentioned above, $^{40}Ar/^{39}Ar$ hornblende and U-Pb sphene for a sample adjacent to 93-NG-17 indicate slow cooling through ~550-450°C from > 77 Ma to < 70 Ma. Amphibole $^{40}Ar/^{39}Ar$ ages of 58-65 Ma from epidote amphibolite mafic schists within the underlying Pelona Schist [Jacobson, 1990] and regional biotite closure at ~65 Ma within higher structural levels in the upper plate [Miller and Morton, 1980] likely constrain the time of initial thrusting. Previous Rb-Sr, K-Ar and $^{40}Ar/^{39}Ar$ results from upper plate mylonites and lower plate micas summarized in [Jacobson, 1990] indicate continued cooling between 60-50 Ma. The initial 5-10% of gas released from both our samples is consistent with slow cooling below 200°C at times later than 50 Ma (Fig. 2d).

Based upon these observations, we have initiated slip at 65 Ma in our numerical models. Isothermal distributions at selected times for two end member cases are shown in Fig. 3a-f. In the first (Model I; Fig. 3a-c), the footwall is maintained stationary while uplift of the hanging wall due to thrusting is balanced by erosion (pure denudation). In the second (Model II; Fig. 3d-f), the hanging wall is fixed (no denudation) while the footwall is subducted. Although we vary the relative vertical separation of the samples in the models to obtain the best agreement with the K-feldspar results, their overall separation is fixed by their present geographic coordinates (10 km; Fig. 1). Because we are most confident in our ability to project surface geology beneath the sample which lies closest to the VT (93-NG-17), we maintain its position at 3.5 km above the fault in both models.

In the following discussion we consider only the solutions to models I and II which produce T-t histories in best agreement with the K-feldspar thermal history results. The results for hanging wall positions indicated in Fig. 3 are shown in Fig. 4a together with contoured T-t histories yielded by K-feldspar thermochronology. Reference to Fig. 4b enables evaluation of T-t conditions for which the K-feldspar thermal history results are able to constrain the numerical models in Fig. 4a. For example, 93-NG-6 provides constraints from 63 to ~55 Ma (Fig. 4b).

The thermal model results shown in Fig. 4a indicate that denudation and cooling of the hanging wall in the manner described by Model I can easily account for the K-feldspar thermal history results. Our best fits are obtained for a relative vertical separation of ~6 km and a net vertical displacement of ~10 km from 65-50 Ma (Fig. 3a-c). Note that the slip rate was increased from ~0.2 to ~0.4 cm/yr between 60-56 Ma to account for the rapid cooling indicated by both samples at this time. Although the model result for 93-NG-6 deviates significantly from the K-feldspar T-t results beyond 55 Ma in Fig. 4a, this portion of the thermal history is constrained only by the first 5% of gas release from the sample (Fig. 4b). Assuming that the fault originally dipped at 15°, the estimated vertical separation of 6 km implies rotation of the VT to a present-day ~50° dip in the subsurface beneath the two samples, possibly due to Neogene doming.

Results from Model II yield somewhat less satisfactory fits to the K-feldspar thermal history results (Fig. 4a-b). Although a closely matched solution is obtained for 93-NG-6 using a subduction rate of 1.5 cm/yr, considerable misfit of 93-NG-17 results (Fig. 4b). More significantly, the best fit to 93-NG-6 requires that the samples be separated vertically by ~8.5 km. This value implies rotation of the VT to a present-day 75° dip in the subsurface beneath the samples. We believe that such a steep inclination of the VT is unsupported by the surface geology. Slip-rates faster than 1.5 cm/yr. require the vertical separation of the samples to exceed the 10 km distance that presently separates them. Alternatively, slower rates of displacement produce steady state temperatures higher than those required to model the K-feldspar thermal history after ~57 Ma.

Although the match to the K-feldspar thermal history results for the case of pure subduction (Model II) is less satisfactory than that obtained for Model I, it appears from Fig. 4a that an improved interpretation would result by considering combined subduction of the lower plate and denudation of the hanging wall. For example, subduction starting at ~65 Ma can account for the slow cooling portion of the thermal histories (>60 Ma and <56 Ma) if the rapid cooling between 60-56 Ma is explained by the superposed effects of subduction plus denudation.

In comparing our results to those obtained in previous studies [Graham and England, 1976; Peacock, 1987], we emphasize that these authors lacked thermochronologic constraints and therefore were concerned only with explaining the formation and preservation of the inverted metamorphic zonation across the VT. Graham and England's preferred model (3 cm/yr displacement for 3 Ma with 0.12 W/m^2 (125 MPa) shear heat source) predicts temperatures that are too high in the VT hanging wall after ~56 Ma to explain the K-feldspar results. Alternatively, Peacock's [1987] model predicts temperatures in the hanging wall that are too low after ~56 Ma.

From the available thermochronology and the simple numerical calculations presented above, we conclude that appreciable (~1 mm/yr.) denudation of the VT hanging wall (particularly between 60-56 Ma; see Model I) likely accompanied modest (<1 cm/yr) underflow of the colder rocks from 65 to <50 Ma. We emphasize that the displacement rates in model I (0.2-0.4 cm/yr) represent

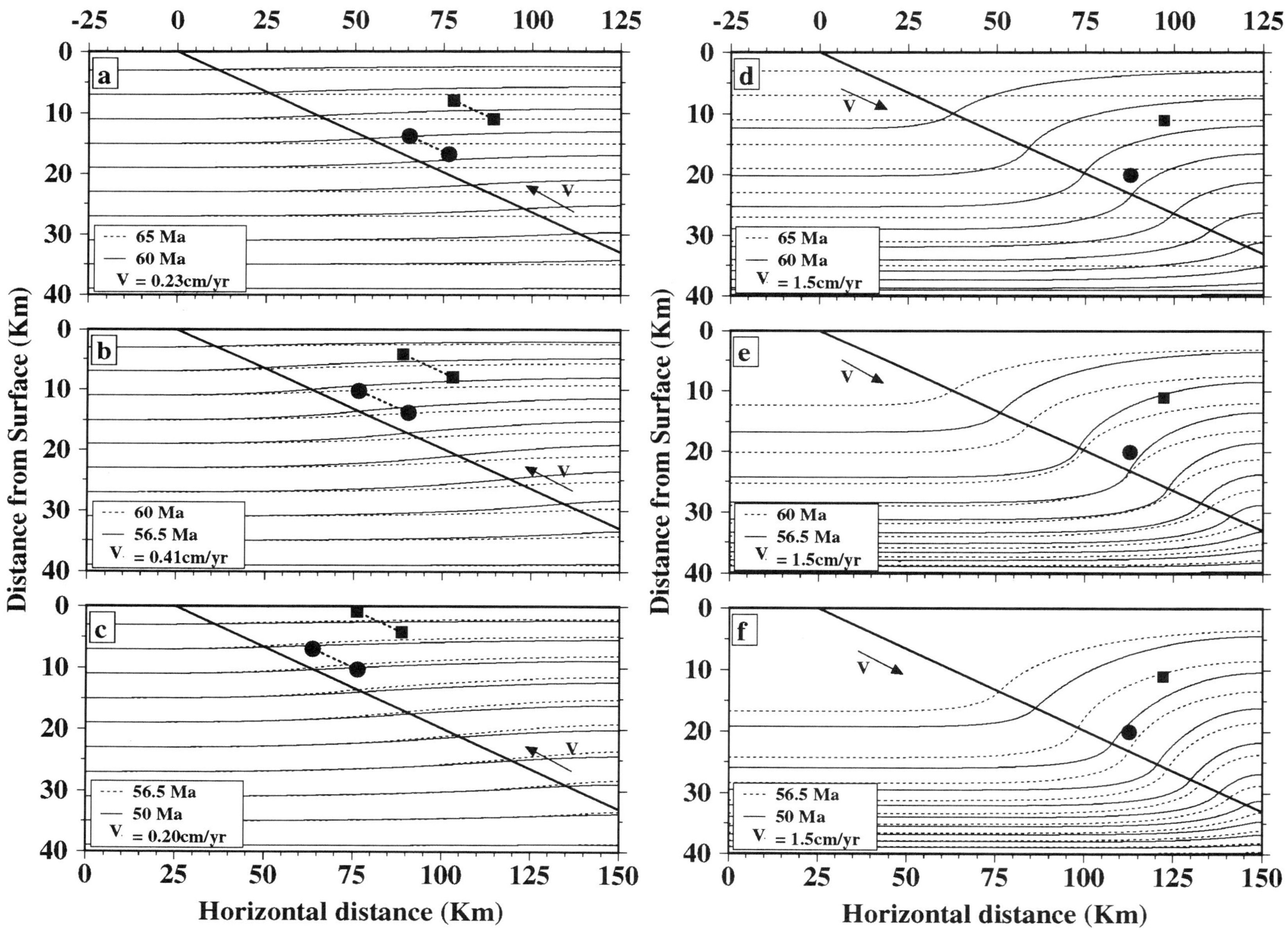

168 SLIP HISTORY OF THE VINCENT THRUST

Fig. 3: Isothermal sections produced from the heat-flow calculations. In model I (a-c), the footwall is fixed while the hanging wall is thrust upwards (denudation and erosion balanced at the surface). In model II (d-f), the hanging wall is stationary while the footwall is subducted. The vertical separation maintained between the samples is 6 km and 8.5 km for models I and II respectively. In both models slip is initiated at 65 Ma. Note that the slip velocities employed between each of the indicated times are provided.

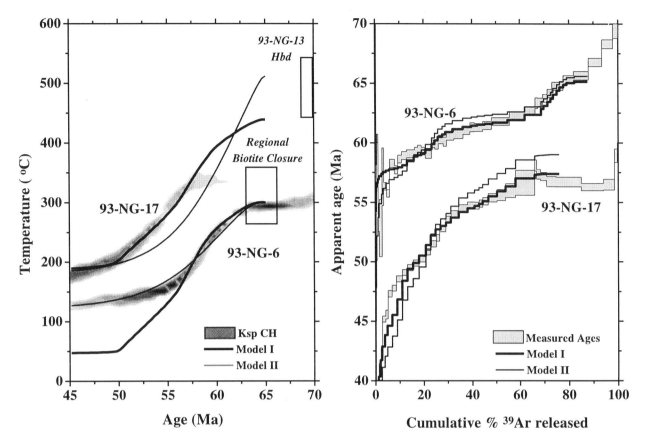

Fig. 4: (a) Thermal evolution of the VT. Thermal histories density contours calculated for 93-NG-17 and 93-NG-6 K-feldspars (45 results each) are represented by shades of gray. The lowest density shown represents a >50% probability that a solution lies within the indicated region.. Open boxes indicate thermochronologic constraints provided by 93-NG-13 hornblende (adjacent to 93-NG-17) and biotite results for the central San Gabriel Mountains (similar to 93-NG-6). Calculated thermal histories from Model I and II for the sample positions indicated in Fig. 3 are shown in bold and light lines respectively. (B) Age spectra calculated from models I and II are compared to the measured K-feldspar age spectra. Note that although Model II yield a satisfactory fit to the measured age spectrum of 93-NG-6, it fails to reproduce that of 93-NG-17 K-feldspar.

relative values that are inversely proportional to the dip angle. If the fault shallowed to become subhorizontal at depth, our model for the slip history of the VT would clearly underestimate the total magnitude of displacement. Similarly, although the thermochronologic data preclude significant underflow of colder rocks beneath the VT hanging wall prior to ~65 Ma, earlier motion along a subhorizontal fault that juxtaposed rocks of equivalent temperature cannot be ruled out.

Acknowledgments. The samples examined in this study were provided by A.P. Barth (funding by the National Geographic Society). We particularly thank C.E. Manning, T.M. Harrison, A.P. Barth, B. Idelman, and A.C. Warnock for providing constructive reviews. Aspects of this research were supported by grants from NSF (Lovera) and DOE (Harrison).

REFERENCES

Barth, A.P., Mid-crustal emplacement of Mesozoic plutons, San Gabriel Mountains, California, and implications for the geologic history of the San Gabriel Terrane, *in The nature and origin of Cordilleran magmatism,* edited by J.L. Anderson, *Geol. Sci. Am.Mem. 174,* 33-45, 1990.

Barth, A.P., Wooden, J.L., Tosdal, R.M., Morrison, J., Crustal contamination in the petrogenesis of a calc-alkalic rock series: Josephine Mountain intrusion, California, , *Geol. Sci. Am. Bull., 107,* 201-212, 1995.

Burchfiel, B.C., and G.A. Davis, Mojave Desert and environs, *in The Geotectonic Developement of California, Rubey Vol. I,* edited by W.G. Ernst, pp. 217-252, Prentice-Hall, Englewood Cliffs, N.J., 1981.

Coney P.J., and S.J. Reynolds, Cordilleran Benioff Zones, *Nature, 270,* 403-406, 1977

Dickinson, W.R., Plate tectonics and the continental margin of California, *in The Geotectonic Developement of California, Rubey Vol. I*, edited by W.G. Ernst, pp. 1-28, Prentice-Hall, Englewood Cliffs, N.J., 1981.

Dillon, J.T., Timing of thrusting and metamorphism along the Vincent-Chocolate Mountain thrust system, southern California, *Geol. Sci. Am. Abstr. Programs, 18*, 101, 1986.

Dumitru, T.A., Effects of subduction parameters on geothermal gradients in foreacs, with an application to Franciscan subduction in California, *JGR, 96* (B1), 621-641, 1991.

Ehlig, P.L., Origin and tectonic history of the basement terrane of the San Gabriel Mountains, central Transverse Ranges, *in The Geotectonic Developement of California, Rubey Vol. I*, edited by W.G. Ernst, pp. 253-283, Prentice-Hall, Englewood Cliffs, N.J., 1981.

Graham, C.M., and England, P.C., Thermal regimes and regional metamorphism in the vicinity of overthrust faults: An example of shear heating and inverted metamorphic zonation from southern California, *Earth Planet. Sci. Lett., 31*, 142-152, 1976.

Graham, C.M., and R. Powell, A garnet-hornblende geothermometer: Calibration, testing, and application to the Pelona Schist, southern California, *J. Metamorph. Geol., 2*, 13-31, 1984

Jacobson, C.E., Relationship of deformation and metamorphism of the Pelona Schist to movement on the Vincent thrust, San Gabriel Mountains, southern California, *Am. J. Sci., 283*, 587-604, 1983.

Jacobson, C.E., The ^{40}Ar/^{39}Ar Geochronology of the Pelona Schist and Related Rocks, Southern California, *JGR, 95* (B1), 509-528, 1990.

Jacobson, C.E., Qualitative thermobarometry of inverted metamorphism in the Pelona and Rand Schists, southern California using calciferous amphibole in mafic schist, *J. Metamorph. Geol., 13*, 79-92, 1995.

Jacobson, C.E., and Dawson, M.R., Structural and metamorphic evolution of the Orocopia Schist and related rocks, sourthern California: Evidence for Late Movement on the Orocopia Fault, *Tectonics, 14*, 733-744, 1995.

Lovera, O.M., F.M. Richter, and T.M. Harrison, The ^{40}Ar/^{39}Ar Thermochronometry for Slowly Cooled Samples Having a Distribution of Diffusion Domain Sizes, *JGR, 94*, 17917-17935, 1989.

Lovera, O.M., M. Grove, T.M. Harrison, and M.T. Heizler, Systematic Analysis of K-feldspar Age and Kinetic Properties, *EOS, 76* (17), 287, 1995.

Mahaffie, M.J, and Dokka, R.K., Thermochronologic evidence for the age and cooling history of the upper plate of the Vincent thrust, California, , *Geol. Sci. Am. Abstr. Programs, 18*, 153, 1986.

May, D.J., and Walker, N.W., Late Cretaceous juxtaposition of metamorphic terranes in the southeastern San Gabriel Mountains, California, *Geol. Sci. Am. Bull. 101*, 1246-1267, 1989.

Miller, F.K., and D.M. Morton, Potassium-argon geochronology of the eastern Transverse Ranges and the southern Mojave Desert, southern California, *U.S. Geol. Surv. Prof. Pap., 1152*, 30 pp, 1980.

Molnar P., and England, P., Temperatures, heat flux, and frictional stress near major thrust faults, *J. Geophys. Res., 95*, 4833-4856, 1990.

Peacock, S.M., Creation and preservation of subduction-related inverted metamorphic gradients, *J. Geophys. Res., 92*, 12763-12781, 1987.

Peacock, S.M., Blueschist-facies metamorphism, shear heating, and P-T-t paths in subduction shear zones, *J. Geophys Res, 97*, 17693-17707, 1992.

Press, W.H., B.P. Flannery, S.A. Teukolsky, and W.T. Vetterling, *Numerical Recipes:The Art of Scientific Computing*, Cambridge University Press, New York, 818p., 1988.

Richter, F.M., O.M. Lovera, T.M. Harrison, and P. Copeland, Tibetan tectonics from ^{40}Ar/^{39}Ar analysis of a single K-feldspar sample, *Earth Planet. Sci. Lett., 105*, 266-278, 1991.

M. Grove and O.M. Lovera, Department of Earth and Space Sciences, Uviversity of California, Los Angeles, Los Angeles, CA 90024.

A Thermotectonic Model for Preservation of Ultrahigh-Pressure Phases in Metamorphosed Continental Crust

W. G. Ernst

Department of Geological and Environmental Sciences, Stanford University, Stanford, California

Simon M. Peacock

Department of Geology, Arizona State University, Tempe, Arizona

Continental rocks subjected to ultrahigh-pressure (UHP) metamorphism contain relict minerals indicating formation at mantle depths approaching or exceeding 100-125 km. Thermobarometric calculations and phase-equilibrium constraints indicate peak metamorphic conditions of approximately 700-900 °C and 28-40 kbar. These extremely high pressures and relatively modest temperatures can be explained through the deep subduction of coherent tracts of continental lithosphere, and reflect the low thermal conductivities of geologic materials. Relatively rapid return to midcrustal levels after detachment from the downgoing slab may result from buoyancy of the UHP continental crust. Incomplete prograde conversion to the ultrahigh-pressure mineral assemblage at depth could enhance the exhumation process. Adiabatic decompression of UHP metamorphic rocks would result in P-T paths that enter the granulite and pyroxene hornfels metamorphic facies; under these conditions, rising, intensely metamorphosed units would remain hot enough to ensure obliteration of precursor UHP phase assemblages. Numerical experiments simulating Pacific-type subduction and continent collision followed by exhumation suggest that survival of traces of UHP metamorphic phases during ascent requires that the terranes migrate toward the surface as relatively thin sheets. Cooling during decompression may occur if (1) subduction continues outboard at tectonically lower levels, allowing heat conduction downward into the refrigerating subducting plate, (2) unroofing occurs along an inboard, extensional fault or shear zone, allowing heat conduction upward into the cooler hanging wall, or (3) a combination of both processes. Documented coesite ± diamond-bearing lithotectonic units from western Europe and central + eastern Asia range in thickness to a maximum of about 10 ± 5 km, but most resurrected ultrahigh-pressure complexes are substantially thinner. Thus, although large, coherent masses of relatively cold material may move down subduction zones resulting in UHP metamorphism at profound depths, ascent toward the surface of thin slabs of such complexes, with cooling across both upper and lower surfaces, may be required for the partial preservation of ultrahigh-pressure mineral assemblages.

INTRODUCTION

A small number of sialic crustal complexes, chiefly located within Eurasia, exhibit rare effects of ultrahigh-pressure (= UHP) recrystallization [*Schreyer*, 1995; *Coleman and Wang, eds.*, 1995]. Figure 1 presents the global distribution of authenticated sections of continental crust subjected to UHP metamorphism. Each complex contains scattered traces of phases indicative of formation at mantle depths approaching, and in some cases considerably exceeding 100 km [*Schreyer*, 1988; *Wang et al.*, 1995]. Mineralogic evidence for ultrahigh pressures includes the preservation of trace amounts of minerals and relict phase assemblages such as coesite, diamond, K-rich clinopyroxene, pyropic garnet, and volatile-bearing associations including magnesite + diopside, coesite + dolomite, talc + kyanite and/or phengite, ellenbergerite, lawsonite, zoisite, and Na-clinoamphibole. Judging from phase relations and

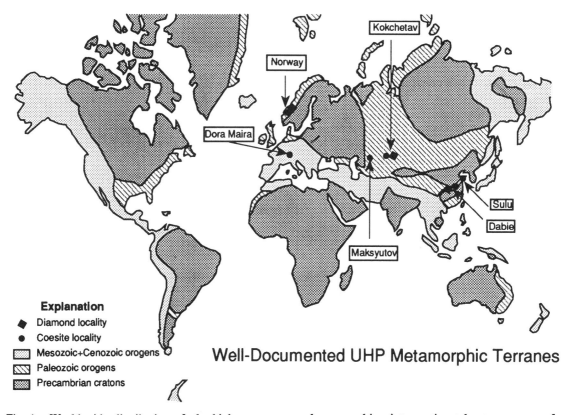

Fig. 1. World-wide distribution of ultrahigh-pressure complexes, marking intracontinental suture zones, after Coleman and Wang, eds. (1995) and Liou et al. (1994).

thermobarometric estimates, temperatures during recrystallization were about 700-900 °C at confining pressures approaching 28-40 kbar.

As with the more widespread high-pressure (= HP) metamorphic terranes, these remarkably high pressures and relatively low temperatures can be generated through the deep subduction of coherent tracts of ancient, cold, sialic-crust-capped lithosphere, and due to the fact that geologic materials are poor thermal conductors [*Ernst*, 1971; *Peacock*, 1995]. Return to shallow depth is also explicable, based on the buoyancy of UHP-metamorphosed continental crust, once it has separated from the downgoing lithospheric slab [*Cloos*, 1993; *Ernst and Liou*, 1995]. Because adiabatic decompression would result in the transit of decoupled subduction complexes through P-T regimes (700-800 °C, 2-10 kbar) appropriate to the granulite, high-rank amphibolite, and pyroxene hornfels metamorphic facies, complete overprinting of the earlier UHP assemblages would be expected. The very rare occurrences of ultra-high-pressure relics suggest that the rocks back reacted almost completely on return toward the Earth's surface. On the other hand, because of the apparent short duration of individual UHP events (see next section), it is conceivable that metastable lower pressure mineral assemblages failed to react to produce the stable prograde UHP configuration [*Erambert and Austrheim*, 1993; *Harley and Carswell*, 1995] except in kinetically favorable sites, such as regimes characterized by trace amounts of fluid.

How is it possible to preserve any relict phases from the original UHP metamorphism? This paper attempts to answer that question. A comparative petrotectonic description of five relatively well-known UHP tracts including the Qinling-Dabie-Sulu belt of east-central China, the Kokchetav Massif of northern Kazakhstan, the Maksyutov Complex of the south Urals, the Dora Maira Massif of the Western Alps, and the Western Gneiss Region of Norway was recently presented by *Ernst et al.* [1995]. Based on these geologic constraints, inferred physical conditions of metamorphism, recovery, and crustal evolution are summarized in tabular form. Insofar as possible, the dimensions of the UHP complexes and timing of events are also approximated. Lastly, thermal modelling of steady-state subduction and of continental collision (burial of sialic crust followed by tectonic return to midcrustal levels) is utilized to demonstrate the following conclusion: Successful retention of UHP mineralogic relics probably depends on the ascent of relatively small blocks and thin slabs—from which heat can be conducted relatively efficiently—between the still descending, outboard, cold lithospheric footwall plate below, and the inboard, cool, hanging wall plate above; accordingly, both plates thereby function as heat sinks [*Ernst*, 1977; *Hacker and Peacock*, 1995].

BRIEF SUMMARY OF ULTRAHIGH-PRESSURE COMPLEXES

Geologic/mineralogic, radiometric, maximum P-T, and tectonic data are briefly summarized in Table 1 for the five intracratonic Eurasian ultrahigh-pressure metamorphic terranes noted above. Physical conditions of recrystallization were determined employing phase-equilibrium constraints (*e. g.*, field of stability for coesite) combined with thermobarometric measurements involving exchange equilibria (*e. g.*, Fe<—>Mg partitioning between coexisting garnet and clinopyroxene). In addition, times and rates of exhumation to midcrustal levels, and the relative areal dimensions + thicknesses of the UHP units are approximated. The overprinting stage presumably reflects midcrustal achievement of neutral buoyancy of the no-longer-ascending complexes. Few firm constraints are available, but it seems clear that decompression of the most profoundly subducted lithotectonic sections was geologically rapid, that the exhumed UHP assemblies involve predominantly old sialic crust (the apparently anomalous mafic UHP rocks of the Zermatt-Saas Fee and Alpe Arami areas of the Western Alps are immersed in low-density serpentinite melange and/or surrounded by quartzofeldspathic schists), and that their dimensions are modest and generally sheetlike. It should be noted that the tabulated thicknesses for the UHP complexes are maxima (except for the very thin UHP sheet exposed in the vicinity of the Dora Maira Massif, where control is especially good), assuming steep tectonic terrane boundaries; most are probably thinner than values listed in Table 1. In addition, nearly all supracrustal rocks subjected to UHP metamorphism appear to have been portions of pre-existing, relatively cold cratons, rather than young, relatively warm oceanic crust or calc-alkaline arcs.

In terms of mineral parageneses, lower temperature terranes such as the Dora Maira, the Maksyutov, and the Dabie Shan are associated with blueschists, whereas the higher T conditions attending UHP metamorphism of the Western Gneiss Region and the Kokchetav Massif evidently destroyed such assemblages on decompression—if, in fact, they existed earlier as prograde lithologies. Moreover, because Kokchetav unit I rocks record the highest temperatures of all, it seems conceivable that the rarity of coesite in this assemblage is a consequence of elevated reaction rates in the system SiO_2 during unloading at ~900 °C compared to kinetics of the diamond —> graphite transformation. Finally, it is worth noting that, among the group considered, the Maksyutov Complex which evidently rose most slowly to midcrustal levels seems to have retained no coesite or other unambiguous UHP relics, whereas the Kokchetav Massif which may have ascended the fastest has retained the most abundant diamonds.

DECOMPRESSION P-T TRAJECTORIES OF UHP COMPLEXES INFERRED FROM MINERALOGY

The quantitative significance of individual relict phases and mineral assemblages summarized by *Ernst et al.* [1995] and listed in Table 1 has been documented by numerous authors [*e. g., Coleman and Wang, eds.*, 1995; *Schreyer*, 1995; *Harley and Carswell*, 1995]. Extremely low P-T gradients, approximately 5-10 °C/km, are required for deeply buried rocks to enter the stability fields of some of the reported minerals and phase associations. Such terrestrial HP and UHP environments are only found in the vicinity of downgoing lithospheric plates. Moreover, the systematic progression in relict subduction-zone minerals and inferred mineral associations support the hypothesis that the parageneses represented close approaches to chemical equilibrium at their times of formation. Of course, the relatively high temperatures attending profound upper mantle depths of underflow would be expected to promote equilibration.

Thermobarometric estimates of the physical conditions attending the HP and UHP metamorphic events summarized in Table 1 are illustrated in Figure 2. Clearly, a range of temperatures and pressures are preserved in the several complexes, reflecting stages of partial re-equilibration on ascent towards midcrustal levels. It also seems likely that different UHP complexes decoupled from the subducting lithosphere at different depths and underwent a period of heating prior to return toward midcrustal levels. The actual exhumation process undoubtedly involved a continuum of changing P-T conditions. Incompleteness of back reaction apparently testifies, at least in part, to the rapidity of exhumation and/or the absence of a rate-enhancing aqueous fluid phase.

Nonetheless, for sections undergoing essentially isothermal decompression at 800 ± 100 °C along an adiabatic decompression path, residence times outside

TABLE 1. Comparison of Ultrahigh-Pressure Metamorphic Complexes [*after Ernst et al.*, 1995]

	Qinling-Dabie-Sulu belt, coesite-eclogite	Kokchetav Massif, unit I	Maksyutov Complex, lower unit	Dora Maira Massif, lower Venasca unit	Western Gneiss Region, Fjordane complex
protolith formation age	1.3-2.9 Ga	2.2-2.3 Ga	1.2 Ga	303 Ma	1.6-1.8 Ga
temp of UHP metamorphism	750 ± 50 °C	950 ± 50 °C	625 ± 50 °C	725 ± 50 °C	825 ± 75 °C
depth of UHP metamorphism	90-120 km	125 km	90 km	90-120 km	90-125 km
time of UHP metamorphism	210-220 Ma	530-540 Ma	589 ± 69 Ma	35-40 Ma	420-440 Ma
midcrustal annealing	180-200 Ma	515-517 Ma	400-500 Ma	15-25 Ma	375 Ma
rise time to midcrust	25 ± 10 Ma	20 ± 5	90-190 Ma	20 ± 5 Ma	55 ± 5 Ma
exhumation rate*	4-5 mm/yr	6 mm/yr	0.5-1 mm/yr	4-6 mm/yr	2 mm/yr
coesite incl.	rel. abundant	rare	very rare	rel. abundant	very rare
diamond incl.	very rare	rel. abundant	absent	absent	very rare
blueschists	present	absent (?)	present	present	absent
areal extent	450 x 75 km	50 x 15 km	120 x 8 km	225 x 60 km	350 x 70 km
max thickness of complex	10 km	5-10 km	4-6 km	1-2 km	10-15 km (?)

* Descriptions and references to primary data are summarized by *Ernst et al.* [1995]. Exhumation rates were estimated by dividing depth of UHP metamorphism by rise time [*see also Coleman and Wang, eds.*, 1995; *Harley and Carswell*, 1995; *Hacker et al.*, in press].

the stability fields of UHP minerals for tens of millions of years ought to have caused the total obliteration of such phases. It therefore seems plausible that deeply subducted sections that retain traces of UHP precursors may have cooled during initial stages of exhumation; lithologic packages evidently more-or-less retraced the prograde P-T trajectory in reverse (= retrograde refrigerated path) by migrating up the subduction channel during continued lithospheric plate descent. For instance, in the Dora Maira UHP Massif, the assemblage talc + phengite remained stable during decompression from peak metamorphic conditions of 700-800 °C at depths of 90-120 km. The assemblage talc + phengite dehydrates to phlogopite + kyanite + quartz + H_2O at ~600 °C and 10 kbar [*Schertl et al.*, 1991], and if it is similar to other devolatilization reactions, is relatively rapid-running. This reaction constrains the maximum temperature of the Dora Maira terrane during decompression. Temperatures evidently declined from 700-800 °C at 35 kbar to 500-600 °C at 7-9 kbar accompanying the low-rank amphibolite to greenschist facies overprint. Because H_2O is evolved by the reaction, the temperatures cited above represent maximum values, and lowered activities of this volatile component would constrain the retrograde P-T trajectory of the talc + phengite pair to even lower temperatures for a given lithostatic pressure. The Dora Maira Massif apparently did not pass through conditions of the pyroxene hornfels, high-rank amphibolite, or granulite facies during exhumation.

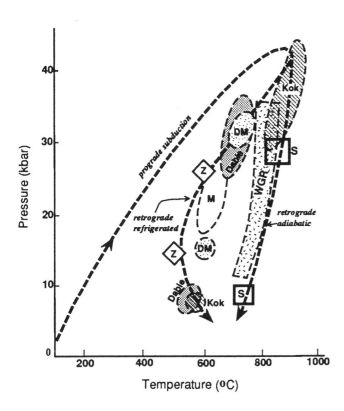

Fig. 2. Pressure-temperature conditions of metamorphism for UHP terranes summarized in Ernst et al. (1995) deduced from phase-equilibrium relations and thermobarometric computations by various workers (see Table 1 for petrotectonic comparisons). Abbreviations are: Dabie = Dabie block of Qinling - Dabie-Sulu belt; DM = Dora Maira Massif; Kok = Kokchetav Massif; M = Maksyutov Complex; S = Sulu block of Qinling - Dabie-Sulu belt; WGR = Western Gneiss Region; and Z = Zermatt-Saas Fee ophiolitic melange. Speculative prograde subduction and retrograde (refrigeration versus approximately adiabatic) paths are also illustrated; tops and bottoms of slabs are not distinguished, and contrasting patterns are for recognition purposes only.

Even in favorable circumstances, only those UHP complexes of appropriate dimensions to be relatively efficiently cooled would be able to preserve evidence of their earlier, deep upper mantle history. The hypothesized mechanism involving ascent of relatively thin tectonic slices is numerically modelled below. It is more-or-less similar to the general mechanism advanced by *Dobretsov* [1991] and *Dobretsov and Kirdyashkin* [1994], as well as the "tectonic wedge extrusion" scenario proposed for the Dabie-Sulu UHP belt by *Maruyama et al.* [1994].

THERMAL MODELLING OF UHP TERRANES—THE IMPORTANCE OF BEING THIN

PROGRADE P-T PATHS

As documented in Table 1, ultrahigh-pressure metamorphic rocks record temperatures of 700-900 °C at pressures of 28-40 kbar, corresponding to depths of 100-125 km. These peak metamorphic temperatures are significantly lower than temperatures of 1000-1450 °C at ~100 km depth estimated for continental and oceanic geotherms far removed from plate margins [*e.g., Pollack and Chapman*, 1977; *Stein and Stein*, 1992].

The low temperatures recorded by UHP metamorphic rocks strongly suggest formation in a convergent plate margin where isotherms are depressed on a regional scale by the descent of cool lithosphere. During subduction, downward conduction of heat warms the top of the sinking lithosphere and cools the overlying hanging wall lithosphere and asthenosphere. For initial stages of underthrusting at typical plate-tectonic rates of 10-100 mm/yr, the top of the footwall is heated to a temperature equal to approximately one-half of the initial prethrusting temperature at any given depth [*Molnar and England*, 1980]. Thus, during the early stages of subduction, rocks underthrust to depths of ~100 km should reach temperatures of 500-725 °C. Continued underthrusting drains heat from the hanging wall and results in even cooler temperatures in the downgoing slab at a given depth, with the final steady-state temperature depending on the geometry and rate of underflow, the rate of shear heating, and the rate and distribution of radioactive heat-producing elements. Thus, temperatures at 100 km depth in convergent margins could be even lower than those recorded by UHP metamorphic mineral assemblages.

The P-T conditions recorded by UHP metamorphic rocks are somewhat warmer than steady-state subduction-zone conditions predicted by two-dimensional thermal models. For example, as illustrated in Figure 3, steady-state temperatures within a subduction shear zone characterized by underflow of oceanic lithosphere may be as cool as 450-650 °C at ~100 km depth, for convergence rates of 10 to 100 mm/yr [*Peacock et al.*, 1994; *Peacock*, 1995]. Temperatures predicted by such subduction-zone thermal models should not be strictly compared to UHP metamorphic conditions because the geometry of convergence and the rate and distribution of radioactive heating in continent collision zones differs from those in Pacific-type subduction zones. The higher temperatures probably reflect one or more of the following possibilities: (1) UHP rocks formed in a cold,

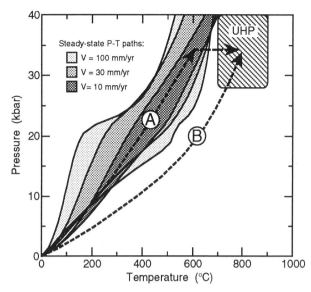

Fig. 3. Steady-state subduction-zone P-T paths and possible prograde P-T paths followed by UHP metamorphic terranes. Shaded regions represent range in calculated P-T conditions within a steady-state subduction shear zone for V = 10, 30, and 100 mm/yr and two different shear heating models, assuming no shear heating and high shear heating where shear stresses increase at 5% P until the brittle-ductile transition at 500 °C (Peacock et al., 1994). Higher peak metamorphic temperatures recorded by UHP terranes may be achieved by post-subduction heating after detachement from the subducting slab (path A), subduction at slow (<10 mm/yr) convergence rates (path B), subduction prior to the achievement of thermal steady state (path B), or heightened heat flow produced in subducted sialic crust (path B).

steady-state subduction zone, but heated up 200-300 °C at ~100 km depth after detachment from the downgoing plate and prior to ascent back toward the Earth's surface (path A, Figure 3); (2) UHP metamorphic rocks were tectonically buried at convergence rates <10 mm/yr, consistent with a slowing of convergence when buoyant continental crust entered the subduction zone (path B, Figure 3); (3) UHP metamorphism occurred prior to the achievement of steady-state conditions in the subduction zone (path B, Figure 3); or (4), the computed Pacific-type thermal structure was modified by the contribution of radiogenic heat from the subducted sialic crust (path B, Figure 3).

RETROGRADE P-T PATHS

The general lack of high-temperature, lower pressure granulite-facies overprinting (except for rocks of the Western Gneiss Region of Norway, and the Sulu block of easternmost China) suggests that UHP terranes cooled during decompression and that thermal relaxation took place at considerable depth. Mineralogic evidence previously presented demonstrates that UHP metamorphic terranes do not follow typical clockwise P-T paths, characteristic of regional metamorphism of sialic crust, which generally reach peak temperatures at about 10 kbar. In contrast, ultrahigh-pressure metamorphic complexes appear to have lost significant amounts of heat as they were exhumed from mantle depths of ~100 km to crustal depths of ~30 km. Such assemblages must have lost heat by conduction to cooler rocks encountered during ascent.

Ultrahigh-pressure metamorphism occurs at temperatures that are unusually cool for depths of 100 km. How is it possible for such rocks to cool even more during ascent from 100 to 30 km depth, where the effects of surface cooling are insignificant? Deep-seated cooling of UHP metamorphic terranes may occur if the rocks are exhumed in the footwall of an extensional fault/shear zone, if exhumation takes place while subduction continues, or, most likely, by a combination of both processes. The volume of orogenic sediments and the lack of mantle detritus strongly suggest that erosional unroofing accounts for only a part of the exhumation of UHP metamorphic rocks [*e.g., Platt*, 1986]. Exhumation of UHP metamorphic terranes from great depth, therefore, requires removal—or lateral displacement—of most of the overlying rocks by extensional unroofing. During normal faulting, footwall rocks are juxtaposed against cooler hanging wall rocks. For moderate displacement rates, the upper parts of the footwall cool as heat is conducted upward into the cooler hanging wall. Accordingly, UHP metamorphic complexes may cool in a similar fashion if they are exhumed in the footwall of extensional structures.

Exhumation of rocks by syncollisional extensional processes has recently been documented in many orogenic belts [*e.g., Platt*, 1993]. In subduction zones, thickening of an accretionary complex by underplating promotes extension in the upper parts of the prism [*Platt*, 1986]. Similarly, during continent collision, crustal thickening can cause surface elevation differences that can drive extension in the upper crust (*e.g*, the South Tibetan detachment system in the High Himalayas). In both settings, exhumation of metamorphic rocks occurs in higher parts of the system while subduction (or crustal convergence) continues below. UHP metamorphic terranes are less dense than mantle rocks, and buoyancy forces must drive exhumation from depths of ~100 km to crustal levels [*Platt*, 1993; *Ernst and Liou*, 1995]. Some UHP metamorphic

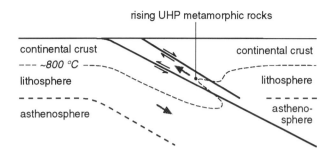

Fig. 4. Schematic cross section attempting to explain the decompression cooling of a rising sheet of UHP rock, assuming steady-state subduction. During uplift of a thin UHP terrane, cooling of the upper margin of the sheet takes place where it is juxtaposed against the lower T hanging wall; cooling along the lower margin of the sheet takes place where it is juxtaposed against the lower T subducting (refrigerating) plate. As the sheet thickness decreases, cooling of the entire UHP complex is effected.

terranes, such as the Dora Maira, evidently represent thin slices of continental crust that were subducted during ocean basin destruction prior to final continent collision. If such a buoyant fragment were detached from the downgoing plate and underplated at ~100 km, continued lithospheric underflow beneath the complex could remove heat from the base of the UHP terrane during its ascent.

Synsubduction exhumation of the footwall of an extensional structure would effectively cool an UHP metamorphic terrane during exhumation, as illustrated schematically in Figure 4. In the illustrated scenario, heat is efficiently withdrawn from both upper and lower surfaces of the UHP complex: (a) downward across the thrust-fault contact with the relatively cold, sinking, outboard lithospheric plate; and (b), upward across the normal-fault contact with the cool, inboard hanging wall plate. The decompression P-T path followed by an UHP metamorphic terrane depends on the rate and geometry of exhumation. Recent radiometric data, summarized in Table 1, suggest that most UHP complexes rise from depths of ~100 km to the midcrust in 20-50 million years at rates of a few mm/yr. Unfortunately, the geometry of exhumation is so poorly known that detailed thermal models of the exhumation process are ill constrained at present.

Results of several simple burial-exhumation calculations are depicted in Figure 5. The purpose of these computations is to demonstrate that substantial cooling of a terrane can occur during exhumation in the footwall of an extensional structure. In these two-dimensional heat-transfer calculations, rocks are tectonically buried

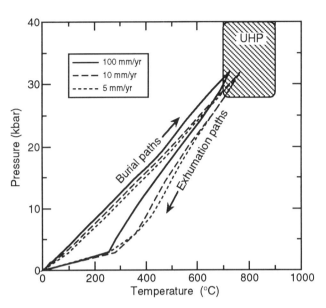

Fig. 5. Calculated burial and exhumation P-T paths for a rock tectonically buried to 100 km depth along a thrust fault dipping 27°, then immediately exhumed by tectonic extension along the same feature reactivated as a normal fault. Convergence and extension take place at the same constant rate; P-T paths are depicted for V = 5, 10, and 100 mm/yr. Prograde burial P-T paths are warmer than those depicted in Fig. 3 because the system does not reach thermal steady-state during convergence. Retrograde exhumation P-T paths depend slightly on the rate of extension, but all paths clearly show that heat is lost to the hanging wall during exhumation. Heat would also be conducted into the tectonically lower subducting lithosphere (not illustrated).

(subducted) to 100 km depth and then immediately exhumed by tectonic extension along the same structure reactivated as a normal fault. The computed exhumation paths are slightly warmer than the burial paths, but clearly, the rising UHP slabs cool during decompression due to loss of heat upward into the cooler hanging wall. The retrograde P-T trajectory depends slightly on the exhumation rate.

Both prograde and retrograde pressure-temperature paths depicted in Figure 5 are for rocks constituting the top of the footwall. Material sited deeper within the lower plate (farther from the fault) will follow different descent and ascent P-T trajectories. During burial, interior rocks will follow cooler P-T paths than rocks close to the thrust fault. In contrast, during exhumation, interior rocks will remain warmer than materials adjacent to the normal fault. For this reason, the inferred cooling of UHP complexes on return toward the surface requires a relatively thin, slab-like geometry. For a time scale of 20 million years (typical rise times to the midcrust, according to Table 1), the characteristic length scale of heat conduction into or out of a slab bounded by faults is approximately 40 km. Each UHP metamorphic terrane summarized in Table 1 is less, probably far less, than 10-15 km thick; accordingly, sufficient time exists during exhumation for these complexes to lose heat, provided that they are juxtaposed against cooler rocks during ascent.

Acknowledgments. This study was supported by Stanford University, Arizona State University, and the National Science Foundation through grants EAR-93-04480/Coleman and EAR-94-17711/Peacock. We thank R. G. Coleman, N. L. Dobretsov, J. G. Liou, B. R. Hacker, and Tracy Rushmer for fruitful discussions and constructive reviews of the first-draft manuscript.

REFERENCES

Cloos, M., Lithospheric buoyancy and collisional oro-genesis: Subduction of oceanic plateaus, continental margins, island arcs, spreading ridges, and sea-mounts, *Geol. Soc. America Bull. 105*, 715-737, 1993.

Coleman, R. G., and X. Wang, editors, Ultrahigh Pressure Metamorphism, *Cambridge University Press*, New York, 528p, 1995.

Dobretsov, N. L., Blueschists and eclogites: A possible plate tectonic mechanism for the emplacement from the upper mantle: *Tectonophysics, 186*, 253-268, 1991.

Dobretsov, N. L., and A. G. Kirdyashkin, Blueschists of North Asia and models of subduction-accretion wedge, in R. G. Coleman, editor, *Reconstruction of the Paleo-Asian Ocean*, 29th Int. Geol. Congress, Kyoto, Japan, Proc., Part B, 99-114, 1994.

Erambert, M., and H. Austrheim, The effect of fluid and deformation on zoning and inclusion patterns in polymetamorphic garnets, *Contr. Min. Petrology, 115*, 204-214, 1993.

Ernst, W. G., Metamorphic zonations on presumably subducted lithospheric plates from Japan, California, and the Alps, *Contr. Min. Petrology, 34*, 43-59, 1971.

Ernst, W.G., Mineral parageneses and plate tectonic settings of relatively high-pressure metamorphic belts, *Fortschr.Min., 54*, 192-222, 1977.

Ernst, W. G., and J. G. Liou, Contrasting plate-tectonic styles of the Qinling-Dabie-Sulu and Franciscan metamorphic belts, *Geology, 23*, 353-356, 1995.

Ernst, W. G., J. G. Liou, and R. G. Coleman, Comparative petrotectonic study of five Eurasian ultrahigh-pressure metamorphic complexes, *Int. Geol. Rev., 37*, 191-211, 1995.

Hacker, B. R., and S. M. Peacock, Creation, preserv-ation, and exhumation of UHPM rocks, 159-181 in R. G. Coleman, and X. Wang, editors, *Ultrahigh Pressure Metamorphism*, Cambridge University Press, New York, 528p, 1995.

Hacker, B. R., X. Wang, and E. A. Eide, Geochronology and structure of the Qinling-Dabie ultrahigh-pressure orogen in China, in T. M. Harrison, and A. Yin, editors, *Rubey Volume VIII: The Tectonic Develop-ment of Asia*, Cambridge University Press, New York, in press.

Harley, S. L., and D. A. Carswell, Ultradeep crustal metamorphism: A prospective review, *Jour. Geophys. Res., 100*, 8367-8380, 1995.

Liou, J.G., R. Zhang, and W. G. Ernst, An introduction to ultrahigh-pressure metamorphism, *The Island Arc, 3*, 1-24, 1994.

Maruyama, S., J. G. Liou, and R. Zhang, Tectonic evolution of the ultrahigh-pressure (UHP) and high-pressure (HP) metamorphic belts from central China, *The Island Arc, 3*, 112-121, 1994.

Molnar, P., and P. England, Temperatures, heat flux, and frictional stress near major thrust faults, *Jour. Geophys. Res., 95*, 4833-4856, 1990.

Peacock, S. M., 1995, Ultrahigh-pressure metamorphic rocks and the thermal evolution of continent collision belts, *The Island Arc, 4*, 376-383.

Peacock, S.M., T. Rushmer, and A. B. Thompson, Partial melting of subducting oceanic crust, *Earth Planet. Sci. Let., 121*, 227-244, 1995.

Platt, J. P., 1986, Dynamics of orogenic wedges and the uplift of high-pressure metamorphic rocks, *Geol. Soc. America Bull., 97*, 1037-1053.

Platt, J. P., 1993, Exhumation of high-pressure rocks: a review of concepts and processes, *Terra Nova, 5*, 199-133.

Pollack, H. N., and D. S. Chapman, On the regional variation of heat flow, geotherms and the lithospheric thickness, *Tectonophysics, 138*, 279-296, 1977.

Schertl, H. P., W. Schreyer, and C. Chopin, The pyrope-coesite rocks and their country rocks at Parigi, Dora-Maira Massif, Western Alps; detailed petrography, mineral chemistry and PT-path, *Contr. Min. Petrology, 108*, 1-21, 1991.

Schreyer, W., 1988, Experimental studies on meta-morphism of crustal rocks under mantle pressures, *Min. Mag., 53*, 1-26.

Schreyer, W., 1995, Ultradeep metamorphic rocks: The retrospective viewpoint, *Jour. Geophys. Res., 100*, 8353-8366.

Stein, C. A., and S. Stein, A model for the global vari-ation in oceanic depth and heat flow with lithospheric age, *Nature, 359*, 123-129 , 1992.

Wang, X., R. Zhang, and J. G. Liou, UHPM terrane in east central China, 356-390 in R. G. Coleman, and X. Wang, editors, *Ultrahigh Pressure Metamorphism*, Cambridge University Press, New York, 528p., 1995.

W. G. Ernst, Department of Geological and Environmental Sciences, Stanford University, Stanford, CA 94305-2115

Simon M. Peacock, Department of Geology, Arizona State University, Tempe, Arizona 85287-1404

Volatile Transfer and Recycling at Convergent Margins: Mass-Balance and Insights from High-P/T Metamorphic Rocks

Gray E. Bebout

Department of Earth and Environmental Sciences, Lehigh University

The efficiency with which volatiles are deeply subducted is governed by devolatilization histories and the geometries and mechanisms of fluid transport deep in subduction zones. Metamorphism along the forearc slab-mantle interface may prevent the deep subduction of many volatile components (e.g., H_2O, Cs, B, N, perhaps As, Sb, and U) and result in their transport in fluids toward shallower reservoirs. The release, by devolatilization, and transport of such components toward the seafloor or into the forearc mantle wedge, could in part explain the imbalances between the estimated amounts of subducted volatiles and the amounts returned to Earth's surface. The proportion of the initially subducted volatile component that is retained in rocks subducted to depths greater than those beneath magmatic arcs (> 100 km) is largely unknown, complicating assessments of deep mantle volatile budgets.

Isotopic and trace element data and volatile contents for the Catalina Schist, the Franciscan Complex, and eclogite-facies complexes in the Alps (and elsewhere) provide insight into the nature and magnitude of fluid production and transport deep in subduction zones and into the possible effects of metamorphism on the compositions of subducting rocks. Compatibilities of the compositions of the subduction-related rocks and fluids with the isotopic and trace element compositions of various mantle-derived materials (igneous rocks, xenoliths, serpentinite seamounts) indicate the potential to trace the recycling of rock and fluid reservoirs chemically and isotopically fractionated during subduction-zone metamorphism.

1. INTRODUCTION

Knowledge of the flux of volatiles at convergent margins is fundamental to our understanding of crust, ocean, and atmosphere evolution [e.g., *Javoy et al.*, 1982; *Berner and Lasaga*, 1989; *Zhang and Zindler*, 1993]. Subduction transfers hydrothermally altered igneous rocks produced at mid-ocean ridges and sediments deposited on the seafloor abyssal plain and in trenches to the upper mantle (Fig. 1). Interstitial pore fluids make up a large fraction of the volatiles initially subducted (> 50 vol. % initially in some subducting sediments; *Moore and Vrolijk* [1992]). *Rea and Ruff* [in press] estimate that seawater constitutes ~40% of the mass initially subducted in sediments, with subduction (globally) of 1.4×10^{15} grams/year of sediment and 0.9×10^{15} grams/year of seawater; these estimates do not account for the pore fluids contained in the sialic sediment contributed to trenches from the continents [*von Huene and Scholl*, 1991]. Large amounts of other major fluid-forming volatiles (e.g., reduced and oxidized C; ammonium and organic N; sulfide and sulfate S) and relatively fluid-mobile trace elements (e.g., B, Cs) are also subducted, impacting their global cycling [*Bebout*, 1995].

Much of the pore fluid initially subducted (perhaps > 90%; *von Huene and Scholl* [1991]) is expelled by compaction processes at shallow levels and escapes toward the surface [*Moore and Vrolijk*, 1992]. The volatiles that persist to greater depths (> 15 km) in subduction zones are primarily those bound in minerals in the subducting rocks (e.g., H as hydroxyl, C in carbonates and graphite, N as ammonium in clays and micas). The volatile components of the subducting rocks are then released through metamorphic devolatilization (and melting) reactions deeper in subduction zones. Fluids released by devolatilization reactions

180 CONVERGENT MARGIN VOLATILE RECYCLING

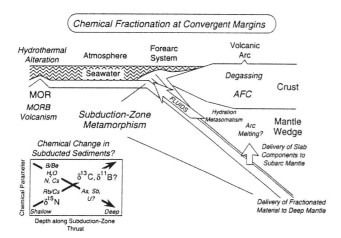

Fig. 1. Schematic illustration of a continental margin-type subduction zone, showing possible sites of trace element fractionation. The large arrow along the slab-mantle interface indicates transfer, toward the surface, of volatile components released by subduction-zone metamorphism. The inset plot demonstrates the likely evolution of chemical parameters with progressive metamorphism of sedimentary rocks during subduction. The concentrations of B, N, Cs, As, Sb and possibly U, the B/Be ratios, and H_2O contents are expected to be decreased dramatically by this metamorphism, whereas the Rb/Cs ratios, $\delta^{15}N$, $\delta^{13}C$ of graphite, and perhaps, the $\delta^{11}B$ values of the sedimentary rocks are expected to be increased by the devolatilization process [Bebout, 1995]. This shift in composition is accomplished by the separation, from the rocks, of reservoirs (e.g., C-O-H-S-N fluids and silicate melts) with complementary, fractionated compositions (e.g., high B/Be, low Rb/Cs, N_2 with low $\delta^{15}N$).

may metasomatize the mantle wedge and flux partial melting leading to arc magmatism, contributing chemical components to arc magma source regions [e.g., Ryan et al., this volume]. Imbalances exist in the amounts of volatiles believed to be subducted and those returned to surficial and shallow crustal reservoirs by arc magmas [e.g., Bebout, 1995]. Ito et al. [1983] suggested that only ~10% of the deeply subducted water is returned in arc magmas; these calculations do not include pore fluids. The majority of the fluids expelled by mechanical processes and devolatilization may be returned to the oceans by updip transport [Bebout, 1991b], perhaps contributing to apparent recycling imbalances. Seafloor fluid venting occurs along trenches, and studies of modern and ancient accretionary complexes document updip shallow fluid flow [Fisher, this volume]. However, the magnitude of the flux of deeper-sourced fluid into shallow parts of accretionary complexes is unknown

[Kastner et al., 1991]. Another significant fraction of the volatiles released during devolatilization could enter the forearc mantle wedge, resulting in widespread hydration and metasomatism [e.g., Ryan et al., this volume]. The retention of volatiles in subducting rocks to depths greater than those beneath arcs [e.g., Thompson, 1992] could also strongly impact the crust-mantle volatile mass-balance.

High-P/T metamorphic rocks cannot be directly used to constrain volatile flux at paleo-convergent margins because of the great complexity of the tectonic processes leading to their ultimate surficial exposure and the potentially biased lithological record inherent with these processes. However, study of the volatile contents and chemical and isotopic compositions of these rocks can potentially yield insight into the magnitudes and mechanisms of devolatilization as functions of varying thermal history and the possible chemical and isotopic fractionations resulting from this devolatilization. Knowledge of this fractionation can improve our efforts to trace the possible additions of "slab signatures" through study of such features as serpentinite seamounts, forearc accretionary wedge fluids, arc volcanic rocks, mantle xenoliths, and ocean-island basalts (OIB; Moran et al. [1992]; Bebout [1995]). Study of such suites can also afford estimates of metamorphic fluid flux and the scales and mechanisms of fluid transport deep in subduction zones [Bebout, 1991a, b; Getty and Selverstone, 1994].

In this paper, I present calculations of subduction-zone inputs and outputs of volatile components in a consideration of the possible effect of subduction-zone metamorphism on convergent margin volatile recycling. I review our knowledge of the mechanisms and magnitude of deep subduction-zone fluid production and transport based on study of high-P/T metamorphic rocks.

2. THE FLUX OF VOLATILE COMPONENTS AT CONVERGENT MARGINS

In this section, I discuss constraints on the magnitudes of volatile flux at convergent margins, comparing the estimated flux of volatiles (H_2O, C, N) into subduction zones with the estimated return fluxes of volatiles from the mantle to the surface (Table 1). Similar calculations for B by Moran et al. [1992] are summarized in Table 1. For considerations of volatile subduction in sediments, I use estimates of Rea and Ruff [in press] and von Huene and Scholl [1991] which consider the volume and composition of sediments in ocean basins seaward of trenches, the addition of

Table 1. Comparisons of Likely Subduction Volatile Fluxes and Outputs from the Mantle (all by weight).

Reservoir/Flux	Subduction Rate (x 10^{15} g/year)	Input Variables for Sediments	H_2O (x 10^{13} g/year)	C (x 10^{13} g/year)	N (x 10^{10} g/year)	B[a] (x 10^9 g/year)
Flux from Mantle [references][b]			10-14 [1, 2]	2.7 [3, 4, 5]	2-8.96 [3, 4, 6]	5-135 [7]
Fluxes into Mantle by Subduction						
Sediments	1.33 to 3.5[c]	100% sh, no CO_2[d]	[3 %] 3.99-10.5[e] [4 %] 5.32-14.0	[1 %] 1.33-3.5	[600ppm] 80-210 [70ppm] 9.3-24.5	[159ppm] 211-557
		100% sh, 0.5 % CO_2[d]	[3 %] 3.99-10.5 [4 %] 5.32-14.0	[1.14 %] 1.52-4.0		
		75% sh [4 %], 25% ls[f]	[3 %] 3.99-10.5	[3.86 %] 5.1-13.5[g]	[450ppm] 60-158g,[h]	[119ppm] 158-417[h]
		50% sh [4 %], 50% ls	[2 %] 2.66-7.00	[6.57 %] 8.74-23.0[g]	[300ppm] 40-105g,[h]	[80ppm] 106-280[h]
Oceanic Crust	58.9 to 60[c]		[1.5] 88.4-90.0 [3 %] 177-180	[.081 %] 4.77-4.86	[10ppm] 59-60 [5ppm] 29.5-30 [1ppm] 5.9-6.0	[24/0.2] 112-114[a]
Ranges of Total Fluxes into Mantle[i]			min=91.1 max=194	min=6.1 max=27.9	min=15.2 max=270	min=218 max=671
Minimum and Maximum % Return to Surface (flux from mantle/flux into mantle)			min=5 max=15	min=10 max=44	min=1 max=59	min=1 max=62

a Values are largely from *Moran et al.* [1992].

b Sources of esimates of flux from the mantle are *Ito et al.* [1993; 1]; *Peacock* [1990; 2]; *Javoy et al.* [1982, 1986; 3]; *Zhang and Zindler* [1993; 4]; *Marty and Jambon* [1987; 5]; *Zhang* [1988; 6]; and *Moran et al.* [1992; 7].

c Range of values from *Ito et al.* [1993] and *Peacock* [1990]. Estimates based on *Rea and Ruff* [in press], *von Huene and Scholl* [1991], and D. Scholl [personal communication, 1995] are intermediate, falling into the range of 1.9×10^{15} to 2.7×10^{15} grams/year.

d Sandstone and shale indicated by, "sh". "CO_2" indicates the presence of small amounts of finely disseminated calcite.

e Values in brackets are the concentrations incorporated into each calculation. Ranges in flux reflect the range of values for subduction rate (e.g., for sediments, 1.33 to 3.5×10^{15} g/year).

f Indicates a 3:1 (by weight) mixture of shale/sandstone and limestone, the latter consisting of 100% calcite. Assumes that shale/sandstone contains 4 wt. % H_2O.

g Shale/sandstone lithology is assumed to contain 1.14 wt. % C and 600 ppm N.

h Limestone (100% calcite) lithology is assumed to contain no N or B.

i Minimum and maximum combinations of sediment and oceanic crust volatile subduction (e.g., for H_2O, 2.66 + 88.4 = 91.1; 14.0 + 180 = 194).

terrigenous sediment in trenches, and subduction erosion of the hanging wall. Of particular interest is the proportion of the sediment subducted to depths beyond accretionary complexes and available for addition to the mantle.

Ito et al. [1983] used a global sediment subduction rate of 3.5×10^{15} grams/year [cf. *Rea and Ruff*, in press] and estimated a rate of H_2O subduction in sediments of ~1×10^{14} grams/year, assuming 3 ± 1 wt. % H_2O in the subducted sediments (Table 1). Data for low-grade Catalina Schist metasedimentary rocks are consistent with somewhat higher H_2O contents of 4 ± 1 wt. % (*Bebout*, 1995]. *Ito et al.* [1983] assumed 1-2 wt. % H_2O content for subducted oceanic crust, deriving a H_2O subduction rate of ~9×10^{14} grams/year (value of 8.84×10^{14} grams/year in Table 1 based on H_2O content of 1.5 wt. %). Pore fluids were not considered by the mass balance calculations of *Ito et al.* [1983] and are not directly considered here. However, such fluids may play a key role in fluid-related processes at shallow levels [*von Huene and Scholl*, 1991; *Moore and Vrolijk*, 1992] if they hydrate mafic (and ultramafic) rocks or are retained in small amounts of porosity. If additional hydration occurs in forearcs [see *Bebout*, 1995], the estimates of deep H_2O subduction in oceanic crust based on analyses of unsubducted seafloor basalts and gabbros (ave. 1-2 wt. % H_2O; *Ito et al.* [1983]; *Peacock* [1990]) may be low. The use of 3.0 wt. % results in a H_2O subduction rate of ~1.8×10^{15} grams/year in oceanic crust (Table 1). Based on their estimates of H_2O subduction and constraints on rates of H_2O expulsion in arc magmatism (they estimated ~1×10^{14} grams/year; *Peacock* [1990] estimated 1.4×10^{14} grams/year), *Ito et al.* [1983] concluded that ~90% of the H_2O is either returned to the mantle or to the crust by processes other than arc magmatism. The higher estimates of H_2O content in subducted rocks based on study of the Catalina Schist would indicate possible greater imbalances (~5-15% return, incorporating ranges in estimates for subduction zone inputs and outputs; see Table 1).

Much attention has been paid to the magnitude of recycling of C into the upper mantle [*Berner and Lasaga*, 1989; *Javoy et al.*, 1982]. *Javoy et al.* [1982, 1986] estimated that the C fluxes from the mantle are >2.7×10^{13} grams/year [cf. *Zhang and Zindler*, 1993; *Marty and Jambon*, 1987]. Those authors suggested the possibility of recycling of large amounts of C with average $\delta^{13}C_{PDB}$ near that of the "average mantle" compositions of ~ -5‰ [*Agrinier et al.*, 1985; cf. *Mattey*, 1987]. If the sediment subduction rate of 3.5×10^{15} grams/year is used, together with a reduced C content of 1 wt. % [*Bebout*, 1995], a C subduction rate of 3.5×10^{13} grams/year is obtained for the reduced sedimentary C alone (Table 1). Consideration of varying carbonate subduction significantly increases estimates of total C subduction. If sediments contain 0.5 wt. % CO_2 in relict diagenetic carbonate cement, then this estimate is increased to 4.0×10^{13} grams/year, again for a sediment subduction rate of 3.5×10^{15} grams/year. The subduction of calcareous oozes dramatically increases the estimates of C flux at convergent margins. If 25% of the 3.5×10^{15} grams/year of subducted sediment is 100% calcite, the C flux estimate is increased to 1.35×10^{14} grams/year. Estimates of the proportion of biogenic carbonate subducted globally (individual margins vary dramatically) by *Rea and Ruff* [in press] and *von Huene and Scholl* [1991] fall between 3 and 4% after consideration of the dilution of ocean sediment by terrigenous sediment deposited in trenches and the contributions of material by subduction erosion. Using estimates of *Ito et al.* [1983] and *Peacock* [1990] for rates of subduction of oceanic lithosphere (58.9-60×10^{15} grams/year) and the estimate of CO_2 content of oceanic crust from *Staudigel et al.* [1989] (average in "bulk oceanic crust" is 0.081 wt. %), a C subduction rate of ~4.8×10^{13} grams/year is inferred for subduction of oceanic crust alone. Thus, depending on the types and amount of sediment subducted, and again recognizing the uncertainties in all estimates, the C flux from the mantle could amount to ~10-45% of the amount of C subducted.

The recycling budget of N, which shows a far more heterogeneous isotopic composition in the mantle [*Javoy et al.*, 1986; *Boyd and Pillinger*, 1994], and for which fewer data exist, is more difficult to evaluate. However, subduction of N is expected where large amounts of sediment are recycled. If the sediment subduction rate of 3.5×10^{15} grams/year is used, along with the N concentration data for the low-grade Catalina Schist metasedimentary rocks (250-1000 ppm N; ave. 600 ppm; *Bebout and Fogel* [1992]), the subduction of ~ 2.1×10^{12} grams/year of N is indicated. This estimate is higher than *Zhang's* [1988] estimate of 2.1×10^{11} grams/year; this estimate was based on a sediment subduction rate of 3.0×10^{15} grams/year and an average sediment N content of 70 ppm which accounts for loss during devolatilization. A conservative estimate using the N content of the high-grade rocks of the Catalina Schist (ave. 125 ppm) yields a N subduction rate of 4.4×10^{11} grams/year. *Zhang* [1988] estimated N expelled by subduction-related magmatism to be $2-4 \times 10^{10}$ grams/year; *Javoy et al.*

[1986] estimated a N flux of 8.5 x 10^{10} grams/year out of the mantle (8.96 x 10^{10} grams/year estimated by *Zhang and Zindler* [1993]). The N imbalance would be more dramatic if seafloor basalts contain appreciable N [*Hall*, 1989].

3. RECORD OF FLUID PROCESSES AND VOLATILE RECYCLING IN SUBDUCTION-RELATED METAMORPHIC ROCKS

3.1. Summary of Evidence from the Catalina Schist

The Catalina Schist, which preserves a record of Early Cretaceous accretion and high-P/T metamorphism, is well-suited for studies of volatile content and stable isotope compositions as it contains metamorphosed sandstones, shales, basalts, and gabbros ranging in grade from lawsonite-albite to amphibolite facies (inferred peak P-T conditions of ~275°-750°C, 0.5-1.1 GPa; *Sorensen and Barton* [1987]; *Grove and Bebout* [1995]). The rock types are characteristic of those in many subduction-zone metamorphic terranes [cf. *Bailey et al.*, 1964; *Ernst et al.*, 1970], largely representing seafloor basaltic and trench and pelagic sedimentary protoliths. The range in grade allows comparison of the volatile contents and isotopic compositions of rocks of similar compositions which have been devolatilized to varying degrees. Such study has yielded a record of progressive high-P/T devolatilization [*Bebout and Fogel*, 1992; *Bebout et al.*, in press] and migmatization of sedimentary and mafic rocks [*Sorensen and Barton*, 1987].

In the Catalina Schist, heterogeneous rheology strongly impacted fluid transfer mechanisms during metamorphism. Mélange zones are homogenized in their stable isotope compositions and show other evidence for metasomatism [*Bebout*, 1991b] consistent with their having served as zones of enhanced fluid flow (relative to more coherent sequences of metasedimentary and metamafic rocks). Stable isotope compositions of mélange and veins from throughout the complex record fluid-rock interactions involving high-P/T, H_2O-rich C-O-H-S-N fluids with relatively uniform O- and H-isotope compositions, indicating possible km-scale fluid flow [*Bebout and Barton*, 1993].

The lowest-grade metasedimentary rocks in the Catalina Schist (peak P-T of 275°C, 0.5 GPa; *Grove and Bebout* [1995]) bear a striking chemical and isotopic resemblance to their seafloor protoliths (e.g., in C content, C/N, $\delta^{15}N$, and trace element contents; *Bebout* [1995]; *Bebout and Fogel* [1992]; Fig. 2). Higher-grade devolatilization of the metasedimentary rocks resulted in the dramatic removal of trace elements such as B, N, Cs, As and Sb by the C-O-H-S-N fluids (*Bebout et al.* [in press]; Fig. 2). Varying degrees of removal of trace elements resulted in changes in the ratios of particular trace elements (B/La, B/Be, Rb/Cs, Cs/Th) with increasing metamorphic grade; these changes have been used to estimate the trace element ratios in the metamorphic fluids [*Bebout et al.*, in press]. Veins, metasomatized mélange, and isotopic alteration throughout the complex reflect extensive chemical redistribution during the high-P/T metamorphism [*Bebout and Barton*, 1993].

Metabasaltic rocks in the Catalina Schist preserve chemical alteration attributable to seafloor processes (e.g., enrichments in large-ion lithophile elements, LILE, elevated $^{87}Sr/^{86}Sr_i$ and $\delta^{18}O$, $\delta^{13}C$ of carbonate breccia fillings; *Bebout and Barton* [1993]; *Bebout* [1995]; *Barnicoat and Cartwright* [1995]). The systematics among the LILE strongly overlap those observed in seafloor altered basalts (Fig. 7 in *Bebout* [1995]), complicating attempts to identify LILE alteration due to subduction-zone processes.

3.2. Comparison of the Catalina Schist with Other High-P/T Metamorphic Suites

Many similarities can be demonstrated in the lithologic characteristics and, where suitable data exist, the stable isotopic characteristics and volatile contents of the metamorphic rocks of the Catalina Schist and other metamorphic suites produced in similar non-collisional settings (i.e., not involving ultimate uplift due to continental collision, as is the case in the Alps; see *Ernst* [1988]). A striking resemblance in structural character, lithology, and geochemistry is observed between the low-grade parts (of lawsonite-blueschist and lower grades) of the Catalina Schist and large parts of the Franciscan Complex [*Bailey et al.*, 1964; *Ernst et al.*, 1970; *Taylor and Coleman*, 1968; *Magaritz and Taylor*, 1976; *Bebout and Barton*, 1993]. Comparison of the Catalina Schist with other rocks representing shallower parts of circum-Pacific accretionary complexes (depths of < 15 km; see *Fisher*, this volume) provides a continuum in inferred depth and temperature over which to examine the progressive evolution of permeability during subduction and its relationship to changing rheology of the subducted sedimentary rocks. However, notable differences exist between the documented histories of fluid and rock compositional evolution in Franciscan-like suites and those for suites in the Alps (and elsewhere) representing prolonged

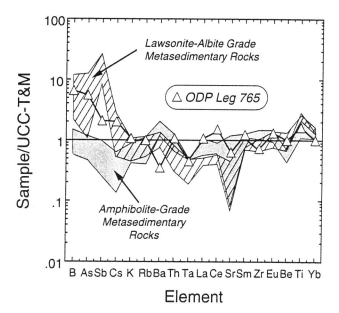

Fig. 2. Comparison of the abundances of selected elements in the Catalina Schist metasedimentary rocks of varying metamorphic grades with samples of deep-sea sediment from trench environments (averaged composition of noncalcareous sediment from ODP Site 765; *Plank and Ludden* [1992]. Data are normalized to the concentrations in the upper continental crust composite of *Taylor and McLennan* [1985]. The ranges of compositions of the lowest-grade rocks (lawsonite-albite) and their amphibolite-grade equivalents are indicated by the diagonal patterned region and the shaded region, respectively.

consumption of oceanic lithosphere with termination of subduction by continental collision. The likely complex effects of varying thermal regimes, lithology, and depth on the resulting fluid release and transport mechanisms deep in subduction zones remain uncertain.

Tables 2 and 3 summarize the reported evidence from subduction-related metamorphic suites relevant to consideration of fluid production and transfer and volatile recycling. Although the suites from collisional settings (Table 2) in many cases contain metamorphosed seafloor basaltic rocks, some contain abundant carbonate rocks (e.g., in Norway, Cyclades, Tauern Window, Alps; see *Agrinier et al.* [1985]; *Matthews and Schliestedt* [1983]; *Getty and Selverstone* [1994]) generally absent in the Franciscan-like suites (see Table 3). Some have argued that the large accumulations of carbonate-rich rocks in Cenozoic orogens such as the Alps and the Himalaya reflect enhanced carbonate accumulation and subduction which affected the Mesozoic to Cenozoic global CO_2 budget and, thus, the global surface temperatures [see *Kerrick and Caldeira*, 1993; *Selverstone and Gutzler*, 1993]. Petrologic studies of the high-P/T complexes in collisional settings have indicated peak metamorphic pressures higher than those experienced by the Franciscan-like suites, with the most extreme inferred metamorphic pressures from the collisional settings (30-35 kbars) corresponding to coesite-bearing rocks from the Alps [*Nadeau et al.*, 1993]. Documented fluid compositional histories in the suites from collisional settings are in general complex, involving significant variations in fluid composition during prograde, high-P/T stages and uplift-related, lower-P/T and retrograde stages of their metamorphic histories. Whereas fluid compositions inferred from petrologic, fluid inclusion, and stable isotope study of Franciscan-like complexes are in general extremely H_2O-rich (with varying CH_4) and low-salinity [*Bebout and Barton*, 1993; *Giaramita and Sorensen*, 1994], fluids during some metamorphic stages in the suites from collisional belts are CO_2 and N_2-rich and relatively saline [*Selverstone et al.*, 1992]. Immiscible low- and high-salinity fluids have been documented in some of these suites [*Selverstone et al.*, 1992].

In Franciscan-like suites, the study of rocks which have experienced peak metamorphic pressures and temperatures similar to those of the Alpine and other eclogitic rocks is in general possible only through study of isolated "exotic" blocks of eclogitic material in lower-grade mélange. *Giaramita and Sorensen* [1994] conducted a detailed study of fluid inclusion populations in eclogitic blocks in mélange in the Franciscan Complex and in the Dominican Republic and documented that fluids during the eclogite-facies metamorphism were extremely H_2O-rich and low-salinity. Thus, the differences in inferred fluid compositional evolution (described above and evident in Tables 2 and 3) are not obviously attributable to differing peak metamorphic P-T conditions, but appear to relate to some combination of varying lithology and P-T-time histories (including uplift-related P-T paths). The fluid evolution in thick accretionary complexes (at depths ≤ 50 km), represented by the Franciscan-like complexes, and fluid processes along the slab-mantle interface and within the slab at greater depths (i.e., as represented by some of the Alpine and other eclogitic suites) may differ significantly [e.g., *Giaramita and Sorensen*, 1994], further complicating the comparisons of the results from the differing settings.

An additional difficulty in comparing the records of volatile production and transport among the eclogite and circum-Pacific suites relates to the disparity in the field and

Table 2. Summary of Field-Based Evidence for High-P/T Fluid Production and Transport (Collisional Settings).

Region	Locality[1]	Lithologies[2]	Fluid Compositions	Sample Textures[2] [approach][3]	Other Information (fluid sources, transport) peak P-T conditions	Reference
Caledonides	Norway	ecl in gneiss	H_2O+N_2 N_2, N_2+CO_2 low to high salinity	qtz seg [inclusions]	progression in fluid chemistry; 15-25 kbar, 600 to >750°C retrograde immiscibility involving aqueous brines	Andersen et al., 1989
	Norway W. Gneiss R.	ecl, amphibolite, marble	?	w.r./veins [st. isot.]	stable isotopic shifts due to fluid-rock interaction	Agrinier et al., 1985
	Norway Bergen Arcs	ecl, granulite	CO_2, Ar, N_2 st. isot.	ecl shear zone [st. isot., inclusions]	infiltration during eclogitization of granulite in shear zones	Mattey et al., 1994
Bohemian Massif	Germany	ecl metasediment	$CO_2+N_2+CH_4$ N_2+CH_4, H_2O-rich varying salinity	w.r. [inclusions]	progression in fluid chemistry 630°C, 17 to 24 kbar	Klemd et al., 1992
Alps	Monviso Massif	ecl-blueschist in metagabbro	$H_2O\pm CO_2$ mod. to high salinity st. isot.	mylonite ecl veins/textures [st. isot., inclusions]	fluid compositional variation at mm-scale shallower fluid loss volatile retention/immobility ≥ 10 kbars, 450-550°C	Nadeau et al., 1993 Philippot, 1993
	Dora Maira	metasediment pyrope quartzites (meta-evaporites?)	$CO_2\pm H_2O$ variable salinity	qtz-rich boudins [st. isot., inclusions]	dehydration and partial melting closed system to fluids at peak P-T 30-35 kbar, 700-800°C	Nadeau et al., 1993
	Zermatt-Val d'Aosta	metabasalt	?	oxygen isotopes	limited fluid-rock interactions along structurally weak zones 17-20 kbar, 550-600°C	Barnicoat and Cartwright, 1995
	Tauern Window	mafic ecl	H_2O+CO_2-NaCl saline brines and carbonic fluids	mm- to cm-scale bands and veins C-O isotopes	limited fluid mixing at mm- to cm-scales; fluid immiscibility and contrasting transport efficiency [inclusions]; 20 kbar, 625°C	Selverstone et al., 1992 Getty and Selverstone, 1994
Greece— Cyclades	Syros	blueschist marbles, pelites	H_2O-rich, saline	w.r., qtz segregations [inclusions; fluid equilibria]		Barr, 1990
	Sifnos, Naxos	blueschist, ecl. marbles		wr, veins [st. isot.]	isot. shifts due to localized infiltration	Matthews and Schliestedt, 1983

[1] abbreviations: W. Gneiss R. = Western Gneiss Region
[2] abbreviations: ecl = eclogite; metasediment = metasedimentary rocks; qtz = quartz
[3] st. isot. = stable isotope; inclusions = fluid inclusions

Table 3. Summary of Field-Based Evidence for High-P/T Fluid Production and Transport (Non-collisional Settings).

Region	Locality	Lithologies[1]	Fluid Compositions	Sample Textures[1] [approach][2]	Other Information (fluid sources, transport) peak P-T conditions)	Reference(s)
Franciscan Complex, CA	Ward Creek	sedimentary, mafic ultramafic	H_2O-rich?	[st. isot.]	localized isotopic reequilibration due to fluid-rock interactions 200-450°C, likely > 6kbar	*Taylor and Coleman, 1968*
	Central Belt	mafic blocks in melange	H_2O-rich?	Nd-Sr isotope	possible km-scale transfer of sedimentary isotope signatures	*Nelson, 1995*
	Central Belt	mafic blocks in melange	H_2O-rich low salinity	[inclusions]	hydrous fluid regime 8-11 kbar, 500-800°C	*Giaramita and Sorensen, 1994*
	various localities	sedimentary, mafic ultramafic	H_2O-rich	[st. isot.]	sediment-derived fluid localized reequilibration with metamorphic fluids	*Magaritz and Taylor, 1978*
	San Simeon	sedimentary, mafic melange	H_2O-rich ($\pm CH_4$) low salinity	[inclusions] vein mineralogy cathodoluminescence	large-scale advection of heat in fluids 100-250°C, 1-5 kbar	*Mashburn and Cloos, pers. comm.*
	Central Belt	sedimentary, mafic	H_2O-rich?	[st. isot.]	oxygen-isotope equilibration across lithologies, indicating fluid exchange at cm-scale	*Rumble and Spear, 1983*
Catalina Schist, CA	Santa Catalina Island	sedimentary, mafic ultramafic, melange	H_2O-rich ($\pm CH_4 \pm N_2$) low salinity	w.r., veins, melange [mapping, st. isot., inclusions]	sediment-derived fluid large-scale fluid transport and metasomatism devolatilization reactions and trace element behavior 5-15 kbar, 275-750°C	*Bebout, 1991a, b Bebout and Barton, 1993 Sorensen and Barton, 1987; Sorensen and Grossman, 1989*
Alaska	Kodiak Formation	sedimentary	H_2O-rich ($\pm CH_4$) low salinity	w.r., veins, cements [mapping, st. isot., inclusions] cathodoluminescence	mechanisms of fluid transport sediment/organic fluid diagenetic cementation 150-300°C, < 5 kbar	*Vrolijk et al., 1988 Fisher, this volume*
Dominican Republic	Samana Peninsula	ecl in melange	H_2O-rich low-salinity	w.r., veins [inclusions]	hydrous fluid regime 8-11 kbar, 500-800°C	*Giaramita and Sorensen, 1994*

[1] abbreviations: ecl = eclogite; metasediment = metasedimentary rocks; qtz = quartz
[2] st. isot. = stable isotope; inclusions = fluid inclusions

analytical approaches taken by researchers in each area. These differences in approach are in many cases obviously related to the quality, extent, and nature of the exposures. As an example of such a difference, whereas it has become apparent from study of the Catalina Schist that km-scale fluid flow was largely localized along zones of structural weakness (i.e., mélange and fractures; *Bebout*, 1991a, b), similar observations regarding the presence or absence of possible fluid transfer zones are for the most part lacking in studies of the Alpine suites (Table 2). In part due to limitations in the extent of the exposures, these other studies have focused on the documentation of open- vs. closed-system chemical behavior at relatively small scales (meter-scale) or on detailed analysis of fluid compositions using the fluid inclusion record [*Selverstone and Munoz*, 1987; *Philippot and Selverstone*, 1993; *Getty and Selverstone*, 1994]. Unfortunately, the studies of the Franciscan-like complexes have devoted less energy to study at this scale and using fluid inclusions. Finally, few studies have been conducted of progressive subduction-zone devolatilization and its effects on volatile content (concentrations of B, LILE, H_2O, etc.). The work on the Catalina Schist and by *Moran et al.* [1992] has focussed on the behavior, during high-P/T metamorphism, of trace elements used to trace slab-mantle transfer in studies of arc magmatism and forearc processes [*Ryan et al.*, this volume; *Bebout et al.*, 1993, in press]. However, few other similar data sets exist, limiting broader considerations of the recycling of these elements.

4. DO HIGH-P/T METAMORPHIC ROCKS PROVIDE A USEFUL RECORD OF VOLATILE CYCLING AT CONVERGENT MARGINS?

The extent to which the characteristics of fluid loss (i.e., P-T conditions, magnitudes, isotope fractionations) in the Catalina Schist or any other subduction-related metamorphic suite reflect the global subduction process is not known. The Catalina Schist metamorphism, particularly of its higher-grade units, occurred in a relatively hot subduction environment, probably during early stages of subduction [*Grove and Bebout*, 1995]. Metamorphism under these conditions may be more typical of the relatively warm subduction proposed for early stages of Earth history or for modern cases of either early-stage subduction or the subduction of young oceanic lithosphere [*Defant and Drummond*, 1990; *Kirby et al.*, this volume]. More rapid subduction of older, cooler oceanic lithosphere results in cooler thermal structures and promotes the retention of volatiles to greater depths [e.g., *Bebout*, 1991a].

A combination of the results from the Catalina Schist and similar rocks and from eclogitic suites reflecting deeper metamorphism may ultimately constrain the degree to which fluid loss occurs at shallow levels of subduction zones as a function of thermal structure. Based on study of the Monviso and Tauern Window eclogites [*Nadeau et al.*, 1993; *Selverstone et al.*, 1992; *Philippot and Selverstone*, 1993; *Getty and Selverstone*, 1994] and the Dora Maira Massif [*Philippot*, 1993], it has been concluded that these Alpine eclogitic rocks representing peak metamorphism at depths >40 km experienced prior dramatic fluid loss during prograde devolatilization at shallower parts of the various subduction zones. Thus, despite dissimilarities in the tectonic settings and the approaches taken in the various studies, there appears to be agreement that, in even the cooler subduction zones, loss of fluid and large-scale fluid transport may be dramatic at relatively shallow levels of subduction zones (< 50 km). Some eclogitic materials reflect complex melting histories [e.g., *Philippot et al.*, 1995] and may provide information regarding trace element behavior during deeper fluid-melt-rock interactions.

A recent detailed petrologic and thermochronological study of the Catalina Schist by *Grove and Bebout* [1995] places better tectonic constraints on the metamorphism and resulting devolatilization and offers insight into the relevance of Catalina fluid-related processes to devolatilization in differing subduction-zone thermal regimes. In this complex, the fortuitous preservation of a relatively coherent set of tectonometamorphic units representing metamorphism at similar depths in a rapidly cooling subduction zone allows evaluation of the magnitude of forearc devolatilization as a function of prograde thermal history (Fig. 3). Such information may be applied to understanding the chemical evolution of present-day subduction zones of varied thermal structures, in particular, in considering the likely degrees of hydrous fluid- and/or melt-mediated transfer contributing to magmatic arcs in these diverse settings [*Bebout et al.*, in press]. The tectonometamorphic units of the Catalina Schist, excluding the lawsonite-albite facies unit, reflect peak metamorphism at similar pressures of 0.8 to 1.1 GPa in a rapidly cooling, newly initiated, Early Cretaceous subduction zone. The amphibolite-facies rocks were metamorphosed and accreted to the hanging-wall in the earliest stages of subduction (i.e., experiencing the highest-temperature prograde thermal histories; Fig. 3), whereas lower-

188 CONVERGENT MARGIN VOLATILE RECYCLING

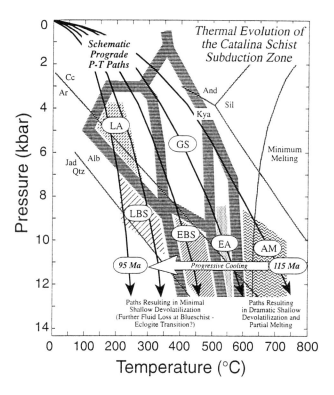

Fig. 3. Pressure-temperature diagram illustrating the varying prograde thermal histories and timing of peak metamorphism and cooling of the tectonometamorphic units of the Catalina Schist. Schematic P-T stability fields of the prograde metamorphism in each of the units (patterned regions) and generalized phase equilibria for relevant volcanic and volcaniclastic rocks are discussed in *Grove and Bebout* [1995]. The dark lines with arrows represent schematic prograde rock P-T paths in a newly formed subduction zone as a function of time [cf. *Peacock*, 1990], with the highest-T path representing the early stages of subduction (labelled "115 Ma") and other progressively lower-T paths depicting the evolution toward an overall cooler subduction zone thermal structure (dark line labelled "95 Ma"). Overall P-T-time loops for each of the units would appear clockwise on this diagram (Fig. 9 in *Grove and Bebout* [1995]). Abbreviations for metamorphic facies are: "LA" = lawsonite-albite; "LBS" = lawsonite-blueschist; "EBS" = epidote-blueschist; "EA" = epidote-amphibolite; "GS" = greenschist; "AM" = amphibolite. Mineral abbreviations are: Jd = jadeite; Ab = albite; Qtz = quartz; And = andalusite; Sil = sillimanite; Kya = kyanite; Cc = calcite; Ar = aragonite.

grade units were metamorphosed and accreted successively during the progressive cooling of the subduction zone (see schematic, progressively cooling, prograde P-T paths for the lower-grade units in Fig. 3). Thus, the units of the Catalina Schist provide a record of devolatilization and trace

element losses over a wide range of prograde thermal histories (P-T paths; *Peacock* [1990]) representing extremely warm regimes (producing extremely devolatilized epidote-amphibolite- and amphibolite-facies rocks) to relatively cool regimes more similar to those expected under "normal" conditions (producing less devolatilized epidote-blueschist-, lawsonite-blueschist-, and lawsonite-albite-facies rocks). The lawsonite-albite-facies metasedimentary rocks, which are similar in their volatile and trace element contents and stable isotope compositions to unmetamorphosed equivalents (see Fig. 2; *Bebout* [1995]; *Bebout et al.* [in press]), are believed to have been subducted to shallower levels than the other units (perhaps 10-15 km) at later stages after cooling of the subduction zone ($^{40}Ar/^{39}Ar$ data indicate that these rocks could be ~20 Ma younger than the other units).

The shallow devolatilization (and melting) experienced by the higher-grade units of the Catalina Schist (i.e., epidote-amphibolite- and amphibolite-facies units) may be analogous to the processes attending forearc metamorphism in relatively warm subduction zones such as the Cascadia plate margin. A comparison of the volatile and trace element contents of the epidote-amphibolite- and amphibolite-facies metasedimentary rocks with those of the lawsonite-albite-facies rocks (representing unmetamorphosed to low-grade equivalents) suggests that dramatic devolatilization occurs in such subduction zones. This shallow devolatilization (at depths < 50 km) should result in smaller hydrous fluid fluxes beneath arcs and the production of fewer arc volcanoes [*Kirby et al.*, this volume] which erupt lavas lacking the distinctive trace element signatures of hydrous fluid additions but possessing trace element abundances consistent with additions of slab sediment melts [*Leeman et al.*, 1994]. Later subduction produced the relatively cooler thermal regimes reflected by the lower-grade units (i.e., epidote-blueschist-, lawsonite-blueschist-, and lawsonite-albite facies units; cooler P-T paths on Fig. 3) and resulted in less dramatic volatile and trace element losses at the same depths. The lawsonite-blueschist-facies metasedimentary rocks are extremely similar in trace element composition to the lawsonite-albite-facies (and unmetamorphosed) equivalents [*Bebout et al.*, in press]. This similarity reflects the retention of volatiles to far greater depths at later stages in the Catalina subduction zone, perhaps to the depths of the blueschist-to-eclogite transition (*Peacock* [1993]; Fig. 3) to be affected by processes of fluid-melt-rock interaction similar to those recorded by the studies of more deeply subducted rocks in collision-related settings. The later-stage thermal regime of the subduction zone producing the Catalina Schist was perhaps more similar to that of the present-day Kurile subduction zone [cf. *Ryan et al.*, 1995]: that is, relatively cool, affording additions of hydrous fluids with distinctive trace element signatures (high B/Be, B/La, B/Zr, Cs/Th, Cs/Rb, As/Ce, and Sb/Ce) to the subarc mantle wedge [*Ryan et al.*, this volume; *Noll et al.*, 1996].

Making the necessary qualifications regarding these likely varied mechanisms of slab-to-mantle transfer (e.g., in C-O-H-S-N fluids or melts; by assimilation), the uncertain degrees of fluid-melt-solid isotopic fractionation within the mantle [*Javoy et al.*, 1986; *Boyd and Pillinger*, 1994; *Boyd et al.*, 1994], and the varying extents of stable isotope fractionation during magma degassing [*Javoy et al.*, 1986; *Taylor*, 1986], it is interesting to make preliminary comparisons of the stable isotopic compositions of the Catalina rocks and fluids with the compositions of various mantle-derived materials (see Fig. 4). These comparisons demonstrate that the diverse isotopic compositions expected for the subducted reservoirs are compatible with the ranges in the stable isotope compositions of certain mantle-derived materials [cf. *Kyser*, 1986; *Boyd and Pillinger*, 1994]. Many have considered the extent to which high-$\delta^{18}O$ mantle xenolith materials (particularly eclogitic xenoliths) could reflect the subduction of hydrothermally altered oceanic lithosphere such as that represented in high-P/T metamorphic suites [*Bebout*, 1995; *Nadeau et al.*, 1993], seafloor cores, and ophiolites [*Muehlenbachs*, 1986]. *Woodhead et al.* [1993] and *Harmon and Hoefs* [1995] suggested that the $\delta^{18}O$ of subduction-related basalts and some EM-type ocean island basalts may reflect the presence of subducted crustal components in their sources.

Evaluations of the bulk $\delta^{13}C$ of the C delivered to the mantle by subduction [*Agrinier et al.*, 1985; *Javoy et al.*, 1986] and the possible effect of subducted organic C on mantle $\delta^{13}C$ [e.g., *Mattey*, 1987; *Mattey et al.*, 1989] have fueled much recent debate regarding the magnitudes of crust-mantle volatile recycling. The calculated CO_2 $\delta^{13}C$ (-14 to +5‰; mean of ~ -5‰) using data for carbonate and carbonaceous matter in the Catalina Schist greatly overlaps the range of -10 to -4‰ (mean near -6.5‰) commonly inferred for "mantle" or "igneous" values [*Taylor*, 1986]. CO_2 $\delta^{13}C$ of back-arc basin basalts ranges from -15 to -9 ‰, somewhat lower than values for mid-ocean ridge basalt (MORB; mean ~ -6.5‰); this difference has been attributed to the addition of "sedimentary carbon" [*Mattey*, 1987]. *Mattey et al.* [1989] attributed low $\delta^{13}C$ in mantle xeno-

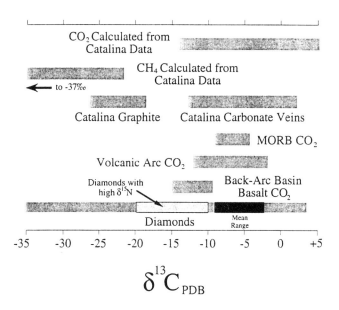

Fig. 4. Comparisons of calculated fluid C-isotopic compositions for the Catalina Schist, and the $\delta^{13}C$ of reduced and oxidized C reservoirs in the Catalina Schist with isotopic compositions of mantle-derived rocks (xenoliths, diamonds, igneous rocks). The range for Catalina Schist CH_4 is calculated using the data for carbonaceous matter; the range for Catalina Schist CO_2 is calculated using the ranges in $\delta^{13}C$ of carbonaceous matter and carbonate at all grades (both ranges calculated for the temperature range of 300-650°C; *Bebout* [1995]). Indicated are the ranges for high-$\delta^{15}N$ diamonds of *Boyd and Pillinger* [1994] and the mean $\delta^{13}C$ of diamonds [*Mattey*, 1987]. *Nadeau et al.* [1993] reported similar $\delta^{13}C$ values of -24.2 ± 1.2‰ in metagabbros from the Monviso ophiolitic complex that could reflect organic C added to oceanic crust.

lith diopsides to contributions by crustal recycling. Additions of fluids with lower $\delta^{13}C$ similar to that of the Catalina fluids (particularly CH_4-rich fluids; Fig. 4) could presumably produce such shifts. The range of calculated CH_4 $\delta^{13}C$ values, together with the $\delta^{13}C$ values of carbonaceous matter in the metasedimentary rocks (total range of -37 to -19‰), is similar to that of relatively ^{13}C-depleted diamonds (see *Mattey* [1987]; *Boyd and Pillinger* [1994]; Fig. 4). The extremely wide range in $\delta^{13}C$ for diamonds may represent some fractionation in the mantle but may also be partly explained by (1) additions of C with low $\delta^{13}C$ in CH_4 derived from the slab or in metamorphosed organic matter [*Mattey*, 1987; *Boyd and Pillinger*, 1994], and (2) the varying influence of relatively higher-$\delta^{13}C$ signatures contributed through subduction of pelagic carbonate or vein/cement carbonate or CO_2-rich fluids equilibrated with carbonate or carbonaceous matter [e.g., *Agrinier et al.*, 1985]. In metamorphic suites with large amounts of carbonate rocks (e.g., Alps, Norway, Cyclades; *Matthews and Schliestedt* [1984]; *Agrinier et al.* [1985]; *Getty and Selverstone* [1994]), carbonate (and fluid) $\delta^{13}C$ differs significantly from that in the Catalina Schist.

Differences in the H-, N-, and S-isotopic compositions of subduction-related basalts and andesites from MORB values are similarly compatible with the addition of one or more of the rock-fluid isotope signatures observed in, or inferred for, the Catalina Schist. Some subduction-related basalts have δD_{SMOW} of -46 to -32‰ [*Poreda*, 1985]; these values are intermediate to the δD of the Catalina Schist metamorphic fluids (-15±15‰; *Bebout* [1991a]) and MORB values, the latter of which are believed to be more negative, near -80‰ [*Kyser*, 1986]. Similarly, values of $\delta^{15}N_{air}$ of +3 to +8‰ were obtained by *Zhang* [1988] for subduction-related volcanic rocks. These $\delta^{15}N$ values are intermediate to values for the Catalina Schist metasedimentary rocks (+1 to +5‰) and fluids (-1.5 to +5.5‰; *Bebout and Fogel* [1992]) and MORB values which appear to approach +10‰ (*Zhang* [1988]; see discussion of uncertainty in mantle $\delta^{15}N$ by *Boyd and Pillinger* [1995]). $\delta^{34}S_{CDT}$ values of 0 to +21‰ have been obtained for subduction-related volcanic rocks [*Woodhead et al.*, 1987; *Alt et al.*, 1993]; these values appear to be shifted from MORB values near 0‰ (in more altered sections ranging to +5‰; *Kyser* [1986], *Alt et al.* [1993]) toward values for marine sulfate (>20‰) and the Catalina Schist metamorphic fluids (+5 to +9‰) and rocks and veins (-1 to +9‰; see *Bebout* [1995]).

5. CONCLUSIONS AND SUGGESTED DIRECTIONS FOR FUTURE RESEARCH

Significant imbalances appear to exist in the amounts of volatiles (H_2O, C, N, and B) subducted and the amounts obviously returned by arc magmatism. These imbalances could in part reflect the release of volatiles from subducting rocks during forearc devolatilization and their transfer into reservoirs other than arc magmas. Evidence for massive, large-scale fluid flow in the Catalina Schist [*Bebout*, 1991a] is consistent with potential contributions of H_2O-rich fluids from subducting materials to the mantle wedge, particularly in forearc regions, or with the large-scale updip movement of such fluids along the slab-mantle interface. Great uncertainty remains regarding the proportion of the volatiles retained in subducting rocks into the deep mantle.

Field-based studies of fluid processes during high-P/T metamorphism demonstrate that devolatilization and other fluid-rock interactions can result in significant chemical and isotopic fractionation in subducting rocks and fluids, thus potentially impacting the "subduction signature" in arc magmas and other mantle-sourced materials. Compatibilities exist between the rock compositions and calculated fluid compositions for the metamorphic suites and the isotopic compositions of various mantle-derived materials (igneous rocks and xenoliths). Further assessments of fractionation during mantle and magma degassing processes and of the volatile contents and stable isotope compositions of mantle-derived materials [*Kyser*, 1986; *Marty*, 1995] are necessary to improve our understanding of volatile recycling. Future work should endeavor to identify geochemical signatures of deep-sourced fluids in shallow parts of accretionary wedges (décollement zones, serpentinite seamounts). Their identification is complicated by mixture of deep-sourced fluids with fluids derived by mechanical expulsion and devolatilization at shallower levels [*Kastner et al.*, 1991], by fluid immiscibility relations [*Selverstone et al.*, 1992], and by the effects of fluid-rock exchange along fluid flow paths [*Bebout and Barton*, 1993].

Finally, calculations of crust-mantle volatile fluxes are generally performed on a global basis, averaging significant differences in the nature of both the subducted sediment section and the volatile returns (e.g., via arc magmatism) among margins. Future study should expand upon these global considerations in assessments of the mass-balance of volatile components, including trace element and isotopic signatures, in individual subduction zones.

Acknowledgments. This research was supported by the National Science Foundation (grants EAR-9206679, EAR-9220691, and EAR-9405625). I acknowledge the support provided by the Santa Catalina Island Conservancy and I thank S. R. Getty and an anonymous reviewer for their reviews.

REFERENCES

Agrinier, P., M. Javoy, D.C. Smith, and F. Pineau, Carbon and oxygen isotopes in eclogites, amphibolites, veins and marbles from the Western Gneiss Region, Norway, *Chem. Geol., 52,* 145-162, 1985.

Alt, J. C., W. C. Shanks III, and M. C. Jackson, Cycling of sulfur in subduction zones: the geochemistry of sulfur in the Mariana Island Arc and back trough, *Ear. Planet. Sci. Lett., 119,* 477-494, 1993.

Andersen, T., E. A. J. Burke, and H. Austrheim, Nitrogen-bearing, aqueous fluid inclusions in some eclogites from the

Western Gneiss Region of the Norwegian Caledonides, *Contrib. Mineral. Petrol., 103*, 153-165, 1989.

Bailey, E. H., W. P. Irwin, and D. L. Jones, Franciscan and related rocks. *Calif. Dept. Mines Geol. Bull., 183,* 1964.

Barnicoat, A. C., and I. Cartwright, Focused fluid flow during subduction: Oxygen isotope data from high-pressure ophiolites of the western Alps, *Ear. Planet. Sci. Lett., 132*, 53-61, 1995.

Barr, H., Preliminary fluid inclusion studies in a high-grade blueschist terrain, Syros, Greece, *Mineral. Mag., 54*, 159-168, 1990.

Bebout, G. E., Field-based evidence for devolatilization in subduction zones: implications for arc magmatism, *Science, 251*, 413-416, 1991a.

Bebout, G. E., Geometry and mechanisms of fluid flow at 15 to 45 kilometer depths of an early Cretaceous accretionary complex, *Geophy. Res. Lett., 18*, 923-926, 1991b.

Bebout, G. E., The impact of subduction-zone metamorphism on mantle-ocean chemical cycling, *Chem. Geol., 126*, 191-218, 1995.

Bebout, G. E., and M. D. Barton, Metasomatism during subduction: products and possible paths in the Catalina Schist, California, *Chem. Geol., 108*, 61-92, 1993.

Bebout, G. E., and M. L. Fogel, Nitrogen-isotope compositions of metasedimentary rocks in the Catalina Schist, California: implications for metamorphic devolatilization history, *Geochim. Cosmochim. Acta., 56*, 2139-2149, 1992.

Bebout, G. E., J. G. Ryan, and W. P. Leeman, B-Be systematics in subduction-related metamorphic rocks: characterization of the subducted component, *Geochim. Cosmochim. Acta, 57*, 2227-2237, 1993.

Bebout, G. E., J. G. Ryan, W. P. Leeman, and A. E. Bebout, Fractionation of trace elements by subduction-zone metamorphism: significance for models of crust-mantle mixing, *Geochim. Cosmochim. Acta,* in press, 1996.

Berner, R. A., and A. C. Lasaga, Modeling the geochemical carbon cycle, *Sci. Amer., 260*, 74-81, 1989.

Boyd, S. R., and C. T. Pillinger, A preliminary study of $^{15}N/^{14}N$ in octahedral growth form diamonds, *Chem. Geol., 116*, 43-59, 1994.

Boyd, S. R., F. Pineau, and M. Javoy, Modelling the growth of natural diamonds, *Chem. Geol., 116,* 29-42, 1994.

Defant, M. F., and M. S. Drummond, Derivation of some modern arc magmas by melting of young subducted lithosphere, *Nature, 347,* 662-665, 1990.

Ernst, W. G., Tectonic history of subduction zones inferred from retrograde blueschist P-T paths, *Geology, 16,* 1081-1084, 1988.

Ernst, W. G., Y. Seki, H. Onuki, and M. C. Gilbert, Comparative study of low-grade metamorphism in the California Coast Ranges and the Outer Metamorphic Belt of Japan, *Geol. Soc. Amer. Mem., 124,* 276 p., 1970.

Fisher, D., Fabrics and veins in the forearc: a record of cyclic fluid flow at depths of < 15 km, in G. E. Bebout, D. Scholl, S. Kirby, and J. P. Platt, eds., *Am. `Geophys. Union, Geophys. Monogr.,* this volume, 1996.

Getty, S. R., and J. Selverstone, Stable isotopic and trace element evidence for restricted fluid migration in 2 GPa eclogites, *Jour. Meta. Geol., 12*, 747-760, 1994.

Giaramita, M. J., and S. S. Sorensen, Primary fluids in low-temperature eclogites: evidence from two subduction complexes (Dominican Republic, and California, USA), *Contrib. Mineral. Petrol., 117*, 279-292, 1994.

Grove, M. and Bebout, G. E., Cretaceous tectonic evolution of coastal southern California: insights from the Catalina Schist. *Tectonics, 14*, 1290-1308, 1995.

Hall, A., Ammonium in spilitized basalts of southwest England and its implications for the recycling of nitrogen, *Geochem. Jour., 23*, 19-23, 1989.

Harmon, R. S., and J. Hoefs, Oxygen isotope heterogeneity of the mantle deduced from global ^{18}O systematics of basalts from different geotectonic settings, *Contrib. Mineral. Petrol., 120*, 95-114, 1995.

Ito, E., D. M. Harris, and A. T. Anderson Jr., Alteration of oceanic crust and geologic cycling of chlorine and water, *Geochim. Cosmochim. Acta, 47*, 1613-1624, 1983.

Javoy, M., F. Pineau, and C. J. Allegre, Carbon geodynamic cycle, *Nature, 300*, 171-173, 1982.

Javoy, M., F. Pineau, F., and H. Delorme, Carbon and nitrogen isotopes in the mantle, *Chem. Geol., 57*, 41-62, 1986.

Kastner, M., H. Elderfield, and J. B. Martin, Fluids in convergent margins: What do we know about their composition, origin, role in diagenesis and importance for oceanic chemical fluxes? *Philos. Trans. R. Soc. London, Ser. A., 335*, 275-288, 1991.

Kerrick, D. M., and K. Caldeira, Paleoatmospheric consequences of CO_2 released during early Cenozoic regional metamorphism in the Tethyan orogen, *Chem. Geol., 108*, 201-230, 1993.

Kirby, S., E. R. Engdahl, and R. Denlinger, Intermediate-depth intraslab earthquakes and arc volcanism as physical expressions of crustal and uppermost mantle metamorphism in subducting slabs. in G. E. Bebout, D. S. Scholl, S. Kirby and J. P. Platt (eds.), *Amer. Geophys. Un. Geophys. Monogr.,* this volume, 1996.

Klemd, R., A. M. van den Kerkhof, and E. E. Horn, High-density CO_2-N_2 inclusions in eclogite-facies metasediments of the Munchberg gneiss complex, SE Germany, *Contrib. Mineral. Petrol., 111*, 409-419, 1992.

Kyser, T. K., Stable isotope variations in the mantle, *Mineral. Soc. Amer. Rev. Mineral., 16*, 141-164, 1986.

Leeman, W. P., M. J. Carr, and J. D. Morris, Boron geochemistry of the Central American Volcanic Arc: constraints on the genesis of subduction-related magmas, *Geochim. Cosmochim. Acta, 58*, 149-168, 1994.

Magaritz, M., and H. P. Taylor Jr., Oxygen, hydrogen and carbon isotope studies of the Franciscan formation, Coast Ranges, California, *Geochim. Cosmochim. Acta, 40*, 215-234, 1976.

Marty, B., Nitrogen content of the mantle inferred from N_2-Ar correlation in oceanic basalts, *Nature, 377*, 326-329, 1995.

Marty, B., and A. Jambon, C/³He in volatile fluxes from the solid Earth: implications for carbon geodynamics, *Ear. Planet. Sci. Lett., 83,* 16-26, 1987.

Mattey, D., D. H. Jackson, N. B. W. Harris, and S. Kelley, Isotopic constraints on fluid infiltration from an eclogite facies shear zone, Holsenoy, Norway, *J. Meta. Geol., 12,* 311-325, 1994.

Mattey, D. P., Carbon isotopes in the mantle, *Ter. Cog., 7,* 31-37, 1987.

Mattey, D. P., R. A. Exley, C. T. Pillinger, M. A. Menzies, D. R. Porcelli, S. Galer, and R. K. O'Nions, Relationships between C, He, Sr and Nd isotopes in mantle diopsides, *Fourth Intn'l Kimberlite Conf., Geol. Soc. Australia Spec. Publ., 14,* 913-921, 1989.

Matthews, A., and M. Schliestedt, Evolution of blueschist and greenschist facies rocks of Sifnos, Cyclades, Greece: a stable isotope study of subduction-related metamorphism, *Contrib. Mineral. Petrol., 88,* 150-163, 1984.

Moore, J. C., and P. Vrolijk, Fluids in accretionary prisms, *Rev. Geophys., 30,* 113-135, 1992.

Moran, A. E., V. B. Sisson, and W. P. Leeman, Boron depletion during progressive metamorphism: implications for subduction processes, *Ear. Planet. Sci. Lett., 111,* 331-349, 1992.

Muehlenbachs, K., Alteration of the oceanic crust and the ^{18}O history of seawater, *Mineral. Soc. Amer. Rev. Mineral., 16,* 425-444, 1986.

Nadeau, S., P. Philippot, and F. Pineau, Fluid inclusion and mineral isotopic compositions (H-C-O) in eclogitic rocks as tracers of local fluid migration during high-pressure metamorphism, *Earth Planet. Sci. Lett., 114,* 431-448, 1993.

Nelson, B. K., Fluid flow in subduction zones: evidence from Nd- and Sr-isotope variations in metabasalts of the Franciscan complex, California, *Contrib. Mineral. Petrol., 119,* 247-262, 1995.

Noll, P. D., Jr., H. E. Newsom, W. P. Leeman, and J. G. Ryan, The role of hydrothermal fluids in the production of subduction zone magmas: evidence from siderophile and chalcophile trace elements and boron, *Geochim. Cosmochim. Acta, 60,* 587-611, 1996.

Peacock, S. M., Fluid processes in subduction zones, *Science, 248,* 329-337, 1990.

Peacock, S. M., Metamorphism, dehydration, and importance of the blueschist -> eclogite transition in subducting oceanic crust, *Geol. Soc. Amer. Bull., 105,* 684-694, 1993.

Philippot, P., Fluid-melt-rock interaction in mafic eclogites and coesite-bearing metasediments: constraints on volatile recycling during subduction, *Chem. Geol., 108,* 93-112, 1993.

Philippot, P., and J. Selverstone, Trace element-rich brines in eclogitic veins: implications for fluid composition and transport during subduction, *Contrib. Mineral. Petrol., 112,* 341-357, 1992.

Philippot, P., P. Chevallier, C. Chopin, and J. Dubessy, Fluid composition and evolution of coesite-bearing rocks (Dora-Maira massif, Western Alps): implications for element recycling during subduction, *Contrib. Mineral. Petrol., 121,* 29-44, 1995.

Plank, T., and J. Ludden, Geochemistry of sediments in the Argo Abyssal Plain at ODP Site 765: a continental margin reference section for sediment recycling in subduction zones, *Proc. ODP Sci. Res., 123,* 167-189, 1992.

Poreda, R., Helium-3 and deuterium in back-arc basalts: Lau Basin and Mariana Trough, *Earth Planet. Sci. Lett., 73,* 244-254, 1985.

Rea, D. K., and L. J. Ruff, Composition and mass flux of sediment entering the World's subduction zones: implications for global sediment budgets, great earthquakes, and volcanism, *Ear. Planet. Sci. Lett.,* in press, 1996.

Rumble, D., III, and F. S. Spear, Oxygen-isotope equilibration and permeability enhancement during regional metamorphism, *J. Geol. Soc. London, 140,* 619-628, 1983.

Ryan, J. G., J. D. Morris, F. Tera, W. P. Leeman, and A. Tsvetkov, Cross-arc geochemical variations in the Kurile island arc as a function of slab depth, *Science, 270,* 625-628, 1995.

Ryan, J., J. Morris, G. Bebout, B. Leeman, and F. Tera, Describing chemical fluxes in subduction zones: insights from "depth-profiling" studies of arc and forearc rocks. in G. E. Bebout, D. S. Scholl, S. Kirby and J. P. Platt (eds.), *Amer. Geophys. Un. Geophys. Monogr.,* this volume, 1996.

Selverstone, J., G. Franz, S. Thomas, S., and S. Getty, Fluid variability in 2 GPa eclogites as an indicator of fluid behavior during subduction, *Contrib. Mineral. Petrol., 112,* 341-357, 1992.

Selverstone, J., and D. S. Gutzler, Post-125 Ma carbon storage associated with continent-continent collision, *Geology, 21,* 885-888, 1993.

Selverstone, J., and J. L. Munoz, Fluid heterogeneities and hornblende stability in interlayered graphitic and non-graphitic schists (Tauern Window, Eastern Alps), *Contrib. Mineral. Petrol., 96,* 426-440, 1987.

Sorensen, S. S., and M. D. Barton, Metasomatism and partial melting in a subduction complex, Catalina Schist, southern California, *Geology, 15,* 115-118, 1987.

Sorensen, S. S., and J. N. Grossman, Enrichment of trace elements in garnet amphibolites from a paleo-subduction zone: Catalina Schist, southern California, *Geochim. Cosmochim. Acta, 53,* 3155-3177, 1989.

Staudigel, H., S. R. Hart, H. -U. Schminke, and B. M. Smith, Cretaceous ocean crust at DSDP Sites 417 and 418: carbon uptake from weathering versus loss by magmatic outgassing, *Geochim. Cosmochim. Acta, 53,* 3091-3094, 1989.

Taylor, B. E., Magmatic volatiles: isotopic variation of C, H, and S, *Mineral. Soc. Amer. Rev. Mineral. 16,* 185-225, 1986.

Taylor, H. P., Jr., and R.G. Coleman, O^{18}/O^{16} ratios of coexisting minerals in glaucophane-bearing metamorphic rocks, *Geol. Soc. Amer. Bull., 79,* 1727-1756, 1968.

Taylor, S. R., and S. M. McLennan, *The Continental Crust: its Composition and Evolution*, Blackwell Scientific Publications, 312 pp., 1985.

Thompson, A. B., Water in the Earth's mantle, *Nature (London), 358*, 295-302, 1992.

von Huene, R., and D. W. Scholl, Observations at convergent margins concerning sediment subduction, subduction erosion, and the growth of continental crust, *Rev. Geophys., 29*, 279-316, 1991.

Vrolijk, P., G. Myers, and J. C. Moore, Warm fluid migration along tectonic melanges in the Kodiak accretionary complex, Alaska, *Jour. Geophys. Res., 93*, 10313-10324, 1988.

Woodhead, J. D., P. Greenwood, R. S. Harmon, and P. Stoffers, Oxygen isotope evidence for recycled crust in the source of EM-type ocean island basalts, *Nature, 362*, 809-813, 1993.

Woodhead, J. D., R. S. Harmon, and D. G. Fraser, O, S, Sr, and Pb isotope variations in volcanic rocks from the Northern Mariana Islands: implications for crustal recycling in intra-ocean arcs, *Earth Planet. Sci. Lett., 83*, 39-52, 1987.

Zhang, D., *Nitrogen Concentrations and Isotopic Compositions of Some Terrestrial Rocks* (Ph.D. Dissertation), The University of Chicago, 157 p., 1988.

Zhang, Y., and A. Zindler, Distribution and evolution of carbon and nitrogen in Earth, *Ear. Planet. Sci. Lett., 117*, 331-345, 1993.

G. E. Bebout, Department of Earth and Environmental Sciences, 31 Williams Drive, Lehigh University, Bethlehem, Pennsylvania 18015.
(e-mail: geb0@lehigh.edu)

Intermediate-Depth Intraslab Earthquakes and Arc Volcanism as Physical Expressions of Crustal and Uppermost Mantle Metamorphism in Subducting Slabs

Stephen Kirby[1], E. Robert Engdahl[2], and Roger Denlinger[3]

U.S. Geological Survey

We elaborate on the well-known spatial association between arc volcanoes and Wadati-Benioff zones and explore in detail their genetic relationships as dual physical expressions of slab metamorphism of the oceanic crust and uppermost mantle. At hypocentral depths less than 200 km intraslab Wadati-Benioff earthquakes tend to occur near the top surfaces of slabs. Subduction of very young lithosphere (age < 15–25 Ma) with high heat flow (> 75 mW/m^2) produces mainly shallow earthquakes and sparse or absent arc volcanism. Subduction of older crust with normal heat flow (50–65 mW/m^2) produces markedly deeper intraslab earthquakes and generally normal volcanic vigor. Seismological observations show that the low-seismic-velocity gabbroic mineralogy of the crust may persist to depths of as much as 150 km in old, cold lithosphere but only to depths of 50–60 km in young, warm slabs. Metamorphic processes in the crust and shallow upper mantle of subducting slabs and the reactivation of faults originally created at shallow depths in the ocean basins probably control the occurrence of intraslab earthquakes to depths of as much as 350 km. A conceptual model for this metamorphism incorporates the likely effects of water liberated by dehydration. Such dehydration facilitates both brittle faulting by fault reactivation and promotes the kinetics of the transformation of the anhydrous gabbro component of the crust to eclogite. Finite-element modelling shows that densification to eclogite is expected to produce extensional stresses in transformed crust and a smaller compression in the underlying mantle. This model helps explain why most intermediate-depth intraslab earthquakes occur just below the top surfaces of slabs and why many have focal mechanisms indicating down-dip extension. Young, warm slabs have mostly shallow intraslab earthquakes and sparse arc volcanoes because dehydration and eclogite formation largely cease before such slabs are in contact with asthenosphere. These processes are evidently delayed by kinetic hindrance and the high-pressure stability of hydrous phases at low temperatures in older, colder subducting crust, and thus earthquake activity and asthenospheric-wedge melting tend to be focused at depths of 100–170 km. Anomalous behavior correlated with the subduction of island-and-seamount chains appears to be associated with anomalous shallow intraplate faulting and with perturbations of slab metamorphism by these chains.

INTRODUCTION

The geographic coincidence of the graceful volcanic arcs of the Circum-Pacific and Alpine-Indonesian orogenic zones with global earthquake belts was widely recognized in the 19th century [e.g., *Mallet*, 1859; *Milne*, 1886]. It was not until well into the era of global instrumental seismology, however, that it was established that inclined bands of earthquakes, now called Wadati-Benioff zones, generally extend under volcanic arcs at depths of about 100 km [*Wadati*, 1935; *Visser*, 1936; *Berlage*, 1937; *Benioff*, 1949; *Sykes*, 1966].

With the development of plate tectonics, convergent plate boundaries were identified as settings where the shallow materials of the oceanic lithosphere are transported to great depths by the subduction process. Such materials slowly heat up by conduction during descent [*McKenzie*, 1969; *Toksöz et al.*, 1971; *Griggs*, 1972]. This heating produces arc magmas that somehow ascend to the surface [*Coats*, 1962; *Ringwood and Green*, 1966; *Ringwood*, 1969]. Wadati-Benioff zones typically are at depths of about 110 km beneath volcanic fronts (the oceanward limit of stratovolcanoes) [*Tatsumi*, 1986; see also *Isacks and Barazangi*, 1977; *Gill*, 1981; *Chiu et al.*, 1991]. Focal mechanisms indicate that earthquakes shallower than about 50 km represent both interplate thrust motion and shallow intraplate failure; events deeper than about 50 km predominately represent failure within descending slabs (intraslab).

Hydrous phases are initially stable at shallow depths in oceanic lithosphere, but they liberate water when that region is heated by conduction during slab descent. It is now generally agreed that most arc magmas result from this release of water by dehydration of the subducting crust or uppermost mantle [e.g., *Gill*, 1981; *Kushiro*, 1983; *Tatsumi et al.*, 1983; *Tatsumi*, 1989; *Davies and Stevenson*, 1992; *Kincaid*, 1995; *Ulmer and Trommsdorf*, 1995; *Tatsumi and Eggins*, 1995]. This water is generally thought to flux peridotite melting in the overlying mantle asthenospheric wedge and produce arc magmas. Densification of mafic crust to eclogite also accompanies dehydration of subducting crust during slab descent.

It is the purpose of this paper to show how many of the characteristics of Wadati-Benioff zones and volcanic arcs are related to each other and that some of these relationships can be understood in terms of a simple unified conceptual model for slab metamorphism that we propose. We will not review the large literature on the metamorphism of mafic oceanic crust based on field petrology and laboratory observations. These are topics discussed elsewhere in this volume. Instead, our approach is to consider the plate-scale consequences of such metamorphism.

We begin by documenting how oceanic plates are mechanically flawed and also mineralogically and thermally heterogeneous as they enter trenches.

These complexities are important in governing the seismological and magmatic expressions of subducting slabs. We then cite seismological observations that place constraints on the depth distributions of large-scale dehydration and eclogite formation, and then show that there is a dichotomy in the depth range of crustal metamorphism among subduction zones: young, slowly-descending slabs metamorphose at shallow depths. In contrast, crustal metamorphism in older, more rapidly descending slabs is probably delayed to depths of 100-170 km or more. We show that this dichotomy in subduction zones with plate age is also expressed as differences in the vigor of arc magmatic activity and that some magmatic arcs and the underlying intraslab seismicity are perturbed by the subduction of the volcanic edifices of island-and-seamount chains. These features of arc volcanism are broadly consistent with a conceptual model proposed here. Lastly, we briefly consider some earthquake- and volcano-hazard implications of this model of metamorphism near the top surfaces of slabs.

WHAT GOES INTO TRENCHES: MECHANICALLY FLAWED AND HETEROGENEOUS LITHOSPHERE

Oceanic plates, created by seafloor spreading and modified by their tenure below ocean basins, are mineralogically and structurally complex and heterogeneous objects. Their subsequent physical evolution with descent as slabs probably reflects this internal complexity (Figure 1a). This section emphasizes three first-order aspects of that complexity that have obvious importance for slab metamorphism and consequent effects on arc volcanism and intraslab earthquakes [*Kirby*, 1995]: (a) *Oceanic lithosphere is flawed* principally by normal faulting near midocean ridges and in trench/outer-rise systems. These mechanical flaws can serve both as channels for hydrothermal fluids and associated localization of hydrous minerals before subduction and as zones of weakness that can be exploited by reactivation accompanying dehydration during slab descent [*Savage*, 1969; *Kirby*, 1995; *Silver et al.*, 1995]. Fracture zones evidently are not as effective as flaws for later reactivation, either seismically in the ocean basins [*M. Wysession*, personal communication 1996] or at depth in slabs. This may be related to the initial vertical orientation of such faults and to differences in the hydrothermal regimes of fracture zones compared to normal faults. (b) *Oceanic crust and uppermost mantle are heterogeneous in the distribution of hydrous phases.* As will be discussed shortly, liberation of water by dehydration of hydrous minerals along faults and fractures in slabs not only can induce melting in

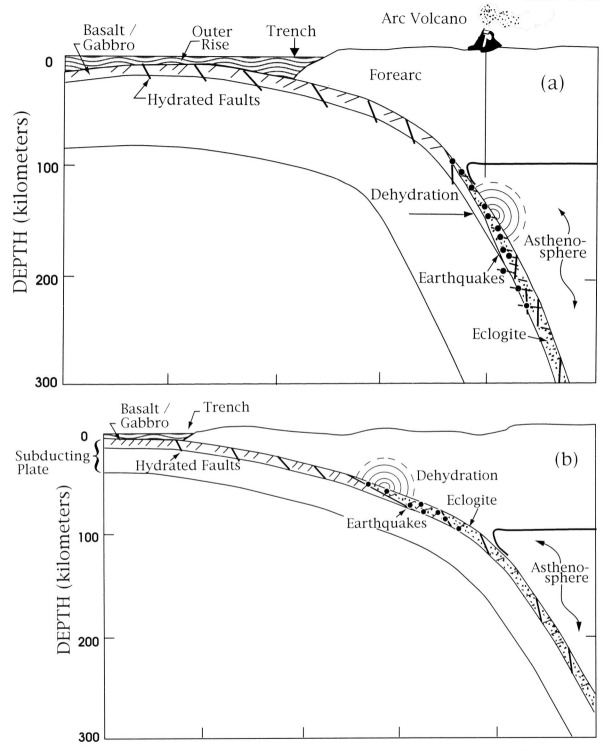

Fig. 1. Schematic diagrams of slab metamorphism associated earthquakes and arc magmatism (a) Thermally-mature slabs [after *Kirby*, 1995]. Oceanic lithosphere layered in mineralogy and flawed principally by hydrated normal faults descends to the depths near the roots of arc volcanoes where it dehydrates and densifies to ecolgite, becomes prone to reactivation of fossil faults by dehydration embrittlement and produces earthquakes. Slab dehydration also fluxes melting in the asthenospheric mantle wedge, producing arc magmas. (b) Young and/or slowly sinking slabs that are warm sustain similar processes as in (a) but largely at shallow depths beneath the forearc, a condition causing shallow intraslab earthquakes and feeble or absent arc volcanism.

the asthenospheric wedge, but can also cause localized failure by faulting and promote eclogite formation.

(c) *Oceanic lithosphere is also mineralogically layered.* The fundamentally mafic mineralogy of the crust and the contrast in its pressure-temperature phase stability compared to the ultramafic slab mantle will produce a heterogeneity in slab volume compression when the oceanic crust dehydrates and forms eclogite during descent. As we discuss later, this heterogeneity can produce stresses in the transformed crust and uppermost mantle that can cause seismic failure by reactivating fossil faults.

Recent high-resolution sonar images of the ocean floor and seismic reflection profiles show that normal faults produced by stretching deformation near midocean ridges [e.g., *Mutter and Karson*, 1992; *Carbotte and Macdonald*, 1994] and by bending at trench-rise systems [e.g., *Masson*, 1991] are nearly ubiquitous features of oceanic lithosphere, confirming earlier investigations. Masson [1991] also showed that normal faults created at midocean ridges are commonly reactivated at trench/outer rise systems if the original fault traces are within 20–30° of the trench axis. Some individual fault scarps at outer trench walls are as much as 100 kilometers long and have throws of as much as 500 m [*Masson*, 1991]. These dimensions and the occurrence of normal-faulting earthquakes at depths of up to a few km at MORs [e.g., *Huang et al.*, 1986] and up to several tens of km at trench-rise systems [e.g., *Chapple and Forsyth*, 1979; *Christensen and Ruff*, 1988; *Seno and Gonzales*, 1987; *Seno and Yamanaka*, this volume] indicate that such faults penetrate well into the tops of descending slabs. As explained more fully below, such flaws may be reactivated during slab descent.

Faults in the top layers of oceanic plates can also serve as channels for hydrothermal fluids, especially near midocean ridges where hydrothermal circulation is important [*Mevel and Cannat*, 1991; *Dick et al.*, 1991; *Agar*, 1994; *Nicolas*, 1995]. It should be noted, however, that the depths that rocks of such fault zones may be hydrated depends upon the depth over which they are cold enough for hydrous minerals to be stable. Very near midocean ridges, this hydration zone may only extend to depths of a few km or less. It is not known how deeply fault-zone hydration may occur in trench-rise systems. As discussed below, this question of how deeply hydrous minerals occur along such faults will likely determine how deep seismic faulting may penetrate into the tops of descending slabs.

Thus in addition to the first-order layering of the oceanic lithosphere into crust and mantle [e.g., *White et al.*, 1992] and the layering of the crust by magmatic and volcanic processes [e.g., *Nicolas*, 1995], the mechanical history of oceanic plates imprints mechanical flaws and produces an associated localized circulation of hydrothermal fluids, and hence a heterogeneous distribution of hydrous minerals.

WHAT GOES INTO TRENCHES: WARM TO COLD CRUST

Lithosphere entering trenches varies in thermal structure, primarily with the time elapsed since the plate is created at midocean ridges [e.g., *Stein and Stein*, 1992; *Stein et al.*, this volume]. This is evident both from the decay in seafloor depths with seafloor age (which reflects the decrease in average plate temperature caused primarily by conduction of heat from plate interiors to the seafloor) and heat flow (which reflects the near-surface thermal gradient). Heat flow is dependent on both normal thermal conduction and on heat losses by hydrothermal circulation [e.g., *Stein et al.*, 1994; 1995]. That heat loss by hydrothermal advection from the shallow lithosphere is important is evident from the fact that observed conductive heat flow is consistently lower than that predicted by purely conduction models for plate cooling to ages of about 50 Ma [e.g., *Stein et al.*, 1995]. Heat flow decreases rapidly with age to 20 Ma old, and is insensitive to age for older lithosphere (Figure 2a).

Slabs are initially coldest at their top surfaces, and they slowly heat up with time by conduction of heat from the plate interface and from the overriding plate and asthenosphere. This heating, in addition to the pressurization that occurs during slab descent, governs where internal metamorphic processes take place [e.g., *Peacock*, this volume]. Earthquake depth distribution varies from subduction zone to subduction zone. Plate age, A, and vertical descent rate, V_z, are the principal parameters that control the thermal structure of descending slabs [e.g., *Molnar et al.*, 1979]. The slab thermal parameter, $\phi = AV_z$ has the units of length and is proportional to the depth that the "v" shaped slab isotherms are advected [*Kostoglodov*, 1989; *Kirby et al.*, 1991; *Gorbatov et al.*, 1996]. As such, it is a measure of how cold slabs are relative to the surrounding mantle. For subduction zones with thermal parameters less than about 1000 km, maximum earthquake depths increase with increasing thermal parameter (Figure 3a). Intermediate-depth earthquake depths are insensitive to variations in thermal parameter beyond $\phi = 1000$ km. Comparison of the variation of these depths and heat flow in the ocean basins with the age of lithosphere entering trenches (Figure 2) shows that heat flow also decreases steeply with increasing age to a critical age of about 20 Ma. Both heat flow and maximum intraslab earthquake depth are insensitive to increasing age in older lithosphere. This transition in behavior probably reflects the effect of hydrothermal circulation in cooling the crust and uppermost mantle in the ocean basins to a nearly-equilibrium thermal gradient for $A \geq 20$ Ma and $\phi \geq 1000$ km. Because heat flow reflects most directly the 'temperature of the oceanic crust and uppermost mantle entering trenches, the parallel behavior shown in Figure 2 suggests that it

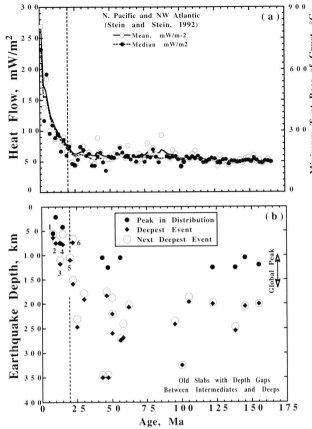

Fig. 2. Parallel variations heat flow and intraslab earthquake depths in oceanic lithosphere with lithosphere age. (a) Mean and median heat flow versus age of the ocean floor from Stein and Stein [1992]. Note the precipitous drop in heat flow with age to about 20 Ma and the insensitivity to variations in age beyond that. (b) Maximum earthquake depth, and depth to the first relative maximum in earthquake numbers versus plate age at the trench. Age from Mueller et al. [1993] and other sources and depths from relocated hypocenters obtained in this study. Subduction of young lithosphere with high heat flow [as shown in (a)] produces mostly shallow earthquakes. Key to young slabs: 1) Cascadia (Puget lowlands); 2) S. Chile (40 to 47°S); 3) S. Mexico; 4) S. America Ecuador near 2°S; 5) Luzon near 16°S; 6) Shikoku/Nankaido, Japan. Intermediate-depth earthquake distribution in older lithosphere is insensitive to variations in age, as is heat flow. Subduction zones with maximum earthquake depths between 300 and 350 km represent those with clusters thought to represent subduction of anomalous crust.

is the thermal regime *near the top surfaces* of slabs that governs the depths of processes controlling this earthquake distribution. Other evidence discussed later also supports this interpretation. Earthquake depth distribution in the Cocos plate beneath Middle America shows this marked variation with plate age entering the Middle America trench (Figure 3b). Earthquakes in the oldest (> 20 Ma) Central-American sector of the Cocos plate persist to depths of > 250 km, whereas events in the younger sectors beneath southern Mexico to the NW and Eastern Costa Rica and Panama to the SE rarely are deeper than 100 km.

Later we show how the parallel variation of heat flow (and hence shallow crustal thermal structure) and depth distribution of intraslab earthquakes with age discussed above also extends to the properties of arc volcanism and to other geophysical properties sensitive to slab metamorphism. We now turn to the seismological observations that can give independent information about slab metamorphism in subducted oceanic crust.

TOP-SIDE INTRASLAB SEISMICITY AND SEISMIC VELOCITY STRUCTURE: IMPLICATIONS FOR CRUSTAL METAMORPHISM

The detailed spatial relationships between intraslab hypocenters and the locations of the top surfaces of slabs provide crucial information regarding the physical environment and the internal processes there. Moreover, intraslab earthquakes may be used as sources of seismic waves that probe the seismic-wave velocity structures of slabs, from which inferences can be drawn as to their mineralogical and thermal structures.

Since the primary expressions of subducting slabs are the Wadati-Benioff-zone earthquakes themselves, how can we independently determine the locations of the top surfaces of slabs? The results detailed below suggest that most intraslab events at intermediate depths occur just below the top surfaces of slabs, an environment we term the "topside" slab setting. This finding is important because this seismic activity occurs just where the thermal gradients and hence the rates of conductive slab heating are greatest, where hydrous phases are most abundant and therefore where prograde metamorphism is expected to be focused.

Some subduction zones also locally have an additional inclined zone of seismic activity at depths of 10-40 km beneath the upper zone [See reviews by *Abers* and by *Seno and Yamanaka*, in this volume]. These events clearly do not fit the phenomenology summarized here. Recent studies of this phenomenon evoke the effects of magmatic processes at mid-plate depths, either at mid-ocean ridges [*Abers*, this volume] or above hot-spot plumes [*Kirby*, 1995; *Seno and Yamanaka*, this volume] in explaining the activity in the lower zone.

Several methods have been used to locate the top surfaces of slabs relative to earthquakes that occur

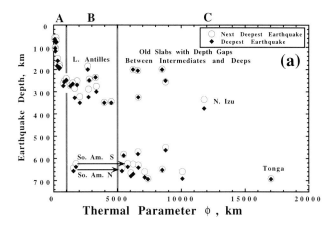

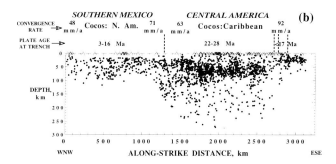

Figure 3. (a) Variation of maximum earthquake depth with thermal parameter ϕ [after *Kirby*, 1995]. ϕ is a measure of how deeply isotherms advect in slabs, i.e., how cold slabs are compared to normal mantle. Arrows show approximate corrections for the thermal parameters of South America (SA N of 15°S and SA S of 15°S) to account for a likely composite age and thermal parameter for the Nazca slab [*Engebreson and Kirby*, 1992]. Note the rapid deepening of earthquakes with increasing ϕ up to $\phi = 1000$ km (Region A) which we associate with deepening of topside slab metamorphic processes. Region B shows no particular trends and corresponds to approximately the age range 20–75 Ma where deepening with age is not observed (Fig. 2b). The deepening near $\phi = 5000$ into Region C is thought to represent the seismic effects of solid-solid phase changes in anhydrous mantle [see reviews in *Kirby* [1995] and *Kirby et al.* [1996b]. (b) Earthquake depth variation with along-trench distance, Middle American subduction zone. Variations in the convergence rates (Cocos : North America and Cocos : Caribbean) and in the age of the lithosphere at the trench are shown (dotted lines represent age discontinuities at fracture zones). Note the factor of about 2.5× increase in maximum earthquake depths with increasing age toward the Central American subduction zone, consistent with pattern seen in Figures 2b and 3a.

within them. Earthquakes with focal mechanisms consistent with interplate thrust motion represent the motion between the overriding and subducting plate; intraslab earthquakes immediately downdip of such events are therefore also near the top surface of the slab (Figure 4c).

Approximate locations of slab top surfaces may also be made using high-resolution seismic tomography. Most of the general seismic-velocity anomaly of slabs is caused by low slab temperatures [*Spakman et al.*, 1989]. Tomographic inversions of regional and teleseismic travel times place most of the intermediate-depth intraslab seismic activity near the tops of these gross slab anomalies (Figure 4) [*Engdahl and Gubbins*, 1987; *Kissling and Lahr*, 1991; *van der Hilst et al.*, 1991; *Hasegawa et al.*, 1994; *Zhao et al.*, 1994; *Engdahl et al.*, 1995].

Other methods of locating the top surfaces of slabs rely upon the velocity discontinuity between the subducting oceanic crust and the overriding plate or the asthenospheric mantle, a discontinuity that is too large and sharply defined to be caused just by the slab thermal anomaly [see reviews by *Helffrich et al.*, 1989; *Helffrich and Stein*, 1993; *Hasegawa et al.*, 1994; *Helffrich*, this volume]. This velocity discontinuity can convert S waves from earthquake sources to P waves or P waves to S waves by refraction [e.g., *Matsuzawa et al.*, 1990]. Again, intermediate-depth intraslab earthquakes tend to be just below the surface that describes those conversion points.

Lastly, the oceanic crust not only can convert seismic body waves but also can serve as a low-velocity seismic wave guide and produce late-arriving body-wave phases at stations updip from the source. Careful identification and timing of these phases allows estimation of the seismic velocities of these "trapped" waves and the thickness of the waveguide. *Fukao et al.* [1983], *Hori et al.* [1985], *Hori* [1990], and *Oda et al.* [1990] studied this crustal waveguide in the young (12–22 Ma) Philippine plate subducting beneath SW Japan and concluded that the waveguide was less than 10 km thick and that $V_p = 7$ km/s and $V_s = 4$ km/s for earthquake sources in the waveguide down to depths of 50–60 km, below which no trapped phases were observed. As these velocities were indistinguishable from those for oceanic crust beneath ocean basins, they concluded that the waveguide was gabbroic crust that survived untransformed to depths of 50–60 km. These Japanese investigations of the crustal waveguide in the Philippine plate also concluded that either the crust transformed to eclogite at the 50–60 km critical depth or that deeper earthquakes occurred outside untransformed crust. In the older and colder Pacific plate subducting beneath Northeast Japan, the low-velocity crustal channel persists to depths of up to 75

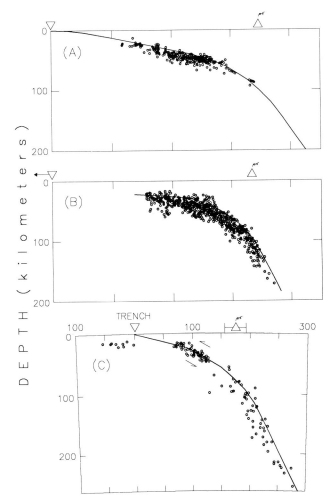

Fig. 4. Profiles through Wadati-Benioff zones with independent placement of the top surfaces of the slabs using seismic tomography. Upper-plate seismicity masked out for clarity. Apparent thicknesses of zones partly reflect the errors in hypocenter determination and also the shortcomings of 2-D projections in describing these 3-D objects. (a) Cascadia subduction zone [*Dragert et al.*, 1994; *Chiao*, 1991]. (b) Southern Alaska [*Kissling and Lahr*, 1991]. (c) Central Aleutians [*Engdahl and Gubbins*, 1987].

to 150 km depending on the locality and the technique, and its thickness is estimated to be about 5 km [*Matsuzawa et al.*, 1986; 1987; *Hurukawa and Imoto*, 1992; *Iidaka and Obara*, 1993]. The foregoing model for these topside velocity anomalies in slabs seems to be a more straightforward interpretation than identifying the low-velocity layer as one of partially serpentinized mantle right above the slab [e.g., *Tatsumi and Eggins*, 1995].

Similar seismic body-wave phases that are evidently produced by velocity anomalies near the top surfaces of slabs have been reported in the Vanuatu subduction zone [*Chiu et al.*, 1985] and in South America [*Snoke et al.*, 1977]. While generally less comprehensive, their results are largely consistent with those from the Japanese studies.

In contrast, *Ansell and Gubbins* [1986], *Gubbins and Sneider* [1991] and *Gubbins et al.* [1994] report the arrivals of anomalously-*fast*, high-frequency *P* waves from earthquakes in the Tonga-Kermadec subduction zone and recorded in New Zealand, ray paths mostly along strike of that subduction zone. Gubbins and his colleagues interpret these early arrivals as representing wave propagation in a thin, high-velocity, low-attenuation layer. They tentatively identify this structure as a continuous eclogite layer extending from 50 km depth to the bottom of the Wadati-Benioff zone. Although attempts have been made at reconciling these observations with those made in Japan [*Gubbins et al.*, 1994], an apparently shallow depth to onset of a continuous high-velocity eclogite layer in the Tonga-Kermadec subduction zone seems incompatible with the more extensive Japanese work. Moreover, an earlier study of velocity anomalies in the northern Tonga slab by *Mitronovas and Isacks* [1971] identified *P*-to-*S* converted phases at Tonga stations like those later observed in Japan and these investigators argued that details of their seismograms were consistent with a low-velocity crustal waveguide existing to depths of about 100 km and roughly coincident with the Wadati-Benioff zone.

We conclude from the above studies that most intermediate-depth intraslab earthquakes occur near the top surfaces of slabs and many lie within subducting crust. This is where plates entering trenches are initially coldest, where brittle normal faults are expected to be concentrated, and where hydrous phases are expected to be initially stable. This "top-side" slab environment should be the locus of rapid slab conductive heating, associated slab metamorphism, and dewatering. Untransformed crust can evidently survive to depths that vary markedly depending upon whether the plate entering trenches is young or old and hence depending upon the slab crustal temperature. Next we outline some of the likely physical connnections between topside slab metamorphism and intraslab earthquakes.

SLAB METAMORPHISM, INTRASLAB EARTHQUAKES AND ARC VOLCANISM: A CONCEPTUAL MODEL

The evidence summarized in the foregoing sections points to intermediate-depth intraslab earthquakes occurring where rapid conductive heating and associated metamorphic processes should also be occurring in subducting crust and uppermost mantle. Why is there this spatial correlation?

Slab metamorphism is important for intraslab earthquakes for several reasons. First, densification of crust by eclogite formation can alter slab stresses by increasing negative buoyancy [e.g., *Ruff and Kanamori*, 1983; *Sacks*, 1983] and thereby increasing the sinking force tending to bend slabs [*Jones et al.*, 1978; *Chapple and Forsyth*, 1979] and to extend them by direct slab pull [e.g., *Spence*, 1987]. *McGarr* [1977] and *Pennington* [1983] also emphasize the general importance of metamorphic volume changes in altering the states of stress in slabs. Eclogite formation should produce a stretching deformation in the crust and a smaller compression in the underlying mantle caused by the strains accompanying the heterogeneous volume change (Figure 1a) [*Denlinger and Kirby*, 1991], provided that transformed crust does not delaminate from the underlying mantle. We examined this effect of eclogite formation by finite-element modelling with a simple rheology that illustrates the localization of deformation to transformed crust (Figure 5). A full description of our computations will be published elsewhere. The stresses and deformations are a consequence of the dimensional mismatch between subducting crust and mantle that occurs when the crust transforms to eclogite. A stretching deformation occurs in the crust, and a smaller compressional deformation occurs in the underlying mantle (Figure 5b). This stress state would be generated in slabs because of the constraint imposed by a "no-slip" condition along the subducting crust-mantle boundary during eclogite formation and would occur even if the crust and uppermost mantle of the slab is segmented by normal faulting, since patches of intact crust-mantle could still provide the "no-slip" constraint. The seismological response of the crust to these stresses depends upon the influence of other sources of stress in slabs and on slab rheologies, as explained below.

Dehydration can lead to high pore pressures, reduce effective normal stresses, and hence promote faulting by dehydration embrittlement [*Raleigh and Paterson*, 1965; *Raleigh*, 1967; *Paterson*, 1978; *Meade and Jeanloz*, 1991; *Kirby*, 1995]. This may be especially important where dewatering occurs along pre-existing normal faults, because such fossil faults may be readily reactivated if water is released just where it is most effective in reducing the local normal stress by elevating pore pressure along the fault [*Kirby*, 1995; *Silver et al.*, 1995]. We suggest that reactivation by dehydration embrittlement of previously hydrated fossil normal faults is the primary rupture process for intraslab earthquakes shallower than about 325 km. Such reactivation may be driven by stresses associated with eclogite formation as explained above and/or by stresses that derive from plate-scale forces [e.g., *Vassiliou and Hager*, 1988]. Stretching deformation along reactivated normal faults in the crust would also facilitate fluid migration out of slabs and into the asthenosphere.

These effects of metamorphism on intraslab faulting and associated earthquakes outlined above would be difficult to isolate from each other by seismological observations, because dehydration reactions in hydrous mafic rocks are themselves key steps in the formation of eclogite from hydrous mafic rocks. In a mineralogically complex and mechanically-flawed crust, such processes may well occur in the same depth range for a given subducting slab with a given thermal structure. Water released by dehydration of hydrous phases along fractures and faults could facilitate transformation in adjacent regions of anhydrous gabbro and basalt. The abrupt change in the decay law in earthquake numbers near 90 km may reflect the average global onset of large-scale crustal metamorphism of thermally mature slabs (Figure 6).

Many previous studies of slab metamorphism in the context of the origin of arc magmas have assumed that metamorphic reactions take place at near-equilibrium conditions. If this were correct, then dehydration of the crust and eclogite formation would take place mainly at shallow depths (<< 70–80 km). However, recent studies of simple polymorphic mineral systems and a model system for Na-pyroxene forming reaction that is a key step in eclogite formation (albite → jadeite + quartz) suggest that if slabs are cold enough, kinetic hindrance of eclogite-forming reactions will be significant [*Hacker*, 1992; *Hacker*, 1995]. The presence of free water is vitally important in governing reaction rates because under comparable pressure oversteps of the albite breakdown reaction, rates of reaction with excess water present may be attained at 200–300°C colder than for nominally dry experiments. This means that eclogite reaction in anhydrous gabbro could be effectively delayed to depths where large-scale dehydration takes place in adjacent hydrous metagabbros. This hypothesis has been largely confirmed in the Japanese seismological studies of the subducting Pacific Plate beneath NE Japan, cited earlier, indicating that a crustal low-velocity waveguide exists in the old, cold Pacific plate to depths of up to 150 km. At greater depth, seismological evidence for the low-velocity crustal layer disappears. This change suggests that the crust transforms to eclogite and its seismic velocity becomes comparable to the mantle in the slab below the crust.

The depth distribution of earthquake activity in thermally mature slabs (thermal parameter greater than about 1000 km) indicates that there is a reversal in the rate of decline in earthquake numbers with increasing depth at about 90 km and that numbers

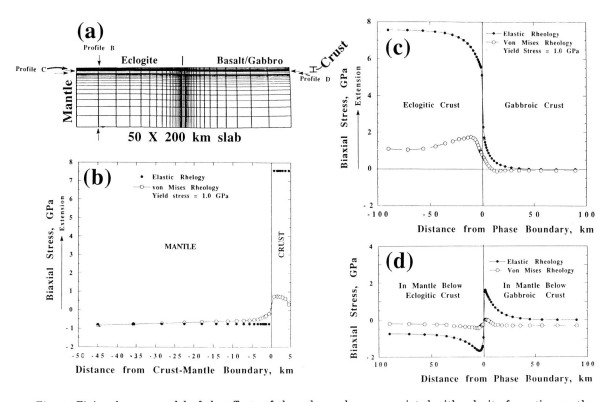

Fig. 5. Finite-element model of the effects of the volume change associated with eclogite formation on the stresses in slabs. (a) Deformed finite-element grid showing the grid geometry and positions of the profiles. Stress profiles in (b), (c) and (d) for a purely elastic rheology (filled circles) and a critical-yield-stress (von Mises) rheology (open circles). Choice of the numerical value of the yield stress was arbitrary and selected for illustrative purposes only. The general pattern of deformation is independent of the choice of yield stress over the range of 10 MPa to 1 GPa. The biaxial stress is the difference between the slab-parallel normal stress and the slab-perpendicular stress. Note the convention that extensional stresses are considered positive. Differences between the stress profiles for the two rheologies are proportional to the amount of inelastic deformation. (b) Profile perpendicular to the slab shows that the deformation in the crust is markedly stretching and in the underlying mantle is mildly compressional. (c) Profile parallel to crust-mantle boundary but within the crust. Note that the rapid falloff of the extensional stress and deformation with distance from the gabbro-eclogite boundary. (d) Profile parallel to crust-mantle boundary but within the mantle. Note the change in sign of the stresses and deformation across the gabbro-eclogite boundary and the markedly lower stresses than in the crust.

peak near 100–170 km. This depth range corresponds to the global range of depths of Wadati–Benioff zones below the volcanic fronts of modern magmatic arcs (Figure 6). This coincidence suggests to us that the focus of crustal metamorphism and associated slab seismicity in thermally-mature slabs is at the roots of the volcanic arcs (Figure 1a). This critical depth range of 100–170 km seems to be insensitive to variations in plate age for age > 20 Ma, in parallel with the lack of sensitivity of heat flow with crustal age (Figure 2a). It should be noted that although many individual subduction zones show peaks in intraslab earthquake activity in this depth range (Figure 2b), some show only a change in slope or show nothing at all distinctive at that depth.

The crust of young lithosphere (< 20 Ma) entering trenches may be as much as 300°C hotter than that for thermally mature subduction zones [*Peacock*, 1996, this volume; *Hyndman and Wang*, 1993; *Wang et al.*, 1995] (Figure 2a). This suggests that dehydration and eclogite formation may take place at shallower depths at more nearly equilibrium conditions (Figure 1b). This conclusion is supported by the Japanese observations cited earlier. Consistent with this interpretation is the observation that intraslab earthquakes in young slabs are generally limited to depths of less than 50–100 km.

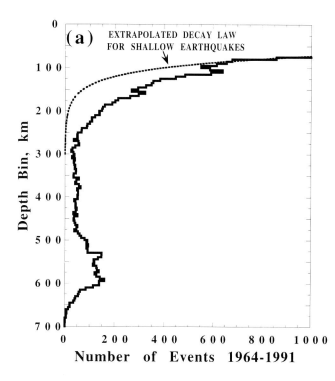

Other examples of shallow seismicity in young slabs include the Cascadia subduction zone [*Ludwin et al.*, 1991], southern Mexico [*Pardo and Suarez*, 1993; 1995; *Kostoglodov and Bandy*, 1995], SE Costa Rica/NW Panama [*Protti et al.*, 1994], in the Nazca plate in southern Chile between the Chile Ridge and 40°S, in the Antarctic plate south of the Chile Ridge [*Engdahl*, unpublished catalogue data], southwest Japan (Nankai-Shikoku) [*Ukawa*, 1982], southern Luzon/Manila trench at 14.5–18°N, and the South Shetlands subduction zone [*Engdahl*, unpublished data]. The lack of appreciable seismicity at greater depths in these young slabs may well reflect the near completion of metamorphic activity in the subducting crust and hence the lack of a mechanism for faulting at greater depths.

IMPLICATIONS FOR ARC MAGMATISM

The fundamental seismological differences we have discussed between subduction zones involving young lithosphere subducting at low to moderate rates and subduction zones that involve older, more rapidly descending plates has significance in helping us interpret trends in arc volcanism. The parallel topology of the variations of heat flow and intraslab earthquake depths with age (and particularly the marked transition in both quantities at an age of about 20 ± 5 Ma (Figure 2a and b) suggests to us that young slabs are hot enough to dehydrate largely at depths shallower than about 100 km. Hence, dehydration would occur before such slabs commence contact with asthenospheric mantle. Other independent

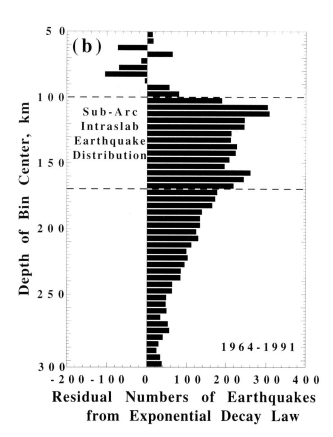

Fig. 6. (a) Histogram showing the depth distribution of subduction-zone earthquakes in the Circumpacific/Indonesian/South Scotia/Caribbean subduction zones. Events are relocated from ISC arrival-time data 1964-1991 and filtered to remove hypocenters with large depth and location uncertainties. Also shown is a fit of the data from 50–90 km to an exponential decay law where earthquake numbers in 5-km depth bins (N) vary as $N = A \exp(az)$ where z is the center of the depth bin ($A = 11,850$ and $a = -0.03395$ km^{-1}). Note the departure of the observed numbers from the decay law at depths greater than about 90 km. (b) Residual earthquake numbers (observed minus predicted numbers from the shallow decay law) versus depth. The scatter in the residuals at depths less than 90 km is due to the errors connected with subtracting large numbers. The onset of the sub-arc population is believed to represent the global average onset depth of large-scale dehydration and eclogite formation in thermally mature subducting crust (Figure 1a).

evidence exists for slab metamorphism localizing at shallow depths in such slabs. *Wannamaker et al.* [1989] note an anomalously-low electrical resistivity region in the Cascadia subduction zone and suggest that large-scale dehydration of the subducting oceanic lithosphere takes place before the slab descends beneath the Cascade volcanic arc. Likewise, *Zhao et al.* [1995] interpret regions of low seismic velocity that coincide with zones of high electrical conductivity establishes where the Juan de Fuca slab is dehydrating beneath the northern Cascadia forearc. These observations provide independent evidence that crustal metamorphism in very young slabs occurs largely beneath forearcs before the crust develops contact with the overlying asthenosphere.

If the widely-held conceptual model for arc magmatism is correct that fluids liberated by dehydration of slabs flux melt the asthenospheric wedge, then premature loss of the water of hydration during the descent of young lithosphere should reduce the rates of asthenospheric melting and the vigor of arc magmatism (Figure 1b). *Simkin et al.* [1981] and *Simkin and Siebert* [1984] emphasized that subduction zones with plate ages less than 20 Ma and/or descending at slow rates often have relatively sparse or absent young arc volcanism. Subduction zones with slabs older than 20 Ma show no particular trends in volcanic vigor with plate age. Simkin and his colleagues cite examples of young slabs lacking vigorous arc volcanism: southern Chile, southern Mexico, and Cascadia. We would add to that list the following: (1) the northern part of the Philippine plate subducting beneath SW Honshu (Japan), (2) the South Shetlands subduction zone (Antarctica) [*Maldano et al.*, 1994], (3) SE Costa Rica/NW Panama in Central America, and (4) possibly northern Luzon near 14.5–18°N where a young but inactive spreading ridge is now being subducted at the Manila trench [*Briais et al.*, 1993]. As noted earlier, the above subduction zones generally do not have significant numbers of intraslab earthquakes appreciably deeper than 100 km. This correspondence between the population of young slabs with mainly shallow intraslab earthquakes and those that lack arc magmatic vigor reinforces the hypothesis that they both reflect a focus of shallow metamorphism that takes place largely beneath forearcs.

Also strengthening this conclusion is that *some* of the arc magmas produced in those settings (S. Chile, Costa Rica/Panama; SW Honshu, Cascadia) have magmas of unusual composition (high in *Al* and possessing rare earth element patterns that are anomalous for arc magmas). Such magmas are also found in some arcs where a plate is obliquely subducting, the slab vertical descent rate is therefore very slow and hence the slab is warm (e.g., W. Aleutians [*Creager and Boyd*, 1991]). It has been argued that "adakites", as they are sometimes called, represent melting of hot subducted crust at conditions near the amphibolite-eclogite transformation with garnet as a residual phase [*Futa and Stern*, 1988; *Defant and Drummond*, 1990; *Defant et al.*, 1991; *Kay et al.*, 1993; *Morris*, 1995]. It should be emphasized that adakites may represent only a minor component of the magmatic rocks of some arcs involving young slabs.

The thermal structure of lithosphere entering trenches and the rates of convergence can vary with time such that marked temporal variations may occur in the slab thermal structure of a given subduction system. The system may cross the critical transition from thermally-mature to warm-slab conditions. How do these changes affect arc volcanism? The Cascadia subduction zone provides an excellent place to study this question. The Juan de Fuca (Farallon): North America motion has markedly decreased during the Cenozoic [*Verplanck and Duncan*, 1987; *Duncan and Kulm*, 1989; *Wilson*, 1988; 1993], and the average age of lithosphere has decreased slightly as the Pacific: Juan de Fuca (Farallon) spreading center slowly approached the North American plate [*Wilson*, 1988; *Atwater*, 1989; *McCrory et al.*, 1995; P. A. McCrory, pers. comm., 1995]. Accompanying this decrease in thermal parameter was a dramatic decrease in the production rates of arc lavas in the Cascadia subduction zone in the last 30 Ma [see reviews by *Muffler and Tamanyu*, 1995; *Verplanck and Duncan*, 1987] and a systematic shift in the volcanic arc to the east with time [*Duncan and Kulm* 1989], suggesting a decrease in average slab dip. These changes in the Cascade volcanic arc are consistent with the expected change from a normal subduction zone with cold crust and uppermost mantle (Figure 1a) to a subduction zone presently involving a warm, buoyant slab and hence shallow metamorphism, shallow earthquakes, shallow slab dip and more feeble arc volcanism (Figure 1b).

EFFECTS OF SUBDUCTING VOLCANIC CHAINS AND RIDGES

Another factor besides plate age and convergence rates that can potentially alter the progress of slab metamorphism is the subduction of anomalous crust, such as is produced in the ocean basins by earlier arc volcanism or hot-spot volcanism. The effects of entrance of such features into trenches has been widely discussed in the literature, mainly in the context of collisional accretion of terranes and their effects on forearc tectonic erosion, slab buoyancy, slab dip, arc magmatism and other expressions of subduction [e.g., *Nur and Ben-Avraham*, 1981; *Greene and Wong*, 1989; *Ballance et al.*, 1989; *Masson et al.*, 1990; *von Huene* and *Scholl*, 1991; *Cloos*, 1993]. This is a complex subject with a huge literature, and a thorough review is beyond the scope of this paper. Instead, we focus on some recent observations that are relevant to this paper.

Evaluating the possible seismological expression of the subduction of aseismic volcanic ridges and volcanic chains at greater depths is made difficult by the lack of knowledge of whether a given structure originally persisted or survived "downstream" of the trench and, if so, what its geometry might be. The seismological expressions of straightforward along-trend projections of large, prominent volcanic ridges can be surprisingly subtle. For example, intermediate-depth events are not unusually intense downstream of the Nazca and Carnegie Ridges in South America, the D'Entrecasteaux Ridge in Vanuatu and the Emperor seamount chain off Kamchatka. In fact, *Marthelot et al.* [1985] argue that the descent of the D'Entrecasteaux Ridge beneath the North Fiji Basin produces a gap in the intermediate-depth seismicity of the Vanuatu subduction zone. Although a number of large and mainly shallow shocks have occurred down-trend of the prominent and long-lived Louisville Ridge [*Christensen and Lay*, 1988], there is surprisingly little, if any, expression of this structure at intermediate depths.

Down the trends of some narrower and generally less continuous volcanic chains, however, intermediate-depth seismicity can be intense. An outstanding example is the Juan Fernandez chain off central Chile, where intermediate-depth clusters occur along a seismic belt that extends nearly 700 km east of the Chile trench [Figure 7a and 7b]. The trend of this seismic belt aligns well with the Valpariso embayment in the Chilean shoreline and with a remarkable tectonic excavation of the inner wall of the Peru–Chile trench where the volcanic chain has collided with the shallow forearc [*von Huene*, 1995; *von Huene et al.*, 1996].

Further north, *Kirby and Engdahl* [1993] have also identified other ENE to NE-trending earthquake clusters that parallel trends in Cenozoic offshore island and seamount chains. Such alignments suggest that some of these clusters are the seismic expressions of these subducted volcanic chains. In the present context, we note that some of the more prominent clusters in this group have young arc volcanism above them in the South American plate that is deflected eastward of the main trend of the Central Volcanic Zone of the Andes (compare Figure 7 with Fig. A1, p. 194 in *de Silva and Francis*, 1991).

To interpret these relationships between subducting volcanic chains, on the one hand and intraslab earthquakes in the Nazca slab and the Andean volcanism, on the other, we consider five possible physical effects of subduction of these volcanic structures on slab metamorphism and fault reactivation in the context of the present hypothesis:

(1) Effects of volcanic ridges in reducing slab buoyancy, decreasing slab dip and thereby eliminating slab contact with an asthenospheric wedge in the critical depth range 100–150 km (see review by *Cahill and Isacks* [1992]). Such a condition evidently extinguishes arc volcanism by excluding partial melting of an asthenospheric wedge [*Barazangi and Isacks*, 1977]. Reduced slab sinking forces should also alter slab stresses. These effects are discussed in the literature cited above and will not be further considered here.

(2) Effects on localized shallow normal faulting in trench/outer rise settings. High-resolution sonar images of the seabed in trench-rise systems where volcanic chains are subducting often reveal intense normal faulting that tectonically dissects the seamounts themselves and the seafloor around them [USGS Gloria Atlas of the Aleutian Exclusive Economic Zone, unpublished; *Kobayashi et al.*, 1987; *Fisher et al.*, 1991; see review in *von Huene and Scholl*, 1991; *von Huene et al.*, 1996]. Intense shallow seismicity commonly occurs where such structures are flexed [*Christensen and Lay*, 1988; *Wysession et al.*, 1991; *Zhao et al.*, 1996].

Kirby et al. [1996] investigated the relocated historical and modern seismicity of the Juan Fernandez earthquake zone and suggest that the fine structure of the Juan Fernandez seismic belt is composed of two parallel zones of earthquakes that are aligned with two clusters of shallow earthquakes near the trench. These shallow clusters are near the intersection of the Juan Fernandez chain with the trench and outer rise and correspond to the gravity highs where the forebulge of the chain associated with seamount loads coincides the outer-rise of the subduction zone. If faults created at near-trench-outer-rise settings in ordinary crust and mantle are reactivated at intermediate depths as described earlier, faults localized near seamounts should also be reactivated and perhaps lead to clustering at intermediate depths. This reactivation hypothesis may help explain why the Juan Fernandez chain produces two zones of intermediate-depth earthquakes.

(3) Effects of differences in the internal structures of the crust compared to normal oceanic crust. One of the biggest internal structures produced in large volcanic edifices are internal rift zones [e.g., *Clague et al.*, 1989]. These structures evidently develop in connection with lateral spreading and volcanic edifice building. Should these rift structures survive the subduction process, they could also serve as flaws that could be reactivated during descent.

(4) Perturbations of the thermal structure of the crust derived from the original volcanic heat trapped during edifice building. The thermal effects of volcanic chains on the thermal structure of crust and mantle entering

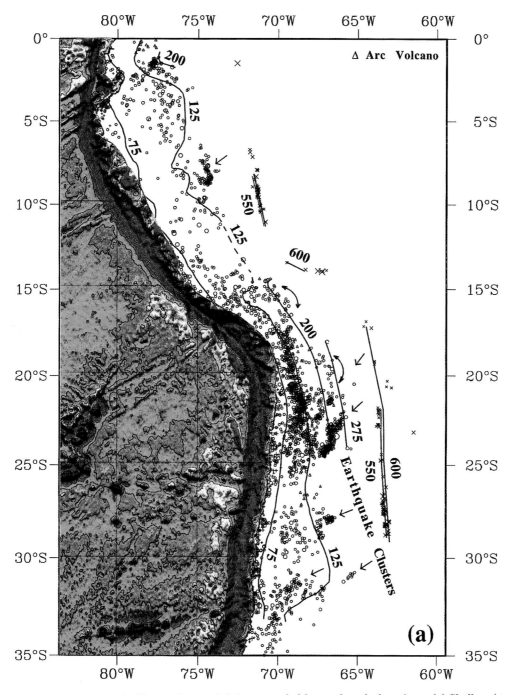

Fig. 7a. Heterogeneities in the Nazca plate and slab as revealed by earthquake locations. (a) Shallow (+: 0–50 km), intermediate (O: 50–350 km) and deep intraslab earthquakes (×: > 350 km) in the Nazca slab beneath South America. Contours are drawn as depth in km to the Wadati–Benioff zone. Shallow earthquakes are stripped from map east of the 75 km contour to emphasize Nazca slab seismicity. Note clustering of intermediate-depth events in roughly ENE- to NE-trending bands (arrows), especially at latitudes between 15 and 33 S. These bands are frequently aligned with offshore volcanic chains and ridges and with shoreline embayments and collisional features as shown by offshore satellite gravity [*Sandwell et al.*, 1995]. Note the alignment of some of the offshore volcanic chains and ridges with coastal embayments and earthquake clusters also shown in Figure 7b.

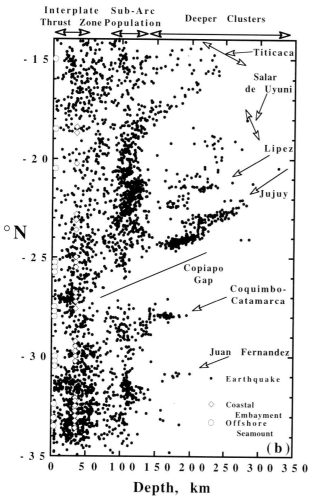

Fig. 7b. North-South cross section of Nazca intraslab seismicity. Note general tendency for earthquake clusters with labeled names to trend in the ENE to NE direction and plunge toward the north and east, corresponding to the range of Nazca plate motion relative to hot-spots [*Gordon and Jurdy*, 1986; *Gripp and Gordon*, 1990] and the offshore trends of volcanic chains shown in (b).

trenches should depend upon the age of volcanic activity and the manner in which they are built and lose their volcanic heat. If they are young enough to be significantly hotter than the surrounding oceanic crust, then crustal metamorphic processes might occur at shallow depths as they evidently do in young crust created at MORs.

(5) Effects of crustal thickness on the thermal time constant of the crust (how fast it takes to heat up by conduction during descent). If a seamount or volcanic ridge is old and cold, the normal oceanic crust under the volcanic edifice will be thermally blanketed and hence may take significantly longer to heat up by conduction during descent than normal oceanic crust. For example, to a good approximation, a crust that is twice as thick as normal has four times the thermal time constant than one of normal thickness. Thus the kinetics of eclogite-forming reactions that are sensitive to temperature should delay eclogite formation and dehydration at the base of the crust to significantly greater depths in the cold, thick crust of fossil hot-spot volcanic edifices.

EFFECTS ON ARC MAGMATISM

Returning to the broader connections between subducted volcanic chains created in the ocean basins and the volcanic arcs above them, it should be noted that the oblique subduction of the NW-trending Louisville Ridge has perturbed the Tonga-Kermadec volcanic arc. Although the Louisville Ridge does not have a prominent effect on intraslab intermediate-depth seismicity, it evidently produced a 450-km gap in Holocene and historical eruptions north of its intersection with the Tonga-Kermadec trench [*Simkin and Siebert*, 1994]. Geological evidence suggests that this gap has been migrating to the south at about the rate that the trench-volcanic chain intersection is migrating south [*D. Scholl*, personal communication]. In other words, subduction of the island and seamount chain "extinguishes" the volcanic arc and "rekindles" it when the arc reaches 400–500 km from the migrating ridge-trench intersection. One possible explanation of this progression is that the rocks of the volcanic ridge thermally blanket the underlying oceanic crust, and this insulating effect delays heating the normal oceanic crust underneath as it descends. By the time dehydration does occur, the melts produced are at sufficient depth that they become entrained in the asthenospheric circulation of back-arc spreading and volcanism rather than fluid circulation to the island arc. Some back-arc lavas in the western Pacific have compositions suggesting a component of island-arc chemistry [Lau Basin: *Vallier et al.*, 1991; Marianas trough: *Stolper and Newman*, 1994]. If a subduction zone lacks active back-arc spreading, water liberated at greater-than-normal depths by dehydration of cold volcanic chains may produce arc volcanism displaced toward the direction of slab motion from the normal magmatic arc. This may explain the outboard deflections of some arc volcanoes in subduction zones that have island- and seamount chains being subducted and that lack present-day back-arc spreading, such as in Kamchatka and South America.

SUMMARY AND CONCLUSIONS

Our conceptual model for the effects of metamorphism of the crust and uppermost mantle of slabs, although highly simplified, recognizes the fundamental roles of temperature and free aqueous fluids in promoting reaction rates and the dual effects of such liberated fluids in facilitating both seismogenic intraslab faulting and melting in mantle asthenospheric wedges. The model is successful in explaining why intermediate-depth earthquakes tend to occur just below the top surfaces of slabs. It also explains the first-order dichotomy of subduction zones: (1) Young lithosphere subducting at low to moderate rates (with thermal parameter less than 500 km) produces largely shallow intraslab earthquakes, and feeble, if any, arc volcanism. Some arc magmas in this setting have bulk chemistries consistent with melting of eclogitic oceanic crust. Dewatering and eclogite formation probably occurs primarily at shallow depth, and hence aqueous fluids are released largely into the overlying forearc. When the dewatered slab descends and develops contact with the overlying asthenosphere, free water is not abundant enough to produce large-scale flux melting, and hence arc volcanism is not vigorous. (2) The crust and uppermost mantle of older lithosphere descending at moderate to high rates is very cold and this thermal structure is essentially independent of plate age. Dehydration of thermally mature slabs not only produces normal arc volcanism but also permits intraslab earthquakes to occur in greatest abundance at depths of 100–170 km at the roots of volcanic arcs and continue to maximum depths of typically 200–325 km. We interpret the peak in seismic activity as indicative of the primary focus of dewatering and associated eclogite formation in that depth interval.

A number of factors may complicate this classification of subduction zones and the conceptual models used to explain their differences. First, plate age, rates of convergence and slab dip may vary with time. It is not known how rapidly magmatic arcs respond to changes in their thermal regime at depth. Second, volcanism in the ocean basins alters to some degree the thermal and mechanical structures of subducting crust and uppermost mantle.

Some of these ideas about shallow metamorphism in young and/or slowly descending slabs may have relevance to the assessment of earthquake and volcano hazards in such settings. If earthquakes in young slabs are localized just at shallow depths, then the falloff in strong motions associated with body waves and surface-wave motion for such events are expected to be less than for intermediate-depth shocks (depths > 70 km) and hence ground motion should be more intense. Examples of such destructive shallow intraslab events are the M_w 7.1 event of 23 April 1949 and the M_w 6.5 shock of 29 April 65 in the Juan de Fuca slab of the Cascadia subduction zone and the M_w 7.5 shock of 15 January 31 in the young Cocos plate beneath southern Mexico (M_w is the earthquake magnitude calculated from the seismic moment). Lastly, the concept of normal faults produced by flexure at shallow depths in trench-outer-rise settings being reactivated at greater depths may be useful in assessing the hazards associated with large interplate thrust earthquakes. Many very large interplate events in strongly coupled subduction zones display shallow intraplate earthquakes seaward of trenches, before and after large underthrusting earthquakes. These intraplate events have been used to evaluate the potential for large interplate earthquakes [e.g., *Christensen and Ruff*, 1988]. If such shallow intra*plate* faults are subsequently reactivated as intra*slab* events, then a fuller seismological record exists of prior shallow trench/outer-rise intraplate activity in the form of deeper intraslab earthquake activity [*Kirby et al.*, 1996a].

This model for shallow metamorphism in young subducting plates also provides a framework for understanding why subduction of young slabs tends to produce feeble arc volcanism and how temporal changes in the slab thermal parameter can produce systematic shifts in volcanic vigor and arc position with time. While arc-wide volcanism may be feeble in such settings, explosive volcanism at individual volcanoes may still be very hazardous and destructive, such as has occurred in the Cascadia subduction zone of the NW U.S during the Holocene [*Simkin and Siebert*, 1994].

Acknowledgments. We thank Tracy Rushmer, Brad Hacker, Mike Clynne, Steve Bohlen, David Gubbins, Emile Okal, Doug Wiens, Akira Hasegawa, Seth Stein, Sue Agar, Tetsuzo Seno, David Scholl, David Clague, Bob Tilling, Rollie von Huene, Diana Comte, Mario Pardo, Sergio Barrientos, and Carol Stein for their helpful discussions, preprints and comments on the manuscript. Laura Stern, Wayne Thatcher, Pat Muffler, George Helffrich and Dave Scholl helped improve the presentation of this paper.

REFERENCES

Agar, S. M., Rheological evolution of the ocean crust: A microstructural view, *J. Geophys. Res.*, 99, 3175–3200, 1994.

Ansell, J. H., and D. Gubbins, Anomalous high-frequency wave propagation from the Tonga–Kermadec seismic zone to New Zealand, *Geophys. J. Roy. Astron. Soc.*, 85, 93–106, 1986.

Ballance, P. F., D. W. Scholl, T. L. Vallier, A. J. Stevenson, H. Ryan, and R. H. Herzer, Subduction of a Late Cretaceous seamount of the Louisville ridge at the Tonga trench, *Tectonics*, 8, 953–962, 1989.

Benioff, H., Seismic evidence for the fault origin of oceanic deeps, *Bull. Geol. Soc. Am.*, 60, 1837–1856, 1949.

Berlage, H. P., A provisional catalogue of deep-focus earthquakes in the Netherlands East Indies 1918–1936, *Gerlands Beiträge zur Geophysik, 50*, 7–17, 1937.

Briais, A., P. Patriat, and P. Tapponnier, Updated interpretation of magnetic anomalies and seafloor spreading stages in the South China Sea, *J. Geophys. Res., 98*, 6299–6328, 1993.

Burnley, P. C., H. W. Green, and D. Prior, Faulting associated with the olivine to spinel transformation in Mg_2GeO_4 and its implications for deep-focus earthquakes, *J. Geophys. Res., 96*, 425–443, 1991.

Cahill, T., and B. L. Isacks, Seismicity and shape of the subducted Nazca plate, *J. Geophys. Res., 97*, 17,503–17,529, 1992.

Carbotte, S. M., and K. C. Macdonald, Comparison of seafloor tectonic fabric at intermediate, fast, and super fast spreading ridges, *J. Geophys. Res., 99*, 13,609–13,631, 1994.

Chapple, W. M., and D. W. Forsyth, Earthquakes and bending of plates at trenches, *J. Geophys. Res., 84*, 6729–6749, 1979.

Chiao, L.-Y., Membrane deformation rate and geometry of subducting slabs, Ph.D. thesis, Geophysics Program, University of Washington, Seattle, WA, 1991.

Chiu, J.-M., B. L. Isacks, and R. K. Cardwell, Propagation of high-frequency seismic waves inside the subducted lithosphere from intermediate-depth earthquakes recorded in the Vanuatu arc, *J. Geophys. Res., 90*, 12,741–12,754, 1985.

Chiu, J.-M., B. L. Isacks, and R. K. Cardwell, 3-D configuration of subducted lithosphere in the western Pacific, *Geophys. J. Int., 106*, 99–111, 1991.

Christensen, D. H., and T. Lay, Large earthquakes in the Tonga region associated with subduction of the Louisville Ridge, *J. Geophys. Res., 93*, 13,367–13,389, 1988.

Christensen, D. H., and L. J. Ruff, Seismic coupling and outer rise earthquakes, *J. Geophys. Res., 93*, 13,421–13,444, 1988.

Clague, D. A., and six others, Chapter 12: The Hawaiian-Emperor Chain, in *The Eastern Pacific Ocean and Hawaii*, edited by E. L. Winterer, D. M. Hussong, and R. W. Decker, vol. N, pp. 187–287, GSA, Decade of Geology of North America, Boulder, Colorado, 1989.

Cloos, M., Lithospheric buoyancy and collisional orogenesis, *Geol. Soc. Am. Bull., 105*, 715–737, 1993.

Coats, R. R., Magma type and crustal structure in the Aleutian arc, in *The Crust of the Pacific Basin*, edited by G. A. Macdonald and H. Kuno, *Geophysical Monograph 6*, 92–109, 1962.

Davies, J. H., and D. Stevenson, Physical model of source region of subduction zone volcanics, *J. Geophys. Res., 97*, 2037–2070, 1992.

Defant, M. J., and others, Andesite and dacite genesis via contrasting processes: The geology and geochemistry of El Valle volcano, Panama, *Contrib. Mineral Petrol., 106*, 309–324, 1991.

Defant, M. J., and M. S. Drummond, Derivation of some modern arc magmas by melting of young subducted lithosphere, *Nature, 347*, 662–665, 1990.

Denlinger, R. P., and S. H. Kirby, Stresses imposed by the basalt-eclogite transformation in descending lithosphere, *Eos Trans. AGU, 72*, 481–482, 1991.

de Silva, S., and P. Francis, *Volcanoes of the Central Andes*, Springer, 216 pp., 1991.

Dick, H. J. B., P. S. Meyer, S. Bloomer, S. Kirby, D. S. Stakes, and C. Mawer, Lithostratigraphic evolution of an in-situ section of oceanic layer 3., in *Proceedings of the Ocean Drilling Program, 118*, 439–538, 1991.

Dragert, H., R. D. Hyndman, G. C. Rogers, and K. Wang, Current deformation and the width of the seismogenic zone of the northern Cascadia subduction thrust, *J. Geophys. Res., 99*, 653–668, 1994.

Duncan, R. A., and L. V. D. Kulm, Plate tectonic evolution of the Cascades arc-subduction complex, in *The Eastern Pacific Ocean and Hawaii*, edited by E. L. Winterer, D. M. Hussong, and R. W. Decker, vol. N, pp. 413–438, GSA, Boulder, CO, 1989.

Engdahl, E. R., and D. Gubbins, Simultaneous travel time inversion for earthquake location and subduction zone structure in the Central Aleutian Islands, *J. Geophys. Res., 92*, 13,855–13,862, 1987.

Engdahl, E. R., R. van der Hilst, and J. Berrocal, Imaging of subducted lithosphere beneath South America, *Geophys. Res. Lett., 22*, 2317–2320, 1995.

Engebretson, D., and S. Kirby, Deep Nazca slab seismicity: Why is it so anomalous?, *Eos, Trans. Am. Geophys. Union, 73*, 379, 1992.

Fisher, M. A., J.-Y. Collot, and E. L. Geist, Structure of the collision zone between Bougainville Guyot and the accretionary wedge of the New Hebrides Island arc, southwest Pacific, *Tectonics, 10*, 887–903, 1991.

Fukao, Y., S. Hori, and M. Ukawa, A seismological constraint on the depth of basalt-eclogite transition in a subducting oceanic crust, *Nature, 303*, 413–415, 1983.

Futa, K., and C. R. Stern, S_r and N_d isotopic and trace element compositions of Quaternary volcanic centers of the southern Andes, *Earth and Planet. Sci. Lett., 88*, 253–262, 1988.

Gill, J. B., *Orogenic Andesites and Plate Tectonics*, 390 pp., Springer-Verlag, Berlin, 1981.

Gorbatov, A., V. Kostoglodov, and E. Burov, Maximum seismic depth versus thermal parameter of subducted slab: Application to deep earthquakes in Chile and Bolivia, *Geofisica Internacional*, *35*, 41–50, 1996.

Gordon, R. G., and D. Jurdy, Cenozoic global plate motions, *J. Geophys. Res.*, *91*, 12,389–12,406, 1986.

Greene, H. G., and F. L. Wong, Ridge collisions along the plate margins of South America compared with those in the southwest Pacific, in *Geology of the Andes and its Relationship to Hydrocarbon and Mineral Resources*, edited by G. D. Erickson, M. T. Canas-Pinochet, and J. A. Reinemund, vol. 11, pp. 39–57, Circum-Pacific Council for Energy and Mineral Resources Earth Science Series, Houston, Texas, 1989.

Griggs, D. T., The sinking lithosphere and the focal mechanism of deep earthquakes, in *The Nature of the Solid Earth*, edited by E. C. Robertson, pp. 361–384, McGraw-Hill, New York, 1972.

Gripp, A. E., and R. G. Gordon, Current plate velocities relative to hotspots incorporating the NUVEL-1 global plate motion model, *Geophys. Res. Lett.*, *17*, 1109–1112, 1990.

Gubbins, D., A. Barnicoat, and J. Cann, Seismological constraints on the gabbro-eclogite transition in subducted oceanic crust, *Earth and Planet. Sci. Lett.*, *122*, 89–101, 1994.

Gubbins, D., and R. Snieder, Dispersion of P waves in subducted lithosphere evidence for an eclogite layer, *J. Geophys. Res.*, *96*, 6321–6333, 1991.

Hacker, B. R., Eclogitization and the rheology, buoyancy, seismicity, and H_2O content of oceanic crust, *J. Geophys. Res.*, submitted, 1995.

Hacker, B., S. Bohlen, and S. Kirby, Time and metamorphic petrology: Calcite to aragonite experiments, *Science*, *258*, 110–112, 1992.

Hasegawa, A., S. Horiuchi, and N. Umino, Seismic structure of the northeastern Japan convergent margin: A synthesis, *J. Geophys. Res.*, *99*, 22,295–22,311, 1994.

Helffrich, G., and S. Stein, Study of the structure of the slab-mantle interface using reflected and converted seismic waves, *Geophys. J. Int.*, *115*, 14–40, 1993.

Helffrich, G. R., S. Stein, and B. J. Wood, Subduction zone thermal structure and mineralogy and their relationship to seismic wave reflections and conversions at the slab/mantle interface, *J. Geophys. Res.*, *94*, 753–763, 1989.

Hori, S., Seismic waves guided by untransformed oceanic crust subducting into the mantle: The case of the Kanto district, central, Japan, *Tectonophysics*, *176*, 355–376, 1990.

Hori, S., H. Inoue, Y. Fukao, and M. Ukawa, Seismic detection of the untransformed "basaltic" oceanic crust subducting into the mantle, *Geophys. J. Royal Astron. Soc.*, *83*, 169–197, 1985.

Huang, P. Y., S. C. Solomon, E. A. Bergman, and J. L. Nabelek, Focal depths and mechanisms of mid-Atlantic ridge earthquakes from body waveform inversion, *J. Geophys. Res.*, *91*, 579–598, 1986.

Hurukawa, N., and M. Imoto, Subducting oceanic crusts of the Philippine Sea and Pacific Plates, *Geophys. J. Int.*, *109*, 639–652, 1992.

Hyndman, R. D., and K. Wang, Thermal constraints on the zone of major thrust earthquake failure: The Cascadia subduction zone, *J. Geophys. Res.*, *98*, 2039–2060, 1993.

Iidaka, T., and K. Obara, The upper boundary of the subducting Pacific plate estimated from S_cS_p waves beneath the Kanto Region, Japan, *J. Phys. Earth*, *41*, 103–108, 1993.

Isacks, B. L., and M. Barazangi, Geometry of Benioff zones: Lateral segmentation and downwards bending of the subducted lithosphere, *AGU Ewing Ser.*, *1*, 99–114, 1977.

Kay, M. S., V. A. Ramos, and M. Marquez, Evidence in Cerro Pampa volcanic rocks for slab-melting prior to ridge-trench collision in southern South America, *J. Geology*, *101*, 703–714, 1993.

Kincaid, C., Subduction dynamics: From the trench to the core-mantle boundary, *Rev. Geophys. Suppl. U.S. National Report to International Un. of Geodesy and Geophys*, *1991–1994*, 401–412, 1995.

Kirby, S. H., Intraslab earthquakes and phase changes in subducting lithosphere, *Rev. Geophys. Supplement, U.S. National Report to the I.U.G.G. 1990–1994, Dynamics of the Solid Earth and Other Planets*, 287–297, 1995.

Kirby, S., and R. Engdahl, Curvilinear belts of intermediate-depth earthquakes in the Nazca slab: An expression of the thermomechanical effects of deeply subducted volcanic island-and-seamount chains?, *Eos, Trans. Am. Geophys. Un.*, *74*, 92, 1993.

Kirby, S. H., W. B. Durham, and L. A. Stern, Mantle phase changes and deep-earthquake faulting in subducting lithosphere, *Science*, *252*, 216–225, 1991.

Kirby, S., E. Engdahl, and E. Okal, The Juan Fernandez earthquake zone of central Chile and Argentina at 34 to 30.5°S: Fine structure and evidence for reactivation of shallow intraplate faults at intermediate depths, *Trans. Amer. Geophys. Union*, Spring 1996 Meeting Abstr., S275, 1996a.

Kirby, S., S. Stein, E. Okal, and D. Rubie, Metastable mantle phase changes and deep earthquakes in subducting oceanic lithosphere, *Rev. Geophys.*, in press, 1996b.

Kissling, E., and J. C. Lahr, Tomographic image of the Pacific slab under southern Alaska, *Eclogae Geologicae Helvetiae, 84*, 297-315, 1991.

Kobayashi, K., and others, Normal faulting of the Daiichi-Kashima Seamount in the Japan trench revealed by the Kaiko I cruise, Leg. 3, *Earth Planet. Sci. Lett., 83*, 257-266, 1987.

Kostoglodov, V., Maximum depth of earthquake and phase transformation within the lithospheric slab descending in the mantle, in *Physics and Interior Structure of the Earth*, edited by V. A. Magnitsky, pp. 52-57, Nauka, Moscow, 1989.

Kostoglodov, G., and W. Bandy, Seismotectonic constraints on the convergence rate between the Rivera and North American plates, *J. Geophys. Res., 100*, 17,977-17,989, 1995.

Kushiro, I., On the lateral variation in chemical composition and volume of Quaternary volcanic rocks across Japanese arcs, *J. Volcanol. Geotherm. Res., 18*, 435-447, 1983.

Ludwin, R. S., C. S. Weaver, and R. S. Crosson, Seismicity of Washington and Oregon, in *Neotectonics of North America*, edited vol. Decade of North American Geology Map Volume, pp. 77-97, GSA, Boulder, CO, 1991.

Maldonado, A., R. D. Larter, and F. Aldaya, Forearc tectonic evolution of the South Shetland margin, Antarctic Peninsula, *Tectonics, 13*, 1345-1370, 1994.

Mallet, R., Seismographic map of the world showing the surface distribution in space as discussed from the British Association catalogue, Plate 11 in Fourth report on the facts of earthquake phenomena, Report of the 28th Meeting (1858) of the British Association for the Advancement of Science, pp. 1-136, 1859.

Masson, D. G., Fault patterns at outer trench walls, *Marine Geophysical Researches, 13*, 209-225, 1991.

Masson, D. G., L. M. Parson, J. Milsom, G. Nichols, N. Sikumbang, B. Dwiyanto, and H. Kallagher, Subduction of seamounts at the Java Trench: A view with long-range sidescan sonar, *Tectonophysics, 185*, 51-65, 1990.

Matsuzawa, T., T. Kono, A. Hasegawa, and A. Takagi, Subducting plate boundary beneath the northern Japan arc estimated from SP converted waves, in *Tectonics of Eastern Asia and Western Pacific Continental Margin*, edited by M. Kono, and B. C. Burchfiel, vol. 181, pp. 123-133, Tectonophysics, 1990.

Matsuzawa, T., N. Umino, A. Hasegawa and A. Takagi, Normal fault type events in the upper plane of the double-planed deep seismic zone beneath the northeastern Japan arc, *J. Phys. Earth, 34*, 85-94, 1986.

Matsuzawa, T., N. Umino, A. Hasegawa, and A. Takagi, Estimation of thickness of a low-velocity layer at the surface of the descending oceanic plate beneath the Northeastern Japan arc by using synthesized PS-wave, *Tohoku Geophys. J., 31*, 19-28, 1987.

McCrory, P. A., D. S. Wilson, and M. H. Murray, Modern plate motions in the Mendocino Triple Junction region: Implications for partitioning of strain, Fall 1995 Meeting Supplement, *Eos, Trans. Am. Geophys. Union, F630*, 1995.

McGarr, A., Seismic moments of earthquakes beneath island arcs, phase changes and subduction velocities, *J. Geophys. Res., 82*, 256-264, 1977.

McKenzie, D. P., Speculations on the consequences and causes of plate motions, *Geophys. J. Royal Astro. Soc., 18*, 1-32, 1969.

McKenzie, D. P., Temperature and potential temperature beneath island arcs, *Tectonophysics, 10*, 357-366, 1970.

Meade, C., and R. Jeanloz, Deep-focus earthquakes and recycling of water into the Earth's mantle, *Science, 252*, 68-72, 1991.

Mevel, C., and M. Cannat, Lithospheric stretching and hydrothermal processes in oceanic gabbros from slow-spreading ridges, in *Ophiolite Genesis and Evolution of the Oceanic Lithosphere*, edited by T. J. Peters, A. Nicolas, and R. G. Coleman, pp. 293-312, Kluwer Academic Publishers, 1991.

Merthelot, J.-M., J.-L. Chatelain, B. L. Isacks, R. K. Cardwell, and E. Coudert, Seismicity and attenuation of the central Vanuatu (New Hebrides) Islands, *J. Geophys. Res., 90*, 8641-8650, 1985.

Milne, J., *Earthquakes and Other Earth Movements*, Appleton and Company International Scientific Series, map insert and pp. 227-228, 1886.

Mitronova, W., and B. L. Isacks, Seismic velocity anomalies in the upper mantle beneath the Tonga-Kermadec island arc, *J. Geophys. Res., 76*, 7154-7180, 1971.

Molnar, P., D. Freedman, and J. S. F. Shih, Lengths of intermediate and deep seismic zones and temperatures in downgoing slabs of lithosphere, *Geophys. J. Roy. Astron. Soc., 56*, 41-54, 1979.

Morris, P. A., Slab melting as an explanation of Quaternary volcanism and aseismicity in southwest Japan, *Geology, 23*, 395-398, 1995.

Mueller, R.D., and four others, A digital age map of the ocean floor, *Scripps Institute of Oceanography Map 93-30*, 1993.

Muffler, L. J. P., and S. Tamanyu, Tectonic, volcanic, and geothermal comparison of the Tohoku volcanic arc (Japan) and the Cascade volcanic arc (USA), in *Proc. World Geothermal Congress*, Florence, Italy, pp. 725-730, International Geothermal Assoc., 1995.

Mutter, J. C., and J. A. Karson, Structural processes at

slow-spreading ridges, *Science, 257*, 627–634, 1992.

Nicolas, A., *Mid-Ocean Ridges: Mountains Below Sea Level*, 200 pp., Springer-Verlag, 1995.

Nur, A., and Z. Ben-Avraham, Volcanic Gaps and the consumption of aseismic ridges in South America, in *Nazca Plate: Crustal Formation and Andean Convergence*, edited by L. D. Kulm, J. Dymond, E. J. Dasch, D. M. Hussong and R. Roderick, vol. Memoir 154, pp. 729–740, GSA, Boulder, CO, 1981.

Oda, H., T. Tanaka, and K. Seya, Subducting oceanic crust on the Philippine Sea plate in southwest Japan, *Tectonophysics, 172*, 175–189, 1990.

Pardo, M., and G. Suarez, Steep subduction geometry of the river plate beneath the Jalisco block in western Mexico, *Geophys. Res. Lett., 20*, 2391–2394, 1993.

Pardo, M., and G. Suarez, Shape of the subducted Rivera and Cocos plates in southern Mexico: Seismic and tectonic implications, *J. Geophys. Res., 100*, 12,357–12,373, 1995.

Paterson, M. S., *Experimental Rock Deformation—The Brittle Field*, 254 p., Springer-Verlag, Berlin, 1978.

Peacock, S., Blueschist-facies metamorphism, shear heating and the P-T-t paths in subduction shear zones, *J. Geophys. Res., 97*, 17,693–17,707, 1992.

Pennington, W. D., Role of shallow phase changes in the subduction of oceanic crust, *Science, 220*, 1045–1046, 1983.

Protti, M., F. Gundel and K. McNally, The geometry of the Wadati-Benioff zone under southern Central America and its tectonic significance: Results from a high-resolution local seismographic network, *Physics of the Earth and Planetary Interiors, 84*, 271–287, 1994.

Raleigh, C. B., Tectonic implications of serpentinite weakening, *Geophys. J. Roy. Astro. Soc., 14*, 113–118, 1967.

Raleigh, C. B., and M. S. Paterson, Experimental deformation of serpentine and its tectonic implications, *J. Geophys. Res., 70*, 3965–3985, 1965.

Ringwood, A. E., Composition and evolution of the upper mantle, *Geophysical Monograph, 13*, 1–17, AGU, 1969.

Ringwood, A. E., and D. H. Green, An experimental investigation of the gabbro-eclogite transformation and some geophysical implications, *Tectonophysics, 3*, 383–427, 1966.

Ruff, L., and H. Kanamori, Seismic coupling and uncoupling at subduction zones, *Tectonophysics, 99*, 99–117, 1983.

Sacks, I. S., The subduction of young lithosphere, *J. Geophys. Res., 88*, 3355–3366, 1983.

Sandwell, D. T., W. H. F. Smith, and M. M. Yale, Gravity Anomaly Map from ERS-1, Topex and Geosat Altimetry, *Scripps Institute of Oceanography*, 1995.

Savage, J., The mechanics of deep-focus faulting, *Tectonophysics, 8*, 115–127, 1969.

Seno, T., and D. G. Gonzalez, Faulting caused by earthquakes beneath the outer slope of the Japan trench, *Journal of Physics of the Earth, 35*, 381–407, 1987.

Silver, P., and six others, Rupture characterisitics of the deep Bolivian earthquake of 9 June 1994 and the mechanism of deep-focus earthquakes, *Science, 268*, 69–73, 1995.

Simkin, T., and L. Siebert, Explosive eruptions in space and time: Durations, intervals, and a comparison of the world's active volcanic belts, in *Explosive Volcanism: Inception, Evolution, and Hazards*, pp. 110–121, National Academy Press, Wash., D.C., 1984.

Simkin, T., and L. Siebert, *Volcanoes of the World*, 349 pp., Geoscience Press, Tucson, Arizona, 1994.

Simkin, T., L. Siebert, L. McClelland, D. Bridge, C. Newhall, and J. H. Latter, *Volcanoes of the World*, 240 pp., Hutchinson Ross, Stroudsburg, PA, 1981.

Snoke, J., Sacks, I. S., and H. Okada, Determination of the subducting lithosphere boundary by use of converted phases, *Bull. Seismol. Soc. Am., 67*, 1051–1060, 1977.

Spakman, W., S. Stein, R. van der Hilst, and R. Wortel, Resolution experiments for NW Pacific subduction zone tomography, *Geophys. Res. Lett., 16*, 1097–1100, 1989.

Spence, W., Slab pull and the seismotectonics of subducting lithosphere, *Rev. Geophy., 25*, 55–69, 1987.

Stein, C. A., and S. Stein, A model for the global variation in oceanic depth and heat flow with lithospheric age, *Nature, 359*, 123–129, 1992.

Stein, C. A., and S. Stein, Comparison of plate and asthenosphere flow models for the thermal evolution of oceanic lithosphere, *Geophys. Res. Lett., 21*, 709–712, 1994.

Stein, C. A., S. Stein, and A. M. Pelayo, Heat flow and hydrothermal circulation, in *Seafloor Hydrothermal Systems*, Geophys. Monogr. 91, edited by S.E. Humphris, R. A. Zierenberg, L.S. Mullineaux, and R.E. Thomson, pp. 425–445, AGU, 1995.

Stolper, E., and S. Newman, The role of water in the petrogenesis of Mariana trough magmas, *Earth and Planet. Sci. Lett., 121*, 293–325, 1994.

Sykes, L. R., The seismicity and deep structure of island arcs, *J. Geophys. Res., 71*, 2981–3006, 1966.

Tatsumi, Y., Formation of the volcanic front in subduction zones, *Geophys. Res. Lett., 13*, 717–720, 1986.

Tatsumi, Y., Migration of fluid phases and genesis of basalt magmas in subduction zones, *J. Geophys. Res., 94*, 4697–4707, 1989.

Tatsumi, Y., and S. Eggins, *Subduction Zone Magmatism*,

Blackwell Science, Frontiers in Earth Science Series, 211 p., 1995.

Tatsumi, Y., H. Sakuyama, H. Fukuyama, and I. Kushiro, Generation of arc basalt magmas and thermal structure of mantle wedge in subduction zones, *J. Geophys. Res., 88*, 5815-5825, 1983.

Toksöz, M. N., J. W. Minear, and B. R. Julian, Temperature field and geophysical effects of a downgoing slab, *J. Geophys. Res., 76*, 1113-1138, 1971.

Ukawa, M., Lateral stretching of the Philippine Sea plate subducting along the Nankai-Suruga Trough, *Tectonics, 1*, 543-571, 1982.

Ulmer, P., and V. Trommsdorff, Serpentine stability to mantle depths and subduction-related magmatism, *Science, 268*, 858-859, 1995.

Vallier, T., and nine others, Subalkcaline andesite from the Valu Fa Ridge, a back-arc spreading center in the southern Lau Basin, *Chemical Geology, 91*, 227-256, 1991.

van der Hilst, R., R. Engdahl, W. Spakman, and G. Nolet, Tomographic imaging of subducted lithosphere below northwest Pacific island arcs, *Nature, 353*, 37-43, 1991.

Vassiliou, M. S., and B. H. Hager, Subduction zone earthquakes and stress in slabs, *PAGEOPH, 128*, 547-624, 1988.

Visser, S. W., Some remarks on the deep-focus earthquakes in the International Seismological Summary, *Gerlands Beiträge zur Geophysik, 48*, 254-267, 1936.

Verplanck, E. P., and R. A. Duncan, Temporal variations in plate convergence and eruption rates in the Western Cascades, Oregon, *Tectonics, 6*, 197-209, 1987.

von Huene, R., Subduction of the Juan Fernandez Ridge beneath the Chile continental margin—results of Cruise SO-101, *Eos, Trans. Am. Geophys. Union*, Fall 1995 Meeting Supplement, F374, 1995.

von Huene, R., and D. W. Scholl, Observations at convergent margins concerning sediment subduction, subduction erosion and the growth of continental crust, *Rev. Geophysics, 29*, 279-316, 1991.

von Huene, R., and six others, Subduction of the Juan Fernandez Ridge beneath the Chile continental margin, *Tectonics*, (under review), 1996.

Wadati, K., On the activity of deep-focus earthquakes in the Japanese Islands and neighborhoods, *Geophysical Magazine, 8*, 305-325, 1935.

Wang, K., R. D. Hyndman, and M. Yamano, Thermal regime of the Southwest Japan subduction zone: Effects of age history of the subducting plate, *Tectonophysics, 248*, 53-69, 1995.

Wannamaker, P. E., J. R. Booker, J. H. Filloux, and others, Magnetotelluric observations across the Juan de Fuca subduction system in the EMSLAB project, *J. Geophys. Res., 94*, 14,111-14,125, 1989.

White, R. S., D. McKenzie, and R. K. O'Nions, Oceanic crustal thickness from seismic measurements and rare earth element inversions, *J. Geophys. Res., 97*, 19,683-19,715, 1992.

Wilson, D. S., Tectonic history of the Juan de Fuca ridge over the last 40 million years, *J. Geophys. Res., 93*, 11,863-11,876, 1988.

Wilson, D. S., Confidence intervals for motion and deformation of the Juan de Fuca plate, *J. Geophys. Res., 98*, 16,053-16,071, 1993.

Wysession, M. E., E. A. Okal, and K. L. Miller, Intraplate seismicity of the Pacific Basin, 1913-1988, *Pure Appl. Geophys., 135*, 261-359, 1991.

Zhao, D., D. Christensen, and H. Pulpan, Tomographic imaging of the Alaska subduction zone, *J. Geophys. Res., 100*, 6487-6504, 1995.

Zhao, D., K. Wang, and G. Rogers, Mapping seismic velocity anomalies in the northern Cascadia subduction zone, *Geophysical J. International*, in press, 1996.

Zhao, D., A. Hasegawa, and H. Kanamori, Deep structure of Japan subduction zone as derived from local, regional, and teleseismic events, *J. Geophys. Res., 99*, 22,313-22,329, 1994.

---

[1]S. Kirby, U.S. Geological Survey, 345 Middlefield Road, MS/977, Menlo Park, CA 94025. e-mail: skirby@isdmnl.wr.usgs.gov

[2]E. R. Engdahl, U.S. Geological Survey, Denver Federal Center, MS/967, Box 25046, Denver, CO 80225

[3]R. Denlinger, U.S. Geological Survey, Hawaii Volcano Observatory, P.O. Box 51, Hilo, Hawaii 96718

Subducted Lithospheric Slab Velocity Structure: Observations and Mineralogical Inferences

George Helffrich

Geology Dept., U. Bristol, Wills Memorial Building, Queens Road, Bristol BS8 1RJ UK

A variety of observational data bearing on slab seismic wave speed structure yields slab anomalies between 4 and 10 per cent relative to ambient mantle. There is also evidence for dip-parallel layering of velocities in slabs, whose thickness is ≤10 km. After a review of the observations, the mineralogical composition and the transformations in the subducted lithosphere are related to the observations. It appears that 1) transformations are taking place in the slab, because slab/mantle velocity contrasts do not monotonically decrease as a thermal origin for these would predict; and 2) velocity layering in the slab must persist to depths in excess of 200 km. Layering persistence implies that wholesale delamination or underplating of slab material does not occur in all subduction zones, and suggests that the oceanic lithosphere subducts coherently. Velocity contrasts in hydrous metabasalts appear large enough to explain low velocity layers in the shallow slabs, implying that the metastable anhydrous gabbro→eclogite reaction may not be a good model for changes in shallow slab mineralogy.

INTRODUCTION

Though subducted slabs are delineated by their seismicity, they also influence the travel times and paths taken by seismic waves from the earthquake source to the seismic station that records their arrival. That slabs affect travel times was an early realization in the plate tectonic paradigm [*Davies and McKenzie*, 1969]. It took a bit longer to recognize the alternative paths that seismic waves could take on account of slabs [*Mitronovas and Isacks*, 1971] and to relate these to the material properties of subducted lithosphere [*Sleep*, 1971]. However, the seismic wave anomalies due to slabs were ultimately appreciated for the information they bear concerning slab mineralogy [*Solomon and U*, 1975], which is their main contemporary interest.

This information is carried in three different ways. Most basic is the travel time of the wave, which, if the path is known, can be turned into an average path velocity or velocity anomaly relative to some reference velocity. A typical choice is a whole-earth seismic velocity model. The second way to garner slab information from seismic waves is to search for secondary arrivals that would not otherwise exist except for the presence of a slab. Discontinuities in a rock's elastic properties generate S waves from P and vice-versa, and slab properties differ greatly from mantle ones. Thus, by inspecting seismograms after P arrivals (when slab-generated P→S arrivals would be expected) and before S arrivals (ditto for S→P), the size of the discontinuity and its location may be inferred. Finally, frequency-dependent effects arise when seismic waves travel along or across layers. If a seismic wave crosses a layer stack, wavelengths that are multiples of the layer spacing will be either enhanced or suppressed, changing the form of the wave emerging from the stack. Running parallel through the stack, longer wavelengths will see an average velocity through the stack (or even outside of it), whereas shorter wavelengths will be confined to narrower regions of the stack and prefer its faster layers, with corresponding changes to its waveform.

All three observational types trade off advantages with disadvantages. Travel times are fundamental, simple to recognize and measure, but are along-path averages and thus smooth out speed variations, which are of central interest. On the other hand, wave conversions are not always seen in one location, nor are they even seen everywhere. Thus the information they give is spotty and arguably biased, representing only an exceptional coincidence of favorable factors. Frequency dependent effects, being intrinsically noisy, are difficult to quantify in detail and only broad features in the spectrum are interpretable. Layer-parallel paths also suffer the same smoothing penalty mentioned earlier.

What seismological investigation provides is a velocity (or an anomaly) in some part of the slab, or a layer thickness. Mineralogical inferences may only be made by comparison with computed velocities. Therefore some idea of slab mineralogies must be to hand for the final step in interpreting the seismic observations mineralogically, which is the goal of this work. This will be deferred until the observational data is discussed.

SEISMOLOGICAL OBSERVATIONS

All the geometries discussed here involve stations sited above subduction zones. Earthquakes either in the slab below the station or at some distant location emit waves that interact with the slab as they ascend to the surface. The interactions are generally one of either reflection from the face of the subducted slab, or refraction at its surface, or lengthy along-strike or up-dip paths taken in the slab, with the slab acting as a wave guide.

All of Japan overlies active subduction, so Japanese subduction zones provide most of the recorded instances of the slab interactions. Slab face reflection is the rarest reported type (Fig. 1a). *Fukao et al.* [1978] observed a distinctive pattern of later arrivals following the direct P wave from shallow earthquakes between central Japan and Taiwan. To be visible, the wave must graze the slab face so as to cause total internal reflection and no transmission through the slab. Thus the observations provide a constraint on the material property contrast between the mantle and the subducted slab, in this case 4-10% change in wave speeds in a region 10-20 km thick. This estimate depends on knowing the interaction geometry, but the source of the constraint is essentially geometric.

More common are observations of refracted waves converted from S→P or P→S at the slab interface. Both the geometry and the relative amplitudes of the P and S waves constrain the properties of the conversion interface. *Mitronovas and Isacks* [1971] inferred P and S velocity contrasts of ~7% at the interface from P→S conversions of deep earthquakes in Tonga (Fig 1b). *Matsuzawa et al.* [1986; 1987], used these to map out a low velocity layer below the surface of the slab in northern Honshu, Japan (Tohoku district). The layer is about 6% slower than the

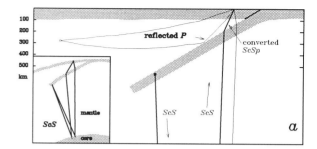

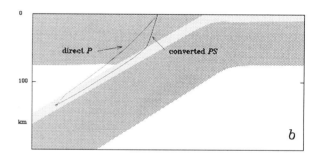

Figure 1. Ray geometries for studies discussed here. (a) P reflector and ScSp geometries. (b) PS conversion geometry.

non-slab mantle at the same depth, 12% slower than the slab, and is about 7 km thick. S→P conversions at the surface of the subducted slab of core-reflected S waves from nearby earthquakes (ScSp) are common and yield local estimates of velocity contrast at the slab interface [*Okada*, 1977; *Snoke et al.*, 1978; *Nakanishi et al.*, 1981; *Iidaka and Obara*, 1993; *Helffrich and Stein*, 1993]. The P amplitude, compared to the amplitude of the S wave whose refraction generated P, implies contrasts between 5-10% at the interface (Fig. 1a). A valuable characteristic of ScSp observations stems from the interaction geometry: at stations farther from the trench along its down-dip side, the S→P conversion occurs deeper in the subduction zone. Thus stepping away from the trench, the change in character with depth of the converter may be examined. In the NE Japan subduction zone, amplitudes vary non-monotonically. Shallow amplitudes are high, intermediate are lower, and deeper ones increase, but not to values seen at the shallowest conversion depths (Figure 2).

Along-strike or up-dip paths provide the longest interaction times with the slab, yielding potentially large slab-related signals, which can be frequency dependent. Observing secondary arrivals devoid of low frequency components in Tonga, *Barazangi et al.* [1972] attributed them to waves confined to the high-velocity subducted lithosphere. Similar anomalous high-frequency arrivals are also observed in S. America [*Snoke et al.*, 1974],

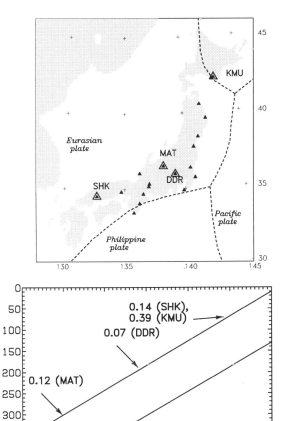

Figure 2. Sites where ScSp is observed in Japan and observed ScSp/ScS amplitude ratios at various depths along the slab-mantle interface. ScS is converted to P at ~ 60 km depth at both KMU and SHK; at DDR the conversion depth is ~ 180 km, and at MAT the conversion depth is ~ 300 km. Amplitude ratios measured from seismograms and corrected for slab focusing effects and differential attenuation [*Helffrich and Stein*, 1993].

than ambient mantle, whereas in Japan and Alaska it is slower. In all cases, the layering is on a scale ≤10 km.

Travel through a layer of low velocity material explains shallow earthquake travel times along strike-parallel paths < 60 km deep in central Japan [*Hori et al*, 1985], with the layer being 8% slower than mantle at the same depth. The up-dip travel times observed by *Suyehiro and Sacks* [1979] in the upper 350 km of the slab led them to propose a two-layer slab model with an upper layer 40-50 km thick and ~1% slower in P speed than the remainder of the slab (a finding in detail at odds with the Matsuzawa studies cited earlier, but possibly a consequence of extending deeper in the slab). The slab/mantle velocity contrast they found to be 5±1% in P and 7±2% in S.

MECHANISMS

The slab of subducted lithosphere differs in temperature, bulk composition and detailed mineralogy from the mantle, any of which may affect slab wave speeds. Slab thermal gradients are largest perpendicular to its surface, differing from mantle temperatures by as much as 400°C 10 km into the slab interior [*Helffrich et al.*, 1989]. In this direction, a succession of mineralogies is encountered. For the moment confining our attention to anhydrous bulk compositions, these dominantly contain mixtures of the phases olivine (ol), orthopyroxene (opx), clinopyroxene (cpx), plagioclase (pl), spinel (sp) and garnet (gt). Outside the slab, the mantle is garnet or spinel lherzolite (ol+opx+cpx±sp±garnet; ± here indicates a phase that may or may not be present). If the top of the subducted plate resembles the oceanic lithosphere at all, it should have a layer of gabbro (basalt) overlying a harzburgite residuum from the mid-ocean ridge differentiation event. Below this lies fertile (undifferentiated) mantle, resembling what overlies the slab. Thus, differentiation generates bulk-compositional and mineralogical differences in the oceanic lithosphere which are carried into the mantle via subduction.

With depth, gabbro (ol+pl+cpx) will convert to eclogite as increasing pressure destabilizes plagioclase and a succession of aluminum bearing phases (spinel and then garnet+clinopyroxene) accommodate this element [*Ringwood*, 1972; *Ahrens and Schubert*, 1975]. Harzburgite (ol+opx±cpx) has little aluminum which is borne by clinopyroxene. With increasing depth, both opx and cpx are incorporated into garnet and disappear by about 450 km [*Bina and Wood*, 1984]. Meanwhile, olivine undergoes a succession of isochemical phase transformations from α-olivine through β-modified-spinel to γ-spinel [*Akimoto and Fujisawa*, 1968]. These transformations are both pressure and temperature dependent and occur at shal-

apparently associated with deep earthquakes but not shallow ones in both regions. In Tonga, high frequency energy arrives early along strike-parallel paths from earthquakes in N. Tonga recorded in New Zealand [*Ansell and Gubbins*, 1986]. The waves' lower frequency components presumably travel more slowly because the longer wavelength's speeds are influenced by the slower mantle outside of the slab. They suggest 8-10 km thick layering in the slab [*Gubbins and Snieder*, 1991; *Gubbins et al.*, 1994]. In Alaska low frequency energy arrives before higher from earthquakes 100-150 km depth, suggesting 2-10 km layer thicknesses [*Abers and Sarker*, 1995]. Thus, the layer speed in the Tonga slab is faster lower depths where the slab is cooler [*Turcotte and Schubert*, 1971].

By calculating seismic wave speeds in these different mineralogies it can be shown that neither temperature nor bulk-compositional difference alone or in combination can generate contrasts as large as those observed [*Helffrich et al.*, 1989]. A purely thermal origin for the contrasts is ruled out by the observed ScSp variation with depth summarized earlier. Thus, the mineralogical changes appear to dominate velocity contrasts observed seismically.

Calculated velocities in a gabbroic layer in the shallow parts of subduction zones yield contrasts in agreement with both the low velocity layers and the amplitudes of ScSp generated at these depths [*Helffrich and Stein*, 1993]. The derived layer thicknesses summarized earlier are virtually all ≤10 km, corresponding to oceanic crustal thickness estimates [*Fox and Stroup*, 1981], strongly implying that the crust's initial layering persists to significant depths, at least 200 km and probably deeper. If underplating of slab material to the overriding plate occurs, it must dominantly involve sediments on top of the oceanic crust, or must operate discontinuously during subduction. This is consistent with inferences drawn from exposure patterns of high-pressure metamorphic rocks [*Platt*, 1993]. Being present in lavas erupted above subduction zones as well as in near-trench sediments approaching them, short-lived cosmogenic isotopes require sediment subduction [*Morris*, 1991], further suggesting that the oceanic crust is subducted intact.

THE GABBRO-ECLOGITE TRANSITION IN SLABS AT SHALLOW DEPTHS

The succession of mineralogical transformations in slab mineralogies not only affect velocities but also density, and thus the forces driving subduction. The shallowest of these is the so-called gabbro to eclogite transformation. There is about a 15% density increase during this process, roughly modeled as $(Mg,Fe)_2SiO_4$ (olivine) + $CaAl_2Si_3O_8$ (anorthite) → $Ca(Mg,Fe)_2Al_2Si_4O_{12}$ (garnet) [*Ahrens and Schubert*, 1975]. The slab buoyancy change this entails may affect slab coupling to the overriding plate and thus the maximum depth extent of earthquakes caused by frictional sliding on faults [*Tiechlaar and Ruff*, 1993] by physically drawing down the slab away from the overriding plate, and may also be a source of seismicity in slabs [*Kirby et al.*, 1995], but thermodynamically predicted transformation depths for this reaction are quite shallow, about 20 km [*Wood*, 1987]. There are good reasons for this reaction to proceed slowly at the low temperatures in subduction zones [*Ahrens and Schubert*, 1975], permitting the low pressure, and low seismic velocity, mineralogy to persist to greater depths during subduction. (This assumption underlies the gabbro layer computation described above.) But while temperatures at shallow subduction levels are indeed low, oceanic crust samples recovered from dredges and drill cores are hydrated and serpentinized [*Gubbins et al.*, 1994; *Cannat et al.*, 1995]. One thus may question the assumption that dry conditions prevail during the gabbro-eclogite transformation, and indeed whether this is even the appropriate model reaction to consider when seeking an explanation for the velocity layering in shallow slabs, since the initially subducted material contains hydrous minerals.

SEISMIC VELOCITIES IN METABASALTS

Peacock [1993] examined the phase relations in the hydrated metabasalt system which represents subducted oceanic crust. He opted for a facies approach to the problem of characterizing the mineralogy in this system (Table 1) on account of the abundance of specific mineralogic reactions present within it. I will adopt these facies boundaries and mineral proportions to compute seismic velocities along plausible slab P-T trajectories at shallow levels. A disadvantage of a facies-based approach is that facies classify repeatedly encountered mineral assemblages [*Turner*, 1968] which are by their nature rare for high-pressure, deeply metamorphosed rocks. This leads to an absence of recognized facies at pressures higher than about 20kb, and limits quantitative discussion of the seismic velocities to depths of about 65 km.

The elastic data and sources are listed in Table 2 (available via WWW URL http://sun1.gly.bris.ac.uk/~george/subcon.tab2.html). *Bina and Helffrich*'s [1992] methodology is used to compute the seismic wave speeds from this data. A guide to the relative velocities of the constituent minerals is given in Figure 3, which includes the hydrous phases lawsonite (Lw), glaucophane (Gp), tremolite (Tr), zoisite (Zo), the zeolites prenhite and pumpellyite (Pr, Pu), hornblende (Hbl) and chlorite (Chl). The P wave speeds in hydrous minerals span the range found in anhydrous minerals. This suggests that the seismic velocity increases in changes to facies bearing significant Zo, Gp and Lw may be as large as the one in the gabbro to eclogite transformation in anhydrous compositions. Keeping with the facies concept, the range of calculated velocities within each is shown in Figure 4. The low-P high-T facies all have velocities between 6.6 and 7 km/sec. Velocities are greater than 7.5 km/sec (maximum ~8.3 km/sec) in the blueschist and eclogite facies except in the epidote blueschist facies, where it is transitional, 7.2-7.3 km/sec. Above pressures of 10 kb, ~5% increases in P wave speeds are typical between the hydrous facies progres-

TABLE 1. Mineral Proportions

Facies	Vol. %	Mineral	Facies	Vol. %	Mineral	Facies	Vol. %	Mineral
Albite Epidote Amphibolite (EA)	58.1	Hbl	High-T Amphibolite (AM)	58.9	Hbl	Low-T Amphibolite (AM)	60.0	Hbl
	11.9	Zo		5.2	$Py_{20}Gr_{80}$		7.9	Zo
	6.0	Chl		2.1	Di		4.6	Chl
	17.1	Ab		28.6	$Ab_{40}An_{60}$		20.8	$An_{20}Ab_{80}$
	7.0	Qtz		5.2	Qtz		6.6	Qtz
High P-T Eclogite (EC)	18.5	Di	Low P-T Eclogite (EC)	25.6	Hbl	Epidote Blueschist (EB)	48.3	$Tr_{50}Gp_{50}$
	19.6	Jd		14.8	Zo		29.1	Zo
	17.7	Gr		17.7	Py		11.8	Chl
	33.5	Py		33.0	$Jd_{50}Di_{50}$		10.0	Ab
	10.8	Qtz		8.8	Qtz		0.7	Qtz
High-T Granulite (GN)	10.8	Fo	Low-T Granulite (GN)	51.8	Hbl	Greenschist (GS)	29.6	Tr
	8.7	En		6.2	En		25.2	Zo
	15.1	Di		2.5	Di		17.9	Chl
	27.8	Ab		36.5	$An_{50}Ab_{50}$		27.2	Ab
	37.6	An		3.1	Qtz		0.1	Qtz
Lawsonite Blueschist (LB)	18.1	Gp	Prenhite Actinolite	15.9	Tr	Garnet Harzburgite	65	$Fo_{90}Fa_{10}$
	28.1	Lw		30.6	Pr		23	$En_{66}Di_{29}Fs_5$
	19.1	Chl		25.9	Chl		4	$Jd_{41}Di_6En_{48}Fs_3$
	29.3	$Jd_{30}Di_{70}$		25.6	Ab		8	Py
	5.4	Qtz		2.1	Qtz			

Source: *Peacock* [1993] except for Garnet Harzburgite [*Helffrich et al.*, 1989].

sively encountered atop subducted slabs, though up to 15% variations may arise at lower pressures.

The trajectory of P-T conditions encountered at the surface of the subducted slab and the base of the oceanic crust is of interest because it controls the velocity contrast across the progressive dehydration fronts represented by metabasalt facies boundaries. Temperatures are obtained from a thermal model developed for throughgoing lithospheric thrusts [*Peacock*, 1993; *Molnar and England*, 1990], which applies to shallow subduction levels before slab heating is dominated by conduction from overlying mantle, again limiting quantitative analysis. Using a 5 km crustal thickness, an average of the layer thickness bounds yielded by observation, a thermal profile is shown for a moderate shear stress τ=0.6 kb on the plate boundary fault (Fig. 5). In order for the slab face to leave the lawsonite blueschist (LB) field above 6 kb pressure, shear stresses must be > 0.5 kb, yielding a P speed increase of 2 to 6% between the top and the bottom of the facies layering. The high end of this range involves the epidote blueschist (EB) facies, and would be even larger if increased fault shear stresses moved the slab face into the greenschist (GS) and amphibolite (EA and AM) fields. Thus contrasts in slab facies are capable of creating velocity increases in slab layers at shallow subduction levels, but require moderate heating at the fault. This may explain low velocity layering down to ~ 60 km [*Hori et al*, 1985; *Matsuzawa et al.*, 1986; 1987], but no deeper because the epidote blueschist facies pinches out (Fig. 5).

Overall, lawsonite blueschist should be the most common facies found during shallow subduction, present perhaps even at the slab surface if shear strain heating is low [*Peacock*, 1993]. Another mineralogy to consider for contrasting velocities, therefore, is the mantle peridotite (garnet harzburgite, Table 1), overlying the slab surface at shallow levels. Relative to ambient mantle at 700°C and 20 kb (65 km depth, a guess of the point of contact with the mantle wedge along the Fig. 5 slab face trajectory), lawsonite blueschist is 7% slower, and thus would constitute a low velocity layer at the top of the slab, again consistent with seismic observations. Thus, a low velocity layer would persist to the depth where blueschist transforms to eclogite by dehydrating. Lawsonite, one of the principal phases in this facies, *Pawley and Holloway* [1993] showed to be stable in metabasalt compositions to high pressures, 30 kb or 100 km depth. Also present in these experiments were a tremolitic amphibole (barroisite) and a zoisite, both relatively low velocity minerals (Figure 3), so that the contrast may be larger than 7%.

There is effectively no difference between wave speeds between eclogites and mantle peridotites [*Helffrich et al.*, 1989], so the same mineralogical composition may be

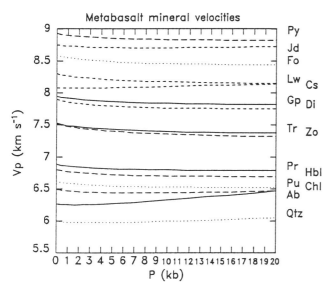
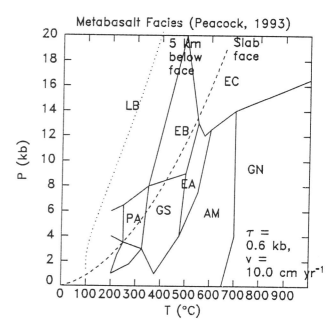

Figure 3. P wave speeds computed in metabasalt minerals along a slab face geotherm [*Peacock*, 1993], convergence rate 10 cm yr^{-1} and shear stress τ 0.66 kb. This convergence rate is characteristic of western Pacific subduction zones (Fig. 2). Phases are albite (Ab), chlorite (Chl), coesite (Cs), diopside (Di), forsterite (Fo), glaucophane (Gp), hornblende (Hbl), jadeite (Jd), lawsonite (Lw), prenhite (Pr), pumpellyite (Pu), pyrope (Py), α-quartz (Qtz), tremolite (Tr), zoisite (Zo). Velocities generally decrease and then increase because temperature and pressure both increase along the path, with competing effects on P speeds.

Figure 4. Facies boundary diagram after *Peacock* [1993], showing range of velocities encountered in each. Facies names are LB lawsonite blueschist, EB epidote blueschist, EC eclogite, GN granulite, AM amphibolite, EA epidote-amphibolite, GS greenschist, PA pumpellyite actinolite.

used to compare wave speeds between the anhydrous slab material below the hydrated metabasalt and the hydrated layer. Here there is an 8% velocity increase from lawsonite blueschist to garnet harzburgite, comparable to the difference between the hydrous metabasalt facies. This makes the upper surface of the slab slower than either ambient mantle or eclogite, constituting a low velocity layer until blueschist ceases to persist. The model reaction separating the lawsonite blueschist - eclogite facies is

$$4 \text{ glaucophane} + 3 \text{ lawsonite} = 8 \text{ jadeite} + 3 \text{ diopside} \\ + 3 \text{ garnet} + 7 \text{ quartz/coesite} + 10 \text{ water} \quad (1)$$

[*Evans*, 1990]. There is an ~6% increase in P speed as eclogite is produced (neglecting the evolved water, which would leave the reaction site). Glaucophane+lawsonite comprise about half the modal mineralogy by volume so the consequence of the reaction would be to increase speeds in the metabasalt by 3%. Nevertheless, this coesite eclogite would still be 4% slower than mantle peridotite.

Some significant uncertainties confound an attempt to extend this analysis to even higher pressures, all mineralogical. At higher pressures, sodic amphibole becomes unstable [*Pawley and Holloway*, 1993], but how this reaction relates to (1) is speculative, perhaps involving

$$\text{lawsonite} + \text{glaucophane} \rightarrow \text{jadeite} + \text{diopside} \\ + \text{chloritoid} \quad (2)$$

Secondly, the high-pressure stability limit for lawsonite itself is unknown. Finally, many dense hydrated magnesium silicate (DHMS) phases appear at low temperatures above 50 kb pressure [*Liu*, 1986] whose equilibrium phase relations are as yet unknown. While the slab face would be too hot for their liking, DHMS phases may be stable in the lowermost hydrated metabasalt layer and influence slab velocity contrasts.

Looking outside of the slab for layered structures, one again meets hydrous mineralogies. *Tatsumi et al.* [1994] attribute deep (>80 km) low-velocity layering above the slab to phlogopite and potassium amphibole (K-richterite) created by reaction of peridotite with fluids liberated by the slab. A shallow, in-slab low velocity layer progressively shifting with depth to one above the slab-mantle interface is in principle possible, and may explain the minimum in velocity contrast with depth inferred from slab refractions (Fig. 2). Ultimately, K-richterite stability controls the depth through which this might work, and experiments to establish this limit would be of interest.

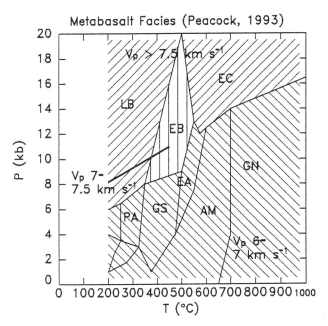

Figure 5. Facies diagram (solid lines), slab-face P-T trajectory (dashed) and similar trajectory for a point 5 km below the slab surface (dotted) for a fault shear stress of 0.6 kb, 10 cm yr^{-1} convergence rate, 0.05 W m^{-2} heat flow at base of lithosphere, and fault dip 20°, parameters describing W. Pacific subduction of ~80 Ma lithosphere [*Peacock*, 1993]. Facies diagram as in Figure 4. For τ<0.5 kb, the slab face P-T path is to the left of the EB field, whereas for τ>0.65 kb it passes through EA+GS. 5 km into the slab remains in LB, independent of velocity, unless τ is large, >0.8 kb. Even for τ=1.0 kb, a path 6 km into the slab still lies in LB. Velocity contrasts between the top of the slab and the bottom of the metabasalt layer involve velocities calculated at corresponding points along the dotted and dashed lines, offset by 1.4 kb in pressure. Contrasts between metabasalt and slab are calculated along the dotted line, and contrasts between the slab and the mantle are calculated at the upper end of the dashed line. Mantle mineralogy is garnet harzburgite (Table 1).

DISCUSSION

The seismic velocity contrast between the subducted slab and the overriding mantle wedge varies non-monotonically with depth. This suggests that the velocity contrast does not arise on account of the temperature difference between the cold slab and the hot mantle because the slab should only warm with continued subduction and the velocity contrast should only decrease. Changes in slab mineralogy within the layering developed prior to subduction satisfactorily explains the weakening of seismic velocity contrast with depth.

A variety of seismological observations from Tonga, Japan and Alaska suggest that the subducted slab is layered. The evidence for layering is in observations of shallow, low-velocity waveguides as well as deep high velocity ones, both about the thickness of the oceanic crust. A straightforward interpretation of these observations would be that the oceanic crust is typically subducted whole.

Oceanic crust returns to the mantle during subduction as gabbro/basalt. At some stage, its reversion to a higher pressure mineral assemblage is expected. While the primary candidate for this phase change has been the gabbro to eclogite transition, it appears improbable that this, or any other anhydrous reaction approximates reality during subduction. Hydrous phases most likely dominate the elastic and chemical evolution of the slab as it subducts. The computations presented here indicate that the seismic velocity contrasts observed at shallow levels of subduction are consistent with those computed in metabasalt mineralogies. If there is significant shear heating at the slab surface, the juxtaposition of amphibolite facies rocks with lawsonite blueschist can yield velocity contrasts up to 14%. This mechanism can only operate to pressures of ~14 kb (~45 km depth) however. With less shear heating, contrasts of 6-8% between metabasalt mineralogies sandwiched between the mantle and the anhydrous slab can be developed, and they would decay with depth to 3%. Thus, there is no need to appeal to dry or cool conditions in the shallow part of the slab to hinder the phase transformations there, which is a central element in some ideas concerning slab seismicity at intermediate (70-350 km) depths [*Kirby et al.*, 1995]. Moreover, major element differences in oceanic lithosphere are not required in eclogite if it does not represent the observed shallow low-velocity layers seen seismically [*Gubbins et al.*, 1994].

Acknowledgements. I thank Satoshi Kaneshima and Takashi Iidaka for their help in retrieving seismograms from the archives of the Earthquake Research Institute, Tokyo, and Selwyn Sacks for encouragement and patience. The manuscript benefited from thoughtful reviews by Tetsuzo Seno and Geoff Abers. Part of this work was done while a postdoctoral fellow at the Carnegie Institution of Washington.

REFERENCES

Abers, G., and G Sarker, Frequency dependent propagation of regional body waves and the nature of subducted crust at 100-150 km depths beneath Alaska (abstract), *XXI IUGG Gen. Ass. Abst.*, A470, 1995.

Ahrens, T. J., and G. Schubert, Gabbro-eclogite reaction rate and its geophysical significance, *Rev. Geophys.*, 13, 383–400, 1975.

Akimoto, S., and H. Fujisawa, Olivine-spinel solid solution equilibria in the system Mg_2SiO_4–Fe_2SiO_4, *J. Geophys. Res.*, 73, 1467–1479, 1968.

Ansell, J. H., and D. Gubbins, Anamalous high-frequency wave propagation from the Tonga-Kermadec seismic zone to New Zealand, *Geophys. J. R. astron. Soc., 85*, 93–106, 1986.

Barazangi, M., B. Isacks, and J. Oliver, Propagation of seismic waves through and beneath the lithosphere that descends under the Tonga island arc, *J. Geophys. Res., 77*, 952–958, 1972.

Bina, C. R., and G. R. Helffrich, Calculation of elastic properties from thermodynamic equation of state principles, *Ann. Rev. Earth Planet. Sci., 20*, 527–552, 1992.

Bina, C. R., and B. J. Wood, The eclogite to garnetite transition -- experimental and thermodynamic constraints, *Geophys. Res. Lett., 11*, 955–958, 1984.

Cannat, M., C. Mevel, M. Maia, C. Deplus, C. Durand, P. Genge, P. Agrinier, A. Belarouchi, G. Dubuisson, E. Humler, and J. Reynolds, Thin crust, ultramafic exposures, and rugged faulting patterns at the Mid-Atlantic Ridge (22°-24°N), *Geology, 23*, 49–52, 1995.

Davies, D., and D. P. McKenzie, Seismic travel-time residuals and plates, *Geophys. J. R. astron. Soc., 18*, 51–63, 1969.

Evans, B. W., Phase relations of epidote-blueschists, *Lithos, 25*, 3–23, 1990.

Fox, P. J., and J. B. Stroup, The plutonic foundation of the oceanic crust, in *The Sea, 7*, edited by C. Emiliani, pp. 119–218, John Wiley, New York, 1981.

Fukao, Y., K. Kanjo, and I. Nakamura, Deep seismic zone as an upper mantle reflector of body waves, *Nature, 272*, 606–608, 1978.

Gubbins, D., and R. Snieder, Dispersion of P waves in subducted lithosphere: evidence for an eclogite layer, *J. Geophys. Res.*, 6321–6333, 1991.

Gubbins, D., A. Barnicoat, and J. Cann, Seismological constraints on the gabbro-eclogite transition in subducted oceanic crust, *Earth Planet. Sci. Lett., 122*, 89–101, 1994.

Helffrich, G. R., and Seth Stein, Study of the structure of the slab-mantle interface using reflected and converted seismic waves, *Geophys. J. Int., 115*, 14–40, 1993.

Helffrich, G. R., Seth Stein, and B. J. Wood, Subduction zone thermal structure and mineralogy and their relationship to seismic wave reflections and conversions at the slab/mantle interface, *J. Geophys. Res., 94*, 753–763, 1989.

Hori, S., H. Inoue, Y. Fukao, and M. Ukawa, Seismic detection of the untransformed "basaltic" oceanic crust subducting into the mantle, *Geophys. J. R. astron. Soc., 83*, 169–197, 1985.

Iidaka, T, and K Obara, The upper boundary of the subducting Pacific plate estimated from ScSp waves beneath the Kanto region, Japan, *J. Phys. Earth, 41*, 103–108, 1993.

Kirby, S., E. R. Engdahl, and R. Denlinger, Crustal slab metamorphism, intraslab earthquakes and arc volcanism (abstract), *XXI IUGG Gen. Ass. Abst.*, A470, 1995.

Liu, L, Phase transformations in serpentine at high pressures and temperatures and implications for subducting lithosphere, *Phys. Earth Planet. Inter., 42*, 255–262, 1986.

Matsuzawa, T., N. Umino, A. Hasegawa, and A. Takagi, Upper mantle velocity structure estimated from PS-converted wave beneath the north-eastern Japan Arc, *Geophys. J. R. astron. Soc., 86*, 767–787, 1986.

Matsuzawa, T., N. Umino, A. Hasegawa, and A. Takagi, Estimation of the thickness of a low-velocity layer at the surface of the descending oceanic plate beneath the Northeastern Japan arc by using synthesized PS-wave, *Tohoku Geophys. J., 31*, 19–28, 1987.

Mitronovas, W., and B. L. Isacks, Seismic velocity anomalies in the upper mantle beneath the Tonga-Kermadec island arc, *J. Geophys. Res., 76*, 7154–7180, 1971.

Molnar, P., and P. England, Temperatures, heat flux and frictional stress near major thrust faults, *J. Geophys. Res., 95*, 4833–4856, 1990.

Morris, J. D., Applications of cosmogenic ^{10}Be to problems in the earth sciences, *Ann. Rev. Earth Planet. Sci., 19*, 313–350, 1991.

Nakanishi, I., K. Suyehiro, and T. Yokota, Regional variations of amplitudes of *ScSp* phases observed in the Japanese Islands, *Geophys. J. R. astron. Soc., 67*, 615–634, 1981.

Okada, H., Fine structure of the upper mantle beneath Japanese island arcs as revealed from body wave analyses, D.Sc. thesis, Dep. of Geophys., Hokkaido Univ., 1977.

Pawley, A., and J. R. Holloway, Water source for subduction zone volcanism: New experimental constraints, *Science, 260*, 664–667, 1993.

Peacock, S. M., The importance of blueschist → eclogite dehydration reactions in subducting oceanic crust, *Geol. Soc. Am. Bull., 105*, 684–694, 1993.

Platt, J. P., Exhumation of high-pressure rocks: A review of concepts and processes, *Terra Nova, 5*, 1993.

Ringwood, A., Phase transformations and mantle dynamics, *Earth Planet. Sci. Lett., 14*, 233–241, 1972.

Snoke, J. A., I. S. Sacks, and H. Okada, A model not requiring continuous lithosphere for anomalous high-frequency arrivals from deep-focus South American earthquakes, *Phys. Earth Planet. Inter., 9*, 199–206, 1974.

Snoke, J. A., I. S. Sacks, and H. Okada, Determination of the subducting lithosphere boundary by use of converted phases, *Bull. Seismol. Soc. Am., 67*, 1051–1060, 1978.

Solomon, S. C., and K. T. P. U, Elevation of the olivine-spinel transition in subducted lithosphere: seismic evidence, *Phys. Earth Planet. Inter., 11*, 97–108, 1975.

Suyehiro, K., and I. S. Sacks, , *Bull. Seismol. Soc. Am., 69*, 97–114, 1979.

Tatsumi, Y., K. Ito, and A. Goto, Elastic wave velocities in an isochemical granulite and amphibolite: Origin of a low-velocity layer at the slab/mantle interface, *Geophys. Res. Lett., 21*, 17–20, 1994.

Tiechlaar, B., and L. Ruff, Depth of seismic coupling along subduction zones, *J. Geophys. Res., 98*, 2017–2038, 1993.

Turcotte, D., and G. Schubert, Structure of the olivine-spinel phase boundary in the descending lithosphere, *J. Geophys. Res., 76*, 7980–7987, 1971.

Turner, F., *Metamorphic Petrology: Metamorphic and field aspects*, xi+403pp, McGraw-Hill, New York, 1968.

Wood, B. J., Computation of multiphase multicomponent equilibrium, in *Reviews in Mineralogy, 17*, edited by I. S. E. Carmichael and H. P. Eugster, pp. 71–95, Min. Soc. of America, Washington, DC USA, 1987.

G. Helffrich, Geology Department, University of Bristol, Wills Memorial Building, Queens Road, Bristol BS8 1RJ, UK. (email: george@geology.bristol.ac.uk)

Plate Structure and the Origin of Double Seismic Zones

Geoffrey A. Abers

Department of Geology, University of Kansas, Lawrence, Kansas

Double seismic zones are seen in several but not all subduction zones. Two of the best-studied such zones are beneath Honshu and beneath the Alaska Peninsula; their regional microearthquake observations are re-examined here. Two subparallel planes of seismicity occur at depths between 70 and 150 km and are separated by 20-40 km. The upper plane seems to follow the top of the subducted plate, while the lower plane is within subducted mantle. Several observations seen here and elsewhere are difficult to reconcile with the popular hypothesis that double zones represent the upper and lower bending fibers of a flexing plate. These include the lack of noticeable curvature change at 70-150 km depth and focal mechanisms in the Alaska Peninsula zone that show extension in both planes. As an alternative, I speculate that compositional anomalies within subducting lithosphere are somehow regulating the occurrence of lower-zone earthquakes much as subduction of crust may be responsible for the upper zone. For example, melt-rich regions may form near mid-ocean ridges at the base of a thermal boundary layer; this melt may occasionally crystallize into the plate without ascending to the surface. Although the presence of such buried mafic zones is speculative, compositional irregularities provide a viable alternative to purely mechanical explanations for double zone seismogenesis.

INTRODUCTION

Recent documentation of transformational faulting in the olivine-spinel system [*Kirby*, 1987; *Green and Burnley*, 1989] has demonstrated a feasible mechanism for producing earthquakes deeper than 350 km by shear failure. At intermediate depths (50-300 km), only the gabbro-to-eclogite transformations show significant volume changes (~16%) which could produce locally high deviatoric stresses, but the transition is probably too complex for transformational faulting to be viable [*Burnley et al.*, 1991; *Kirby et al.*, 1991]. Still the associated density changes may provide the driving stresses for shear failure at intermediate depths [*Kirby et al.*, this volume], either directly or through dehydration processes [e.g. *Meade and Jeanloz*, 1991].

Both dehydration and gabbro phase changes only explain intermediate-depth seismicity if earthquakes are near subducted crust at the top of the downgoing slab, as is observed in most places where constraints are good [e.g., *Matsuzawa et al.*, 1986; *Engdahl and Gubbins*, 1987; *Abers*, 1992]. However, observations of double seismic zones (DSZ's), subparallel planes of seismicity 20-40 km apart, require that at least some intermediate-depth earthquakes occur well outside the subducted crust, posing a challenge to most theories of intermediate-depth seismogenesis. One possibility, suggested by *Abers* [1992] and elucidated here, is that the second zone reveals a region of anomalous composition within subducting mantle lithosphere where earthquakes are favored.

DSZ's in Japan [*Hasegawa et al.*, 1978] and elsewhere [e.g., *Kawakatsu*, 1986] are found between 70 and 250 km depth (Figure 1). Often the upper plane is contiguous with the shallower region of interplate thrusting, and the lower plane is subparallel and 20-40 km deeper into the slab [e.g., *Hasegawa et al.*, 1978; *Reyners and Coles*, 1982]. Several DSZ's have been confirmed while monitoring elsewhere has shown only one seismicity plane [*Bevis and Isacks*, 1984; *Kawakatsu*, 1986]. Thus DSZ's share a number of common characteristics where they are found.

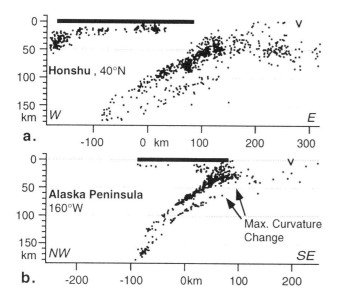

Fig. 1. Cross sections of double seismic zones, from regional network observations. (a.) Seismic zone beneath North Honshu, Japan, from Earthquake Research Institute of Japan's University Network Catalog, 1985-1990 [*Tsuboi*, 1992]. Events selection described in text. Section centered 40°N, 141°E; events are projected ±0.5° along strike from section. (b.) Alaska Peninsula seismic zone, reprocessed from East Aleutian Seismic Network data 1982-1989 [*Abers*, 1992]. Section centered 55°N, 161.5°W. In both regions, solid bars show region spanned by local seismic networks, and V shows location of trench. Events near trench are not beneath network so their depth constraints are relatively weak.

Recent improvements in teleseismic arrival time and waveform matching have led to a number of new DSZ's being identified [e.g., *Comte and Suárez*, 1994; *Kao and Chen*, 1994; *McGuire and Wiens*, 1995]. Unfortunately the teleseismically-delineated DSZ's are nearly all defined by less than 20 events in at least one of the planes over several hundred km along strike, some by as few as one or two events, so it is difficult to assess the spatial completeness of the zones (presumably, stringent requirements on data quality limit the number of events). Also some known DSZ's such as the East Aleutian zone have not yet been detected teleseismically so these techniques do not yet provide complete sampling. Teleseismic techniques are improving, though, and the near future an accurate and more complete catalog of DSZ's is likely.

SEISMIC NETWORK OBSERVATIONS

Data. By examining accurate microearthquake data sets, the distribution of DSZ earthquakes is evaluated in greater detail. New regional catalogs and recent re-analyses are used here to re-examine double zones beneath Honshu [*Tsuboi*, 1992] and the Alaska Peninsula [*Abers*, 1992, 1994].

The Japanese JUNEC catalog for Honshu, provided by the Earthquake Research Institute, has been compiled from the numerous University-operated networks in Japan and is analyzed for the 66-month period of data availability (July 1985 to December 1990). These events have been relocated after combining arrival times from the different networks [*Tsuboi*, 1992]. A subset of the events are used here, those located by eleven or more stations, along a 600 km long arc section centered on 39°N, 141°E, between depths of 70 and 150 km (shallower events are included on Figure 1). A plunging cylindrical shell is fit to seismicity in order to define the surface separating the two seismic zones (see figure captions). Event distribution perpendicular to the surface (Figure 2a) shows two populations corresponding to the two planes of the DSZ. In map view (Figure 3a) the lower zone includes a few anomalous events far trenchward (east) of the island (e.g., Figure 1a). Ocean bottom seismograph observations [*Suyehiro and Nishizawa*, 1994] show that forearc events such as these are often mislocated by on-land networks and most are actually thrust-zone events near 20 km depth, although a handful may be related to bending seaward of the trench [*Seno and Gonzalez*, 1987]. These near-trench events are not discussed further.

The Shumagin or East Aleutian network was operated between 1973 and 1992, recording digitally after 1982, and consisted of 15-20 stations at 25-50 km spacing in the East Aleutian region [*Abers*, 1992; *Reyners and Coles*, 1982]. The digitally-recorded events have been re-evaluated and relocated in a series of inversions for three-dimensional velocity structure and hypocenters [*Abers*, 1992; 1994]. A series of numerical experiments, together with observations of event clustering and formal error analysis, suggests hypocentral uncertainties are less than 5 km for most East Aleutian events discussed here. Similar procedures as used for the Honshu hypocenters were used to distinguish upper zone from lower zone events (Figures 2b, 3c and 3d).

Results. Both regions show an upper plane of intermediate-depth seismicity that abuts the zone of interplate thrust earthquakes, at depths near 40-50 km (Figures 1). The two planes are separated by 25 km in Alaska and 40 km in Japan (Figure 2). The updip end of the lower zone is near 70 km deep in Alaska and 90 km deep in Japan (Figure 1). In Alaska, where scatter is smallest, earthquakes in both zones are in planes no thicker than 10 km (Figure 2b) and probably 5 km thick locally [*Abers*, 1992]. In both regions the DSZ's are present at depths where slabs are planar and curvature changes little; slab curvature changes are largest near the trench and where the seismic zone is 35-50 km deep. The two zones appear to merge below 150-200 km

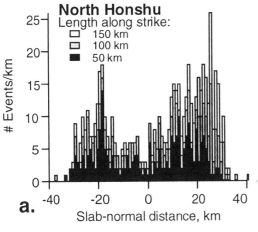

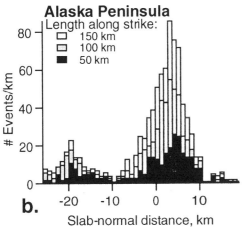

Fig. 2. Abundance histograms of intermediate-depth events with distance normal to slab, for the two zones studied. Increased scatter for Honshu relative to Alaska is expected because Honshu hypocenters are from catalog while Alaska hypocenters are relocated. (a.) North Honshu; slab is defined as in Figure 3. (b.) East Aleutians. Abundances are calculated in 1 km increments. Events are all between 70 and 150 km depth for the two arcs. Positive slab normal distance is up. Three different along-strike sample widths are shown, all beneath the centers of monitoring networks, to illustrate heterogeneity.

depth beneath both arcs where the seismic zones appear to thicken. It is unclear if this is a real feature or an artifact of decreasing hypocenter quality at greater depth. All of these attributes are observed in other regions, indicating that the range of seismic zone spacing, depth extent of DSZ behavior, and a perhaps the lack of curvature are common DSZ characteristics.

In plan view (Figure 2) seismicity is relatively uniform along the upper zone while the seismicity in the lower zone shows heterogeneity. In Alaska the lower zone is completely absent beneath the eastern 100 km of the seismic network in the Alaska Peninsula (Figure 2b) even though monitoring is good [*Hudnut and Taber*, 1987]. Beneath Honshu the lower zone shows gaps 50-100 km across (Figure 3a). The time span of available data is only 6 years for Japan but the gaps are observable for each 2-year time span within the available data (not shown), so may be stationary. The upper plane in Honshu shows no such features (Figure 3b). Such heterogeneity, along with the global observation that DSZ's are not always present, suggests that the conditions necessary for existence of DSZ's can vary over length scales of 100 km or less.

DISCUSSION

Bending hypotheses. Several process have been proposed to explain two zones, often associating them with the top and bottom of the subducted mechanical plate that is unbending, sagging, or contracting [*Engdahl and Scholz*, 1977; *Sleep*, 1979; *Hasegawa et al.*, 1978; *House and Jacob*, 1982; *Kawakatsu*, 1986]. All these processes induce DSZ's by a force couple, that creates high stresses of opposing signs at the tops and bottoms of flexing plates and low stresses in the aseismic core. This stress state accounts for the observation (in most west-Pacific subduction zones) of downdip P axes in the upper plane and downdip T axes in the lower plane [e.g. *Kawakatsu*, 1986; *Kao and Chen*, 1994].

However the flexure models do not explain the geometry of fault plane solutions everywhere. Focal mechanisms from the East Aleutians [*Abers*, 1992] show both planes in downdip extension, suggesting no stress reversal (also those from North Chile show downdip T axes in the upper plane and downdip P axes in the lower plane [*Comte and Suárez*, 1994]). Also, the largest flexural bending stresses (and hence the most seismicity) are expected where curvature changes most rapidly, while most DSZ's are associated with little if any curvature change (Figure 1). Unbending stresses may be generated up dip from DSZ's but appear to be insufficient to trigger seismicity there; if unbending stresses generate DSZ earthquakes then the triggering mechanism must be more complicated than simple elastic loading. Although unbending stresses may be significant, other factors seem necessary to trigger DSZ seismicity. (A different situation prevails near the trench, where curvature changes are well correlated with outer-rise seismicity.) Finally, focal mechanisms of intermediate-depth earthquakes show extreme variability over distances of a few kilometers, difficult to explain if seismicity is driven by a uniform stress field [e.g., *Abers*, 1992]. Hence even if flexing forces are important contributors to the stress state

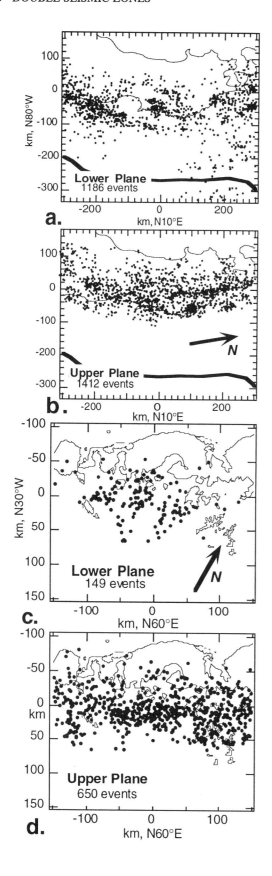

Fig. 3. Map view of upper and lower planes of intermediate-depth seismicity (70-150 km depth). Upper plane seismicity appears uniform in distribution compared to patchy lower-plane distribution. (a.) Lower seismic zone of North Honshu, July 1985 and December 1990. Surface separating the two seismic zones is a section of a canted cylinder, with a 1500 km radius of curvature and a local dip of 25°. (b.) Upper plane of North Honshu. (c.) Lower plane of Alaska Peninsula [*Abers*, 1992]. (d.) Upper plane of Alaska Peninsula. Thin line shows coastline, thick line shows trench (in 3c and 3d trench is just off plot, between 150 and 200 km S30°E). Note difference in map scale between two regions.

in slabs they seem insufficient in themselves to account for DSZ's.

Failure criteria. Flexure models only describe stress orientation and do not explain how these intermediate-depth earthquakes can exist. High pressures at intermediate depths make simple frictional failure unlikely because the needed differential stresses are many times the ductile rock strength [*Jeffreys*, 1929]. An increasing body of evidence suggests that single seismic zones and the upper planes of double zones correspond to the top of the subducting plate, rather than its cold core [e.g., *Engdahl and Gubbins*, 1987; *Abers*, 1992]. As a result the subduction of oceanic crust is suspected of playing an important role in seismogenesis. The large volume changes associated with basalt → eclogite transformations may produce locally high deviatoric stresses [*Pennington*, 1983; *Kirby et al.*, this volume], and dehydration reactions may produce embrittlement and large transients in pore fluid pressures [e.g., *Rayleigh*, 1967; *Kirby et al.*, this volume]. Such phase changes cannot account for a second seismic zone that lies 20-40 km deeper into the subducted mantle lithosphere, as oceanic crust is only 7 km thick.

Speculations on compositional stratification. By analogy with processes envisioned to occur upon subduction of crust, I explore the possibility that a compositional aspect of the subducting slab is inducing the lower zone of DSZ seismicity. Elsewhere in this volume the effects of volatile enrichment by plumes are considered. Here, the possibility is explored that DSZ's are instead products of normal variability in mid-ocean ridge environments (Figure 4). Complexities of melt ascent near ridges may produce the necessary lithospheric heterogeneity without appealing to later modification by plumes. For example, mafic magmas could accumulate and crystallize below the surface within oceanic lithosphere, perhaps in packets, which would then behave like crust upon subduction. Were such material present, it would undergo large density changes over the depths where double zones are found, and similar to subducted crust may be a likely site for producing earthquakes. This

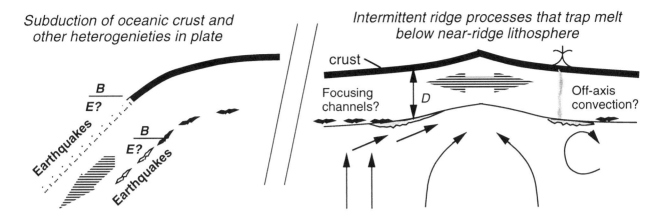

Fig. 4. Some possible ways in which compositional heterogeneity generated near ridges could generate double seismic zones. At spreading centers, some melt ponding may occur as melt is focused toward the ridge crest. Excess melt near ridges might tend to form preferentially at a single depth (D), an isotherm defining the base of a less permeable lithosphere. Heterogeneities created where excess melt cools may then affect the lower planes in double seismic zones when the plate later subducts. The depth of the onset of double-zone activity may represent phase transformations, such as those involved in the basalt-to-eclogite transition (B/E).

particular scenario is not necessarily right, but it demonstrates how material heterogeneity generated at ridges may be significant.

Studies of oceanic crust chemistry suggest that the melts that form it are funneled at depth from a large region [e.g., *Plank and Langmuir*, 1992], and much lateral melt transport is needed to supply ridge crests. A variety of mechanisms have been proposed to transport melt laterally through convection of partly molten rock [e.g. *Buck and Su*, 1989; *Scott and Stevenson*, 1989]. Among the most effective at focusing are the formation of high-porosity channels where ascending melt collects just below an impermeable lithosphere [*Sparks and Parmentier*, 1991]. Because melt ponds at a freezing boundary, a large fraction can crystallize into the mantle lithosphere rather than ascend to the ridge crest. Some melt may remain trapped at the base of the lithosphere off the ridge axis and either crystallize at depth or act as a source for off-axis seamount volcanism. Alternatively, off-axis convection of partly molten mantle may lead to much accumulation of melt above upwelling limbs at the base of the lithosphere which cannot escape to the ridge [*Tackley and Stevenson*, 1993]. Because the melt concentrations produced by either process are highly sensitive to permeability and rheology, and the spatial scales of the processes are on the order of plate thickness, either process could have spatial heterogeneity comparable to that of lower planes of DSZ's (Figures 3a, 3c).

Both of the melt concentration mechanisms discussed above lead to the pooling of melt in a high-porosity region, just below a freezing boundary that roughly follows an isotherm. Such a boundary would be a few tens of km below the sea floor and relatively flat except very near the ridge, closer to the sea floor at faster plate spreading rates. Isolated remnants of frozen melt could be later available to later influence the plate upon subduction. Hence the potential region of frozen melt is at depths similar to those where lower planes of DSZ's are found.

How earthquakes would be then produced upon later subduction is not clear, but lenticular pods or filled cracks of basalt are likely to produce high stresses at their edges as they transform to eclogite [e.g., *McGarr*, 1977; *Pennington*, 1983]. The ~70 km depth where DSZ's are first observed is somewhat greater than depths where eclogite formation is expected. The phase changes are sluggish [*Ahrens and Schubert*, 1975], so such earthquakes could be distributed in depth. The melts may be sufficiently rich in volatiles that dehydration reactions may be important as is suggested for earthquakes in subducted crust [*Kirby et al.*, this volume]. Volatile-rich early melts are among the first likely to reach the freezing front. In any case, off-axis melt collection near mid-ocean ridges may play an important role in influencing the generation of double seismic zones.

Unfortunately, little observational evidence exists to confirm or refute the existence of anomalous regions 20-50 km deep in oceanic mantle. As more DSZ's become better documented it may be possible to compare DSZ spacing with spreading rates, which should reflect ridge thermal structure, independent of other variables. Intriguingly, seismic observations are occasionally made of reflectors or refractors 30-60 km beneath sea floor [e.g. *Shimamura et al.*, 1983]; these may reflect sites of potential future DSZ formation.

Acknowledgments. These ideas grew out of discussions of ridge structure with M. Spiegelman, D. Sparks, and T. Plank. I thank S. Tsuboi for help with the ERI catalog, and the Lamont Alaska group for collecting the Shumagin data. The manuscript was substantially improved by comments from S. Kirby, E. Kissling and an anonymous reviewer.

REFERENCES

Abers, G.A., Relationship between shallow- and intermediate-depth seismicity in the eastern Aleutian subduction zone, *Geophys. Res. Lett., 19*, 2019-2022, 1992.

Abers, G.A., Three-dimensional inversion of regional P and S arrival times in the East Aleutians and sources of subduction zone gravity highs, *J. Geophys. Res., 99*, 4395-4412, 1994.

Ahrens, T. J. and G. Schubert, Gabbro-eclogite reaction rate and its geophysical significance, *Rev. Geophys. Space Phys., 13*, 383-400, 1975.

Bevis, M., and B.L. Isacks, Hypocentral trend surface analysis: Probing the geometry of Benioff zones *J. Geophys. Res., 89*, 6153-6170, 1984.

Buck, W.R., and W. Su, Focused mantle upwelling below mid-ocean ridges due to feedback between viscosity and melting, *Geophys. Res. Lett., 16*, 641-644, 1989.

Burnley, P.C., H.W. Green II, and D.J. Prior, Faulting associated with the olivine to spinel transformation in Mg_2GeO_4 and its implications for deep-focus earthquakes, *J. Geophys. Res., 96*, 425-444, 1991.

Comte, D., and G. Suárez, An inverted double seismic zone in Chile: evidence of phase transformation in the subducted slab, *Science, 263*, 212-215, 1994.

Engdahl, E.R., and D. Gubbins, Simultaneous travel time inversion for earthquake location and subduction zone structure in the Central Aleutian Islands, *J. Geophys. Res., 92*, 13,855-13,862, 1987.

Engdahl, E.R. and C.H. Scholz, A double Benioff zone beneath the Central Aleutians: An unbending of the lithosphere, *Geophys. Res. Lett., 4*, 473-476, 1977.

Green, H.W., II, and P.C. Burnley, A new self-organizing mechanism for deep-focus earthquakes, *Nature, 341*, 733-737, 1989.

Hasegawa, A., N. Umino and A. Tokagi, Double-planed deep seismic zone and upper-mantle structure in the northeastern Japan arc, *Geophys. J. R. Astron. Soc., 54*, 281-296, 1978.

House, L.S., and K.H. Jacob, Thermal stresses in subducting lithosphere can explain double seismic zones, *Nature, 295*, 587-589, 1982.

Hudnut, K.W., and J.J. Taber, Transition from double to single Wadati-Benioff seismic zone in the Shumagin Islands, Alaska, *Geophys. Res. Lett., 14*, 143-146, 1987.

Jeffreys, H., *The Earth*, 2nd ed., Cambridge University Press, New York, 1929.

Kao H., and W.-P. Chen, The double seismic zone in Kuril-Kamchatka: The tale of two overlapping single zones, *J. Geophys. Res., 99*, 6913-6930, 1994.

Kawakatsu, H., Double seismic zones: Kinematics, *J. Geophys. Res., 91*, 4811-4825, 1986.

Kirby, S.H., Localized polymorphic phase transformations in high-pressure faults and applications to the physical mechanism of deep earthquakes, *J. Geophys. Res., 92*, 13,789-13,800, 1987.

Kirby, S.H., W.B. Durham, and L.A. Stern, Mantle phase changes and deep-earthquake faulting in subducting lithosphere, *Science, 252*, 216-225, 1991.

Kirby, S., E.R. Engdahl, and R. Denlinger, Intraslab earthquakes and arc volcanism: Dual physical expressions of crustal and uppermost mantle metamorphism in subducting slabs, this volume.

Matsuzawa, T., N. Umino, A. Hasegawa, and A. Takagi, Upper mantle velocity structure estimated from PS-converted wave beneath the north-eastern Japan Arc, *Geophys. J. R. Astr. Soc., 86*, 767-787, 1986.

McGarr, A., Seismic moments of earthquakes beneath island arcs, phase changes and subduction velocities, *J. Geophys. Res., 82*, 256-264, 1977.

McGuire, J.J. and D.A. Wiens, A double seismic zone in New Britain and the morphology of the Solomon Plate at intermediate depths, *Geophys. Res. Lett., 22*, 1965-1968, 1995.

Meade, C., and R. Jeanloz, Deep-focus earthquakes and recycling of water into the Earth's mantle, *Science, 252*, 68-72, 1991.

Pennington, W.D., Role of shallow phase changes in the subduction of oceanic crust, *Science, 220*, 1045-1047, 1983.

Plank, T., and C.H. Langmuir, Effects of the melting regime on the composition of the oceanic crust, *J. Geophys. Res., 97*, 19,749-19,770, 1992.

Raleigh, C.B., Tectonic implications of serpentinite weakening, *Geophys. J. R Astr. Soc., 14*, 113-118, 1967.

Reyners, M., and K. Coles, Fine structure of the dipping seismic zone and subduction mechanics in the Shumagin Islands, Alaska, *J. Geophys. Res., 87*, 356-366, 1982.

Scott, D.R. and D.J. Stevenson, A self-consistent model of melting, magma migration and buoyancy-driven circulation beneath mid-ocean ridges, *J. Geophys. Res., 94*, 2973-2988, 1989.

Seno, T. and D.G. Gonzalez, Faulting caused by earthquakes beneath the outer slope of the Japan trench, *J. Phys. Earth, 35*, 381-407, 1987.

Seno, T., and Yamanaka, Double seismic zone and superplume, this volume.

Shimamura, H., T. Asada, K. Suyehiro, T. Yamada, and H. Inatani, Longshot experiments to study velocity anisotropy in the oceanic lithosphere of the northwestern Pacific, *Phys. Earth and Planet. Int., 31*, 348-362, 1983.

Sleep, N.H., The double seismic zone in downgoing slabs and the viscosity of the mesosphere, *J. Geophys. Res., 84*, 4565-4571, 1979.

Sparks, D.W., and E.M. Parmentier, Melt extraction from the mantle beneath spreading centers, *Earth and Planet. Sci. Lett., 105*, 368-377, 1991.

Suyehiro, K, and A. Nishizawa, Crustal structure and seismicity beneath the forearc off northeastern Japan, *J. Geophys. Res., 99*, 22,331-22,347, 1994.

Tackley, P.J. and D.J. Stevenson, A mechanism for spontaneous self-perpetuating volcanism on the terrestrial planets, in *Flow and Creep in the Solar System: Observations, Modeling and Theory* (eds. D.B. Stone and S.K. Runkorn), 307-321, Kluwer Academic Publishing, The Netherlands, 1993.

Tsuboi, S., Japan University network earthquake catalog, *EOS Trans. AGU, 73*, 344, 1992.

G.A. Abers, Department of Geology, 120 Lindley Hall, University of Kansas, Lawrence, KS 66045.

Phase Equilibria Constraints on Models of Subduction Zone Magmatism

James D. Myers and A. Dana Johnston

Department of Geology and Geophysics, University of Wyoming, Laramie, WY

Department of Geological Sciences, University of Oregon, Eugene, OR

Petrologic models of subduction zone magmatism can be grouped into three broad classes: (1) predominantly slab-derived, (2) mainly mantle-derived, and (3) multi-source. Slab-derived models assume high-alumina basalt (HAB) approximates primary magma and is derived by partial fusion of the subducting slab. Such melts must, therefore, be saturated with some combination of eclogite phases, e.g. cpx, garnet, qtz, at the pressures, temperatures and water contents of magma generation. In contrast, mantle-dominated models suggest partial melting of the mantle wedge produces primary high-magnesia basalts (HMB) which fractionate to yield derivative HAB magmas. In this context, HMB melts should be saturated with a combination of peridotite phases, i.e. ol, cpx and opx, and have liquid-lines-of-descent that produce high-alumina basalts. HAB generated in this manner must be saturated with a mafic phase assemblage at the intensive conditions of fractionation. Multi-source models combine slab and mantle components in varying proportions to generate the four main lava types (HMB, HAB, high-magnesia andesites (HMA) and evolved lavas) characteristic of subduction zones. The mechanism of mass transfer from slab to wedge as well as the nature and fate of primary magmas vary considerably among these models. Because of their complexity, these models imply a wide range of phase equilibria. Although the experiments conducted on calc-alkaline lavas are limited, they place the following limitations on arc petrologic models: (1) HAB cannot be derived from HMB by crystal fractionation at the intensive conditions thus far investigated, (2) HAB could be produced by anhydrous partial fusion of eclogite at high pressure, (3) HMB liquids can be produced by peridotite partial fusion 50-60 km above the slab-mantle interface, (4) HMA cannot be primary magmas derived by partial melting of the subducted slab, but could have formed by slab melt-peridotite interaction, and (5) many evolved calc-alkaline lavas could have been formed by crystal fractionation at a range of crustal pressures.

INTRODUCTION

In the broadest sense, petrologic research on subduction zone magmatism is designed to answer two fundamental questions: (1) What is the origin of primary arc magmas?, and 2) How is the variety of magmas erupted in arcs produced? One means of addressing these questions is to use lavas as probes of magma sources and the physical processes of magmatic evolution, i.e. the inverse approach. The majority of inverse arc petrologic studies have been geochemical investigations using various combinations of major, trace and rare earth element and isotopic data. Such studies have been particularly successful in identifying the contributions of minor source components [e.g. *Tera et al.*, 1986].

Experimental studies of phase equilibria represent an alternative test of arc magmatic models. Inverse phase equilibria investigations of near-primary samples provide first order constraints on the phases controlling primary

Table 1: Averages of Natural High-magnesia Basalts and Corresponding Experimental Starting Compositions

	Subduction Zone					Experimental Starting Compositions							
	Aleutians	Kermadec	Tonga	Mariana	Scotia	7935g	RZ6	ID16	SD438	TM0	MK15	AKT12b	AKT12a
SiO_2	48.50(2.07)			48.83(1.06)	45.88(4.36)	47.70	49.09	49.32	49.39	49.76	50.57	51.20	50.63	51.33
TiO_2	0.75(0.21)			0.69(0.08)	0.33(0.15)	0.65	0.65	0.60	0.85	1.02	0.51	0.75	0.90	0.96
Al_2O_3	14.83(1.69)			15.21(0.91)	18.75(4.45)	18.60	15.09	15.84	15.70	15.48	14.48	15.69	15.05	16.13
FeO^t	9.54(1.04)			9.75(0.80)	8.10(0.87)	8.53	8.97	8.81	9.76	8.87	8.36	9.21	8.51	8.43
MnO	0.17(0.01)			0.24(0.08)	0.13(0.05)	0.15	0.19	0.16	0.15	0.16	0.17	0.16	0.13	0.14
MgO	11.67(2.66)			10.11(1.18)	11.01(0.25)	9.92	13.34	11.35	12.05	11.68	12.64	9.64	13.23	10.62
CaO	11.18(1.38)			12.17(0.72)	13.13(0.16)	12.10	10.60	10.67	9.43	8.91	10.57	10.12	8.45	9.06
Na_2O	2.04(0.50)			1.68(0.04)	1.20(0.15)	2.25	1.77	2.10	2.33	2.60	1.95	2.77	2.44	2.61
K_2O	0.65(0.24)			0.58(0.36)	0.06(0.04)	0.07	0.20	0.51	0.34	1.28	0.65	0.92	0.49	0.53
P_2O_5	0.15(0.05)			0.10(0.06)	0.01(0.01)	0.05	0.09	0.10	-	0.22	0.10	0.21	0.17	0.19
Total						100.02	100.00	99.21	100.00	99.98	100.00	100.67	100.00	100.00
n	23	0	0	3	2									
# basalts	366	20	6	68	67									
reference	1	1	1	1	1	2,3,4,5	6	7	8	9	6	10	6	6

1 - references can be found by viewing WWW page http://www.uwyo.wdu/a&s/faculty/myers.htm; 2 - *Grove et al.* [1982]; 3 - *Bartels et al.* [1991]; 4 - *Sisson and* Grove [1993a]; 5 - *Sisson and Grove* [1993b]; 6 - Tatsumi *et al.* [1994]; 7 - *Draper and Johnston* [1992]; 8 - *Tatusmi et al.*, [1983]; 9- *Tatsumi* [1982]; 10 - *Gust and Perfit* [1987]

magma composition and the depths of source-melt equilibration as well as the intensive parameters of magma generation. At the same time, forward experiments using likely protoliths delineate the compositional range of potential primary liquids as well as the nature of the solid residuals in equilibrium with these liquids. If samples representing evolved magmas are used, inverse experimental investigations can be utilized to: (1) determine the phases controlling derivative liquid composition, (2) test proposed fractionation schemes, and (3) identify the magmatic processes important in intermediate to shallow depth magma chambers. Conversely, forward experiments can be used to determine liquid-lines-of-descent for different parental liquids and combinations of intensive parameters. Because every petrologic model dictates a particular set of crystal-liquid relations, phase equilibria studies provide a direct means of evaluating arc magmatic models.

HMB, HAB and HMA

Typically, subduction zone magmatic suites are characterized by a compositional range extending from basalt to rhyolite. Within this compositional spectrum, petrologic discussions generally focus on four important lava types: (1) high-magnesia basalts, (2) high-alumina basalts, (3) high-magnesia andesites, and (4) evolved lavas such as basaltic andesites, dacites and rhyolites. High-magnesia basalts (HMB) are typified by high MgO ($\geq$ 9 wt %), Ni and Cr and generally low Al_2O_3 (Table 1). These rocks often contain only ol phenocrysts [*Nye and Reid*, 1986]. In contrast, high-alumina basalts (HAB) are highly porphyritic (< 25-40 % crystals) with phenocrysts of plag, ol, cpx, ±sp.

Chemically, these lavas are marked by high Al_2O_3 (18-22 wt %) and low MgO (< 6 %) and compatible element abundances (Table 2). High-magnesia andesites (HMA) are more siliceous (55-60 % SiO_2) but with unusually high concentrations of MgO (5-8 wt %) and compatible elements. Petrographically, these lavas are generally characterized by a disequilibrium phenocryst assemblage often including xenocrysts and xenoliths. Although quantitative estimates are difficult to make, HAB are volumetrically much more abundant than either HMB or HMA (Table 3).

EXPERIMENTAL APPROACHES AND THEIR LIMITATIONS

As with any geologic problem, questions of subduction zone magmatism can be addressed using either an inverse or forward approach. Using an inverse approach, one starts with a sample and utilizes its characteristics to identify the processes by which it was formed. In contrast, a forward investigation begins with a sample and determines what can be generated from that sample by different petrologic processes. Depending upon the starting material and study objectives, the same phase equilibria experiment may represent an inverse study in one situation but a forward investigation in another.

A forward phase equilibria experiment begins with a presumed source material and determines the nature of liquids produced from it under different intensive conditions. The starting material represents the beginning point of the igneous process simulated by the experimental design. Depending upon the study objectives, the source, i.e. the starting material, may be a presumed protolith or derivative

Table 2: Comparison of Natural HAB and Experimental Starting Materials

	Subduction Zone					Experimental Starting Compositions				
	Aleutians	Kermadec	Tonga	Mariana	Scotia	AT-112	82-62	AT-1	SSS.1.4	82-66
SiO_2	50.13(1.43)	49.87(1.26)	51.90(0.00)	50.15(1.59)	50.30(0.76)	49.00	49.30	49.89	50.26	51.20
TiO_2	0.97(0.26)	0.74(0.23)	0.69(0.08)	0.81(0.10)	0.58(0.09)	1.02	0.95	1.02	0.54	1.01
Al_2O_3	19.34(0.96)	19.23(1.07)	18.30(0.00)	19.34(0.91)	19.46(1.08)	19.00	17.70	19.43	18.46	17.30
FeO^t	9.03(1.08)	9.70(1.66)	9.55(0.99)	9.34(0.93)	8.60(0.59)	9.47	9.02	9.45	7.61	8.66
MnO	0.17(0.02)	0.19(0.06)	0.18(0.03)	0.16(0.05)	0.17(0.02)	0.21	0.17	0.28	0.21	0.16
MgO	5.00(1.16)	5.17(0.79)	6.49(1.34)	4.51(1.00)	5.96(1.12)	6.18	8.61	4.79	7.32	7.47
CaO	10.51(0.93)	11.64(1.94)	12.10(0.42)	11.11(1.08)	12.04(0.62)	11.4	11.30	9.17	11.81	10.2
Na_2O	2.88(0.43)	1.83(0.38)	1.98(0.02)	2.47(0.41)	1.78(0.28)	2.56	2.58	3.39	1.80	3.11
K_2O	0.71(0.26)	0.32(0.10)	0.53(0.03)	0.72(0.43)	0.13(0.06)	0.65	0.30	0.76	0.25	0.70
P_2O_5	0.18(0.07)	0.09(0.10)	0.62(0.69)	0.15(0.10)	0.05(0.07)	0.11	0.11	-	-	0.16
Total						99.60	100.00	98.18	98.62	99.97
n	228	10	2	28	34					
# basalts	366	20	6	68	67					
reference	1	1	1	1	1	2	3,4	5	6	3,4

1 - references can be found by viewing WWW page http://www.uwyo.wdu/a&s/faculty/myers.htm; 2 - *Baker and Eggler* [1987]; 3 - *Sisson and Grove* [1993a]; 4 - *Sisson and Grove* [1993b]; 5 - *Baker and Eggler* [1983]; 6 - *Johnston* [1986]

liquid. For the protolith, a solid source is postulated to have partially melted to produce a primary magma. These experiments generally focus on the p-T region near the solidus and consist of melting experiments at different intensive parameters and degrees of melting. Compositional similarity of experimental melts and natural samples representing primary magmas provides evidence that the conditions of primary magma generation may have been identified. In the other type of forward phase equilibria experiment, the starting material is a derivative lava assumed parental to evolved lavas through differentiation. These forward experiments consist of crystallization experiments concentrated on the near-liquidus region (Figure 1). Compositional similarity between experimental liquids and natural lavas is permissive, but not conclusive, evidence that the latter could have been formed by the process simulated in the experiments. As these examples show, the factor determining whether an experiment employs a forward or inverse approach is not its physical nature, e.g. crystallization vs. melting, but the objective of the experiment.

For an inverse experiment, the choice of a starting material is not protolith versus derivative liquid but primary versus derivative magma. In this instance, the starting material represents the end product of an igneous process, i.e., a primary magma produced by partial melting or an evolved liquid generated by differentiation. Unlike with the forward experiments, the experimental setup is the same for both starting materials. In both cases, experiments near the liquidus define crystallization sequences over a range of intensive parameters (Figure 2). A proposed petrologic model is deemed feasible if the experimentally determined phase equilibria agree with those predicted by the model. Such agreement further constrains the intensive conditions under which the process may have occurred.

In most instances, the interpretation of inverse experiments is straightforward. For example, the liquidus appearance of a phase indicates the sample is saturated in the observed phase. Conversely, the absence of a phase indicates either the sample was not saturated or it was saturated but the phase was dissolving, not crystallizing, i.e. it was reacting with the liquid (Figure 3). Inverse experiments on derivative liquids lacking reaction relations will always correctly predict potential evolutionary trends. In contrast, failure to recognize a reaction relation between liquid and solid can result in serious misinterpretation of the petrologic importance of experimental phase equilibria (Figure 3). When interpreting inverse experimental results, the possibility of reaction relations must, therefore, always be considered. Experimental tests to identify a reaction relationship can be performed by conducting a series of mineral-addition experiments [*Green et al.*, 1979; Basaltic Volcanism 1981, sec. 3.1.4]. In such experiments, the potential reacting phase is added to the starting material. Because experiments are generally conducted isothermally, i.e. they represent a single point along a phase boundary curve (Figure 3), the added phase will persist through the experiment if the liquid is saturated, but dissolve if it is not. For arc magmas, ol is the most likely phase to exhibit a reaction relation.

A more rigorous test of petrologic models can be made using inverse and forward experiments simultaneously. This procedure utilizes p-T diagrams for the presumed source as well as derivative lavas (Figure 4). Assuming equilibrium between solids and liquid during melting,

Table 3: High-magnesia Andesites - Natural Examples and Experimental Starting Compositions

	Subduction Zone					Experimental Starting Compositions				
	Aleutians	Kermadec	Tonga	Mariana	Scotia	KMA	SD-261	KMA2	TGI	
SiO_2	56.50(1.04)			55.83(0.64)		57.04	57.13	57.32	59.44	59.59
TiO_2	0.75(0.13)			0.74(0.04)		0.69	0.73	0.69	0.44	0.44
Al_2O_3	16.45(1.05)			17.50(0.75)		15.36	15.83	15.46	13.51	13.55
FeO^t	6.98(0.67)			7.48(0.53)		6.03	6.33	5.96	6.27	6.32
MnO	0.14(0.02)			0.17(0.01)		0.12	0.13	0.13	0.12	0.12
MgO	5.85(1.22)			4.78(0.41)		9.03	7.39	8.90	9.62	9.65
CaO	8.41(0.69)			8.98(0.69)		7.04	7.19	7.08	6.23	6.24
Na_2O	3.14(0.53)			2.88(0.36)		3.11	2.89	3.07	2.65	2.66
K_2O	1.30(0.41)			0.86(0.23)		1.76	2.27	1.84	1.30	1.30
P_2O_5	0.18(0.07)			0.15(0.06)		-	0.14	-	0.13	0.13
Total						100.18	100.03	100.44	99.71	100.00
n	29	0	0	3	0					
# andesites	350	3	12	50	51					
reference	1	1	1	1	1	2	3	2	4	3

1 - references can be found by viewing WWW page http://www.uwyo.wdu/a&s/faculty/myers.htm; 2 - *Kushiro and Sato* [1978]; 3 - *Tatsumi* [1982]; 4 - *Tatsumi* [1981]

phase equilibria defined by inverse experiments must match those of forward experiments. This kind of combined experimental setup is particularly useful because it approaches melt extraction from both sides of the reaction and better constrains the intensive conditions at which a process could have occurred.

Experimental studies provide only permissive evidence for the operation of a particular process. The failure of an experimental study to confirm a proposed petrologic process can be interpreted in at least three ways. First, the petrologic process and associated end members may have been correctly deduced but not the intensive parameters. For example, anhydrous crystallization experiments on a basalt are unlikely to reproduce liquid-lines-of-descent generated from that parent under hydrous conditions. Second, the end members and intensive conditions may be correct but the experiments simulated the incorrect petrologic process. An example would be using an experiment to test a crystal fractionation mechanism when the lava was formed by magma mixing. Finally, identification of the end members, i.e. the starting materials, may have been wrong. For example, crystallization experiments are unlikely to produce the observed fractionation trend if the mafic parent was misidentified. Although discussed as three separate and unrelated factors, the "failure" of an experimental study to confirm a model could result from any combination of these variables.

ARC MAGMATIC MODELS AND PREDICTED PHASE RELATIONS

Models proposed to explain the origin of subduction zone magmas can be grouped into three broad classes: (1) predominantly slab-dominated, (2) mainly mantle-derived, and (3) multi-source models. Because they involve different petrologic mechanisms, each type of model dictates different phase relations. These differences form the basis for phase equilibria tests of arc petrologic models.

Slab-dominated Magmatic Models

These models are conceptually simple and based on the direct, observed link between subduction and magmatism, i.e. when the former is absent, the latter does not occur. They assume high-alumina basalts, the dominant arc lava, represent primary magmas (Figure 5) and their distinctive geochemical character, e.g. high-alumina and low MgO, are most readily explained by partial fusion of the down-going slab, i.e. subducted sediment and oceanic crust [*Myers et al.*, 1986]. Earlier models implicitly assumed partial fusion and melt extraction occurred at the slab-mantle interface. Later models suggested small to moderate degrees of partial melting produces a solid-liquid diapir that begins to ascend buoyantly [*Brophy and Marsh*, 1986]. As the diapir rises, melting increases and a primary HAB magma is extracted when the diapir has ascended several tens of kilometers. The preponderance of plag phenocrysts, absence of hydrous phases and paucity of explosive arc *basaltic* volcanism were interpreted as evidence that primary HAB contain little water thereby making primary magma generation a dominantly anhydrous process. In these models, the origins of the volumetrically minor HMB and HMA were not specifically addressed. Upon entering crustal magma chambers, parental HAB magmas differentiate to form evolved lavas, e.g. basaltic andesites, andesites, dacites, etc. (Figure 5). Recognized crustal processes include crystal fractiona-

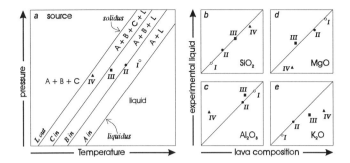

Fig. 1a. Diagrammatic representation of a forward experimental study designed to determine if an assumed parent could have produced a given series of derivative liquids, i.e. lavas. The p-T diagram for the presumed parent is determined by a series of crystallization experiments near the liquidus.

Fig. 1b. Experimental liquids at positions *I-IV* on the source p-T diagram have silica contents similar to some of the natural lavas, i.e. they fall on or within analytical error of the 1:1 line on the experimental liquid versus lava SiO$_2$ plot. These similarities suggest the natural suite might be the product of crystal fractionation of the presumed parent at the experimental conditions. This possibility is evaluated further by considering other major element oxides (c-e). For all oxides, the close compositional agreement for liquids *I* and *II* is consistent with their proposed fractionation origin. The markedly different alumina, MgO and K$_2$O between experimental liquid *IV* and its possible natural analog suggest the latter was not produced by crystal fractionation at the intensive parameters of the experiments. It could, however, have been produced by fractionation at intensive conditions other than those investigated experimentally. The compositional agreement between experimental liquid and lava is not as good for liquid *III* but is near analytical limits. For this sample, determining if the proposed petrogenetic scheme is valid is not straightforward. An approach similar to that described here could be used for a forward experimental study relating a protolith and its presumed primary magma. The major difference would be the appropriate p-T diagram would be determined by a series of melting experiments near the solidus.

tion [e.g. *Stern*, 1979; *Perfit et al.*, 1980; *Kay et al.*, 1982, 1983; *Singer et al.*, 1992], magma mixing [e.g. *Conrad et al.*, 1983; *Myers et al.*, 1995] as well as complex combinations of crystal retention and magma mixing [e.g. *Brophy*, 1990]. For the evolved lavas, the absence of hydrous phases and great abundance of plag have been interpreted as indicators that crustal processes also occurred at low water contents [*Marsh*, 1982; *Singer et al.*, 1992; *Fournelle et al.*, 1994].

These magmatic models are characterized by two stages of evolution: (1) primary HAB magma generation by slab melting, and (2) differentiation of these parents to form

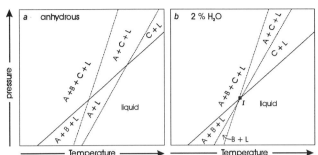

Fig. 2a. Illustration of the inverse experimental procedure to test the fractionation origin of a derivative lava. If geochemical modeling, which suggests the lava was formed by fractionation of phases A, B and C, is correct, this phase assemblage must appear on the lava's liquidus. The anhydrous phase diagram reveals a liquidus marked by A at low pressures and C at higher pressures but no point of multiple saturation. Consequently, this lava could not have been formed by fractionation of A + B + C under anhydrous conditions.

Fig. 2b. The addition of a small amount of water depresses the stability of phase A more than B or C and produces a point of multiple saturation at point *I* (b). In systems with 2 % H$_2$O, the pressure and temperature of this point represent the intensive parameters at which the lava could have been formed by the proposed fractionation scheme.

evolved liquids. To test this model of magma genesis, forward experiments must determine if eclogite melting can produce HAB liquids (Table 4). The matching inverse crystallization experiments must also show that the high pressure HAB liquidus is defined by the same solid assemblage found in the forward experiments. A saturation assemblage of garnet and cpx at ~30 kb would indicate primary melt extraction near the slab-mantle interface after small to moderate degrees of partial fusion. In contrast, if cpx alone occurs on the HAB liquidus, melting must have been sufficient to eliminate garnet or the melt was extracted from a rising diapir at depths shallower than the stability field of garnet [*Brophy and Marsh*, 1986].

To understand evolved lava generation, experiments at low-pressure (≤ 10 kb) and variable water contents are necessary (Table 4). Potential starting materials for inverse experiments include basaltic andesites, andesites or dacites. To be consistent with the slab-melting model, these experiments must show that HAB liquids are multiply-saturated with the phase assemblages used in successful mass-balance based HAB fractionation models (Table 4) [*Kay et al.*, 1982; *Myers et al.*, 1995]. Forward HAB melting experiments must reveal an evolutionary trend similar

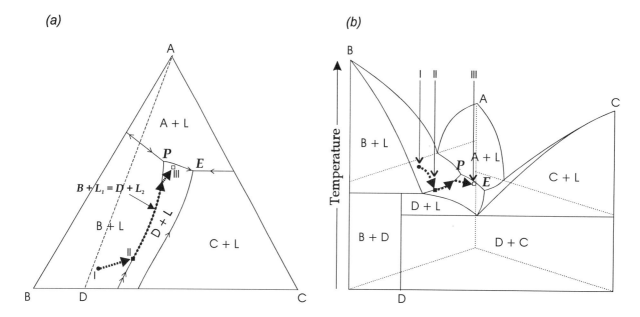

Fig. 3a. Graphical depiction of the effect of a reaction relation on the interpretation of inverse experiments. The A-B-C system is a simple ternary with an intermediate phase, D, that melts incongruently. (a) Projection of the liquidus surface. Liquid *I* initially crystallizes B which moves the residual liquid directly away from apex B. When the liquid reaches *II*, B begins to react with the liquid to produce D and the liquid simultaneously crystallizes D. Continued dissolution of B and crystallization of D moves the liquid down temperature along the reaction curve toward the peritectic at P. When all of B is consumed, the liquid composition leaves the reaction curve and moves onto the D + L surface. Crystallization of a finite amount of D moves the liquid away from D to *III*.

Fig. 3b. Perspective drawing of the ternary liquidus surface illustrating the effect of this reaction relation on the phase equilibria of derivative liquids *I*, *II* and *III*. Inverse experiments on *I* correctly identify B as its liquidus phase and reveal its potential liquid-line-of-descent. Similarly, experiments on liquid *III* reveal D as its liquidus phase and correctly predicts liquid trends produced by fractionation of this phase. In contrast, the interpretation of the inverse experiments for liquid *II* is more complicated. Although experiments on this liquid reveal a liquidus characterized by D, the original liquid was also saturated with B. Because the latter was dissolving not precipitating, it does not appear on the experimental liquidus. Thus, the experiments suggest subsequent liquid evolution was controlled by crystallization of D alone when in fact it was determined by crystallization of D *and* dissolution of B.

to that of calc-alkaline suites. In addition, the solid assemblages coexisting with these liquids must match, in kind and composition, that of the inverse experiments. (Because the mechanisms of evolved lava formation are similar in all the arc petrologic models, the experimental tests described here apply to the other models as well and are therefore not described again.)

Mantle-dominated Petrologic Models

These models assume *a priori* peridotite is the source of arc magmas thereby fixing the mantle wedge as that portion of subduction zones where arc magmatism originates. Because only HMB satisfy the geochemical characteristics dictated by a mantle source [*Tatsumi et al.*, 1983; *Nye and Reid*, 1986; *Gust and Perfit*, 1987], they are necessarily assigned primary magma status (Figure 5). According to these models, water from the dehydrating slab enters the wedge and lowers the peridotite solidus thereby permitting partial fusion and primary HMB generation. Unlike later models, these less complex mantle-dominated models assign only a fluxing role to the slab-derived fluid. Ascent of HMB magmas is implicitly assumed to be isochemical and they do not differentiate until entering subcrustal or crustal magma chambers. Here primary HMB magmas undergo ol, cpx and plag fractionation at moderate to low pressure to produce HAB liquids (Figure 5). This fractionation step requires removal of as much as 70-80 % of the original HMB [e.g. *Kay et al.*, 1982]. Typically, the paucity of HMB and abundance of HAB (Tables 1-2) are

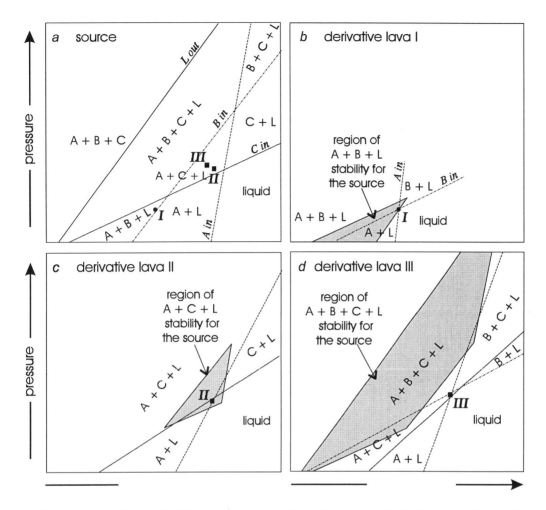

Fig. 4a. Representation of a combined forward-inverse experimental approach. If p-T diagrams are known for a parental magma (a) and its presumed derivative liquids (b-d), a combined experimental approach can be used to test proposed fractionation models. For this example, lava *I* is presumed to have been formed by fractionation of A + B, *II* by removal of A + C and fractionation of A + B + C is believed to have produced lava *III*.

Fig. 4b. The appearance of A and B on the liquidus of *I* indicates it could have been produced by the proposed fractionation scheme. This point of multiple saturation also lies within the stability field of A + B on the source p-T diagram (point *I* on a) thereby indicating the phases relations of the presumed source are also consistent with the proposed fractionation scheme. Compositional similarity between the solids in equilibrium with the source's derivative liquid as well as the phases on the derivative lava's liquidus at these intensive parameters is further evidence that the proposed fractionation scheme is possible.

Fig. 4c. Similar relations for lava *II* also support its proposed fractionation scheme.

Fig. 4d. The phase relations of lava *III* include a point of multiple saturation in A, B and C that is consistent with a fractionation origin (d). However, the pressure and temperature of this point are removed from the A + B + C + L field of the parental magma (a). This observation plus compositional dissimilarity between source and derivative solids and liquids at the common point of multiple saturation indicate the proposed fractionation scheme is not feasible. Although the fractionation scheme suggested by the geochemical modeling may be correct, it must have occurred at conditions other than those simulated in the experiments. Using only an inverse approach, the lack of compatibility between the phase relations of the parent and derivative liquid would have gone unrecognized. Shaded regions on b-d represent the source p-T fields from (a) of the respective fractionating phase assemblages.

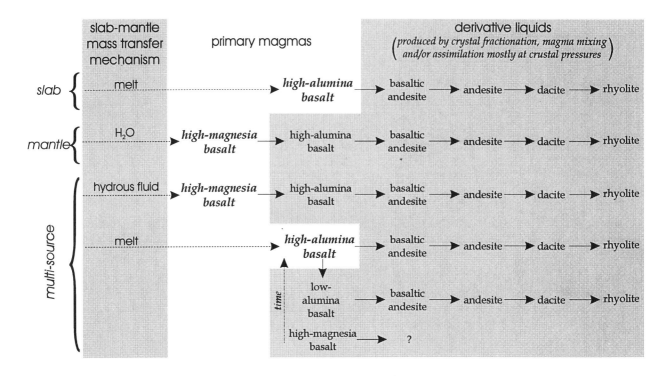

Fig. 5. Diagrammatic representation of idealized liquid-lines-of-descent for the main types of subduction zone magmatic models. In all cases, a significant magmatic component must be derived from the subducting slab. The manner in which this component(s) is transferred from the slab (shaded region on left) to the mantle varies between the three models. In slab-dominated and some multi-source models, this material is transferred via a silicate melt. For mantle-derived models, the transfer is accomplished by water but the nature of the fluid and its geochemical characteristics are rarely explicitly stated. These cases represent primarily earlier magmatic models and have generally evolved over time. Many current multi-source models attribute the slab-mantle transfer to a hydrous fluid that moves a significant number of geochemical components. For all models, there are two possible primary magmas, i.e. high-alumina basalt or high-magnesia basalt (unshaded region in center). In mantle-dominated and some multi-source models, the latter undergo extensive crystal fractionation to produce high-alumina basaltic liquids. In one subclass of multi-source models, the primary high-alumina basalt magma produces high-magnesia basaltic liquid by interaction with the mantle during ascent. In nearly all the different magmatic models, evolved lavas, i.e. those more siliceous than basalt, are produced by some combination of crystal fractionation, magma mixing and/or crustal assimilation (shaded region on right). Typically, these processes occur and subsequent lavas are produced in intermediate to shallow level crustal magma chambers.

attributed to crustal density filtering. The differentiated HAB liquids are parental to the evolved calc-alkaline lavas through a variety of magmatic processes that are broadly similar to those described for the slab-dominated models (Figure 5). As with the slab-dominated models, the origin of HMA is not specifically addressed in these models.

Mantle-dominated arc magmatic models can be divided into three major stages. Two of these stages differ from those of slab-melting models and necessitate different experimental conditions. Generation of primary HMB from the mantle wedge can be investigated using water-bearing experiments conducted at high pressures (Table 4). The forward experiments involve peridotite melting whereas the complementary inverse investigations concentrate on crystallization of HMB. Because water serves simply as a flux in these mantle-dominated models, the forward experiments must yield HMB partial melts in equilibrium with a peridotite residuum. To be mutually consistent, inverse experiments must show that HMB, at the same intensive conditions, are multiply saturated with the solid assemblage of the forward experiments.

The other unique stage of mantle-dominated models involves fractionation of HMB to produce HAB. Experiments designed to reproduce this magmatic stage must be conducted at moderate pressures with variable amounts of water (Table 4). Forward crystallization experiments con-

Table 4: Experimental Phase Equilibria Tests of Arc Petrologic Models

Models	Magmatic Process	Conditions	Forward Experiments		Inverse Experiments	
			Starting Material	Predicted Phases	Starting Material	Predicted Phases
slab-dominated	1. primary magma generation (HAB)	25-30 kb - anhydrous	eclogite	HAB liquid w/ cpx, ±gar	HAB	cpx, ±gar
	2. formation of evolved lavas	≤10 kb - anhydrous to moderate H$_2$O	HAB	evolved liquid w/ pl, ol, cpx, ±sp, ±opx	evolved lavas	pl, ol, cpx, ±sp, ±opx
mantle-dominated	1. primary magma generation (HMB)	20-30 kb - water bearing	peridotite	HMB liquid w/ ol, cpx, ±opx, ±sp	HMB	ol, cpx, ±opx, ±sp
	2. petrologic relation between HMB and HAB	15-25 kb - water bearing	HMB	HAB liquid w/ ol, cpx, ±pl, ±sp	HAB	ol, cpx, ±pl, ±sp
	3. formation of evolved lavas	≤10 kb - water bearing	HAB	evolved liquid w/ pl, ol, cpx, ±sp, ±opx	evolved lavas	pl, ol, cpx, ±sp, ±opx
multi-source	1. HAB generation by slab melting	25-30 kb - anhydrous	eclogite	HAB liquid w/ cpx, ±gar	HAB	cpx, ±gar
	2. HMB origin: peridotite partial melt + slab mantle contamination of HAB	15-30 kb - water bearing and free	-	-	HMB	ol, ±cpx, ±opx
	3. petrologic relation between HMB and HAB	20-30 kb - water bearing and free	HMB	HAB liquid w/ ol, cpx, ±pl, ±sp	HAB	ol, cpx, ±pl, ±sp
	4. HMA generation: slab melting w/ mantle interaction	15-30 kb - water bearing and free	eclogite -	HMA liquid w/ cpx, ±gar -	HMA HMA	cpx, ±gar ol, cpx, ±opx, ±pl, ±sp
	5. formation of evolved lavas	≤10 kb - water bearing and free	HAB	evolved liquid w/ pl, ol, cpx, ±sp, ±opx	evolved lavas	pl, ol, cpx, ±sp, ±opx

sistent with this model should yield HAB derivative liquids from HMB starting materials. The complimentary inverse experiments must document HAB saturation with the solid assemblage of the forward experiments (Table 4). Because production of HAB in nearly all mantle-dominated models requires removal of ol, HAB must be saturated with this phase. The presence of ol on the liquidus proves saturation and would provide permissive evidence for a fractionation relation between HAB and HMB. Because of the possibility of a reaction relation, the lack of ol is inconclusive evidence that ol saturation did not occur. Accordingly, ol addition experiments should be conducted whenever ol is absent from the liquidus [*Nicholls*, 1974; *Draper and Johnston*, 1992]. Confirmation of a reaction relation would provide permissive evidence for a fractionation relation between HMB and HAB.

Multi-source Magmatic Models

Although some early petrologic models [e.g. *Ringwood*, 1974; *Kay*, 1978] envisioned a combination of arc magmatic sources, nearly all recent models incorporate slab *and* mantle wedge components in primary magma formation (Figure 5). The manner in which material is transferred from slab to wedge, the nature of primary magmas and the subsequent fate of these magmas differ markedly. The most common multi-source models transfer slab components in a hydrous fluid of unspecified or unknown physical nature. Three alternative models move material as a silicate melt, but the nature of the melt and its ultimate fate vary substantially.

Fluid transfer models. The most widely accepted models suggest material is transferred in a hydrous fluid produced by slab dehydration (Figure 5). Because the fluid carries a wide range of chemical components, it acts as a mass transfer agent as well as a flux [e.g. *Tera et al.*, 1986]. The physical path by which the fluid enters the wedge differs considerably between models [*Tatsumi*, 1989; *Davies and Stevenson*, 1992]. In the wedge, the fluid lowers the peridotite solidus thereby promoting partial fusion and generation of primary HMB magmas. Nearly all models assume implicitly that HMB magmas reach shallow-level magma chambers without significant differentiation. At crustal levels, fractionation of ol (±other phases) produces HAB [e.g. *Kay et al.*, 1982]. Subsequent differentiation of HAB in crustal chambers produces both calc-alkaline and tholeiitic arc suites.

Magma-mantle interaction models. Two models explaining the origin of arc basalts suggest primary magmas do not ascend isochemically and that the diversity of arc mafic lavas reflects the interaction of primary magma and mantle during ascent. In one of the earliest arc models, *Ringwood* [1974] suggested partial fusion of the downgoing slab produced a siliceous magma that interacted with peridotite during ascent to produce arc basalts. *Myers et al.* [1985] retained the concept of magma-mantle interaction but suggested the primary magma was HAB produced by slab partial melting. Interaction of this primary magma with the mantle produced the volumetrically minor HMB and HMA characteristic of several arcs [*Myers and Frost*, 1994].

High-magnesia andesite models. Slab melting also plays an important role in two models specifically proposed to explain the origin of the rare HMA found in arcs worldwide (Table 3). To account for the origin of an unusual high-MgO Aleutian andesite, *Kay* [1978] suggested partial melting of subducted crust produced a dacitic melt that during ascent reacted with peridotite to produce HMA. *Defant and Drummond* [1990] named this special class of andesite adakite but suggested it represented primary partial melts of the subducting slab. They also suggested such melts could be produced only from unusually hot, and therefore young, slabs.

Multi-source phase equilibria Multi-source model issues to be addressed experimentally can be summarized in five questions: (1) Can primary HAB be generated by slab melting? (2) How are HMB produced? (3) What is the petrologic relation between HMB and HAB? (4) How are HMA generated? and (5) What processes are responsible for evolved arc lavas? Of these questions, only (2) and (4) require new experimental designs.

Because HMB generation from multiple sources is necessarily an open-system process, there is no single starting material with which to conduct forward experiments. For example, peridotite partial melting experiments cannot produce HMB liquids because they lack the components introduced from the slab, e.g. K_2O. Consequently, only inverse experiments can be used to test HMB generation within the context of multi-source models (Table 4). Inverse HMB crystallization over a range of pressures and water contents can show if these liquids ever equilibrated with a peridotite residuum. Whereas the appearance of this assemblage is permissive evidence for melt equilibration with peridotite, it does not distinguish between a melting equilibrium or equilibration of a primary magma with peridotite during ascent.

Testing the two HMA generation models requires different experimental approaches because one model is an open-system process whereas the other is closed. For the model of *Defant and Drummond* [1990], the forward-inverse experimental procedure described for HAB is appropriate. The only modification required is that HMA must be substituted for HAB as the starting material in the inverse experiments (Table 4). Conversely, the HMA model of *Kay* [1978] combines slab and mantle components thereby making it an open-system process and rendering forward experiments ineffective. As with multi-source HMB generation, inverse crystallization experiments on HMA can show if these liquids ever equilibrated with peridotite as predicted by the model (Table 4).

EXPERIMENTAL RESULTS

Forward Experiments

Arc-related forward experiments can be divided into four classes: (1) characterization of fluids produced by slab dehydration, (2) evaluation of sediment melting, (3) investigation of primary magma (HMB or HAB) generation, and (4) examination of evolved lava petrogenesis. Dehydration experiments [*Tatsumi and Nakamura*, 1986; *Tatsumi and Isoyama*, 1988; *Tatsumi et al.*, 1986] are fundamentally different from phase equilibria studies because fluid compositions are not buffered, i.e. fixed, by the phases present, but are dependent upon starting material (solid and fluid) composition and how the fluid is introduced into the experiments, e.g. structurally bound, intergranular, etc. Accordingly, such experiments are not discussed. Because the physical nature of sediment-basalt melting is poorly known, sediment melting experiments [*Stern and Wyllie*, 1973; *Huang and Wyllie*, 1974; *Johnson and Plank*, 1993; *Nichols et al.*, 1994] are also not reviewed.

Slab melting. Forward investigations of slab melting must begin with eclogite or MORB (Table 4). Unfortunately, most MORB melting experiments (anhydrous and hydrous) have been carried out at pressures (≤ 15 kb) below the eclogite stability field [e.g. *Nicholls and Ringwood*, 1973; *Stolper*, 1980; *Wyllie and Wolf*, 1993]. In one of the few high-pressure studies, *Johnston* [1986] partially melted three MORBs at 26-27 kb and 1430-1450°C. At these conditions, the basalts produced three different phase assemblages, liq + gar + cpx, liq + cpx and liq, despite similar bulk compositions. Liquids coexisting with gar + cpx, are broadly similar to HAB but exhibit serious mismatches for

CaO and FeO. Agreement between experimental liquids and HAB was better for those experiments producing only a cpx residuum [*Johnston*, 1986]. High-pressure (8-32 kb) melting experiments between 1000 and 1150°C were carried out by *Rapp et al.* [1991] and *Rapp and Watson* [1995] on three amphibolites compositionally similar to MORB. These experiments contained between 1 and 2 wt % water that was supplied by the breakdown of amphibole. Depending upon pressure and temperature, the experimental liquids ranged from 70 to 51 % SiO_2. Siliceous liquids were produced at the lowest degrees of melting (~10 %) whereas the more extensive partial melts were less evolved (51-55 wt % silica). Alumina contents correlate inversely with silica content and hence degree of melting. At higher pressures, mafic compositions were produced at smaller degrees of melting. For example, andesites were produced at 32 kb after 20-30 % melting whereas 40-50 % partial melting was required to produce similar silica contents at 8 kb. (Because the maximum experimental temperature was 1150°C, the maximum degree of melting attained at high pressure was less than 30 %.) Saturating assemblages ranged from amp + pl + opx + il at low pressures and temperatures to cpx + gar at high pressures and temperatures.

Wedge melting. Forward experiments of wedge melting utilize peridotite as the starting material (Table 4). A large number of peridotite melting experiments have been conducted over a range of experimental conditions [*Kushiro et al.*, 1972; *Green*, 1973; *Kushiro*, 1972, 1973; *Mysen and Boettcher*, 1975; *Mysen and Kushiro*, 1977; *Jaques and Green*, 1979, 1980; *Baker and Stopler*, 1994]. Because most were directed at questions of MORB petrogenesis, many of these experiments were performed at pressures well below those appropriate to arc magma genesis. One peridotite melting experiment was, however, interpreted in terms of arc petrogenesis despite its low pressure. *Nicholls* [1974] produced andesitic liquids at 5 kb with 10 % water. In these types of experiments, the small amounts of melt produced made determining melt compositions extremely difficult. Moreover, loss of iron to Pt capsules almost certainly occurred but was not evaluated. By sandwiching a basalt layer between two peridotite layers, *Takahashi and Kushiro* [1983] generated glass pools large enough for accurate microprobe analysis. Using their new technique, these authors showed that under anhydrous conditions peridotite partial melts near the solidus changed from MORB-like at 10 kb to alkali picrite at 25 kb and did not approximate HMB. In the same study, 20% potassium feldspar was added to peridotite to test multi-source peridotite melting. The added phase lowered the peridotite solidus 70-150°C, but near solidus melts were leucite phonolites, not HMB. *Kushiro* [1990] performed another series of peridotite sandwich experiments at pressures and temperatures appropriate to "... shallow levels of the mantle wedge...." At 12 kb with 5-6 wt % water, HMB-like partial melts were produced at 1250°C but changed to low-MgO HAB at 1150°C and 4.4 wt % H_2O. Increasing water content produced more siliceous liquids. Aside from the 12 kb experiments of *Kushiro* [1990], peridotite melting experiments have failed to produce HMB-like liquids, and the compositional mismatch is considerable for the alkalis.

Inverse experiments

Inverse arc experimental studies can be divided into four categories based on starting material: (1) HAB, (2) HMB, (3) HMA, and (4) evolved magmas. Starting compositions for HMB, HAB and HMA experiments are presented in Tables 1-3 and for evolved lava investigations in Table 5. The experimental conditions for these experiments are listed in Table 6 and points of multiple saturation summarized in Table 7.

High-alumina basalt experiments. Several inverse experimental studies on HAB have been reported (Table 6). These investigations, which employed only five different starting materials (Table 2), were conducted over a range of conditions. In one of the first studies, *Baker and Eggler* [1983] investigated the anhydrous phase relations of an Aleutian HAB. This basalt has an anhydrous liquidus defined by plag below ~ 18 kb, is multiply saturated with gar + cpx + plag near 18 kb (Table 7) and crystallizes garnet at greater pressures. The compositions of glasses coexisting with these phases were not reported. At 2 kb, plag is the liquidus phase up to 5 % water when it was replaced by a Fe-Ti oxide. In a follow-up study from 0 to 10 kb, *Baker and Eggler* [1987] determined the anhydrous phase relations of another Aleutian HAB (AT-112) with 2 wt % more MgO and CaO. This sample was not multiply saturated at the intensive parameters of the experiments (Table 7) and below 8 kb its liquidus phase relations are similar to those of the other Aleutian HAB. In another HAB study, *Johnston* [1986] determined the anhydrous liquidus phase relations of a Scotia arc basalt (Table 6). This basalt has less alumina, Na_2O and K_2O but higher MgO than the Aleutian HAB studied (Table 2). Its liquidus is characterized by garnet above 27 kb, cpx from 27 to 17.5 kb and plag below 17.5 kb. Multiple saturation with gar + cpx occurs at 27 kb and with cpx + plag at 17 kb (Table 7). *Sisson and Grove* [1993a] defined the water-saturated phase equilibria at 2 kb of two calc-alkaline basalts from Medicine Lake Volcano (Table 6). At 2 kb, both liquids are multiply saturated with ol + cpx + pl at 1012°C (Table 7). The hydrous liquids coexisting with these phases are low-alumina basalts (< 18 wt

Table 5: Compositions of Starting Materials used in Inverse Experiments to Understand Evolved Lava Petrogenesis

sam #	79-38b	AT-41	AT-29	AT-25	R-2	79-9c	MHA	T101	FP1652	GRM-4	II-GRM	I-GRM	GRM-1	P_S-G	P_A-G
SiO_2	54.50	54.70	56.80	57.20	58.63	58.90	59.10	59.14	60.50	61.07	64.07	65.40	66.18	68.49	69.05
TiO_2	0.86	0.79	1.01	0.92	1.02	1.29	0.94	0.79	0.91	0.94	0.97	1.02	0.76	0.63	0.66
Al_2O_3	17.10	18.70	16.90	17.10	16.30	16.50	17.80	18.23	17.30	14.38	14.06	13.70	15.32	14.50	14.82
FeO^t	8.21	7.90	8.03	6.98	9.00	7.91	6.43	5.68	4.30	9.01	6.65	6.48	5.14	3.88	3.56
MnO	0.18	0.31	0.17	0.21	0.20	0.17	-	0.11	-	0.22	0.19	0.28	0.08	0.11	0.06
MgO	5.85	3.29	3.09	3.32	2.96	3.04	3.05	2.50	3.80	2.96	2.36	2.00	1.64	0.74	1.06
CaO	8.39	7.96	7.05	7.07	7.37	6.08	6.85	5.92	6.30	5.86	4.36	4.01	4.74	3.24	3.65
Na_2O	2.65	4.10	3.99	3.90	3.39	3.04	4.27	3.81	4.30	3.66	3.87	3.72	3.76	4.26	3.59
K_2O	0.90	0.99	2.05	2.47	0.47	1.24	1.08	2.19	1.69	0.59	2.14	2.12	1.89	2.59	2.27
P_2O_5	-	0.25	0.28	0.16	0.12	-	-	0.30	-	0.17	0.26	-	0.07	-	0.18
Total	98.64	98.99	99.37	99.33	99.46	98.17	99.52	98.67	99.10	98.86	98.93	98.73	99.58	98.44	98.90
reference	1	2	2	2	3	1	4	5	6	3	3	3	3	3	3

1 - Grove et al. [1982]; 2 - Baker and Eggler [1987]; 3 - Sekine et al. [1979]; 4 - Eggler and Burnham [1973]; 5 - Stern and Wyllie [1978]; 6 - Eggler [1972]

%) that when considered on an anhydrous basis are similar to typical arc HAB. For one sample, the same liquidus assemblage was found at 1 kb but at a slightly higher temperature (1050°C) [*Sisson and Grove*, 1993b]. Coexisting liquids were, however, andesitic not basaltic (again on an anhydrous basis). The samples used in these experiments are compositionally unlike typical arc HAB. In particular, they are much less aluminous than typical HAB (Table 2).

High-magnesia basalt studies. Unlike for HAB, a variety of HMB have been investigated experimentally (Table 6). As part of a larger study, *Tatsumi* [1982] investigated the phase relations of a HMB. At low pressures (< 11 kb), ol marks the anhydrous liquidus but is replaced by opx at higher pressures. The occurrence of cpx just below the liquidus at 11 kb led *Tatsumi* [1982] to suggest this HMB was saturated with a three phase assemblage at 11 kb and 1305°C (Table 7). The addition of small amounts of water (3.8 %) moved the point of multiple saturation to 15.5 kb and 1210°C, i.e. lower temperature but higher pressure. In a later study, *Tatsumi et al.* [1983] conducted experiments on what they termed a "...primary high-alumina basalt...(p. 5820)." Examination of their starting composition (Table 1) clearly shows that this material is, in fact, a HMB. Multiple saturation in a lherzolite (ol + cpx + opx) assemblage occurs at 15 kb and 1340°C under anhydrous conditions (Table 7). The addition of 1.5 % water decreases the temperature of saturation to 1325°C but increases the pressure to 17 kb. Unfortunately, glass compositions were not reported for either set of experiments. In another experimental study, *Gust and Perfit* [1987] determined the anhydrous phase relations of an Aleutian HMB (Table 1). This HMB is multiply-saturated with ol + cpx around 9 kb and is the only HMB investigated experimentally saturated with a wehrlitic (ol + cpx) phase assemblage (Table 7). In general, glass compositions were basaltic but did not approximate HAB. The anhydrous phase relations of another Aleutian HMB were determined by *Draper and Johnston* [1992]. This sample is multiply-saturated with ol + plag + cpx + opx + sp near 12 kb (Table 7). Although high-pressure glasses have compositions similar to some HABs, they do not approximate the majority of Aleutian basalts and basaltic andesites [*Draper and Johnston*, 1992]. A reaction relation between ol and melt was found at 10 kb and low temperatures, but the resultant liquids were enriched in alkalis and depleted in CaO relative to typical Aleutian HAB. The 1 atm crystallization sequence of a compositionally unusual Medicine Lake HMB (Table 1) was determined by *Grove et al.* [1982]. Ol and pl first appear together at 1228°C and are the only phases to crystallize for over 50°C. *Bartels et al.* [1991] investigated the phase relations of this sample to 15 kb under anhydrous conditions and suggest it is multiply saturated with ol + cpx + sp + plag at 11 kb and approximately 1285°C (Table 7). Glass compositions at 10 and 12 kb are basaltic but are less aluminous and more magnesian than typical HAB. Under water-saturated conditions, *Sisson and Grove* [1993a] found that at 2 kb the liquidus of this Medicine Lake basalt is marked by the appearance of ol at 1050°C. If considered on an anhydrous basis, the coexisting liquids have high alumina, but are more calcic and less alkalic than most HAB. At 1 kb, the liquidus temperature increases to 1100°C and the phase assemblage remains unchanged (Table 7) [*Sisson and Grove*, 1993b]. Coexisting liquids are crudely similar to HAB when recalculated on anhydrous basis but exhibit the same compositional mismatches as the higher pressure liquids. Compositionally, the sample used in these experiments is not a typical HMB (Table 1). It has less silica, MgO and K_2O but more alumina than most natural HMB and HMB experimental starting materials. In a study of Japanese arc basalts, *Tatsumi et al.* [1994]

Table 6 Summary of Experimental Conditions of Important Inverse Experiments

Study	Starting Material(s)	1 atm T (°C)	1 atm f_{O_2}	high-pressure p (kb)	high-pressure T (°C)	high-pressure f_{O_2}	H_2O (wt %)
High-alumina basalt							
Baker and Eggler [1983]	Aleutian HAB	1080-1150	NNO	2-20	900-1300	unbuffered	0-6
Johnston [1986]	Scotia HAB	-		10-31	1233-1500	unbuffered	0
Baker and Eggler [1987]	Aleutian HAB	1060-1250	NNO	2-10	1000-1250	unbuffered	0-2
Sisson and Grove [1993a]	Medicine Lake high-MgO, low-Al_2O_3 HAB	-		2	965-1012	NNO	saturated
Sisson and Grove [1993b]	Medicine Lake high-MgO, low-Al_2O_3 HAB	-		1	1020-1050	NNO	saturated
High-magnesia basalt							
Tatsumi [1982]	Japanese HMB	-		7-22	980-1340	unbuffered	3.8,7,20
Tatsumi et al. [1983]	synthetic HMB	-		8-25	1295-1380	~NNO	0-1.5
Gust and Perfit [1987]	Aleutian HMB	1175-1263	NNO	5-15	1175-1350	unbuffered?	0
Grove et al. [1982]	Medicine Lake high-Al_2O_3 HMB	1059-1234	QFM			-	
Bartels et al. [1991]	Medicine Lake high-Al_2O_3 HMB	1238-1270	?	10-15	1240-1370	unbuffered	0
Sisson and Grove [1993a]	Medicine Lake high-Al_2O_3 HMB	-		2	1000-1050	NNO	saturated
Sisson and Grove [1993b]	Medicine Lake high-Al_2O_3 HMB	-		1	1082-1100	NNO	saturated
Draper and Johnston [1992]	Aleutian HMB	1150-1257	NNO-1	10-20	1150-1475	unbuffered	0
Tatsumi et al. [1994]	Japanese HMB	-		9-15	1280-1340	unbuffered	0
High-magnesia andesite							
Kushiro and Sato [1978]	Japanese HMA	-		12-18	1000-1100	unbuffered	7,12-16
Tatsumi [1981]	Japanese HMA	-		11-17	1000-1300	unbuffered	8, saturated
Tatsumi [1982]	Japanese HMA	-		7-22	980-1340	unbuffered	3.8,7,20
Evolved Lavas							
Eggler [1972]	Paricutin, Mexico andesite	1088-1216	QFM	0.5-8	900-1140	QFM	2,4.7,saturated
Eggler and Burnhan [1973]	Mount Hood andesite	-		0.5-8	900-1158	QFM	2,4.7,saturated
Stern and Wyllie [1978]	Sierra Nevada tonalite	-		30	850-1250	unbuffered	5-30
Sekine et al. [1979]	Japanese andesites and dacites	1050-1200	range	0.5-1.5	900-1070	NNO	saturated
Grove et al. [1982]	Medicine Lake andesites	1059-1234	QFM			-	
Baker and Eggler [1987]	Aleutian basaltic andesite to andesite	1060-1250	NNO	2.8	1000-1250	unbuffered	0-2

investigated three HMB. Two samples, TM-0 and RZ-6, could have been in equilibrium with mantle ol and were used as starting material (Table 1). Because the third basalt (AKT12) was deemed a fractionated sample, ol and opx were added to create two starting compositions in equilibrium with ol of composition Fo_{89} and Fo_{91}, respectively. These samples are saturated with a harzburgitic (ol + opx) phase assemblage at pressures ranging from 10 to 14 kb and temperatures from 1300 to 1330°C (Table 7).

High-magnesia andesite experiments. Three experimental studies [*Kushiro and Sato*, 1978; *Tatsumi*, 1981, 1982] have examined the phase relations of HMA (Table 6). In the earliest study, *Kushiro and Sato* [1978] found that with 12-16 wt % water the HMA liquidus was defined by ol up to 16 kb but by opx above 18 kb. Based on these relations, they suggested the liquid was multiply saturated with ol and opx between 16 and 18 kb (Table 7). *Tatsumi* [1981] conducted H_2O-saturated and -undersaturated experiments on another Japanese HMA similar in MgO but 2 wt % more siliceous and 2 % less aluminous (Table 3). At water-saturated conditions, this HMA is multiply saturated with a harzburgite (ol + opx) phase assemblage at ~ 16 kb and 1080°C. In water-undersaturated experiments, the saturation point shifted to lower pressures (approximately 12 kb) but higher temperatures (Table 7). Whereas these two HMA have very similar phase relations, those of a HMA with 2 wt % less MgO are markedly different [*Tatsumi*, 1982]. This HMA is saturated with a lherzolitic (ol + opx + cpx) not harzburgitic phase assemblage at 15 kb and 1030° C and water-saturated conditions. The point of multiple saturation shifts to higher pressure and lower temperature under water-undersaturated conditions (Table 7).

Evolved lavas. Historically, evolved calc-alkaline lavas have been the subject of considerable experimental interest [e.g. *Eggler*, 1972; *Eggler and Burnham*, 1973; *Stern and Wyllie*, 1978; *Sekine et al.*, 1979]. *Eggler* [1972] conducted water-saturated and undersaturated experiments on an andesite (Table 5) from Paricutin volcano. Multiple saturation with opx + plag occurs at about 6 kb with 2 % H_2O but does not occur below 10 kb (the maximum experimental pressure) for water contents of 4 %. When the magma is water saturated, opx + plag appearance shifts to about 0.5 kb (Table 7). Experiments on a Mount Hood andesite that is less siliceous and potassic but more iron-rich

Table 7: Conditions of Multiple Saturation Defined by Inverse Experiments

Study	sam #	Composition				Anhydrous			Hydrous			
		SiO_2	Al_2O_3	MgO	CaO	T (°C)	p (kb)	phases	% H_2O	T (°C)	p (kb)	phases
High-alumina basalt												
Baker and Eggler [1983]	AT-1	49.89	19.43	4.79	9.17	1300	18	gar+cpx+pl		-		
Johnston [1986]	SSS.1.4	50.26	18.46	7.32	11.81	1330	17	cpx+pl		-		
						1440	27	cpx+gar		-		
Baker and Eggler [1987]	AT-112	49.00	19.00	6.18	11.40	not multiply saturated < 10 kb			2	not multiply saturated at 2 kb		
Sisson and Grove [1993a]	82-62	49.30	17.70	8.61	11.30	-			saturated	1012	2	ol+pl+cpx
	82-66	51.20	17.30	7.47	10.20	-			saturated	1012	2	ol+pl+cpx
Sisson and Grove [1993b]	82-66	51.20	17.30	7.47	10.20	-			saturated	1050	1	ol+pl+cpx
High-magnesia basalt												
Tatsumi [1982]	SD-438	49.76	15.48	11.68	8.91	1305	11	ol+cpx+opx	3.8	1210	15.5	ol+cpx+opx
Tatsumi et al. [1983]		49.39	15.70	12.05	9.43	1340	15	ol+cpx+opx	1.5	1325	17	ol+cpx+opx
Gust and Perfit [1987]	MK-15	51.20	15.69	9.64	10.12	1310	9.5	ol+cpx		-		
Grove et al. [1982]	79-35g	47.70	18.60	9.92	12.10	1228	0.001	ol+pl		-		
Bartels et al. [1991]	79-35g	47.70	18.60	9.92	12.10	1285	11	ol+cpx+sp+pl		-		
Sisson and Grove [1993a]	79-35g	47.70	18.60	9.92	12.10	-			saturated	not multiply saturated at 2 kb		
Sisson and Grove [1993b]	79-35g	47.70	18.60	9.92	12.10	-			saturated	not multiply saturated at 1 kb		
Draper and Johnston [1992]	ID-16	49.32	15.84	11.35	10.67	1315	12	ol+cpx+opx+pl		-		
Tatsumi et al. [1994]	RZ-6	49.09	15.09	13.34	10.60	1330	14	ol+opx		-		
	TM-0	50.57	14.48	12.64	10.57	1315	12	ol+opx		-		
	AKT12b	50.63	15.05	13.23	8.45	1330	13	ol+opx		-		
	AKT12a	51.33	16.13	10.62	9.06	1300	10	ol+opx		-		
High-magnesia andesite												
Kushiro and Sato [1978]	KMA	57.04	15.36	9.03	7.04	-			12-16	1040	17	ol+opx
Tatsumi [1981]	TGI	59.44	13.51	9.62	6.23	-			8	1120	12	ol+opx
						-			saturated	1080	16	ol+opx
Tatsumi [1982]	SD-261	57.13	15.83	7.39	7.19	-			7	1070	10	ol+cpx+opx
						-			20	1030	15	ol+cpx+opx
Evolved lavas												
Eggler [1972]	FP1652	60.50	17.30	3.80	6.30	-			saturated	1080	0.5	pl+opx
						-			2	1110	5.5	pl+opx
									4	not multiply saturated < 10 kb		
Eggler and Burnham [1973]	MHA	59.10	17.80	3.05	6.85	-			saturated	950	6	pl+opx
						-				950	8	opx+amph
									2	not multiply saturated < 8 kb		
						-			4.7	1120	9	ol+opx
Stern and Wyllie [1978]	T101	59.14	18.23	2.50	5.92	-			5-30	not multiply saturated at 30 kb		
Sekine et al. [1979]	R-2	58.63	16.30	2.96	7.37	-			saturated	1026	0.75	pl+opx+cpx
	GRM-4	61.07	14.38	2.96	5.86	-			saturated	1097	0.13	pl+opx+cpx
	II-GRM	64.07	14.06	2.36	4.36	-			saturated	1080	0.16	pl+opx
	I-GRM	65.40	13.70	2.00	4.01	-			saturated	1012	0.50	pl+opx
	GRM-1	66.18	15.32	1.64	4.74	-			saturated	1050	0.38	pl+opx+cpx
	P_S-G	68.49	14.50	0.74	3.24	-			saturated	950	0.80	pl+opx
	P_A-G	69.05	14.82	1.06	3.65	-			saturated	981	0.88	pl+opx
Grove et al. [1982]	79-38b	54.50	17.10	5.85	8.39	not multiply saturated at 0.001 kb				-		
	79-9c	58.90	16.50	3.04	6.08	not multiply saturated at 0.001 kb				-		
Baker and Eggler [1987]	AT-41	54.70	18.70	3.29	7.96	not multiply saturated < 10 kb			2	not multiply saturated at 2 kb		
	AT-29	56.80	16.90	3.09	7.05	not multiply saturated < 10 kb			2	not multiply saturated at 2 kb		
	AT-25	57.20	17.10	3.32	7.07	not multiply saturated < 10 kb			2	not multiply saturated at 2 kb		

than the Paricutin lava (Table 5) displayed slightly different phase equilibria [*Eggler and Burnham*, 1973]. Under saturated conditions, the point of multiple saturation occurs not at 0.5 kb but at 6 kb and the assemblage is plag + amphibole. In another inverse experimental study, *Stern and Wyllie* [1978] examined the phase relations of a tonalite (Table 5) at 30 kb with 5 to 30 % H_2O. At this pressure and all water contents, garnet is the liquidus phase followed by cpx and the melt does not have a point of multiple saturation on its liquidus. Water-saturated phase equilibria to 1.5 kb of several whole rock and groundmass separate samples were studied by *Sekine et al.* [1979]. Compositionally, these materials ranged from andesite to dacite (Table 5). These authors found that multiple saturation occurred at pressures of less than 1 kb and the solid assemblage was either plag + opx + cpx or plag + opx (Table 7).

In a more recent experimental study, *Grove et al.* [1982] determined the 1 atm crystallization sequences of a Medicine Lake basaltic andesite and an andesite (Table 5). For both samples, plag is the first phase to crystallize and is followed by ol in the basaltic andesite and opx in the andesite. Glass compositions determined in these experiments were also used to locate phase boundaries on pseudoternary diagrams. The higher pressure phase relations of these samples have not been determined. *Baker and Eggler* [1987] conducted inverse experiments on three Aleutian samples ranging from basaltic andesite to andesite (Table 5). Under anhydrous conditions and up to 10 kb (the highest pressure attained experimentally), these samples are not multiply saturated. In all cases, liquidus temperatures are quite high, e.g. ~1200°C [*Baker and Eggler*, 1987], and reaction between ol and andesitic melt saturated with plag produces augite and opx. Addition of 2 % H_2O at 2 kb reduces liquidus temperatures by as much as 100°C and greatly shrinks the crystallization interval but does not change the crystallization sequence or produce a point of multiple saturation (Table 7). The 1 atm phase relations of these Aleutian lavas are markedly different from those of the Medicine Lake samples [*Grove et al.*, 1982]. *Baker and Eggler* [1987] suggested these differences may be due to compositional differences between the starting materials and noted that they have important ramifications for possible liquid-lines-of-descent at low pressures.

Summary. Experiments have shown that HAB liquids are saturated in an eclogite assemblage, cpx and garnet, at high pressure. In addition, ol does not occur as a liquidus phase under the anhydrous conditions investigated. At very low pressures, ol does occur as a liquidus phase in a water-saturated, unusually primitive HAB melt. Inverse experiments on a variety of HMB have defined a range of conditions for multiple saturation but the nature of the saturating assemblage varies (Table 7). Multiple saturation with a lherzolitic or harzburgitic assemblage occurs at 11-15 kb and 1300-1330°C in water-free systems. Adding water shifts multiple saturation to higher pressures but lower temperatures. Only one sample studied experimentally, MK-15 [*Gust and Perfit*, 1987], was saturated with a wehrlite assemblage and this occurred at 9.5 kb and 1310°C. A simple correlation between major element composition and saturating assemblage is not readily apparent (Table 7). As with HMB, high-magnesia andesites are multiply saturated (harzburgite or lherzolite) in the range 10-17 kb and between 1030 and 1120°C under water-saturated and under-saturated conditions.

SIGNIFICANCE OF SUBDUCTION PHASE EQUILIBRIA EXPERIMENTS

Because a given experiment can be interpreted in either a forward or inverse manner, individual phase equilibria studies can often be used to constrain several petrologic components of different arc magmatic models (Table 8). As for most geologic studies, these investigations have provided few definitive answers to the many questions of arc petrogenesis. However, they do impose important restrictions on plausible petrologic models.

High-magnesia Basalt Generation

Inverse experiments on a variety of HMB under different experimental conditions have identified several points of multiple saturation with lherzolite phase assemblages (Table 7). For example, *Tatsumi* [1982], *Tatsumi et al.* [1983], *Bartels et al.* [1991] and *Draper and Johnston* [1992] have found points of multiple saturation under anhydrous conditions in the pressure range 11-12 kb and 1305-1340°C. The addition of small amounts of water shifts multiple saturation to higher pressures (15-17 kb) but lower temperatures [*Tatsumi*, 1982]. *Johnston and Draper* [1992] suggested two interpretations are consistent with these experimental results. The samples could have been generated at 35 km by small degrees of melting that did not exhaust any major phase. If this melting was initiated by fluid from the slab, the latter must be moved from the slab-mantle interface to shallower depths. Alternatively, these samples could have been derived from a melt (of unknown composition) produced at greater depths but which last equilibrated with the mantle at 35 km. As noted above, many forward experiments involving peridotite melting have been conducted at pressures too low for arc magma genesis or analytical problems have rendered analyzed melt compositions suspect. Such experiments at a range of pressures and intensive parameters may provide, however, a means of choosing between the two petrologic possibilities consistent with the inverse experiments.

Other HMB experiments [*Gust and Perfit*, 1987; *Tatsumi et al*, 1994] have found points of multiple saturation with ol + cpx or ol + opx (Table 7). The petrologic significance of these results is uncertain. Assuming the samples investigated truly represent primary magmas, the liquids could have been produced by: (1) large degrees of lherzolite melting, or (2) an unknown degree of partial fusion of depleted peridotite, e.g. harzburgite or wehrlite. Alterna-

Table 8: Summary of Phase Equilibria Tests	
Forward Experiments	Inverse Investigations
High-magnesia Basalt Generation	
	Tatsumi [1982]; *Tatsumi et al.* [1983]; *Gust and Perfit* [1987]; *Grove et al.* [1982]; *Bartels et al.* [1991]; *Sisson and Grove* [1993a,b]; *Draper and Johnston* [1992]; *Tatsumi et al.* [1994]
High-alumina Basalt Generation	
Johnston [1986]; *Rapp et al.* [1991]; *Rapp and Watson* [1995]	*Johnston* [1986]; *Baker and Eggler* [1983, 1987]; *Sisson and Grove* [1993a,b]
HMB/HAB Petrologic Connection	
Tatsumi [1982]; *Tatsumi et al.* [1983]; *Gust and Perfit* [1987]; *Grove et al.* [1982]; *Bartels et al.* [1991]; *Sisson and Grove* [1993a,b]; *Draper and Johnston* [1992]; *Tatsumi et al.* [1994]	*Johnston* [1986]; *Baker and Eggler* [1983, 1987]; *Sisson and Grove* [1993a,b]
High-magnesia Andesite Generation	
Johnston [1986]; *Rapp et al.* [1991]; *Rapp and Watson* [1995]	*Kushiro and Sato* [1978]; *Tatusmi* [1981, 1982]
Evolved lava Evolution	
Johnston [1986]; *Baker and Eggler* [1983, 1987]; *Sisson and Grove* [1993a,b]	*Eggler* [1972]; *Eggler and Burnham* [1973]; *Stern and Wyllie* [1978]; *Sekine et al.* [1979]; *Grove et al.* [1982]; *Baker and Eggler* [1987]

tively, the samples may not represent primary magmas. Rather they may be derivative liquids produced by crystal fractionation or magma-mantle interaction. Again the forward peridotite melting experiments that might provide additional constraints are lacking.

High-alumina Basalt Generation

Because they have high-pressure anhydrous liquidii defined by garnet and cpx (Table 7), the HAB studied experimentally could be primary magmas derived by eclogite partial fusion [*Baker and Eggler*, 1983; *Johnston*, 1986]. There is, however, a significant difference in the pressure of saturation for the different HAB (Table 7). The Scotia sample is saturated at pressures (~ 27 kb) close to those expected at the slab-mantle interface. Based on these results and the observation that most HAB have nearly flat REE patterns, *Johnston* [1986] concluded that HAB "can be extracted from a MORB composition eclogite source at 20-27 kb providing the degrees of melting are sufficient (ca. 50%) to totally consume garnet, leaving only cpx in the residue (p. 381)." Current thermal models of subduction zones do not, however, predict temperatures at the slab interface as high as those suggested by these anhydrous experiments.

In contrast to this sample, the Aleutian HAB studied by *Baker and Eggler* [1983] was multiply saturated at 18 kb, a pressure that would indicate a depth considerably above the slab-mantle interface. These results could be interpreted as evidence for the diapir ascent model of *Brophy and Marsh* [1986]. The differences in experimental results may also suggest that HAB are produced in a variety of ways.

The two forward experiments on MORB/eclogite melting also provide permissive evidence for generation of HAB by slab melting. For example, the presence of cpx ± gar on the eclogite liquidus is consistent with this process [*Johnston*, 1986]. The compositional similarity of experimental liquids and HAB in the cpx only runs suggest large degrees of melting or melt extraction at shallower levels [*Brophy and Marsh*, 1986]. Gar + cpx residuum in the high-pressure melting experiments (27 and 32 kb) experiments of *Rapp et al.* [1991] and *Rapp and Watson* [1995] are also consistent with a slab origin for HAB. None of the liquid compositions produced match HAB compositions but this may be due to the limited degrees of partial fusion attained in the experiments. Compositional trends at all pressures suggest that at higher degrees of melting HAB-like liquids may be produced. These forward experiments, like the inverse ones, suggest HAB could be generated by eclogite melting in the pressure range of 25-35 kb and with little or no water.

The High-magnesia Basalt/High-alumina Basalt Petrologic Connection

Because ol does not occur on their anhydrous liquidii, HAB can be generated by fractionation of ol (± other phases) from HMB in water-free systems only under special conditions [*Johnston*, 1986; *Baker and Eggler*, 1983, 1987]. In particular, a reaction relation must exist between ol and HAB melt. In this case, ol could have been fractionated in earlier stages of magmatic development even though it does not appear on the derivative melt liquidus. Although the existence of such a relationship has never been specifically addressed by ol-addition experiments, the results of *Draper and Johnston* [1992] suggest that an ol-melt reaction may be unlikely.

Given the greatly expanded ol volume in hydrous systems, a fractionation relation between HAB and HMB may be possible in water-bearing systems. Unfortunately, only one water-bearing experimental study has been conducted on typical HAB (Table 7). *Baker and Eggler* [1987] found that at 2 kb the addition of up to 5 % water failed to place ol on the HAB liquidus. The hydrous experiments of *Sisson and Grove* [1993a] did define a point of ol + pl + cpx saturation, the phase assemblage used in most HMB/ HAB fractionation schemes, e.g. *Kay et al.* [1982]. Although coexisting hydrous melts are not HAB-like, when considered on an anhydrous basis they have high alumina contents but are enriched in CaO and depleted in alkalis. Despite these compositional differences, *Sisson and Grove* [1993a,b] suggested HAB were produced from such liquids by degassing at shallow levels. As noted above, the starting materials for these experiments are high-MgO, low alumina "HAB" that are compositionally very much different from either HMB or HAB (Tables 1 and 2). Whether or not HAB are produced by this fractionation scheme, the origin of these unusual low-alumina calc-alkaline basalts is unclear.

As described above, numerous HMB experimental studies have defined points of ol + pl + cpx saturation (Table 7). These temperatures and pressures represent conditions under which HMB could fractionate to produce HAB-like liquids and suggest HAB generation must occur in magma chambers located at depths greater than 30 km (Table 7). In general, liquids coexisting with this phase assemblage do not, however, approximate HAB. Consequently, the forward experiments, like their inverse counterparts, have failed to confirm the proposed HMB/HAB relationship. In light of these results, HAB could be derivative magmas of HMB only at conditions not yet addressed experimentally, e.g. different pressures, higher water contents, different protoliths.

Origin of High-magnesia Andesites

Because the HMA studied are not saturated with an eclogite assemblage, they could not have been generated by slab melting at the intensive parameters of these experiments (Table 7). The HMA are, however, multiply saturated with mantle phases at moderate pressures and high water contents (> 7 wt %). These phase relations are incompatible with the slab-melting model of *Defant and Drummond* [1990], but are consistent with the interaction of magma with peridotite during ascent. Because the inverse experiments reveal nothing about the nature of the reacting magma, either the model of *Kay* [1978] or that of *Myers and Frost* [1994] could explain the observed relations. Both of these models imply the existence of a lithospheric magmatic plumbing system between the magma source and crustal-level magma chambers. An alternative interpretation of the observed phase relations suggests HMA are produced by peridotite partial melting with large amounts of water in the pressure range 10-17 kb. Because peridotite melting is assumed to produce HMB in many arc models, a peridotite melting origin for HMA would require a different melting regime. Differences in the melting processes could include variations in peridotite protolith, degree of partial melting or intensive conditions of melting. In contrast to the other models, a melting origin for HMA suggests the positioning of a critical peridotite melting isotherm and/or a change in the lithologic nature of the mantle at 30-50 km within the wedge. Given the rarity of HMA (Table 3) it must be remembered that their occurrence signals an unusual magmatic regime. Consequently, the physical and thermal conditions suggested by the HMA phase relations are unlikely to play a significant role in the generation of typical calc-alkaline suites.

Two forward eclogite melting experiments provide important constraints on the types of primary magmas that can be produced by slab melting. Large degrees of anhydrous melting at 27 kb yield basaltic liquids coexisting with some combination of cpx and garnet [*Johnston*, 1986]. In contrast, smaller degrees of partial fusion over a range of pressures and with small amounts of water produce compositions ranging from basaltic andesite to rhyodacite [*Rapp et al.*, 1991; *Rapp and Watson*, 1995]. Depending upon pressure, the solid residuum coexisting with these liquids varies from amphibolite to eclogite. Dacitic and andesitic liquids with silica contents similar to HMA (55-65 wt %) have been produced at a variety of melting conditions and small degrees of melting (10-20 %). Despite their similar silica contents, none of the experimental liquids have MgO contents as high as HMA. Consequently, the types of slab melts proposed by *Defant and Drummond* [1990] cannot be produced at the conditions investigated experimentally. Different degrees of melting or intensive parameters might, however, produce the siliceous, high-MgO melts. As with the inverse experiments, the melting investigations are broadly consistent with the models of *Kay* [1978] and *Myers and Frost* [1994]. The phase relations do, however, limit the conditions under which each model would be viable. Production of siliceous liquids [*Kay*, 1978] requires small degrees of melting at high pressures whereas much larger degrees of melting are required for primary and parental HAB melts [*Myers and Frost*, 1994].

Petrogenesis of Evolved Calc-alkaline Lavas

None of the andesites investigated by *Baker and Eggler* [1987] were multiply saturated under anhydrous conditions at pressures less than 10 kb (Table 7). The addition of 2 % water at 2 kb did not produce multiple saturation or alter the order of crystallization. In contrast, the high-SiO_2 andesites crystallize pl + opx between 5 and 8 kb with low water contents [*Eggler*, 1972; *Eggler and Burnham*, 1973] and dacites are multiply saturated with pl + opx + cpx or pl + opx at very low pressures (< 1 kb) when the melts are water-saturated [*Eggler*, 1972; *Sekine et al.* 1979]. Because fractionation schemes for the formation of evolved lavas generally involve removal of pl ± ol/opx ± cpx ± mt [e.g. *Myers et al.*, 1995], these experimental results suggest "crustal" magma chambers may be located over a range of depths. Under anhydrous conditions, calc-alkaline suites must have evolved at pressures greater than 10 kb thereby placing crustal chambers at depths greater than 30 km. Conversely, fractionation from water-saturated magmas would require very shallow (~ 3 km) plumbing systems. Intermediate water contents would place magma chambers at moderate depths. The positioning of these chambers has important implications for crustal growth models as well as thermal and physical models of crustal and subcrustal structure. Additional phase equilibria studies of evolved lavas and HAB are necessary to determine which of these experiments are appropriate for the genesis of the majority of evolved calc-alkaline lavas. Rather than using genetically unrelated samples as in the past, these investigations should focus on individual calc-alkaline suites. In this manner, the phase equilibria results can be combined with other petrologic information to constrain better models of magma evolution at "crustal" depths.

SUMMARY AND SUGGESTIONS

Every subduction zone petrologic model, whether derived from geochemical, isotopic or geophysical constraints, implies a certain set of phase equilibria. These phase relations provide additional and independent tests of the proposed models. For the three major classes of petrologic models, these criteria include:

1. *slab-dominated models*: (a) eclogite partial melts must be HAB-like in composition, (b) HAB must have high-pressure liquidii saturated with some combination of eclogite phases, i.e. cpx and garnet; (c) HAB liquids must produce calc-alkaline liquid-lines-of-descent at low pressure, and (d) evolved calc-alkaline lavas must be saturated with plag, ol, cpx, ±opx. (Because the proposed generation of evolved calc-alkaline lavas is the same for all three models, this last condition also holds for the other two models discussed below.)

2. *mantle-dominated models*: (a) partial fusion of peridotite at high pressure and under hydrous conditions must produce HMB liquids, (b) HMB liquids must be saturated with some combination of ol, cpx and opx under the same conditions, (c) HMB liquid-lines-of-descent must produce HAB-like liquids, and (d) HAB liquidii must be defined by ol (with or without other phases).

3. *multi-source models*: Because there are a variety of multi-source models each producing major arc mafic lava types differently, the phase equilibria predicted by multi-source models are best summarized individually by lava type. Depending upon the model, high-magnesia basalts may be either primary or derivative liquids. Despite the difference in origin, the same phase relations are consistent with both models: (a) peridotite melting will not produce HMB liquids because they were formed by open-system processes, and (b) ol + cpx ± opx must define the HMB liquidii. In this instance, experimental studies cannot differentiate between a primary or differentiated origin. For the more mafic basalts, multi-source models explain HAB lavas as either primary or differentiated magmas but predicted phase equilibria differ between the two models. A primary origin for HAB dictates: (a) eclogite partial melting must produce HAB liquids, and (b) HAB must be saturated with some combination of cpx and garnet. Conversely, if they are derivative liquids: (a) fractionation of HMB liquids must produce HAB magmas, and (b) ol must appear on the HAB liquidus. Phase equilibria predicted for HMA depend upon whether they are presumed to be primary or derivative melts. If they are of primary origin: (a) eclogite melting at high pressure must yield HMA-like liquids, and (b) HMA must be saturated with some combination of cpx and garnet. A differentiated origin suggests: (a) HMA liquids cannot be produced by simple forward experiments because of their open-system origin, and (b) ol + cpx ± opx multiple saturation must define the HMA liquidus. The latter condition could reflect generation of HMA by partial melting within the wedge or interaction of a primary magma of unknown character with the mantle during ascent.

Presently, the number of experimental phase equilibria studies that bear on the major questions of subduction zone magmatism is limited. For example, experimental studies under anhydrous conditions exist for only three HAB. In addition, experimental difficulties, e.g. iron loss or quench crystallization, as well as significant differences between starting materials and typical calc-alkaline lavas limit the general applicability of some experiments. Because most phase equilibria studies have focused on genetically unrelated samples, their potential for unraveling the origin of more evolved arc lavas, e.g. basaltic andesites through dacites, has also not been fully realized.

Despite these limitations, phase equilibria studies have been useful in placing certain constraints on the origin of primary arc magmas and their evolution. These constraints can be summarized as follows:
1. HAB could be generated by eclogite partial melting under anhydrous conditions at high pressure;
2. HMB may be generated by peridotite melting under

Table 9: Phase Equilibria Studies Needed to Test Arc Models

Forward Experiments	Inverse Investigations
High-magnesia Basalt Generation	
• peridotite melting at moderate to high pressures and with variable H_2O	• anhydrous HMB studies
	• hydrous HMB experiments
• peridotite-HAB sandwich experiments	
High-alumina Basalt Generation	
• eclogite melting experiments covering a range of conditions and starting compositions	• anhydrous HAB studies
	• HAB mineral addition investigations
	• hydrous HAB experiments
HMB/HAB Petrologic Connection	
• hydrous HMB crystallization	• anhydrous HAB studies
	• HAB mineral addition investigations
	• hydrous HAB experiments
High-magnesia Andesite Generation	
• anhydrous eclogite melting at high pressure	• anhydrous HMA investigations
	• hydrous HMA experiments
• peridotite melting at low to moderate pressures and with variable H_2O	
• peridotite-HAB sandwich experiments	
• peridotite-siliceous melt sandwich experiments	
Evolved lava Evolution	
• anhydrous HAB studies	• anhydrous crystallization of evolved lavas
• hydrous HAB experiments	
• anhydrous crystallization of evolved lavas	• crystallization of evolved lavas with variable water content
• crystallization of evolved lavas with variable water contents	
(these are low-p experiments)	

anhydrous to H_2O under-saturated conditions in the pressure range 15-25 kb;

3. under anhydrous conditions, HMB cannot produce HAB by fractionation involving ol unless a reaction relationship exists between this phase and the HAB melt;

4. HMA cannot be primary eclogite partial melts in water-bearing systems. They may, however, have been in equilibrium with peridotite in the pressure range 15-20 kb; and

5. under anhydrous or water-undersaturated conditions, evolved calc-alkaline lavas cannot be formed by proposed crystal fractionation schemes at pressures less than 10 kb. If, however, the suites evolved with large amounts of water, pressures of formation are much lower (< 10 kb).

Because of the limited experimental results (Table 8) and the complex nature of arc petrologic models (Figure 5), many opportunities exist for future phase equilibria experiments (Table 9). These experiments can be grouped into four classes each focused on the origin of one of the main lava classes important in arcs worldwide. The utility of these experiments can be greatly enhanced by careful consideration of the consequences of a variety of petrologic models. In this manner, experiments addressing a variety of objectives can be designed (Table 9). Whereas previous investigations have commonly focused on individual samples or starting materials that are not genetically related, future experimental investigations of cogenetic calc-alkaline suites offer a means of addressing arc petrologic questions within a coherent and systematic framework. Such an approach permits a close integration of experimental studies with comprehensive geologic, petrographic and geochemical investigations.

Acknowledgments. Unofficial reviews by Travis McElfresh, Carol Frost, Kirsten Nicolaysen and Susan Swapp and official reviews by Simon Peacock and Jeffrey Ryan materially improved this paper and are gratefully acknowledged. This project was supported, in part, by NSF Grants EAR 91-17809 (J.D. Myers) and EAR 95-06045 (A.D. Johnston).

REFERENCES

Baker, D.R., and D.H. Eggler, Fractionation paths of Atka (Aleutians) high-alumina basalts: constraints from phase relations, *J. Volcanol. Geotherm. Res., 18,* 387-404, 1983.

Baker, D.R., and D.H. Eggler, Compositions of anhydrous and hydrous melts coexisting with plagioclase, augite, and olivine or low-Ca pyroxene from 1 atm to 8 kb: application to the Aleutian volcanic center of Atka, *Am. Mineral., 72,* 12-28, 1987.

Baker, M.B., and E.M. Stolper, Determining the composition of high-pressure mantle melts using diamond aggregates, *Geochim. Cosmoshim Acta,* 58, 2811-2827, 1994.

Bartels, K.S., R.J. Kinzler, and T.L. Grove, High pressure phase relations of primitive high-alumina basalts from Medicine Lake volcano, northern California, *Contrib. Mineral. Petrol., 108,* 253-270, 1991.

Basaltic Volcanism Study Project, Basaltic Volcanism on the Terrestrial planets, Pergamon Press, New York, 1981.

Brophy, J.G., Andesites from northeastern Kanaga Island, Aleutians. implications for calc-alkaline fractionation mechanisms and magma chamber development, *Contrib. Mineral. Petrol., 104,* 568-581, 1990.

Brophy, J.G., and B.D. Marsh, On the origin of high-alumina arc basalt and the mechanics of melt extraction, *J. Petrol., 27,* 763-789, 1986.

Conrad, W.K., S.M. Kay, and R.W. Kay, Magma mixing in the Aleutian arc: evidence from cognate inclusions and composite xenoliths, *J. Volcanol. Geotherm. Res., 18*, 279-295, 1983.

Davies, J.H., and D.J. Stevenson, Physical model of source region of subduction zone volcanics, *J. Geophys. Res., 97*, 2037-2070, 1992.

Defant, M., and M.S. Drummond, Derivation of some modern arc magmas by melting of young subducted lithosphere, *Nature, 347*, 662-665, 1990.

Draper, D.S., and A.D. Johnston, Anhydrous PT phase relations of an Aleutian high-MgO basalt: an investigation of the role of olivine-liquid reaction in the generation of arc high-alumina basalts, *Contrib. Mineral. Petrol., 112*, 501-519, 1992.

Eggler, D.H., Water-saturated and undersaturated melting relations in a Paricutin andesite and an estimate of water content in the natural magma, *Contrib. Mineral. Petrol., 34*, 261-271, 1972.

Eggler, D.H., and C.W. Burnham, Crystallization and fractionation trends in the system andesite-H_2O-CO_2-O_2 at pressures to 10 kb, *Geol. Soc. Am. Bull., 84*, 2517-2532, 1973.

Fournelle, J.H., B.D. Marsh, and J.D. Myers, Age, character and significance of Aleutian arc volcanism, in The Geology of Alaska, vol. G-1, edited by G. Plafker and H.C. Berg, pp. 723-757, , Geol. Soc. Am., Boulder, CO 1994.

Green, D.H., Experimental melting studies on a model upper mantle composition at high pressure under water-saturated and water-undersaturated conditions, *Earth Planet. Sci. Lett., 19*, 37-53, 1973.

Green, D.H., W.O. Hibberson, and A.L. Jaques, Petrogenesis of mid-ocean ridge basalts, in *The Earth: its origin, structure and evolution*, edited by M.W. McElhinny, pp. 265-299, Academic Press, 1979.

Grove, T.L., D.C. Gerlach, T.W. Sando, and M.B. Baker, Origin of calc-alkaline series lavas at Medicine Lake volcano by fractionation, assimilation and mixing, *Contrib. Mineral. Petrol., 80*, 160-182, 1982.

Gust, D.A., and M.R. Perfit, Phase relations of a high-Mg basalt from the Aleutian Island arc: implications for primary island arc basalts and high-Al basalts, *Contrib. Mineral. Petrol., 97*, 7-18, 1987.

Huang, W.L., and P.J. Wyllie, Melting relations of muscovite-granite to 35 kbar as a model for fusion of metamorphosed subducted oceanic sediments, *Contrib. Mineral. Petrol., 42*, 1-14, 1974.

Jaques, A.L., and D.H. Green, Determination of liquid compositions in experimental, high-pressure melting of peridotite, *Am. Mineral., 64*, 1312-1321, 1979.

Jaques, A.L., and D.H. Green, Anhydrous melting of peridotite at 0-15 kb pressure and the genesis of tholeiitic basalts, *Contrib. Mineral. Petrol., 73*, 287-310, 1980.

Johnson, M.C., and T. Plank, Experimental constraints on sediment melting during subduction, *Eos, 74*, 680, 1993.

Johnston, A.D., Anhydrous P-T phase relations of near-primary high-alumina basalt from the South Sandwich Islands: implications for the origin of island arcs and tonalite-trondhjemite series rocks, *Contrib. Mineral. Petrol., 92*, 368-382, 1986.

Johnston, A.D., and D.S. Draper, Near-liquidus phase relations of an anhydrous high-magnesia basalt from the Aleutian Islands: implications for arc magma genesis and ascent, *J. Volcanol. Geotherm. Res., 52*, 27-41, 1992.

Kay, R.W., Aleutian magnesian andesites: melts from subducted Pacific Ocean crust, *J. Volcanol. Geotherm. Res., 4*, 117-132, 1978.

Kay, S.M., R.W. Kay, and G.P. Citron, Tectonic controls on tholeiitic and calc-alkaline magmatism in the Aleutian arc, *J. Geophys. Res., 87*, 4051-4072, 1982.

Kay, S.M., R.W. Kay, H.K. Brueckner, and J.L. Rubenstone, Tholeiitic Aleutian arc plutonism: the Finger Bay Pluton, Adak, Alaska, *Contrib. Mineral. Petrol., 82*, 99-116, 1983.

Kushiro, I., Effect of water on the composition of magmas formed at high pressures, *J. Petrol., 13*, 311-334, 1972.

Kushiro, I., Origin of some magmas in oceanic and circumoceanic regions, *Tectonophysics, 17*, 211-222, 1973.

Kushiro, I., Partial melting of mantle wedge and evolution of island arc crust, *J. Geophys. Res., 95*, 15,929-15,939, 1990.

Kushiro, I., N. Shimizu, Y. Nakamura, and S. Akimoto, Compositions of coexisting liquid and solid phases formed upon melting of natural garnet and spinel lherzolites at high pressures: a preliminary report, *Earth Planet. Sci. Lett., 14*, 19-25, 1972.

Kushiro, I., and H. Sato, Origin of some calc-alkalic andesites in the Japanese Islands, *Bull. Volcanol., 41*, 576-585, 1978.

Marsh, B.D., The Aleutians: in *Andesites: orogenic andesites and related rocks*, edited by R.S. Thorpe, pp. 99-115, John Wiley, New York, 1982.

Myers, J.D., and C.D. Frost, A petrologic re-investigation of the Adak volcanic center, central Aleutian arc, Alaska, *J. Volcanol. Geotherm. Res., 60*, 109-146, 1994.

Myers, J.D., C.D. Frost, and C.L. Angevine, A test of the quartz eclogite source for parental Aleutian magmas: a mass balance approach, *J. Geol., 94*, 811-828, 1986.

Myers, J.D., B.D. Marsh, and A.K. Sinha, Strontium isotopic and selected trace element variations between two Aleutian volcanic centers (Adak and Atka): implications for the development of arc volcanic plumbing systems, *Contrib. Mineral. Petrol., 91*, 221-234, 1985.

Myers, J.D., B.D. Marsh, C.D. Frost, and J.A. Linton, Petrologic constraints on the spatial relations of crustal magma chambers beneath the Atka volcanic center, central Aleutian arc, *Contrib. Mineral. Petrol.*, in review, 1995.

Mysen, B.O., and A.L. Boettcher, Melting of a hydrous upper mantle: II, geochemistry of crystal and liquids formed by anatexsis of mantle peridotite at high pressure and high temperatures as a function of controlled activities of water, hydrogen and carbon dioxide, *J. Petrol., 16*, 549-593, 1975.

Mysen, B.O., and I. Kushiro, Compositional variations of coexisting phases with degree of melting of peridotite in the upper mantle, *Am. Mineral., 62*, 843-865, 1977.

Nichols, G.T., P.J. Wyllie, and C.R. Stern, Subduction zone melting of pelagic sediments constrained by melting experiments, *Nature, 371*, 785-788, 1994.

Nicholls, I.A., Liquids in equilibrium with peridotitic mineral assemblages at high water pressures, *Contrib. Mineral. Petrol., 45*, 289-316, 1974.

Nicholls, I.A., and A.E. Ringwood, Effect of water on olivine stability in tholeiites and the production of silica-saturated magmas in the island-arc environment, *J. Geol., 81*, 285-300, 1973.

Nye, C.J., and M.R. Reid, Geochemistry of primary and least fractionated lavas from Okmok volcano, central Aleutians: implications for magma genesis, *J. Geophys. Res., 91*, 10271-10287, 1986.

Perfit, M.R., H. Brueckner, J.R. Lawrence, and R.W. Kay, Trace element and isotopic variations in a zoned pluton and associated volcanic rocks, Unalaska Island, Alaska: a model for fractionation in the Aleutian calc-alkaline suite, *Contrib. Mineral. Petrol., 73*, 69-87, 1980.

Rapp, R.P., and E.B. Watson, Dehydration melting of metabasalt at 8-12 kbar: implications for continental growth and crust-mantle recycling, *J. Petrol., 36* 891-931, 1995.

Rapp, R.P., E.B. Watson, and C.F. Miller, Partial melting of amphibolite/eclogite and the origin of Archean trondhjemites and tonalites, *Precamb. Res., 51*, 1-25, 1991.

Ringwood, A.E., The petrologicall evolution of island arc systems, *J. Geol. Soc. Lond., 130*, 183-204, 1974.

Sekine, T., T. Katsura, and S. Aramaki, Water saturated phase relations of some andesites with application to the estimation of the initial temperature and water pressure at the time of eruption, *Geochim. Cosmochim. Acta, 43*, 1367-1376, 1979.

Singer, B.S., J.D. Myers, and C.D. Frost, Mid-Pleistocene lavas from the Seguam volcanic center, central Aleutian arc: closed-system fractional crystallization of a basalt to rhyodacite eruptive suite, *Contrib. Mineral. Petrol., 110*, 87-112, 1992.

Sisson, T.W., and T.L. Grove, Experimental investigations of the role of H_2O in calc-alkaline differentiation and subduction zone magmatism, *Contrib. Mineral. Petrol., 113*, 143-166, 1993a.

Sisson, T.W., and T.L. Grove, Temperatures and H_2O contents of low-MgO high-alumina basalts, *Contrib. Mineral. Petrol., 113*, 167-184, 1993b.

Stern, C. and P.J. Wyllie, Melting relations and basalt-andesite-rhyolite-H_2O and a pelagic red clay at 30 kilobars, *Contrib. Mineral. Petrol., 42*, 313-323, 1973.

Stern, C. and P.J. Wyllie, Phase compositions through crystallization intervals in basalt-andesite-H_2O at 30 kbar with implications for subduction zone magmas, *Am. Mineral., 63*, 641-663, 1978.

Stern, R.J., On the origin of andesite in the Northern Mariana island arc: implications from Agrigan, *Contrib. Mineral. Petrol., 68*, 207-219, 1979.

Stolper, E., A phase diagram for mid-ocean ridge basalts: preliminary results and implications for petrogenesis, *Contrib. Mineral. Petrol., 74*, 13-27, 1980.

Takahashi, E., and I. Kushiro, Melting of a dry peridotite at high pressures and basalt magma genesis, *Am. Mineral., 68*, 859-879, 1983.

Tatsumi, Y., Melting experiments on a high-magnesian andesite, *Earth Planet. Sci. Lett., 54*, 357-365, 1981.

Tatsumi, Y., Origin of high-magnesian andesites in the Setouchi volcanic belt, southwest Japan, II. Melting phase relations at high pressures, *Earth Planet. Sci. Lett., 60*, 305-317, 1982.

Tatsumi, Y., Migration of fluid phases and genesis of basalt magmas in subduction zones, *J. Geophys. Res., 94*, 4697-4707, 1989.

Tatsumi, Y., and H. Isoyama, Transportation of beryllium with H_2O at high pressures: implication for magma genesis in subduction zones, *Geophys. Res. Lett., 15*, 180-183, 1988.

Tatsumi, Y., and N. Nakamura, Composition of aqueous fluid from serpentinite in the subducted lithosphere, *Geochem. J., 20*, 191-196, 1986.

Tatsumi, Y., Y. Furukawa, and S. Yamashita, Thermal and geochemical evolution of the mantle wedge in the northeast Japan arc 1. Contribution from experimental petrology, *J. Geophys. Res., 99*, 22,275-22,283, 1994.

Tatsumi, Y., M. Sakuyama, H. Fukuyama, and I. Kushiro, Generation of arc basalt magmas and thermal structure of the mantle wedge in subduction zones, *J. Geophys. Res., 88*, 5815-5825, 1983.

Tatsumi, Y., D.L. Hamilton, and R.W. Nesbitt, Chemical characteristics of fluid phase released from a subducted lithosphere and origin of arc magmas: evidence from high-pressure experiments and natural rocks, *J. Volcanol. Geotherm. Res., 29*, 293-309, 1986.

Tera, F., L. Brown, J. Morris, I.S. Sacks, J. Klein, and R. Middleton, Sediment incorporation in island-arc magmas: inferences from ^{10}Be, *Geochem. Cosmochim. Acta, 50*, 535-550, 1986.

Wyllie, P.J., and M.B. Wolf, Amphibolite dehydration-melting: sorting out the solidus, in *Magmatic Processes and Plate Tectonics*, edited by H.M. Prichard, T. Alabaster, N.B.W. Harris, and C.R. Neary, Geol. Soc. Spec. Publ. 76, pp. 405-416, 1993.

A.D. Johnston, Department of Geological Sciences, University of Oregon, Eugene, OR 97403-1272 (email: adjohn@oregon.uoregon.edu)

J.D. Myers, Department of Geology and Geophysics, University of Wyoming, Laramie, WY 82071 (email: magma@uwyo.edu)

Deciphering Mantle and Crustal Signatures in Subduction Zone Magmatism

Jon P. Davidson

Department of Earth and Space Sciences, UCLA, Los Angeles, CA 90095

A compilation of data for arc volcanic rocks from modern subduction zones shows that they are highly differentiated relative to probable mantle-derived magmas. When examined on a volcano-by-volcano basis, the differentiation of arc magmas can commonly be shown to be open system, so that source geochemical characteristics may be obscured. When such overprinting effects are accounted for, mantle-derived compositions are shown to be enriched in fluid-mobile incompatible elements, consistent with the addition of a slab-derived component to the arc source. The remaining incompatible elements are depleted relative to MORB, and suggest derivation from a wedge from which melt has already been recently extracted.

IN PURSUIT OF ARC MANTLE SOURCES

Fundamental questions that remain unanswered in arc magmagenesis are;
1. What is the (presubduction) composition of the mantle wedge source of arc magmas?
2. To what extent does it melt, and by what process?
3. What is the composition and amount of slab-derived component added to the wedge?

Clearly if we can sample undifferentiated magmas that have come directly from the subduction-modified wedge then we can potentially address the above points. If we can definitively establish the composition of the wedge, then mass balance considerations will enable us to constrain questions 2 and 3. With such constraints we will be able to address questions of the extent of crustal recycling at arcs, the role of arc magmatism in generating the continental crust, and the role of arc magmatism in controlling the evolution of crust and mantle compositions.

Why have the answers to these apparently straightforward questions eluded us for so long? Two principle limitations have confounded us:
1. The scarcity of near-primary magma compositions at arcs, and the realization that magmas have been profoundly modified during ascent through the lithosphere.
2. Arcs are dynamic systems in space and time, such that compositional variations occur in magmas erupted at different edifices *along* arcs, *across* arcs, *through time* at one location, and between one arc and the next.

Some definitions: The term *primitive* is used to describe rocks that show limited effects of differentiation. Note that this is a relative term, roughly equivalent to the term "mafic". The term *primary* is reserved for undifferentiated magmas - in this case those which can be demonstrably in equilibrium with mantle compositions. While primitive magmas are erupted at some arcs, it is debatable whether any truly primary magmas reach the surface [e.g. *Hart and Davis* 1978; *O'Hara*, 1968]. Commonly used acronyms used herein include; LILE = Large Ion Lithophile Elements (large ionic radius, low ionic charge elements, generally alkali and alkali earth metals; groups I and II of the periodic table - Rb, K, Cs, Ba, Sr), REE = Rare Earth Elements (the actinides; a geochemically coherent group with atomic numbers corresponding to La through Lu), HFSE = High Field Strength Elements, small highly charged ionic species - Nb, Ta, Zr, Hf, Ti. MORB and OIB refer to different basalt types - mid ocean ridge and ocean island respectively.

CHARACTERISTICS OF PRIMARY ARC MAGMAS

Experiments

The general consensus among arc petrologists is that magmas are derived from the mantle wedge, with magmatism triggered by lowering of the ambient solidus through the addition of a fluid from the slab. This fundamental assumption forms the basis of the subsequent discussion, but deserves at least brief justification. Two important lines of evidence are offered; (1) high MgO arc basalts, which appear to be parental to many arc suites, cannot be derived by direct melting of lower MgO oceanic crust, and (2) most arc magmas are compositionally distinct from those that can be produced by melting of amphibolite or eclogite [e.g. *Wyllie*, 1982]. In fact high SiO_2 low K rocks (adakites) found at some arcs where the subducted plate is anomalously young and hot, have been interpreted to represent slab melts [*Defant and Drummond*, 1990]. The fact that these are chemically quite distinct from the majority of calc-alkaline and tholeiitic arc magmas (e.g. they are strongly light REE enriched and have high Sr/Y ratios) might be taken as the exception that proves the rule.

It should arguably be a simple matter to melt peridotite with H_2O at about 30 kbar. This pressure corresponds to the 100-150 km depth beneath the arc front at which the slab commonly lies and may be a logical starting estimate for the depth of melting, although consideration of prograde metamorphic reactions in the slab suggest that dehydration probably takes place at shallower depths. Lateral fluid migration by cycles of amphibole formation and break down in the convecting wedge [e.g. *Davies and Stevenson*, 1992] may be a more realistic scenario, implying that the actual depth of melting is shallower. Surprisingly few peridotite + H_2O melting experiments have been performed. Some results are plotted in Fig. 1 and compared with primitive arc data.

Can we take these experimentally produced liquids as representative of primary arc magmas? Perhaps, but with some caveats.

1. Melt composition is dependent on the mineral assemblage in the peridotite. The experiments of *Kushiro* [1990] use a spinel peridotite from Hawaii. The cpx content of peridotite in the wedge beneath arcs may well be considerably less, based on the observed mineralogy of xenolith suites from arcs [*Maury et al.*, 1992; *Kepezhinskas and Defant*, this volume]. Some trace element characteristics suggest the source is even more depleted than that of MORB (see below), so the source peridotite may be more depleted in cpx (and garnet/spinel), thus lower in CaO and Al_2O_3 content.

2. Degree of melting may vary. Do compositions vary widely with % partial melting as in *Jaques and Green's*

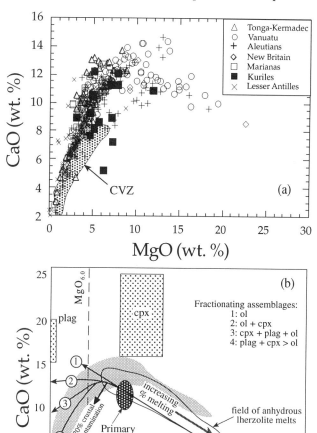

Fig. 1. (a) Compilation of CaO-MgO data for oceanic arcs. Data from the central Andes (central volcanic zone = CVZ), an arc built on thick continental crust, are shown for comparison (from *Davidson et al.*, 1990). Data sources in this and later figures; *Bailey et al.*, 1989; *Barsdell*, 1988; *Barsdell and Berry*, 1990; *Brophy*, 1987; *Castellana and Davidson*, unpublished; *Eggins*, 1993; *Ewart and Hawkesworth*, 1987; *Ewart et. al.*, 1977; *Fournelle et al.*, 1994; *Gorton*, 1977; *Kay and Kay*, 1985, 1994; *Kersting and Arculus*, 1994; *Myers et al.*, 1985, 1986; *Nye and Reid*, 1986; *Romick et al.*, 1990; *Singer et al.*, 1992a, b; *Woodhead*, 1989). (b) Comparison of arc CaO-MgO systematics with primary magmas produced experimentally from peridotite + H_2O [*Kushiro*, 1990], and partial melts of anhydrous peridotite [*Jaques and Green*, 1980]. The estimated primary magmas field includes primitive non-cumulate basalts and calculated primary magmas from *Plank and Langmuir* [1988]. The diagram shows general fields for common arc phenocryst phases; olivine, clinopyroxene and plagioclase, to illustrate the effects of removal/ accumulation of these phases in varying proportions. Also shown is the effect of adding 20% bulk crust to a primitive arc magma. Extrapolating to 6% MgO ($MgO_{6.0}$ line) may be ineffective in distinguishing different primary magma characteristics.

[1980] anhydrous experiments? Control may be through variations in the fertility of the peridotite, variations in the amount of fluid added from the slab [e.g. *Luhr*, 1992; *Stolper and Newman*, 1994], or variations in the temperature structure of the wedge - which itself may be a function of the dynamics of subduction.

3. The nature of the melting process may have a large effect on melt composition. Experiments produce effectively an *in situ* batch melt. In reality small melt fractions could be continually extracted and aggregated, melting may occur over a vertical column in which the controlling factors (P,T, source composition) change, or melts may react with the mantle matrix as they migrate [*e.g. Myers et al.*, 1985; *Navon and Stolper*, 1987; *Keleman et al.*, 1990; *Plank and Langmuir*, 1992].

While simple melting experiments can be used as a guide to evaluate whether natural samples are primitive, they do not help us greatly in constraining the trace element characteristics of primary melts, which, for many elements, will be dominated by the inventory from the slab.

Natural Primitive arc samples

Rare mafic lavas that have been sampled at some arcs, along with experimentally derived liquids from hydrous peridotite are commonly >10% MgO. In contrast, (Figs. 1, 2) the majority of arc lavas are highly differentiated (MgO <5%). Note that few data points correspond to the experimentally produced compositions. Differentiation has, at least in part, modified virtually all of the samples, and the effects of crystal fractionation or accumulation can hardly be denied in most arc suites.

Many attempts have been made to account for the effects of fractionation, such as filtering data to include rocks with > 5 or 6% MgO, or extrapolating fractionation trends to an arbitrary parental value of say 6% MgO [*Plank and Langmuir*, 1988; *Pearce and Parkinson*, 1993]. Although these approaches implicitly recognize the problem of differentiation affecting primary compositions, they do not constitute a realistic solution. This, compounded with the probability that much of the protracted intra-crustal differentiation of arc magmas is open system, renders the majority of arc data bases impotent in the quest for well-constrained primary compositions.

The approach of restricting consideration to samples with >5% MgO excludes some 75% of the arc data base, and does not circumvent the possible effects of olivine accumulation (Fig. 1b). Furthermore, by virtue of characteristics such as relatively low Mg numbers or porphyritic textures, even these samples appear to have undergone considerable differentiation from primary magmas. Correcting for differentiation by extrapolating or interpolating data trends to some nominal primitive composition is also ambiguous. From the CaO-MgO compilation (Fig. 1), it is clear that the data do not project directly back to the experimentally produced melts, perhaps because melting involves the addition of H_2O, or involves a different phase assemblage from that used in experiments to date. The turnover in the CaO-MgO trend may reflect a change in the fractionating assemblage from olivine dominated, to olivine + pyroxene + plagioclase. The relative proportions of these minerals and the onset of fractionation of each phase, will be pressure sensitive [*Grove and Baker*, 1984] and should vary from arc to arc. Note that there is a small but significant displacement to lower CaO at a given MgO in the general field of compositions erupted through thick continental crust (the central Andes) when compared with island arcs. This can be, in part, explained by fractionation at higher pressures with earlier onset of cpx crystallization depleting CaO (Fig. 1b). The alternative explanation, that the primary magma in the central Andes represents a lower degree of partial melting because the mantle column is less in extent [*Plank and Langmuir*, 1988], does not account for the distinctly different isotopic compositions, which reflect crustal contamination and underscore the importance of crustal level processes. In fact, *Plank and Langmuir* [1988] show that high pressure fractionation alone may not adequately explain CaO-MgO systematics, suggesting that additional open system effects may be operating (Figure 1b).

TRACE ELEMENT CHARACTERISTICS

Arcs are characterized by a distinct incompatible trace element distribution, with high LILE/REE and HFSE (Fig. 3). These characteristics (often referred to as the arc "signature") are distinct from the trace element distributions of most other magmas (MORB, OIB) and suggest that there is a connection of causality between the trace element composition and the process of subduction [e.g. *Pearce*, 1982]. The consensus is that this pattern is a reflection of fluid-controlled element partitioning from the slab to the wedge. Theoretical and experimental [*Tatsumi et al.*, 1986; *Keppler*, 1995] considerations predict that LILE are fluid-soluble relative to HFSE, so that in the simplest of cases arc magmas may be interpreted as simply melts of a mantle comprising MORB source + LILE-rich fluid (Fig. 3). Note, however, that many arcs are also characterized by light REE (LREE) enrichment, and it has been claimed that high LREE/HFSE (e.g. La/Nb) are also a fundamental characteristic. Close inspection reveals that, even though the light REE may be more mobile than heavy REE in

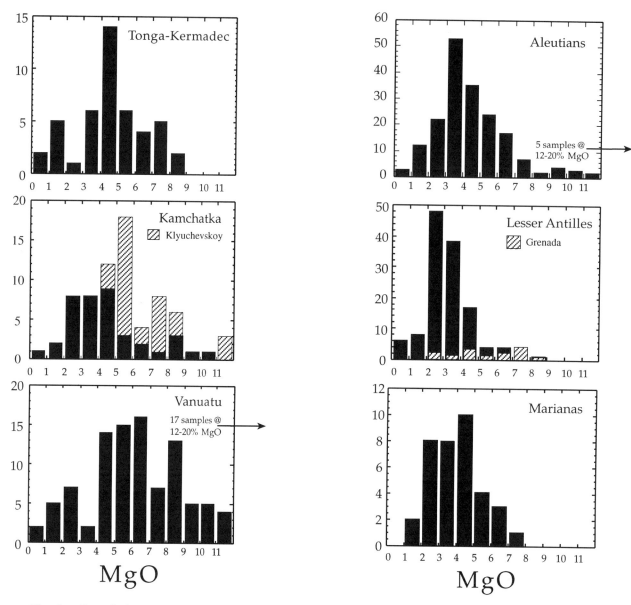

Fig. 2. Compilation of MgO contents for a number of intra-oceanic arcs, illustrating clustering around differentiated compositions (largely <6% MgO). Kamchatka data are included as the arc is developed on accreted oceanic lithosphere, albeit thickened. Kluyechevskoy is distinguished from other Kamchatka samples as it is formed on the edge of the Central Kamchatka Depression, a back-arc rift. There may also be tectonic complications at the southern end of the Antilles arc, such as the juxtaposition of South American lithosphere, that influence magma compositions to produce distinctly more mafic lavas on Grenada.

fluids [*Tatsumi et al.,* 1986], the most primitive arc magmas are actually quite low in La/Nb (although still high in Ba/Nb), and therefore the LREE enrichment observed at some arcs may be a secondary effect (M. Dungan, personal communication, 1992).

Recent studies focusing in particular on B and Be systematics have made considerable progress in identifying the nature of the slab-derived component [e.g. *Morris et al.,* 1990; *Edwards et al.,* 1993; *Leeman et al.,* 1994]. The inventory of both elements at arcs is dominated by the slab-derived flux. Boron has been identified as a particularly slab-fugitive element, released from the subducted slab and concentrated into magmas at the arc front. Correlations with ^{10}Be, which is concentrated in young (less than 5 Ma)

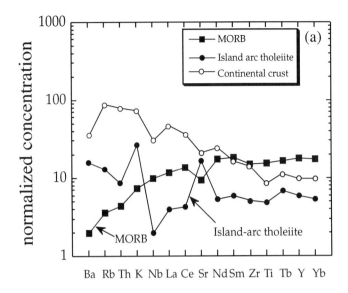

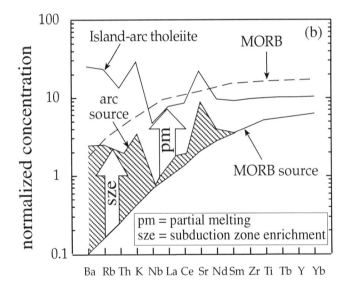

Fig. 3. Trace element distributions in arcs. (a) comparison of island arc, bulk continental crust and MORB distributions (data sources; *Sun*, 1980; *Taylor and McLennan*, 1985. Normalization constants from *Thompson et al.*, 1984). The trace element patterns of island arc tholeiite and bulk crust are similar, with the exception of Ba, K and Sr, which appear as "spikes" in arc lavas, and are subdued in the bulk crust probably as a result of feldspar and amphibole fractionation in the more differentiated continental crust. (b) schematic model for generating arc trace element distribution from a MORB source. Subduction zone enrichment (sze) adds fluid-soluble elements [cf *Tatsumi et al.*, 1986] as shown by the shading. This enriched source is then partially melted (pm) to generate a primitive arc tholeiite.

subducted sediment, indicate that both elements have short mantle residence times and are effectively flushed through the lithosphere without significant modification, thus providing robust indicators of slab contributions. Nevertheless, most other trace elements reflect a more complex inheritance, and cannot be used to determine mantle or slab contributions without judicious consideration of the effects of differentiation

DIFFERENTIATION - OVERPRINTING OF PRIMARY CHARACTERISTICS

This section underscores the importance of recognizing "secondary effects" - namely those of differentiation. It is clear from above that we cannot equate *basaltic* with *undifferentiated* or *uncontaminated*.

A major problem at continental margin arcs, is that the crust shares general trace element characteristics with arc basalts (Fig. 3). Of course, this may simply reflect the fact that much of the crust was formed at subduction zones or in subduction-like environments. Other than to point out the importance of understanding arc magmatism before we can evaluate the origins of the continental crust, the implications of this observation are beyond the scope of this paper. But the similarity in trace element characteristics means that the effects of crustal contamination on arc magmas may be obscured. Trace element signatures such as high Rb/Nb which have been used to fingerprint crustal effects on flood basalts [e.g. *Leeman and Hawkesworth* 1986] are of limited use in monitoring crustal contamination of arc basalts for which high Rb/Nb is a primary characteristic. The *isotopic* signature of continental crust may serve as a simple identifying tracer, but even this may be of little value if the crust is sufficiently young that a radiogenic isotope contrast has not developed [*Davidson et al.*, 1987].

If we restrict consideration to oceanic island arcs alone then surely we can circumvent the potential effects of contamination by continental crust. Unfortunately this is not the case. Continental sediments are found throughout the oceans. The thickest accumulations are near the margins of the oceanic lithosphere where island arcs tend to be located. But can we really expect a relatively thin veneer of sediment to be a potential contaminant to arc magmas? Many models of arc magmagenesis invoke of the order 1-2% bulk subducted sediment to explain many of the trace element and isotopic characteristics of the volcanic rocks. If the mantle melts by 10%, incompatible element abundances will be a factor of ~10 greater than in the source, so 10 times as much sediment (10-20% bulk) will be needed to cause the same effects by contamination during differentiation as 1-2% source contamination. This may not be obvious from the major element data (Fig. 1b).

As an example, consider a suite of samples from a typical island arc volcano, Mt. Pelée in the Lesser Antilles. A standard $^{206}Pb/^{204}Pb$-$^{207}Pb/^{204}Pb$ isotope diagram (Fig. 4) shows samples from Mt. Pelée defining an inclined positive array, that could reasonably be interpreted as reflecting mixing between subducted sediment (represented by the DSDP Hole 543 sediments) and the mantle wedge (represented by ocean floor basalts from the same DSDP hole). However, closer inspection of the data reveals a good correlation between Pb isotope ratios and indices of differentiation - such as SiO_2. Such correlations imply that the isotope ratios are modified *during differentiation*, and a process of simultaneous assimilation and fractional crystallization (AFC) is implicated. Again, this result does *not* preclude initial sediment contamination in the source, such as described by *Plank and Langmuir* [1993], but underscores the leverage that intra-crustal contamination may ultimately have on the compositions of arc rocks. At this stage the point is not to suggest this as an alternative to source contamination, which on the basis of ^{10}Be data alone must be a factor at many arcs [*Morris et al.*, 1990]. The point is to recognize the possibility of crustal contamination during differentiation, which may modify the inventory of elements.

CIRCUMVENTING THE SLAB COMPONENT: HFSE CHARACTERISTICS

Accepting the conventional wisdom that a slab-derived fluid enriches the mantle source of arc magmas (Fig. 3), it will be difficult to constrain the relative contributions of fluid-soluble elements from the slab and wedge respectively. The inventory of such elements in primary arc magmas will be a function of; (1) the mineral-fluid and mineral-melt partition coefficients, (2) the element concentrations in the slab and mantle, and (3) the degree and mechanism of partial melting of the enriched mantle. All of these factors are poorly constrained, and compounded by the possible overprinting effects of crustal contamination for which contaminant compositions, phase assemblages and partition coefficients are equally vague. Limited experimental data on trace element partitioning into fluids indicate that LILE and possibly LREE are fluid mobile compared with heavy REE and HFSE [*e.g. Tatsumi et al.*, 1986; *Keppler*, 1995]. Based on these data, and predictions from known geochemical characteristics, a suite of trace elements can be identified, the distributions of which are largely unaffected by the slab contribution, and are controlled only by melting of the wedge. For our purposes

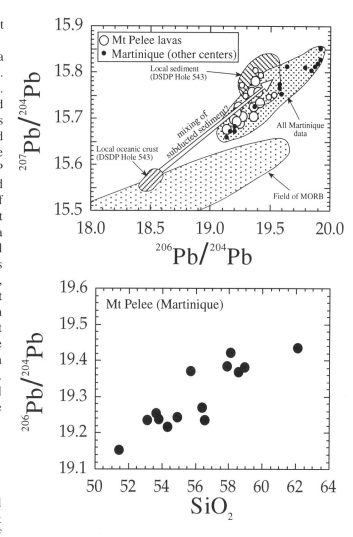

Fig. 4. (a) Pb isotope data for the Lesser Antilles, emphasizing samples from a single volcanic edifice, Mt. Pelée. Arrow illustrates conventional interpretation of Pb isotopes; mixing of subducted sediment (represented by compositions recovered from DSDP hole 543 just outboard of the arc) with the mantle wedge (represented by local oceanic crust from the same hole). (b) Correlation between $^{206}Pb/^{204}Pb$ and SiO_2 for comagmatic Mt. Pelée samples, indicating that significant modification of Pb isotopes, generating a range in $^{206}Pb/^{204}Pb$, occurs during intra-crustal differentiation. All samples < 50,000 years old. Data from *Davidson* [1986], and *Ellisor and Davidson* [1992].

the elements Nb, Ta, Zr, Hf, Y, P, Yb, V and Sc will be referred to as WC elements (**W**edge-**C**ontrolled). If totally incompatible, concentrations in primary magmas would be a function of simply source composition, degree and mechanism of melting. But most of the elements listed

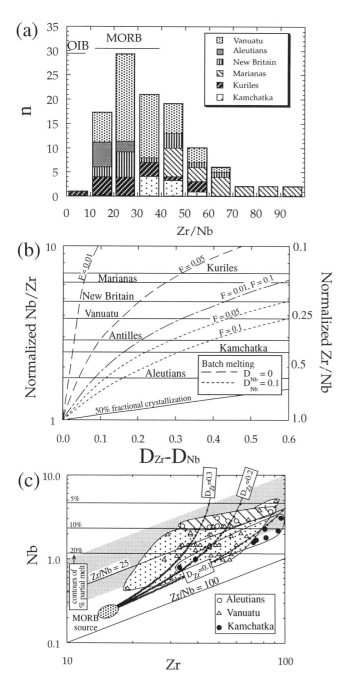

Fig. 5. (a) Histogram of Zr/Nb ratios for various primitive arcs, compared with ranges for OIB and MORB. (b) Model examining feasibility of producing the *range* in arc Nb/Zr (and therefore Zr/Nb as in (a)) ratios from a common source through silicate-melt fractionation processes. The y axis shows Nb/Zr normalized to a source ratio of 1, while the x axis shows the difference in bulk distribution coefficient between Zr and Nb. Model curves are constructed for a normalized Nb/Zr source ratio of 1. Normalization enables comparison of Zr/Nb systematics regardless of source ratio. Horizontal lines marked for each arc indicate the maximum *range* in Nb/Zr, normalized to the most primitive (lowest) Nb/Zr ratio in each arc, which is presumed to be closest to that of the source. In order to explain the large ranges in many arcs either (1) the degree of melting must range to very low (<5%: F= 0.05) and Nb must be perfectly incompatible, and/or (2) the difference in distribution coefficients between Zr and Nb during mantle melting must be large (>0.3). (c) Nb-Zr relationships showing how Zr/Nb may be fractionated during melting of a MORB source, given $D_{Nb} = 0$ and $D_{Zr} = 0.1$, 0.2 or 0.3. Comparison with selected data suggests that Zr/Nb ratios do not vary much with differentiation, and are relatively high (arguing for a large % melting and/or a mantle source with higher Zr/Nb than that of MORB). Range of MORB Zr/Nb given by shaded band.

have non-zero distribution coefficients, and both the mineral assemblage in the source and the fractionating mineral assemblage during subsequent differentiation may significantly affect their distribution in arc magmas. Virtually all of the highly incompatible trace elements are fluid-soluble (e.g. Ba, K, Cs), with the exception of Nb and Ta. Nb and Ta offer the greatest potential as geochemical tracers, but are hampered in their use by analytical difficulties - such that very little high quality Nb and Ta data are available for primitive arc rocks. This deficiency may be redressed soon with the application of ICP-MS technology [*e.g. Plank* 1994].

Recent work by *Woodhead et al.* [1993] and *Pearce and Parkinson* [1993] has shown that the distribution of WC elements in primitive arc rocks is consistent with derivation from a depleted mantle wedge - commonly even more depleted in incompatible elements than the source of MORB. A simple comparison of MORB and arcs can be made using Zr/Nb ratios (Fig. 5a). In general, Zr/Nb ratios in arcs are higher than those in MORB which, given $D_{Zr}>D_{Nb}$, suggests that the arc source is more depleted than MORB source or/and the degree of partial melting at arcs is higher than that at ridges. Furthermore, the actual concentrations of both Zr and Nb are low in primitive arc lavas compared with MORB. This also argues against small degrees of melting or significant effects of dynamic/ fractional melting processes that are commonly called on to produce significant fractionations in incompatible elements.

A first order model can be devised to examine whether variations in Zr/Nb within arcs and between arcs and MORB (and by analogy the overall variations in the distribution of all WC elements) can be simply due to variations in the degree of melting of a common source given $D_{Zr} \neq D_{Nb}$. In Fig. 5b Nb/Zr is calculated as a function of the *difference* in distribution coefficients between the two elements ($D_{Zr}-D_{Nb}$) for various F values (degree of partial melting or residual melt during crystallization). As expected, significant fractionations of Zr from Nb require significant differences in D_i and/or low degrees of partial melting. The *range* of Nb/Zr values for primitive rocks from many arcs is too high to be explained by derivation from a common source. With a value of F = 0.1, arguably a minimum degree of partial melting, (D_{Zr}-D_{Nb}) is required to be ≥0.3 for most arcs. While residual HFSE phases could effect such fractionations in the wedge it is considered unlikely, because primitive suites from many arcs do not show systematic changes in Zr/Nb with Zr or Nb concentration (Fig. 5c). Examination of HFSE systematics relative to other incompatible trace elements has also shown that the relative depletion exists in the wedge before melting, and may be largely a reflection of relative enrichment in elements other than HFSE that are derived from the slab [*Davidson and Wolff*, 1989; *Thirlwall et al.*, 1994]. It should be realized that the same analytical difficulties alluded to above may have compromised Nb data, and make the data shown in Fig. 5 qualitative at best. Similar considerations can be made for element ratios such as Sm/Yb - models comparable with Fig. 5b suggest that the ranges in Sm/Yb within individual arcs are unlikely to result from simple variations in wedge melting at a given locality. The implication is that arc sources do not have the same distribution of WC elements as MORB, and vary within arcs and from arc to arc.

Compilation of WC elements for a few representative primitive arc rocks and comparison with MORB shows depletion in the most incompatible elements relative to MORB. The distribution can be modeled to a first order by melting of a depleted MORB source from which 5% melt has already been extracted (Fig. 6). *Woodhead et al* [1993] suggest that the depletion occurs as a result of melt extraction by back-arc magmatism, and advection of the depleted material into the wedge where it is fluxed by slab-derived fluids to generate arc magmas. The possibility deserves exploration and could be addressed simply by comparing arcs with and without back-arc systems, and examining arc - back-arc systems through time. The most primitive arc magmas do appear to be associated with extensional lithospheric stresses. This in itself may be further support for the role of the upper plate in modifying magma compositions, since lithosphere under extension either through back-arc spreading or local intra-arc extension, would both facilitate unimpeded passage of primitive basalts through the crust [*e.g. Singer and Myers*, 1990] and promote back-arc spreading.

The contention that arc magmas are derived from a melt-depleted wedge which has been subsequently metasomatized by a component from the subducted slab, is consistent with limited data from arc mantle xenolith suites. In general,

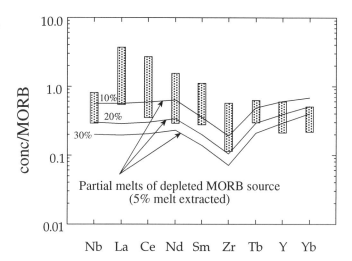

Fig. 6. Incompatible trace element concentrations in primitive arc lavas (from Vanuatu), compared with model melts of a 2-stage depleted MORB source. Shaded bars represent the range in concentrations of the arc lavas. Model melts were calculated for batch melting from a MORB source, from which 5% melt has been previously extracted. Note that WC element concentrations in the arc lavas are reproduced by ~ 20% melts of the depleted source. LREE show slight relative enrichment in the arc lavas, and may include a contribution from the slab-derived fluid.

mantle xenoliths are rare in arc lavas - in keeping with the overall differentiated nature of the lavas. Where they have been recovered, for instance in Kamchatka and the Philippines, they are depleted in basaltic components (low modal clinopyroxene contents), but enriched in fluid-mobile trace elements [*e.g. Maury et al.*, 1992; *Kepezhinskas and Defant*, this volume]. Although it cannot be unequivocally demonstrated that these samples represent the source mantle for arc magmas, the chemical characteristics are certainly similar to those of arc basalts, suggesting that experiments using these natural peridotites might provide a better analog for arc magmagenesis than those presented in Fig. 1b.

USING ISOTOPE SIGNATURES TO DECIPHER SOURCE CHARACTERISTICS.

Isotopes of Sr, Nd and Pb have traditionally been used to identify contributions from mantle sources and the subducting slab (including sediments). Pb is particularly useful in this context because the concentration contrast between subducted sediments (typically > 10 ppm) and the mantle (<0.5 ppm) is enormous, the isotopic contrast is generally significant (continental sediments have relatively high $^{207}Pb/^{204}Pb$) and small sediment contributions are readily detected in Pb isotope systematics from arcs [*e.g. Armstrong*, 1971]. Sr, as a LILE, is expected to be fluid-mobile and the Sr budget in arc magmas should also be dominated by the slab-derived component. Indeed, $^{87}Sr/^{86}Sr$ of primitive arc rocks are uniformly higher than those of MORB. The difference in $^{87}Sr/^{86}Sr$ between primitive arc rocks and MORB source is not as great as would be expected if the slab Sr is derived purely from radiogenic Sr in the sediment, arguing that a significant proportion of the slab Sr comes from variably altered oceanic crust rather than subducted sediment. At the same time $^{143}Nd/^{144}Nd$ ratios in arcs tend to be lower than MORB. This too may reflect Nd from the sediment component of the slab, consistent with limited fluid mobility of the LREE [*Tatsumi et al.*, 1986]. On the other hand, the most primitive arc rocks generally tend not to be LREE enriched, and it is equally plausible that Nd isotope signatures (and therefore also Sr and Pb) may also be modified by contamination in the upper plate (see Fig. 4). The Sr, Nd and Pb isotope signatures of arc magmas are therefore of dubious utility in establishing the characteristics of the wedge source, since they have been modified by additions from the slab and/or crustal contamination.

Since Hf is a WC element, Hf isotope compositions may help to document wedge characteristics. Hf in arc magmas is likely to be derived from the wedge with very little contribution from the slab or arc crust. The limited data for arcs [*Salters and Hart* 1991; *White and Patchett*, 1984] indicate that primitive arc rocks have $^{176}Hf/^{177}Hf$ ratios indistinguishable from those of MORB, consistent with derivation from a common depleted source. But it has been argued above that arc sources are commonly *more* depleted in incompatible elements than MORB sources. $^{176}Hf/^{177}Hf$ ratios may therefore be able to constrain the timing of the depletion. Fig. 7 shows the Hf isotope evolution of a source from which melts have been extracted, fractionating Lu from Hf. The actual fractionation of Lu from Hf will depend on the role of garnet during melting [*Salters and Hart*, 1989], but the overall effect with time, if $D_{Lu} \neq D_{Hf}$, will be to generate different Hf isotopic signatures between arc and MORB sources. The observed similarity in ε_{Hf} suggests that depletion of arc sources was relatively recent and is consistent with back-arc processing of the wedge, as suggested by *Woodhead et al.*, 1993 and *Hochstaedter et al.*, 1995. This conclusion could be modified given a more representative data base and better resolution of Hf isotope data.

CONCLUSIONS AND PROSPECTS

Constraining the petrogenesis of arc magmas is severely impeded by the paucity of primitive arc rocks. Protracted

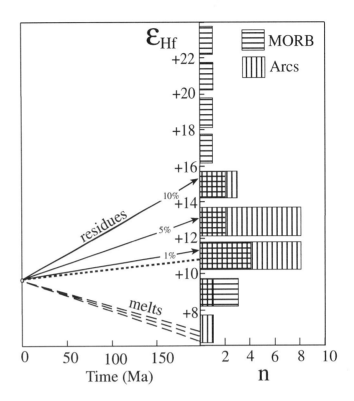

Fig. 7. Schematic Hf isotope evolution of melts and residues from a MORB source mantle over 200 Ma, given $D_{Hf} = 0.1$ and $D_{Lu} = 0.3$. Hf isotope ratios are expressed as deviations relative to a chondritic reservoir as ε_{Hf}, where;

$$\varepsilon_{Hf} = \left(\frac{(^{176}Hf/^{177}Hf)_{measured}}{(^{176}Hf/^{177}Hf)_{CHUR}} - 1 \right) \cdot 10^4$$

CHUR = Chondritic Uniform Reservoir, representing the bulk silicate Earth)

For comparison are range of ε_{Hf} data for arcs (omitting those that may have been affected by crustal contamination) and MORB [from *White and Patchett*, 1984, *Patchett and Tatsumoto*, 1980, *Salters and Hart*, 1991]. Cross-hatching corresponds to overlapping arc and MORB data - for instance there are four MORB and one arc sample with ε_{Hf} between +8 and +10. Based on these limited data, there is no significant difference between MORB and arc ε_{Hf}. If arcs are derived from more depleted sources than MORB then the depletion must have occurred relatively recently. If the depletion event occurred in the upper mantle during MORB extraction >100 Ma ago we might expect ε_{Hf} values to be distinct.

differentiation at arcs, commonly open system, can invalidate the inversion of geochemical data to constrain source contributions. Correcting for the effects of differentiation to determine primary magma compositions can only realistically be attempted at individual volcanic edifices for which a sufficiently diverse suite of rocks has been erupted to enable magmatic evolution to be quantitatively modeled. Comparison between such edifices within and between arcs may help to constrain the effects of variations in subduction zone geometry and thickness/ composition of the upper plate [*e.g. Luhr*, 1992; *Leeman et al.*, 1994].

Qualitatively, primary arc magmas appear to have been derived from extremely depleted sources. The composition of the wedge source and the nature of the melting process, may, in principle be understood by consideration of the distribution of non-fluid-mobile (HFSE and heavy REE) elements. Depletion likely occurs as a result of melt extraction from the wedge in the back-arc, shortly before arc magmatism.

Acknowledgments. This contribution is a crystallization of ideas and personal dogmas accumulated through work on magmatism at a number of arcs. Marge Wilson, John Wolff, Richard Arculus, Chris Hawkesworth, Marc Defant, Mike Dungan, Terry Plank, Bill Leeman, Julian Pearce, Huw Davies, Jon Woodhead, Bill White, Julie Morris, Matthew Thirlwall, Todd Feeley, and many of the authors included in the reference list have significantly influenced my understanding of arcs - but should not be held responsible for any of my heretical claims. Insightful comments and suggestions from B. Singer, M. Defant and an anonymous reviewer greatly improved the manuscript. I am extremely grateful to the organizers of SUBCON for providing us with an engaging and fertile forum for the discussion of subduction zone processes, and to NSF (EAR 9303791, 9405340) for financial support.

REFERENCES

Armstrong, R.L. Isotopic and chemical constraints on models of magma genesis in volcanic arcs, *Earth Planet. Sci. Lett., 12,* 137-142, 1971.

Bailey, J.C., T.I. Frolova, and I.A. Burikova, Mineralogy, geochemistry and petrogenesis of Kurile island-arc basalts, *Contrib. Mineral. Petrol., 102,* 265-280, 1989.

Barsdell, M., Petrology and petrogenesis of clinopyroxene-rich tholeiitic lavas, Merelava Volcano, Vanuatu, *J. Petrol., 29,* 927-964, 1988.

Barsdell, M. and R.F. Berry, Origin and evolution of primitive island arc ankaramites from western Epi, Vanuatu, *J. Petrol., 31,* 747-777, 1990.

Brophy, J.G., The Cold Bay Volcanic Center, Aleutian Volcanic Arc; II, Implications for fractionation and mixing mechanism in calc-alkaline andesite genesis, *Contrib. Mineral. Petrol., 97,* 378-388, 1987.

Davidson, J.P., Isotopic and geochemical constraints on the petrogenesis of subduction - related lavas from Martinique, Lesser Antilles, *J. Geophys. Res., 91,* 5943-5962, 1986.

Davidson, J.P., M.A. Dungan, K. Ferguson and M.A. Colucci, Crust-magma interaction and the evolution of arc magmas from the San Pedro-Pellado volcanic complex, Southern Chilean Andes, *Geology, 15,* 443-446, 1987.

Davidson J.P., N.J. McMillan, S. Moorbath, G. Worner, R.S. Harmon and L. Lopez-Escobar, The Nevados de Payachata volcanic region (18°S, 69°W, N. Chile) II. Evidence for widespread crustal involvement in Andean magmatism, *Contrib. Mineral. Petrol., 105,* 412-432, 1990.

Davidson, J.P. and J.A. Wolff, On the origin of the Nb-Ta "anomaly" in arc magmas (abstract), *Eos Trans. AGU, 70*(43), Fall Meeting Suppl., 1387, 1989.

Davies, J.H and D.J. Stevenson, Physical model of source region of subduction zone volcanics, *J. Geophys. Res., 97,* 2037-2070, 1992.

Defant, M.J. and M.S. Drummond, Derivation of some modern arc magmas by melting of young subducted lithosphere, *Nature, 347,* 662-665, 1990.

Edwards, C.M.H., J.D. Morris and M.F. Thirlwall, Separating mantle from slab signatures in arc lavas using B/Be and radiogenic isotope systematics, *Nature, 362,* 530-533, 1993.

Eggins S.M., Origin and differentiation of picritic arc magmas, Ambae (Aoba), Vanuatu, *Contrib. Mineral. Petrol., 14,* 79-100, 1993.

Ellisor, R. and J.P. Davidson, Crustal contamination at island arc volcanoes: The Quill and Mt Pelée, Lesser Antilles (abstract), *Eos Trans. AGU, 73*(43), Fall Meeting Suppl., 646, 1992.

Ewart, A., R.N. Brothers and A. Mateen, An outline of the geology and geochemistry and the possible petrogenetic evolution of the volcanic rocks of the Tonga-Kermadec-New Zealand island arc, *J. Volcanol. Geotherm. Res., 2,* 205-250, 1977.

Ewart, A. and C.J. Hawkesworth, The Pleistocene to Recent Tonga-Kermadec arc lavas: interpretation of new isotope and rare earth element data in terms of a depleted source model, *J. Petrol., 28,* 495-530, 1987.

Fournelle, J.H., B.D. Marsh, and J.D.Myers, Age, character, and significance of Aleutian arc volcanism, in *The Geology of North America v.G-1, The Geology of Alaska,* edited by G. Plafker and H.C. Berg, pp. 723-757, Geological Society of America, Boulder, CO, 1994.

Gorton, M.P., The geochemistry and origin of Quaternary volcanism in the New Hebrides, *Geochim. Cosmochim. Acta, 41,* 1257-1270, 1977.

Grove, T.L. and M.B. Baker, Phase equilibrium controls on the tholeiitic versus calc-alkaline differentiation trends, *J. Geophys. Res., 89,* 3253-3274, 1984.

Hart S.R., and K.E. Davis, Nickel partitioning between olivine and silicate melt, *Earth Planet. Sci. Lett., 40,* 203-219, 1978.

Hochstaedter, A., P. Kepezhinskas, M. Defant and A. Koloskov, Insights into the volcanic-arc mantle wedge from magnesian lavas from the Kamchatka Arc, *J. Geophys. Res.,* (in press).

Jaques, A.L. and D.H. Green, Anhydrous melting of peridotite at 0-15 kb pressure and the genesis of tholeiitic basalts, *Contrib. Mineral. Petrol., 73,* 287-310, 1980.

Kay, S. M. and R.W. Kay, Aleutian tholeiitic and calc-alkaline magma series: the mafic phenocrysts, *Contrib. Mineral. Petrol., 90,* 276-290, 1985.

Kay, S. M. and R.W. Kay, Aleutian magmas in space and time, in *The Geology of North America v.G-1, The Geology of Alaska,* edited by G. Plafker and H.C. Berg, pp. 687-722, Geological Society of America, Boulder, CO, 1994.

Keleman, P.B., K.T.M. Johnson, R.J. Kinzler and A.J. Irving, High-field strength element depletions in arc basalts due to mantle-magma interaction, *Nature, 345,* 521-524, 1990.

Kepezhinskas, P. and M.J. Defant, Contrasting styles of mantle metasomatism in subduction zones: constraints from ultramafic xenoliths in volcanic arcs, in *Dynamics of subduction,* edited by G.E. Bebout, D. Scholl, S. Kirby and J.P. Platt, Geophysical Monograph Series, AGU, 1996.

Keppler, H., Experimental constraints on trace element transport by fluids in subduction zones (abstract), *Eos Trans. AGU, 76*(46), Fall Meeting Suppl., 655, 1995.

Kersting, A.B. and R.J. Arculus, Kluyechevskoy Volcano, Kamchatka, Russia: the role of high-flux recharged, tapped and fractionated magma chamber(s) in the genesis of high-Al_2O_3 from high-MgO basalt, *J. Petrol., 35,* 1-41, 1994.

Kushiro, I., Partial melting of mantle wedge and evolution of island arc crust, *J. Geophys. Res., 95,* 15929-15939, 1990.

Leeman, W.P., M.J. Carr and J.D. Morris, Boron geochemistry of the Central American Volcanic Arc: constraints on the genesis of subduction-related magmas, *Geochim. Cosmochim. Acta, 58,* 149-168, 1994.

Leeman, W.P. and C.J. Hawkesworth, Open magma systems: trace element and isotopic constraints, *J. Geophys. Res., 91.,* 5901-5912, 1986.

Luhr, J.F., Slab-derived fluids and partial melting in subduction zones: insights from two contrasting Mexican volcanoes (Colima and Ceboruco). *J. Volcanol. Geotherm. Res., 54,* 1-18, 1992.

Maury, R.C., M.J. Defant, and J.-L. Joron, Metasomatism of the sub-arc mantle inferred from trace elements in Philippine xenoliths, *Nature, 360,* 661-663, 1992.

Morris, J.D., W.P. Leeman, and F. Tera, The subducted component in island arc lavas: constraints from Be isotopes and B-Be systematics, *Nature, 344,* 31-36, 1990.

Myers, J.D., B.D. Marsh, and A. Krishna Sinha, Strontium isotopic and selected trace element variations between two Aleutian volcanic centers (Adak and Atka): implications for the development of arc volcanic plumbing systems, *Contrib. Mineral. Petrol., 91,* 221-234, 1985.

Myers, J.D., B.D. Marsh, and A. Krishna Sinha, Geochemical and strontium isotopic characteristics of parental Aleutian Arc magmas: evidence from the basaltic lavas of Atka, *Contrib. Mineral. Petrol., 94,* 1-11, 1986.

Navon, O. and E. Stolper, Geochemical consequences of melt percolation: the upper mantle as a chromatographic column. *J. Geol., 95,* 285-307, 1987.

Nye C.J., and M.R. Reid, Geochemistry of primary and least fractionated lavas from Okmok Volcano, central Aleutians: Implications for arc magmagenesis, *J. Geophys. Res., 91,* 10271-10287, 1986.

O'Hara, M.J., The bearing of phase equilibria studies in synthetic and natural systems on the origin of basic and ultrabasic rocks, *Earth Sci. Rev., 4,* 69-133, 1968.

Patchett, P.J. and M. Tatsumoto, Hafnium isotope variations in oceanic basalts, *Geophys. Res. Lett., 7,* 1077-1080, 1980.

Pearce, J.A., Trace element characteristics of lavas from destructive plate boundaries, in *Andesites: Orogenic Andesites and Related Rocks,* edited by R.S. Thorpe, pp. 525-548, Wiley and Sons, New York, 1982.

Pearce, J.A. and I.J. Parkinson, Trace element models for mantle melting: application to volcanic arc petrogenesis, in Magmatic processes and plate tectonics, edited by H.M. Prichard, T. Alabaster, N.B.W. Harris, and C.R. Neary, *Spec. Pub. Geol. Soc. London, 76,* pp. 373-403, 1993.

Plank, T., Nb anomalies in arc lavas: insights from the Marianas (abstract), *Eos Trans. AGU, 75*(44), Fall Meeting Suppl., 730, 1994.

Plank, T. and C.H. Langmuir, An evaluation of the global variations in the major element chemistry of arc basalts, *Earth Planet. Sci. Lett., 90,* 349-370, 1988.

Plank, T. and C.H. Langmuir, Effects of the melting regime on the composition of oceanic crust, *J. Geophys. Res., 97,* 19749-19770, 1992.

Plank, T. and C. H. Langmuir, Tracing trace elements from sediment input to volcanic output at subduction zones, *Nature, 362,* 739-743, 1993.

Romick, J.D., M.R. Perfit, S.E. Swanson and R.D. Shuster, Magmatism in the eastern Aleutian Arc: temporal characteristics of igneous activity in Akutan Island, *Contrib. Mineral. Petrol., 104,* 700-721, 1990.

Salters, V.J.M. and S.R. Hart, The Hafnium paradox and the role of garnet in mid-ocean ridge basalts, *Nature, 342,* 420-422, 1989.

Salters, V.J.M. and S.R. Hart, The mantle sources of ocean ridges, islands and arcs: the Hf isotope connection, *Earth Planet. Sci. Lett., 104,* 364-380, 1991.

Singer, B.S. and J.D. Myers, Intra-arc extension and magmatic evolution in the central Aleutian arc, Alaska. *Geology, 18,* 1050-1053, 1990.

Singer, B.S., J.D. Myers and C.D. Frost, Mid-Pleistocene basalt from the Seguam volcanic center, central Aleutian arc, Alaska: local lithospheric structures and source variability in the Aleutian arc, *J. Geophys. Res., 97,* 4561-4578, 1992a.

Singer, B.S., J.D. Myers and C.D. Frost, Mid-Pleistocene lavas from the Seguam volcanic center, central Aleutian arc: closed-system fractional crystallization of a basalt to rhyodacite eruptive suite, *Contrib. Mineral. Petrol., 110,* 87-112, 1992b.

Stolper E., and S. Newman, The role of water in the petrogenesis of Mariana Trough magmas, *Earth Planet. Sci. Lett., 121,* 293-326, 1994.

Sun, S.S., Lead isotopic study of young volcanic rocks from mid-ocean ridges, ocean islands and island arcs. *Phil. Trans. R. Soc., A297,* 409-445, 1980.

Tatsumi Y., D.L. Hamilton and R.W. Nesbitt, Chemical characteristics of fluid phase released from a subducted lithosphere and the origin of arc magmas: evidence from high-pressure experiments and natural rocks, *J. Volcanol. Geotherm. Res., 29,* 293-309, 1986.

Taylor S.R., and S.M. McLennan, *The continental crust: its composition and evolution,* 312 pp., Blackwell, Oxford, 1985.

Thirlwall, M.F., T.E. Smith, A.M. Graham, N. Theodorou, P. Hollings, J.P. Davidson, and R.J. Arculus, High Field Strength Element Anomalies in Arc Lavas: Source or Process? *J. Petrol., 35,* 819-838, 1994.

Thompson, R.N., M.A. Morrison, G.L. Hendry, and S.J. Parry, An assessment of the relative roles of crust and mantle in magma genesis: an elemental approach, *Phil. Trans. R. Soc. 310,* 549-590, 1984.

White, W.M. and P.J. Patchett, Hf-Nd-Sr Isotopes and incompatible element abundances in island arcs: Implications for magma origins and crust-mantle evolution, *Earth Planet. Sci. Lett., 67,* 167-185, 1984.

Woodhead, J.D., Geochemistry of the Mariana arc (western Pacific): Source composition and processes, *Chem. Geol., 76,* 1-24, 1989.

Woodhead, J.D., S.E. Eggins, and J.G. Gamble, High field strength and transition element systematics in island arc and back-arc basin basalts: evidence for multi-phase extraction and a depleted mantle wedge, *Earth Planet. Sci. Lett., 114,* 491-504, 1993.

Wyllie, P.J., Subduction products according to experimental prediction, *Bull. Geol. Soc. Amer., 93,* 468-476, 1982.

Jon P. Davidson, Department of Earth and Space Sciences, UCLA, Los Angeles, CA 90095.
(email: davidson@zephyr.ess.ucla.edu)

Describing Chemical Fluxes in Subduction Zones: Insights from "Depth-Profiling" Studies of Arc and Forearc Rocks

Jeff Ryan[1], Julie Morris[2], Gray Bebout[3], and Bill Leeman[4]

Trace element systematics in convergent margin metamorphic and volcanic rocks show that subducting slabs release fluids of changing composition as a function of depth. Volcanic transects across arcs record declines in H_2O-soluble elements (B, Cs, As, and Sb) with increasing depth that parallel declines with increasing metamorphic grade in "subduction complex" associations. These paired, prograde declines point to decreasing inputs of H_2O-rich fluids from the subducting slab. Uniform K, Ba, REE and ^{10}Be levels across arcs suggest that slab-derived fluxes of different compositions persist to greater depths. Slabs returned to the mantle via subduction should have profoundly fractionated chemical signatures, and substantial fluid releases should occur through forearc regions. Serpentinites from the Marianas forearc show elevated B contents, and fractionated trace element signatures suggesting inputs of fluids like those released from metamorphosed slabs at low temperatures.

1. INTRODUCTION

In the past decade, geologists studying subduction zones have come to believe that melting at arcs occurs in the mantle in response to material additions from the subducting plate. Current models view these additions as fluids released from the slab, though it is suggested that melting is triggered when hydrated minerals in the mantle (formed through reactions with slab-derived fluids) decompose [*Tatsumi and Eggins*, 1995; *Davies and Stevenson*, 1992]. Insights into the compositions of such fluids, or into how slab and mantle compositions change with fluid exchange are limited. By contrast, models for the isotopic evolution of the mantle infer a subducted origin for certain chemical signatures in intraplate lavas, implying that the "fingerprint" of subducted materials can be preserved in the mantle [*Hart* 1988; *Hofmann* 1988]. Reconciling intraplate "slab signatures" with the petrologic evolution of subducting plates requires an understanding of the chemistry of subduction: what species leave slabs and enter the mantle, do these inputs change, and how do these additions affect arc magmatism?

We have addressed these questions through "depth profile" geochemical studies of subduction zone magmatic and metamorphic rocks. We have examined the trace element systematics of rock suites reflecting specific depth ranges in the subduction cycle to discover how and where different species are liberated from the slab. From these observations, we can begin to describe the mobile phases released from the slab at different depths. Below, we integrate results from several diffrrent studies of subduction-related volcanic and metamorphic suites. Our combined results suggest that slab inputs to the mantle occur over a wide depth range. Trace element abundance levels change with slab depth, suggesting progressive changes in the composition of the slab flux. Our results also shed light on the controls over melting beneath arcs, and into crust-mantle chemical recycling processes mediated by subduction.

2. CROSS-ARC ELEMENTAL VARIATIONS

The complex nature of slab-mantle exchanges is apparent in the elemental variation patterns observed in lavas from volcanoes aligned perpendicular to the strikes of volcanic arcs. These "cross-arc transects" offer a means to see if trace element enrichments in arc source regions change with progressive subduction. Cross-arc geochemical data are available from several arcs: Bismarck [*Gill et al.* 1993], the Aleutians, Izu-Bonin [*Ryan and Langmuir*,

[1]Department of Geology, University of South Florida, Tampa, FL
[2]Department of Earth and Planetary Sciences, Washington University, St. Louis, MO
[3]Department of Earth and Environmental Sciences, Lehigh University, Bethlehem, PA
[4]Department of Geology and Geophysics, Rice University, Houston, TX

1993; *Morris et al.*, 1990], Central America [*Walker et al.*, 1995] the Cascades [*Leeman et al.* 1990], and most recently the Kurils [*Ryan et al*, 1995]. Of these all but the Cascades and Izu-Bonin lavas contain >10^6 atoms/g ^{10}Be, indicating material inputs from subducted trench sediments to sub-arc source regions [*Tera et al..* 1986]. We focus on results from the Kurils, and to a lesser extent Bismarck and the Aleutians, as in these cases lavas erupt through relatively thin (~15 km) oceanic crust. Crust-magma interactions should thus be minimized, and primitive lavas should preserve elemental signatures that more directly reflect their mantle sources.

Figure 1 plots element ratios versus Benioff zone depth for mafic lavas from the Kuril, Aleutian, and Bismarck arcs. The concentration of a trace element in an arc lava relates both to the mantle source abundance of that element, and to the extents of melting and crystallization involved in its magmatic evolution. To see through crystal/melt fractionation effects, we ratio the elements of interest to others with similar solid/melt distribution behaviors, but much lower apparent solubilities in H_2O-rich fluids, such as might be liberated from a subducting slab. Variations in these ratios will thus reflect the extent of mantle contamination by slab-derived fluid phases.

Two general cross-arc variation patterns appear:

1. High ratios relative to ocean ridge (MORB) or ocean island (OIB) basalt mantle sources at the arc volcanic front (VF), with a progressive decline in ratios behind the front (BTF) toward MORB and OIB values (Figure 1a).

2. No significant changes in element ratios across the arc, though ratios may range from values similar to MORBs to highly elevated values (Figures 1b, c).

These cross-arc elemental enrichment patterns are difficult to reconcile with subduction zone melting models that assume "point source" slab input events associated with specific mineral breakdown reactions [e.g., *Tatsumi* 1986], as some elements show progressive declines with depth, suggesting a progressively "distilled" slab; and others show strong enrichment, but no changes with depth. Elemental fractionations during the transport of slab-derived materials [*Davies and Stevenson* 1992] will be controlled by mantle/fluid or mantle/melt equilibria. However, all of the elements discussed above behave highly incompatibly in arc lavas, and many (Cs, Rb, Ba, K, B?) show mobility in fluids during serpentinite dehydration reactions, such as might occur in the mantle [*Tatsumi et al.* 1986; *Tatsumi and Eggins* 1995]. Models that assume sediment melt inputs from the slab [*Plank* 1992] can explain K, Ba and REE enrichment patttterns, but have difficulties producing variable, depth-dependent enrichments of B and Cs.

3. EVIDENCE FROM SUBDUCTION-RELATED METAMORPHIC ROCKS

Trace element data from metamorphic rocks suggest fractionations that in key ways parallel those seen in arc

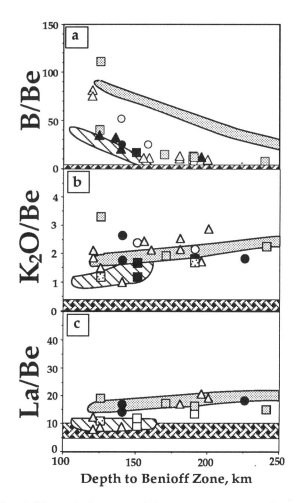

Fig. 1: Plots of element enrichment patterns versus Benioff zone depth for cross-arc transects. Data from *Ryan et al.* [1995]; *Morris et al.* [1990]; and *Ryan and Langmuir* [1993] Kuril arc data: black triangles: Lvinaya transect; light squares: Medvhyzia transect; black circles: Chirpoy transect; white circles: Chirinkotan transect; white triangles: Onekotan transect; black squares: Paramushir-Alaid transect. Shaded trend represents Bismarck transect (Pago-Ulawan, Makalia, Garove), and and striped trend represents Okmok-Bogoslof transect in the Aleutians. Hatched fields represent enrichment levels of MORB and OIB source mantle. Each diagram presents a pcross-arc enrichment pattern shown by a subset of trace elements: a) declining enrichment with depth, shown by B, Cs, As, and Sb; b) constant high enrichment relative to other mantle sources, (K, Ba, Rb, and ^{10}Be); c) little enrichment relative to mantle sources, (LREE, Th, ^{9}Be, and Zr).

lavas. Metamorphic massifs from around the world show declines in B content with increasing grade, even under low pressure and temperature conditions (greenschist facies or lower)[*Moran et al.*, 1992; *Bebout et al.*, 1993; *Truscott et al.*, 1985; *Leeman et al.*, 1992]. This early mobilization of B contrasts with the behavior of K, Rb, and Li during metamorphism, which all show depletions only in granulite

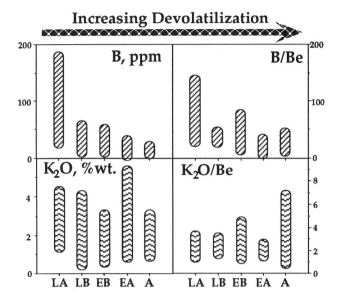

Fig. 2: Element concentration and ratio plots of prograde metasedimentary rocks from the Catalina Schist. Data from *Bebout et al.* [1993; 1995]. Shaded fields represent ranges of sample values at a specific metamorphic facies: each field includes 6-15 samples. X-axis legend are metamorphic facies, arranged in order of increasing P-T conditions. LA: lawsonite-albite; LB: lawsonite-blueschist; EB: epidote blueschist; EA: epidote amphibolite; A: amphibolite. Each pair of diagrams presents typical variations of subsets of elements. a,b) B content and B/Be vs. grade (B, Cs, As and Sb). c,d) K_2O and K_2O/Be vs. grade (K, Ba, Rb, LREE).

facies rocks [*Sighinolfi and Gorgoni*, 1978]. The Catalina Schist, a "subduction complex" metamorphic massif, includes metasedimentary and metamafic rocks which range in grade from lawsonite-albite up to amphibolite facies (P≈0.8-1.2 GPa; T≈300-650°C)[*Bebout et al.*, 1996]. Rocks of all facies in the Catalina are heavily veined, indicating the passage of large volumes of fluid.

Trace element variations with increasing grade in Catalina Schist metasediments and veins show patterns similar to those observed in the Kurils: B and Cs (also As and Sb) contents decline steadily with increasing grade, while K and Ba show little change in concentration save a slight increase in amphibolite facies rocks [*Bebout et al.*, 1996]. Declines in B and Cs contents correlate with declines in H_2O, suggesting an H_2O-rich fluid was the medium in which these species were removed [*Bebout et al.*, 1996].

Pegmatites, which occur only in amphibolite facies Catalina rocks, show trace element signatures similar to their metamorphosed hosts. These pegmatites may represent a higher temperature component of the slab flux in which elements that are relatively insoluble in H_2O-rich fluids (K, Ba, REE?) may be liberated. Although the Catalina Schist reflects hotter metamorphic conditions than are typical of subducting plates, we view the patterns of elemental variation in Catalina rocks as broadly representative of devolatilizing slabs.

4. DISCUSSION

4.1 A Changing Slab Flux

Data from subduction-related igneous and metamorphic rocks strongly suggest that slab-derived fluid phases mobilize elements selectively as a function of changing pressure and temperature conditions on the slab. Elements like B and Cs, which have affinities for lower temperature, H_2O-rich fluids, are mobilized in the early stages of subduction. The inventories of these elements decline rapidly as the slab metamorphoses, so that at the depths of arc source regions, these species are present in diminished abundances. B contents in arc lavas account for <30% of the boron subducted in marine sediments and ocean crust reaching trenches [*Ryan*, 1989; *Moran et al.*, 1992]. This degree of attenuation suggests slab conditions thermally equivalent to amphibolite facies Catalina rocks.

Nonetheless, in many arc lavas, B enrichments are pronounced. Figure 3 plots B/Be versus La/Sm for mafic lavas from the volcanic fronts of several arcs. La/Sm is strongly affected by changes in extents of partial melting, but is little impacted by changes in slab fluid inputs; while B/Be is a sensitive indicator of slab inputs, but is little changed by melting processes. Figure 3 demonstrates a relationship between slab inputs, and the degree of partial melting: the greater the flux from the slab, the more extensive the melting event. While other slab-derived species show similar correlations (i.e., Ba/La in some arcs), none show it as prominently as B, which implies that inputs of H_2O-rich fluids from the slab may act as a primary control over arc melting.

Melting behind the volcanic front are less clearly associated with an H_2O-rich flux, as enrichments of B and like elements are much lower in these lavas. However, consistently high abundances of K and Ba (and in some arcs ^{10}Be) [*Morris and Tera*, 1989] in lavas erupted behind volcanic fronts indicate significant, but different slab inputs to these source regions. Alkaline element contents in arc lavas globally show correlations with abundances in modern trench sediments [*Plank and Langmuir* 1993], suggesting a sediment-dominated subduction component. Melting of slab sediments, induced possibly by muscovite breakdown, or inputs of pegmatite-like fluids generated on slabs at higher metamorphic grades, may serve to add K, Ba, REEs and like species from slab sediments to the mantle without strongly fractionating them, as solid/melt distribution coefficients for many species in high-SiO_2 magmatic systems approach 1 [*Henderson*, 1982; *Plank*, 1992; *Johnson and Plank*, 1993]. Elements soluble in higher temperature, SiO_2-rich fluids/melts are liberated later in the

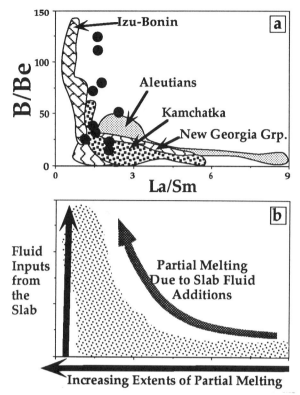

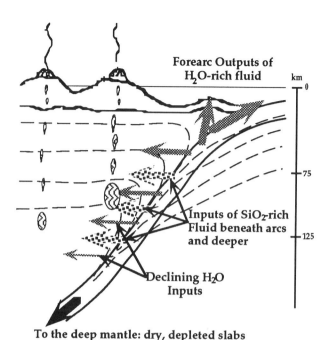

Fig. 3a. B/Be vs. La/Sm for mafic VF lavas from several arcs. Data sources are *Ryan and Langmuir* [1993]; *Ryan et al.* [1995]; and *Austin et al.*, in prep. Individual data points represent VF basalts from the Kuril arc; fields are as marked. 3b. Explanatory schematic of 3a., outlining relationship between slab inputs and degree of melting described in text.

Fig. 4. Profile of a subduction zone depicting model of varying slab-mantle exchanges described in the text. Dotted contours represent typical subduction zone isotherms; shaded and speckled arrows, respectively, represent H_2O- and SiO_2-rich fluids released from the slab at different depths. The source regions of arc lavas are contaminated by a mixed slab flux in which the proportion of H_2O-rich fluids decreases as a function of subduction-related metamorphism.

subduction process, so significant inventories should remain on the slab to contaminate mantle sources behind arc volcanic fronts.

Slab inputs to arc source regions can thus be described as a mixture of at least two components: one rich in H_2O, and another which may be rich in SiO_2. The proportions of these components vary as a function of progressive subduction, such that forearc and volcanic front inputs are H_2O dominated, and behind-the-front sources see a more SiO_2-rich flux (Figure 4). Implicit in this model is the idea that slab-mantle chemical exchanges occur over a wide range of depths, and that elemental fractionations induced by mantle/melt or mantle/fluid equilibria are negligible relative to those produced by the metamorphic evolution of the slab. While high pressure mantle/fluid and mantle/magma interactions may be viable for generating HFSE depletions in arc lavas [e.g., *Kelemen et al.*, 1990], such processes need not be invoked to explain most of the elemental fractionations observed in arcs. Arc lavas preserve trace element systematics consistent with elemental patterns inferred to develop on the slab as a result of subduction-induced metamorphism.

4.2 Implications for Deep Slabs, and Forearc Fluid Fluxes

The model for slab-mantle exchanges outlined above will result in pervasive chemical modification of subducting plates. Deep slabs will be highly devolatilized, and strongly depleted in H_2O-soluble species like B and Cs. The degree to which more "refractory" trace elements like K and Rb are released from the slab is unclear, as these species show little change within the zone of melting in arcs. Back-arc basin lavas may preserve elemental signatures derived from deep slabs, though disentangling the effects of such inputs from those of fluid/mantle interactions may be difficult. *Stolper and Newman* [1994] recognized in the trace element systematics of Marianas Trough basalts additions of a phase high in H_2O, but enrichments in K, Rb, and Ba smaller than those seen in Marianas arc lavas, which they describe as inputs of a slab-derived fluid which has undergone extensive chemical exchange with the mantle. The modest K, Rb, and Ba enrichments of Trough lavas could also reflect inputs from an evolved Mariana slab, as metamorphic processing should diminish slab inventories of these species. Distinguishing slab- and mantle-induced elemental fractionations in the back arc is difficult given current knowledge, and both processes may well be important.

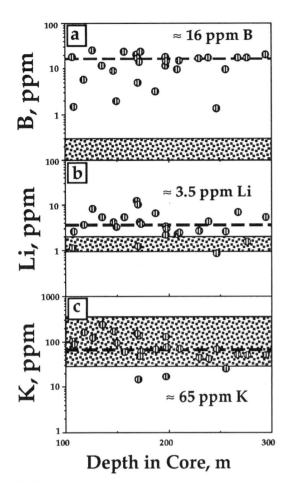

Fig. 5. Plots of element abundance versus core depth for Conical Seamount serpentinite samples from Hole 779A, ODP Leg 125. Data from *Mattie and Ryan*, in prep. Dashed lines refer to mean abundance levels listed in each diagram. Shaded fields represent primitive-to-depleted mantle abundance levels based on data from *Ryan and Langmuir* [1987; 1993] and *McDonough and Sun* [1995].

A second implication of our model is that large amounts of H_2O (and entrained H_2O-soluble elements) must leave the slab in forearc regions. As mentioned above, >60% of the boron subducted at trenches leaves the slab before reaching arcs. However, the only large output flux of B in the forearc, desorptive fluid releases from sediments in accretionary prisms [*Spivack*, 1986; *You et al.*, 1992; 1995], returns only ~10% of the subducted B budget. Large amounts of B (and H_2O) must follow other paths.

Serpentinite seamounts in the Marianas and Izu-Bonin forearcs may reflect this unseen forearc flux. The Conical Seamount in the Marianas forearc is an active mud volcano, venting fluids and serpentinite clasts and muds [*Fryer et al.*, 1992]. Pore fluid compositions from Conical show elevated B and Rb contents, and slightly elevated K contents relative to seawater, which along with low chlorinity, and high levels of volatile hydrocarbons and ammonia, strongly indicate a subducted origin [*Mottl*, 1992]. Conical Seamount core samples collected on ODP Leg 125 are serpentinized dunites and harzburgites with high B contents, modestly elevated Li contents, and K and Ba contents indistinguishable from depleted mantle sources [*Mattie and Ryan*, 1994](Figure 5). While elemental distributions in Leg 125 serpentinites are still being examined, these preliminary results suggest inputs of an H_2O-rich fluid high in B and low in K and Ba, similar to fluids released by low-grade Catalina Schist metasediments.

5. CONCLUSIONS

Through the study of geochemical "depth profiles" across subduction zones, we have come to view the slab input process as a spectrum of slab/mantle exchanges occurring over a range of depths. The compositions of slab-derived fluids change due to progressive metamorphic evolution of slab sediments and ocean crust. Highly H_2O-soluble trace elements (B, Cs) mobilize early, showing depletions in low grade metamorphic rocks and enrichments in forearc fluids; and subduct poorly, showing little enrichment at volcanoes sited above deep slabs. The role of subduction in the chemical evolution of the mantle is still imperfectly understood. Further study of chemical transport processes both deeper and shallower than seen in arcs will help quantify the effects of subduction on both modern and ancient mantle sources.

Aknowledgments. Thanks to A. Tsvetkov for providing samples and data from the Kuriles, and to Steve Shirey, Brady Byrd, and Dave Kuentz for their aid in analyzing Kuriles and Catalina samples on the ICP at DTM. Thanks also to C. Edwards, D. Miller, T. Plank, A.F. Hochstaedter, M. Defant, and P. Kepezhinskas for discussions which led to insights, and for helping to create a ferment of subduction thinking through which the first author has had the pleasure of swimming for the past ten years. This work was supported by NSF Grants EAR90-04389 to Morris and Tera, EAR91-19110 to Leeman, and EAR92-05804 to Ryan, Hochstaedter, and Defant.

REFERENCES

Arculus, R., Arc Volcanism as indicators of slab and mantle wedge evolution. *Dynamics of Subduction, AGU Monograph __*, pp. __ 1996 (THIS VOLUME).

Bebout, G.E., Ryan, J.G., and Leeman, W.P., B-Be systematics in subduction-related metamorphic rocks: characterization of the subducted component. *Geochim. Cosmochim. Acta*, 57 2227-2237, 1993.

Bebout G.E., Ryan, J.G., Leeman, W.P., and Bebout, A.E., Fractionation of trace elements by subduction zone metamorphism: significance for models of crust-mantle mixing *Geochim. Cosmochim. Acta* , 1996, in press.

Davies, J.H. and Stevenson, D.J., Physical model of source region of subduction zone volcanics. *J. Geophys. Res.*, 97, 2037-2070, 1991.

Fryer, P. Saboda, K.L., Johnson, L.E., MacKay, M.E., Moore, G.F. and Stoffers, P., Conical Seamount: SeaMARCII, Alvin submersible, and seismic reflection studies. In *Proc. ODP, Init. Repts.*, *125*, College Station, TX, Ocean Drilling Program, 69-80, 1992.

Gill, J.B., Morris, J.D., and Johnson, R.W., Timescale for producing the geochemical signature of island arc magmas: U-Th-Po and Be-B systematics in recent Papua New Guinea lavas. *Geochim. Cosmochim. Acta*, *57*, 4269-4283, 1993.

Hart, S.R., Heterogeneous mantle domains: signatures, genesis, and mixing chronologies. *Earth Planet. Sci. Lett.*, *90*, 273-296, 1988.

Henderson, P. *Inorganic Geochemistry*. Oxford, Pergamon Press, 1982.

Hofmann, A.W., Chemical differentiation of the Earth: the relationship between mantle, continental crust, and oceanic crust. *Earth Planet. Sci. Lett.* *90*, 297-314, 1988.

Johnson, M.C. and Plank, T., Experimental constraints on sediment melting during subduction. *EOS*, *74*, 680, 1993.

Kelemen, P.B., Johnson, K.T.M, Kinzler, R.J., and Irving, A.J., High-field-strength element depletions in arc basalts due to mantle-magma interaction. *Nature*, *345*, 521-524. 1990.

Leeman, W.P., Sisson, V.B., and Reid, M.R., Boron geochemistry of the lower crust: evidence from granulite terranes and deep crustal xenoliths. *Geochim. Cosmochim. Acta*, *56*, 775-788, 1992.

Leeman, W.P., Smith, D.R., Hildreth, W., Palacz, Z., and Rogers, N., Compositional diversity of Late Cenozoic basalts in a transect across the southern Washington Cascades: implications for subduction zone magmatism. *J. Geophys. Res.*, *95*, 19561-19582, 1990.

Mattie P.D., and Ryan, J.G., Boron and alkaline element systematics in serpentinites from Holes 779A, 780C, and 784A, ODP Leg 125: describing fluid mediated slab additions. *EOS Suppl.*, *75*, 352, 1994.

McDonough, W.F. and Sun, S.-s., The composition of the Earth. *Chemical Geology*, *120*, 223-253.

Moran, A.E., Sisson, V.B., and Leeman W.P., Boron depletion during progressive metamorphism: implications for subduction processes. *Earth Planet. Sci. Lett.*, *111*, 331-349, 1992.

Morris, J.D., Leeman, W.P., and Tera, F., The subducted component in island arc lavas: constraints from Be isotopes and B-Be systematics. *Nature*, *344*, 31-36, 1990.

Mottl, M.J., Pore waters from serpentinite seamounts in the Mariana and Izu-Bonin forearcs, Leg 125: Evidence for volatiles from the subducting slab. In *Proc. ODP, Scientific Results*, *125*, College Station, TX, Ocean Drilling Program, 373-385, 1992.

Plank, T., *Mantle Melting and Crustal Recycling in Subduction Zones*. Ph.D. dissert., Columbia Univ., 1992.

Plank, T. and Langmuir, C.H., Tracing trace elements from sediment input to volcanic output at subduction zones. *Nature*, *362*, 739-742, 1993.

Ryan, J.G. *The Systematics of Lithium, Beryllium, and Boron in Young Volcanic Rocks*, Ph.D. Dissert., Columbia Univ., 313 pp., 1989.

Ryan, J.G., and Langmuir, C.H., The systematics of lithium abundances in young volcanic rocks. *Geochimica et Cosmochimica Acta*, *51*, 1727-1741, 1987.

Ryan, J.G. and Langmuir, C.H., The systematics of boron abundances in young volcanic rocks. *Geochimica et Cosmochimica Acta*,. *57*, 1489-1498, 1993.

Ryan, J.G., Morris, J.D., Leeman, W.P., Tera, F. and Tsvetkov, A., Cross-arc geochemical variations in the Kuril arc as a function of slab depth. *Science*, *270*, 625-627.

Sighinolfi, G.P., and Gorgoni, C., Chemical evolution of high grade metamorphic rocks: anatexis and remotion of material from granulite terranes. *Chem. Geol.*, *22*, 157-176, 1978.

Spivack, A.J., *Boron Isotope Geochemistry*. Ph.D. dissert., Woods Hole-MIT, 1986.

Stolper, E. and Newman, S., The role of water in the petrogenesis of Mariana Trough magmas. *Earth Planet. Sci. Lett.*, *121*, 293-325, 1994.

Tatsumi, Y., Formation of volcanic front in subduction zones. *Geophys. Res. Lett.*, *13*, 717-720, 1986.

Tatsumi, Y., Hamilton, D.L., and Nesbitt, R.W., Chemical characteristics of fluid phase released from a subducted lithosphere and origin of arc magmas: evidence from high-pressure experiments and natural rocks. *J. Volc. Geotherm. Res.*, *29*, 293-309, 1986.

Tatsumi, Y., and Eggins, S., *Subduction Zone Magmatism*, 211pp., Blackwell Science, Cambridge, MA, 1995

Tera, F., Brown, L., Morris, J., Sacks, I.S., Klein, J. and Middleton, R., Sediment incorporation in island-arc magmas: inferences from ^{10}Be. *Geochim. Cosmochim. Acta*, *50*, 535-550, 1986.

Truscott, M.G., Shaw, D.M., and Cramer, J.J., Boron abundance and localization in granulites and the lower continental crust. *Bull. Geol. Soc. Finland* 58, 169-177, 1986.

Walker, J.A., Carr, M.J., Patino, L.C., Johnson, C.M., Feigenson, M.D., and Ward, R.L., Abrupt change in magma generation processes across the Central American arc in southeastern Guatemala: flux dominated melting near the base of the wedge to decompression melting near the top of the wedge. *Contrib. Mineral. Petrol.* *120*, 378-390, 1995.

You, C-F., A.J. Spivack, J.H. Smith, and J.M. Gieskes, Mobilization of boron in convergent margins: implications for the boron geochemical cycle. *Geology*, *21*, 207-210, 1993.

You, C-F., Spivack, A.J., Gieskes, J.M., Rosenbauer, R., and Bischoff, J.L., Experimental study of boron geochemistry: implications for fluid processes in subduction zones. *Geochim. Cosmochim. Acta*, *59*, 2435-2442, 1995.

J.G. Ryan, Department of Geology, University of South Florida, 4202 East Fowler Ave. Tampa, FL 33620

Boron and Other Fluid-mobile Elements in Volcanic Arc Lavas: Implications for Subduction Processes

William P. Leeman

Keith-Wiess Geological Laboratory, Rice University, Houston, Texas

This paper emphasizes the utility of fluid-mobile elements (B, As, Sb, Pb) in evaluating the transfer of material from subducted slabs to sources of arc magmas - which in most arcs are believed to reside in the overlying mantle wedge. Correlated enrichments of such elements, along with ^{10}Be, relative to fluid-immobile elements in mafic arc lavas strongly suggest that the mode of transfer is via an aqueous fluid inasmuch as melting processes alone do not seem to produce the observed elemental fractionations. This line of reasoning leads to the views that [1] enrichment levels of fluid-mobile elements in arc magmas ultimately depend on the efficiency of subduction of hydrated materials, and [2] this efficiency is directly related to physical aspects of subduction that control rates and loci of dehydration reactions - namely subduction rate, slab temperature, etc. A direct linkage between physical conditions and chemical signatures suggests that a spectrum of volcanic arcs exists - ranging from 'hot and dry' (e.g., Cascades) to 'cold and wet' (e.g., Japan) endmembers. Other factors, such as amount and type of sediment subducted, may influence the composition of arc magma sources with respect to some elements or isotopes, but are unlikely to be the dominant control for fluid-mobile element composition.

INTRODUCTION

The origins of subduction-related magmas remain controversial despite many man-years of effort on this problem. In part this uncertainty reflects real diversity in boundary conditions, mass fluxes, etc. from one volcanic arc to another. Also, many characteristics of primitive arc magmas may be overprinted by processes attending magma ascent, storage, and eruption, thus complicating efforts to understand the fundamental melt-forming processes. There are relatively few geochemical 'tracers' that can be interpreted unambiguously as signifying transfer of subducted material into source regions of arc magmas [cf. *Davidson*, 1994]. It is even more difficult to describe and quantify the actual fluxes of such components or to determine how these fluxes are influenced by various physical and kinematic parameters.

One of the most effective tracers is the cosmogenic isotope ^{10}Be, for which the only significant source is the atmosphere. This isotope is concentrated by adsorption onto soil and sediment particles, and its only known means of reaching the mantle is via subduction of oceanic lithosphere [*Tera et al.*, 1986]. Because of its short half-life (1.6 m.y.), there is negligible accumulation of ^{10}Be in the deep mantle other than enrichments associated with subduction. The amount of ^{10}Be that can accumulate in a given subduction zone is limited by a quasi steady-state balance between the rate of supply and radioactive decay. Under optimal conditions, the entire cycle from initial subduction to eruption must occur within about 8-10 m.y. for ^{10}Be to be detected in arc magmas. The fact that ^{10}Be is detected in young lavas from most volcanic arcs, yet is below detection limits for fresh lavas from non-subduction tectonic settings, provides convincing evidence for involvement of a subduction component in arc magmas [*Morris and Tera*, 1989]. Other geochemical parameters that correlate strongly with ^{10}Be in young arc magmas may also be derived from subducted material. The element boron is a good example, but other fluid-mobile elements display analogous correlations with ^{10}Be [Leeman et al., in preparation]. A significant advantage of using B and ^{10}Be as petrogenetic indicators is the fact that 'normal' mantle and lower crust rocks contain very low concentrations of both; thus, where magma-wall rock interactions occur, they likely exert only a small dilution effect on the contents of these components and no effect on their ratio.

BORON IN ARC LAVAS AS AN INDICATOR OF SLAB PROCESSES

Boron is enriched in pelagic sediments and altered oceanic crust, whereas data for oceanic basalts [*Leeman*, unpub.; *Ryan and Langmuir*, 1993] and high grade metamorphic rocks [*Leeman et al.*, 1992] indicate that its abundance is very low in the mantle (away from subduction zones) and in the lower crust (cf. Fig. 1). However, high B contents are observed in basalts from most arcs and, where sufficient data are available for lavas from specific arcs, B enrichment correlates strongly with that of ^{10}Be [*Morris et al.*, 1990]. This relation suggests that B and ^{10}Be enrichments result from addition of a rather homogeneous, yet distinctive subduction component for each arc. B and ^{10}Be enrichment levels depend upon the specific nature of the downgoing plate and the physical conditions (e.g., temperature distribution) attending subduction in each case; sparse data preclude detailed evaluation of along-strike variations for these components at most arcs (but see discussion below). The near uniformity of ^{10}Be/B ratios within each arc suggest that both elements are controlled by the same slab-to-mantle transfer process. A fluid transfer mechanism is virtually required for B [cf. *Leeman et al.*, 1994], but Be is not very soluble in aqueous fluids (at least at near-surface conditions). As an alternative mechanism, melt transfer has been considered to account for the behavior of Be; as discussed below, there are problems with this view.

Another important factor concerns the extent to which subducted materials actually resemble their inventories in oceanic crust and sediments on the incoming plate. Mass balance calculations for the Central America arc imply that the ^{10}Be/B ratio of the subduction component there is similar to the average (time-corrected) composition of the entire incoming sediment pile; that is, these materials appear to be subducted more or less intact with little mechanical fractionation by selective accretion or underplating [*Leeman et al.*, 1994]. At other convergent

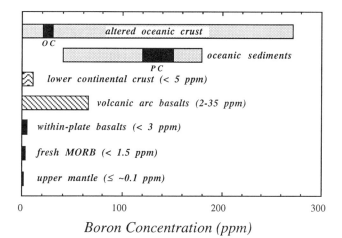

Fig. 1. Ranges in boron concentration in various reservoirs. Data sources: *Leeman et al.* [1992], *Ryan and Langmuir* [1993], *Leeman et al.* [1994], *Chaussidon and Jambon* [1994], Leeman (unpublished data). Average boron content is estimated to be ca. 26 ppm for basaltic oceanic crust and ca. 120 ppm for clay-rich pelagic sediment (Leeman, unpublished data; *Moran et al.*, 1992).

margins, offscraping or underplating processes related to accretionary wedge formation may lead to mechanical fractionation of incoming sediment, hence some uncertainty as to the exact composition of the sedimentary material actually subducted. This is particularly problematic in places where the sediment column is characterized by wide lithologic diversity.

Cross-arc transects for several arcs show that B/Be typically is highest in lavas from the volcanic front and decreases toward backarc regions [*Morris et al.*, 1990; *Ryan et al.*, 1996]. Similar relations are seen for ratios of boron to other incompatible elements (e.g., B/Zr, B/La, B/Nb; [cf. *Ryan et al.*, 1995]). These ratios are expected to vary little during closed system partial melting or magmatic differentiation processes owing to close similarities in bulk solid/melt distribution coefficients. It is inferred that the cross-arc decrease in boron enrichment reflects changing composition of the subducted slab (e.g., decreasing B/Be) as it descends and warms. This seemingly requires operation of a non-igneous process to account for the observed elemental fractionations [cf. *Hart and Reid*, 1991]. It is inferred [*Morris et al.*, 1990; *Moran et al.*, 1992; *Ryan et al.*, 1996] that B is selectively and progressively mobilized and removed from the slab via solution in aqueous fluids released during prograde dehydration reactions involving phyllosilicates, amphiboles, and possibly other hydrous phases [cf. *Pawley*, 1994; *Sorensen*, 1994] - primarily in the sediment and oceanic crust upper portions of subducted slabs. Figure 2 schematically illustrates this concept.

Studies of metamorphic suites reveal that boron is systematically depleted as temperature increases [*Moran et al.*, 1992; *Bebout et al.*, 1993, 1996; *Bebout*, 1994]. Because B is hosted mainly in phyllosilicates (even in some tourmaline-bearing rocks), its abundance is controlled largely by progressive devolatilization reactions involving these phases. Such reactions clearly control the release of aqueous fluids and apparently the distribution of fluid-mobile elements (B, As, Sb, Pb), all of which are strongly depleted in most high grade metasediments [*Bebout et al.*, 1996; W. Leeman and V. Sisson, unpublished data]. The temperature-dependent depletion of these elements from subducting slabs implies that if the uppermost slab exceeds upper amphibolite conditions (ca. 700°C) it is unlikely to retain sufficient amounts of B (and probably the other fluid-mobile elements) to match their estimated outputs at typical ('cool') arcs [*Moran et al.*, 1992]. In most cases, melting of oceanic crust is highly improbable [*Peacock*, 1994; *Peacock et al.*, 1994], and even sediments are unlikely to melt unless water-saturated conditions prevail [*Nichols et al.*, 1994]. In other words, direct slab melting is unlikely to play a significant role in transferring subducted components into the mantle wedge unless an unusually warm slab is involved - for example, where very young lithosphere is subducted. Moreover, significant dehydration of the slab will occur as the result of a series of prograde metamorphic reactions; these will be more pronounced in warmer subduction zones. In sedimentary and upper crustal parts of subducting slabs (i.e., the main reservoirs for B and other FMEs) dehydration and breakdown of host minerals (clays, micas, chlorite) may be quite advanced and FME inventories largely depleted by fluid transfer before solidus temperatures are reached. If melting occurs, it is likely to be at fluid-undersaturated conditions [cf. *Rushmer*, 1994] unless fluids are provided from dehydration of deeper parts of the slab. For example, in serpentinized oceanic mantle, reaction of antigorite to form forsterite and enstatite at ca. 25 kb can release up to 13% water [*Ulmer and Trommsdorff*, 1995]; talc may provide another repository for water in subducted slabs [*Pawley and Wood*, 1995]. Release of such fluids could promote localized water-saturated melting in some subduction zones. But migmatites from the Catalina subduction complex are demonstrably depleted in B and FMEs [*Bebout et al.*, 1997]; if they are representative, it appears that deep subduction fluids do not introduce significant amounts of FMEs. It is fair to say that there remains considerable uncertainty concerning the importance of slab melting with regard to slab-to-mantle transfer of FMEs.

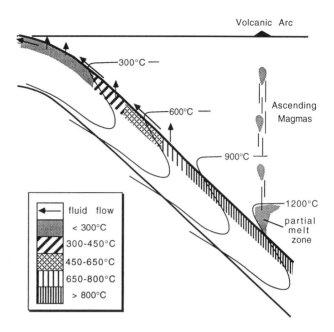

Fig. 2. Schematic cross-section of a subduction zone showing generalized isotherms [cf. *Abbott and Lyle*, 1984]. Sediments and altered oceanic crust (especially the upper few km of the subducted slab, but here exaggerated in thickness for clarity) will contain the bulk of inventories of alkalies, B, As, Sb, Pb, and water; the underlying predominantly ultramafic part of the slab is expected to be a minor reservoir for these components. Prograde metamorphism of the upper layer - from greenschist to eclogite facies - will result in progressive dehydration, recrystallization, and selective losses of water and aqueous-soluble elements [*Moran et al.*, 1992; *Bebout et al.*, 1993, 1995]. At temperatures above ~750°C, the residual upper layer may be extensively dehydrated with only a small fraction of the original water retained in such host phases as amphibole, phlogopite, or lawsonite. In detail, stabilities of phyllosilicates (B, Ba, alkalies), sulfides (As, Sb, Pb), epidote (Sr, REE) with respect to P-T trajectories along the top of the slab will dictate the release patterns of fluid-mobile elements. Actual melting of the slab may rarely occur, and then only in subduction zones with very warm slabs; this is because melting will involve the dehydrated residual upper layer, which is far removed from the water-rich material first subducted.

Along-strike variations in B enrichment seen in some arcs provide strong evidence that the nature and amount of subducted material may significantly influence the composition of arc magmas. In Central America, high B/La is characteristic of most mafic lavas from Nicaragua to Guatemala, whereas low B/La (and negligible ^{10}Be) typifies lavas from Costa Rica [*Leeman et al.*, 1994]. The latter region is characterized by low-angle subduction of the Cocos Ridge and relatively young, warm oceanic lithosphere. Because of tectonic factors, sediment subduction rates are inferred to range significantly along the Middle America trench - from very low off Costa Rica to significantly higher off Guatemala. Subduction contributions to arc magma sources appear to be minimal below Costa Rica and increase in other parts of the arc in proportion to the amount of sediment subducted. In the Aleutians, B/La ratios (also $^{87}Sr/^{86}Sr$ ratios) are strikingly higher in the Seguam-Yunaska sector of the arc, where the Amlia fracture zone has been subducted [*Singer et al.,* 1996]. This observation is consistent with a higher than usual subduction flux of altered oceanic crust and sediment associated with relatively extensive water-rock interaction in the Amlia fracture zone as compared with other sectors in the arc. Thus, some along-strike variations in the composition of arc lavas apparently are inherited from heterogeneities in the subducting plates.

The Cascades arc is one of the least-enriched arc 'end members' recognized so far in that its basalts have no ^{10}Be, low B/La, B/Zr, and similar ratios, and generally resemble many oceanic island basalts [*Leeman et al.,* 1990]. For this arc, sediment offscraping to form the Cascadia accretionary prism probably limits the amount of sediment subducted [*von Huene and Scholl,* 1991]. Also, because the subducted Juan de Fuca plate is young and warm, thermal maturation, metamorphism, and fluid release will deplete the upper slab of B well before it reaches a sub-arc position [cf. *Abbott and Lyle,* 1984; *Jambon and Zimmerman,* 1990; *Peacock,* 1993, 1994]. The western Trans-Mexico Volcanic Belt provides another example of a warm subduction zone; the subducting plate there is young and the arc is in incipient stages of rifting. Calcalkaline lavas from this arc exhibit low B-enrichments and associated alkaline volcanic rocks are negligibly enriched [*Hochstaedter et al.,* 1996].

ENRICHMENTS OF BORON AND OTHER FLUID-MOBILE ELEMENTS IN ARC MAGMAS

Predictably, the chemistry of lavas from 'hot' arcs will provide the most direct compositional representation of mantle wedge materials that have been minimally metasomatized by fluid inputs from the subducted plate. Such lavas exhibit little or no significant enrichment of fluid-mobile elements and can be modelled as being derived from a mixed OIB-MORB source [cf. *Leeman et al.,* 1990, 1994]. At cooler arcs, variable slab-derived fluid fluxes may enhance the wedge in B and other fluid-mobile elements. This view is based on the assumption that B is largely slab-derived - as indicated by the now familiar correlation of B with ^{10}Be [*Morris et al.,* 1990; *Ryan et al.,* 1995] and the likelihood that normal mantle (≤ 0.1 ppm B) and lower crustal rocks (≤ 1 ppm B) are not adequate sources for B-rich arc magmas [cf. *Leeman et al.,* 1992]. The following provides a qualitative discussion of relations between B and other fluid-mobile elements.

Figure 3 summarizes analytical data for Ce, As, Sb, Pb, and B for oceanic basalts, island arc lavas, sediments and metasediments, and several important geochemical 'reservoirs'. Most data are from Newsom et al. [1986], with B determined on many of these same samples (Leeman, unpublished); estimated reservoir compositions are from *Taylor and McLennan* [1985] and *Leeman et al.,* [1992]. Key observations are as follows: (1) As/Ce, B/Ce, Sb/Ce, and Pb/Ce ratios in MORB and OIB samples largely overlap and define restricted ranges (in the case of B, the variation may be largely due to analytical uncertainty); (2) arc lavas, like sediments and average upper crust and pelagic clay, are variably enriched in As, Sb, Pb, and B relative to MORB and OIB. As for many other elements, these data show that subducted sediments provide a potential source for the excess As, Sb, Pb, and B found in arc lavas. How these components are transferred to arc magma sources is a fundamental question.

Assuming (for the moment) that bulk solid/melt distribution coefficients are similarly low for these elements, the Ce-normalized ratios in MORB and OIB samples should be representative of those ratios in the respective mantle source regions; this is generally consistent with the lack of systematic variation in these ratios with concentrations of the numerator elements. Approximate ratios estimated for the suboceanic mantle are as follows: As/Ce (0.01), Sb/Ce (0.001), Pb/Ce (0.04), and B/Ce (0.05). The higher estimates for these ratios in primitive upper mantle (PUM; [*Taylor and McLennan,* 1985]) are based on analyses of several spinel lherzolites which may have been contaminated (W.F. McDonough, personal communication, 1989); in any case they are quite different from values in MORB and OIB and are unlikely to represent viable sources for those magmas. It is also unlikely that arc lavas could be derived from MORB- or OIB-source mantle. This possibility would require that bulk solid/melt distribution coefficients for As, Sb, Pb, and B be significantly smaller than that for Ce. That this clearly is not the case for Pb is well established [cf. *Hofmann,* 1988]. As and Sb partitioning will be influenced by the stability of sulfides and perhaps other host phases, which if present would ensure that these elements would be less incompatible than Ce. Under these conditions it is difficult to reconcile formation

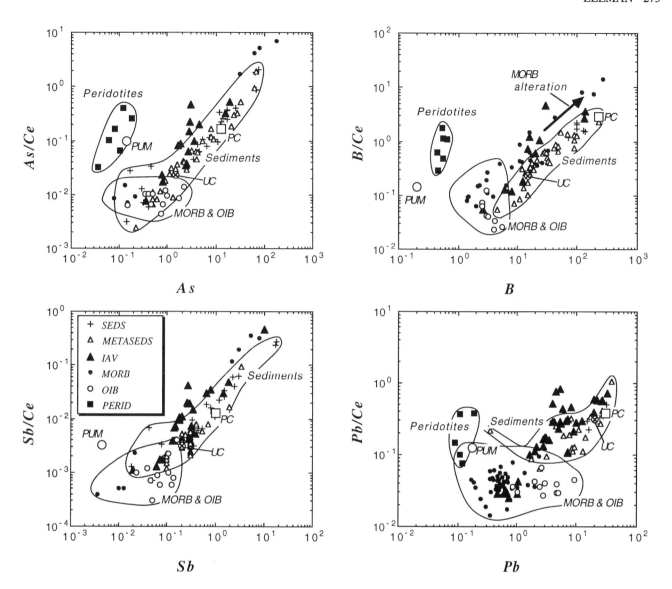

Fig. 3. As-B-Sb-Pb systematics in selected island arc (IAV) and non-arc (fresh and altered MORB; OIB) volcanics, various sediments (SEDS) and metasediments (METASEDS). Averages are shown for pelagic clay (PC), upper continental crust (UC), and estimated primitive upper mantle (PUM) [*Taylor and McLennan*, 1985]; As, Sb, Pb, and Ce data for MORB and OIB samples are from *Newsom et al.* [1986] and Leeman (unpublished data); boron data are analyses of the same samples (Leeman, unpublished data). Data for peridotites are from *Higgins and Shaw* [1988]; note that the Ce-normalized ratios for these samples are too high for these rocks to represent sources of MORB or OIB magmas. Arrows show the effects of seawater alteration in increasing B, As, and Sb contents of MORB.

of arc magma sources with a strictly igneous (e.g., silicate melt-related) transfer process.

As, Sb, Pb, and B have in common the fact that all are highly soluble in aqueous solutions. They are enriched together in many natural geothermal fluids, they are common metasomatic components in alteration haloes around silicic intrusions and hydrothermal veins, and they are enriched together in many epithermal ore deposits [*Weissberg et al.*, 1979; *Shearer et al.*, 1984; *Galbreath et al.*, 1988; *Peucker-Ehrenbrink et al.*, 1994]. In these environments, the aqueous solubility of light rare earth elements (LREE; e.g., Ce) and high field strength elements

(HFSE; e.g., Zr) is usually near or below detection limits [*Michard*, 1989; *Bau*, 1991]. Thus, a more plausible explanation for the enrichments seen in arc lavas involves selective fluid transport of As, Sb, Pb, and B from subducted slabs into overlying mantle wedge domains. Such metasomatic enrichment processes can produce the unique sources needed to generate arc magmas. *Noll et al.* [1996] report that enrichments of these elements in arc lavas also are correlated with ^{10}Be, thus supporting their derivation from subducted materials.

The extent of source metasomatism will strongly influence enrichment levels for fluid-mobile elements in arc magmas. A relative measure of this effect is given by B/Zr ratios in selected arcs as determined for a common B concentration (10 ppm) from regressions of available B-B/Zr data for each arc segment (Table 1). Source enrichment depends on available fluid flux and slab chemical inventories - factors that hinge on thermal state of the subducting plate and on types and amounts of sediment, degree of alteration of oceanic crust, etc. in the subducted slab (Leeman, in preparation). An important corollary of these results is that preferential losses of fluid-mobile elements almost certainly modify the slab chemistry and limit the efficiency of their recycling into deep mantle regions. Quantification of these fluxes is an important goal for future studies.

CONCLUSIONS

Boron and correlated ^{10}Be enrichments in arc lavas are interpreted as the result of direct involvement of subducted sediment and ocean crust components during magma generation. Based on analog studies of metamorphic rocks, this is likely to involve mass transfer of these components from slabs to mantle wedge domains where melting occurs. A major question concerns the mechanism(s) of mass transfer. The following conclusions lead to the view that subduction zone magmatism is largely 'fluid-modulated' in response to several fundamental controlling factors, each of which exhibit some latitude.

1. The role of silicate melts in modifying arc magma sources seems minimal in all but the warmest of subduction zones because, at temperatures approaching melting conditions, volatiles and fluid-mobile element inventories in the slab are likely to be too low to balance their observed enrichments in arc magmas. Direct melting of oceanic crust to form arc *basalts* is not a viable process. If fluid-saturated conditions are attained, sediment melts may form in some subduction zones and act as transfer media for a number of elements that have low solubility in aqueous fluids.

Table 1. B/Zr ratios in selected basalt suites[a]

Volcanic arcs	B/Zr-10[b]
Andes (18-22°S)	0.117
Middle America	
central Costa Rica	0.066
Arenal, Costa Rica	0.175
Mexico	0.059
Cascades	
Medicine Lake	0.055
Mount St. Helens	0.057
Mount Adams	0.063
Aleutians	
eastern	0.135
central	0.182
Kurile-Kamchatka	0.160
NE Japan	0.283
Marianas	0.181
Lesser Antilles	
northern	0.115
central	0.153
S. Sandwich	0.146

Oceanic basalts	Average B/Zr	Range
avg OIB (n=16)	0.012 ± 0.003	0.007 - 0.015
avg MORB (n=49)	0.015 ± 0.005	0.004 - 0.029
altered MORB (n=25)	0.248 - 0.390	0.028 - 1.667

[a] Arc and oceanic basalt data from Leeman (unpublished data) and *Ryan* [1989].
[b] Value interpolated at B = 10 ppm from regression of B versus B/Zr for available basaltic rocks from each arc (Leeman, in preparation).

2. Additional information concerning slab contributions to arc magma sources is provided by the other fluid-mobile trace elements discussed here (As, Sb, Pb). Enrichments of these elements (and B) in arc lavas, compared to elements (LREE, HFSE) that are not readily soluble in aqueous fluids, are easier to reconcile with mass transfer predominantly via aqueous fluids rather than silicate melts. A detailed inventory of how these elements are distributed and details of host mineral stability in subduction zones are not available at present.

3. Exceptionally hot subduction zones (e.g., Cascades, Mexico) produce basaltic magmas that in many respects are indistinguishable from OIBs. These lavas may be derived from domains consisting essentially of mixtures of OIB-

and MORB-source mantle wedge; the minimal overprint by subduction fluids in these places allows more direct assessment of mantle wedge composition. In the Cascades for example, nearly twofold ranges are observed in concentrations of many incompatible elements (e.g., Sr, Ba, La, Th, Ta, Zr, Hf) in basaltic rocks of similar MgO content near 8 wt. % [*Leeman et al., 1990*]. This observation is unlikely to be related simply to variations in flux of subducted sediment as has been proposed by *Plank and Langmuir* [1993]; compositional heterogeneities inherent in the mantle wedge (e.g., due to varied proportions of OIB- and MORB-source material) must also be considered as contributing to this variability.

4. Superimposed on such heterogeneities are the effects of fluid infiltration from subducting slabs which may range from little or none for the Cascades (low B/Zr) to significant for the Japan, Marianas, and South Sandwich arcs (high B/Zr). Local or regional differences in the composition and flux of slab-derived fluids can be ascribed to physical attributes and thermal maturity of the subducting slab and the mechanical efficiency and compositional nature of sediment subduction. Relationships between these factors remain to be established.

5. Finally, extraction of fluid-mobile elements from subducting materials limits their recycling into the deep mantle. Rather, they are concentrated in the mantle wedge at convergent margins, and transferred to shallower levels via fluids or arc magmas. Thus, there may be long-term depletion of these elements from the upper mantle and concentration in the crust.

Acknowledgements. This paper is based on research funded by the National Science Foundation (Grants EAR85-12172, EAR90-14802, EAR90-18996, and EAR91-19110). It builds on collaborative efforts over the years with the following colleagues: Jeff Ryan, Julie Morris, Fouad Tera, Jinny Sisson, Hort Newsom, Phil Noll, Gray Bebout, and Ann Moran-Bebout. This does not imply that they fully subscribe to my views, but they have all contributed to my understanding of subduction processes. Also, I thank Bill White who provided samples for this study that previously had been characterized for As, Sb, and Pb. Some of the As, Sb, and REE data used in Figure 3 were obtained at the Oregon State University Radiation Center through their reactor-sharing program.

W.P. Leeman, Keith-Wiess Geological Laboratories, MS 126, Rice University, Houston, TX 77005; leeman@rice.edu

REFERENCES

Abbott, D., and M. Lyle, Age of oceanic plates at subduction and volatile recycling. *Geophys. Res. Lett. 11*, 951-954, 1984.

Bau M., Rare-earth element mobility during hydrothermal and metamorphic fluid-rock interaction and the significance of the oxidation state of europium. *Earth Planet. Sci. Lett. 93*, 219-230, 1991.

Bebout, G.E., Fluid processes deep in subduction zones - the record in high P/T metamorphic terranes. *SUBCON abstracts*, 175-177, 1994.

Bebout, G.E., J.G. Ryan, and W.P. Leeman, B-Be systematics in subduction-related metamorphic rocks: Characterization of the subducted component. *Geochim. Cosmochim. Acta 57*, 2227-2237, 1993.

Bebout, G.E., Ryan, J.G., Leeman, W.P., and Bebout, A.E., Fractionation of trace elements by subduction-zone metamorphism: significance for models of crust-mantle mixing. *Geochim. Cosmochim. Acta* (in press) 1997.

Chaussidon, M. and A. Jambon, Boron content and isotopic composition of oceanic basalts: geochemical and cosmochemical implications. *Earth Planet. Sci. Lett. 121*, 277-291, 1994.

Davidson, J., Source contribution to arc magmas: Problems and prospects. *SUBCON abstracts*, 253-255, 1994.

Galbreath, K.C., E.F. Duke, J.J. Papike, J.C. Laul, Mass transfer during wall-rock alteration: An example from a quartz-graphite vein, Black Hills, South Dakota. *Geochim. Cosmochim. Acta 52*, 1905-1918, 1988.

Hart, S.R., and M.R. Reid, Rb/Cs fractionation: A link between granulite metamorphism and the S-process. *Geochim Cosmochim Acta 55*, 2379-2383, 1991.

Higgins, M.D., and D.M. Shaw, Boron cosmochemistry interpreted from abundances in mantle xenoliths. *Nature 308*, 172-173.

Hochstaedter, A., J.G. Ryan, J.F. Luhr, and T. Hasenaka, On B/Be ratios in the Mexican Volcanic belt. *Geochim. Cosmochim. Acta* (in press), 1996.

Hofmann, A.W., Chemical differentiation of the Earth: the relationship between mantle, continental crust, and oceanic crust. *Earth Planet. Sci. Lett. 90*, 297-314, 1988.

Jambon, A., and J.L. Zimmerman, Water in oceanic basalts: Evidence for dehydration of recycled crust. *Earth Planet. Sci. Lett, 101*, 323-331, 1990.

Leeman, W.P., D.R. Smith, W. Hildreth, Z. Palacz, and N.W. Rogers, Compositional diversity of Late Cenozoic basalts in a transect across the southern Washington Cascades. *J. Geophys. Res. 95*, 19561-19582, 1990.

Leeman, W.P., V.B. Sisson, and M.R. Reid, Boron geochemistry of the lower crust: Evidence from granulite terranes and deep crustal xenoliths. *Geochim. Cosmochim. Acta 56*, 775-788, 1992.

Leeman, W.P., M.J. Carr, and J.D. Morris, Boron geochemistry of the Central American Volcanic Arc: Constraints on the genesis of subduction-related magmas. *Geochim. Cosmochim. Acta 58*, 149-168, 1994.

Michard A., Rare earth element systematics in hydrothermal fluids. *Geochim. Cosmochim. Acta 53*, 745-750, 1989.

Moran, A.E., V.B. Sisson, and W.P. Leeman, Boron depletion during progressive metamorphism: Implications for subduction processes. *Earth Planet. Sci. Lett. 111*, 331-349, 1992.

Morris, J., and F. Tera, F., ^{10}Be and ^{9}Be in mineral separates and whole rocks from volcanic arcs: Implications for sediment subduction. *Geochim. Cosmochim. Acta 53*, 3197-3206, 1989.

Morris, J., W.P. Leeman, and F. Tera, The subducted component in island arc lavas: Constraints from Be isotopes and B-Be systematics. *Nature 344*, 31-36, 1990.

Newsom, H.E., W.M. White, K.P. Jochum, and A.W. Hofmann, Siderophile and chalcophile element abundances in oceanic basalts, Pb isotopic evolution and growth of the Earth's core. *Earth Planet. Sci. Lett, 80*, 299-313, 1986.

Nichols, G.T., Wyllie, P.J., and Stern, C.J., Subduction zone melting of pelagic sediments constrained by melting experiments. *Nature 371*, 785-788, 1994.

Noll, P.D., Newsom, H., Leeman, W.P., and Ryan, J., The role of hydrothermal fluids in the production of subduction zone magmas: Evidence from siderophile and chalcophile trace elements and boron. *Geochim. Cosmochim. Acta*, (in press) 1996.

Pawley, A.R., Experimental constraints on dehydration of mafic systems. *SUBCON abstracts*, 247-249, 1994.

Pawley, A.R., and B.J. Wood, High pressure stability of talc and 10-Angstrom phase: potential storage sites for H_2O in subduction zones. *Amer. Mineral. 80*, 998-1003. 1995.

Peacock, S.M., Large-scale hydration of the lithosphere above subducting slabs. *Chem. Geol. 108*, 49-59, 1993.

Peacock, S.M., Thermal structures of subduction zones. *SUBCON abstracts*, 172-174, 1994.

Peacock, S.M., T. Rushmer, and A.B. Thompson, Partial melting of subducting oceanic crust. *Earth Planet. Sci. Lett. 121*, 227-244, 1994.

Peucker-Ehrenbrink, B., A.W. Hofmann, and S.R. Hart, Hydrothermal lead transfer from mantle to continental crust: the role of metalliferous sediments. *Earth Planet. Sci. Lett. 125*, 129-142, 1994.

Plank, T., and Langmuir, C.H., Tracing trace elements from sediment to volcanic output at subduction zones. *Nature 362*, 739-742, 1993.

Rushmer, T., The influence of dehydration and partial melting reactions in warm, down-going oceanic crust on slab deformation. *SUBCON abstracts*, 297-299, 1994.

Ryan, J.G., The systematics of lithium, beryllium, and boron in young volcanic rocks. Ph.D. Thesis, Columbia University, 1989.

Ryan, J.G., and C.H. Langmuir, The systematics of boron abundances in young volcanic rocks. *Geochim. Cosmochim. Acta 57*, 1489-1498, 1993.

Ryan, J., J. Morris, F. Tera, B. Leeman, and A. Tsvetkof. The slab effect as a function of depth: evidence from cross-arc geochemical variations in the Kurile Arc. *Science, 270*, 625-628, 1995.

Ryan, J., J. Morris, G. Bebout, B. Leeman, and F. Tera, Describing chemical fluxes in subduction zones: insights from "depth-profiling" studies of arc and forearc rocks. *This volume*, 1996.

Shearer, C.K., J.J. Papike, S.B. Simon, J.C. Laul, and R.P. Christian, Pegmatite/wallrock interactions, Black Hills, South Dakota: progressive boron metasomatism adjacent to the Tip Top pegmatite. *Geochim. Cosmochim. Acta 48*, 2563-2579, 1984.

Singer, B., W.P. Leeman, W.P., M. Thirlwall, and N. Rogers, Does fracture zone subduction increase sediment flux and mantle melting in subduction zones? Trace element evidence from Aleutian arc basalt. *This volume*, 1996.

Sorensen, S.S., Subduction-related K-Rb-Ba metasomatism of high-grade melange blocks from California. *SUBCON abstracts*, 213-215, 1994.

Taylor, S.R., and S.M. McLennan, *The Continental Crust: Its Composition and Evolution*, Blackwell, Oxford, 1985.

Tera, F., L. Brown, J. Morris, I.S. Sacks, J. Klein, and R. Middleton, Sediment incorporation in island arc magmas: Inferences from ^{10}Be. *Geochim. Cosmochim. Acta 50*, 535-550, 1986.

Umner, P., and V. Trommsdorff, Serpentine stability to mantle depths and subduction-related magmatism. *Science 268*, 858-861, 1995.

von Huene, R., and D.W. Scholl, Observations at convergent margins concerning sediment subduction, subduction erosion, and the growth of continental crust. *Rev. Geophys. 29*, 279-316., 1991.

Weissberg, B.G., P.R.L. Browne, and T.M. Seward, Ore metals in active geothermal systems. *In* H.L. Barnes (ed.), *Geochemistry of Hydrothermal Ore Deposits*, 2nd ed., Wiley and Sons, New York, 738-780, 1979.

Effect of Sediments on Aqueous Silica Transport in Subduction Zones

Craig E. Manning

Department of Earth and Space Sciences, University of California, Los Angeles, California

Where sediments are present at the slab-mantle interface in subduction zones, most migration paths of aqueous solutions will result in chemical interaction between sediment and fluid. The ability of sediment to influence fluid composition can be appreciated by examining aqueous Si concentrations with a model chemical system, K_2O-Al_2O_3-SiO_2-H_2O, for three subduction-zone pressure-temperature trajectories (*PT* paths). Analysis of mineral-melt-H_2O phase relations as a function of aqueous Si concentration shows that metasedimentary mineral assemblages fix dissolved silica contents at or near those required by quartz saturation, even in the absence of quartz. Changes in dissolved silica content with pressure and temperature are much greater than those required by the range in plausible mineral assemblages. Sediments thus influence Si transport in subduction zones by buffering fluids at the slab-mantle interface at or near quartz saturation as pressure and temperature change, and, as a result, maximizing aqueous Si concentrations along the slab-mantle interface. Changes in pressure and temperature along model slab trajectories result in increases in Si content of pore fluids by $\sim 10^3$ and $\sim 10^5$ times along the lowest and highest temperature paths, respectively. The spatial gradient in dissolved silica concentration also increases with depth, and is greater for higher temperature *PT* paths. Most Si redistribution will occur in the deep portions of high-temperature subduction zones, and fluid-sediment interaction dictates that the amount of aqueous silica transported from the slab to the mantle wedge at any point along the slab-mantle interface is maximized if equilibrium is attained.

1. INTRODUCTION

Subduction zones can be divided into those that accumulate, or accrete, sediments, and those that do not [*von Heune and Scholl*, 1991]. In non-accreting margins, all ocean-floor sediment transported to the trench is subducted; in accreting margins, a décollement develops in the sedimentary section, below which all material transported to the trench is subducted. The décollement most likely develops near the interface between ocean-floor sediments and the overlying clastic wedge [*Moore*, 1975; *Plank*, 1993; *Plank and Langmuir*, 1993]. Downgoing slabs are therefore mantled by a veneer of ocean-floor sediments, which may be hundreds of meters thick [*Plank and Langmuir*, 1993].

Studies of volcanic arcs and subduction complexes suggest that sediments exert an important control on the petrology, magmatic evolution, and aqueous geochemistry of convergent margins [e.g., *Church*, 1976; *Kay et al.*, 1978; *Sun*, 1980; *Whitford and Jezek*, 1982; *White and Dupré*, 1986; *Tera et al.*, 1986; *Morris et al.*, 1990; *Ernst*, 1990; *Peacock*, 1990; *Bebout and Barton*, 1993; *Philippot*, 1993; *Plank and Langmuir*, 1993]. The physical and chemical mechanisms by which sediments influence subduction processes remain poorly understood, but redistribution of chemical components by aqueous fluids provides a viable hypothesis for material transport in this environment [e.g., *Tatsumi*, 1989; *Bebout and Barton*, 1989, 1993; *Bebout*, 1991; *Davies and Stevenson*, 1992]. As illustrated in Figure 1, sediments may mix mechanically with mantle material along the slab-mantle interface, or with other components of the subducting lithosphere. Fluids will be liberated by metamorphic devolatilization reactions in either sediments or mafic and ultramafic portions of the subducting plate. Wide variations in possible pressure and temperature gradients in subduction zones, coupled with strongly heterogeneous permeability-fields, imply that low-density fluids may have flux vectors with directions ranging from subvertical to updip along the slab. Figure 1 shows that of the potential flux directions, only fluids liberated in, and remaining in, oceanic basement rocks will not interact with sediments. In cases where the slab melts, movement of the resulting magma into the mantle wedge will also require chemical interaction with subducted sedimentary material.

Recent studies have addressed the role played by sediments in the redistribution of chemical components by aqueous fluids during subduction [*Bebout and Barton*, 1989, 1993; *Ernst*, 1990; *Peacock*, 1990; *Bebout*, 1991]. As noted by *Bebout and Barton* [1993], such studies are hindered by the lack of experimental data or theoretical models for mineral solubility at subduction-zone conditions. However, new experimental developments now allow the acquisition of these data [*Ayers and Watson*, 1991, 1993; *Brenan and Watson*, 1991; *Brenan*, 1993; *Manning*, 1994, *Manning and Boettcher*, 1994], which can be incorporated into coupled reaction and fluid-flow models of subducting slabs and the mantle wedge [e.g., *Manning*, 1995]. I have combined this approach with phase equilibrium calculations in the model system K_2O-Al_2O_3-SiO_2-H_2O (KASH) to illustrate how sediment may control the concentration and transport potential of a major element, silicon, in subduction zones.

2. METHODS

As a model for sediment-fluid interaction, I evaluated phase relations in the KASH system assuming the presence of an Al or Al-Si mineral and H_2O. This provides an approximation of subducting Al-rich pelagic sediment, which is dominated by micas, aluminosilicate clays, and quartz, and allows illustration of the first-order consequences of subduction of this sediment on element transport by aqueous solutions.

Analysis of phase relations in a model system is a simplification limiting the bulk compositions addressed by the calculations. For example, subducting sedimentary material may be a mixture of pelagic, terrigenous, and arc-derived sediment [e.g., *von Heune and Scholl*, 1991]; such sediments would be better approximated in the KASH system by assuming the presence of K feldspar instead of an Al or Al-Si mineral. In addition, this analysis ignores biogenic carbonate, which could lead to significant dissolved carbon species. This is reasonable for modern subduction, in which little biogenic carbonate is consumed [e.g., *Plank*, 1993]; but it limits applicability to subduction environments rich in CO_2. Finally, pelagic sediments range widely in com-

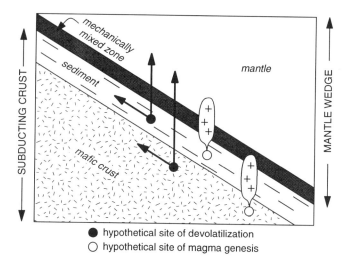

Fig. 1. Schematic illustration of the effect of sediment on material transport in subduction zones. Bold arrows are schematic H_2O flux vectors; crosses denote magma. The only transport path that does not involve interaction with sediment is aqueous fluid liberated from mafic crust and migrating updip within its source. This conceptual model is based on *Bebout and Barton* [1989, 1993] and *Bebout* [1991].

position [e.g., *Kyte et al.*, 1993; *Plank and Ludden*, 1992; *Plank and Langmuir*, 1993]. Important components not considered here include Na and Mg. On average, pelagic sediments have subequal molar concentrations of Na and K, and increasing Mg leads to the formation of chlorite, smectite, and/or phengite rather than aluminosilicate clays and muscovite. Model KASH sediment therefore corresponds to an aluminous, potassic, Mg-poor end member of the spectrum of pelagic sediment compositions. Similar analyses of phase relations in Mg- and Na-bearing model systems (C. E. Manning, unpublished data) shows that varying chemical components does not significantly alter conclusions about Si dissolution, transport, and precipitation.

Phase relations were computed for three model subduction zones from *Peacock* [1993]. The steady-state increase in pressure and temperature at the slab-mantle interface with depth in the subduction zone, or *PT* path, was calculated for subduction at 10 cm yr^{-1} at an angle of 20° for rock densities of 3000 kg m^{-3} (Figure 2). The three *PT* paths are distinct in that temperature at a given depth from 0 to ~70 km decreases from Path 1 through Path 3. They can be viewed as illustrating the range in pressure and temperature conditions resulting from different subduction scenarios. For example, Path 1 represents the *PT* regime to be expected when young oceanic crust is subducted, whereas

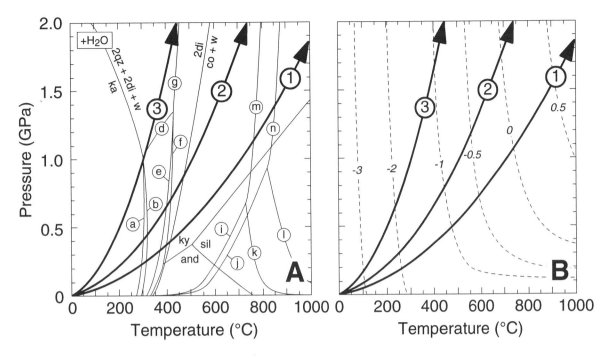

Fig. 2. Pressure-temperature diagrams with *PT* paths 1-3 (bold arrows) from *Peacock* [1993] (see text). (A) Phase relations in the system K_2O-Al_2O_3-SiO_2-H_2O and in the presence of H_2O. All phases assumed to be stoichiometric, with unit activity. Mineral-H_2O equilibria calculated using *Berman* [1988 (1992 extension)] and *Haar et al.* [1984]; mineral-melt-H_2O equilibria are from *Storre and Karotke* [1972] and *Huang and Wyllie* [1974]. Abbreviations: and, andalusite; co, corundum; di, diaspore; ka, kaolinite; ky, kyanite; qz, quartz; sil, sillimanite; w, water. Reaction boundaries labeled with circled letters keyed to Table 1. (B) Isopleths of $\log m_{SiO_2(aq)}$ calculated using the equation of *Manning* [1994].

Path 3 probably best reflects conditions likely in a long-lived, steady state subduction zone. Although different, more complex *PT* paths may result from consideration of the brittle-ductile transition [*Peacock et al.*, 1994] or alternative numerical analyses of heat and mass transfer [e.g., *Davies and Stevenson*, 1992], the paths used here serve as a simple framework to illustrate how changes in pressure and temperature influence aqueous mass transfer of Si from the Earth's surface to 2 GPa.

3. PHASE RELATIONS IN THE MODEL SEDIMENT

Figure 2a shows equilibria relevant to high pressure metamorphism and melting in the KASH system, with the *PT* paths of the three subduction zones. Equilibria in Figure 2a are listed in Table 1. Not shown for simplicity are the α–β quartz transition, low-pressure high-temperature equilibria involving leucite and melt, and low-pressure low-temperature reactions involving silica and clay diagenesis.

Each path traverses different regions of the *PT* projection, which results in contrasting phase relations.

Along Path 1, dehydration of hydrous Al and Al-Si minerals occurs at low pressures in the upper levels of the subduction zone, such that muscovite + kyanite will be the stable mineral assemblage in the presence of quartz and H_2O from ~0.4-1.1 GPa (14-37 km depth). Below this point, a KASH melt forms (equilibrium *m*, Table 1). Note that the assumption of the presence of an aluminosilicate mineral prevents the stable existence of K feldspar with muscovite at high pressures, which leads to the water-saturated solidus geometry defined by equilibria *k* and *m* (Figure 2a, Table 1). Melting in the presence of K feldspar instead of kyanite would produce the more familiar solidus geometry and would occur at 32 km, which is 5 km shallower, or 15 km updip, of the melting position in Figure 2a. Neither case corresponds to the temperature of melting of natural pelagic clay in the presence of H_2O (~650°C, *Nichols et al.* [1994]); but, despite this discrepancy, the effect of melt on metasomatic phase relations can still be evaluated from the equilibria in Figure 2a (see below).

Along Path 2, the maximum depth of dehydration of hydrous Al and Al-Si minerals (equilibria *e*, *f*, and *h*, Table 1)

TABLE 1. Relevant Equilibria in the KASH System

$Al_2Si_2O_5(OH)_4 + 2SiO_2 = Al_2Si_4O_{10}(OH)_2 + H_2O$ (a)
kaolinite   quartz   pyrophyllite

$2Al_2Si_2O_5(OH)_4 = 2AlOOH + Al_2Si_4O_{10}(OH)_2 + 2H_2O$ (b)
kaolinite   diaspore   pyrophyllite

$Al_2Si_2O_5(OH)_4 = 2SiO_2 + 2AlOOH + H_2O$ (c)
kaolinite   quartz   diaspore

$2AlOOH + 4SiO_2 = Al_2Si_4O_{10}(OH)_2$ (d)
diaspore   quartz   pyrophyllite

$6AlOOH + Al_2Si_4O_{10}(OH)_2 = 4Al_2SiO_5 + 4H_2O$ (e)
diaspore   pyrophyllite

$Al_2Si_4O_{10}(OH)_2 = 3SiO_2 + Al_2SiO_5 + H_2O$ (f)
pyrophyllite   quartz

$2AlOOH + SiO_2 = Al_2SiO_5 + H_2O$ (g)
diaspore   quartz

$2AlOOH = Al_2O_3 + H_2O$ (h)
diaspore   corundum

$KAl_3Si_3O_{10}(OH)_2 + SiO_2 = KAlSi_3O_8 + Al_2SiO_5 + H_2O$ (i)
muscovite   quartz   K feldspar

$KAl_3Si_3O_{10}(OH)_2 = KAlSi_3O_8 + Al_2O_3 + H_2O$ (j)
muscovite   K feldspar corundum

$KAlSi_3O_8 + Al_2SiO_5 + SiO_2 + H_2O = $ Melt (k)
K feldspar   quartz

$KAlSi_3O_8 + Al_2O_3 + H_2O = $ Melt (l)
K feldspar corundum

$KAl_3Si_3O_{10}(OH)_2 + SiO_2 + H_2O = Al_2SiO_5 + $ Melt (m)
muscovite   quartz

$KAl_3Si_3O_{10}(OH)_2 + H_2O = KAlSi_3O_8 + Al_2O_3 + $ Melt (n)
muscovite   K feldspar corundum

Phase assemblage on left-hand side of equilibrium is stable at low temperature relative to assemblage on right.

is greater than along Path 1 by a factor of about two. Melting in the model system in the presence of aluminosilicate and H$_2$O occurs at >2 GPa because of the lower temperature intersection of Path 2 and equilibrium *m*. Path 3 is characterized by such low temperatures to 2 GPa that water-rich aluminous minerals (kaolinite and diaspore) are stable to great depths, and melting in the model system will not occur. As noted above, an analysis allowing for compositional and structural variations in white micas and silica polymorphs would include stability fields for illite and amorphous silica at shallow levels along all paths; these would be limited to temperatures of less than ~200°C.

In general, the change in thermal regime from Path 1 to Path 3 makes the path of the slab-mantle interface similar to the Clapeyron slopes of dehydration equilibria (e.g., *e-h*, Table 1). This means H$_2$O may be stored in sediments to greater depths along cooler *PT* paths [*Peacock*, 1990, 1993]. Since H$_2$O is the principal solvent for material transport, this will result in metasomatic reactions proceeding at substantially greater depths when physical conditions favor cool *PT* paths. However, mineral solubilities generally decrease with decreasing temperature at constant pressure, so the extent of metasomatism may be low.

4. PHASE RELATIONS AS A FUNCTION OF DISSOLVED SILICA CONTENT

The concentration of aqueous silica (SiO$_{2(aq)}$) in H$_2$O in equilibrium with quartz has been determined experimentally by *Manning* [1994] at 0.5-2.0 GPa and 500-900°C. These experiments, combined with previous work, allow calculation of the concentration of SiO$_{2(aq)}$ in equilibrium with quartz from 25°C, 1 bar to >2 GPa and ~1000°C. Predicted SiO$_{2(aq)}$ molality ($m_{SiO_{2(aq)}}$ at quartz saturation are shown as a function of pressure and temperature in Figure 2b with the three *PT* paths. Below ~0.6 GPa and at high temperature, $m_{SiO_{2(aq)}}$ is sensitive to small changes in pressure at constant temperature because of correspondingly large changes in the density of H$_2$O with pressure. Above 1 GPa, isopleths of SiO$_{2(aq)}$ concentration have steep negative slopes. This indicates that high-temperature paths, which traverse isopleths at a high angle, will result in larger changes in quartz-saturated SiO$_{2(aq)}$ concentration along the slab than low-temperature paths.

By computing $m_{SiO_{2(aq)}}$ along Paths 1-3, the Si concentrations required by specific mineral assemblages may be determined, even if they do not include quartz. The concentration of SiO$_{2(aq)}$ in equilibrium with quartz (Figure 2b) can be used to constrain such an analysis because equilibrium between pure quartz and aqueous silica,

$$SiO_2 = SiO_{2(aq)} \quad (1)$$
quartz

requires at constant pressure and temperature that

$$\Delta G°_{SiO_{2(aq)}} = \Delta G°_{quartz} - RT \ln a_{SiO_{2(aq)}} \quad (2)$$

where $\Delta G°$ is the standard molal Gibbs free energy difference between a reference state (25°C, 10^5 Pa) and the P and T of interest, R is the gas constant, a is activity. In this study, standard states for minerals and water are unit activities of the pure phases at any pressure and temperature. Thermodynamic properties of the aqueous solution are taken to be those of pure H_2O.

The activity of $SiO_{2(aq)}$ can be equated to its molality because Si forms a neutral hydrated species with an activity coefficient of one over a wide range in pH [*Walther and Helgeson*, 1977]. Equation (2) thus becomes

$$\Delta G°_{SiO_{2(aq)}} = \Delta G°_{quartz} - RT \ln m_{SiO_{2(aq)}} \quad (3)$$

which allows calculation of the standard molal Gibbs free energy of aqueous silica at the P and T of interest from $\Delta G°_{quartz}$ and $m_{SiO_{2(aq)}}$ as given by experiments and compilations of thermodynamic data. The value of $\Delta G°_{SiO_{2(aq)}}$ may then be combined with thermodynamic data for minerals and H_2O to determine metasomatic phase relations as a function of silica concentration in the fluid. For example, the equilibrium between kaolinite and pyrophyllite at a given P and T,

$$Al_2Si_2O_5(OH)_4 + 2SiO_{2(aq)}$$
kaolinite
$$= Al_2Si_4O_{10}(OH)_2 + H_2O \quad (4)$$
pyrophyllite

leads to

$$\begin{aligned}\ln m_{SiO_{2(aq)}} &= (\Delta_r G°(4))/2RT \\ &= (\Delta G°_{py} + \Delta G°_{H_2O} - \Delta G°_{ka} \\ &\quad - 2\Delta G°_{SiO_{2(aq)}})/2RT\end{aligned} \quad (5)$$

where $\Delta_r G°(4)$ is the difference between standard molal Gibbs free energies of reactants and products for equilibrium (4).

Figure 3 shows calculated phase relations along Paths 1-3 assuming the presence of an Al- or Al-Si-bearing mineral. These phase diagrams differ from conventional diagrams describing Si metasomatism [e.g., *Hemley et al.*, 1980] in that pressure and temperature covary along the ordinate. The diagrams illustrate stability boundaries among both K-bearing phases (heavy lines) and K-absent phases (light lines). Thus, fields in Figure 3 represent the range of pressure, temperature, and fluid composition over which a K-bearing and a K-absent phase coexist with aqueous solution of varying Si concentration. Fields in Figure 3 are divariant, because pressure and temperature are not independent, and the system has four components. Similarly, phase boundaries represent the addition of a

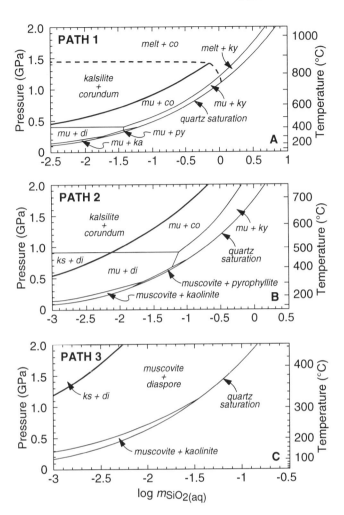

Fig. 3. Phase relations in the system K_2O-Al_2O_3-SiO_2-H_2O as a function of $SiO_{2(aq)}$ concentration and PT path. Light lines represent phase boundaries between stable minerals in the subsystem Al_2O_3-SiO_2-H_2O; heavy lines represent phase boundaries between stable K-bearing phases. The phase boundary involving KASH melt (heavy dashed line) is inferred from *Huang and Wyllie* [Figure 3, 1974]. Abbreviations as in Figure 3, except: mu, muscovite; py, pyrophyllite.

phase to the assemblage, and are therefore univariant. Fluids coexisting at equilibrium with three additional phases must adjust their composition as pressure and temperature change, if they are to remain on the phase boundary. Finally, four phases coexist with aqueous solution at phase-boundary intersections, which are invariant: any change in pressure, temperature, or fluid composition requires loss of one phase from the assemblage.

Melting in the KASH system occurs along Path 1 at <2 GPa (Figure 2a). The stability of a KASH melt as a function of $SiO_{2(aq)}$ concentration along Path 1 is shown schematically in Figure 3a. Si-rich bulk compositions containing quartz have the lowest solidus temperature in this system (reactions m and n, Figure 2a). Melting will therefore occur at the shallowest levels of warm subduction zones where aqueous fluids contain high dissolved silica contents (Figure 3). The depth of melting increases as $m_{SiO_{2(aq)}}$ decreases, and the total melting interval defined by variations in $m_{SiO_{2(aq)}}$ corresponds to a depth range of ~10 km, or 29 km of slab length. Note that, unlike minerals in Figure 3, the KASH liquid has variable composition.

Figure 3 represents a set of chemical maps that show how phase assemblage and fluid composition must change in model metasediments at the slab-mantle interface for different subduction scenarios. Along each path, the maximum molality of $SiO_{2(aq)}$ is defined by quartz saturation. Any metasediment containing quartz must lie on the phase boundaries labeled "quartz saturation," along which muscovite or KASH melt coexist with quartz and a K-absent, Al-bearing mineral at equilibrium (Figure 3). Fluids with greater Si contents are metastably supersaturated with respect to quartz, and must precipitate sufficient quartz to decrease $m_{SiO_{2(aq)}}$ until they lie on the quartz-saturation boundary. Fluids with $m_{SiO_{2(aq)}}$ below quartz saturation can not coexist with quartz at equilibrium. These fluids will be in equilibrium with quartz-absent phase assemblages, such as muscovite + corundum, or kalsilite + diaspore. Note that Figure 3 illustrates the well-known incompatibility certain mineral assemblages at equilibrium; for example, the Si-poor phases kalsilite or corundum will never be in equilibrium with quartz under the conditions considered, as they will react with quartz to form intervening muscovite or kyanite instead.

For any bulk composition, $m_{SiO_{2(aq)}}$ in coexisting fluid increases strongly with increasing depth (and correspondingly, temperature) along each subduction path. For example, $m_{SiO_{2(aq)}}$ in equilibrium with quartz-bearing assemblages along Path 3 will increase by a factor of ~10^3 (10^{-4} to >10^{-1} mol kg H_2O^{-1}) between 0 and 70 km; along Path 1, an even greater increase of ~10^5 times will occur (Figure 3). Below quartz saturation, all univariant phase boundaries buffer $m_{SiO_{2(aq)}}$ along the slab-mantle interface, and their positive slopes indicate that they similarly require increasing $m_{SiO_{2(aq)}}$ with depth for quartz-absent assemblages.

Figure 3 shows that the presence of two or more Si-bearing minerals in the KASH system requires $SiO_{2(aq)}$ concentrations within several tenths of a log unit of quartz saturation. This implies that at a given pressure and temperature, the absence of quartz, or its loss through dissolution, will result in only minor changes in $SiO_{2(aq)}$ concentration compared to those attending variation in pressure and temperature. Si mass transfer in sediments in subduction zones can therefore be reasonably assumed to occur at, or very close to, quartz saturation, which means that $m_{SiO_{2(aq)}}$ will be maximized in the presence of sediment. The rarity of silica undersaturated phases, like corundum and kalsilite, in exposed metasediments from subduction-zone settings supports this conclusion.

5. IMPLICATIONS

Phase relations among model KASH sediment and aqueous fluid (Figure 3) illustrate three important points about material transport and metasomatism in subduction zones. The first is that, for Si redistribution, the dissolved silica content of pore fluids in metasediments, and its change as subduction proceeds, depends strongly on the *PT* path followed by the subduction zone. As shown in Figure 3, $m_{SiO_{2(aq)}}$ in equilibrium with quartz-bearing assemblages increases with temperature. This indicates that conditions which lead to high temperatures in subduction zones (e.g., young oceanic crust, incipient subduction, high shear stress, low slab velocities) maximize the potential for aqueous transport of silica. A particularly well-documented natural example illustrating this observation is described by *Bebout and Barton* [1993] on Santa Catalina Island, California.

The second implication of Figure 3 is that increases in $SiO_{2(aq)}$ must be achieved by dissolving silica from coexisting minerals. Thus, as model KASH metasediments are subducted along different PT paths, their bulk compositions will become progressively depleted in Si with depth in the presence of a static pore fluid. Like the amount of dissolved silica, the magnitude of the shift in bulk composition will be greater for higher temperature paths. In addition, if metasediment is returned to the surface, dissolved silica in static pore fluids must decrease, leading to precipitation of silica-rich phases such as quartz.

Finally, along with lithologically controlled bulk compositional differences, the changes in dissolved silica with path and depth represent the driving potential for Si metasomatism by flowing fluid (Figure 1). As pore fluids migrate independently of the rock matrix in a model KASH metasediment, their compositions must adjust at equilibrium as required by the relevant phase relations [e.g., Figure 3]. This will lead to shifts in bulk composition of varying magnitude in the rocks through which the fluid flows. For fluids migrating from the slab-mantle interface into the mantle wedge, the presence of sediment will maximize the amount of dissolved silica redistributed form slab to mantle, regardless of the depth at which the fluids leave the slab [see *Manning*, 1995].

Fluids may also migrate updip along the slab-mantle interface. The potential for Si transfer back up the slab along a sediment veneer can be assessed by evaluating the change in $m_{SiO_2(aq)}$ at quartz saturation along each path. Figure 4a shows the quantity $dm_{SiO_2(aq)}/dz$, where z is distance in kilometers along the subducting slab for Paths 1-3. This quantity was obtained from finite difference derivatives of quartz solubility as a function of pressure (Figure 3), with pressure transformed to distance using a rock density of 3000 kg m^{-3} and a 20° slab angle. The change in quartz solubility with position in the slab is greater for higher temperature paths, just as absolute solubilities are higher along higher temperature paths (Figure 2b). Because the change in solubility will require quartz precipitation or dissolution at equilibrium, higher temperature paths have a significantly greater capacity to redistribute Si than cooler paths.

The magnitude of this difference can be appreciated by integrating $dm_{SiO_2(aq)}/dz$ over the length of the flow path along the slab. For example, assuming that quartz-saturated H$_2$O begins migrating updip in sediments at a depth corresponding to 185 km along the slab surface, and that this fluid remains in equilibrium with quartz along its flow path, the volume of quartz precipitated during upward flow to the Earth's surface will be ~0.4 cm^3 kg H$_2$O^{-1}, ~0.1 cm^3 kg H$_2$O^{-1}, and ~0.01 cm^3 kg H$_2$O^{-1} along Paths 1, 2, and 3, respectively (Figure 4b). In addition, most of the increase in the amount of quartz precipitated occurs in the deep parts of the slab. Veins of quartz, which are common in subducted metasediments [e.g., *Bebout and Barton*, 1989; *Ernst*, 1990], require removal of Si from the fluid in response to solubility decreases along its flow path. Figure 4 implies that for constant flux, the greatest volume of vein quartz should be expected to originate in the deeper parts of any given subduction zone, and that high-temperature paths will result in larger volumes of precipitated quartz.

In conclusion, this analysis shows that because most flux trajectories carry aqueous fluids through sediment at the slab-mantle interface (Figure 1), chemical interaction between sediment and fluid must be taken into account in considering aqueous mass transfer. For Si, it is likely that sediment-fluid interaction will result in high SiO$_2$(aq) concentrations, at or near quartz saturation. If this fluid migrates back up the slab, the decrease in quartz solubility with pressure and temperature will lead to precipitation of quartz in fluid conduits such as fractures. If the fluid migrates into the low-Si environment of the mantle-wedge, the result may be substantial increases in the mantle SiO$_2$ content near the slab through conversion of olivine and orthopyroxene to, for example, talc and Fe-Mg amphibole [e.g., *Bebout and Barton*, 1993; *Manning*, 1995].

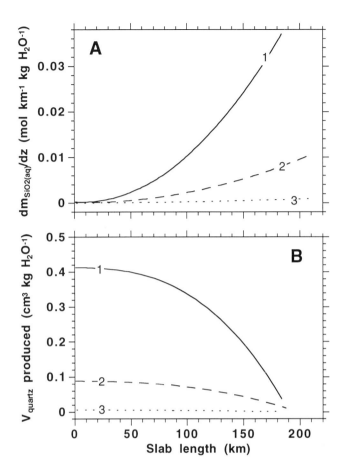

Fig. 4. (A) Change in concentration of aqueous silica in equilibrium with quartz ($dm_{SiO_2(aq)}/dz$) with distance along slab for different PT paths (1-3). (B) Volume of quartz produced by quartz-saturated H$_2$O migrating updip along Paths 1-3 185 km from the trench.

Acknowledgments. This study was funded by NSF EAR-9405999. Thanks to M. Barton, G. Nichols, and P. Vrolijk for insightful reviews. M Grove, K. Knesel, and D. Rothstein read and improved an early draft of the manuscript.

REFERENCES

Ayers, J. C., and E. B. Watson, Solubility of apatite, monazite, zircon, and rutile in supercritical fluids with implications for subduction zone geochemistry, *Philos. Trans. R. Soc. London, Ser. A, 335,* 365-375, 1991.

Ayers, J. C., and E. B. Watson, Rutile solubility and mobility in supercritical aqueous fluids, *Contrib. Mineral. Petrol., 114,* 321-330, 1993.

Bebout, G. E., Geometry and mechanisms of fluid flow at 15 to 45 kilometer depths in an Early Cretaceous accretionary complex, *Geophys. Res. Lett., 18,* 923-926, 1991.

Bebout, G. E., and M. D. Barton, Fluid flow and metasomatism in a subduction zone hydrothermal system: Catalina Schist terrane, California, *Geology, 17*, 976-980, 1989.

Bebout, G. E., and M. D. Barton, Metasomatism during subduction: Products and possible paths in the Catalina Schist, California, *Chem. Geol., 108*, 61-92, 1993.

Berman, R. G., Internally-consistent thermodynamic data for minerals in the system $Na_2O-K_2O-CaO-MgO-FeO-Fe_2O_3-Al_2O_3-SiO_2-TiO_2-H_2O-CO_2$, *J. Petrol., 29*, 445-522, 1988.

Brenan, J. M., 1993, Partitioning of fluorine and chlorine between apatite and aqueous fluids at high pressure and temperature: implications for the F and Cl content of high P-T fluids, *Earth Planet. Sci. Lett., 117*, 251-263, 1993.

Brenan, J. M., and E. B. Watson, Partitioning of trace elements between olivine and aqueous fluids at high P-T conditions: Implications for the effect of fluid composition on trace element transport, *Earth Planet. Sci. Lett., 107*, 672-688, 1991.

Church, S. E., The Cascade mountains revisited: a re-evaluation in light of new lead isotopic data: *Earth Planet. Sci. Lett., 29*, 175-188, 1976.

Davies, J. H., and D. J. Stevenson, Physical model of source region of subduction zone volcanics, *J. Geophys. Res., 97*, 2037-2070, 1992.

Ernst, W. G., Thermobarometric and fluid expulsion history of subduction zones, *J. Geophys. Res., 95*, 9047-9053, 1990.

Haar, L., J. S. Gallagher, and G. S. Kell, *NBS/NRC Steam Tables*, 320 pp., Hemisphere, New York, NY, 1984.

Hemley, J. J., J. W. Montoya, J. W. Marinenko, and R. W. Luce, Equilibria in the system $Al_2O_3-SiO_2-H_2O$ and some general implications for alteration/mineralization processes, *Econ. Geol., 75*, 210-228, 1980.

Huang, W. L., and P. J. Wyllie, Melting relations of muscovite with quartz and sanidine in the $K_2O-Al_2O_3-SiO_2-H_2O$ system to 30 kilobars and an outline of paragonite melting relations, *Am. J. Sci., 274*, 378-395, 1974.

Kay, R. W., S.-S. Sun, and S.-N. Lee-Hu, Pb and Sr isotopes in volcanic rocks from the Aleutian Islands and Pribilof Islands, Alaska, *Geochim. Cosmochim. Acta, 42*, 263-273, 1978.

Kyte, F. T., M. Leinen, G. R. Heath, and L. Zhou, Cenozoic sedimentation history of the central North Pacific: inferences from the elemental geochemistry of core LL44-GPC3. *Geochim. Cosmochim. Acta, 57*, 1719-1740, 1993.

Manning, C. E., The solubility of quartz in the lower crust and upper mantle, *Geochim. Cosmochim. Acta, 58*, 4831-4839, 1994.

Manning, C. E., Coupled reaction and flow in subduction zones: Si metasomatism in the mantle wedge, in *Fluid Flow and Transport in Rocks*, edited by B. Jamtveit and B. W. D. Yardley, Chapman and Hall, in press, 1995.

Manning, C. E., and S. L. Boettcher, Rapid-quench hydrothermal experiments at mantle pressures and temperatures, *Am. Mineral., 79*, 1153-1158, 1994

Moore, J. C., Selective subduction, *Geology, 3*, 530-532, 1975.

Morris, J. D., W. P. Leeman, and F. Tera, The subducted component in island arc lavas: Constraints from Be isotopes and B-Be systematics, *Nature, 344*, 31-36, 1990.

Nichols, G. T., P. J. Wyllie, and C. R. Stern, Subduction zone melting of pelagic sediments constrained by melting experiments, *Nature, 371*, 785-788, 1994.

Peacock, S. M., Fluid processes in subduction zones, *Science, 248*, 329-337, 1990.

Peacock, S. M., The importance of blueschist → eclogite dehydration reactions in subducting oceanic slabs, *Geol. Soc. Am. Bull., 105*, 684-694, 1993.

Peacock, S. M., T. Rushmer, and A. B. Thompson, Partial melting of subducting oceanic crust, *Earth Planet. Sci. Letters, 121*, 227-244, 1994.

Philippot, P., Fluid-melt-rock interaction in mafic eclogites and coesite-bearing metasediments: constraints on volatile recycling during subduction, *Chem. Geol., 108*, 93-112, 1993.

Plank, T., Mantle melting and crustal recycling in subduction zones, Ph.D. Thesis, 444 pp., Columbia University, New York, 1993.

Plank, T., and J. N. Ludden, Geochemistry of sediments in the Argo abyssal plain at Site 765: a continental margin reference section for sediment recycling in subduction zones, *Proc. Ocean Drill. Program: Sci. Results, 123*, 167-189, 1992.

Plank, T., and C. H. Langmuir, Tracing trace elements from sediment input to volcanic output at subduction zones, *Nature, 362*, 739-743, 1993.

Storre, B., and E. Karotke, Experimental data on melting reactions of muscovite + quartz in the system $K_2O-Al_2O_3-SiO_2-H_2O$ to 20 Kb water pressure, *Contrib. Mineral. Petrol., 36*, 343-345, 1972.

Sun, S.-S., Lead isotopic study of young volcanic rocks from mid-ocean ridges, ocean islands and island arcs: *Philos. Trans. R. Soc. London, Ser. A, 297*, 409-445, 1980.

Tatsumi, Y., Migration of fluid phases and genesis of basalt magmas in subduction zones, *J. Geophys. Res., 94*, 4697-4707, 1989.

Tera, F., L. Brown, J. Morris, I. S. Sacks, J. Klein, and R. Middleton, Sediment incorporation in island-arc magmas: Inferences from [10]Be, *Geochim. Cosmochim. Acta, 50*, 535-550, 1986.

von Heune, R., and D. W. Scholl, Observations at convergent margins concerning sediment subduction, subduction erosion, and the growth of continental crust, *Rev. Geophys., 29*, 279-316, 1991.

Walther, J. V., and H. C. Helgeson, Calculation of the thermodynamic properties of aqueous silica and the solubility of quartz and its polymorphs at high pressures and temperatures, *Am. J. Sci., 277*, 1315-1351, 1977.

White, W. M., and B. Dupré, Sediment subduction and magma genesis in the Lesser Antilles: Isotopic and trace element constraints, *J. Geophys. Res., 91*, 5927-5941, 1986.

Whitford, D. J., and P. A. Jezek, Isotopic constraints on the role of subducted sialic material in Indonesian island-arc magmatism. *Geol. Soc. Am. Bull., 93*, 504-513, 1982.

C. E. Manning, Department of Earth and Space Sciences, University of California, Los Angeles, CA, 90024-1567.

Does Fracture Zone Subduction Increase Sediment Flux and Mantle Melting in Subduction Zones? Trace Element Evidence from Aleutian Arc Basalt

Bradley S. Singer[1], William P. Leeman[2], Matthew F. Thirlwall[3], Nicholas W. Rogers[4]

New trace element data from 57 Aleutian arc basalt samples, including B, Be, Cs, Nb, Zr, La and Yb concentrations are used to examine along-arc variability in magma source compositions and processes. The large ranges in ratios of Nb/Zr and La/Yb are consistent with an origin of most basalts through different degrees of partial melting of a broadly similar mantle wedge. Concentrations of the fluid-mobile element B (>20 ppm) and ratios of B/La (>3), B/Be (>25) Cs/La (>0.16) and $^{87}Sr/^{86}Sr$ (>0.7035) are distinctly elevated in those basalts for which we infer the highest degrees of mantle melting. These samples are from Seguam Island in the central Aleutian arc, beneath which the Amlia fracture zone in the Pacific plate was subducted in the past 1 Ma. Strongly correlated B/Be and B/La ratios are modeled as a mixture of mantle and slab-derived fluid; the mantle beneath Seguam Island may reflect a two- to five-fold increase in the flux of subducted sediment relative to regions unaffected by fracture zone subduction. We propose that the Amlia fracture zone provided a channelway for subduction of exceptionally large quantities of sediment and water into the subarc mantle. Release of a large volume of B- and Cs-rich, high $^{87}Sr/^{86}Sr$ fluid from the slab may have lowered the melting temperature of the mantle, thereby promoting more extensive partial melting.

INTRODUCTION

Since *Coats* [1962] suggested in a pre- plate tectonic paper that dehydration of the underthrust Pacific Ocean crust initiated mantle melting beneath the Aleutian arc, the Aleutians have been important to understanding chemical fluxes and magma generation at subduction zones. Although the potential role of overriding lithosphere in modifying ascending arc magmas was emphasized globally by *Plank and Langmuir* [1988] and *Keleman et al.* [1990] and in the Aleutians by *Kay et al.* [1982], *Myers et al.* [1985], and *Singer et al.* [1992a], few studies have explored the effects of structural or compositional variations in the downgoing plate on the generation of arc magma. Quantifying elemental fluxes through subduction zones will, however, require moving beyond simple models [*e.g., Plank and Langmuir*, 1993] that tend to obscure the variability observed in both the column of crust plus sediment subducted and the lavas erupted along many arcs [*e.g., Varne*, 1994].

Some arcs show extreme topography on the subducting plate where oceanic plateaus or young ocean ridge systems are being subducted, for example subduction of the Cocos Ridge beneath the Central American volcanic arc [*Leeman et al.*, 1994] or the Chile Rise beneath the southern Andean arc [*Kay et al.*, 1993]. Although such topographic features are rare, fracture zone structures influence 10-20% of Pacific ocean crust [*Mammerickx*, 1989; *Atwater and Severinghaus*, 1989] and by virtue of their unusual structure, sediment distribution, basement lithology, and alteration may exert a disproportionate influence on the chemical flux of subducted crustal components into the subarc mantle.

---

[1]Département de Minéralogie, Université de Genève, Switzerland.
[2]Department of Geology and Geophysics, Rice University, Houston, Texas, U.S.A.
[3]Geology Department, Royal Holloway and Bedford New College, University of London, U.K.
[4]Department of Earth Sciences, The Open University, Milton Keynes, U.K.

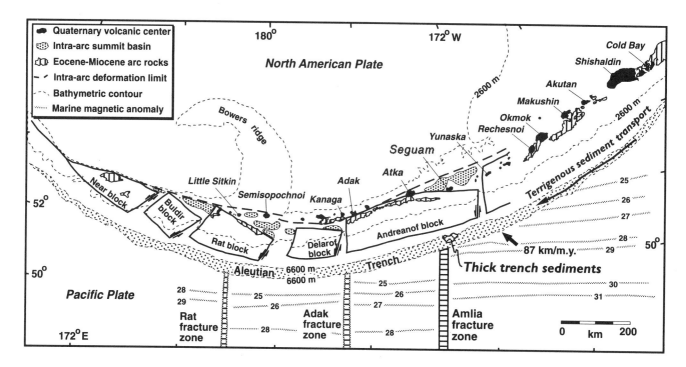

Fig. 1. Structure of the Aleutian arc after *Geist et al.* [1988]. The narrow island crested ridge is fragmented into five structural blocks; in the wake of each rotated block is an intra-arc basin. The Amlia fracture zone offsets marine magnetic anomalies 220 km [*Hayes and Heirtzler*, 1968], and has a 0.5 -1.0 km escarpment defining its western wall [*Scholl et al.*, 1982]. Turbidity currents carry sediment from the Gulf of Alaska into the Aleutian trench south of the Seguam volcanic center. This terrigenous sediment overlies 200-300 m of pelagic sediment and is typically 2-2.5 km thick in the trench floor. Where the Amlia fracture zone intersects the trench, the sediment is, however, 3.7-4.0 km thick [*Scholl et al.*, 1982]. Arrow gives relative plate convergence direction and velocity. Since subduction is highly oblique, the Amlia fracture zone was below Yunaska 3 Ma and beneath Seguam 1 Ma.

Fracture zone walls can expose large volumes of water-rich (13 wt%) serpentine [*Bonatti and Crane*, 1984] that may be stable to depths of 150-200 km [*Ulmer and Trommsdorff, 1995*]. Moreover, intense faulting and hydrothermal alteration in fracture zones will increase the surface area upon which devolatilization reactions occur [*Tatsumi et al.*, 1986; *Bebout et al.*, 1993]. These factors may greatly enhance fluid transfer from subducted crust into the mantle wedge below arcs. Thus, the global importance of fracture zone subduction to element recycling is probably seriously neglected. Recent geophysical work in the Aleutian arc [*Scholl et al.*, 1982; *Geist et al.*, 1988; *Geist and Scholl*, 1992] provides an exceptional opportunity to test the effects of fracture zone subduction on arc magma generation and elemental fluxes (Figure 1).

Despite many geochemical studies, concentration data for key trace elements in Aleutian basalts are either lacking or fragmentary at best. To address this problem plus questions concerning the relative roles of the mantle, crust, and fluids, and to test the hypothesis that fracture zone subduction may impact arc-wide chemical variations, we have undertaken concentration measurements of rare-earth elements, Zr, Hf, Ta, Nb, plus Rb, Ba, Cs, U, Th, Pb, Be, and B in 57 representative basalt samples (45-53% SiO_2; 4-16% MgO) from ten volcanic centers along the Aleutian island arc. Of particular importance are new B, Be, and Cs concentration determinations. Our results suggest that lateral variations in the trace element, isotopic, and possibly the major element compositions of Aleutian basalt reflect along-arc differences in the quantity of sediment and water subducted beneath the arc.

TRACE ELEMENTS IN ALEUTIAN BASALTS

In attempting to use basalt compositions to explore arc magma sources, the absolute abundances of incompatible trace elements can be misleading as they are strongly affected by crystallization or contamination during ascent of the magma through the lithosphere. Concentration *ratios* of incompatible elements with similar or slightly different distribution coefficients, in conjunction with isotopic data, can however, provide insight to variations in magma source composition, degree of mantle melting, and additions of subducted crustal components to the magma source prior to melting. Our results are therefore discussed mainly in terms of salient trace element ratios.

The rare earth elements La and Yb and high field strength elements Nb and Zr are relatively immobile under hydrothermal conditions and are strongly fractionated only during melting or magma mixing processes. Different partition coefficients for the otherwise geochemically similar La and Yb or Nb and Zr in clinopyroxene and garnet permit the La/Yb and Nb/Zr ratios to be used as monitors of the degree of partial melting of the mantle [*e.g., Thirlwall et al.*, 1994]. Chondrite normalized La/Yb ratios generally decrease from about 4 to 1 as La decreases from 11 to <3 ppm. Similarly, Nb/Zr ratios decrease from 0.040 to 0.015 as Nb decreases from 3 to 0.5 ppm (Figures 2a and b). These geochemical trends suggest that most Aleutian basalts can be related to a broadly common mantle source composition by varying degrees of partial melting of this mantle. Samples from the central Aleutian arc volcanic center of Seguam (Figure 1) are among those for which we infer an origin via the highest degrees of melting (Figure 2).

In contrast to rare earth or high field strength elements, experimental data [*You et al.*, 1995] indicate that B is strongly partitioned into aqueous fluids during dehydration of minerals at temperatures below 350°C, thus it records the progressive devolatilization attending metamorphism of subducting ocean crust [*Moran et al.*, 1992; *Bebout et al.*, 1993; *You et al.*, 1993]. Similar partition coefficients for the highly incompatible elements B, Be, and La [*Ryan and Langmuir*, 1988; 1993], coupled with the very low concentration and short residence time of B in the mantle and lower arc crust [*Moran et al.*, 1992; *Leeman et al.*, 1992] predict that B/La and B/Be in arc basalts are largely unaffected by crystallization or contamination of the ascending magmas. Cesium is a fluid-mobile element that should exhibit geochemical behavior similar to B [*Leeman et al.*, 1992], hence Cs/La ratios are used here to strengthen arguments that rely mainly on B.

Ratios of B/La and B/Be are strongly correlated with $^{10}Be/^9Be$ ratios and thus serve as useful indicators of subducted sediment in arc magma sources [*Morris et al.*, 1990; *Edwards et al.*, 1993; *Leeman et al.*, 1994]. Boron concentrations in Aleutian basalt samples vary nearly ten-fold from 3 to 28 ppm; their B/La ratios range from 0.6 to >7.0 (Figure 3a). Samples from the central Aleutian arc volcanic centers of Seguam and Yunaska have the highest B contents (>20 ppm) and B/La (>3), B/Be (>25), and Cs/La (>0.16) ratios observed (Figures 3a, 3c and 4). The Seguam basalt samples also have the highest $^{87}Sr/^{86}Sr$ ratios measured in the Aleutian arc [*Singer et al.*, 1992b] (Figure 3b). Assuming an initial subarc mantle wedge that was broadly uniform in composition, the relations between B, B/La, B/Be, Cs/La, and $^{87}Sr/^{86}Sr$ suggest that addition of greater quantities of a fluid rich in B, Cs, and ^{87}Sr may have imprinted a larger signature of subducted sediment on the source of Seguam and Yunaska basalts than at other centers (Figures 3c and 4).

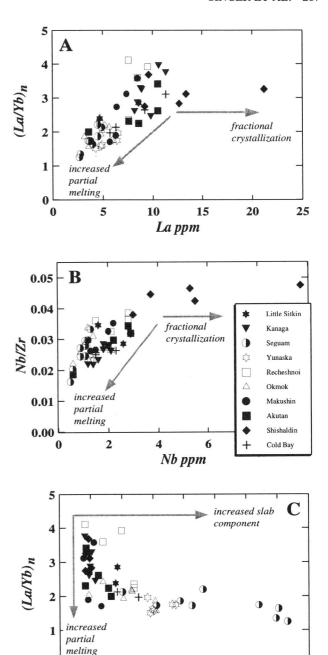

Fig. 2. Trace element plots of Aleutian basaltic lava samples illustrating effects of partial melting. (a) chondrite-normalized La/Yb vs. La. (b) Nb/Zr vs. Nb. (c) chondrite-normalized La/Yb vs. B/La. La and Yb (and Cs) were determined by INAA at the Open University, Nb and Zr by XRF analysis at the University of London, B by prompt gamma neutron activation at McMaster University. With the exception of high Nb and La samples from Shishaldin, the samples in (a) and (b) define steep arrays suggestive of relationships via partial melting. Samples from Seguam are consistent with the highest degree of melting.

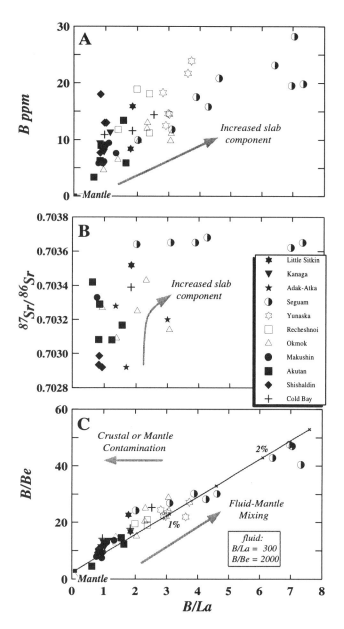

Fig. 3. Trace element and Sr isotope ratio plots of Aleutian basaltic lava samples illustrating effects of adding a subducted crustal component to the mantle wedge. (a) Boron concentration vs. B/La ratio. (b) $^{87}Sr/^{86}Sr$ vs. B/La ratio. (c) B/Be ratio vs. B/La ratio. Beryllium measured by ICP-MS at the University of South Florida. The mixing line in (c) assumes mantle and fluid end members similar to those in *Leeman et al.* [1994] and *Morris et al.* [1990]. Tick marks show 0.5% increments of fluid in the mixture. Data sources for $^{87}Sr/^{86}Sr$ ratios and the La values used to calculate B/La ratios for Adak and Atka samples are given in *Singer et al.* [1992b].

DISCUSSION

Basalt samples for which the quantity of slab-derived crustal component added to the mantle wedge was apparently largest correspond to those that we infer to have originated via the highest percentages of mantle melting (Figure 2c). These basalts are from the Seguam and Yunaska volcanic centers (Figure 4). The Aleutian arc crust west of Shishaldin volcano (Figure 1) consists of deformed Eocene-Pliocene arc rocks plus older oceanic crust and probably varies little in thickness [*Geist and Scholl*, 1992].

It is unlikely that contamination processes during magma ascent obscured the imprint of source characteristics in Seguam basalt for three reasons. First, combined Sr, Nd, Pb, and O isotope data from basaltic through rhyodacitic (50-71% SiO_2) lavas at Seguam strongly suggest that basaltic magmas crystallized with little contamination during ascent through the arc crust [*Singer et al.*, 1992a and 1992c]. Second, since the B and Be contents of lower crustal and mantle rocks are vanishingly low [*Leeman et al.*, 1992; *Bebout et al.*, 1993], magma-mantle or magma-lower crust interactions would lower the B/La ratio with a negligible affect on the B/Be ratio (Figure 3c). Finally, assimilation of enormous quantities of mantle or lower crust would be required to explain the arc-wide range of $^{87}Sr/^{86}Sr$ ratios [*Singer et al.*, 1992a] and impossibly large quantities to explain the ten-fold range in B concentration (Figure 3a).

The Seguam and Yunaska volcanic centers overlie a region of the mantle into which the Amlia fracture zone was subducted obliquely over the past 3 Ma. As the Amlia fracture zone offsets magnetic lineations by at least 220 km, it is the only one of the three fracture zones subducting beneath the Aleutian arc that has a significant topographic expression as a 30 km wide, 1 km deep trough in the Pacific ocean crust (Figure 1). Seismic reflection profiles [*Scholl et al.*, 1982] show that an extra kilometer or more of terrigenous sediment overlies 200-300 m of pelagic sediment in the Aleutian trench east of the Amlia fracture zone. The east-facing escarpments of the Amlia fracture zone reflect older Pacific ocean crust to the east and have acted to pond west-flowing terrigenous sediment in the trench (Figure 1). This escarpment would almost certainly prevent much of the pelagic and terrigenous sediment from being scraped off below the accretionary wedge and may juxtapose abundant water-rich serpentinite [*Ulmer and Trommsdorff*, 1995] with these sediments at mantle depths.

We propose that the larger quantity of slab-derived components involved in magma sources for Seguam (and possibly Yunaska) basalts reflects efficient focussing of terrigenous and pelagic sediment into the mantle via subduction of the Amlia fracture zone. Extra fluid released into the mantle wedge from the excess sediment and serpentinite in the Amlia fracture zone may have (1) locally increased by two- to five-fold the amount of crustally derived elements (*e.g*, B, Cs, radiogenic Sr and other large

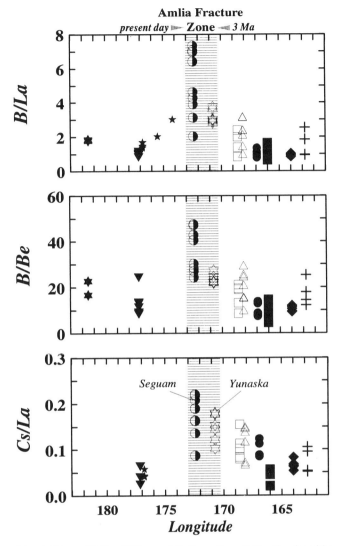

Fig. 4. Plots of B/La, B/Be, and Cs/La ratios of Aleutian basaltic lava samples vs. longitude. The ruled area shows position of the Amlia fracture zone between 3 Ma and the present day (Figure 1). The volcanic centers of Seguam and Yunaska overlie subarc mantle that was in contact with this fracture zone and have the highest ratios. Data sources for La and Cs values used to calculate B/La and Cs/La ratios for Adak, Atka, and Kasatochi samples are given in *Singer et al.* [1992b]. Symbols as in Figure 3.

ion lithophile elements) transferred from the slab to the mantle wedge, and (2) lowered the melting temperature of the mantle, thus increasing the extent to which partial melting proceeded [*e.g., Luhr*, 1992]. Intra-arc extension (Figure 1) may also have promoted more extensive mantle melting [*Singer et al.*, 1992b], however the amount of lithospheric attenuation is probably small [*Geist and Scholl*, 1992] and extension alone cannot explain the high B contents or B/La, B/Be, Cs/La and $^{87}Sr/^{86}Sr$ ratios observed at Seguam.

IMPLICATIONS

Seguam basalt is distinctively lower in K_2O, TiO_2, and total FeO than other Aleutian basalts [*Singer et al.*, 1992b], consistent with an origin via higher degrees of mantle melting. If the degree of partial melting reflects the quantity of slab-derived fluid added to the mantle wedge, then, as suggested by *Luhr* [1992], the small but variable quantities of fluid liberated from the subducted plate can have a profound effect on bulk magma compositions. Our conclusions that the flux of subducted sediment is quite variable and that the degree of mantle melting may reflect differences in sediment flux are at odds with *Plank and Langmuir's* [1988] conclusion that bulk magma composition and degree of partial melting in arcs are determined primarily by crustal thickness.

Trace element data from the central Aleutian arc imply a strong link between slab-mantle chemical exchanges and the products of arc volcanism. Single stage models of melting directly either the slab or mantle wedge are ruled out as these would lead to (1) very low B contents in the melt, owing to severe B-depletion of any protolith as it approached its melting temperature, and (2) decreases of K *and* B contents in magmas produced by increased percentages of melting. Neither case is observed. Increased availability of water in the presence of a thickened column of subducted B-rich sediment can potentially stabilize larger quantites of hydrous minerals in the shallow reaches of the mantle wedge [*e.g., Tatsumi et al.*, 1986]. Selective partitioning into these hydrous mantle minerals will accomodate only a fraction of the subducted B (and Cs); upon transport to the depths of melt generation [*e.g., Davies and Stevenson*, 1992] the higher modal concentration of hydrous minerals will release this water to the mantle which then melts to a greater extent than otherwise possible in the absence of the fracture zone. Simple two component mixing calculations (Figure 3c) suggest that the the flux of hydrous minerals to the depth of melting in the mantle may vary by a factor of two to five along the Aleutian arc.

Despite the complex pathway of subducted fluid to the depth of melting and the large apparent variation in sediment flux, the B/La-B/Be correlation (Figure 3c) indicates that the fluid composition must be remarkably uniform below 1400 km of the arc. This seems to be true in all arcs where similar data exist [*Morris et al.*, 1990; *Edwards et al.*, 1993; *Leeman et al.*, 1994] and implies, with respect to B, Be, La, and Cs at least, effective homogenization of the subducted sediment component during evolution of the fluid.

The short residence times postulated for B (and Cs) in the mantle wedge [*Morris et al.*, 1990] are confirmed by the lower B/La, B/Be, and Cs/La ratios at Yunaska, where the Amlia fracture zone interacted with the mantle wedge 3 million years ago. Changes in the flux of subducted sediment may produce subtle transient signals in the arc volcanoes, in this case reflecting oblique subduction of the Pacific plate.

Acknowledgements. Supported by Swiss NSF grants 21-36509.92 and 20-42124.94 (Singer) and U.S. NSF grant EAR 91-19110 (Leeman). Samples upon which this study is based were generously provided to Singer by Jim Brophy, John Fournelle, James Myers, Chris Nye, Mike Perfit, and George Snyder. Jeff Ryan kindly provided the Be analyses discussed here. We thank Mike Dungan for his enthusiastic support of this project. Dave Scholl, Gray Bebout, and Steve Kirby are applauded for envisioning the *SUBCON* forum and stimulating exchange of ideas across disciplines. Insightful reviews by John Davidson, Geoff Nichols, and Jeff Ryan are greatly appreciated as they helped to clarify many points and to improve the presentation.

REFERENCES

Atwater, T., and J. Severinghaus, Tectonic maps of the northeast Pacific, in: Winterer, E.L., D.M. Hussong, and R.W. Decker, eds., The Eastern Pacific Ocean and Hawaii. Geol. Soc. Amer., *The Geology of North America, v. N*, pp. 15-20, 1989.

Bebout, G.E., J.G. Ryan, and W.P. Leeman, B-Be systematics in subduction-related metamorphic rocks: characterization of the subducted component. *Geochim. Cosmochim. Acta, 57*, 2227-2237, 1993.

Bonatti, E., and K. Crane, Oceanic fracture zones. *Scientific American, 250*, 36-47, 1984.

Coats, R.R., Magma type and crustal structure in the Aleutian Arc. *Am. Geophys. Union Monograph 6*, G.A. MacDonald and, H. Kuno, eds., pp. 92-109, 1962.

Davies, J.H., and D.J. Stevenson, Physical model of source region of subduction zone volcanics. *J. Geophys. Res., 97*, 2037-2070, 1992.

Edwards, C.M.H., J.D. Morris, and M.F. Thirlwall, Separating slab from mantle signatures in arc lavas using B/Be and radiogenic isotope systematics. *Nature, 362*, 530-533, 1993.

Geist, E.L., J.R. Childs, and D.W. Scholl, The origin of summit basins of the Aleutian ridge: Implications for block rotation of an arc massif. *Tectonics, 7*, 327-341, 1988.

Geist, E.L., and D.W. Scholl, Application of continuum models to deformation of the Aleutian Island arc. *J. Geophys. Res. 97*, 4953-4967, 1992.

Hayes, D.E., and J.R. Heirtzler, Magnetic anomalies and their relation to the Aleutian arc. *J. Geophys. Res., 73*, 4637-4646, 1968.

Kay, SM, R.W. Kay, and G.P. Citron, Tectonic controls on tholeiitic and calc-alkaline magmatism in the Aleutian arc, *J. Geophys. Res. 87*, 4051-4072, 1982.

Kay, S.M., V.A. Ramos, and M., Marques, Evidence in Cerro Pampa volcanic rocks for slab melting prior to ridge-trench collision in southern South America: *J. Geology, 101*, 703-714., 1993.

Keleman, P.B., K.T.M. Johnson, R.J. Kinzler, and A.J. Irving, High-field strength element depletions in arc basalts due to mantle-magma interaction, *Nature, 345*, 521-524, 1990.

Leeman, W.P., V.B. Sisson, and M.R. Reid, Boron geochemistry of the lower crust: Evidence from granulite terranes and deep crustal xenoliths. *Geochim. Cosmochim. Acta, 56*, 775-788, 1992.

Leeman, W.P., M.J. Carr, and J.D. Morris, Boron geochemistry of the Central American Volcanic Arc: constraints on the genesis of subduction-related magmas. *Geochim. Cosmochim. Acta, 58*, 149-168, 1994.

Luhr, J.F., Slab-derived fluids and partial melting in subduction zones: insights from two contrasting Mexican volcanoes (Colima and Ceboruco). *J. Volcanol. Geotherm Res., 54*, 1-18, 1992.

Mammerickx, J., Large scale undersea features of the northeast Pacific, in: Winterer, E.L., D.M. Hussong, and R.W. Decker, eds., The Eastern Pacific Ocean and Hawaii. Geol. Soc. Amer. *The Geology of North America, v. N*, pp. 5-13, 1989.

Moran, A.E., V.B. Sisson, and W.P. Leeman, Boron in subducted oceanic crust and sediments: effects of metamorphism and implications for arc magmatism. *Earth Planet Sci. Lett., 111*, 331-349, 1992.

Morris, J.D., W.P. Leeman, and F. Tera, The subducted component in island arc lavas: constraints from Be isotopes and B-Be systematics. *Nature, 344*, 31-36, 1990.

Myers, J.D., B.D. Marsh, and A.K. Sinha, Strontium isotopic and selected trace element variations between two Aleutian volcanic centers (Adak and Atka): implications for the development of arc volcanic plumbing systems. *Contrib. Mineral. Petrol., 91*, 221-234. 1985.

Plank, T., and C.H. Langmuir, An evaluation of the global variations in the chemistry of arc basalts, *Earth Planet. Sci. Lett., 90*, 349-370, 1988.

Plank, T., and C.H. Langmuir, Tracing trace elements from sediment input to volcanic output at subduction zones. *Nature, 362*, 739-742, 1993.

Ryan, J., and C.H., Langmuir, Beryllium systematics in young volcanic rocks: implications for ^{10}Be. *Geochim. Cosmochim. Acta, 52*, 237-244, 1988.

Ryan, J., and C.H., Langmuir, The systematics of boron abundances in young volcanic rocks: *Geochim. Cosmochim. Acta, 57*, 1489-1498, 1993.

Scholl, D.W., T.L. Vallier, and A.J. Stevenson, Sedimentation and deformation in the Amlia fracture zone sector of the Aleutian trench. *Marine Geol., 48*, 105-134, 1982.

Singer, B.S., J.D. Myers, and C.D. Frost, Mid-Pleistocene lavas from the Seguam volcanic center, central Aleutian arc: closed-system fractional crystallization of a basalt to rhyodacite eruptive suite. *Contrib. Mineral. Petrol., 110*, 87-112, 1992a.

Singer, B.S., J.D. Myers, and C.D. Frost, Mid-Pleistocene basalt from the Seguam volcanic center, central Aleutian arc, Alaska: local lithospheric structures and source variability in the Aleutian arc. *J. Geophys. Res., 97*, 4561-4578, 1992b.

Singer, B.S., J.G. Brophy, and J.R. O'Neil, Oxygen isotope constraints on the petrogenesis of Aleutian arc magmas. *Geology, 20*, 367-370, 1992c.

Tatsumi Y., D.L. Hamilton, and R.W. Nesbitt, Chemical characteristics of fluid phase released from a subducted lithosphere and origin of arc magmas: evidence from high-pressure experiments and natural rocks. *J. Volcanol. Geotherm. Res., 29*, 293-309, 1986.

Thirlwall, M.F., T.E. Smith, A.M. Graham, N. Theodorou, P. Hollings, J.P. Davidson, and R.J. Arculus, High field strength element anomalies in arc lavas: source or process? *J. Petrol, 35*, 819-838, 1994.

Ulmer, P., and Trommsdorff, V., Serpentine stability to mantle depths and subduction-related magmatism. *Science*, 268, 858-861.

Varne, R., A view from the Sunda arc, *Nature, 367*, 224, 1994.

You, C.-F., A.J. Spivack, J.M. Gieskes, R. Rosenbauer R., and J.L. Bischoff, Experimental study of boron geochemistry: implications for fluid processes in subduction zones. *Geochim. Cosmochim. Acta, 59,* 2435-2442, 1995.

You, C.-F., A.J. Spivack, J.H. Smith, and J.M. Gieskes, Mobilization of boron in convergent margins: implications for the boron geochemical cycle, *Geology, 21*, 207-210, 1993.

B.S. Singer, Département de Minéralogie, Université de Genève, 13 rue des Maraîchers, 1211 Genève 4, Switzerland (email: singer@sc2a.unige.ch).

W.P. Leeman, Department of Geology and Geophysics, Rice University, Houston, TX 77251, U.S.A.

M.F. Thirlwall, Geology Department, Royal Holloway and Bedford New College, University of London, Egham Hill, Egham, Surrey TW20 0EX, U.K.

N.W. Rogers, Department of Earth Sciences, The Open University, Milton Keynes, MK7 6AA, U.K.

Experimental Melting of Pelagic Sediment, Constraints Relevant to Subduction

Geoffrey T. Nichols

Macquarie University, GEMOC, School of Earth Sciences, Sydney, New South Wales, 2109, Australia, and California Institute of Technology, Division of Geological and Planetary Sciences, Pasadena, CA. 91125, U.S.A.

Peter J. Wyllie

California Institute of Technology, Division of Geological and Planetary Sciences, Pasadena, CA. 91125, U.S.A.

Charles R. Stern

University of Colorado, Department of Geological Sciences, Boulder, Colorado, CO. 80309, U.S.A.

The experimental melting relations of pelagic red clay with water indicate a low temperature solidus, close to 650°C. This is significant as the red clay water solidus is lower than the solidus of gabbro with water, to depths of at least 140 km. Such a solidi configuration allows for melting of sediments, whilst gabbroic crust dehydrates, in moderate-temperature steady-state subduction regimes. This new experimental evidence lends support to (1) the sediment melting - gabbro dehydration hypothesis, recently proposed on independent geochemical criteria, and (2) may place relatively narrow limits on temperatures of the upper slab-mantle boundary, over a considerable depth interval, an apropos constraint for thermal modelling of steady-state subduction regimes.

1. INTRODUCTION

Sedimentary input into subduction zones as proposed in the 1960's by *Coats* [1962], *Armstrong* [1968], and *Oxburgh and Turcotte* [1968], but was later disputed by *Pankhurst* [1969], and *Oversby and Ewart* [1972]. *Stern* [1973] studied the melting relationships of pelagic red clay to 30 kbar, and published the 30 kbar results in *Stern and Wyllie* [1973]. *Huang and Wyllie* [1973] used the phase relationships of a muscovite-granite as a model for the partial melting of pelagic sediments in subduction zones, drawing attention to the prospect of two stages of melting, one at shallower depths associated with pore fluids, followed by melting at deeper levels and higher temperatures where residual mica experienced vapor-absent dehydration-melting. Many calculations for the thermal structures of subduction zones during the 1970's presented isotherms requiring partial melting of subducted sediments (see *Gill* [1981]), however, geochemists in general failed to find evidence for the involvement of sediments in the generation of arc-magmas. The prevailing opinion through the 1980's was that the main material source for arc magmas was peridotite in the mantle-wedge, but that fluids of uncertain source and character had transferred specific isotope and trace-element signatures from the subducted slab to the melting region in the mantle-wedge. By the 1990's, evaluations of the thermal structure of subduction zones (e.g. *Peacock* [1991]) were emphasizing that temperatures were normally too low for melting of oceanic crust (e.g. *Davies and Stevenson* [1992]), and that aqueous vapors were the dominant transfer agents for the characteristic trace-element signatures of arc magmas. However, in 1990, *Drummond and Defant* presented evidence

294 SUBDUCTION RELATED SEDIMENT MELTING EXPERIMENTS

Table 1. Comparison of Red Clay Compositions and Normative Mineral Proportions

	Red Clay #256	volatile free	Average Ocean Clay[a]	volatile free	Red Clay[b]	volatile free
SiO_2	48.79	57.31	53.70	57.60	52.77	57.54
TiO_2	0.83	0.97	1.00	1.10	0.84	0.92
Al_2O_3	17.90	21.02	17.40	18.70	14.67	16.00
FeO[c]	5.62	6.60	0.5 / 8.5	0.5 / 9.1	7.39	8.06
MnO	0.58	0.68	0.80	0.90	1.24	1.35
MgO	3.05	3.58	4.60	4.90	3.29	3.59
CaO	2.06	2.42	1.60	1.70	2.40	2.62
Na_2O	2.61	3.07	1.30	1.40	4.33	4.72
K_2O	3.49	4.10	3.70	4.00	3.58	3.90
P_2O_5	0.21	0.25	0.10	0.10	1.20	1.31
Cl	1.90	-	-	-	-	-
H_2O+	7.79		6.30			
H_2O-	1.05					
CO_2	1.70		0.40			
S	0.06		-			
LOI			0.40		8.30	
Total	97.64	100.00	100.00	100.00	91.71	100.01
Norms						
Qz		14.6		39.3		1.1
Ab		44.2		20.3		62.7
Or		41.2		40.4		36.2
Qz		12.4		34.7		0.9
Ab + An		52.6		29.8		65.2
Or		35.0		35.6		33.9

[a] *Seibold and Berger*, 1982
[b] composition used by *Johnson and Plank* [1993]
[c] Fe expressed as FeO or FeO / Fe_2O_3

for the derivation of silicic volcanics (adakites) from the partial melting of subducted hydrous gabbro, using trace-element considerations.

Renewed support for the involvement of sediments has arisen with evidence from Sr, Pb and Be isotopes [*Kay et al.*, 1978; *White and Dupré*, 1986; *Tera et al.*, 1986], as well as from trace-elements *Hawkesworth et al.*, [1991], and studies combining this evidence with mass flux calculations [*Plank and Langmuir*, 1993]. Abundances of Be and Th in arc-derived magmas have led several workers [*Ryan and Langmuir*, 1988; *Plank and Langmuir*, 1992; *Reagan et al.*, 1994] to conclude that not only are these elements recycled efficiently, but that the recycling agent is probably a sediment-melt (with or without additional hydrous fluids) since Be and Th are not known to partition into hydrous fluids.

Experiments described here, are aimed at understanding the phase relations of a typical ocean pelagic sediment, at conditions relevant to subduction, and investigating the chemistry of melts generated through the melting interval. *Johnson and Plank* [1993] are engaged in similar experiments at higher pressures.

2. APPARATUS AND STARTING MATERIAL

Our experiments were conducted in 1/2 and 3/4 inch piston-cylinder apparatus with an Atlantic Ocean pelagic red clay, collected from a water depth of 5949 m, approximately 1560 km east of Miami. This red clay has a composition close to that of many ocean-pelagic red clays (Table 1) and has an initial mineralogy of quartz, muscovite, illite, albite, halite, minor kaolinite, and probably orthoclase. We conducted the experiments in gold capsules using the red clay (which has 8% H_2O bound in hydrous minerals) and in separate runs added 20% H_2O; in one experiment we added only 3% water. Our data are therefore relevant to red clay with 8%, 11%, and 28% total water, over the range of pressures from 2-30 kbar, and temperatures from 550-1200°C.

3. EXPERIMENTAL PHASE RELATIONS

3.1. Sub-solidus and Solidus Relations

The red clay reacts sub-solidus to produce predominantly quartz and micas (muscovite, illite, and minor biotite); with mafic minerals, including cordierite, amphibole, garnet and clinopyroxene, depending on the pressure. Our experiments both with added and without added water (Figs 1a and b), define solidi close to that of the granite-H_2O system [e.g. *Boettcher and Wyllie*, 1968], between 630 - 675°C at 10 and 30 kbar. *Johnson and Plank* [1993] report a higher temperature solidus at 30 kbar close to our Bi-out curve. This discrepancy in solidi temperatures may, in part, be caused by the differing normative quartz contents of the clays used, with *Johnson and Plank's* [1993] clay having only ~1% calculated normative quartz compared with ~13% normative quartz for the red clay used in these experiments (Table 1).

3.2. Melting Interval Relations

Above the solidus the primary mineralogy is dominated by quartz and biotite, with positions of the phase-fields of clinopyroxene, amphibole, garnet and plagioclase varying with pressure [*Nichols et al.*, 1993]. Through the melting interval, both quartz and biotite occur as large crystals (~40 - 60 μm) along the lower capsule margins. At 650°C, immediately above the solidus, fine-grained quartz occurs throughout the lower half of the capsules, and ~5 μm garnet occurs in the upper half.

3.3. Superliquidus Relations

In runs interpreted to be within the superliquidus field, above 850°C (at 20 kbar), experimental-melts quench to wide glass rims (~400 μm) which surround a core of glass crowded with ~3-5 μm garnets. Occurring between the central garnet-rich region and the glass rims, is a transitional zone of glass containing atoll-shaped and skeletal garnets. The garnets change form gradationally from atolls nearest the glass rims, to euhedral habits nearest the centre. Along with the euhedral garnets in the central region, are rare areas of fine-grained spinifex-textured quenched kyanite. In these experiments no crystal-settling is observed. Combining the observations of gradational changes in garnet morphology,

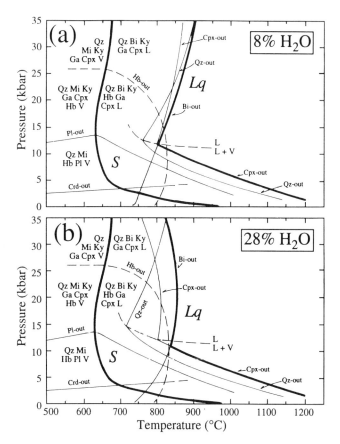

Figure 1. Experimental phase diagrams showing the melting relations of red clay in experiments run with a total of 8% water (a), and 28% total water (b). Interpretation for the figures is based on the combined data from 64 experiments conducted over the P-T range of the plot. The lower temperature bold curves labelled *S* are the red clay solidi, and are identical for both 8% and 28% water experiments; higher temperature bold curves labelled *Lq* are the liquidi surfaces. Mineral abbreviations: **Bi** biotite, **Cpx** clinopyroxene, **Crd** cordierite, **Ga** garnet, **Hb** hornblende, **Ky** kyanite, **L** liquid, **Mi** mica, **Pl** plagioclase, **Qz** quartz, **V** vapor.

the lack of garnet-settling (compared with quartz and biotite settling in lower temperature runs), and the associated spinifex quenched kyanite, leads us to the interpretation that these garnets grew from liquid during quench [*Nichols et al.*, 1994]. In further support of this conclusion, we compared the compositions of garnets within each experiment. Garnets do not display the characteristic bipartite compositional groupings that are usually useful to discriminate quench from primary minerals. Garnets do, however, display the usual variations in Fe, Mg and Ca characteristic of the temperatures at which they began to form.

Figures 1a and b compare the phase diagrams of red clay with 8% and 28% total water. They show the final interpretations combining early experiments which were crushed and examined under immersion oils, with the latest polished runs examined by SEM and analysed using a microprobe. The solidus for red clay with 8% water (bold, Fig. 1a) is identical to that determined for red clay with 28% water (Fig. 1b), indicating that with 8% H_2O there is free vapor remaining after saturation of the partial melt near the solidus. The liquidi surfaces (in bold) are similar at low pressure, but deviate at higher pressures where the Bi-out curve (liquidus) in the red clay-28% H_2O system curves to lower temperatures at pressures above 20 kbar. At pressures above 15 kbar, both quartz and clinopyroxene melt at lower temperatures in the red clay-28% system, than in the red clay 8% system (cf. Figs 1a and b).

3.4. Glass Compositions

The experimental glasses display large and systematic compositional variations throughout the melting interval, with variations strongly dependent on run temperatures. In experiments immediately above the solidus (650°C, 20 kbar), glass rims are very narrow and are compositionally heterogeneous; these glasses have variable K_2O, and low SiO_2 (~50 wt%, Fig. 2), and therefore probably represent disequilibrium melts produced from the partial melting of micas. At higher temperatures, 750 - 800°C, glasses are rhyolitic (SiO_2 ~75 wt%), peraluminous (Al / Na+Ca+K ~1 - 2.7), and have low Cl ~0.1 - 0.4. From 825 - 950°C, glasses become progressively less silicic (SiO_2 ~75 - 60), less peraluminous (~1.5) and become increasingly Cl rich (0.5 - 1.7 wt%, Fig. 2).

Compositions of experimental glasses also depend on the degree of vapor saturation. Figure 2 displays data from experiments run at 15-20 kbar with total water contents ranging from 8% to 28%. The data define two trends, one from vapor saturated experiments (28% total water) and one trend with glasses from vapor present runs (8 and 11% total water). The glasses produced in experiments with less vapor contain more Cl and, at a given temperature, are more silicic (Fig. 2). Excess vapor thus effectively enhances Cl partitioning into the vapor, depleting the coexisting glass.

Chlorine strongly partitions into the vapor phase (in agreement with the work of *Webster and Holloway* [1988]) at lower experimental temperatures along with the alkalis, whereas at higher temperatures Cl (and the alkalis) are progressively redissolved into the coexisting melts thereby reducing the melt's peraluminosity. *Webster and Holloway* [1988], in addition to determining that Cl favors H_2O and H_2O+CO_2 vapors more than coexisting melts, found that this partitioning behavior increased in more peraluminous melts, and also demonstrated that Cl partitioning into vapor increased with temperature and pressure.

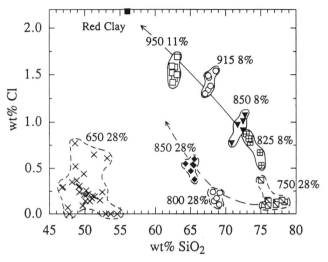

Figure 2. Silica and chlorine compositional variations of red clay glasses from experiments between 15 and 20 kbar. Data are subdivided into two trends; one trend displays data from experiments with excess vapor (28% total water, dashed line), and the other trend shows data from runs with 8% and, for one run 11%, water (solid line). Except for the disequilibrium glass compositions from a 650°C run, data define trends which approach the red clay starting composition (black square), with increasing temperature.

4. SIGNIFICANCE OF EXPERIMENTAL RESULTS

4.1. Thermal Regimes

Davies and Stevenson [1992] reviewed current thermal models for subduction zones, concluding that shear heating does not contribute significantly to the heat budget (contrary to *Molnar and England* [1990]), and therefore despite the large calculated range in possible thermal regimes, viable models were generally cool. *Peacock* [1991] and *Peacock et al.* [1994] determined that for similar subduction parameters (slab velocity, dip angle, age and no shear heating) P-T trajectories were also cool. These thermal models thus predict no melting of the oceanic crust during steady-state subduction, except in rare circumstances where shear heating is significant, or when young, hot oceanic crust is subducted slowly.

Figure 3 compares the current thermal models of *Davies and Stevenson* [1992] (labelled *DS*), and one of the P-T paths from *Peacock et al.* [1994] (labelled *PRT*), with the solidus of red clay-H_2O as well as the solidi and dissociation curves of other subduction zone lithologies. For both calculated paths (*DS* and *PRT*), only slight temperature

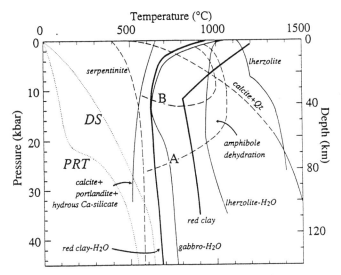

Figure 3. Pressure-temperature diagram comparing recent thermal models for the top of subducting oceanic crust from *Davies and Stevenson* [1992] (*DS*), (slab dipping 30°, velocity 7.2 cm yr-1, no shear heating), and *Peacock et al.* [1994] (*PRT*), (slab dipping 26.6°, velocity 7.2 cm yr-1, no shear heating), with solidi and dissociation curves for subduction zone lithologies. The lower temperature curve in bold is the solidus for red clay-H_2O, and the higher temperature curve (in bold) is the deduced vapor-absent solidus of red clay (vapor present liquidus surface). Between the points labelled B and A amphibolite undergoes dehydration melting. Curves from the following references: serpentinite, *Wyllie* [1979]. The recent antigorite stability experiments of Ulmer *and Trommsdorff* [1995] extend the serpentinite stability field to temperatures higher by ~100°C; calcite+portlandite+hydrous Ca-silicate, *Boettcher and Wyllie* [1969]; gabbro-H_2O and amphibole dehydration, *Wyllie and Wolf* [1993]; calcite+quartz, *Huang et al.* [1980]; lherzolite, *Takahashi* [1986]; and lherzolite-H_2O, *Green* [1973].

increases are required before they intersect the red clay-H_2O solidus. Such temperature increases would produce a situation where pelagic sediments melt for a significant depth interval, whilst the underlying gabbroic crust dehydrates. These thermal models, combined with the red clay-H_2O experimental results, support the findings of *Plank and Langmuir* [1992] who concluded, on the basis of geochemical mass balances for various trace elements, that a mantle-wedge contribution was derived from both melted sediments and aqueous fluids originating in the underlying basaltic crust. In this sediment, the extended stability field of biotite (which defines the liquidus at pressures above ~12 kbar) would significantly influence not only the trace-element abundances in the sediment-derived melts, but assuming that these melts migrate away from the slab/mantle interface, also the trace-element ratios imparted to the overlying mantle-wedge.

Also depicted on Figure 3 is the deduced vapor-absent solidus for red clay, equivalent to the vapor-present liquidus surface of red clay with water. The relatively high temperature position of this curve (~850-900°C at pressures >12 kbar) suggests that sediments would not melt in many moderate temperature subduction regimes, if such regimes are entirely vapor absent. For vapor-present subduction conditions, without sufficient fluid for saturation (<8% total water), the actual solidus of red clay will be bracketed by the red clay-H_2O solidus and the deduced vapor-absent red clay solidus. For realistic conditions however, some H_2O vapor would be present, produced by the dehydration of hydrous minerals such as mica, amphibole and serpentine. For example, up to pressures of ~20 kbar muscovite dehydration occurs at temperatures below (but close to) the inferred vapor-absent solidus of red clay [*Huang and Wyllie*, 1973]. Serpentine and amphibole also undergo dehydration at temperatures below the deduced vapor-absent solidus of red clay.

These experiments provide further support for *Plank and Langmuir's* [1992] deduction that sediments melt, while gabbro dehydrates. This analysis may provide a unique temperature constraint on the upper slab-mantle boundary, a regime difficult to model, and indicates that during steady state subduction the interface remains between ~650 and ~750°C, for a significant depth interval [*Nichols et al.*, 1994].

Acknowledgments: We thank Dr. Emiliani, University of Miami, for the Red Clay used in our experiments. We thank James Myers, Terry Plank and Tracy Rushmer for helpful reviews. This work has been funded by the National Science Foundation, grant EAR-9303967. Publication 59 in the Key Centre for GEMOC.

REFERENCES

Armstrong, R. L., A model for the evolution of Strontium and Lead isotopes in a dynamic earth, *Rev. Geophys.*, 6, 175-199, 1968.

Boettcher, A. L., and P. J. Wyllie, Melting of granite with excess water to 30 kilobars pressure, *J. Geol.*, 76, 235-244, 1968.

Boettcher, A. L., and P. J. Wyllie, The system $CaO-SiO_2-CO_2-H_2O$—III. Second critical end-point on the melting curve, *Geochim. Cosmochim. Acta*, 33, 611-632, 1969.

Coats, R. R., Magma type and crustal structure in the Aleutian arc, *Geophys. Mono.*, 6, 92-109, 1962.

Davies, J. H., and D. J. Stevenson, Physical model of source region of subduction zone volcanics, *J. Geophys. Res.*, 97, 2037-2070, 1992.

Drummond, M. S., and M. J. Defant, A model for trondhjemite-tonalite-dacite genesis and crustal growth via slab melting: Archean to modern comparisons, *J. Geophys. Res., 95*, 21503-21521, 1990.

Gill, J. B., *Orogenic Andesites and Plate Tectonics*, 390 pp., Springer-Verlag, Berlin, Heidelberg, New York, 1981.

Green, D. H., Experimental melting studies on a model upper mantle composition at high pressure under water-saturated and water-undersaturated conditions, *Earth Planet. Sci. Lett., 19*, 37-53, 1973.

Hawkesworth, C. J., J. M. Hergt, R. M. Ellam, and F. McDermott, Element fluxes associated with subduction related magmatism. *Philos. Trans. R. Soc. London, Ser. A, 335*, 393-405, 1991.

Huang, W. -L., and P. J. Wyllie, Melting relations of muscovite-granite to 35 kbars as a model for fusion of metamorphosed subducted oceanic sediments, *Contrib. Mineral. Petrol., 42*, 1-14, 1973.

Huang, W. -L., P. J. Wyllie, and C. E. Nehru, Subsolidus and liquidus phase relationships in the system $CaO-SiO_2-CO_2$ to 30 kbar with geological applications, *Am. Miner. 65*, 285-301, 1980.

Johnson, M. C., and T. Plank, Experimental constraints on sediment melting during subduction (abstract), *Eos Trans. AGU, 74*(43), Fall Meeting Suppl., 680, 1993.

Kay, R. W., S. -S. Sun, and C. -N. Lee-Hu, Pb and Sr isotopes in volcanic rocks from the Aleutian Islands and Pribilof Islands, Alaska, *Geochim. Cosmochim. Acta, 42*, 263-273, 1978.

Molnar, P., and P. J. England, Temperature, heat flux, and frictional stress near major thrust faults. *J. Geophys. Res., 95*, 4833-4856, 1990.

Nichols, G. T., C. R. Stern, and P. J. Wyllie, Experimental metamorphism and melting of pelagic red clay: on the formation of granitoid magmas in subduction zones (abstract), *Eos Trans. AGU, 74*(43), Fall Meeting Suppl., 657-658, 1993.

Nichols, G. T., P. J. Wyllie, and C. R. Stern, Subduction zone melting of pelagic sediments constrained by melting experiments, *Nature, 371*, 785-788, 1994.

Oversby, V. M., and A. Ewart, Lead isotopic composition of Tonga-Kermadec volcanics and their petrogenetic significance, *Contrib. Mineral. Petrol., 37*, 181-210, 1972.

Oxburgh, E. R., and D. L. Turcotte, Problem of high heat flow and volcanism associated with zones of descending mantle convection flow, *Nature, 218*, 1041-1043, 1968.

Pankhurst, R. J., Strontium isotope studies related to petrogenesis in the Caledonian basic igneous province of NE Scotland, *J. Petrol., 10*, 115-143, 1969.

Peacock, S. M., Numerical simulation of subduction zone pressure-temperature-time paths: constraints on fluid production and arc magmatism. *Philos. Trans. R. Soc. London, 335*, 341-353, 1991.

Peacock, S. M., T. Rushmer, and A. B. Thompson, Partial melting of subducting oceanic crust. *Earth Planet. Sci. Lett., 121*, 227-244, 1994.

Plank, T., and C. H. Langmuir, Sediments melt and basaltic crust dehydrates at subduction zones (abstract), *Eos Trans. AGU, 72*, Fall Meeting Suppl., 637, 1992.

Plank, T., and C. H. Langmuir, Tracing trace elements from sediment input to volcanic output at subduction zones, *Nature, 362*, 739-742, 1993.

Reagan, M. K., J. D. Morris, E. A. Herrstrom, and M. T. Murrell, Uranium series and beryllium isotope evidence for an extended history of subduction modification of the mantle below Nicaragua, *Geochim. Cosmochim. Acta, 58*, 4199-4212, 1994.

Ryan, J. G., and C. H. Langmuir, Beryllium systematics in young volcanic rocks: Implications for ^{10}Be, *Geochim. Cosmochim. Acta, 52*, 237-244, 1988.

Seibold, E., and W. H. Berger, *The Sea Floor: An Introduction to Marine Geology.* 288 pp., Springer-Verlag, Berlin, Heidelberg, New York, 1982.

Stern, C. R., Melting relations of Gabbro-Tonalite-Granite-Red Clay with H_2O at 30 kb: the implications for melting in subduction zones, Ph.D. thesis, Uni. of Chicago, Illinois, December 1973.

Stern, C. R., and P. J. Wyllie, Melting relations of basalt-andesite-rhyolite-H_2O and a pelagic red clay at 30 kb, *Contrib. Mineral. Petrol., 42*, 313-323, 1973.

Takahashi, E. J., Melting of a dry peridotite KLB-1 up to 14 GPa: implications on the origin of peridotitic upper mantle, *J. Geophys. Res. 91*, 9367-9382, 1986.

Tera, F., L. Brown, J. Morris, I. S. Sacks, J. Klein, and R. Middleton, Sediment incorporation in island-arc magmas: Inferences from ^{10}Be, *Geochim. Cosmochim. Acta, 50*, 535-550, 1986.

Ulmer, P., and V. Trommsdorff, Serpentine stability to mantle depths and subduction-related magmatism, *Science, 268*, 858-861, 1995.

Webster, J. D., and J. R. Holloway, Experimental constraints on the partitioning of Cl between topaz rhyolite melt and H_2O and $H_2O + CO_2$ fluids: New implications for granitic differentiation and ore deposition, *Geochim. Cosmochim. Acta, 52*, 2091-2105, 1988.

White, W. M., and B. Dupré, Sediment subduction and magma genesis in the Lesser Antilles: Isotopic and trace element constraints, *J. Geophys. Res., 91*, 5927-5941, 1986.

Wyllie, P. J., Magmas and volatile components, *Am. Miner. 64*, 469-500, 1979.

Wyllie, P. J., and M. B. Wolf, Amphibolite dehydration-melting: sorting out the solidus, in *Magmatic Processes and Plate Tectonics,* edited by H. M. Prichard, T. Alabaster, N. B. W. Harris, and C. R. Neary, *Geol. soc. sp. Pub., 76*, 405-416, 1993.

G. T. Nichols, Macquarie University, GEMOC, School of Earth Sciences, Sydney, New South Wales, 2109, Australia.

P. J. Wyllie, California Institute of Technology, Division of Geological and Planetary Sciences, Pasadena, CA. 91125, U.S.A.

C. R. Stern, University of Colorado, Department of Geological Sciences, Boulder, Colorado, CO. 80309, U.S.A.

The Influence of Dehydration and Partial Melting Reactions on the Seismicity and Deformation in Warm Subducting Crust

Tracy Rushmer

Department of Geology, Perkins Hall, University of Vermont, Burlington, Vermont, 05405-0122

In subduction zones involving young, warm lithosphere ($\leq$ 20 Ma) most all the observed seismicity is shallow and located in the upper part of the downgoing slab, or, seismicity is absent. The top portion of the slab is composed of mainly basaltic oceanic crust and as it subducts, dehydration and, in some cases, partial melting takes place. In this paper, the results from an experimental deformation study on basaltic amphibolite which investigated the effects of mineral reactions on deformation, are applied here to young subducting oceanic lithosphere and the role of metamorphic reactions in accommodating deformation in the slab is explored. Pressure-temperature-time (PTt) paths which calculate the changing pressure and temperature conditions of the uppermost portion of the slab as it subducts are also presented, together with experimental results. Between 15 and 75 km, most warm PTt paths intersect mineral reaction curves of not only hornblende, but of other hydrous minerals such as zoisite, lawsonite and chloritoid. The specific depth at which these reactions will occur increases with age of the slab. By combining the calculated PTt paths with the experimentally determined stability curves for high-pressure hydrous phases and the experimental deformation results, potential effects of dehydration/hydration and partial melting reactions on deformation within the upper portion of the downgoing slab can be proposed. The formation of micron-sized product phases (both hydrous and anhydrous) from dehydration reactions can localize ductile deformation into broad shear zones by mass transfer and grain-boundary sliding. This is one potential mechanism (a reaction-enhanced ductility process) by which stress can be accommodated in warm slabs. This mechanism would suppress seismicity and may be active in subduction zones which are characterized by relatively aseismic behavior (e.g. Cascadia). At high temperatures, or during unusual "hot" subduction, melting of the slab may produce local brittle failure. Punctuated shallow intraslab seismic events in subduction zones involving young oceanic lithosphere could be brought on by rapid dehydration- or melt-enhanced embrittlement.

INTRODUCTION

One of the major consequences of subduction is the transformation of the downgoing oceanic crust into blueschist, amphibolite and eclogite by prograde reactions [*Peacock*, 1993]. It is considered that these reactions are the likely source of water which triggers arc magmatism in the overlying mantle wedge and can contribute to seismic activity [*Raleigh and Paterson*, 1965; *Gill*, 1981]. Under conditions where the downgoing oceanic crust is "hot", such as during the subduction of very young oceanic crust or at the initiation of subduction, partial melting of the slab may occur in addition to dehydration [*Drummond and Defant*, 1990; *Defant and Drummond*, 1990; *Kay et al.*, 1993; *Peacock et al.*, 1994; *Molnar and England*, 1995]. These prograde mineral reactions not only have important geochemical consequences, but as subduction zones are one of the tectonic environments where strain rates are relatively high, mineral reactions may also influence the rheological behavior of the uppermost portion of the slab, including triggering intermediate-depth intraplate earthquakes [*Kirby et al.*, this volume].

In more typical subduction regimes, where the slab is older than 20 Ma, *Shimamoto et al.* [1993] have suggested that the relatively aseismic zone that is common in many subduction zones along the subducting plate boundary down to ~ 30 km is due to the release of large quantities of free H_2O during dehydration thus promoting ductile deformation by solution-transfer processes. *Shimamoto et al.* [1993] further suggest that the intermediate-depth earthquakes, which begin at greater than ~30 km, are caused by dehydration embrittlement processes because circulation of free H_2O is not likely at these greater depths and the ductile deformation promoted by the availability of free fluid is dampened. *Kirby et al.* [1991, 1994] and *Kirby et al.* [this volume] also propose that dehydration embrittlement may be responsible for intermediate-depth seismic events but include reactivating fossil faults by mineral reaction, combined with the stress caused by the total volume change during the basalt to eclogite transition.

During warm and hot subduction (where partial melting of the slab may be possible) seismic patterns are different than those observed in subduction zones involving old, cold lithosphere. Dominantly shallow seismic activity (<100 km) is observed, with earthquakes located just below the top surfaces of the slabs [*Kao and Chen*, 1991; *Kirby et al.*, 1991; *Kirby et al.*, this volume]. In addition to the shallow seismicity, some warm subduction zones are also characterized by aseismic behavior all along the slab-mantle interface (e.g. Southernmost Chile; or Cascadia, Pacific northwest) [*Acharya*, 1992]. Aseismic deformation of the warm slab may be due to massive dehydration keeping slab deformation ductile by solution-transfer processes but this will only be true at very shallow levels. What mechanisms could be active at deeper depths? When seismic activity does occur in the top portion of these warm slabs, is it due to an embrittlement process? Observations made in an experimental investigation focused on the rheological changes in basaltic amphibolite as it undergoes partial melting and to lesser extent, dehydration/hydration reactions, have been used to explore potential deformation mechanisms active during mineral reactions [*Rushmer*, 1995]. Some of these results, which are presented in a condensed form below, provide potential answers to the questions regarding seismicity and deformation in the upper portion of warm slabs. By combining the PTt paths calculated in *Peacock et al.* [1994] for warm downgoing oceanic crust (0-20 Ma) with experimentally determined melting and dehydration reactions in basalt, it is possible to show which potential dehydration/hydration reactions can occur at a given pressure and temperature along the subducting plate interface. These reactions together with the experimental deformation results show how deformation in the slab may be accommodated in some known warm, subduction zone regimes.

EXPERIMENTAL DEFORMATION INVESTIGATION

An experimental deformation study of a natural basaltic amphibolite has been carried out to explore the effect of mineral reactions on active deformation mechanisms. An in-depth discussion of the deformation experimental procedure and results is presented in *Rushmer* [1995]. Therefore, the deformation behavior of the amphibolite at the lowest temperature investigated, where no reaction occurs, and two experiments (N 768 and N 711) performed at higher temperatures will be the only experiments discussed here. The amphibolite is a metamorphosed alkali basalt and is composed of mainly hornblende and plagioclase with minor quartz. The grain size of the samples range between 200 and 250 µm. The experiments were all performed in a solid-media Griggs' apparatus at 1.8 GPa confining pressure (equal to approximately 60 km depth) with deformation strain rates equal to 10^{-5} seconds^{-1}. There is no added free water in the experiments, so the breakdown of hornblende in the presence of plagioclase +/- quartz is the solidus of the sample (fluid-absent or dehydration melting). Experiments were performed between 650°C and 1000°C so the amphibolite could be deformed at conditions much below and above the fluid-absent solidus (between 800°C and 850°C).

Summary of results

At 650°C (N 725) the amphibolite did not undergo any mineral reactions. Under these conditions, the deformation observed on the macroscopic scale is ductile. The sample deforms mainly by homogeneous flattening of plagioclase and quartz and both phases show strong undulatory extinction. Slip along cleavage planes in hornblende and plagioclase grains is also observed, but no localized fracturing is present. In the experiments which do not undergo mineral reactions, hornblende is always more brittle than the plagioclase and often grains have small cracks oriented near-parallel to the compression direction. Grain boundaries are not broken, and no through-going fractures or cracks are present.

In experiment N 768 at 850°C, the beginning of the hornblende-breakdown reaction is observed. Maximum yield strength is 1/4 of the subsolidus experiment (N 725) described above and most of the deformation in this sample is taken up in a single ductile shear zone (Figure 1). No melt is observed in the shear zone itself, but is present in dilatant cracks in the hornblende (<5 vol %) which are oriented near-parallel to the shortening direction. Within the shear zone, very fine grained (0.1-10.0 µm) clinozoisite + albite aggregates are found in a matrix of anorthitic

Fig. 1. SEM backscatter photomicrograph of experiment N 768 at 850°C. A single ductile shear zone dominates the deformation in the entire sample. Extensive reaction of anorthitic plagioclase (plag) to albite + zoisite is observed within the shear zone (see Figure 2) and deformation is likely accommodated by a combination of grain-boundary sliding and mass transfer. Hornblende (hbd) contains melt (g) in dilatant fractures parallel to compression direction (oriented top to bottom in this photomicrograph). Reflective phases are titanite.

plagioclase (Figure 2). Neither hornblende nor melt appear to be involved in this reaction. The zoisite and albite phases are attributed to the reaction of anorthite and H_2O, where the H_2O has been produced during the dehydration of altered grain-boundary phases or of sericite present in some plagioclase grains. A reaction-enhanced deformation process (*Rubie*, 1983) appears to have caused a pronounced weakening of this sample, not the presence of melt.

At a higher temperature (N 711, at 935°C), melt fractions of ~10-15 vol% are achieved and melt-enhanced embrittlement has localized deformation. A conjugate set of broad shear zones (300-500 μm in width) oriented at an approximate angle of 45° to the compression direction is the main deformation observed. The increased melt fraction is likely due to the more extensive breakdown of hornblende + plagioclase at this temperature. Abundant evidence for reaction is found in the shear zones and hornblende is strongly, cataclastically deformed which may have in turn promoted more reaction due to the decrease in grain size and the increase in surface area. Overall, the deformation observed within the shear zones is considered to be mainly brittle. Outside of the sheared areas, garnet ($\leq$ 5 vol %), various quantities of melt (5-12 vol %), accessory titanite (3 vol %), zoisite (~1 vol %), and hornblende altered to clinopyroxene (50-54 vol %) are observed. Melt is usually found at grain boundaries between plagioclase and hornblende (Figure 3).

The results from the higher temperature experiments, N 768 at 850°C and N 711 at 935°C, emphasize the importance of mineral reactions on deformation style. Below the solidus (at 650°C) no reaction is observed and the samples deform macroscopically ductiley, mainly by the flattening and flow of quartz and plagioclase grains perpendicular to the compression direction. As mineral reactions occur however, deformation becomes localized. In experiment N 768, deformation is focused along a shear zone partly composed of very-fine grained reaction products. The very-fine grained aggregates deform ductiley, most likely by both grain-boundary sliding and mass transfer processes [*Rushmer & Stünitz*, 1993]. The combination of the dehydration of partly altered plagioclase (by either grain-boundary phases or sericite) and the subsequent hydration reaction which forms the zoisite + albite aggregates does not produce large quantities of free fluid. There is some melt present which locally fractures the hornblende, but overall deformation is not by embrittlement, but by a combination of ductile processes. Embrittlement is not observed until melt fractions are higher, as in experiment N 711. Here the sample exhibits localized brittle deformation and cataclastic textures are

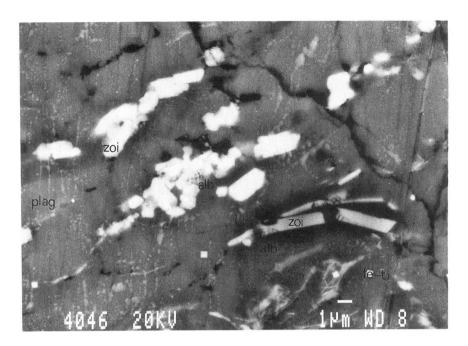

Fig. 2. SEM backscatter photomicrograph showing detail of very fine-grained zoisite (zoi) and albite (alb) forming from the reaction anorthitic plagioclase (plag) + H_2O. Free H_2O may have been released during the experiment by the breakdown of altered phases on grain boundaries. Fe-Ti (fe-ti) oxides are also observed in the shear zone.

developed in the shear zones. The development of melt overpressure which fractures the rock is termed melt-enhanced embrittlement and in these experiments, at melt fractions between 5-15 vol.%, it appears to be the main mode of deformation. This type of embrittlement, induced by increasing pore fluid pressure, can also occur during dehydration reactions which produce large quantities of free fluid as shown in earlier experiments [*Raleigh and Paterson*, 1965].

DISCUSSION

Strain rate, pressure and temperature, type of mineral reaction (solid-solid, dehydration/hydration, partial melting) and kinetics are clearly major factors in determining how mineral reactions will influence deformation. Application of these experimental deformation results to a specific deformation style in warm, subducting oceanic lithosphere requires a knowledge of strain rates at the subducting slab interface, the pressure and temperature conditions along the upper portion of the slab and the approximate stability range of hydrous phases present in the basaltic crust. Figure 4 is a compilation P-T diagram which shows the current experimentally-determined stability of several important hydrous phases such as lawsonite and chloritoid, in addition to amphibole. It also shows the PTt paths followed by subducting oceanic crust between the ages of 0 and 20 Ma converging at a relatively slow rate of 3 cm/yr from *Peacock et al.* [1994]. Figure 4 allows the location of several specific hydrous mineral reactions to be determined in P-T space for a given slab age. It is difficult to establish strain rates along the down-going slab interface. Shear stresses are certainly among the highest of Earth's active tectonic environments and have led to the proposal of frictional heating along the slab interface (e.g. *Molnar and England*, 1990). The natural strain rates are certainly slower than the experimental strain rate of 10^{-5} second^{-1}, however the relatively high natural strain rates may allow deformation processes observed experimentally to readily occur in this environment.

Potential Hydrous Mineral Reactions In Young Oceanic Lithosphere

Pressure-temperature-time (PTt) paths calculated for the top of young downgoing basaltic oceanic crust (5 - 20 Ma) will intersect the stability curves of several high-pressure hydrous phases [*Peacock et al.*, 1994, Figure 4]. Hornblende, zoisite, Mg-chloritoid and lawsonite are all potential hydrous phases found in basaltic compositions which may survive at depth during subduction [*Poli and Schmid*, 1995]. Currently the stability of high-pressure phases in downgoing oceanic crust is under debate. Recent experimental work on basalt compositions by *Tatsumi*

Fig. 3. At 935°C, melt fraction increases to 10-15 vol%. SEM backscatter photomicrograph shows how hornblende grains (hbd + cpx) become cracked and deformed by cataclasis. Shear zones have formed at approximately 45° angle to the compression direction, which is top to bottom in the photograph. The bottom right-hand corner is the top portion of one shear zone and garnet (gar, the more reflective phases), melt (g) and zoisite (very fine-grained reflective phases) are found in more abundance and hornblende grains-size is reduced in this area. Deformation is dominantly brittle in this sample.

[1989]; *Pawley and Holloway* [1993]; *Poli* [1993]; *Poli and Schmid* [1995]; *Schmid and Poli* [CASH-system, 1994]; *Pawley* [CASH-system, 1994] and *Thompson and Ellis* [1994] have suggested that hydrous phases such as lawsonite and zoisite may be stable to pressures greater than 7.0 GPa. However, *Bohlen et al.*, [1994] and *Liu et al.*, [1995] suggest that these phases may be only metastable, as the equilibrium pressure-temperature conditions needed to experimentally duplicate conditions in the downgoing slab are difficult to achieve. It is considered here, as shown in Figure 4, that there are potential high-pressure hydrous phases stable in addition to amphibole.

Figure 4 also illustrates how the age of the slab will determine at which depth a specific reaction will occur in the uppermost portion as it descends into the mantle. The PTt paths (A and B) of two warm aseismic subduction zones are given here as examples to illustrate the potential dehydration/hydration reactions that may occur in the top portion of the slab and their possible role in suppressing seismicity. Data from *Kao and Chen* [1991] and *Acharya* [1992] show that the subduction zone in Southern Chile which involves the Antarctic-Scotia plate at Tierra del Fuego is relatively aseismic. The age of the oceanic lithosphere is 10-14 Ma and it converges at a slow rate of ~2.0 cm/yr. The Cascadia subduction zone (Juan de Fuca plate) is 3-10 Ma and is also converging at a slow rate of ~ 2.0 cm/yr. This subduction zone is currently considered relatively aseismic with, possibly, a very narrow portion of the downdip subduction thrust fault locked [*Hyndman and Wang*, 1995]. Figure 4 shows that the surfaces of the two plates will approximately follow paths A (10 Ma) and B (5 Ma).

For Southernmost Chile, path A will parallel and then intersect (marked by filled circles), the lawsonite-in curve experimentally determined in a natural basalt from *Poli and Schmidt* [1995] at 1.8 GPa, 475°C. Then the path will cross the zoisite-out curve at 2.3 GPa, 550°C [*Green*, 1982] and the lawsonite-out curve at 3.0 GPa and 700°C [*Poli and Schmidt*, 1995]. The basalt wet solidus will be intersected at ~3.3 GPa. The depth of reaction will also depend, of course, on variations in the starting basalt chemistry but the overall path shows that between ~1.8 GPa and 3.0 GPa several dehydration/hydration reactions can take place, but no partial melting of the uppermost portion of the slab will occur until ~$\geq$3.3 GPa (approximately at 100 km).

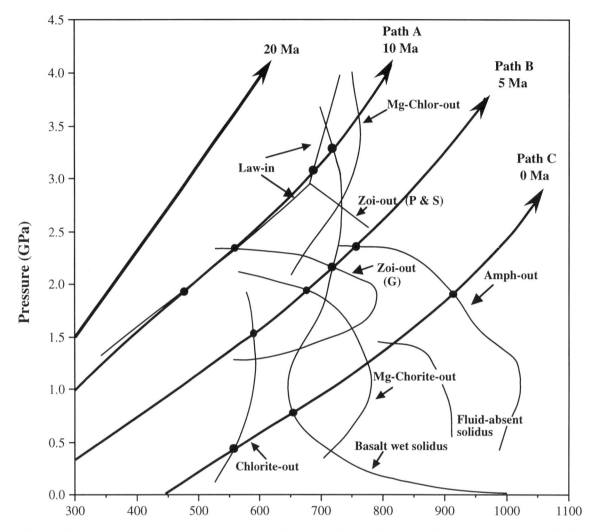

Fig. 4. This pressure-temperature phase diagram which shows the stability curves determined experimentally for several high-pressure hydrous phases in basaltic compositions. The basalt wet solidus and hornblende fluid-absent melting curve are also given. Paths A-C are 10 Ma, 5 Ma and 0 Ma old subducting oceanic lithosphere, respectively. 20 Ma is also marked on the diagram (from *Peacock et al., 1994*). Filled circles show where the different ages in subducting oceanic crust will intersect potential dehydration, hydration and partial melting reactions (see text). References for the experimental data: Amph-out, amphibolite-out [*Rushmer*, 1993]; Zoi-out (G), zoisite-out [*Green*, 1982]; Zoi-out (P & S), zoisite-out [*Poli and Schmidt*, 1995]; Law-in, lawsonite-in [*Poli and Schmidt*, 1995] Mg-Chlor-out, Mg-chloritoid-out [*Schreyer*, 1988]; Mg-Chlorite-out [*Tatsumi*, 1989].

The Cascadia subduction zone will follow a slighter hotter path (B, 5 Ma) and will first intersect (filled circles) chlorite-out [*Poli*, 1993] at ~1.2 GPa, 575°C and zoisite will be a stable hydrous phase according to the data from *Green* [1982]. At 1.8 GPa and 675°C, the Mg-chlorite-out [*Tatsumi*, 1989] curve will be crossed, then the zoisite-out reaction [*Green*, 1982] at ~2.3 GPa, 700°C. The basalt wet solidus is reached at approximately the same pressure and temperature for this path. This path will also intersect the amphibole-out [*Rushmer*, 1993] curve at ~2.3-2.5 GPa, 750°C. Hornblende most likely composes a large portion of the basalt so this reaction (even above the basalt wet solidus) will produce a pulse of additional fluid (melt). For this path, most reactions will occur between ~1.0 and 2.5 GPa (approximately 75 km deep).

As the slabs descend and react, deformation may be initially accommodated by a reaction-enhanced ductility mechanism because both dehydration and hydration reactions can occur in this pressure-temperature regime. As H_2O is released, some may be used to crystallize other hydrous phases still stable at pressures between 0.5-3.0 GPa. The formation of the very-fined grained zoisite + albite aggregates observed in the experimental study show that these reaction products can be very effective in focusing deformation, ductilely. Reaction-enhanced ductility has been described by *Rubie* [1983] who has shown that localization of deformation can be caused by prograde mineral reactions which produce fine-grained reaction products. He suggests that it is the presence of these new phases which significantly effects the mechanical properties of the rocks by enhancing their ductility and ability to deform. In an experimental deformation study, *Rutter and Brodie* [1987] show that the slow dehydration under controlled pore-pressure of serpentine forms very-fined grained olivine + talc + H_2O. Shear zones observed in the samples are lined with the fine-grained olivine which, in turn, deform by a diffusion-accommodated grain-boundary-sliding mechanism. The authors suggest that the this reaction of serpentine may suppress seismicity below the 400°C isotherm in oceanic transform faults.

The formation of micron-sized reaction products during subduction of young oceanic lithosphere may also suppress seismicity in some of these warm subduction zones. The example of the Southernmost Chile and Cascadia shows that below 100 km (~3.3 GPa) and 75 km (~2.5 GPa) respectively, several dehydration/hydration reactions are encountered during subduction. If new, hydrous reaction products such as zoisite and lawsonite are formed subsequent to dehydration, then deformation in the slab to intermediate depths may be accommodated by reaction-enhanced ductility processes rather than the free H_2O inducing embrittlement. Massive rapid dehydration which releases large quantities of water will, more than likely, overwhelm the ductility processes and induce embrittlement and may produce seismicity [*Kirby et al.*, this volume]. However, steady progressive slow dehydration accompanied by rehydration and ductile deformation may be a major mechanism active in some of these warm aseismic subduction zones.

Partial Melting in Young Subducting Oceanic Lithosphere

Previous work has shown that partial melting of downgoing oceanic lithosphere is possible only when the slab is very young, ≤ 5 Ma [*Peacock et al.*, 1994, *Molnar and England*, 1995]. Path C on Figure 4 is an example of a "hot" path, equivalent to ridge subduction. During ridge subduction, the oceanic crust is melting and the age of the lithosphere is 0 Ma. For this path, chlorite-out is crossed at very shallow levels, 0.3 GPa and 550°C [*Poli*, 1993], then the basalt wet solidus [*Green*, 1982] is reached at 0.5 GPa, at 750°C. Again the amphibolite-out reaction is encountered at higher temperatures, which will produce a pulse of fluid.

Ridge subduction is observed in Southern Chile, where the Chilean Ridge is currently thrusting under the South American plate. Melting of the oceanic crust is likely occurring in this environment and syntectonic deformation may induce fracturing along upper portion of the hot slab. Whether this will induce seismic activity is not clear, as the presence of melt may attenuate seismicity. However, microseismicity data with earthquake magnitudes ranging between 0 and 4 have been reported at this site by *Muride et al.* [1993]. Their suggestion is that the shallow seismicity is the response of continued oceanic spreading of the subducted ridge-transform system. Syntectonic melting in the upper portion of the slab may also contribute to shallow brittle deformation of the downgoing slab by a melt-enhanced embrittlement process similar to that observed in the experimental study at melt fractions between 5 and 15 vol.%.

CONCLUSIONS

The thermal structure in warm subduction zones (involving oceanic crust less than 20 ma) will cause the upper part of the down-going slab to intersect many of the dehydration/hydration reactions including partial melting of basalt between 0.5 and 3.0 GPa. Interaction of these reactions with on-going deformation may ultimately trigger some of the shallow seismicity observed. However, the focusing of deformation into ductile shear zones by dehydration followed by hydration reactions may allow much of the deformation to be accommodated by ductile shearing, without inducing fracture. A reaction-enhanced ductility process may be an important deformation mechanism operative in warm downgoing oceanic lithosphere. This would accommodate stress in the slab, but by ductile shearing and therefore aseismically.

Acknowledgments. This work was supported by Schweizerischer Nationalfonds project 2-77-590-92 to K. Hsü and R. Schmid/ R. Schmid and Jean-Pierre Burg. I wish to thank Steve Kirby, Simon Peacock for many helpful discussions and the SUBCON committee who generously provided the support to attend the SUBCON conference in June, 1994.

REFERENCES

Acharya, H., Comparison of seismicity parameters in different subduction zones and its implications for the Cascadia subduction zone, *J. Geophys. Res. 97*, 8831-8842, 1992.

Drummond, M. J. and M. S. Defant, A model for trondhjemite-tonalite-dacite genesis and crustal growth via slab melting: Archean to modern comparisons, *J. Geophys. Res. 95*, 21503-21521, 1990.

Defant, M. S., and M. J. Drummond, Subducted lithosphere-derived andesitic and dacitic rocks in young volcanic arc setting, *Nature, 347*, 662-5, 1990.

Bohlen, S. R., J. Liu, W. G. Ernst and J. G. Liou, Stability of hydrous phases and source of water in a deep subduction zone, *EOS Transactions, Amer. Geophys. Union, 75/44*, 748, 1994.

Gill J., *Orogenic Andesites and Plate Tectonics*, 390 pp., ed. Springer-Verlag, New York, 1981.

Green, T. H., Anatexis of mafic crust and high pressure crystallization of andesite, in *Andesites--Orogenic Andesites and Related Rocks*, edited by R. S. Thorpe, pp. 465-487, John Wiley, New York, 1982.

Hyndman, R. D. and K. Wang, The rupture zone of Cascadia great earthquakes from current deformation and thermal regime, *J. Geophys. Res. 100*, 22,133-22,154, 1995.

Kao, H. and W.-P. Chen, Earthquakes along the Ryukyu-Kyushu arc: strain segmentation, lateral compression, and the thermomechanical state of the plate interface, *J. Geophys. Res. 96*, 21443-21485, 1991.

Kirby, S. H., E. R. Engdahl, B. Hacker, R. Denlinger and S. Bohlen, Metastable phase transformations in subducting oceanic crust and their possible roles in the physics of intermediate-depth earthquakes, in *SUBCON abstract volume*, 236-239, 1994.

Kirby, S. H., and W. B. Durham, and L. A. Stern, Mantle phase changes and deep-focus earthquake faulting in subducting lithosphere, *Science, 252*, 216-225, 1991.

Kirby, S. H., E. R Engdahl and R. Denlinger, Intraslab earthquakes and arc volcanism: Dual physical expressions of crustal and uppermost mantle metamorphism in subducting slabs, in *SUBCON: An interdisciplinary conference on the subduction process, Santa Catalina Island, Geophys. Monogr. Ser.*, edited by G. Bebout, D. Scholl and S. H. Kirby, this volume, AGU, Washington, D. C., in press.

Liu, J., W. G. Ernst, J. G. Liou and S. R. Bohlen, Experimental constraint on stability of hydrous phases in subduction oceanic crust, *EOS Transactions, Amer. Geophys. Union, 76/17*, S298, 1995.

Mahlburg Kay, S., V. A. Ramos, and M. Marquez, Evidence in Cerro Pampa volcanic rocks for slab-melting prior to ridge-trench collision in southern South America, *J. Geology*, in press.

Molnar, P. and P. England, Temperatures, heat flux, and frictional stress near major thrust faults, *J. Geophys. Res. 95*, 4833-4856, 1990.

Molnar, P. and P. England, Temperatures in zones of steady-state underthrusting of young oceanic lithosphere, *Earth and Planet. Sci. Lett., 131*, 57-70, 1995.

Murdie, R. E., D. J. Prior, P. Styles, S. S. Flint, R. G. Pearce and S. M. Agar, Seismic responses to ridge-transform subduction: Chile triple junction, *Geology, 21/12*, 1095-1098, 1993.

Pawley, A. R., The pressure and temperature stability limits of lawsonite: implications for H_2O recycling in subduction zones, *Contrib. Mineral. Petrol., 118*, 99-108, 1994.

Pawley, A. R., and Holloway, J. R., Water sources for subduction zone volcanism: new experimental constraints, *Science, 248*, 329-337, 1993.

Peacock, S. M., Metamorphism, dehydration, and the importance of the blueschist -> eclogite transition in subducting oceanic crust, *Geol. Soc. Am. Bull. 105*, 684-694, 1993.

Peacock, S. M., T. Rushmer, and A. B. Thompson, Partial melting of subducting oceanic crust. *Earth and Planet. Sci. Lett., 121*, 227-243, 1994.

Poli, S., Amphibole behavior through amphibolite-eclogite transition: An experimental study on phase relationships in basalts, *Amer. Jour. Sci., 293*, 1061-1077, 1993.

Poli, S. and M. W. Schmidt, H_2O transport and release in subduction zones: Experimental constraints on basaltic and andesitic systems, *J. Geophys. Res. 100*, 22,299-22,214, 1995.

Raleigh, C. B. and M. S. Paterson, Experimental deformation of serpentinite and its tectonic implications, *J. Geophys. Res., 70*, 3965-3985, 1965.

Rubie, D.C., Reaction-enhance ductility: the role of solid-solid univariant reactions in deformation of the crust and mantle, *Tectonophysics 96*, 331-352, 1983.

Rushmer, T., Experimental high pressure granulites: Some implications to natural mafic xenolith suites and Archean granulite terranes, *Geology, 21/5*, 411-414, 1993.

Rushmer, T., An experimental deformation study of partially molten amphibolite: Application to low-melt fraction segregation, *J. Geophys. Res., 100*, 15,681-15,696, 1995.

Rushmer, T., and H. Stünitz, Experimental deformation of amphibolite at high temperature and pressure: melt or reaction weakening? (abs) *Swiss Tectonic Studies Group Annual Meeting*, Zürich, February, 1993.

Rutter, E. H. and K. H. Brodie, On the mechanical properties of oceanic transform faults, *Annales Tectonicae, 1/2*, 87-96, 1987.

Schmidt, M. W., and S. Poli, The stability of lawsonite and zoisite at high pressures: experiments in CASH to 92 kbar and implications for the presence of hydrous phases in subducted lithosphere, *Earth and Planet. Sci. Lett., 124*, 105-118, 1994.

Schreyer, W., Experimental studies on metamorphism of crustal rocks under mantle pressures, *Mineral. Mag.*, 52, 1-26, 1988.

Shimamoto, T., T. Seno, and S. Uyeda, A simple rheological framework for comparative subductology, *Relating Geophysical Structures and Processes: The Jefferies Volume, Geophysical Monograph 76, IUGG Volume 16*, 39-51, 1993.

Tatsumi, Y., Migration of fluid phases and genesis of basalt magmas in subduction zones, *J. Geophys. Res., 94*, 4697-4707, 1989.

Thompson, A B. and D. Ellis, $CaO + MgO + Al_2O_3 + SiO_2 + H_2O$ to 35 kbar: Amphibole, talc and zoisite dehydration and melting reactions in the silica-excess part of the system and their possible significance in subduction zones, amphibolite melting and magma fractionation, *Amer. Jour. Sci., 294*, 1229-1289, 1994.

T. Rushmer, Department of Geology, Perkins Hall, University of Vermont, Burlington, Vermont, 05405-0122.

Contrasting Styles of Mantle Metasomatism Above Subduction Zones: Constraints From Ultramafic Xenoliths in Kamchatka

Pavel Kepezhinskas and Marc J. Defant

Department of Geology, University of South Florida, Tampa, Florida

Whole-rock and mineral-trace element data from a suite of ultramafic xenoliths collected from several young volcanoes along the Kamchatka arc suggest at least two stages of mantle-wedge metasomatism. Fluid-induced cryptic metasomatism (stage I) was caused by shallow-level slab devolatilization which introduced several fluid-dependent trace elements (light rare earth elements - LREE, Ba, U, platinum group elements - PGEs) into the sub-arc mantle wedge without changing the composition of the mineral phases. This metasomatism is mainly governed by metamorphic reactions and trace element partitioning in the downgoing slab. Melt-induced modal metasomatism (stage II) resulted in the formation of new, trace element enriched mineral assemblages and glasses under mantle wedge conditions. These metasomatic changes are strongly dependent on the age of the subducting slab and are caused by hydrous siliceous melts derived from young and hot slab or carbonated, alkaline melts derived by partial melting of old slab (carbonated basalt and/or carbonate sediment). The ultimate result of multi-stage slab-induced metasomatism is the creation of a hybridized veined mantle wedge capable of generating a variety of arc magmas.

1. INTRODUCTION

It is a generally accepted paradigm in subduction-zone geology that supercritical aqueous fluids expelled from the downgoing slab transport selected incompatible elements into the overlying mantle wedge [*Gill*, 1981; *Morris et al.*, 1990]. Subsequent metasomatism of the island-arc mantle by slab-derived fluids creates a mantle source capable of generating primitive arc magmas [*Maury et al.*, 1992; *Hawkesworth et al.*, 1993; *Bebout et al.*, 1993]. However, because mantle xenoliths are extremely rare in arcs, modelling of the composition of the arc mantle wedge and source processes have relied primarily on inferences made from arc lavas. We have collected mantle wedge-derived ultramafic xenoliths from several localities in the Kamchatka volcanic arc (Fig. 1) that give us the opportunity to test various island-arc models.

2. ISLAND ARC MANTLE PRIOR TO METASOMATISM

Pre-metasomatic clinopyroxenes occur as primary-textured porphyroclasts associated with deformed olivine and orthopyroxene. These clinopyroxenes have low Ti, Na, Al, and high Mg and Cr concentrations coupled with rather uniform LREE depletions (Table 1). Pre-metasomatic clinopyroxenes display negative high-field strength element (HFSE - e.g., Ti, Zr) anomalies on chondrite-normalized graphs similar to clinopyroxenes from depleted (MORB-type) residual mantle (Fig. 2). Primary clinopyroxene geochemistry suggests that the mantle wedge is depleted with respect to HFSE and LREE throughout the 1000-km length of the Kamchatka arc and, therefore, any enrichment signatures in Kamchatka xenoliths can be attributed to metasomatism.

3. FLUID-INDUCED (CRYPTIC) SUB-ARC METASOMATISM - STAGE I

Modal mineralogy (very low modal clinopyroxene contents - all peridotite nodules are harzburgites or dunites)

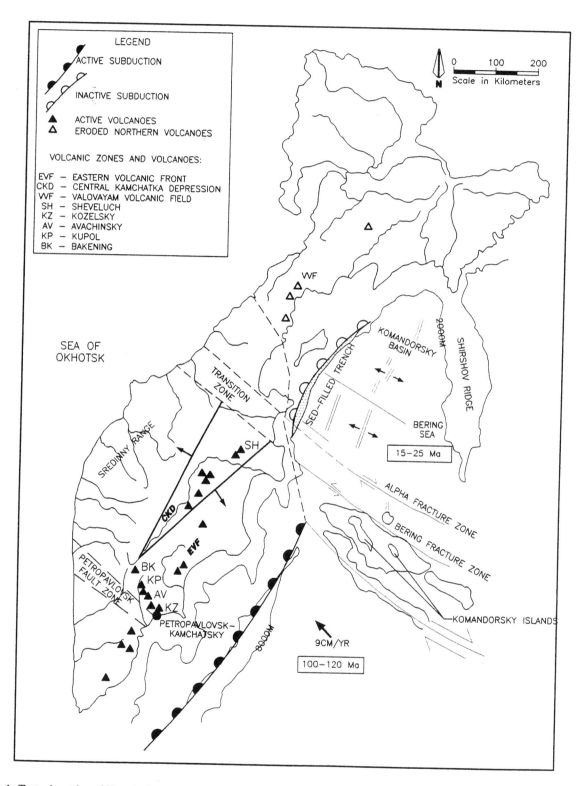

Fig. 1. Tectonic setting of Kamchatka arc along with mantle xenolith locations, position of active and fossil subduction zones and trenches. Ages for the Komandorsky Basin crust and Pacific lithosphere are adopted from Baranov et al. [1991]. Boundary between northern and southern segments is manifested by major transcurrent faults and crustal discontinuities and is adopted from Hochstaedter et al. [1994].

Table 1. Trace-element concentrations (ppm) in clinopyroxenes and glasses from selected Kamchatka xenoliths

Segment	**Northern**							
Xen. suite	*Valovayam*							
Sample	Val 58/2	Val 32/8	Val 58/1	Val 55/7	Val 55/7	Val 55/12	Val 55/12	Val 55/4
Phase	Prim. Cpx	Prim. Cpx	Prim. Cpx	Met. Cpx	Met. Cpx	Met. Cpx	Met. Cpx	Glass
Ti	1761.8	2878.7	1580.2	5636.1	10037.4	7045.5	13686.3	1855.3
V	259.4	327.8	207.9	258.3	389.9	355.7	413.4	8.9
Cr	1012.8	1089.1	1403.3	2203.9	2657.9	875.3	258.7	348.6
Sr	38	40.1	35	99.5	86	65.2	122.1	1736.1
Y	6.2	7.5	5.2	18.7	25	23.9	29.4	6.1
Zr	10.2	15.2	9.9	90.2	151.4	128.4	84.4	30.4
Nb	BDL	BDL	BDL	2.2	2.2	1.1	2.2	13.2
La	0.41	0.52	0.61	7.94	12.68	6.78	3.78	66.58
Ce	2.36	2.91	2.22	19.99	32.95	20.81	13.93	114.2
Nd	2.68	3.72	2.69	15.14	19.48	14.07	12.21	37.61
Sm	1.6	2.18	1.48	5.29	6.52	4.23	5.21	8.29
Eu	0.43	0.63	0.47	1.71	1.84	1.61	1.67	2.67
Dy	1.29	2.03	1.52	4.36	5.27	4.61	5.31	4.99
Er	0.88	1.08	0.82	2.26	3.15	2.48	2.51	3.17
Yb	0.91	0.87	0.61	2.09	2.58	1.87	2.21	2.92

Segment	**Southern**							
Xen. suite	*Bakening*							
Sample	Bak 48-50	Bak 48-74	Bak 48-74	Bak 48-109	Bak 48-60	Bak 48-60	Bak 48-109	Bak 48-109
Phase	Prim. Cpx	Prim. Cpx	Prim. Cpx	Met. Cpx	Met. Cpx	Met. Cpx	Glass	Glass
Ti	2652.6	2176.7	1821.1	4408.7	1218.9	1014.6	516.8	683.6
V	332.5	310.6	293.2	404.8	210.1	203.1	18.3	148.6
Cr	2705.7	2689.9	3070.6	367.6	3288.8	2964.2	113.2	50.9
Sr	46.6	29.4	29.6	89.5	63	67.9	1171.3	1040.6
Y	11.8	9.4	8	16.5	6.8	6.7	17	18
Zr	16	11.9	9.8	44.2	45.3	45.6	233	270.1
Nb	BDL	BDL	BDL	1.1	BDL	1.1	23.1	35.1
La	0.46	0.81	0.29	3.87	3.32	6.06	98.88	125.25
Ce	1.79	2.66	1.42	13.37	6.39	12.24	198.32	253.18
Nd	2.41	3.2	2.1	11.14	4.73	6.12	95.87	107.75
Sm	0.96	1.49	0.96	3.52	1.49	1.56	25.42	21.61
Eu	0.37	0.52	0.32	1.51	0.34	0.58	7.23	5.33
Dy	1.43	1.96	1.81	4.05	1.87	1.57	14.32	12.36
Er	0.75	1.15	1.29	2.17	1.31	1.21	5.91	5.18
Yb	0.69	0.89	1.31	1.98	0.66	0.81	4.94	4.21

Note. Prim. cpx - primary porphyroclastic clinopyroxene, met. cpx - metasomatic disseminated clinopyroxene. Trace elements in mineral phases and glass from the ultramafic xenoliths were analyzed using a Cameca IMS 3f ion microprobe at Woods Hole Oceanographic Institution. The accuracy and precision of the results are within 10% for Ti, V, Cr, Sr, Y, Zr, Dy, Er and Yb, 20% for Ce, Nd, Sm and Eu, and 40% for La. BDL - below detection limit.

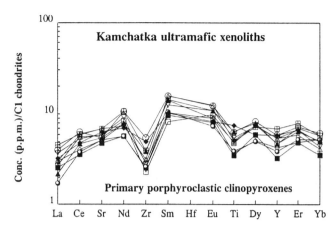

Fig. 2. Chondrite-normalized trace element patterns for primary (pre-metasomatic) porphyroclastic clinopyroxenes from Kamchatka ultramafic xenoliths. Chondrite normalizing values are from Anders and Grevesse [1989].

and refractory mineral chemistry (high Cr/(Cr + Al) ratios of spinels, high Mg# of olivines and orthopyroxenes) of Kamchatka peridotite xenoliths suggest their overall depletion in mafic components and that they represent mantle residual after melt extraction. However, the bulk-rock geochemistry of these xenoliths indicates the presence of an enriched component represented by relatively high bulk LREE/HREE ratios and positive Ba and U anomalies on chondrite-normalized graphs (Fig. 3). This cryptic enriched component is further manifested by high bulk-xenolith contents of Pt, Pd, and Rh (10-35 ppm) suggesting that considerable bulk addition of these fluid-mobile platinum group elements took place prior to any modal metasomatic event in the Kamchatka mantle [*Kepezhinskas and Defant*, 1994]. Ba, U, LREE, and platinum group elements (especially Rh, Pd, and Pt) are the elements commonly transported by water [*Kay*, 1980; *Gill*, 1981; *Crocket*, 1981; *Morris et al.*, 1990]. Consequently, this cryptic enrichment of modally depleted peridotite xenoliths in fluid-controlled trace elements is attributed to slab devolatilization below the Kamchatka arc. This is further supported by high boron concentrations of 0.16 to 0.61 ppm in Kamchatka ultramafic xenoliths reported by Austin [1995]. These xenoliths also display elevated B/Be ratios of 16 which suggest that these nodules underwent a recent subduction-related boron enrichment. Ryan et al. [1996] estimated boron contents in depleted and primitive mantle reservoirs to be 0.08 ppm and 0.15 ppm respectively. Elevated boron concentrations in Kamchatka xenoliths indicate that the metasomatic events in sub-arc mantle are likely related to the recent subduction processes.

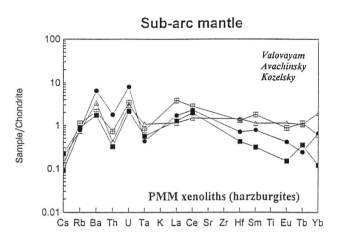

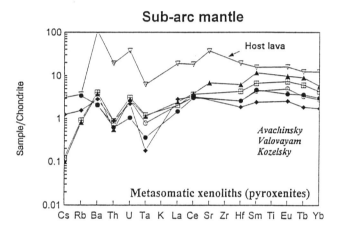

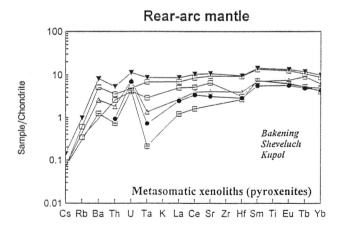

Fig. 3. Chondrite-normalized whole-rock rare earth patterns for primitive and metasomatic xenoliths from Kamchatka sub-arc and rear-arc mantle. Chondrite normalizing values are from Sun and McDonough [1989].

4. MELT-INDUCED (MODAL) SUB-ARC METASOMATISM - STAGE II

4.1. Mantle Metasomatism Associated with the Young Slab

The northern segment of the Kamchatka arc was produced in response to subduction of the young Komandorsky Basin lithosphere [*Baranov et al.*, 1991; *Hochstaedter et al.*, 1994] (Fig. 1). Ultramafic xenoliths from this segment display abundant textural and chemical signatures of melt-induced mantle metasomatism [*Kepezhinskas et al.*, 1995]. Felsic glass veins and associated metasomatic clinopyroxenes in these xenoliths exhibit high Na, Si, Al, Sr, and low Y and Yb concentrations coupled with high La/Yb and Zr/Sm ratios (Table 1). Felsic veins are compositionally similar to siliceous melts derived by melting of young subducted lithosphere, e.g. adakites [*Defant and Drummond*, 1990].

4.2. Mantle Metasomatism Associated with the Old Slab

The southern segment of the Kamchatka arc is associated with the subduction of old Pacific lithosphere (Fig. 1). Avachinsky harzburgites commonly host amphibole pyroxenite veins which, based on their textures, precipitated by reaction with mafic melts. Metasomatic clinopyroxenes exhibit LREE and HFSE depletions similar to primary clinopyroxenes.

Bakening Volcano xenoliths contain melt pockets with apatite, amphibole, and phlogopite. Clinopyroxenes associated with these glass pockets exhibit enrichments in Cr, Mg, Al, Ti, and Na and diverge from the typical metasomatic trends caused by silicate melts (Table 1). Ion-probe analyses of glasses reveal high Sr, Nb, Zr and REE concentrations (Table 1) coupled with high La/Yb, Zr/Hf, Sr/Sm, and Nb/La and low Ti/Eu and Y/Er ratios (Fig. 4) similar to carbonate-rich melts [*Rudnick et al.*, 1993]. These glasses either directly represent carbonate-rich melt introduced into the lithospheric mantle or have originated through decomposition of primary mantle carbonates during xenolith transport. Elevated Na contents of clinopyroxenes and amphiboles suggest that the metasomatizing melt was similar to a sodic carbonate-rich liquid rather than to typical low-Na calcitic or dolomitic carbonatite.

5. MANTLE WEDGE METASOMATISM: A SUBDUCTING SLAB CONNECTION

Two stages of metasomatism are recognized in the mantle wedge below Kamchatka. Stage I is probably caused by

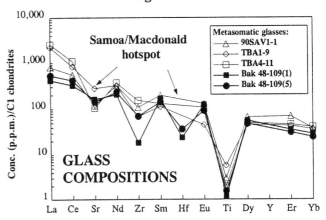

Fig. 4. Chondrite-normalized trace-element patterns for southern Kamchatka xenolith glasses (open symbols) compared to metasomatic glasses from the Samoa-Macdonald hot spot xenoliths (filled symbols). The latter are believed to represent carbonatite metasomatism in the oceanic upper mantle [*Hauri et al.*, 1993]. Chondrite-normalizing values are from Anders and Grevesse [1989].

H_2O-CO_2 fluids produced through slab devolatilization which resulted in pervasive cryptic metasomatism of the down-dragged mantle wedge peridotite along the slab-wedge interface. This stage I metasomatism is detected in all analyzed peridotite xenoliths independently of their location along the Kamchatka arc (Fig. 1) or geometry of subduction and is therefore independent from the subducted slab age. Since primary mineral phases present in peridotite xenoliths do not display any chemical enrichments, we believe that the slab-derived, fluid-induced component resides in the intergranular interfaces or fluid inclusions trapped in primary mantle minerals. The latter is confirmed by a direct laser ablation ICP-MS analysis of primary H_2O-CO_2 fluid inclusions trapped in deformed olivine porphyroclasts in Kamchatka peridotite xenoliths which revealed significant concentrations of La, Ce, Pr, Nd (light REE) and Sr and Rb (large-ion lithophiles) in a trapped fluid [*Clague et al.*, 1995]. These are the elements which are believed to be introduced into the mantle wedge by a subduction fluid component derived from the downgoing slab [*Gill*, 1981; *Hawkesworth et al.*, 1993; *Ryan et al.*, 1995].

Further enrichment of the sub-arc mantle below Kamchatka (stage II) occurred through melt-induced crystallization of new mineral phases (modal metasomatism) in the mantle wedge. The composition of the new mineral assemblages in the modally metasomatized sub-arc mantle along with associated felsic glasses (frozen

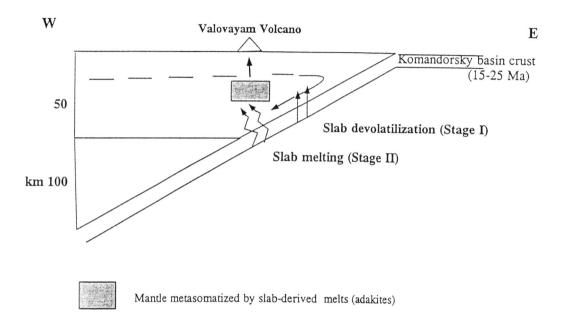

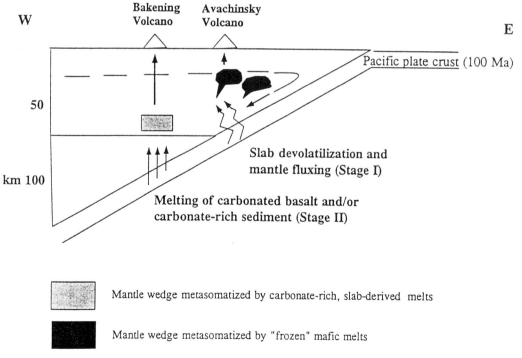

Fig. 5a and b. A schematic cartoon showing proposed models of contrasting mantle wedge metasomatism above the Kamchatka subduction zone as a function of the age of subducted slab. Ages for the Komandorsky Basin crust and Pacific plate crust are adopted from Baranov et al. [1991].

metasomatizing melts) is strongly governed by the age of the downgoing slab.

In arcs (or arc segments) related to young slab subduction (northern Kamchatka), slab melting at depths of 60-70 km will introduce REE and LILE-enriched, water-saturated adakite magmas causing pervasive modal Na-metasomatism in the overlying mantle wedge (Fig. 5A). Metasomatism appears to be shallow (35-45 km) within the plagioclase stability field in the sub-arc mantle and the presence of Na-plagioclase stable under these P-T-conditions will govern trace-element partitioning among metasomatic mineral phases (Fig. 5A).

Hydrous fluids expelled from the old slab below southern Kamchatka are capable of introducing LREE and LIL elements into the wedge. These fluids are also likely to cause local mantle fluxing and melting [*Arculus and Powell*, 1986] (Fig. 5B). Interaction between these melts and the mantle will potentially result in the creation of a variably veined mantle, like the one below Avachinsky volcano (Fig. 5B). Melting of carbonated basalt and/or carbonate-rich sediment at depths of approximately 100-120 km [*McInnes and Cameron*, 1994] will produce a carbonated siliceous melt (Fig. 5B). Since melting of the carbonated basalt occurs deeper than slab devolatilization, a portion of the mantle wedge metasomatized by carbonate-rich felsic melts will underlie the rear-arc (Bakening) rather than frontal-arc (Avachinsky) volcanoes (Fig. 5B). This is consistent with the observed location of mantle xenolith suites in the southern Kamchatka arc (Fig. 1).

6. CONCLUSIONS

1. Metasomatism in the mantle wedge above subduction zones is caused by slab-derived H_2O-CO_2-Cl fluids (first stage) and siliceous melts (second stage). Fluids mainly cause cryptic metasomatism and introduce large ion lithophile, rare earth, and volatile platinum group elements which are concentrated in intergranular films or fluid inclusions trapped in mantle minerals. Siliceous melts are responsible for modal metasomatism indicated by crystallization of new phases, e.g. clinopyroxenes and amphiboles.

2. Fluid-induced metasomatism is pervasive in the sub-arc mantle wedge and is not dependent on the slab age. Slab devolatilization mainly introduces fluid-mobile elements into the sub-arc mantle, e.g. Ba, Sr, Rb, U, light rare earth elements.

3. Melt-induced sub-arc metasomatism is strongly dependent on the subducting slab age. In arcs (or arc segments) related to young slab subduction, slab-derived adakitic melts cause pervasive Na-metasomatism in sub-arc mantle. Arcs associated with subduction of old slab can experience melting of carbonated basalt and/or carbonate-rich sediment which will result in production of a carbonated, Ne-normative melt. Both types of siliceous melts introduce melt-mobile elements into the sub-arc wedge, such as heavy rare earths. Slab melt-mantle interaction may also result in selective enrichment of sub-arc peridotites in high field strength elements (Nb and Ta) resulting in creation of local "OIB-like" mantle domains. It remains to be seen whether these diverse mantle compositions give rise to the suite of arc magmas erupted at Kamchatkan volcanoes.

Acknowledgements. This paper benefitted from numerous discussions with Mark Drummond, Rene Maury, Julie Morris and Jeff Ryan. Detailed constructive reviews by Jon Davidson and Brad Singer significantly improved the manuscript. This work was supported by NSF grant EAR-9304105 to Marc Defant, Mark Drummond, Julie Morris, and Pavel Kepezhinskas.

REFERENCES

Anders, E. and N.Crevesse, Abundance of the elements: meteoritic and Solar, *Geochim.Cosmochim. Acta, 53,* 197-214, 1989.

Arculus, R.J. and R.Powell, Source component mixing in the regions of arc magma generation, *J.Geophys.Res., 91,* 5913-5926, 1986.

Austin, Ph., Boron and Beryllium of the Kamchatka arc: the roles of rifting and assimilation in continental arcs, M.Sc. thesis, University of South Florida, Tampa, 1995.

Baranov, B.V., N.I. Seliverstov, A.V.Muraviev, and E.L.Muzurov, Commander Basin: product of a back-transform spreading, *Tectonophysics, 199,* 237-269, 1991.

Bebout, G.E., J.G.Ryan, and W.P.Leeman, B-Be systematics in subduction-related metamorphic rocks: characterization of the subducted component, *Geochim. Cosmochim. Acta, 57,* 2227-2237, 1993.

Clague, A.J., P.Kepezhinskas, M.Defant, R.W.Nesbitt, and J.A.Milton, Laser ablation ICP-MS study of fluid inclusions in mantle xenoliths from Kamchatka, Russia: preliminary results, *EOS, 76,* 538, 1995.

Crocket, J.H., Geochemistry of the platinum-group elements, *Canadian Inst.Min.Metall.Spec.Issue, 23,* 47- 64, 1981.

Defant, M.J., and M.S.Drummond, Derivation of some modern arc magmas by melting of young subducted lithosphere, *Nature, 347,* 662-665, 1990.

Gill, J.B., *Orogenic Andesites and Plate Tectonics,* Springer-Verlag, New York, 1981.

Hauri, E.H., N.Shimizu, J.J.Dieu and S.R.Hart, Evidence for hotspot-related carbonatite metasomatism in the oceanic upper mantle, *Nature, 365,* 221-227, 1993.

Hawkesworth, C.J., K.Gallagher, J.M.Hergt, and F.McDermott, Mantle and slab contributions in arc magmas, *Ann.Rev.Earth Planet. Sci., 21,* 175-204, 1993.

Kepezhinskas P.K., and M.J.Defant, Trace element systematics in island-arc mantle xenoliths: evidence for a depleted mantle wedge metasomatized by volatile-rich felsic melts, *EOS, 75,* 730, 1994.

Kepezhinskas P.K., M.J.Defant and M.S.Drummond, Na metasomatism in the island arc mantle by slab melt-peridotite interaction: evidence from mantle xenoliths in theNorth Kamchatka arc, *J. Petrol., 36,* 1250-1267, 1995.

Maury, R.C., M.J.Defant and J.-L.Joron, Metasomatism of the sub-arc mantle inferred from trace elements in Philippine xenoliths, *Nature, 360,* 661-663, 1992.

McInnes, B.I.A. and E.M.Cameron, Carbonated, alkaline hybridizing melts from a sub-arc environment: mantle wedge samples from the Tabar-Lihir-Tanga-Feni arc, Papua New Guinea, *Earth Planet.Sci.Lett., 122,* 125-141, 1994.

Morris, J.D., W.P.Leeman, and F.Tera, The subducted component in island arc lavas: constraints from Be isotopes and B-Be systematics, *Nature, 344,* 31-36, 1990.

Rudnick, R.L., W.F.McDonough, and B.W.Chappell, Carbonatite metasomatism in the northern Tanzanian mantle: petrographic and geochemical characteristics, *EarthPlanet.Sci. Lett., 114,* 463-475, 1993.

Ryan, J.G., J.D.Morris, F.Tera, W.P.Leeman, and A.Tsvetkov, Cross-arc geochemical variations in the Kurile arc as a function of slab depth, *Science, 270,* 625-627, 1995.

Ryan,J.G., W.P.Leeman, J.D.Morris, and C.H.Langmuir, The Boron systematics of intraplate lavas: implications for crust and mantle evolution, *Geochim. Cosmochim. Acta, 60,* 501-514, 1996.

Sun, S.-s., and W.F.McDonough, Chemical and isotopic systematics of oceanic basalts: Implications for mantle composition and processes, In: Magmatism in the ocean basins,Sanders, A.D., and M.J.Norry (eds.), *Geol.Soc. London Spec.Publ.,* 313-345, 1989.

P.K.Kepezhinskas and M.J.Defant, Department of Geology, University of South Florida, Tampa, FL 33620-5200.

Suprasubduction Mineralization: Metallo-tectonic Terranes of the Southernmost Andes

Eric P. Nelson

Department of Geology and Geological Engineering, Colorado School of Mines, Golden, Colorado

The suprasubduction environment, here defined as any crust above a subduction zone, is one of the most prolific tectonic environments for the formation of ore deposits. Mineralization in this environment is controlled by three principal factors: (1) suprasubduction magmatism (forearc, arc, and backarc), (2) major faults, and (3) the plate tectonic regime. Ore deposits in the suprasubduction environment are classified here into forearc, arc, and backarc types. The southernmost Andes has been a suprasubduction orogen at least since the late Paleozoic, and steady-state subduction has been punctuated at times by diachronous and episodic orogenic events. Although major mineral resources are known in the central Andes, mineral exploration has been neglected in the southernmost Andes because of real and perceived logistical problems including difficult weather, vegetation cover, and poor infrastructure. Six metallo-tectonic terranes are defined on the basis of petrological character and tectonic history, and are characterized by associations of known or inferred mineral resources: I Paleozoic-early Mesozoic forearc; II Triassic?-Jurassic volcano-tectonic rift and Jurassic-Cretaceous intra- and backarc basins; III Jurassic to early Tertiary Patagonian batholith; IV Mesozoic marginal basin; V Tertiary intra-arc transtensional basins; VI Tertiary-Recent volcanic arc. Recognition of suprasubduction environments and definition of metallo-tectonic terranes can be useful in designing exploration programs in such frontier regions.

INTRODUCTION

Many ore deposits, including some of the largest known, formed, or were tectonically emplaced, in what is here termed the suprasubduction environment. The term suprasubduction was first used in the ophiolite literature to refer to ophiolites with geochemical characteristics indicating an origin above a subduction zone (originally "supra-subduction", *Alabaster et al.,* 1982). Here the suprasubduction environment is broadly defined as any crustal region that was, at some time during its evolution, above a subduction zone. This environment is thus associated with convergent plate boundary orogenic belts [*Burchfiel*, 1980], and includes forearc, arc, and backarc regions.

Although many suprasubduction orogens have a history of successful mineral exploration and production, a number of frontier regions in this tectonic setting remain to be explored. For such regions, where few resources are known, a first approach in exploration planning is to predict the types of deposits that could exist. One approach is to compare tectonic evolution models with deposit classifications based on tectonic environment. For example, Cyprus-type massive sulfide deposits, which are classified as forming in exhalative hot-spring systems on oceanic spreading ridges, are predicted to exist in ophiolites.

The southernmost Andes is an example of an orogenic belt with a tectonic evolution dominated by convergent boundary tectonics, and with a historical record of little exploration. To analyze the mineral potential of this poorly explored, but potentially mineral-rich orogenic belt, the orogen is here divided into metallo-tectonic terranes. Such terranes are similar to tectono-stratigraphic terranes [*Coney et al.,*

1980; *Howell et al.*, 1985], but are defined here as regions characterized by distinctive associations of known or inferred mineral resources, predicted on the basis of the tectonic evolution of each terrane. This paper first outlines important factors controlling the occurrence of ore deposits in the suprasubduction environment, then proposes a classification of ore deposits based on tectonic setting within this environment. The tectonic evolution of the southern-most Andes is outlined, and six metallo-tectonic terranes are described. The mineral potential of each terrane is then given in terms of the deposits types predicted from the deposit classification.

FACTORS IN SUPRASUBDUCTION MINERALIZATION

Mineralization in the suprasubduction environment is controlled by three principal interrelated factors: (1) suprasubduction magmatism (forearc, arc, and back-arc); (2) major structures; and (3) the plate tectonic regime. Magmas are a source of heat, fluids, and probably metals. Metal sources associated with magmas may include subducted oceanic crust, the mantle wedge above the subducting slab, and crustal wall rocks along the path of rising magmas and circulating hydrothermal fluids. The petrochemical characteristics of magmas have been correlated with some ore deposits. For example, Sillitoe [1981] used the I- and S-type granitoid classification [Chappell and *White*, 1974; *White et al.*, 1977], and the magnetite- and ilmenite-series granitoid classification [*Ishihara*, 1980, 1981], to correlate petrochemical/petrological features and metallogenic character of granitoid rocks in magmatic arcs. For example, tin and tungsten deposits are typically associated with S-type granitoids and I- or S-type ilmenite series granitoids. Also, Cu-Fe-Mo-Au-Ag deposits in magmatic arcs are typically associated with I-type granitoids.

In the upper crust, major brittle to brittle-ductile structures can act as important fluid conduits, as permeable zones for mineral deposition, and as controls on magma intrusion [*Sibson et al.*, 1988; *Bursnall*, 1989]. In the lower crust, ductile shear zones may act as inverted, elongate funnels along which incompatible elements and metals migrate upward. Such migration is accomplished by fluid flow combined with diffusion along chemical potential gradients at a high angle to isothermal surfaces, and is assisted by crystal plastic and pressure solution deformation mechanisms, and by seismic pumping within the brittle-plastic transition zone [*Sibson et al.*, 1988]. Major structures include suture zones, arc-parallel strike-slip faults, transverse segmentation structures, arc-parallel normal faults, thrust faults, low-angle normal faults [*Doblas et al.*, 1988; *Roddy et al.*, 1988; *Beaudoin et al.*, 1991], and caldera collapse structures.

The plate tectonic regime of the suprasubduction environment is controlled by tectonic parameters that are potentially important in mineralization. These include slab dip, convergence vector, age of subducting plate, forearc accretionary prism development vs. tectonic erosion, and collision of oceanic bathymetric highs (ridges, plateaus, seamounts, etc.). Changes in any of these parameters may be especially important.

PLATE TECTONIC CLASSIFICATION OF ORE DEPOSITS

A number of ore deposit classifications based on, or incorporating, plate tectonic theory have been proposed [e.g., *Pereira and Dixon*, 1971; *Guilbert*, 1981; *Mitchell and Garson*, 1981; *Sillitoe*, 1981; *Hutchison*, 1983; *Sawkins*, 1984; *Cox and Singer*, 1987]. Some of the main classifications are summarized in Table 1. Note that, although there are some similarities between these classifications, there exists little correspondence between the major categories. One of the most extensive classifications was that proposed by *Guilbert* [1981; revised in *Guilbert and Park*, 1986], which includes approximately 109 deposit types in five major tectonic settings (Table 1). Guilbert's classification includes even those deposits which lack immediate plate tectonic influence, such as placer deposits on stable cratonic interiors. Sawkins' classification [1984, updated 1990], which contains eight main categories, is a mixture of deposits related to both tectonic setting and tectonic events (Table 1). Cox and Singer [1987] made an extensive "lithotectonic" classification with approximately 90 deposit types (Table 1). Although their main categories do not easily fit into a plate tectonic framework, each of their descriptive deposit models includes a depositional environment and tectonic setting.

Although classification of ore deposits in a plate tectonic context is an important and logical step, a number of problems are inherent with this approach. Such problems include juxtaposition and superposition of terranes and associated deposits, the evolution of tectonic and metallogenic processes in time, and the non-steady-state nature of some tectonic processes at convergent boundaries.

TABLE 1. Selected Plate Tectonic Classifications of Ore Deposits: Main Categories.

Guilbert (1981)	Sawkins (1984)	Mitchell and Garson (1981)	Hutchison (1983)
Mid-ocean ridges/ocean floor	Oceanic-type crust	Oceanic settings	Oceanic lithosphere
Consuming, subducting margins	Principal arcs	Subduction-related settings	Sea-floor sulfides
Ensialic-ensimatic back-arc basins	Inner sides of principal arcs	Collision-related settings	Intracratonic basins
Cratonic openings	Collisional events	Passive continental margins and interior basins	Other epicontinental sea
Cratons	Intracontinental hotspots, anorogenic magmatism	Continental hot spots, rifts, aulacogens	Intrusive bodies in stable cratonic terrain
	Continental rifting; early vs. late stages	Transform faults and lineaments in continental crust	Batholith-associated
	Arc-related rifts		Epigenetic volcanic and epizonal plutonic associations
	Other arc-related deposits		Archean-style
			Proterozoic-style
			Surficial continental

Metallogenic juxtaposition and superposition

Classification of ore deposits is compounded by the complex history of convergent boundary orogens. Tectonic provinces can be juxtaposed and superimposed, as can the deposits that typify them. Sillitoe [1981] noted that "metallogenic juxtaposition" can result from accretion of ophiolites, arc terranes, or other oceanic or continental fragments. Thus, ore deposits formed in allochthonous terranes can be attached to a continent by terrane suturing and the distinction between forearc, arc, and backarc thereby obscured. In addition, some ore-forming processes may be associated with suturing events. For example, metamorphic fluids, segregated during such accretion events, may act as epigenetic ore-forming agents [*Sillitoe*, 1979; *Kerrich and Wyman*, 1990]. If collision and crustal thickening are great enough, anatexis may occur, forming lithophile suite deposits. Also, the migration of magmatic arcs in response to slab dip changes will cause superposition of different magmatic and related metallogenic systems. For example, the well documented sweep of volcanism in the western United States and northern Mexico during the Cretaceous-Tertiary [*Coney and Reynolds*, 1977; *Clark et al.*, 1982] caused arc-type magmatism to occur over vast regions that previously had been in a backarc setting.

Evolution of plate tectonics in time

As pointed out by Guilbert [1981], one weakness of plate tectonic classifications that incorporate all deposit types is in projecting plate tectonic processes, as interpreted today, back into the Archean. The principle of uniformitarianism can most likely be applied to tectonics in the Phanerozoic and much of the Proterozoic. However, Archean and possibly early Proterozoic tectonic systems may not have operated in an analogous fashion to the tectonics of today [*Sangster*, 1979; *Barley and Groves*, 1992; *Hutchinson*, 1992; *Abbott et al.*, 1994]. Some differences in tectonic processes hypothesized for the Archean include a higher geothermal gradient [*Bickle*, 1978], greater oceanic ridge length [Burke et al., 1976; *Hargraves*, 1986], increased spreading rates [*Burke and Kidd*, 1978], thicker oceanic crust, and shallower subduction dip [*Abbott et al.*, 1994]. Hutchinson [1992] pointed out that mineral deposits have evolved along with tectonic processes. He notes a remarkable decline of certain earlier deposit types, such as komatiite-hosted Ni and Superior-type Fe, and the appearance, proliferation, and diversification of certain new deposit types (such as porphyry-type deposits).

Although plate tectonic processes can operate in steady-state over long periods, some processes are episodic and/or diachronous, and therefore affect the temporal and spatial distribution of related deposits. For example, lode gold deposits are most abundant in the late Archean, lower Paleozoic and Mesozoic, because these are times of extensive Cordilleran-style terrane accretion and associated transpressive deformation [*Kerrich and Wyman*, 1990; *Kerrich*, 1993]. In another example, Bradley et al. [1993] and Haeussler et al. [1995] have documented the diachronous development of gold-quartz vein mineralization in the forearc of southern Alaska in response to ridge collision and lateral migration of the related triple junction along the continental margin.

TABLE 2. Plate Tectonic Classifications of Ore Deposits: Convergent Plate Boundaries.

Guilbert (1981)	Sawkins (1984)	Mitchell and Garson (1981)	Sillitoe (1981)
Consuming, subducting margins • obduction • ocean/ocean -island arc-'eugeosynclines' • ocean/continent, trench/arc, Cordilleran orogenics • ocean/continent extension (unique[?] to SW N. America) Ensialic-ensimatic back-arc basins • volcanic tendencies - outboard (arc) side • sedimentary tendencies - inboard (continent) side	Principal arcs Inner side of principal arcs Arc-related rifts Other arc-related deposits	Submarine trenches and outer arcs Magmatic arcs Outer arc troughs Back-arc magmatic belts and thrust belts Back-arc compressive cratonic basins Back-arc extensional cratonic basins Back-arc marginal basins and inter-arc troughs	Principal arcs Inner side of principal arcs: Pb-Zn-Ag-Cu-(Mo-Fe) Back-arc basins Arc-related rifts, lithophile suite: Mo-F-U-Be-(Sn) Arc-related rifts, base/precious-metal suite Forearcs: Sn-W-(Cu) Non-magmatic back-arcs: U-V-Cu-(Pb-Zn-Ag)

CLASSIFICATION OF ORE DEPOSITS IN SUPRASUBDUCTION SETTINGS

Most tectonic classifications of ore deposits incorporate deposit types found at convergent plate margins (Table 2). Mitchell [1976] subdivided ore deposits related to Andean and island arc magmatism into pre-, syn-, and post-collision types, and distinguished calc-alkaline from alkaline/peralkaline associations. He further subdivided subduction-related ore deposits based on upper vs. lower plate, on emplacement location relative to continent, and on the nature of the host rocks and major faults. Sillitoe [1981] specifically classified ore deposits formed at convergent margins and noted that subduction style influences the magmatic and metallogenic characteristics of resultant Andean and island arcs. The interrelation of subduction style, petrochemical character of granitoid rocks, and metallogeny has been noted by Ishihara [1980, 1981], Keith and Swan [1987], Uyeda and Nishiwaki [1980], Sillitoe [1981], and others. Three fundamental end-member classes of subduction style are recognized [*Uyeda and Kanamori*, 1979; *Dewey*, 1980]: (1) typically intraoceanic, extensional arcs with backarc basins, underlain by steep Benioff zones, (2) typically continental, contractional arcs with backarc thrust belts, underlain by flat Benioff zones, and (3) neutral arcs. Sillitoe [1981], however, emphasized that certain deposits are associated with transitional stages between these end-member classes (Figure 1). This is best illustrated by deposit types in the Basin and Range province of the western United States, many of which formed in the Tertiary during the transition between a contractional arc-backarc setting and a neutral or extensional setting. Many of these deposit types are somewhat unique, and are placed in special categories in some classifications (e.g., Table 2; 'Other arc related deposits', *Sawkins*, 1984).

Proposed classification

The following simplified classification (Table 3) is partly based on a number of previous classifications [*Guilbert*, 1981; *Mitchell and Garson*, 1981; *Sillitoe*, 1981; *Sawkins*, 1984], and focuses on deposits found in the suprasubduction environment. I emphasize *found* in this environment, as some deposits may not have *formed* in this environment, but may have been emplaced there tectonically. For example, podiform chromite is usually formed by some magmatic concentration process during the formation of oceanic lithosphere, but is often found in ophiolites in forearc or suture zone settings.

In the proposed classification, deposits are classified into forearc, magmatic arc, and backarc subdivisions of the suprasubduction environment, and oceanic (island arc) and continental (Andean arc) settings are distinguished (Table 3, Figure 2). This classification is useful as it is based on recognition in the rock record of calc-alkaline magmatic rocks of arc affinity. However, recognition of these tectonic environments can be complicated by many factors (discussed below). The proposed classification omits deposits primarily

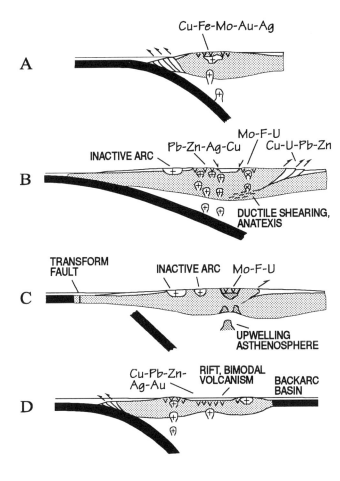

Fig. 1. Schematic relationship between styles of subduction and metallogeny (from Sillitoe, 1981). Black = oceanic crust; light gray = continental crust; v - volcanic rocks; + = I-type magma; x = S-type magma; dark gray = A-type magma. A) Moderate to steep subduction; neutral stress regime. B) Shallow subduction; contractional regime. C) Steepening of detached slab following ridge collision (with transform fault possibly formed); extensional regime. D) Steep subduction and commencement of intra-arc rifting; extensional regime.

restricted to Archean and early Proterozoic rocks because, as noted earlier, tectonic processes have evolved since the early history of the earth. The proposed classification also omits deposits formed as a result of continental collision, because continental collision shuts off subduction. Note, however, that most collisional orogenic belts had a history of convergent margin tectonics prior to collision, and will therefore contain some of the deposits included in the classification presented here.

A number of benefits result from this new classification. The main tectonic subdivisions are simplified over previous classifications and are consistent with modern tectonic terminology. In geologically less known frontier regions, recognition of forearc, arc, and backarc geology should be relatively easy in most cases. In addition, the deposit classses encompass most major ore deposit types; many subclasses or minor deposits [cf., *Cox and Singer*, 1987] are omitted. Thus the classification will be useful in designing exploration programs in frontier regions within suprasubduction orogens.

The proposed classification is not perfect in a number of aspects. First, ore deposit terminology is historically complex, and tends to be a mixture of descriptive and genetic terms. For example, "massive magnetite" is a descriptive designation, whereas "exhalative Mn" is a genetic designation. Terms referring to a type locality, such as "Kuroko-type deposits", although descriptive, may carry a genetic implication. The proposed classification includes both genetic and descriptive terminology, but is fundamentally descriptive, as it groups deposit types by where they occur. Nonetheless, deposits in the classification include those formed by magmatic, hydrothermal, metamorphic, metasomatic, and sedimentary processes, as well as those emplaced by tectonic processes. For example, most porphyry deposits are found in a magmatic arc setting, and are assumed to form in a magmatic-hydrothermal system. In addition, a suite of other deposit types, including epithermal veins, mantos, skarns, and pipes or chimneys, are associated broadly with magmatic and hydrothermal processes in the magmatic arc [e.g., *Arribas et al.*, 1995].

Second, the origin of some deposit types is controversial. For example, a number of deposit types ascribed to an epigenetic replacement origin, such as skarn, manto, and disseminated Carlin-type, also have been interpreted as having a modified syngenetic origin [*Hutchinson and Burlington*, 1984; *Kerrich*, 1993].

Third, the classification of deposits in forearc, arc, and backarc environments is compounded by the variability of such environments, and by a number of factors which affect the evolution of such environments. Examples of variability include: (1) some modern forearc regions have an extensive accretionary prism (e.g., Alaska), whereas others have no accretionary prism (e.g., Peru), and (2) some backarc regions are contractional (e.g., central Andes), whereas others are extensional (e.g., western Pacific arcs) [*Uyeda*, 1983]. Factors affecting the evolution of

TABLE 3. Classification of Ore Deposits in Suprasubduction Setting.

FOREARC RELATED ORE DEPOSITS

1a   accreted oceanic terranes: Cyprus-type exhalative deposits, distal exhalative Mn, ophiolite-related podiform Cr and Ni laterites

1b   epithermal Au, Hg veins, and Sn-W-Cu veins and skarns associated with intrusions

MAGMATIC ARC RELATED ORE DEPOSITS

2a   epizonal calc-alkaline systems (Cu, Mo, Au*, ± Pb, Zn, Ag), including porphyry and related breccia pipes, epithermal veins and hot springs, skarns, and mantos

2b   batholith-related systems (pegmatites, granitic U)

2c*   Kuroko-type exhalative deposits in submarine intra-arc basins

2d   lode gold, shear zone associated

2e   massive magnetite

BACKARC RELATED ORE DEPOSITS

3a   alkaline igneous systems (e.g., Au-tellurides, Mo-porphyries, and other lithophile suites)

3b   tin-tungsten systems (porphyries, veins, pipes, skarns, greisens)

3c   disseminated sediment-hosted Au (Carlin-type)

3d*   Besshi-type exhalative deposits in backarc basins

3e   metamorphic core complex ('detachment-hosted')

* = island arc affinity, otherwise Andean arc affinity

suprasubduction environments include: (1) state of stress in the upper plate, (2) nature of crust in the upper plate, (3) relative and absolute plate motions, (4) age of subducting lithosphere, and (5) prior tectonic history. Because of these factors, the distribution of arc and backarc regions can be difficult to determine. For example, during the period between 70-55 Ma, subduction-related magmas in the western United States erupted over a vast area in what had previously been the backarc [*Burchfiel and Davis*, 1975; *Coney and Reynolds*, 1977]. Subsequently, in the Oligocene, magmatism swept westward in a widespread, nearly simultaneous volcanic event. Although many of the magmas produced during this "flare-up" were fundamentally calc-alkaline arc-type magmas, the dispersed nature of the magmatism over a relatively short period of time, and the transitional to extensional tectonic regime, affected the nature of mineralization in the region.

Although suprasubduction settings normally evolve through processes associated with steady-state subduction, most are affected by episodic events such as collision of seamounts, aseismic ridges, and actively spreading ridges [e.g., *Nelson and Forsythe*, 1989]. These episodic events can affect the metallogeny of suprasubduction settings.

SOUTHERNMOST ANDES: EXAMPLE OF A SUPRASUBDUCTION OROGEN

The Andes south of about 42°S latitude (Figure 3) have been a suprasubduction orogen semi-continually at least since the late Paleozoic [*Mpodozis and Ramos*, 1990], although steady-state subduction has been punctuated at times by diachronous and episodic orogenic events (Figure 4). Tectonic development of the southernmost Andes began in the late Paleozoic with formation of a forearc accretionary complex along the margin of Gondwana between the Late Devonian(?) and Triassic to Early Jurassic [*Forsythe*, 1982; *Dalziel and Forsythe*, 1985; *Davidson et al.*, 1987; *Mpodozis and Ramos*, 1990]. This complex, exposed mostly in a Pacific coastal belt, formed west of a magmatic arc and includes submarine fan turbidites, pelagic chert, broken formation (sheared sandstone, mudstone, and conglomerate), foraminiferal limestone, pillow lava, ultramafic rocks, mica schist, and greenschist. Metamorphic grade is generally zeolite to middle greenschist facies, with local exposures of blueschist and amphibolite facies rocks. This complex is interpreted to represent accretion of oceanic terranes mixed with continental and arc detritus along the convergent margin of Gondwana.

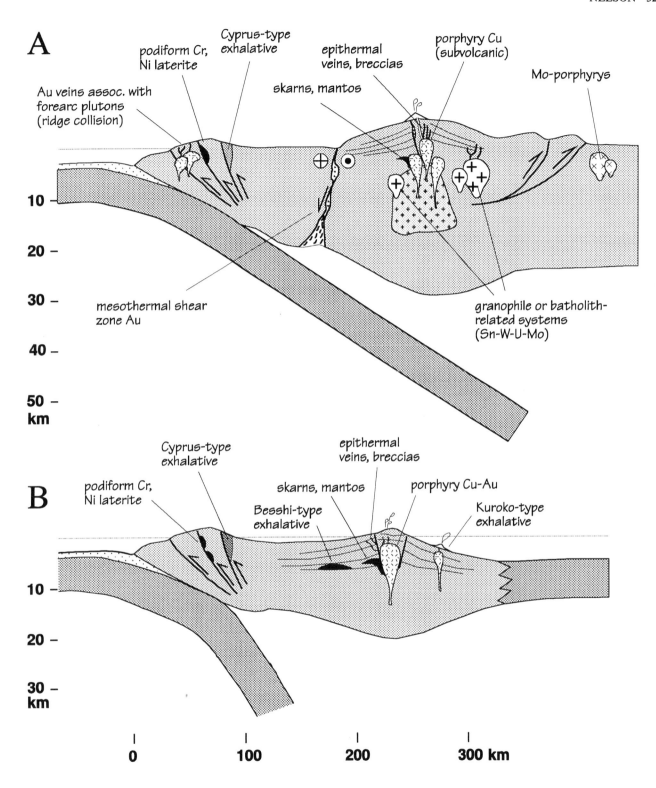

Fig. 2. Schematic cross sections of suprasubduction settings showing major deposit types that form, or are emplaced tectonically in, A) Andean-type orogens and B) oceanic island arc-type orogens.

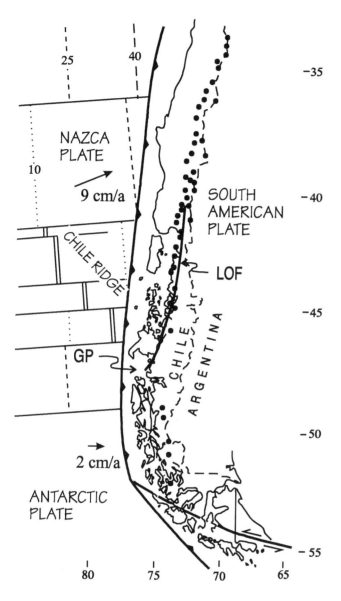

Fig. 3. Tectonic setting of southern Andes. Black dots = active volcanoes. GP = Golfo de Penas; LOF = Liquiñe-Ofqui fault system. Dotted and dashed lines on oceanic plates labeled with age of magnetic anomaly in Ma. Arrows on oceanic plates show relative convergence vector.

Following a period of uplift and erosion, and coincident with early rifting of the South Atlantic in the Late Jurassic [Rabinowitz and La Brecque, 1979], a widespread volcano-tectonic rift province developed with voluminous bimodal, but mainly silicic, volcanism and minor S-type granitoids [Bruhn et al., 1978; Nelson and Elthon, 1983; Gust et al., 1985]. This event coincided with a major paleogeographic change in the Middle to Late Jurassic as the magmatic arc shifted westward to intrude the late Paleozoic accretionary complex. With resumption of normal subduction, the calc-alkaline Patagonian batholith developed between latest Jurassic and Tertiary time [Hervé et al., 1984; Nelson et al., 1986; Bruce et al., 1991].

North of 49° S, ephemeral marine transgressions in intra-arc basins deposited sandstone, shale, and limestone interbedded with andesitic, dacitic, and rhyolitic rocks [Haller and Lapido, 1982; Ramos et al., 1982; Niemeyer et al., 1984]. Weak deformation, represented by open folds and normal faults, occurred in the backarc zone during the Mesozoic. The Liquiñe-Ofqui fault system, a major arc-parallel lineament, developed by the Miocene in response to oblique subduction, and appears to have localized Miocene and Quaternary eruptive centers [Hervé, 1976; Skarmeta, 1976; Bartholomew and Tarney, 1984, Forsythe and Nelson, 1985] during transtensional phases.

South of 49°S during the Early Cretaceous, pyritic black shale was deposited on a stable shelf cratonward of an ophiolite-floored, flysch-filled marginal basin [Dalziel et al., 1974; Suárez and Pettigrew, 1976]. The Andean orogeny began in the mid-Cretaceous and involved shortening and cratonward thrusting of the marginal basin [Bruhn and Dalziel, 1977; Nelson et al., 1980; Dalziel, 1986] and development of the Magallanes foreland basin [Winslow, 1979]. At approximately 15 Ma the Chile Rise spreading center collided with the margin forming two triple junctions [Herron et al., 1977; Herron et al., 1981]. The southern triple junction is today a diffuse, complex zone of faulting near Tierra del Fuego, and involves the Magallanes fault zone [Fuenzalida, 1976], a major lineament and possible strike-slip structure which cuts diagonally across the main cordillera. The northern triple junction has migrated north along the continental margin [Herron et al., 1981; Cande et al., 1987] to its present position at 46.5°S latitude where the actively spreading Chile Rise is being subducted [Forsythe and Nelson, 1985; Nelson et al., 1994].

Recent tectonics of the southernmost Andes

Today, the southern Andes can be subdivided longitudinally into two zones separated by the triple junction region at 46.5°S (Figure 3). North of the triple junction, oblique convergence occurs at approximately 9 cm/a and the orogen contains a well defined

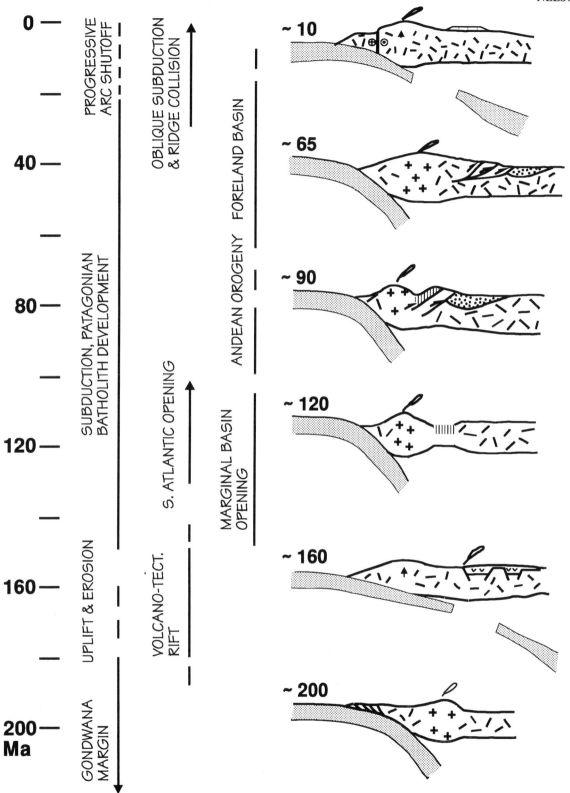

Fig. 4. Tectonic history of southern Andes, mostly taken from south of 49° S. Gray = oceanic crust; dashed = older continental crust; + = magmatic additions to continental crust; vertical lines = ophiolitic marginal basin crust; dots = sedimentary rocks in foreland basin. Plateau basalts shown in backarc region in ~10 Ma cross section.

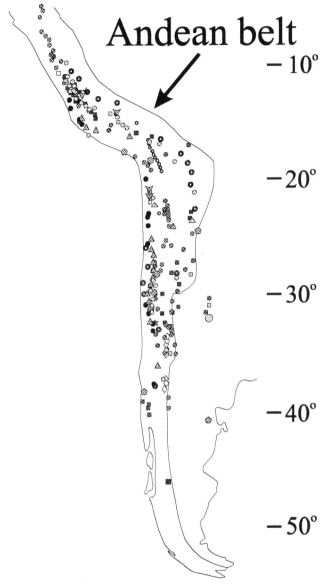

Fig. 5. Ore deposits in the central and southern Andes (compiled from Petersen, 1990). Note the paucity of deposits in the southernmost Andes. Symbols represent various types of ore deposits.

forearc, arc, and backarc (generally contractional). Near the triple junction, forearc tectonics are affected by ridge collision and oblique convergence [*Forsythe and Nelson*, 1985; *Nelson and Forsythe*, 1989]. Two major tectonic features accommodate movement of an elongate forearc sliver: (1) normal faults in the Golfo de Penas forearc basin (GP, Figure 3), and (2) the Liquiñe-Ofqui right-normal fault system (LOF; Figure 3). Magmatic effects of ridge collision include silicic magmatism and emplacement of the Pliocene Taitao ophiolite in the forearc [*Nelson et al.*, 1994]. South of the triple junction, subduction has slowed to less than 2 cm/a and arc volcanism is nearly extinct (Figure 3) [*Stern et al.*, 1976].

METALLO-TECTONIC TERRANES AND MINERAL POTENTIAL OF THE SOUTHERNMOST ANDES

Although major mineral resources are known in the central Andes, mineral exploration has been neglected in the southernmost Andes because of real and perceived logistical problems including difficult weather, vegetation cover, and poor infrastructure. Nearly all maps showing the distribution of ore deposits in Chile are truncated south of about 37°-40° S latitude [e.g., *Ruiz*, 1990; *Davidson and Mpodozis*, 1991]. The few maps that do include the southern region indicate very few known ore deposits (Figure 5) [e.g., *Ruiz*, 1965; *Frutos and Pincheira*, 1986a; *Petersen*, 1990]. Nonetheless, analysis of the tectonic development of the region suggests that it has potential for major deposits.

Indeed, some resources are known in the southernmost Andes [*Ruiz*, 1965; *Frutos and Pincheira*, 1986b], and exploration activity has increased in the past few years. Known resources include stratiform Zn-Pb in the Toqui district [*Wellmer et al.*, 1983], epithermal Ag-Au in the Fachinal district [*Little*, 1993], the Cutter Cove Cu-sulfide deposit [*Thomas*, 1973; *Arias*, 1985], massive sulfides in bimodal volcanics in Tierra del Fuego [*Anonymous*, 1994], and gold placers in Chiloé [*Frutos and Pincheira*, 1986b; *Emparan et al.*, 1985], and in Tierra del Fuego [*Heylmun*, 1993].

In order to aid exploration design in the region, six metallo-tectonic terranes are recognized (Table 4). Metallo-tectonic terranes, which are similar to tectono-stratigraphic terranes [*Coney et al.*, 1980; *Howell et al.*, 1985], are here defined as having distinct petrological features and tectonic histories. They are further defined as regions characterized by associations of known or inferred mineral resources, predicted on the basis of the tectonic evolution of each terrane. Some terranes are allochthonous (e.g., much of terrane I), although most have formed essentially in place or have been affected by subduction-related processes. Terrane characteristics are summarized in Table 4 and terranes are located in Figure 6. Predicted mineral deposit types for each terrane, taken from the classification in Table 3, are also given in Table 4.

TABLE 4. Metallo-tectonic Terranes in the Southernmost Andes. Deposit types in Table 3.

#	TERRANE DESCRIPTION	DEPOSIT TYPES
I	Late Paleozoic-early Mesozoic accreted forearc terrane (includes Quaternary ridge-collision related rocks)	1a, 1b, 2c?, 3d?
II	Triassic?-Jurassic volcano-tectonic rift terrane (bimodal volcanics), and Jurassic-Cretaceous volcanic-sedimentary intra-arc and backarc terranes (includes early Cretaceous shale basin south of 49°S)	2a, 2c, 3a, 3b, 3c, 3d?, 3e
III	Jurassic to early Tertiary Patagonian batholith and volcanic roof pendants	2a, 2b, 2c, 3c
IV	Mesozoic marginal basin terrane: ophiolites and volcaniclastic flysch	1a, 2c?, 3d
V	Tertiary intra-arc transtensional basins: mafic pillow lavas, ultramafics (includes sheared rocks along Liquiñe-Ofqui fault zone)	1a, 2c, 2d, 3d
VI	Tertiary-Recent volcanic arc	2a, 2e?

With this approach, exploration programs for certain deposit types can best be directed to specific terranes. I give two examples of areas in the southern Andes which show mineral potential in the field consistent with the potential predicted by the terrane analysis. First, within terrane II in the Cordillera Darwin region of Tierra del Fuego (~69.4°W, 54.4°S; A in Figure 6), a belt of altered, pyritic Jurassic rhyolitic rock is exposed for a strike length of over 30 km. The geology of this belt suggests a Kuroko-type setting (deposit type 2c), however the volcanic sequence, known as the Tobifera Formation, is poorly mapped and no information on mineral content has been published. Nonetheless, the exposure of intensely altered rock is vast and an obvious target for exploration. Second, within terrane V in the Llancaqué area (~72.6°W, 42.1°S; B in Figure 6) a belt of sheared felsic and mafic volcanic rocks is associated with fuchsite-ankerite alteration, an association indicative of lode Au deposits (deposit type 2d). Again, although much of this area is poorly mapped, the geology along portions of the Liquiñe-Ofqui fault zone (Figures 3 and 6) is a likely target for exploration.

The apparent paucity of some deposit types in the southernmost Andes may be related to geological factors, not just lack of exploration. For example, erosion level may be a factor as the southernmost Andes generally expose more deeply eroded levels toward the south. This is certainly true for the predominantly Cretaceous Patagonian batholith, from which most epithermal systems probably have been eroded. The erosion may reflect a combination of increased Pleistocene glaciation toward the south as well as increased uplift caused by subduction of young oceanic crust and ridge collision to the south over the past 14 Ma or so [*Cande et al.*, 1987; *Delong and Fox*, 1977]. Nonetheless, a number of epithermal systems in Jurassic and Cretaceous volcanic and subvolcanic rocks in the less-eroded backarc region have shown indications of mineralization. Another factor could be the likely lack of Precambrian basement rock below the southernmost Andes, which could limit the source of some metals. Note, however, that the presence of large Tertiary epithermal deposits in the western Pacific magmatic arcs [e.g., Lihir (Au), *Moyle et al.*, 1990; Vatukoula (Au), *Anderson and Eaton*, 1990; Bougainville (Cu-Au), *Clark*, 1990] indicates that an ancient basement is not required as a source for younger deposits.

Despite the apparent paucity of ore deposits in the Andes south of 42°S, definition of metallo-tectonic terranes, using tectonic analysis of the orogen and tectonic classification of ore deposits, shows potential for exploration. This potential is supported by the existence of some deposits in the southernmost Andes, and the increased exploration activity in the past few years. Indeed, given the long history of convergent margin tectonics, punctuated by episodic extensional, transpressional, and transtensional phases, it is predicted that the southernmost Andes should house a wealth of mineral resources.

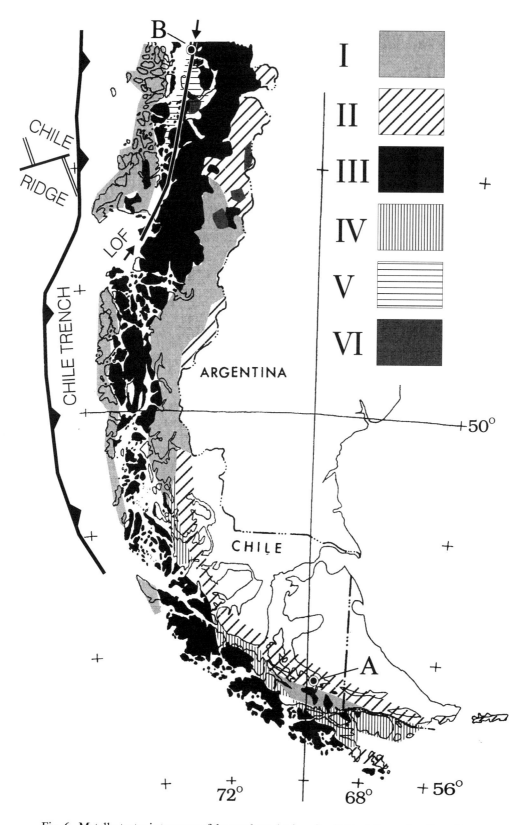

Fig. 6. Metallo-tectonic terranes of the southern Andes. See Table 4 for explanation of terranes.

Acknowledgments. Dick Hutchinson, Cliff Taylor, and Steve Turner were extremely helpful in discussions concerning ore deposit classification and formation. Study of numerous ore deposits over the years on field trips led by the energetic and enthusiastic Dick Hutchinson was invaluable. I thank Bradley Hacker, Craig Manning, Steve Turner, and an unknown reviewer for thoughtful and helpful reviews.

REFERENCES

Abbott, D., R. Drury, and W.H.F. Smith, Flat to steep transition in subduction style, *Geology, 22*, 937-940, 1994.

Alabaster, T., J.A. Pearce, and J. Malpas, The volcanic stratigraphy and petrogenesis of the Oman ophiolite complex, *Contrib. Mineral. Petrol., 81*, 168-183, 1982.

Anderson, W.B. and P.C. Eaton, Gold mineralisation at the Emperor Mine, Vatukoula, Fiji, in *Epithermal Gold Mineralization of the Circum-Pacific, II*, Association of Exploration Geochemists Special Publication No. 16b, edited by J.W. Hedenquist, N.C. White, and G. Siddeley, pp. 267-296, 1990.

Anonymous, Exploration efforts pick up steam in Argentina camp, *Northern Miner, 80*, no. 41, p.11, 1994.

Arias, J., Exploración geoquimica en los fiordos del Seno de Otway (XII Region, Chile), *IV Congreso Geol. Chileno, 3*, 452-478, 1985.

Arribas, A., Jr., J.W. Hedenquist, T. Itaya, T. Okada, R.A. Concepción, and J.S. Garcia, Jr., Contemporaneous formation of adjacent porphyry and epithermal Cu-Au deposits over 300 ka in northern Luzon, Philippines, *Geology, 23*, 337-340, 1995.

Barley, M.E., and D.I. Groves, Supercontinent cycles and the distribution of metal deposits through time, *Geology, 20*, 291-294, 1992.

Bartholomew, D.S., and J. Tarney, Geochemical characteristics of magmatism in the southern Andes (45-46°S), in *Andean Magmatism: Chemical and Isotopic Constraints*, Shiva Geological Series, edited by R. Harmon and B.A. Barreiro, pp. 220-250, 1984.

Beaudoin, G., B.E. Taylor, and D.F. Sangster, Silver-lead-zinc veins, metamorphic core complexes, and hydrologic regimes during crustal extension, *Geology, 19*, 1217-1220, 1991.

Bickle, M.J., Heat loss from the earth: a constraint on Archaean tectonics from the relation between geothermal gradients and the rate of plate production, *Earth Planet. Sci. Lett., 40*, 301-315, 1978.

Bradley, D.C., P.J. Haeussler and T.M. Kusky, Timing of early Tertiary ridge subduction in southern Alaska, *U.S. Geol. Surv. Bull. 2068*, 163-177, 1993.

Bruce, R.M., E.P. Nelson, S.G. Weaver, and D.R. Lux, Temporal and spatial variations in the Southern Patagonian batholith: Constraints on magmatic arc development, *Geol. Soc. Am. Spec. Paper 265*, 1-12, 1991.

Bruhn, R.L., and Dalziel, I.W.D., Destruction of the Early Cretaceous marginal basin in the Andes of Tierra del Fuego, in, *Island Arcs, Deep Sea trenches and Back-arc Basins*, Maurice Ewing Series, vol. I, edited by M. Talwani and W.C. Pitman, pp. 395-405, Am. Geophys. Union, Washington, D.C., 1977.

Bruhn, R.L., C.R. Stern, and M.J. DeWitt, Field and geochemical data bearing on the development of a Mesozoic volcano-tectonic and back arc basin in southernmost South America, *Earth Planet. Sci. Lett., 41*, 32-46, 1978.

Burchfiel, B.C., Tectonics of noncollisional regimes, in *Continental Tectonics*, National Academy of Sciences, Washington, D.C., pp. 65-72, 1980.

Burchfiel, B.C., and G.A. Davis, Nature and controls of Cordilleran orogenesis, western United States: extensions of an earlier synthesis, *Am. J. Sci., 275-A*, 363-396, 1975.

Burke, K., and W.S.F. Kidd, Were Archaean continental gradients much steeper than those today?, *Nature, 272*, 240-241, 1978.

Burke, K., J.F. Dewey, and W.S.F. Kidd, Dominance of horizontal movements, arc and microcontinental collisions during the later permobile regime, in *The Early History of the Earth*, edited by B.F. Windley, pp. 113-129, Wiley and Sons, London, 1976.

Bursnall, J.T. (editor), *Mineralization and Shear Zones*, Geological Association of Canada, Short Course Notes, 6, 299 pp., 1989.

Cande, S., R.B. Leslie, J.C. Parra, and M. Hobart, Interaction between the Chile Ridge and Chile Trench: Geophysical and geothermal evidence, *J. Geophys. Res., 92*, 495-520, 1987.

Carlile, J.C., and A.H.G. Mitchell, Magmatic arcs and associated gold and copper mineralization in Indonesia, *J. Geochem. Exploration, 50*, 91-142, 1994.

Chappell, B.W., and A.J.R., White, Two contrasting granite types, *Pacific Geol., 8*, 173-174, 1974.

Clark, G.H., Panguna copper-gold deposit, in *Geology of the Mineral Deposits of Australia and Papua New Guinea*, The Australian Institute of Mining and Metallurgy, Melbourne, edited by F.E. Hughes, pp. 1807-1816, 1990.

Clark, K.F., C.T. Foster and P.E. Damon, Cenozoic mineral deposits and subduction related magmatic arcs in Mexico. *Geol. Soc. Am. Bull., 93*, 533-544, 1982.

Coney, P.J., and S.J. Reynolds, Cordilleran Benioff zones. *Nature, 270*, 403-406, 1977.

Coney, P.J., D.L. Jones, and J.W.H. Monger, Cordilleran suspect terranes, *Nature, 288*, 329-333.

Cox, D.P., and D.A. Singer (editors), Mineral deposit models, *U.S. Geol. Surv. Bull. 1693*, 379pp., 1987.

Cuadra, W.A., and P.M. Dunkerley, A history of gold in Chile, *Econ. Geol., 86*, 1155-1173, 1991.

Dalziel, I.W.D., Collision and Cordilleran orogenesis: an Andean perspective, in *Collisional Tectonics*, edited by M.P. Coward and A.C. Ries, Geol. Soc. Spec. Pub. 19, 389-404, 1986.

Dalziel, I.W.D., and Forsythe, R.D., Andean evolution and the terrane concept, in *Tectonostratigraphic terranes of the circum-Pacific region*, edited by D.G. Howell, Circum-Pacific Council for Energy and Mineral Resources, Earth Science Series, no. 1, pp. 565-581, 1985.

Dalziel, I.W.D., M.J. DeWitt, and K. Palmer, Fossil marginal basin in the southern Andes, *Nature, 250,* 291-294, 1974.

Davidson, J., and C. Mpodozis, Regional geologic setting of epithermal gold deposits, Chile. *Econ. Geol., 86,* 1174-1186, 1991.

Davidson, J., C. Mpodozis, E. Godoy, F. Hervé, R. Pankhurst, and M. Brook, Late Paleozoic accretionary complexes on the Gondwana margin of southern Chile: evidence from the Chonos Archipelago, in: *Gondwana Six: Structure, Tectonics, and Geophysics,* Geophys. Monograph, vol. 40, edited by G.D. McKenzie, pp. 221-227, Am. Geophys. Union, Washington, D.C., 1987.

DeLong, S.E., and P.J. Fox, Geological Consequences of Ridge Subduction. in: *Island Arcs, Deep Sea Trenches and Back-arc Basins*, Maurice Ewing Series, vol. I, edited by M. Talwani and W.C. Pitman III, pp. 221-228, Am. Geophys. Union, Washington, D.C., 1977

Dewey, J., Episodicity, sequence and style at convergent plate boundaries, in *The Continental Crust and its Mineral Deposits*, Geol. Assoc. Can. Spec. Paper, 20, edited by D.W. Strangway, pp. 553-573, 1980.

Doblas, M., R. Oyarzun, R. Lunar, N. Mayor, and J. Martinez, Detachment faulting and late Paleozoic epithermal Ag-base metal mineralization in the Spanish central system, *Geology, 16,* 800--803, 1988.

Emparan, C., and C. Portigliati, Exploración de placeres auriferos en el area del Río Queualt, Comuna de Puerto Cisnes, XI Region, *IV Congreso Geol. Chileno, 3,* 579-602, 1985.

Forsythe, R.D., The late Paleozoic to early Mesozoic evolution of southern South America: a plate tectonic interpretation, *Quarterly J. Geol. Soc. London, 139,* 671-682, 1982.

Forsythe, R.D., and E.P. Nelson, Geological manifestations of ridge collision: evidence from the Golfo de Penas-Taitao basin, southern Chile, *Tectonics 4,* 447-495, 1985.

Frutos, J., and M. Pincheira, Metallogénesis y yacimientos metalíferos chilenos, in *Geología y Recursos Minerales de Chile*, edited by J. Frutos, R. Oyarzún, and M. Pincheira, pp. 469-487, Univ. Concepción, Concepción, 1986a.

Frutos, J., and M. Pincheira, Fichas metalogénicas de yacimientos minerales metálicos Chilenos, in *Geología y Recursos Minerales de Chile*, edited by J. Frutos, R. Oyarzún, and M. Pincheira, pp. 839-908, Univ. Concepción, Concepción, 1986b.

Fuenzalida, P.R., The Magellan fault zone, in Proceedings, Symposium of Andean Antarctic problems, Naples, Int. Assoc. Volc. Geochem. of the Earth Interior, 373-391, 1976.

Guilbert, J.M., A plate tectonic-lithotectonic classification of ore deposits, *Arizona Geol. Soc. Digest, XIV,* 1-10, 1981.

Guilbert, J.M., and C.F. Park, Jr., *The Geology of Ore Deposits*, 985 pp., W.H. Freeman and Company, New York, 1986.

Gust, D.A., K.T. Biddle, D.W. Phelps, and M.A. Uliana, Associated Middle to Late Jurassic volcanism and extension in southern South America, *Tectonophysics, 116,* 223-253, 1985.

Haeussler, P.J., D.C. Bradley, R.J. Goldfarb and L.W. Snee, A link between ridge subduction and turbidite-hosted gold mineralization in southern Alaska, *Geol. Soc. Am., Abstracts with Programs,* 27, no. 5, 1995.

Haller, M.J., and O.R. Lapido, The Jurassic-Cretaceous volcanism in the Septentrional Patagonian Andes, *Earth Sci. Rev., 18,* 395-410, 1982.

Hargraves, R.B., Faster spreading or greater ridge length in the Archean?, *Geology, 14,* 750-752, 1986.

Herron, E.M., R. Bruhn, M. Winslow, and L. Chuaqui, Post Miocene tectonics of the margin of southern Chile, in *Island Arcs, Deep Sea Trenches and Back-arc Basins*, Maurice Ewing Series, vol. I, edited by M. Talwani and W.C. Pitman III, pp. 273-283, Am. Geophys. Union, Washington, D.C., 1977.

Herron, E.M., S.C. Cande and B.R. Hall, An active spreading center collides with a subduction zone: A geophysical survey of the Chile Margin Triple Junction, *Geol. Soc. Amer., Memoir 154,* 683-701, 1981.

Hervé, F., J. Davidson, E. Godoy, C. Mpodozis, and V. Covacevich, The Late Paleozoic in Chile: Stratigraphy, structure and possible tectonic framework, *Rev. Acad. brazil. Ciênc., 53,* 361-373, 1981.

Hervé, M., Estudio geológico de la falla Liquiñe-Reloncavi en el area de Liquñe; Antecedentes de un movimiento transcurrente (Provincia de Valdivia), *I Cong. Geol. Chileno Actas,* Santiago, B38- B65, 1976.

Hervé, M., M. Suárez, and A. Puig, The Patagonian batholith south of Tierra del Fuego, Chile: Timing and tectonic implications, *J. Geol. Soc. London, 14,* 909-917, 1984.

Heylmun, E.B., Gold in Tierra del Fuego, *California Mining J.,* 40-42, November, 1993.

Howell, D.G., D.L. Jones and E.R. Schermer, Tectonostratigraphic terranes of the circum-Pacific region, in *Tectonostratigraphic terranes of the circum-Pacific region*, Circum-Pacific Council for Energy and Mineral Resources, Earth Science Series, vol. 1, edited by D.G. Howell, pp. 3-30, 1985.

Hutchison, C.F., *Economic Deposits and Their Tectonic Setting*, 365pp., John Wiley and Sons, New York, 1983.

Hutchinson, R.W., Mineral deposits and metallogeny: Indicators of Earth's evolution, in *Early Organic Evolution: Implications for Mineral and Energy Resources*, edited by M. Schidlowski et al., pp.521-545, Springer-Verlag, Berlin Heidelberg, 1992.

Hutchinson, R.W., and J.L. Burlington, Some broad characteristics of greenstone belt gold lodes, in: *Geology, geochemistry and genesis of gold deposits*, Proceedings Symposium Gold'82, edited by R.P. Foster, pp. 339-372, Balkema Rotterdam, Netherlands, 1984.

Ishihara, S., Significance of the magnetite-series and ilmenite-series of granitoids in mineral exploration, in Proc. 5th Quadrennial IAGOD Symposium, vol. I, edited by J.D. Ridge, pp. 309-312, Stuttgarg, Schweizerbartsche, 1980.

Ishihara, S., The granitoid series and mineralization, *Econ. Geol. 75th Anniversary volume*, 458-484, 1981.

Keith, S.B., and M.M. Swan, Oxidation state of magma series in the southwestern US: implications for geographic distribution of base, precious, and lithophile metal metallogeny, (abstr.), Geol. Soc. Amer, Annual Meeting Program, Phoenix, 723-724, 1987.

Kerrich, R., Perspectives on genetic models for lode gold deposits, *Mineralium Deposita, 28*, 362-365, 1993.

Kerrich, R., and D. Wyman, Geodynamic setting of mesothermal gold deposits: An association with accretionary tectonic regimes, *Geology, 18*, 882-885, 1990.

Koski, R.A., R.C. Lamons, J.A. Dumoulin and R.M. Bouse, Massive sulfide metallogenesis at a late Mesozoic sediment-covered spreading axis: Evidence from the Franciscan complex and contemporary analogues, *Geology, 12*, 137-140, 1993.

Little, M.L. (editor), South America exploration review, Chile, *SEG Newsletter, 15*, 14-15, 1993.

Mitchell, A.H.G., Tectonic settings for emplacement of subduction-related magmas and associated mineral deposits. *Geol. Assoc. Canada,* Special Paper 14, 3-21, 1976.

Mitchell, A.H.G., and M.S. Garson, *Mineral Deposits and Global Tectonic Settings*, Academic Press, London, 405pp., 1981.

Moyle, A.J., B.J. Doyle, H. Hoogvliet, and A.R. Ware, Ladolam gold deposit, Lihir Island, in *Geology of the Mineral Deposits of Australia and Papua New Guinea*, The Australian Institute of Mining and Metallurgy, Melbourne, edited by F.E. Hughes, pp. 1793-1805, 1990.

Mpodozis, C., and Ramos, V., The Andes of Chile and Argentina, in *Geology of the Andes and its Relation to Hydrocarbon and Mineral Resources*, Circum-Pacific Council for Energy and Mineral Resources Earth Science Series 11, edited by G.E. Ericksen, M.T. Cañas P. and J.A. Reinemund, pp. 59-90, 1990.

Nelson, E., and D. Elthon, Petrology, geochemistry and origin of I- and S-type 'granites' in the southern Andes, in *Antarctic Earth Science*, Proc. 4th Int. Symp. on Antarctic Earth Sci., edited by R.L. Oliver, P.R. James, and J.B. Jago, p.372, Australian Academy of Science, Canberra, 1983. (note error in abstract: I-type and S-type should be reversed).

Nelson, E.P., and R.D. Forsythe, Ridge collision at convergent margins: implications for Archean and post-Archean crustal growth, *Tectonophysics, 161*, 307-315, 1989.

Nelson, E.P., I.W.D. Dalziel, and A.G. Milnes, Structural geology of the Cordillera Darwin: Collisional-style orogenesis in the southernmost Chilean Andes, *Ecol. Geol. Helv., 73*, 727-751, 1980.

Nelson, E., D. Elthon, S. Weaver, D. Kammer, and B. Bruce, Regional lithologic variations in the Patagonian batholith, *J. South Am. Earth Sci., 1*, 239-247, 1986.

Nelson, E.P., R. Forsythe, J. Diemer, M. Allen, and O. Urbino, Taitao ophiolite: a ridge collision ophiolite in the forearc of southern Chile (46°S), *Revista Geológica de Chile, 20*, 137-165, 1993.

Nelson, E., R. Forsythe, and I. Arit, Ridge collision tectonics in terrane development, *J. South Amer. Earth Sci., 7*, 271-278, 1994.

Niemeyer, H., J. Skarmeta, R. Fuenzalida, and W. Espinoza, Hoja Península de Taitao y Puerto Aysén, Carta Geológica de Chile, no. 60-61, Servicio Nacional de Geología y Minería, Santiago, 80pp., 1984.

Panteleyev, A., A Canadian Cordillera model for epithermal gold-silver deposits, in *Ore Deposit Models*, Geoscience Canada Reprint Series 3, edited by R.G. Roberts and P.A. Sheahan, pp. 31-43, 1985.

Pereira, J., and C.J. Dixon, Mineralization and plate tectonics, *Mineralium Deposita, 6*, 253-259, 1971.

Petersen, U., Geological framework of Andean mineral resources, in *Geology of the Andes and its Relation to Hydrocarbon and Mineral Resources*, Circum-Pacific Council for Energy and Mineral Resources Earth Science Series 11, edited by G.E. Ericksen, M.T. Cañas P. and J.A. Reinemund, pp. 213-232, 1990.

Rabinowitz, P.D., and J. La Brecque, The Mesozoic South Atlantic Ocean and evolution of its continental margins, *J. Geophys. Res., 84*, 5973-6002, 1979.

Ramos, V., Patagonia: un continente paleazoico a la deriva?, IX Congreso Geológico Argentino Actas 2, 311-325, 1984.

Ramos, V.A., H. Niemeyer, J. Skarmeta, and J. Muñoz, The magmatic evolution of the Austral Patagonian Andes, *Earth Sci. Reviews, 18*, 411-443, 1982.

Roddy, M.S., S.J. Reynolds, B.M. Smith, and J. Ruiz, K-metasomatism and detachment-related mineralization, Harcuvar Mountains, Arizona, *Geol. Soc. Am. Bull., 100*, 1627-1639, 1988.

Ruiz F.C., (and collaborators), *Geología y yacimientos metalíferos de Chile*, 305pp., Inst. Invest. Geológicas, Santiago, 1965.

Ruiz F.C., Distribution and characteristics of Chilean copper deposits, in Geology of the Andes and its Relation to Hydrocarbon and Mineral Resources, edited by G.E. Ericksen, M.T. Cañas P. and J.A. Reinemund, *Circum-Pacific Council for Energy and Mineral Resources Earth Science Series 11*, 245-256, 1990.

Sangster, D.F., Plate tectonics and mineral deposits, *Geoscience Canada 6*, 185-189, 1979.

Sawkins, F.J., *Metal Deposits in Relation to Plate Tectonics*, 461pp., Springer-Verlag, New York, 1984 (second edition 1990).

Sibson, R.H., F. Robert, and H. Poulsen, High angle faults, fluid pressure cycling and mesothermal gold deposits, *Geology, 16*, 551-555, 1988.

Sillitoe, R. H., Andean mineralization: a model for the metallogeny of convergent plate margins, in *Metallogeny and Plate Tectonics*, Geological Association of Canada, Special Paper 14, edited by D.F. Strong, pp.59-100, Waterloo, Ontario, 1976.

Sillitoe, R. H., Speculations on Himalayan metallogeny based on evidence from Pakistan, in *Geodynamics of Pakistan*, edited by A. Farah and K.A. deJong, pp.167-179, Geol. Surv. Pakistan, 1979.

Sillitoe, R. H., Ore Deposits in Cordilleran Settings. *Arizona Geological Society Digest, XIV*, 49-65, 1981.

Skarmeta, J., Evolucion tectónica y paleográfica de los Andes Patagónicos de Aysen (Chile) durante el Neocomiano, *I Congreso Geológico Chileno Actas 1*, B1-B15, 1976.

Solomon, M., Subduction, arc reversal, and the origin of porphyry copper-gold deposits in island arcs, *Geology, 18*, 630-633, 1990.

Stern, C., M.A. Skewes, and A.M. Durán, Volcanismo orogénico in Chile austral, I Congreso Geol. Chileno, Actas 2, F195-F212, 1976.

Strong, D.F. (editor), *Metallogeny and Plate Tectonics*, Geological Association of Canada, Special Paper 14, 660pp., Waterloo, Ontario, 1976.

Strong, D.F., A model for granophile mineral deposits, in *Ore Deposit Models*, Geoscience Canada Reprint Series 3, edited by R.G. Roberts and P.A. Sheahan, pp.59-66, 1985.

Suárez, M., and T.H. Pettigrew, 1976, An Upper Mesozoic island-arc-basin-arc system in the southern Andes and South Georgia, *Geol. Mag., 113*, 305-328, 1976.

Thomas, A., Geología y perspectiva económica del yacimiento polimetálico Cutter-Cove, Provincia de Magallanes, ENAMI, unpublished report, 15pp., 1973.

Wellmer, F.W., E.J. Reeve, E. Wentzlau, and H. Westenberger, Geology and ore deposits of the Toqui District, Aysen, Chile, *Econ. Geol., 78*, 1119-1143, 1983.

White, A.J.R., S.D. Beams, and J.J. Cramer, Granitoid types and mineralization with special reference to tin, in *Plutonism in relation to volcanism and metamorphism*, Proc. 7th CPPP Meeting, edited by N. Yamada, pp.89-100, Toyama, 1977.

Winslow, M., Mechanism for basement shortening in the Andean foreland fold belt of southern South America, in *Thrust and Nappe Tectonics*, Geol. Soc. London Spec. Publ. 9, edited by K.R. McClay and N.J. Price, pp.513-526, 1979.

Uyeda, S., Comparative subductology, *Episodes 2*, 19-24, 1983.

Uyeda, S., and H. Kanamori, Back-arc opening and the mode of subduction, *J. Geophys. Res., 84*, 1049-1061, 1979.

Uyeda, S., and C. Nishiwaki, Stress field, metallogenesis and mode of subduction, in *The Continental Crust and its Mineral Deposits*, Geol. Assoc. Can. Spec. Paper 20, edited by D.W. Strangway, pp.323-340, 1980.

E. P. Nelson, Department of Geology and Geological Engineering, Colorado School of Mines, Golden, CO 80401

Hazards and Climatic Impact of Subduction-Zone Volcanism: A Global and Historical Perspective

Robert I. Tilling

Volcano Hazards Team, U.S. Geological Survey, Menlo Park, California

Subduction-zone volcanoes account for more than 80 percent of the documented eruptions in recorded history, even though volcanism--deep and, hence, unobserved--along the global oceanic ridge systems overwhelmingly dominates in eruptive output. Because subduction-zone eruptions can be highly explosive, they pose some of the greatest natural hazards to society if the eruptions occur in densely populated regions. Of the six worst volcanic disasters since A.D. 1600, five have occurred at subduction-zone volcanoes: Unzen, Japan (1792); Tambora, Indonesia (1815); Krakatau, Indonesia (1883); Mont Pelée, Martinique (1902); and Nevado del Ruiz, Colombia (1985). Sulfuric acid droplets in stratospheric volcanic clouds produced by voluminous explosive eruptions can influence global climate. The 1815 Tambora eruption caused in 1816 a decrease of several Celsius degrees in average summer temperature in Europe and the eastern United States and Canada, resulting in the well-known "Year Without Summer." Similarly, the eruptions of El Chichón (Mexico) in 1982 and of Mount Pinatubo (Philippines) in 1991 lowered average temperatures for the northern hemisphere by as much as 0.2 to 0.5 °C, respectively. However, eruption-induced climatic effects of historical eruptions appear to be short-lived, lasting at most for only a few years.

1. INTRODUCTION

Powerful explosive eruptions are awesome culminations of the magmatic and tectonic processes operative in subduction zones, and they can have severe adverse effects on civilizations. Although subduction-zone volcanism produces only about 15 percent of the averaged global volcanic output (~ 4 km^3/yr), it accounts for more than 80 percent of documented historical eruptions (Figure 1). This is simply because the vast majority of eruptions occur sight unseen along the global oceanic ridge systems that crisscross the deep ocean floor. To date, none of these deep-sea eruptions have been observed in real time, whereas subduction-zone and other subaerial eruptive activity rarely goes unreported.

Beginning in 1980, a number of eruptions at subduction-zone volcanoes--such as Mount St., Helens (U.S.A., 1980), El Chichón (Mexico, 1982), Galunggung (Java, Indonesia, 1982), Nevado del Ruiz (Colombia, 1985), Redoubt (Alaska, U.S.A., 1989), Mount Pinatubo (Philippines, 1991), and Unzen (Japan, 1991)--have caused fatalities and (or) substantial economic loss, gaining notoriety and capturing the attention of scientists, public officials, and the populace worldwide [*Tilling*, 1989; *Yanagi et al.*, 1992; *Wolfe*, 1992; *Chester*, 1993; *Tilling and Lipman*, 1993]. In addition to wreaking devastation locally, some of these eruptions may have influenced global climate [*Simarski*, 1992; *Johnson*, 1993].

2. VOLCANIC HAZARDS: DIRECT AND INDIRECT

Since A.D. 1600, more than 260, 000 people have died from eruptions (Figure 2), overwhelmingly involving volcanoes in subduction zones. In the 17th-19th centuries, most deaths were from *indirect* volcanic hazards, most

Subduction: Top to Bottom
Geophysical Monograph 96
Copyright 1996 by the American Geophysical Union

332 HAZARDS AND IMPACTS OF SUBDUCTION-ZONE VOLCANISM

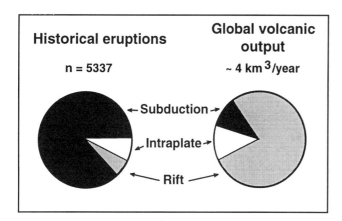

Fig. 1. Comparison of historical eruptions with global volcanic output apportioned to tectonic regimes: subduction zone, intraplate (hotspot), and rift (including spreading centers). [Data sources: *Crisp*, 1984; *Simkin*, 1993].

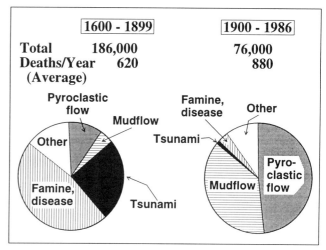

Fig. 2. Eruption-related fatalities apportioned to principal cause for the periods 1600-1899 and 1900-1986. [Data sources: *Blong*, 1984; *Tilling*, 1989].

notably tsunami and post-eruption famine and disease. However, in the 20th century, the principal killers were *direct* hazards commonly associated with subduction-zone volcanoes: namely, pyroclastic flows and surges and volcanic debris avalanches and flows (mudflows). The large reduction in fatalities from indirect hazards in this century reflects two factors: 1) the existence of modern, rapid communications and disaster-relief delivery systems; and 2) the non-occurrence (to date) of any large volcanogenic tsunami. Most of the deaths in the 20th century resulted from only two events, the 1902 eruption of Mont Pelée, Martinique, and the 1985 eruption of Nevado del Ruiz, Colombia (Figure 3). A comprehensive review of volcano-hazards mitigation, including relevant demographic and socio-economic factors, is given by *Tilling* [1989].

Of the six deadliest eruptions since A.D. 1600 (Figure 4), only the Lakagígar (also called Laki) in Iceland did not involve an explosive eruption of a subduction-zone volcano. The Ruiz eruption, which produced only about 0.03 km³ of magma, also tragically demonstrated that even a very small eruption can cause many fatalities (> 25,000), especially if there is potential for significant interaction between eruptive activity and ice and snow. By contrast, the 1912 eruption of Novarupta (Katmai, Alaska)--the most voluminous to date in this century (~ 13 km³ of magma)--produced no known deaths and only minimal property damage because it occurred in a sparsely populated region.

Beginning in the 1980s, another hazard of subduction-zone volcanism--beyond imagination at the beginning of the 20th century--became a cause for concern: encounters between volcanic ash and jet aircraft and airport facilities [*Casadevall*, 1992, 1993, 1994]. This modern-day volcanic hazard will inexorably become more severe with the continuing increase in air travel, because many international air routes cross zones of active volcanism [*Casadevall and Thompson*, 1995]. Since the early 1970s, more than 60 airplanes, mostly commercial jetliners, have been damaged by such encounters, with several of them experiencing total power loss, necessitating emergency landings. During the 15 December 1989 eruption of Redoubt Volcano, Alaska, a new Boeing 747-400 was heavily damaged by in-flight encounter with ash and incurred repair costs, which included the replacement of all four engines, exceeding 80 million dollars! [*Casadevall*, 1992].

3. CLIMATIC IMPACT OF EXPLOSIVE ERUPTIONS

A connection between volcanism and climate change is now well demonstrated and generally accepted [e.g., *Robock*, 1991; *Simarski*, 1992; *Johnson*, 1993]. The prevailing view is that the amount of gas (principally SO_2)--not the tephra--injected into the stratosphere largely determines whether an explosive eruption might affect climate. The SO_2 forms an aerosol layer of sulfuric acid droplets; this layer tends to cool the troposphere by reflecting solar radiation, and to warm the stratosphere by absorbing radiated Earth heat. With the single exception of the Laki fissure eruption in 1783, which occurred in a divergent plate-boundary regime, all other known historical climate-influencing eruptions have been from subduction-zone volcanoes, including Tambora (Indonesia, 1815), Krakatau (Indonesia, 1883), Santa María (Guatemala, 1902),

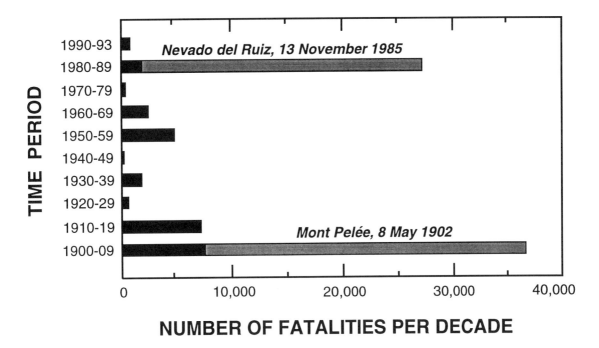

Fig. 3. Volcano-related fatalities by decade for the 20th century through May 1993; shaded bars show fatalities from the two deadliest events. [From *Tilling and Lipman*, 1993, Figure 1].

Novarupta (Katmai, U.S.A 1912), Agung (Indonesia, 1963), El Chichón (Mexico, 1982), and Mount Pinatubo (Philippines, 1991). The stratospheric volcanic cloud associated with the 1815 Tambora eruption--the largest in recorded history--resulted in the well-documented "Year Without Summer" (Figure 5) in the following year [*Stommel and Stommel*, 1983].

The 1980 Mount St. Helens eruption apparently had minimal climatic impact, producing only negligible change (0 to -0.1 °C) in average surface temperature for the northern hemisphere. In contrast, the 1982 El Chichón and the 1991 Pinatubo eruptions lowered surface temperature by 0.2 to 0.5 °C [*Simarski*, 1992]. Significantly, the Mount St. Helens magma contains much less sulfur than those of El Chichón or Pinatubo, which contain *anhydrite* ($CaSO_4$) as a primary phase (Figure 6). Interestingly, the 1985 Ruiz eruption, despite its very small magma volume (0.03 km^3), was exceptionally high in released SO_2 [*Williams et al.*, 1986].

While the 1991 Pinatubo eruption clearly perturbed the atmosphere, the effect seems to be relatively short-lived, as demonstrated by measurement of ozone levels in the northern hemisphere. The greatly accelerated erosion of the ozone level in the spring of 1993 caused by Pinatubo had largely dissipated by spring of 1994 [*Stone*, 1994]. However, the effects of Pinatubo are not clearly understood by atmospheric scientists. For example, the global level of carbon monoxide, which had been steadily increasing, began to level off and downturn with the Pinatubo eruption (Figure 7); similar patterns have also been noted for carbon dioxide and other closely monitored "greenhouse" gases.

4. REDUCING VOLCANO RISK IN SUBDUCTION-ZONE REGIONS

Not only are the subduction-zone regions highly prone to earthquake and volcano hazards, they also are among the most densely populated parts of the world. By the year 2000, the population at risk from volcanic hazards is likely to increase to at least 500 million; about 90 percent of these people live along the subduction zones of the circum-Pacific region [*Tilling*, 1989; *Tilling and Lipman*, 1993]. It is sobering to note that the people ($\geq$ 500 million) currently at risk represents the entire world population at the start of the 17th century.

Although eruption frequency and hazards severity have not increased in recent centuries [*Simkin*, 1993], the problem of reducing volcano risk inexorably becomes more acute with continuing explosive population growth, especially for many developing countries rimming the Pacific Ocean. Against this daunting scenario, *Tilling and*

VOLCANO	YEAR	EJECTA VOLUME (km^3)	DEATHS
Ruiz, Colombia	1985	$< 10^{-2}$	25,000
Novarupta, U.S.A.	1912	10^1	0
Mont Pelée, Martinique	1902	10^{-1}	30,000
Krakatau, Indonesia	1883	10^1	36,000
Tambora, Indonesia	1815	$> 10^2$	92,000
Unzen, Japan	1792	10^{-1}	15,000
Lakagígar, Iceland (Laki)	1783	10^1	9,000

> 90% Post-eruption starvation

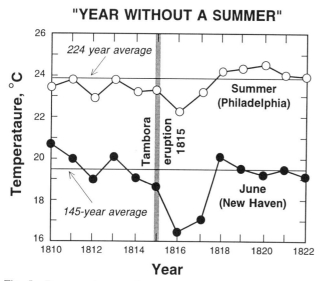

Fig. 4. The six deadliest eruptions since A.D. 1600; the 1912 Novarupta (Katmai) eruption is also shown to emphasize that eruption size (ejecta volume) is not primary factor in number of deaths. [Data sources: *Blong*, 1984; *Tilling*, 1989].

Fig. 5. Decrease in annual summer temperature in the eastern United States caused by the 1815 eruption of Tambora Volcano, Indonesia. [Modified from *Johnson*, 1993].

Lipman [1993, 279-280] have identified the following critical issues in reducing volcano risk:

1. need for additional reliable, real-time monitoring systems;
2. new approaches to eruption prediction;
3. study of more volcanoes;
4. more effective interaction with civil authorities and the public; and

ERUPTION (Month,Year)	Erupted volume, km^3 (DRE)	Sulfur in magma, (wt. % SO_3)	SO_2 released, megatons	Primary anhydrite ($CaSO_4$)	SO_2 released / Erupted volume
Mount St. Helens (May, 1980)	~ 0.2	< 0.01	0.8	NO	~ 4
El Chichón (March-April, 1982)	~ 0.4	1 - 4	~ 7	YES	~ 18
Nevado del Ruiz (November, 1985)	~ 0.02	< 0.01	~ 2	YES ??	~ 100
Mount Pinatubo (June, 1991)	3 - 5	0.3 - 0.5	20	YES	~ 5

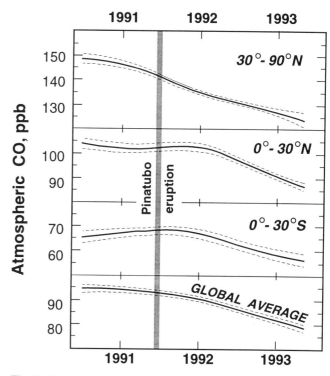

Fig. 6. Comparison of magma volumes (dense-rock equivalent, DRE), release of SO_2, and the ratio SO_2 released/magma volume for selected explosive eruptions in the period 1980-1991. [Data sources: *Bluth et al.*, 1993; *Calvache V.*, 1990; *Lipman et al.*, 1981; *Rose et al.*, 1984; and *Wolfe*, 1992].

Fig. 7. Reversal of increasing level of carbon monoxide attributed to atmospheric effects of the 15 June 1991 eruption of Mount Pinatubo, Luzon, Philippines. The upper three curves reflect differences in concentration according to latitude range; the bottom curve gives the global average. Thin dashed lines bracket 1-σ confidence intervals. [Modified from *Novelli et al.*, 1994, Figure 4].

5. more effective international cooperation.

These issues can begin to be addressed with present-day technology and modestly funded, but stable, national or international programs, but many countries lack the resources and (or) the political will to initiate or adequately sustain such programs. To increase governmental and public support for programs to reduce volcano risk, volcanologists must play a more active role to enhance awareness of volcano hazards and to develop rapport and effective communications with civil authorities and the affected populace [*Peterson and Tilling*, 1993].

Acknowledgments. This brief paper is both a scientific and philosophical statement, and I wish to thank my colleagues in the volcanologic community for helping me to frame it. I also greatly appreciate the careful and thoughtful reviews of earlier drafts of the paper by Robert L. Christiansen, Michael Clynne, Donald W. Peterson, William I. Rose, and Stanley N. Williams.

REFERENCES

Blong, R.J., *Volcanic Hazards: A Sourcebook on the Effects of Eruptions*, Academic Press, San Diego, Calif., 424 pp., 1984.

Bluth. G.J.S., C.C. Schnetzler. A.J. Krueger. and L.S. Walter, The contribution of explosive volcanism to global atmospheric sulfur dioxide concentrations, *Nature, 366,* 327-329, 1993.

Calvache V., M.L., Pyroclastic deposits of the November 13, 1985 eruption of Nevado del Ruiz volcano. Colombia, *J. Volcanol. Geotherm. Res., 41,* 67-78, 1990.

Casadevall. T.J. (Ed.), Volcanic ash and aviation safety: Proceedings of the First International Symposium on Volcanic Ash and Aviation Safety: *U.S. Geol. Survey Bulletin 2047,* 450 pp., 1994.

Casadevall, T.J., Volcanic hazards and aviation safety: Lessons of the past decade: *FAA Aviation Safety Journal, 2,* 3-11, 1992.

Casadevall, T.J., Volcanic ash and airports: Discussions and recommendations from the Workshop on Impacts of Volcanic Ash on Airport Facilities. Seattle, Washington. April 26-28, 1993, *U.S. Geol. Survey Open-File Report 93-518,* 52 pp., 1993.

Casadevall, T.J., and T.B. Thompson, World map of volcanoes and principal aeronautical features: *U. S. Geol. Survey Geophysical Investigations Map GP-1011,* 1995.

Chester, David, *Volcanoes and Society,* Edward Arnold (a Division of Hodder & Stoughton), London, 351 pp., 1993.

Crisp, J.A., Rates of magma emplacement and volcanic output, *J. Volcanol. Geotherm. Res., 20,* 177-211, 1984.

Johnson, R.W., *Volcanic eruptions & atmosphere change,* AGSO Issues Paper. Australian Geological Survey Organization, Canberra, 36 pp., 1993.

Lipman, P.W., D.R. Norton, J. E. Taggart, Jr., E.L. Brandt, and E.E. Engleman, Compositional variations in 1980 magmatic deposits, *in* Lipman, P.W. and D.R. Mullineaux (Eds.), The 1980 eruptions of Mount St. Helens. Washington, *U.S. Geol. Survey Prof. Paper 1250,* 631-640, 1981.

Novelli, P.C., K.A. Masarie, P.P. Tans, and P.M. Lang, Recent changes in atmospheric carbon monoxide, *Science, 263,* 1587-1590, 1994.

Peterson, D.W., and Tilling, R.I., .Interactions between scientists, civil authorities and the public at hazardous volcanoes, in *Active Volcanoes: Monitoring and Modelling*, edited by C.R.J. Kilburn and G. Luongo, pp. 339-365, UCL Press, London, 1993.

Robock, Alan, The volcanic contribution to climate change of the past 100 years, in *Greenhouse-Gas-Induced Climatic Change: A Critical Appraisal of Simulations and Observations,* edited by M.E. Schlesinger, pp. 429-443, Elsevier Science Publishers B.V., Amsterdam, 1991.

Rose, W.I., Jr., T.J. Bornhorst, S.P. Halsor, W.A. Capaul, and P.J. Plumley, Volcán El Chichón. Mexico: pre-1982 S-rich eruptive activity, *J. Volcanol. Geotherm. Res., 23,* 147-167, 1984.

Simarski, L.T., *Volcanism and climate change,* Special Report, Amer. Geophys. Union Washington, D.C., 27 pp., 1992.

Simkin, Tom, Terrestrial volcanism in space and time, *Annu. Rev. Earth Planet. Sci., 21,* 427-452, 1993.

Stommel, H., and E. Stommel, *Volcano Weather: The Story of 1816, The Year Without a Summer,* Seven Seas Press, Newport, Rhode Island, U.S.A., 177 pp., 1983.

Stone, Richard, editor, Ozone has recovered from Pinatubo's jolt *(in* Random Samples) *Science, 264,* 1078, 1994.

Tilling, R I., Volcanic hazards and their mitigation: Progress and problems, *Rev. Geophys., 27,* 237-269, 1989.

Tilling, R.I., and P.W. Lipman, Lessons in reducing volcano risk, *Nature, 364,* 277-280, 1993.

Williams, S.N., R.E. Stoiber, G.P. Garcia, A. Londoño C., J.B. Gemmell, D.R. Lowe, and C.B. Connor, Eruption of the Nevado del Ruiz Volcano, Colombia, on 13 November 1985: Gas flux and fluid geochemistry, *Science,* **233,** 964-967, 1986.

Wolfe, E.W., The 1991 eruptions of Mount Pinatubo, Philippines, *Earthquakes and Volcanoes, 23,* 5-35, 1992.

Yanagi, T., H. Okada, and K. Ohta, *Unzen Volcano: The 1990-1992 eruption,* The Nishinippon & Kyushu University Press, Fukuoka, Japan, 137 pp., 1992.

R.I. Tilling, Volcano Hazards Team, U.S. Geological Survey, MS-910, 345 Middlefield Road, Menlo Park, CA 94025-3591.

Eclogite Formation and the Rheology, Buoyancy, Seismicity, and H_2O Content of Oceanic Crust

Bradley R. Hacker[1]

Department of Geological and Environmental Sciences, Stanford University, Stanford, California

A broad spectrum of variably altered igneous rocks with a wide range of grain sizes are compressed and heated over a wide range of pressure-temperature paths in subduction zones. Although experimental kinetic data cannot be extrapolated to predict the rates of blueschist and eclogite formation in nature, textural data from rocks indicate that transformation below temperatures of 150°C is minimal. Complete transformation of volcanic rocks occurs by ~250°C, but incomplete transformation of gabbroic rocks heated to 800°C has been observed. There are important consequences to the rapid transformation of volcanic rocks and the metastable persistence of gabbroic rocks into the blueschist and eclogite stability fields. Fast seismic velocities should be evident first in the upper oceanic crust and may be substantially retarded in the lower oceanic crust. The upper oceanic crust will be denser than asthenosphere before the lower oceanic crust. Early in the process of eclogite formation, volcanic rocks will be placed in deviatoric tension and the underlying coarser grained rocks in compression; with further reaction, the state of stress in gabbroic rocks will change from compressive to tensile. Earthquakes at shallow depths should be extensional in basalt and contractional in gabbro, changing at deeper levels to extensional throughout the crust.

INTRODUCTION

Formation of eclogite is a key factor in plate tectonics, influencing the size and shape of plates as well as their age and rate of disappearance from the Earth's surface. Subduction zones have long been of interest because they are the builders of continents, collecting material from the ocean basins and constructing magmatic arcs. In spite of the dynamic nature of subduction zones, much of our understanding of their behavior stems from equilibrium thermodynamics. The equilibrium viewpoint has proven a valuable first means of addressing processes in subduction zones, but a more in-depth comprehension requires consideration of the rates and mechanisms by which phase changes occur during subduction.

[1] Also at U.S. Geological Survey, Menlo Park, California

Subduction: Top to Bottom
Geophysical Monograph 96
This paper is not subject to U.S. copyright.
Published in 1996 by the American Geophysical Union

This paper summarizes the rates and textures of densification reactions in subduction zones by examining experimental and field studies, and then considers the effects of these findings on the rheology, buoyancy, seismicity, and H_2O content of subducting oceanic crust.

WHAT ARE BLUESCHIST AND ECLOGITE?

Blueschist facies mafic rocks are characterized by the coexistence of sodic amphibole and lawsonite at low temperature or epidote at high temperature [*Ernst*, 1963]. They may also contain garnet or omphacite. Eclogite *sensu stricto* is mafic rock consisting of garnet and omphacitic clinopyroxene with or without additional minor phases [*Coleman et al.*, 1965]. The geophysical importance of eclogite is that it represents the highest density commonly attained by crustal rocks. Omphacite and garnet have densities of 3.2–3.4 gm/cm^3 and 3.6–4.0 gm/cm^3, a marked elevation beyond the density of a typical basalt at 2.9 gm/cm^3 [*Carmichael*, 1989]. This shift from positive to negative buoyancy relative to asthenosphere, which has a density of roughly 3.23 gm/cm^3 [*Cloos*, 1993], is as important a factor in controlling plate subduction as are conductive cooling of the lithosphere and the olivine → spinel transformation in subducting mantle [*Ahrens and Schubert*, 1975].

PRINCIPAL DENSIFICATION REACTIONS

A broad spectrum of reactions cause rocks to become more dense in subduction zones. At the high temperatures found in very hot subduction zones the transformation of basaltic rocks to garnet granulite and then eclogite involves the breakdown of plagioclase, addition of Na_2O to cliopyroxene and the growth of garnet (Figure 1). PT conditions for the appearance of garnet have been determined by experimental study [*Green and Ringwood*, 1967; *Ito and Kennedy*, 1971; *Liu et al.*, 1993; *Poli*, 1993], and occursat lower pressures/higher temperatures in SiO_2-undersaturated [*Ito and Kennedy*, 1971] and MgO-rich rocks [*Green and Ringwood*, 1967]. The disappearance of plagioclase is less well defined by the same studies and depends more strongly on rock composition.

The transformation of basaltic rocks to blueschist and then eclogite at the low temperatures that prevail in most subduction zones is more complicated because of the prevalence of disequilibrium in nature and experiment. One reaction inferred to have transformed blueschist to eclogite at several localities worldwide is epidote + glaucophane → garnet + omphacite ± paragaonite + quartz + H_2O [*El-Shazly et al.*, 1990; *Ridley and Dixon*, 1984; *Schliestedt*, 1986].

Eclogite facies conditions exist at temperatures >500°C and pressures >1.2 GPa (Figure 1). Eclogite facies *rocks* do not necessarily exist wherever such conditions prevail in the Earth because of sluggish kinetics. All reactions require overstepping of equilibrium to overcome local free energy increases related to attachment or detachment of atoms at interfaces, and all reactions involve diffusion or atom–atom bond rearrangement. Perceptible reaction only occurs at sufficiently high diffusivities and reaction free energies.

HOW FAST ARE REACTIONS IN THE EARTH?

Experimental and Theoretical Inferences

How long can rocks remain out of equilibrium at elevated pressures and temperatures? Specifically, how long can rocks remain within the eclogite stability field before transforming partially or completely to the stable phase assemblage? Information relevant to this question derives from two sources: experiment and study of natural rocks [*Rubie*, 1990]. Experimental kinetic data collected on

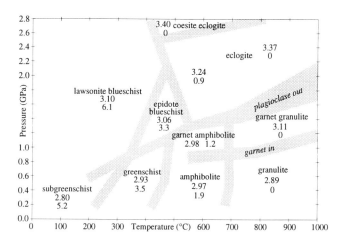

Fig. 1. Metamorphic facies and reactions pertinent to eclogite formation, after [*Liou*, 1971], [*Maresch*, 1977; *Brown*, 1977; *Brown and Ghent*, 1983; *Evans*, 1990], [*Green and Ringwood*, 1967; *Ito and Kennedy*, 1971; *Liu et al.*, 1993; *Oh and Liou*, 1990; *Poli*, 1993]. Three- and two-digit number for each facies indicate density (g/cm^3) and H_2O content (wt%) calculated using NCMASH model of Peacock [1993] assuming equilibrium (i.e., no kinetic hindrance).

powdered materials may not be directly applicable to the majority of geologic situations where rocks are tightly packed crystalline aggregates—principally because the interfacial free energy of a rock is much less than that of a powder. For extrapolation to geologic conditions, data on phase transformations in unpowdered materials are required—information that is exceedingly sparse.

Recent experiments on reactions known to occur in subduction zones demonstrate that most kinetic data cannot be extrapolated to the Earth and that H_2O has major effects on reaction rate and texture. For instance, *Holland* [1980] observed jadeite growth in 24 hr from powdered albite + quartz at 900°C and a pressure overstep of 50 MPa. In contrast, albite → jadeite + quartz experiments by *Hacker et al.* [1993] on albite rock showed no transformation in 24 hours at 1100°C at a pressure overstep of 500 MPa!

Quantitative transformation-rate measurements are available for the calcite ↔ aragonite reaction. As with albite, the calcite → aragonite transformation in Carrara marble [e.g., *Hacker et al.*, 1992] is much slower than in calcite powder. For example, at 600°C and 2.0 GPa confining pressure, powdered calcite transforms 50% to aragonite in <1 hour [*Brar and Schloessin*, 1979, p. 1409], whereas 192 hours were required for similar conversion in unpowdered marble. *Carlson and Rosenfeld* [1981] and *Liu and Yund* [1993] measured the rate of the aragonite → calcite reaction; extrapolated to geologic timescales and grain sizes, their data imply that 1 mm aragonite grains will revert completely to calcite in less than 1 m.y. at temperatures as low as 200°C. *Hacker et al.* [1992] explored the reverse transformation in marble and found a similar result, that the calcite → aragonite transformation occurs within less than 1 m.y. at 200°C for oversteps of <100 MPa. The reaction depends strongly on temperature, such that even several GPa overstep at temperatures of <100°C are unlikely to promote transformation.

The catalytic action of H_2O is well known from kinetic studies of powders and rocks. *Hacker and Kirby* [1993] deformed polycrystalline, vacuum-dried albite rocks at strain rates of 10^{-4}–10^{-6} s^{-1} at temperatures of 600–800°C and confining pressures of 1.0–2.0 GPa. In spite of extreme pressure overstepping and extreme sample strain, deformed anhydrous samples contained no jadeite. Differential stresses as high as 2000 MPa produced maximum normal stresses up to 4.0 GPa—all at temperatures where the equilibrium pressure is <2.0 GPa. Some of these samples were strained 70%, and individual grains attained aspect ratios of greater than 10:1. In marked contrast to this, the addition of 1 wt% H_2O produced partial reaction at temperatures as low as 600°C, even in undeformed samples.

More complicated experiments measured the olivine → spinel transformation in hot-pressed Mg_2GeO_4 [*Burnley*, 1990] and Ni_2SiO_4 [*Rubie et al.*, 1990] at pressures and temperatures of 800–1900°C and 1–15 GPa. When extrapolated to geologic conditions these investigations suggest that transformation may be suppressed to depths 150–200 km below the equilibrium boundary in rapidly subducting slabs.

In summary, existing experimental data indicate that H_2O has a major catalytic affect on reaction rates and that some reactions are geologically fast (e.g., calcite→aragonite) whereas others are probably slow (olivine → spinel). We have essentially no experimental information on reactions forming blueschist and eclogite.

Metamorphic Textures

Observations on rocks show that transformation to blueschist and eclogite may be suppressed long after rocks have passed into the respective stability fields of these assemblages. Figure 2 was constructed from studies worldwide that reported not only peak metamorphic pressures and temperatures but also textural information about the degree of reaction. Rocks reported as feebly recrystallized or whose high-pressure phases were so minor as to require detection by electron microscopy are shown in Figure 2 as "feeble, 25%." Rocks containing abundant relict phases (usually clinopyroxene) or partially developed

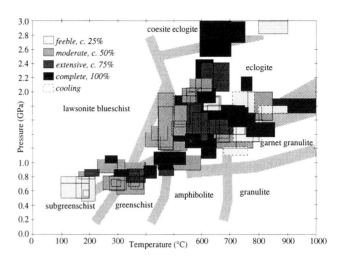

Fig. 2. Reported extents of transformation, pressures, and temperatures, for blueschist and eclogite facies metamorphism worldwide. References and data used to compile figure are available from the author on request.

reaction coronas are denoted as "moderate, 50%", and those with rare relict phases are labeled "extensive, 75%". Metamafic rocks specifically described as lacking relict phases are shown as "complete, 100%".

There are some important limitations to Figure 2. 1) Reported peak pressures and temperatures may not have occurred at the same time and, at best, represent only a single way-station on the PT path experienced by the rocks. 2) How long the rocks were at the reported pressures and temperatures is unknown. 3) Pressure estimates for many rocks are *minima* because of the high variance of the phase assemblages; these are denoted by U-shaped lines rather than rectangles. 4) Many high-pressure rocks experienced substantial post-high pressure alteration that likely consumed existing relict phases; studies reporting severe alteration of this type were excluded from consideration. 5) Fine-grained rocks react faster than coarse-grained rocks because of increased interfacial free energy; gabbroic rocks are only partially transformed at conditions where volcanics are completely recrystallized. Numerous papers that do not specifically state whether relict phases are present or absent had to be excluded from Figure 2, but this may simply mean that the absence of relics was not significant for the purposes of the study. Thus Figure 2 should be taken as an illustration of the transformation behavior of coarse-grained, i.e., gabbroic rocks.

With these limitations in mind, Figure 2 indicates that feebly metamorphosed rocks generally persist to temperatures of 200–250°C. Moderate transformation begins at 200–250°C and is found in rocks that were heated to just above 600°C. There are a few cases of extremely limited reaction at high temperature shown by dashed boxes in Figure 2. Incomplete conversion of gabbro to eclogite is known to have happened at temperatures >700°C and pressures of 1.5–2.0 GPa in the presence of aqueous fluid [*Mørk*, 1985]. Suppression of transformation of gabbro to eclogite was reported from rocks that reached 800–900°C and pressures of >2.8 GPa in the absence of fluid [*Zhang et al.*, 1995]. Oddly enough, there is little difference in the PT range over which extensive to complete transformation have been found. Both have been observed from rocks heated to less than 300°C and yet partial transformation of gabbro to eclogite has occurred at 760–850°C and 1.6–2.0 GPa [*Indares and Rivers*, 1995]. In general, 500–600°C marks the transition to extensive or complete reaction for most rocks. A few cases, however, indicate that the transformation of gabbro to eclogite may be suppressed to much higher temperatures.

Fe-rich rocks transform to eclogite along high P/T trajectories more readily than Mg-rich rocks because reactions involving Fe-rich phases generally occur at lower temperatures [*Bohlen et al.*, 1983]. Field evidence of this— MgO-rich rocks with abundant igneous relics adjacent to Fe-rich rocks with rare relict phases—has been noted at Bardoney Valley in the Alps [*Reynard and Ballevre*, 1988], at Flemsøy, Norway [*Mørk*, 1985], and in Piedmont ophiolites [*Pognante*, 1991; *Sandrone et al.*, 1986].

Eclogite formation is faster in finer grained rocks. For example, in the Bardoney Valley of the Alps, abundant relict augites are present in gabbros, but absent in metavolcanics [*Reynard and Ballevre*, 1988]. In the Sesia Lanzo zone farther east, amphibolite transformed to eclogite contains relics in coarse rocks but none in fine-grained rocks [*Lardeaux and Spalla*, 1991].

Not only is H_2O proven to accelerate reactions in the laboratory, numerous localities demonstrate that this is true in nature as well. Mafic granulite in the Musgrave Ranges of Australia cooled into the eclogite stability field and, though the bulk of the rock is unaltered, local shear zones contain omphacite + garnet + minor zoisite [*Ellis and Maboko*, 1992]; the presence of zoisite implies that the reaction was fluxed by H_2O. At Holsnøy in the Bergen Arcs, zoisite-bearing eclogite haloes in undeformed rock surrounding deformed veins also indicate that fluid drove reaction [*Boundy et al.*, 1992; *Klaper*, 1990]. Deformation is involved in these fluid-mediated transformations inasmuch as the fluid must make its way to the reaction site via cracks.

COMPOSITION OF IGNEOUS OCEANIC CRUST

To understand the transformation of oceanic crust to blueschist and eclogite we must know the composition and structure of oceanic crust. Oceanic crust has been studied by dredging and drilling in ocean basins, and by the examination of ophiolites. The Samail ophiolite in Oman is the best-exposed, largest, least-deformed, and perhaps most-studied ophiolite in the world. Plutonic rocks of the Samail ophiolite are divisible into a lower layered sequence and an upper isotropic section. The lower 2.3 km (avg.) is well-foliated and lineated, layered ultramafic (chiefly wehrlite, dunite, and clinopyroxenite) through mafic cumulates [*Pallister and Hopson*, 1981]. Olivine predominates at the base of the section, whereas plagioclase is dominant at the top [*Juteau et al.*, 1988b]. Three-quarters of the section is gabbro—typically olivine–clinopyroxene or clinopyroxene gabbro with minor hornblende and FeTi oxides—but at rare localities it is gabbronorite [*Juteau et al.*, 1988a]. The gabbro consists of mm-size grains of 55% plagioclase, 35% clinopyroxene, 10% olivine, and rarely Ti-pargasite [*Browning*, 1984]. Overlying the layered section are 200–900 m of isotropic high-level mafic rocks, chiefly hornblende-clinopyroxene gabbro [*Pallister and Hopson*, 1981]. Grain sizes range from mm to pegmatitic [*Juteau et al.*, 1988a]. Sheeted dikes contain abundant to sparse plagioclase and augite [*Hopson et al.*, 1981]. The volcanic rocks are non- to moderately vesicular, almost aphyric pillows and rare brecciated flows and massive flows [*Ernewein et al.*, 1988].

Mid-ocean-ridge basalts (MORBs) typically are glassy or contain plagioclase with less olivine [*Hekinian*, 1982], while rocks crystallized in intraoceanic magmatic arcs generally contain plagioclase and augite [*Ewart*, 1982]. Mid-ocean-ridge basalts (MORBs) are subalkaline tholeiites that typically contain plagioclase with less olivine [*Hekinian*, 1982]. Oceanic plateaus, which make up a sizeable proportion of oceanic crust, are similar in their major element chemistry to MORB, but are 10–40 km thick [*Cloos*, 1993]. Hotspot lavas and propagating rifts are enriched in Fe and Ti relative to MORB [*Flower*, 1991]. Lavas produced at fast spreading ridges such as the Pacific have lower Mg/Fe ratios than MORB erupted at slow spreading ridges [*Batiza*, 1991].

ALTERATION OF IGNEOUS OCEANIC CRUST

Because phase transformations are strongly influenced by the availability of H_2O and the extent and type of pre-existing alteration, understanding the alteration of oceanic crust is critical to evaluating when and where phase transformations occur. Variable alteration of oceanic crust to lower grade mineral assemblages occurs by active hydrothermal circulation at spreading centers and by later weathering on the seafloor [*Alt et al.*, 1986; *Humphris and Thompson*, 1978]. Some rocks escape alteration entirely, while others are wholly recrystallized to new minerals. Most typically, lavas are variably metamorphosed to subgreenschist facies, dikes change partially to greenschist facies, and gabbro recrystallizes weakly to amphibolite-facies minerals. The pressures at which these metamorphic assemblages form are relatively low (≤300 MPa; determined by ocean depth and the thickness of oceanic crust) and the formation temperatures may range up to ~650°C.

Our best information about the alteration of oceanic lavas and dikes comes from DSDP hole 504B [*Alt et al.*, 1989; *Alt et al.*, 1986]. There the lavas contain ~10% alteration haloes around veins and cracks, leading to about 1 wt% H_2O in the bulk rock. Veins and vugs contain Fe-hydroxides, clay, celadonite, smectite, zeolites, and carbonate; groundmass minerals are altered to Fe-hydroxides and clay; plagioclase is partly altered to smectite; olivine is partly to totally replaced by smectite and minor calcite; glass is partly to completely replaced by clay; and pyroxenes are unaltered. Alteration in the dikes increases down section to a maximum of ~50 vol%. Olivine is replaced by chlorite, clay, and talc; plagioclase is partially to totally replaced by albite + zeolites, chlorite, or clay; and augite is unaltered high in the section and rimmed by actinolite deeper in the section. The H_2O content of the dikes averages 2 wt% [*Alt et al.*, 1989; *Alt et al.*, 1986].

Good information on the alteration of plutonic rocks comes from the Samail ophiolite and ODP core holes. Alteration of the Samail layered plutonic rocks is limited to rare replacement of clinopyroxene by actinolite [*Lippard et al.*, 1986] and alteration of the isotropic plutonic rocks is limited to the growth of up to 10% actinolite and FeTi oxides, and the infilling of cavities with epidote + sphene + prehnite + quartz [*Ernewein et al.*, 1988]. In contrast, gabbroic rocks drilled from ODP hole 735B in the Indian Ocean show widespread amphibolite-facies alteration from 0–100%, with an average of perhaps 20–30% [*Dick et al.*, 1991]. Olivine and pyroxene are replaced by hornblende, there is partial to complete alteration of olivine to colorless amphibole and talc, and 2.4 vol% of the rock consists of hydrothermal and late magmatic veins composed of hornblende or oligoclase + diopside + epidote. H_2O contents of the 735B gabbros are mostly 0.5–1.5 wt% [*Robinson et al.*, 1991].

EFFECTS OF SLOW TRANSFORMATION ON DENSITY AND H_2O CONTENT

When oceanic crust is subducted it is compressed and heated along one of a wide variety of possible *PT* paths (Fig 3). The highest-temperature path into the eclogite

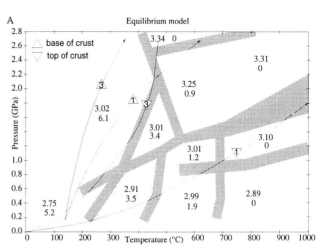
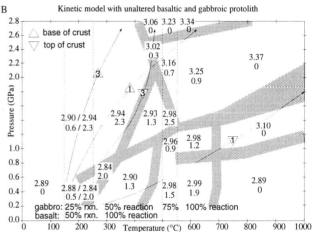
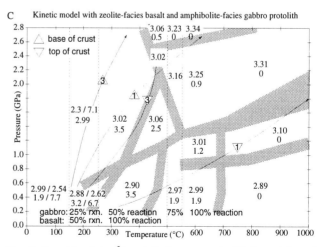

Fig. 3. Densities (g/cm³) and H_2O contents (wt%) for subducted oceanic crust calculated using NCMASH model of Peacock and kinetic data discussed in text. A) Equilibrium model. B) Kinetic model with unaltered basalt/gabbro protolith. C) Kinetic model with subgreenschist-facies basaltic protolith and amphibolite-facies gabbroic protolith. Where two numbers are given, the first refers to gabbro and the second to basalt.

stability field is from the granulite facies. In the hottest of subduction zones involving lithosphere less than a few m.y. old, it is possible for oceanic igneous rocks to pass into the eclogite stability field by this path [*Hacker*, 1991]. Rocks that recrystallized in the subgreenschist facies during hydrothermal alteration at the mid-ocean ridge will pass through the greenschist and amphibolite facies prior to entering granulite conditions, and may transform variably to greenschist, amphibolite, or granulite facies assemblages prior to entering the stability field of eclogite. At the opposite end of the temperature spectrum, rocks that previously recrystallized in the prehnite-pumpellyite facies will pass through the blueschist facies prior to entering eclogite conditions [*Peacock*, 1990], and may thus transform variably to blueschist prior to entering the eclogite stability field.

Peacock [1993] used the restricted compositional system NCMASH (Na-Ca-Mg-Al-Si-H-O) to calculate the mineral modes and maximum H_2O contents of subducting mafic rocks. He demonstrated that moderately altered oceanic crust containing 1–2 wt% H_2O begins to dehydrate at the onset of eclogite or amphibolite facies metamorphism, and suggested that the transition from blueschist to eclogite facies, associated with the breakdown of lawsonite or clinozoisite, releases the most H_2O during subduction.

Peacock calculated a range of paths that traverse most of the region of PT space relevant to subduction zones. Figure 3 illustrates two paths for the uppermost and lowermost parts of subducted 7-km thick oceanic crust—calculated for a subduction velocity of 50 mm/a and shear stresses of 100 and 33 MPa. Most of the PT paths followed by the uppermost layer of the crust (basalt) are limited by the paths identified by upright triangles, and most of paths followed by the lowermost layer of the crust (gabbro) are limited by the paths with inverted triangles. Intermediate levels in the crust follow intermediate PT paths. For path 1, the upper crust passes progressively through the greenschist to amphibolite to eclogite facies conditions, while the lower crust evolves from blueschist to eclogite facies. Along path 3, all the crust remains at blueschist facies conditions to pressures of 2.5 GPa. For paths intermediate between 1 and 3, most of the crust passes through the blueschist and eclogite facies.

In the spirit of Peacock's calculations and using the information presented in Figure 2, assume that coarse-grained rocks (i.e., gabbro) transform 25%, 50%, 75%, and 100% at 150°C, 250°C, 500°C and 550°C, respectively, in accordance with field observations described earlier. Further assume that fine-grained rocks (i.e., basalt) transform twice as fast—such that they reach 100% transformation at 250°C. Assume also that oceanic crust is made of 3 km of basalt and 4 km of gabbro. Given these assumptions, what conclusions can be drawn about the rates of eclogite formation in subduction zones and its effect on rock physical properties? Possible variables to be considered include the amount of H_2O and the composition of the crust.

Figure 3A and 4A illustrate the predicted densities and H_2O contents of NCMASH assemblages determined by *Peacock* [1993] for the various metamorphic facies. Figures 3B and 4B repeat this theme, using the kinetic inferences discussed above. Figures 3C and 4C use the same kinetic hindrance considerations but begin with altered protoliths. Because these figures are based on modeling real assemblages with a subset defined by the NCMASH system, the predicted densities and H_2O contents cannot be considered exact; however, several interesting features are apparent.

Figures 3B and 4B predict the behavior of subducted basalt and gabbro using the mineral assemblage olivine + orthopyroxene + clinopyroxene + plagioclase as a protolith. In the upper, volcanic crust, 50% reaction is assumed to occur at 150°C and complete reaction at 250°C; in the lower plutonic crust, 25, 50, 75, and 100% reaction is modeled as happening at 150, 250, 500, and 550°C, respectively. For example, the density of 2.97 g/cm^3 for lawsonite-blueschist facies gabbro in Figure 3B was calculated using unaltered gabbro (2.89 g/cm^3) transformed 50% to lawsonite blueschist (3.10 g/cm^3). The H_2O required for hydration in Figures 3B and 4B is assumed to derive from sediments or hydrated mantle. In the absence of sufficient H_2O, all facies other than CE, EC, GG, and GN have a density of 2.89 g/cm^3 and 0 wt% H_2O.

The results for volcanic rocks are close to the predictions of Peacock's equilibrium model (Figure 1) because the transformation rate is rapid. Basalts traveling via path 1 undergo only slight changes in density prior to the appearance of garnet and the disappearance of plagioclase. In contrast, Path 3 indicates that 5–10% volume loss occurs in basalts upon entering both the blueschist and eclogite facies. As in the equilibrium case (Figures 3A and 4A), the biggest step in dehydration occurs at the blueschist/eclogite facies boundary; dehydration along path 1 is subequally between the greenschist/amphibolite and amphibolite/granulite transitions. Basalts moving along path 3 or slightly warmer trajectories lose roughly equal amounts of H_2O when entering and leaving the epidote blueschist stability field. Compared to basalt, where volume loss occurs fairly evenly over the 150–450°C temperature range, half the volume loss of gabbro occurs in the 500–550° interval. In contrast to basalt, which may undergo substantial hydration by 250°C, gabbro is predicted to

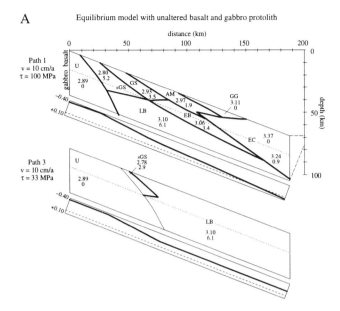

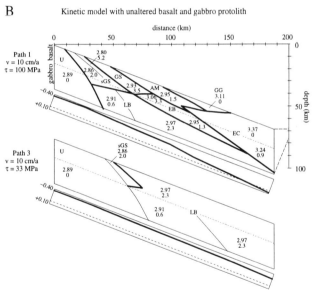

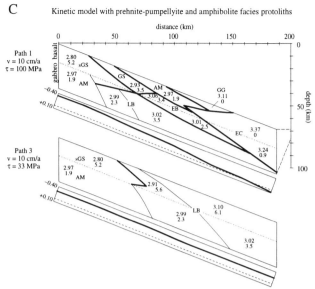

Fig. 4. Densities (g/cm³) and H₂O contents (wt%) of subducted crust, based on Figure 3. A) Equilibrium model with essentially unaltered crust. B) Kinetic model with unaltered crust. C) Kinetic model with crust altered to subgreenschist-facies and amphibolite-facies minerals. Compare with [*Peacock*, 1993] Figure 4. Thicker lines are facies boundaries and thinner lines represent changes from 25% to 50% to 75% to complete transformation. The thickness of the crust is stretched by a factor of 5 for clarity. Elongate rectangle at bottom of each panel shows density difference relative to asthenosphere (3.23 gm/cm³).

DISCUSSION: GEODYNAMIC IMPLICATIONS

The retardation of reaction during subduction illustrated in Figures 3 and 4 will affect 1) how the structure of subducted crust is interpreted from seismic data, 2) the buoyancy forces that produce sinking and bending of the slab, 3) the distribution of stresses induced by volume changes, 4) seismicity in the slab induced by volume-change-related stresses, 5) the mechanical behavior of the slab, 6) the distribution of H₂O in the slab, and 7) the generation of arc magmatism. All these factors differ from the equilibrium state because the densification reactions in descending slabs depend on grain size, H₂O content, bulk composition, and temperature. Relative to the equilibrium model, transformation to blueschist and eclogite is always delayed to greater depths in the kinetic model.

The first rocks to transform will be glassy basalts, followed by holocrystalline basalts, then diabases, and finally gabbros. Note that this derives from thermal as well as kinetic causes, as the upper crust is warmer than the lower crust as well as being finer grained. The seismic velocity signature of dense rock should be evident first in the upper crust and may be substantially retarded in the

undergo hydration to temperatures up to 550°C—although at any given condition the amount of H₂O in gabbro is less than in basalt. Again, as in the equilibrium model, the bulk of the dehydration occurs at the blueschist/eclogite transition.

Figures 3C and 4C predict the behavior of subducted altered basalt and gabbro using the prehnite-pumpellyite facies mineral assemblage chlorite + albite + pumpellyite + quartz + calcite for altered basalt and the amphibolite-facies assemblage hornblende + clinozoisite + chlorite + quartz for altered gabbro. The predictions are qualitatively similar to Figures 3B and 4B, except that the H₂O content at any chosen P and T is higher.

lower crust. The absence of fast velocities at eclogite facies conditions cannot be taken to indicate the absence of oceanic crust. High densities will appear first in the upper crust and may be considerably suppressed in the lower crust. In Figure 4, the lower crust is not only less dense than asthenosphere, but also less dense than the basaltic upper crust. In these situations the upper crust may sink in the asthenosphere but the lower crust cannot. Moreover, *Kirby et al.* [1996] calculated that densification reactions in subducting crust can place the crust in deviatoric tension and the mantle lithosphere in deviatoric compression. The difference in reaction rate between basalt and gabbro means that their calculations are oversimplified. The initiation of densification in the volcanic rocks will initially place those rocks in tension and the underlying coarser grained rocks in compression. As reaction progresses into the coarser lower crust, the state of deviatoric stress in the gabbroic rocks will change from compressive to tensile. If there were no other forces at work in subduction zones, earthquakes at shallow depths would be extensional in basalt and contractional in gabbro, changing as the rocks descend deeper in subduction zones to extensional throughout the oceanic crust.

Little is known about the mechanical behavior of blueschist facies minerals, but clinopyroxene and garnet are two of the strongest phases in the crust [*Ji and Martignole*, 1994; *Kolle and Blacic*, 1983] and eclogite should be strong compared to altered basalt. The replacement of fine-grained volcanic phases by blueschist or eclogite facies minerals may result in strengthening, whereas the alteration of coarse-grained minerals in gabbro to blueschist or eclogite phases may enhance deformation [*Rubie*, 1983]. The volcanic layer, though hotter, may actually become stronger than the gabbroic layer as a result of eclogite formation. It is conceivable that such rheological layering could result in a downward jump of the subduction decollement from the volcanic into the gabbroic layer, resulting in underplating of eclogitized upper crust to the mantle hanging wall and continued subduction of the gabbroic layer.

Crust that undergoes slow transformation is less capable of containing H_2O. Figure 4 shows that a slowly transforming gabbro layer has a maximum H_2O content ~1/2 that of the equilibrium model. The flux of H_2O carried into the mantle beyond a depth of 70 km is 1.3×10^8 grams per meter per year in the equilibrium model, but only 0.48×10^8 grams per meter per year in the kinetic model with unaltered protolith. Note that if the only source of H_2O in a subduction zone is the igneous crust, situations such as Figure 4C, where the H_2O content increases with depth, are impossible. Only if there is another source of H_2O in the subduction zone, from subducted sediments or mantle, can the H_2O content of igneous crust increase in subduction zones.

Kirby and Hacker [1991] proposed a link between arc volcanism, intermediate depth earthquakes, and eclogite formation. They noted that hypocenters of subduction zone earthquakes in the depth range 50–250 km frequently coincide with arc volcanoes, and proposed that these intermediate-depth earthquakes might be caused by delayed eclogite formation within subducting crust. This hypothesis requires that eclogite does not form at shallower depths because low temperatures and low H_2O activity hinder transformation, and that the devolatilization of amphibole in the 50–250 km depth range triggers eclogite formation and releases volatiles that lead to the overlying volcanoes. Figure 2 suggests that transformation to eclogite occurs at much shallower depths in most subduction zones. For Kirby and Hacker's hypothesis to be correct, the subducting crust must be essentially anhydrous.

Calculations and conclusions presented above apply principally to normal mid-ocean ridge crust. Ocean crust that is anomalously thick or of unusual composition will transform differently. Oceanic plateaus, which are up to 6 times thicker than mid-ocean crust, but compositionally similar, will have somewhat retarded transformation because of the additional time required for the subducting material to heat; this may be partly offset however by the greater thickness of volcanic rocks, which will transform more rapidly. Field observations discussed above indicate that Fe-rich rocks transform notably faster than MgO-rich rocks. Hydrothermal alteration at spreading centers often produces marked MgO enrichment in lavas [*Humphris and Thompson*, 1978], which would be expected to slow transformation. Oceanic islands built by hotspots and crust from propagating rifts are both enriched in Fe relative to MORB, and will transform more rapidly than standard mid-ocean-ridge crust. Crust erupted at fast spreading ridges (>60 mm/a) is also richer in Fe than crust erupted at slow spreading (<50 mm/a) ridges [*Niu and Batiza*, 1993; *Sinton and Detrick*, 1992], and is expected to transform more rapidly.

CONCLUSIONS

The transformation of mafic rocks to blueschist and eclogite in subduction zones is complex, encompassing a wide variety of igneous mineral assemblages and grain sizes, variably developed and equilibrated metamorphic mineral assemblages, all pressurized and heated via a broad range of PT paths. Existing experimental kinetic data cannot be extrapolated to actual petrotectonic settings in order to predict where rocks transform to blueschist and eclogite. However, textural data from exhumed subduction zones worldwide indicate that little transformation occurs at temperatures below 150°C. Volcanic rocks are completely transformed by perhaps 250°C. Coarser gabbroic rocks rarely avoid complete eclogitization at temperatures above

550°C, although examples of incomplete transformation are known from rocks heated at temperatures as high as 800°C.

The rapid transformation of volcanic rocks and the metastable persistence of gabbroic rocks into the blueschist and eclogite stability fields has several implications. The seismic velocity signature of dense rocks should be evident first in the upper crust and may be substantially retarded in the lower crust. The buoyancy forces causing the slab to sink will appear first in the upper crust and may be considerably suppressed in the lower crust. Earlier formation of garnet in the upper crust may cause the upper crust to become stronger than the gabbroic layer, and may lead to underplating of eclogitized upper crust and continued subduction of the lower crust. The initiation of densification in the volcanic rocks will initially place those rocks in deviatoric tension and the underlying coarser grained rocks in compression. As reactions progress into the coarser lower crust, the state of stress in the gabbroic rocks will change from compressive to tensile. If there were no other forces at work in subduction zones, earthquakes at shallow depths would be extensional in the basalt and contractional in the gabbro, changing at deeper levels to extensional throughout the crust. All these effects will be minimized in Fe-rich crust produced at oceanic islands and fast spreading ridges and most pronounced in Mg-rich crust formed at slow spreading centers.

Acknowledgments. Reviewed by S.H. Bloomer, S.R. Bohlen, S. DeBari, W.G. Ernst, J.G. Liou. A particularly thorough review was given by J.H. Natland.

REFERENCES

Ahrens, T.J., and G. Schubert, Gabbro-eclogite reaction rate and its geophysical significance, Reviews of Geophysics and Space Physics, 13, 383–400, 1975.

Alt, J.C., T.F. Anderson, L. Bonnell, and K. Muehlenbachs, Mineralogy, chemistry and stable isotopic compositions of hydrothermally altered sheeted dikes; ODP Hole 504B, Leg 111, *Proceedings of the Ocean Drilling Program, Scientific Results*, *111*, 27-40, 1989.

Alt, J.C., J. Honnorez, C. Laverne, and R. Emmermann, Hydrothermal alteration of a 1-km section through the upper oceanic crust, Deep Sea Drilling Project Hole 504B; mineralogy, chemistry, and evolution of seawater-basalt interactions, *Journal of Geophysical Research*, 91, 10,309-10,335, 1986.

Batiza, R., Pacific Ocean crust, in *Oceanic basalts*, edited by P.A. Floyd, pp. 264-288, Blackie and Son, Glasgow, 1991.

Bohlen, S.R., V.J. Wall, and A.L. Boettcher, Geobarometry in granulites, in *Kinetics and equilibrium in mineral reactions*, edited by S.K. Saxena, pp. 141-171, Springer-Verlag, New York, 1983.

Boundy, T.M., D.M. Fountain, and H. Austrheim, Structural development and petrofabrics of eclogite facies shear zones, Bergen arcs, western Norway; implications for deep crustal deformational processes, *Journal of Metamorphic Geology*, 10, 127-146, 1992.

Brown, E.H., Phase equilibria among pumpellyite, lawsonite, epidote and associated minerals in low grade metamorphic rocks, *Contributions to Mineralogy and Petrology*, 64, 123-136, 1977.

Brown, E.H., and E.D. Ghent, Mineralogy and phase relations in the blueschist facies of the Black Butte and Ball Rock areas, Northern California Coast Ranges, *American Mineralogist*, 68, 365-372, 1983.

Browning, P., Cryptic variation within the cumulate sequence of the Oman ophiolite: magma chamber depth and petrological implications, in *Ophiolites and oceanic lithosphere, Geological Society Special Publications*, edited by I.G. Gass, S.J. Lippard, and A.W. Shelton, pp. 71-82, Geological Society of London, London, 1984.

Carlson, W.D., and J.L. Rosenfeld, Optical determination of topotactic aragonite-calcite growth kinetics: metamorphic implications, *Journal of Geology*, 89, 615–638, 1981.

Carmichael, R.S., *CRC practical handbook of physical properties of rocks and minerals*, 741 pp., CRC Press, Boca Raton, FL, 1989.

Cloos, M., Lithospheric buoyancy and collisional orogenesis: subduction of oceanic plateaus, continental margins, island arcs, spreading ridges, and seamounts, *Geological Society of America Bulletin*, 105, 715–737, 1993.

Coleman, R.G., D.E. Lee, L.B. Beatty, and W.W. Brannock, Eclogites and eclogites--Their differences and similarities, *Geol. Soc. America Bull.*, 76, 483-508, 1965.

Dick, H.J.B., P.S. Meyer, S.H. Bloomer, S.H. Kirby, D.S. Stakes, and C. Mawer, Lithostratigraphic evolution of an in-situ section of oceanic layer 3, *Proceedings of the Ocean Drilling Program, Scientific Results*, 118, 439-538, 1991.

El-Shazly, A.E.-D., R.G. Coleman, and J.G. Liou, Eclogites and blueschists from northeastern Oman: petrology and P-T evolution, *Journal of Petrology*, 31, 629–666, 1990.

Ellis, D.J., and M.A.H. Maboko, Precambrian tectonics and the physicochemical evolution of the continental crust. I. The gabbro–eclogite transition revisited, *Precambrian Research*, 55, 491–506, 1992.

Ernewein, M., C. Pflumio, and H. Whitechurch, The death of an accretion zone as evidenced by the magmatic history of the Sumail ophiolite (Oman), *Tectonophysics*, 151, 247-274, 1988.

Ernst, W.G., Petrogenesis of glaucophane schists, *J. Petrology*, 4, 1-30, 1963.

Evans, B.W., Phase relations of epidote-blueschists, *Lithos*, 25, 3-23, 1990.

Ewart, A., The mineralogy and petrology of Tertiary–Recent orogenic volcanic rocks: with special reference to the andesite–basaltic andesite compositional range, in *Andesites: orogenic andesites and related rocks*, edited by R.S. Thorpe, pp. 26–87, Wiley, Chichester, 1982.

Flower, M., Magmatic processes in oceanic ridge and intraplate settings, in *Oceanic basalts*, edited by P.A. Floyd, pp. 116-147, Blackie and Son, Glasgow, 1991.

Green, D.H., and A.E. Ringwood, An experimental investigation of the gabbro-eclogite transformation and its petrological implications, *Geochimica Cosmochimica et Acta*, 31, 767–833, 1967.

Hacker, B.R., The role of deformation in the formation of metamorphic field gradients: Ridge subduction beneath the Oman ophiolite, *Tectonics*, 10, 455-473, 1991.

Hacker, B.R., S.R. Bohlen, and S.H. Kirby, Albite → jadeite + quartz transformation in albitite, *Eos, Transactions American Geophysical Union*, 74, 611, 1993.

Hacker, B.R., S.H. Kirby, and S.R. Bohlen, Time and metamorphic petrology: the calcite → aragonite transformation, *Science*, 258, 110–113, 1992.

Hekinian, R., *Petrology of the ocean floor*, 393 pp., Elsevier, Amsterdam, 1982.

Holland, T.J.B., The reaction albite = jadeite + quartz determined experimentally in the range 600-1200°C, *American Mineralogist*, 65, 129–134, 1980.

Hopson, C.A., R.G. Coleman, R.T. Gregory, J.S. Pallister, and E.H. Bailey, Geologic section through the Samail Ophiolite and associated rocks along a Miscat-Ibra transect, southeastern Oman Mountains, *Journal of Geophysical Research*, 86, 2527-2544, 1981.

Humphris, S.E., and G. Thompson, Hydrothermal alteration of oceanic basalts by seawater, *Geochimica Cosmochimica et Acta*, 42, 107-125, 1978.

Indares, A., and T. Rivers, Textures, metamorphic reactions and thermobarometry of eclogitized metagabbros; a Proterozoic example, *European Journal of Mineralogy*, 7, 43-56, 1995.

Ito, K., and G.C. Kennedy, An experimental study of the basalt–garnet granulite–eclogite transition, in *The Structure and Physical Properties of the Earth's Crust*, edited by J.G. Heacock, pp. 303–314, American Geophysical Union, Washington D.C., 1971.

Ji, S., and J. Martignole, Ductility of garnet as an indicator of extremely high temperature deformation, *Journal of Structural Geology*, 16, 985-996, 1994.

Juteau, T., M. Beurrier, R. Dahl, and P. Nehlig, Segmentation at a fossil spreading axis: the plutonic sequence of the Wadi Haymiliyah area (Haylayn block, Sumail nappe, Oman), *Tectonophysics*, 1881, 167–197, 1988a.

Juteau, T., M. Ernewein, I. Reuber, H. Whitechurch, and R. Dahl, Duality of magmatism in the plutonic sequence of the Sumail nappe, Oman, *Tectonophysics*, 151, 107–136, 1988b.

Kirby, S.H., E.R. Engdahl, and R. Denlinger, Intraslab earthquakes and arc volcanism: dual physical expressions of crustal and uppermost mantle metamorphism in subducting slabs, in *Dynamics of Subduction, Geophysical Monograph*, edited by G.E. Bebout, D. Scholl, and S. Kirby, AGU, Washington, D.C., 1996.

Kirby, S.H., and B.R. Hacker, Intermediate-depth earthquakes, crustal phase changes and the roots of arc volcanoes, *Eos, Transactions American Geophysical Union*, 72, 481, 1991.

Klaper, E.M., Reaction-enhanced formation of eclogite-facies shear zones in granulite-facies anorthosites, in *Deformation Mechanisms, Rheology and Tectonics, Geological Society of London Special Publication*, edited by R.J. Knipe, and E.H. Rutter, pp. 167–173, London, 1990.

Kolle, J.J., and J.D. Blacic, Deformation of single-crystal clinopyroxenes; 2, Dislocation-controlled flow processes in hedenbergite, *Journal of Geophysical Research*, 88, 2381-2393, 1983.

Lardeaux, J.-M., and M.I. Spalla, From granulites to eclogites in the Sesia Zone (Italian Western Alps); a record of the opening and closure of the Piedmont ocean, *Journal of Metamorphic Geology*, 9, 35-59, 1991.

Liou, J.G., P-T stabilities of laumontite, wairakite, lawsonite, and related minerals in the system $CaAl_2Si_2O_8$-SiO_2-H_2O, *Journal of Petrology*, 12, 379-411, 1971.

Lippard, S.J., A.W. Shelton, and I.G. Gass, *The Ophiolites of Northern Oman*, 178 pp., 1986.

Liu, J., W.G. Ernst, J.G. Liou, and S.R. Bohlen, Experimental constraints on the amphibolite-eclogite transition, *Geological Society of America Abstracts with Programs*, 25, 213, 1993.

Liu, M., and R.A. Yund, Transformation kinetics of polycrystalline aragonite to calcite: new experimental data, modelling, and implications, *Contributions to Mineralogy and Petrology*, 114, 465–478, 1993.

Maresch, W.V., Experimental studies on glaucophane; an analysis of present knowledge, *Tectonophysics*, 43, 109-125, 1977.

Mørk, M.B.E., A gabbro to eclogite transition on Flemsøy, Sunnmøre, western Norway, *Chemical Geology*, 50, 283–310, 1985.

Niu, Y., and R. Batiza, Chemical variation trends at fast and slow spreading mid-ocean ridges, *Journal of Geophysical Research*, 98, 7887-7902, 1993.

Oh, C.W., and J.G. Liou, Metamorphic evolution of two different eclogites in the Franciscan Complex, California, USA, *Lithos*, 25, 41-53, 1990.

Pallister, J.S., and C.A. Hopson, Samail ophiolite plutonic suite: field relations, phase variation, cryptic variation and layering, and a model of a spreading ridge magma chamber, *Journal of Geophysical Research*, 86, 2673-2645, 1981.

Peacock, S.M., Numerical simulation of metamorphic pressure-temperature-time paths and fluid production in subducting slabs, *Tectonics*, 9, 1197, 1990.

Peacock, S.M., The importance of blueschist → eclogite dehydration reactions in subducting oceanic crust, *Geological Society of America Bulletin*, 105, 684-694, 1993.

Pognante, U., Petrological constraints on the eclogite- and blueschist-facies metamorphism and P-T-t paths in the Western Alps, *Journal of Metamorphic Geology*, 9, 5-17, 1991.

Poli, S., The amphibolite-eclogite transformation; an experimental study on basalt, *American Journal of Science*, 293, 1061-1107, 1993.

Reynard, B., and M. Ballevre, Coexisting amphiboles in an eclogite from the Western Alps; new constraints on the miscibility gap between sodic and calcic amphiboles, *Journal of Metamorphic Geology*, 6, 333-350, 1988.

Ridley, J., and J.E. Dixon, Reaction pathways during the progressive deformation of a blueschist metabasite; the role of chemical disequilibrium and restricted range equilibrium, *Journal of Metamorphic Geology*, 2, 115-128, 1984.

Robinson, P.T., H.J.B. Dick, V. Herzen, and R. Pierre, Metamorphism and alteration in oceanic layer 3; Hole 735B, *Proceedings of the Ocean Drilling Program, Scientific Results*, 118, 541-562, 1991.

Rubie, D.C., Reaction-enhanced ductility: the role of solid-solid univariant reactions in deformation of the crust and mantle, *Tectonophysics*, 96, 331–352, 1983.

Rubie, D.C., Role of kinetics in the formation and preservation of eclogites, in *Eclogite Facies Rocks*, edited by D.A. Carswell, pp. 111–140, Blackie, Glasgow, 1990.

Sandrone, R., L. Leardi, P. Rossetti, and R. Compagnoni, P-T conditions for the eclogitic re-equilibration of the metaophiolites from Val d'Ala di Lanzo (internal Piemontese zone, Western Alps), *Journal of Metamorphic Geology*, 4, 161-178, 1986.

Schliestedt, M., Eclogite-blueschist relationships as evidenced by mineral equilibria in the high-pressure metabasic rocks of Sifnos (Cycladic Islands), Greece, *Journal of Petrology*, 27, 1437-1459, 1986.

Sinton, J.M., and R.S. Detrick, Mid-ocean ridge magma chambers, *Journal of Geophysical Research*, 97, 197-216, 1992.

Zhang, R.Y., J.G. Liou, T.F. Yui, and D. Rumble, Transformation of gabbro and granitoid to coesite-bearing eclogite facies rock from Yangkou, the Sulu Terrane, eastern China, *International Geological Congress*, in press, 1995.

B. R. Hacker, Department of Geological and Environmental Sciences, Stanford University, Stanford, California 94305-2115. (e-mail: hacker@geo.stanford.edu)

Double Seismic Zones, Compressional Deep Trench-Outer Rise Events, and Superplumes

Tetsuzo Seno and Yoshiko Yamanaka

Earthquake Research Institute, University of Tokyo

Subduction zones having double seismic zones (DSZs) are limited. They are generally characterized by the occurrence of compressional deep events (CDE) at the trench-outer rise region. Since shallow normal faulting is also common there, this pairing constitutes a kind of double zone at the trench-outer rise region, suggesting that a DSZ and CDEs are causally related. We propose that the lower zone of a DSZ and CDEs are both caused by dehydration embrittlement at the mid-plate depths in the subducting oceanic lithosphere, which might have been originally hydrated by the magmatic effects of superplumes or plumes.

INTRODUCTION

Some subduction zones show a double seismic zone (DSZ) at intermediate depth of the Wadati-Benioff zone and others do not. Figure 1 shows the subduction zones which have a DSZ; data are from various sources in the literature [Kuril-Kamchatka: *Veith*, 1974; *Kao and Chen*, 1994; *Gorbatov et al.*, 1994; Japan: *Hasegawa et al.*, 1978; *Barazangi and Isacks*, 1979; *Kawakatsu and Seno*, 1983; New Britain: *McGuire and Wiens*, 1995; Tonga: *Kawakatsu*, 1986a; N. Chile: *Comte and Suarez*, 1994; Peru: *Isacks and Barazangi*, 1977; E. Aleutians-Alaska Peninsula: *Reyners and Coles*, 1982; *Abers*, 1992]. DSZs listed here are those found at intermediate depth and the deep double zones of a different origin [*Wiens et al.*, 1993; *Iidaka and Furukawa*, 1994] are not included. It has been often disputed whether there is a DSZ in a particular area; sometimes a DSZ is an artifact due to the projection or mislocations. Even if it exists, it sometimes is a faint feature. We thus listed only the subduction zones having a conspicuous DSZ, though we included Tonga where the lower zone, made up of down-dip tensional events, is far less active than the upper zone [*Kawakatsu*, 1986a] and Peru where the DSZ was only found in the aftershock activity of the May 31, 1970 Peru earthquake [*Barazangi and Isacks*, 1977].

It has been thought that down-dip compressional and tensional stresses in DSZ slabs are caused by an unbending or thermal stresses [*Engdahl and Scholz*, 1977; *Sleep*, 1979; *House and Jacob*, 1982; *Goto et al.*, 1985; *Kawakatsu*, 1986b]. On the other hand, it has been difficult to explain why DSZs are not universal features of slabs when unbending or thermal stresses are expected to be common to any slab [e.g., *Fujita and Kanamori*, 1981; *Kawakatsu*, 1986b]. It also should be noted that under the high pressures expected at intermediate depth, mantle material will not fracture in a brittle manner under differential stresses [*Paterson*, 1978]. The strength estimated from the Coulomb-Navier failure criterion amounts to no less than a few GPa at 100-200 km depth. Flow laws also predict large strength at this depth and temperature [*Brodholt and Stein*, 1988]. Therefore special weakening mechanisms are required for the occurrence of intermediate-depth earthquakes [e.g., *McGarr*, 1977; *Pennington*, 1983; *Liu*, 1983; *Green and Burnley*, 1989; *Kirby*, 1995; *Abers*, 1995; *Kirby et al.*, 1996; *Kao and Liu*, 1996; See also the Nature News and Views article by *Frohlich*, 1994].

In this paper, we show that many compressional events in the trench-outer rise regions occur in the deeper portion of the oceanic plate (we call these events CDEs, abbreviating Compressional Deep Events), and that they

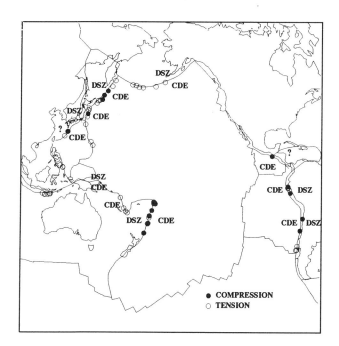

Fig. 1. Distribution of trench-outer rise earthquakes with well-constrained depths and focal mechanisms. Compression is denoted by the closed circle and tension by the open circle. Data of the trench-outer rise events are listed in Table 1. The location of double seismic zones is indicated by DSZ and that of compressional deep trench-outer rise events is indicated by CDE. DSZs and CDEs are associated with each other except for N. Ryukyu, New Hebrides, and Middle America.

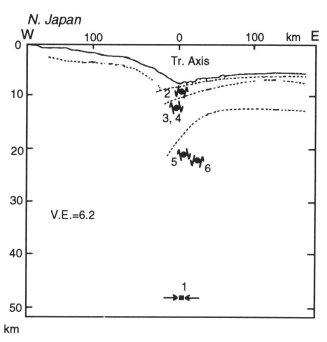

Fig. 2. E-W cross-section of trench-outer rise earthquakes in northern Honshu [modified from *Seno and Gonzalez*, 1987]. The number attached shows the order of occurrence (Table 1). The compressional event (event 1) occurred deeper than the tensional events (events 2-6).

occur in subduction zones that have DSZs. Since shallow tensional trench-outer rise events are common, this implies that tensional-compressional event pairing in the trench-outer rise region might be the shallow counterparts of DSZs. We then propose that this DSZ-CDE association could be caused by dehydration embrittlement [*Raleigh and Paterson*, 1965; *Raleigh*, 1967; *Green and Burnley*, 1989; *Meade and Jeanloz*, 1991; *Nishiyama*, 1992; *Kirby*, 1995] at mid-plate depths in oceanic lithosphere when it enters the subduction zone. We further will suggest a possibility that hydration deep inside of oceanic plates might occur by the magmatic effects of superplumes or other smaller plumes in the Pacific Basin.

COMPRESSIONAL TRENCH-OUTER RISE EVENTS

In the trench-outer rise regions seaward of subduction zones, seismic activity occurs within oceanic plates prior to subduction [e.g., *Stauder*, 1975; *Chapple and Forsyth*, 1979; *Christensen and Ruff*, 1988]. Figure 2 shows a vertical section of such trench-outer rise events in the northern Honshu arc [*Seno and Gonzalez*, 1987]. These events have nearly horizontal T-axes (tension type) or P-axes (compression type) that are perpendicular to the trench axis. We notice that the compressional event occurred at depth below the shallow tensional events; their depths are well-constrained by waveform analyses [*Seno and Gonzalez*, 1987]. Down dip at the 60-150 km depth range, a DSZ exists in the Pacific plate subducting beneath northern Honshu [*Hasegawa et al.*, 1975; *Barazangi and Isacks*, 1979; *Kawakatsu and Seno*, 1983].

We searched for a geographical relationship between DSZ and CDE by making a global compilation of trench-outer rise events with focal depths well-constrained by waveform analyses (Table 1; See examples of such depth analyses in *Stein and Wiens*, 1986; *Christensen and Ruff*, 1985; *Seno and Gonzalez*, 1987; *Honda and Seno*, 1989; *Honda et al.*, 1990; *Seno and Honda*, 1990). Figure 3 is a depth-versus-plate age plot of these events; the abscissa is the plate age near the epicenter of each event. For very large events, we showed the probable depth range of rupture [*Seno and Honda*, 1990; *Seno et al.*, 1992] by the thick vertical bar. The isotherms calculated from the plate model of *Parsons and Sclater* [1977] are also plotted. The depth-versus-age plot of the trench-outer rise events is similar to the depth-age relationship seen for intra-oceanic earthquakes [*Wiens and Stein*, 1983].

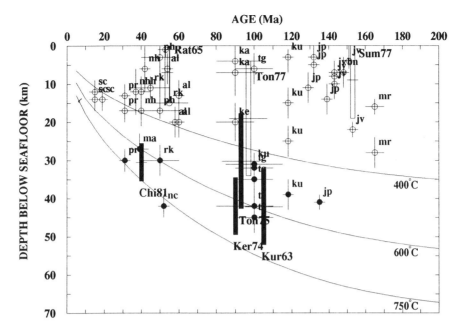

Fig. 3. Depth-versus-age plot for the trench-outer rise earthquakes listed in Table 1. Only well-constrained depths are included in the data set. Bars indicate the depth ranges of the rupture for large events. The abscissa is the plate age where the events are located. The isotherms are calculated from a plate model using the parameters of Parsons and Sclater (1977). Tensional events (open circles) occur shallower than compressional events (closed circles). The label beside each event abbreviates as follows: al, Aleutian; ka, Kamchatka; ku, Kuril; jp, North Japan; bn, Izu-Bonin; mr, Mariana; rk, Ryukyu; ph, Philippines; jv, Java; nh, New Hebrides; tg, Tonga; ke, Kermadec; sc, South Chile; nc, North Chile; pr, Peru; ma, Middle America; Rat65, 3/30/65 Rat Is. (M_s=7.5); Kur63, 3/16/63 Kuril (m_B=7.7); Sum77, 8/19/77 Sumba (M_s=7.9). Ton75, 10/11/75 Tonga (M_s=7.8); Ton77, 10/10/77 Tonga (M_s=7.2); Ker74, 7/02/74 Kermadec (M_s=7.2); Chi81, 10/16/81 Chile (M_s=7.2).

More important is the fact that the compressional events (closed circles) have deeper foci than those of the tensional events (open circles), suggesting that these trench-outer rise events are caused by bending of oceanic plates prior to subduction. The neutral-surface between tension and compression is almost fixed along the 450°C isotherm, which implies that the bending stress is large relative to regional tectonic stresses or the cyclic stresses associated with seismic coupling at the thrust zone. These regional stress perturbations might trigger the trench-outer rise or intraslab seismicity, however [*Dmowska et al.*, 1988; *Lay et al.*, 1989]. This constant neutral-surface interpretation is in contrast with that of *Christensen and Ruff* [1983, 1988] who claimed that the depth of the compressional 1981 Chilean trench-outer rise event is shallow, and that shallow compression was produced by seismic coupling prior to a subsequent large thrust earthquake. However, we questioned their depth estimation for that event [See *Honda and Seno*, 1989; *Honda et al.*, 1990; *Seno and Honda*, 1990]. Christensen and Ruff now acknowledge that their hypocentral depth was too shallow and that its centroid depth is consistent with bending of the Nazca plate prior to subduction [*Tichelaar et al.*, 1992]. *Ward* [1983] and *Liu and McNally* [1993], on the other hand, noted that the neutral surface in some regions might shoal as much as 10-15 km above the neutral surface due to tectonic stresses or thrust zone coupling. However, with the uncertainties of the depth determination in their studies, this does not contradict our constant neutral-surface viewpoint.

In Figure 1, we plot epicenters of all the trench-outer rise earthquakes listed in Table 1. The closed circles represent CDEs. We notice that CDEs are not always associated with every subduction zone. Trenches having CDEs are labeled by "CDE" and include Kuril-Kamchatka, N. Japan, N. Ryukyu, Tonga, N. Chile, Peru, and Middle America. New Britain, New Hebrides, and the E. Aleutians are also labeled by "CDE", because compressional events are listed by *Christensen and Ruff* [1988], *Christensen* [1995] and *Liu and McNally* [1993], though they are not listed in our Table 1 or shown in our Figure 1 because their focal depths are unconstrained.

Notice that the subduction zones which have a DSZ generally have CDEs at the trench. There are few excep-

TABLE 1. Trench-Outer Rise Events of Which Depths Are Well-Constrained

Y	M	D	H	M	Lat. (°N)	Long. (°E)	Type	Depth (km)	Age (Ma)	Reference
Aleutians (al)										
65	03	30	02	27	50.32	177.93	t	0-20	54	Seno et al. [1992]
66	06	02	03	27	51.01	175.98	t	13	60	Forsyth [1982]
66	08	07	02	13	50.57	-171.22	t	20	58	Herrmann [1976]
70	02	27	07	07	50.13	-179.59	t	6	54	Forsyth [1982]
70	03	19	23	33	51.34	173.75	t	20	60	Forsyth [1982]
81	06	05	07	09	52.34	-165.21	t	1	53	Ward [1983]
Kamchatka (ka)										
81	02	01	22	43	53.02	162.41	t	7	90	Ward [1983]
81	10	01	17	04	50.72	160.40	t	4	90	Ward [1983]
Kuril (ku)										
63	03	16	08	44	46.79	154.83	c	32-52	105	Seno et al. [1992]
65	04	05	13	52	44.51	151.90	t	25	118	Forsyth [1982]
71	09	09	23	01	44.34	150.85	t	15	118	Forsyth [1982]
71	12	02	17	18	44.77	153.55	c	39	118	Seno el. [1992]
81	04	30	14	41	43.23	149.94	t	3	118	Ward [1983]
81	08	23	12	00	48.71	157.37	c	31	100	Ward [1983]
N. Japan (jp)										
67	07	08	19	18	37.74	143.88	c	41	135	Seno and Gonzalez [1987]
69	08	23	02	54	39.70	144.38	t	3	132	Seno and Gonzalez [1987]
69	08	23	06	30	39.72	144.29	t	5	132	Seno and Gonzalez [1987]
69	08	24	22	03	39.80	144.30	t	5	132	Seno and Gonzalez [1987]
69	12	04	08	50	40.74	144.69	t	11	129	Seno and Gonzalez [1987]
75	06	14	23	36	36.31	143.40	t	14	139	Seno and Gonzalez [1987]
Izu-Bonin (bn)										
74	08	25	01	18	32.18	142.37	t	7	148	Forsyth [1982]
Mariana (mr)										
66	10	27	14	21	22.11	145.90	t	28	165	Seno et al. [1992]
67	04	05	02	47	19.97	147.28	t	16	165	Seno et al. [1992]
Ryukyu (rk)										
68	08	03	04	54	25.73	128.50	t	15	55	Ward [1979]
76	12	14	16	06	28.27	130.67	t	11	45	Seno et al. [1992]
76	12	14	19	35	28.26	130.65	c	30	50	Seno et al. [1992]
Philippine (ph)										
73	03	09	10	06	6.32	127.38	t	17	50	Seno et al. [1992]
81	05	26	06	47	6.14	127.50	t	3	50	Ward [1983]
Java (jv)										
72	05	04	04	11	-10.73	113.65	t	10	143	Forsyth [1982]
77	08	19	06	08	-11.19	118.41	t	0-19	153	Seno et al. [1992]
82	02	28	17	52	-11.46	117.24	t	4	151	Ward [1983]
83	04	23	09	20	-11.21	118.92	t	22	153	Seno et al. [1992]
83	09	29	02	06	-11.37	115.32	t	8	143	Seno et al. [1992]
83	11	15	10	38	-11.60	115.20	t	7	143	Seno et al. [1992]
New Hebrides (nh)										
64	01	22	23	59	-13.64	165.96	t	17	40	Chinn and Isacks [1983]
66	09	12	11	29	-23.00	170.60	t	12	37	Chinn and Isacks [1983]
81	11	16	13	53	-22.11	169.52	t	12	40	Ward [1983]
82	05	20	21	29	-20.24	168.20	t	6	42	Ward [1983]
Tonga (tg)										
67	11	12	10	36	-17.20	-172.00	c	42	100	Forsyth [1982]
69	01	29	17	44	-17.15	-171.57	c	32	100	Seno et al. [1992]
72	08	07	09	24	-16.66	-172.01	c	45	100	Forsyth [1982]
72	09	27	09	01	-16.47	-172.17	t	6	100	Forsyth [1982]
75	10	11	14	35	-24.91	-175.16	c	18-43	93	Seno et al. [1992]
77	10	10	11	53	-25.87	-175.37	t	4-34	97	Seno et al. [1992]
82	02	28	17	00	-21.65	-173.51	c	35	100	Ward [1983]
Kermadec (ke)										
74	07	02	23	26	-29.22	-175.94	c	35-50	90	Seno et al. [1992]
74	07	03	23	25	-29.37	-176.13	t	20	90	Forsyth [1982]

TABLE 1. Continued

Y	M	D	H	M	Lat. (°N)	Long. (°E)	Type	Depth (km)	Age (Ma)	Reference
Chile (nc, and sc)										
62	11	11	22	14	-43.06	-75.82	t	12	15	*Chinn and Isacks* [1983]
64	08	05	22	23	-41.13	-74.99	t	14	19	*Chinn and Isacks* [1983]
64	08	18	04	44	-26.40	-71.50	c	42	52	*Chinn and Isacks* [1983]
65	10	03	16	14	-42.90	-75.13	t	14	15	*Chinn and Isacks* [1983]
81	10	16	03	25	-33.15	-73.10	c	25-35	40	*Seno and Honda* [1990]
Peru (pr)										
63	08	29	15	30	-7.10	-81.60	t	17	31	*Chinn and Isacks* [1983]
65	08	03	02	01	-7.40	-81.30	t	13	31	*Chinn and Isacks* [1983]
67	09	03	21	07	-10.60	-79.80	c	30	31	*Chinn and Isacks* [1983]
Middle America (ma)										
71	08	20	21	36	13.30	-92.41	c	27	39	*Forsyth* [1982]

Plate ages are read from Plate-tectonic map of the circum-Pacific region [*Halbouty et al.*, 1981]. The letters in the parentheses which abbreviate each region's name correspond to those shown in Figure 3.

tions: In N. Ryukyu, New Hebrides, and Middle America, CDEs are found without DSZs. However, existence of a DSZ has been in dispute in northern Ryukyu [*Ishikawa*, 1985; *Takahashi*, 1987]. In Middle America, we examined all the CMT solutions by Harvard University [*Dziewonski and Woodhouse*, 1983 and subsequent papers]. We found a number of events at around 80 km depth with P-axes oriented along the slab dip among numerous events with shallower dipping T-axes, of which geometry suggests possible existence of a DSZ here. Therefore the association between a DSZ and CDEs in Figure 1 is fairly good.

ROCK MECHANICS FOR THE DOUBLE SEISMIC ZONE-COMPRESSIONAL DEEP TRENCH-OUTER RISE EVENTS ASSOCIATION

The difficulty in explaining the occurrence of intermediate-depth earthquakes in the Wadati-Benioff zone mentioned in the Introduction also applies to CDEs in the trench-outer rise region. The depth and temperature ranges of these events are 25 to 50 km and 450 to 750°C (Figure 3), corresponding to the semi-brittle to ductile regime of the strength envelope for the oceanic plates [*Kirby*, 1980; *Wiens and Stein*, 1983]. If the events are in the semi-brittle regime, it is hard for them to occur because the strength becomes very high due to large confining pressures at depth. On the other hand, if they are in the ductile regime, it is again hard for them to occur because ductile deformation prevails.

To avoid this difficulty for the intermediate-depth events of the Wadati-Benioff zone, dehydration embrittlement has been proposed as an attractive physical mechanism of failure [See *Kirby*, 1995 and *Frohlich*, 1994 for review]. *Raleigh and Paterson* [1965] showed that serpentine becomes brittle under anomalously high temperature (500-600°C) due to dehydration if it is loaded under a differential stress with a confining pressure below 0.5 GPa. *Raleigh* [1967] applied this dehydration embrittlement to a failure mechanism for intermediate and deep events. *Meade and Jeanloz* [1991] also showed in their experiments that applying differential stress under confining pressures corresponding to the depth range of intermediate earthquakes (2-9 GPa), acoustic emissions associated with dehydration are observed at 630±100°C. Dehydration embrittlement has been usually presented, however, as a failure mechanism for intermediate-depth events just below the upper surface of the subducting slab, since this portion of the slab, affected by the normal faulting and associated hydrothermal activity at the mid-ocean ridges and in the trench-outer rise regions, can be reactivated easily by dehydration during heating [See *Kirby*, 1995]. In DSZs, the lower zone is usually located at a few tens kilometers below the slab upper surface, and thus has not been believed prone to hydration or dehydration.

In this paper, we are going to extend the above dehydration embrittlement mechanism to DSZs and CDEs. If the middle portion of the subducting plate is somehow hydrated, along with the hydrated shallow portion of the plate, such a structure could produce a DSZ if the slab passes through the pressure-temperature conditions for dehydration, which corresponds to roughly 600°C. *Nishiyama* [1992] first proposed this mechanism and we subsequently conceived it independently. We will discuss the hydration mechanism for the middle portion of the plate in the next section. The loading due to an unbending will cause faulting where dehydration raises pore pressures or cause mechanical instabilities where serpentine is metastable. Figure 4a shows the estimated dehydration loci within a slab where antigorite dehydrates under the

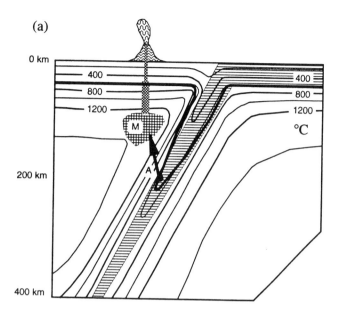

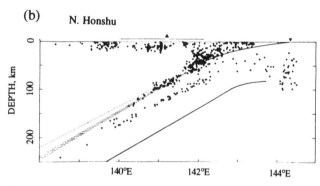

Fig. 4. (a) The dehydration loci estimated from the antigorite breakdown curve (thick solid line) [Figure 2 of *Ulmer and Trommsdorff*, 1995]. The experimental pressure-temperature diagram for the breakdown of antigorite was applied to the slab subducting with a velocity of 4.5 cm/yr at a dip angle of 60°; the fluids produced at point A reach the source region for calc-alkaline magma in the mantle wedge [*Ulmer and Trommsdorff*, 1995]. (b) The double seismic zone found beneath northern Honshu [*Hasegawa et al.*, 1978]. The upper and lower zones of the double seismic zone merge at depth, as does the boundary where serpentine dehydrates.

confining pressure of 1-8 GPa [*Ulmer and Trommsdorff*, 1995]. It mimics the shape of a DSZ seen in northern Honshu (Figure 4b). Dehydration from the mantle part is also important for the upper zone activity (Figure 4a) because dehydration from crust and sediments will be completed mostly at depths shallower than 100 km [*Fyfe et al.*, 1978; *Shimamoto et al.*, 1993; *Poli and Schmidt*, 1995; See also *Ulmer and Trommsdorff*, 1995].

Furthermore this embrittlement mechanism may be applicable to CDEs at the trench-outer rise region. The phase boundary extends seaward beyond the trench (Figure 4a) and the bending moment applied at the trench-outer rise region pressurizes the deeper half of the plate making serpentine unstable there. Thus if the middle portion of the oceanic plate is hydrated, it can produce CDEs and a DSZ simultaneously in the same subduction zone as associated phenomena.

HYDRATION MECHANISM

We suggest a possibility here that hydration of the middle to deep portion of oceanic plates occurs when some plates pass over plumes and superplumes. This is similar to the hypothesis of CO_2 encapsulation by plumes proposed by *Wilshire and Kirby* [1988] for intraplate events near Hawaii and by *Kirby* [1995] for the lower zone of a DSZ, except that the volatile species involved is different. Consider that an oceanic plate, which has thickened as it ages to a few tens of million years and then passes over a huge upwelling plume like the Mesozoic superplume in the south Pacific [*Larson*, 1991](Figure 5). Magma having dissolved water accumulates through a partial melting zone in the superplume head. Some of this magma would solidify within the plate when it is injected, releasing water that results in hydration of surrounding harzburgite and produces serpentine and other hydrous minerals by metasomatism [*Menzies et al.*, 1987]. However, most of the magma will solidify within the partially molten asthenosphere at the plume head beneath the plate, which will accrete to the oceanic plate as it cools and thickens [*Parker and Oldenberg*, 1973]. This accretion brings water into the ambient mantle as the partial melts solidify and release water, resulting in hydration of the middle part of the lithosphere in the stability field of serpentine-talc below 700°C at ~1 GPa [*Raleigh and Paterson*, 1965; *Ulmer and Trommsdorf*, 1995]. This hydration mechanism would also operate for plates passing over smaller plumes, but to a lesser extent. Though part of this serpentinized mantle could be erupted as serpentine diapirs, some could survive. As the plate cools with age, the region over which serpentine is stable enlarges. The exact extent of hydration by this mechanism is not well constrained at present due to lack of knowledge of the volume and degree of partial melting.

It is very difficult to prove that the slabs having CDE/DSZ pairings (Figure 1) were actually hydrated by magmatic effects of superplumes or plumes, however, since prior magmatic anomalies within or under lithosphere has already subducted. We feel it is plausible that a significant fraction of the Pacific and Nazca plates subducting over the last few million years passed over

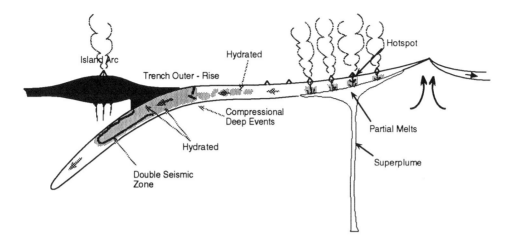

Fig. 5. A schematic illustration showing the posssible hydration mechanism of the middle portion of the plate over an active superplume by solidification of the partial melts. The hydrated plate causes a dehydration instability at depth in the trench-outer rise region, resulting in CDEs. At depths in the subduction zone, dehydration causes DSZs to develop.

plumes or a superplume sometime in their early histories because there are numerous volcanic ridges, plateaus and seamount chains in the Pacific Basin.

DISCUSSION AND CONCLUSIONS

The dehydration embrittlement mechanism has been referred to particularly for the upper zone seismic activity in the crust and uppermost mantle [Kirby, 1995; Kirby et al., 1996]. In our dehydration embrittlement hypothesis for DSZs, the upper zone activity at depth is expected to occur mostly in the mantle (Figure 4a). There is some evidence which suggests that there is seismic activity within the mantle part in the upper zone. As shown in Figure 4b, the upper zone seems to merge with the lower zone at 200 km depth. If this is true, there should be upper zone activity within the mantle at least in the deepest portion of a DSZ. In accord with this, Zhao and Sacks [1994] suggested that the upper zone activity beneath northern Honshu is best fitted by the inclined plane 4 degrees steeper than the slab surface using converted seismic phases. Kao and Chen [1995] showed that some events in the Kuril slab deeper than 100 km probably occurred within the mantle beneath the crust based on waveform analyses.

It would be difficult to explain the DSZ geometry merging at its deepest end by devolatilization of CO_2 for the lower zone and dehydration embrittlement for the upper zone [Kirby, 1995]. This merging would be a mere coincidence if these zones have different origins. On the other hand, Kao and Liu [1996] proposed that the phase change of the metastable Al-rich enstatite to Al-poor enstatite and garnet can produce a DSZ. This mechanism has an advantage similar to that proposed here in explaining the observed shape of a DSZ because the phase change is controlled by a P/T boundary. However kinetics of this metastable phase change is not known. Moreover, it would be difficult to explain the occurrence of CDEs by this mechanism.

We have shown that when CDEs are found in the trench-outer rise regions, they accompany DSZs that occur down dip. The key fact is that CDEs occur in the deep portion of the plate, an observation not easily explained within the framework of the failure mechanism of shallow earthquakes. This is also true for intermediate-depth earthquakes in DSZs. We proposed that the mechanism for the CDE/DSZ pairing is dehydration embrittlement at mid-plate depths associated with loading by bending-unbending. The hydration at the mid-plate depths of lithosphere might occur by magmatic effects of superplumes or plumes. The hypothesis presented here explains why some arcs have CDE/DSZ pairings and why others do not. Its application to the observed CDE/DSZ or non CDE/DSZ associations is still premature, and it should be substantiated by direct observations in the trench-outer rise regions or by other lines of evidence. If there are abundant hydrous minerals at mid-plate depths, we would expect lower seismic velocities and lower densities than those for normal lithosphere, according to the effect of serpentinization [Hess, 1962].

Acknowledgments. We thank Steve Kirby for his critical reading, advice and discussion. We also thanks John Brodholt, and two anonymous reviewers for their critical reviews. We also benefited from discussions with Tadao Nishiyama, Taku Koyaguchi, Yoji Kobayashi, and Masao Nakanishi, and from preprints provided by Steve Kirby and Hon Kao.

REFERENCES

Abers, G. A., Relationship between shallow- and intermediate-depth seismicity in the eastern Aleutian subduction zone, *Geophys. Res. Lett., 19*, 2019-2022, 1992.

Abers, G. A., Buy one, get one free: Structure and origins of double seismic zones, *Proc. SUBCON: An Interdisciplinary Conference on the Subduction Process, June 12-17, 1994, Santa Catalina Island, California, U. S. Geolo. Surv., Open File Rep., 95-21*, in press, 1995.

Barazangi, M., and B. L. Isacks, A comparison of the spatial distribution of mantle earthquakes determined from data produced by local and by teleseismic networks for the Japan and Aleutian arc, *Bull. Seism. Soc. Am., 69*, 1763-1770, 1979.

Brodholt, J., and S. Stein, Rheological control of Wadati-Benioff Zone seismicity, *Geophys. Res. Lett., 15*, 1081-1084, 1988.

Chapple, W. M., and D. W. Forsyth, Earthquakes and bending of plates at trenches, *J. Geophys. Res., 84*, 6729-6749, 1979.

Chinn, D. S., and B. L. Isacks, Accurate source depths and focal mechanisms of shallow earthquakes in western South America and in the New Hebrides island arc, *Tectonics, 2*, 529-563, 1983.

Christensen, D., Shallow seismicity at the outer rise and trench, *Proc. SUBCON: An Interdisciplinary Conference on the Subduction Process, June 12-17, 1994, Santa Catalina Island, California, U. S. Geol. Surv., Open File Rep., 95-21*, in press, 1995.

Christensen, D. H., and L. J. Ruff, Outer-rise earthquakes and seismic coupling, *Geophys. Res. Lett., 10*, 697-700, 1983.

Christensen, D. H., and L. J. Ruff, Analysis of the trade-off between hypocentral depth and source time function, *Bull. Seism. Soc. Am., 75*, 1637-1656, 1985.

Christensen, D. H., and L. J. Ruff, Seismic coupling and outer-rise earthquakes, *J. Geophys. Res., 93*, 13421-13444, 1988.

Comte, D., and G. Suarez, An inverted double seismic zone in Chile: Evidence of phase transformation in the subducted slab, *Science, 263*, 212-215, 1994.

Dmowska, R. J., R. Rice, L. C. Lovinson and D. Josell, Stress transfer and seismic phenomena in coupled subduction zones during earthquake cycle, *J. Geophys. Res., 93*, 7869-7884, 1988.

Dziewonski, A. M., and J. H. Woodhouse, An experiment in systematic study of global seismicity: Centroid-moment tensor solutions for 201 moderate and large earthquakes of 1981, *J. Geophys. Res., 88*, 3247-3271, 1983.

Engdahl, E. R., and C. H. Scholz, A double Benioff zone beneath the central Aleutians: An unbending of the lithosphere, *Geophys. Res. Lett., 4*, 473-476, 1977.

Forsyth, D. W., Determination of focal depths of earthquakes associated with the bending of oceanic plates at trenches, *Phys. Earth Planet. Inter., 28*, 141-160, 1982.

Frohlich, C., A break in the deep, *Nature, 368*, 100-101, 1994.

Fujita, K., and H. Kanamori, Double seismic zones and stresses of intermediate depth earthquakes, *Geophys. J. R. astron. Soc., 66*, 131-156, 1981.

Fyfe, W. S., N. J. Price, and A. B. Thompson, *Fluids in the Earth's Crust*, Elsevier, Amsterdam, 383 pp., 1978.

Gorbatov, A., G. Suarez, V. Kostoglodov, and E. Gordeev, A double-planed seismic zone in Kamchatka from local and teleseismic data, *Geophys. Res. Lett., 21*, 1675-1678, 1994.

Goto, K., Hamaguchi, H., and Z. Suzuki, Earthquake generating stresses in a descending slab, *Tectonophysics, 112*, 111-128, 1985.

Green II, H. W., and P. C. Burnley, A new self-organizing mechanism for deep-focus earthquakes, *Nature, 341*, 733-737, 1989.

Halbouty, M. T., J. A. Reinemund, and M. J. Terman, Plate-tectonic map of the circum-Pacific region, *Circum-Pacific Council for Energy and Mineral Resources*, 1981.

Hasegawa, A., Umino, N., and A. Takagi, Double-planed structure of the deep seismic zone in the northeastern Japan arc, *Tectonophysics, 47*, 43-58, 1978.

Herrmann, R. B., Focal depth determination from the signal character of long-period P waves, *Bull. Seism. Soc. Am., 66*, 1221-1232, 1976.

Hess, H. H., History of ocean basins, in *Petrological Studies: A Volume in Honor of A. F. Buddington*, edited by A. E. Engel, H. L. James, and B. F. Leonard, GSA, 599-620, 1962.

Honda, S., and T. Seno, Seismic moment tensors and source depths determined by the simultaneous inversion of body and surface waves, *Phys. Earth Planet. Inter., 57*, 311-329, 1989.

Honda, S., H. Kawakatsu, and T. Seno, Centroid depth of the October 1981 off Chile outer-rise earthquake (M_s=7.2) determined by a comparison of several waveform inversion methods, *Bull. Seismol. Soc. Am., 80*, 69-87, 1990.

House, L. S., and K. H. Jacob, Thermal stress in subducting lithosphere can explain double seismic zone, *Nature, 295*, 587-589, 1982.

Iidaka, T., and Y. Furukawa, Double seismic zone for deep earthquakes in the Izu-Bonin subduction zone, *Science, 263*, 1116-1118, 1994.

Isacks, B. L., and M. Barazangi, Geometry of Benioff zones: Lateral segmentation and downwards bending of the subducted lithosphere, in *Island Arcs, Deep Sea Trenches and Back Arc Basins, Maurice Ewing Series, 1*, editd by M. Talwani and W. C. Pitman III, AGU, Washington, D. C., 99-114, 1977.

Ishikawa, Y., Double seismic zone beneath Kyushu, *Jisin, 38*, 265-269, 1985 (in Japanese).

Kao, H., and W.-P. Chen, Double seismic zone in Kuril-Kamchatka: The tale of two overlapping single seismic zone, *J. Geophys. Res., 99*, 6913-6930, 1994.

Kao, H., and W.-P. Chen, Transition from interplate slip to double seismic zone along the Kuril-Kamchatka arc, *J. Geophys. Res, 100*, 9881-9903, 1995.

Kao, H., and L.-G. Liu, A hypothesis for the seismogenesis of double seismic zone, *Geophys. J. Inter., 123*, 71-84, 1996.

Kawakatsu, H., and T. Seno, Triple seismic zone and the regional variation of seismicity along the northern Honshu arc, *J. Geophys. Res., 88*, 4215-4230, 1983.

Kawakatsu, H., Downdip tensional earthquakes beneath the Tonga arc: A double seismic zone?, *J. Geophys. Res., 91*, 6432-6440, 1986a.

Kawakatsu, H., Double seismic zones: kinematics, *J. Geophys. Res., 91*, 4811-4825, 1986b.

Kirby, S. H., Tectonic stresses in the lithosphere: Constraints provided by the experimental deformation of rocks, *J. Geophys. Res., 85*, 6353-6363, 1980.

Kirby, S. H., W. B. Durham, and L. A. Stern, Mantle phase changes and deep-earthquake faulting in subducting lithosphere, *Science, 252,* 216-225, 1991.

Kirby, S., Intraslab earthquakes and phase changes in subducting lithosphere, *Rev. Geophys., Suppl.,* 287-297, 1995.

Kirby, S., E. R. Engdhal, and R. Denlinger, Intraslab earthquakes and arc volcanism: Dual physical expressions of crustal and uppermost mantle metamorphism in subducting slabs, *in this volume, Geophys. Monogr.,* edited by G. Bebout, D. Scholl, and S. Kirby, 1996.

Larson, R. L., Latest pulse of earth: evidence for a mid-Cretaceous superplume, *Geology, 19,* 547-550, 1991.

Lay, T., D. H. Christensen, L. Astiz, and H. Kanamori, Temporal variation of large intraplate earthquakes in coupled subduction zones, *Phys. Earth Planet. Inter., 54,* 258-312, 1989.

Liu, L.-G., phase transformations, earthquakes and the descending lithosphere, *Phys. Earth Planet. Inter., 32,* 226-240, 1983.

Mcgarr, A., Seismic moments of earthquakes beneath island arcs, phase changes, and subduction velocities, *J. Geophys. Res., 82,* 256-264, 1977.

McGuire, J. J., and D. A. Wiens, A double seismic zone in New Britain and the morphology of the Solomon plate at intermediate depth, *Geophys. Res. Lett., 22,* 1965-1968, 1995.

Meade, C. and R. Jeanloz, Deep-focus earthquakes and recycling of water into the earth's mantle, *Science, 252,* 68-72, 1991.

Menzies, M., N. Rogers, A. Tindle, and C. Hawkesworth, Metasomatic and enrichment processes in lithospheric peridotites, an effect of athenosphere-lithosphere interaction, *in Mantle Metasomatism,* edited by M. A. Menzies and C. J. Hawkesworth, Academic press, 313-361, 1987.

Nishiyama, S., Mantle hydrology in a subduction zone: A key to episodic geologic events, double Wadati-Benioff zones and magma genesis, *in Mathematical Seismology VII, Rep. Stat. Math. Inst., 34,* 31-67, 1992.

Parker, L., and D. W. Oldenburg, Thermal model of oceanic ridges, *Nature Phys. Sci., 242,* 137-139, 1973.

Parsons, B., and J. G. Sclater, An analysis of the variation of ocean floor bathymetry and heat flow with age, *J. Geophys. Res., 82,* 803-827, 1977.

Paterson, M. S., *Experimental Rock Deformation - The Brittle Field,* Springer-Verlag, Berlin, 254 pp., 1978.

Pennington, W. D., Role of shallow phase changes in the subduction of oceanic crust, *Science, 220,* 1045-1047, 1983.

Poli, S., and M. W. Schmidt, H_2O transpot and release in subduction zones: Experimental constraints on basaltic and andesitic systems, *J. Geophys. Res., 100,* 22299-22314, 1995.

Raleigh, C. B., Tectonic implications of serpentinite weakening, *Geophys. J. R. astron. Soc., 14,* 113-118, 1967.

Raleigh, C. B., and M. S. Paterson, Experimental deformation of serpentinite and its tectonic implications, *J. Geophys. Res., 70,* 3965-3985, 1965.

Reyners, M. and K. S. Coles, Fine structure of the dipping seismic zone and subduction mechanics in the Shumagin islands, Alaska, *J. Geophys. Res., 87,* 356-366, 1982.

Seno, T., and D. G. Gonzalez, Faulting caused by earthquakes beneath the outer slope of the Japan Trench, *J. Phys. Earth, 35,* 381-407, 1987.

Seno, T., and S. Honda, Depth extent analysis of the 1981 October 16 Chile earthquake, *Bull. Earthq. Res. Inst., 65,* 1-32, 1990.

Seno, T., S. Honda, E. T. Peterson, and T. Okamoto, Trench-outer rise earthquakes: implications to the rheology of oceanic lithosphere, *Abstr. 29th IGC,* 659, 1992.

Shimamoto, T., T. Seno, and S. Uyeda, Rheological framework for comparative subductology, *in Relating Geophysical Structures and Processes: The Jeffreys Volume, Geophys. Monogr., 76,* edited by K. Aki and R. Dmowska, IUGG and AGU, 39-52, 1993.

Sleep, N. H., The double seismic zone in downgoing slabs and the viscosity of the mesosphere, *J. Geophys. Res., 84,* 4565-4571, 1979.

Stauder, W., Subduction of the Nazca plate under Peru as evidenced by focal mechanisms and by seismicity, *J. Geophys. Res., 80,* 1053-1064, 1975.

Stein, S., and D. A. Wiens, Depth determination for shallow teleseismic earthquakes: Method and results, *Rev. Geophys., 24,* 806-832, 1986.

Takahashi, M., Non-double seismic zone beneath Kyushu, *Jisin, 40,* 115-117, 1987 (in Japanese).

Tichelaar, B. W., D. H. Christensen, and L. J. Ruff, Depth extent of rupture of the 1981 Chilean outer-rise earthquake as inferred from long-period body waves, *Bull. Seism. Soc. Am., 82,* 1236-1252, 1992.

Turcotte, D. L., and G. Schubert, *Geodynamics,* John Wiley & Sons, New York, 450 pp., 1982.

Ulmer, P., and V. Trommsdorff, Serpentine stability to mantle depths and subduction-related magmatism, *Science, 268,* 858-859, 1995.

Veith, K. F., The nature of the dual zone of seismicity in the kuriles arc, *EOS, Trans. Am. Geophys. Union, 58,* 1232, 1977.

Ward, S. N., Ringing P waves and submarine faulting, *J. Geophys. Res., 84,* 3057-3062, 1979.

Ward, S. N., Body wave inversion: moment tensors and depths of oceanic intraplate bending earthquakes, *J. Geophys. Res., 88,* 9315-9330, 1983.

Wiens, D. A., and S, Stein, Age dependence of oceanic intraplate seismicity and implications for lithospheric evolution, *J. Geophys. Res., 88,* 6455-6468, 1983.

Wiens, D. A., J. J. McGuire, and P. J. Shore, Evidence for transformational faulting from a deep double seismic zone in Tonga, *Nature, 364,* 790-793, 1993.

Wilshire, H. G., and S. H. Kirby, Dikes, joints, and faults in the upper mantle, *Tectonophysics, 157,* 1988.

Zhao, D., and I. S. Sacks, Metastable phase change mechanism for deep earthquakes: A seismological constraint, *EOS, Trans. Am. Geophys. Union, 75,* 234, 1994.

T. Seno and Y. Yamanaka, Earthquake Research Institute, University of Tokyo, Bunkyo-ku, Tokyo 113, Japan.

Characteristics of Multiple Ruptures During Large Deep-Focus Earthquakes

Wang-Ping Chen and Li-Ru Wu

Department of Geology, University of Illinois, Urbana, IL

Mary Ann Glennon

Energy and Environmental Systems Division, Argonne National Laboratory, Argonne, IL

Based on the inversion of broad-band (high-resolution) *P* and *SH* waveforms, we summarize the configuration and propagation of ruptures during several recent, large deep-focus earthquakes. In all cases, bursts of seismic moment seem to have been released as sub-horizontally propagating ruptures over elongate, narrow regions. New results for the event of July 21, 1994 (M_W=7.3, depth 474 km) in the Japan slab show that the earthquake is likely to have occurred as two sub-horizontal, *en echelon* ruptures, separated in depth by approximately 7 km. Given an extremely consistent downdip orientation of *P* axes for all large to moderate-sized earthquakes in this slab in the past 30 years, such a rupture geometry is discussed in the context of transformational faulting that may trigger deep-focus seismicity. Seismic moment release over multiple sub-horizontal regions has also been documented for the great Bolivian deep-focus earthquake of June 9, 1994 and a large deep-focus earthquake along the Izu–Bonin subduction zone. Such a configuration seems to be common for many large deep-focus earthquakes, and thus accommodates high moment release over a limited seismogenic volume within subducted lithosphere. This mode of moment release also allows for uneven rupture speeds during a single large earthquake: while the apparent rupture speed between subevents can be only 1–2 km/s, the speed is as high as 4 km/s within large subevents. In sum, several characteristics of the multiple rupture model dovetail with predictions based on the transformational faulting mechanisms for deep-focus seismicity. Whatever mechanisms that extend a large, multiple rupture, there is no conclusive evidence for a slow rupture speed during such a process. One outstanding question is whether transformational faults can propagate far beyond the expected thickness of metastable olivine wedge.

INTRODUCTION

With the advent of plate tectonics, deep-focus seismicity (300–680 km) is considered to originate in the interior of cold, sinking oceanic lithosphere. It follows that the distribution of deep-focus seismicity provides a direct way of mapping out the geometry of subducted lithosphere in the upper mantle. Furthermore, fault plane solutions of deep earthquakes constrain the state of stress in subducted lithosphere as it interacts with the surrounding asthenosphere and mesosphere [e.g., *Isacks and Molnar*, 1971]. However, the physical mechanism for generating deep-focus earthquakes has been enigmatic. Unlike shallow earthquakes, confining pressure is so large at depths greater than approximately 100 km that the shear stress necessary to drive frictional sliding is expected to exceed the ductile strength of geologic materials [e.g., *Scholz*, 1990]. On the other hand, there is no clear evidence that the kinematics of earthquake faulting changes significantly with focal depth [e.g., *Frohlich*, 1989].

Recently, results from experimental rock mechanics suggested some specific models for the origin of deep-focus earthquakes. For instance, *Meade and Jeanloz* [1991] proposed that amorphization of hydrothermally altered oceanic crust is a viable mechanism for sudden release of strain at great depths within subducted oceanic lithosphere. Meanwhile, other researchers have suggested that transformational faulting induced by the presence of metastable olivine is the cause of deep-focus earthquakes [e.g., *Kirby*, 1987; *Green and Burnley*, 1989; *Kirby et al.*, 1991]. Metastable olivine is expected to be the dominant mineral only in a cold wedge in the interior of subducted lithosphere, at depths below approximately 350 km [e.g., *Sung and Burns*, 1976; *Goto et al.*, 1985]. Both models imply that deep-focus earthquakes originate within a restricted region of subducted lithosphere.

Given that very large deep-focus earthquakes are frequently observed [e.g., *Frohlich*, 1989], two issues arise. First, are ruptures during large deep-focus earthquakes confined within a specific region, as implied by these recent models of seismogenesis discussed above? Second, given a thin subducted lithosphere in general, how is the large source region associated with a large earthquake accommodated at depth? The latter is a geometric consideration, not necessarily tied to specific physical mechanisms for the generation of deep-focus earthquakes.

The extent of ruptures during large deep-focus earthquakes also has important implications on the fate of subducted lithosphere in the transition zone of the mantle. For instance, *Giardini and Woodhouse* [1986] suggested that since some of the deepest earthquakes along the Tonga subduction zone seem to involve faulting on sub-horizontal planes, horizontal shear flow exists in the transition zone. In other words, wholesale slab penetration into the lower mantle is unlikely. Such an argument, however, is valid only if the entire thickness of the slab is broken during large earthquakes [*Jackson and McKenzie*, 1988].

Starting in 1987, abundant broad-band seismic data has been accumulating. In particular, 1994 has been an extraordinary year during which three very large deep-focus earthquakes occurred on March 9 along the Tonga subduction zone (M_W=7.6), on June 9 beneath Bolivia (M_W=8.3), and on July 21 along the Japan subduction zone (M_W=7.3). The unprecedented high resolution of broad-band seismic data recorded during these earthquakes provides an opportunity to carry out detailed investigation on the kinematics of rupture during deep-focus earthquakes. The orientation, extent, and amount of faulting during deep-focus earthquakes constrain the geometry and volume of the source region which, in turn, are useful for testing models of seismogenesis and for understanding how subducted slab deforms near the base of the upper mantle.

Using broad-band body waves, we first summarize the kinematics of rupture during several large deep-focus earthquakes that occurred between 1987 and 1994, including new results from the 1994 event along the Japan subduction zone and a comparison of different source models for the Bolivian earthquake of 1994. We then discuss issues concerning the seismogenic volume at depth. The results suggest that during large deep-focus earthquakes, long, narrow ruptures, accompanied by uneven rupture speeds, facilitate the release of large seismic moments over narrow source regions.

BACKGROUND AND DATA ANALYSIS

Except for the obvious difference in focal depths, seismic waves radiated from deep-focus earthquakes resemble those from shallow events. Generally speaking, these waves can be adequately explained by slip across a fault plane, the double-couple source model [e.g., *Isacks and Molnar*, 1971]. For a simple point source, or a circular rupture, the far-field displacement field of *P* and *S* phases is a pulse of variable amplitude but constant width (Figure 1a). In this case, one cannot obtain independent estimates of fault radius and rupture speed. Systematic variations in amplitudes observed over a wide range of azimuths and epicentral distances (i.e., the radiation pattern), however, allow a precise estimate for the orientation of the rupture plane. The area under the pulse is proportional to the seismic moment which, in turn, equals the product of the amount of slip, the rupture area, and the shear modulus in the source region. The shape of the pulse, somewhat modified by anelastic attenuation of the Earth, largely reflects the time derivative of the history of displacement across the fault, known as the source-time function.

The rupture of some earthquakes can be approximated by a line source on a long, narrow fault, with a constant speed of rupture propagation along the length of the fault [*Haskell*, 1964]. The width of rupture must be less than its length but the precise width is unknown (Figure 1b). This rupture will show directivity akin to the Doppler effect. Along source-to-station azimuths close to the leading edge of the rupture, observed pulses have short durations but high amplitudes. Along azimuths near the trailing edge of the rupture, the opposite effect on waveforms is expected. Thus observations at several different azimuths and take-off angles allow estimates of both rupture velocity and fault length. (The take-off angle is the angle between a ray path leaving the source and the downward vertical.)

For a complex rupture, discrete bursts of seismic moment release can be approximated by a sequence of subevents distributed over space and time, with each subevent constrained to be a double-couple point source. In this case, body-wave signals are a sequence of pulses

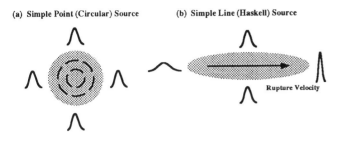

Fig. 1. Schematic map showing the effect of rupture propagation on pulse-width and absolute amplitude of body waves in the far-field. Direct arrivals of body-wave pulses at four distinct azimuths, each 90° apart, are shown for each case. For simplicity, variations in amplitudes due to the radiation pattern of a double-couple source have been removed. Shaded regions show areas of fault slip; dashed circular curves show successive rupture fronts; and the arrow shows the direction of rupture propagation for a line source. (a) Simple point source (circular rupture). The pulses have identical amplitude and frequency content at all azimuths. (b) Simple line source (Haskell model). The Doppler effect, affecting both amplitude and frequency content, is most evident at azimuths near the leading and trailing edges of the rupture.

whose relative timing varies with azimuth and take-off angle. Such an effect has also been described as "directivity" of the source by some researchers. If for some of the individual subevents, the Doppler effect is discernible in observed waveforms, one can model the source as a mixed sequence of point and line sources. Either way, the overall extent of the source region can be estimated.

We used an inversion technique that simultaneously models P and SH wave trains by minimizing the differences between observed and synthetic seismograms in a least-squares sense. For each subevent, the unknown source parameters determined in the inversion are the focal mechanism, the depth, the scalar seismic moment, the duration and the shape of the far-field source-time function, as well as the spatial and temporal separation among subevents. The source-time function is composed of a sequence of triangular elements whose amplitudes are determined by the inversion. For each subevent, the source is allowed to be a line source, if data can resolve the Doppler effect. In such cases, the reported source-time function duration is the pulse-width measured normal to the fault. Details of the inversion procedure and criteria for estimating uncertainties are discussed by *Nábelek* [1984] and *Glennon and Chen* [1993, 1995].

We collected broad-band seismograms from the Incorporated Research Institutions for Seismology (IRIS) and from the GEOSCOPE and the Pre-POSEIDON projects. Technical aspects of data selection and processing are essentially identical to the procedures described by *Glennon and Chen* [1993, 1995] and *Chen* [1995].

MULTIPLE RUPTURES

Because the azimuthal coverage of seismic stations is excellent for the northwestern Pacific, both the configuration of the Wadati-Benioff zone and source kinematics of large earthquakes are well constrained in this region. In this case, new results presented below for the Japan slab and the work of *Glennon and Chen* [1995] represent a comprehensive study of large earthquakes deeper than 400 km since 1987 when abundant broad-band seismic data became available.

It so happens that observations at close-in distances above the source regions are available for all the three largest deep-focus earthquakes in 1994. Thus we shall concentrate our discussion on these most recent deep-focus earthquakes for which good constraints are available. Among these large earthquakes in 1994, the event along the Tonga subduction zone seems to have the least amount of directivity [*Antolik et al.*, 1994; *McGuire et al.*, 1994] and will receive only cursory discussion here.

En Echelon Ruptures: The Event of July 21, 1994 Along the Japan Wadati-Benioff Zone

The Japan Wadati-Benioff zone is unique in that a nearly continuous belt of seismicity, reaching a depth of almost 600 km, dips at a shallow angle of only approximately 30° (Figure 2). Based on detailed analysis of focal depths and source kinematics of large to moderate-sized earthquakes that occurred in this region for the past 30 years, the Wadati-Benioff zone appears simple, with no evidence for large-scale deviation from a planar configuration (Figure 2b).

The large event of July 21, 1994 occurred at a depth of approximately 474 km. The azimuthal coverage of both P and SH waveforms is excellent. Observed waveforms have high signal-to-noise ratios and show a clear bimodal moment release (Figure 3).

Since ray paths for direct arrivals to teleseismic distances dive downward at steep angles when leaving the source, up-going phases such as pP and sP are required to resolve any vertical component of directivity. Using the entire body-wave train containing arrivals as late as sS, we inverted for focal depths by the relative timing between direct arrivals (P and SH) and phases reflected off the free surface (pP, sP, and sSH). Generally speaking, the time interval between the two prominent peaks of pP phase is larger than that of the direct P phase by approximately 1 s,

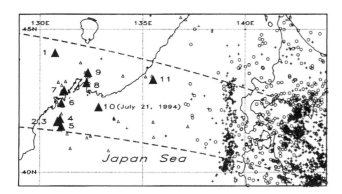
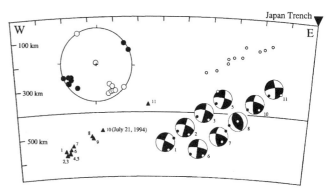

Fig. 2a. Map showing background seismicity along the Japan (Honshu) Wadati-Benioff zone. Events that occurred between 1964 and 1994 with $m_b \geq 5.0$ are plotted. Different symbols correspond to ranges in focal depths: crosses for shallow depths (0–70 km), circles for intermediate-depths (70–300 km), and triangles for deep-foci (>300 km). Large triangles are events whose source parameters have been investigated by the inversion of body-waveform data (Wu, Li-Ru, unpublished data, 1996). Large to moderate-sized events ($m_b \geq 5.5$) that occurred at depths greater than 100 km and fell within the dashed curves (approximately region N10 of *Zhou* [1990]) are included in Figure 2b. Sources of data are from the International Seismological Center (ISC, 1964–1987) and the Preliminary Determination of Epicenters (PDE, 1988–1994).

Fig. 2b. An east–west cross section (without vertical exaggeration) of the northern portion of the Japan subduction zone. Viewing from the south, large symbols show orientations of nodal planes, P (solid circle), and T axes (open circles) in equal-area projections of the back hemispheres of the focal spheres. The height of the triangles (showing hypocenters) is 10 km, roughly comparable to the uncertainties in our estimates of focal depths. Notice the consistent downdip orientation of all the P axes (see stereogram insert in upper-left corner).

indicating that the centroid of the second subevent is deeper than that of the first subevent. Modeling of the pP phase shows that the difference in depth is approximately 7±4 km (Figure 3a, Table 1).

The timing between the onset of the direct P phase and that of the second subevent also shows an azimuthal variation of up to 1 s, with stations to the east-southeast of the epicenter having the shortest time interval. This observation suggests that in addition to a deeper depth, the centroid of the second subevent is displaced to the east-southeast with respect to that of the first subevent (Figures 3b and 4).

Meanwhile, azimuthal variations in both pulse-width and absolute amplitude suggest a sub-horizontal Doppler effect in each subevent. The best case is observed for the large second subevent. Figure 3b compares the results of waveform inversion between our preferred solution (EER, Table 1) and a best-fitting solution with two point sources separated in space and time. The data constrain the azimuth of rupture propagation to be 180° +25°/-15°. Due to its smaller seismic moment, the northward Doppler effect of the first subevent can only be constrained to lie between azimuths of -30° and +60° and we have assumed that the two rupture azimuths are anti-parallel in our interpretation (Figure 4).

The strike of the steep dipping nodal plane lies on one extreme of the uncertainty for the direction of rupture during the second subevent. Unless the direction of rupture propagation is exactly parallel to the strike, given the near vertical dip of this plane, a sub-horizontally propagating rupture over a horizontal distance of ten's of kilometers must be accompanied by an even greater amount of change in focal depth. Consequently, we identify the shallow dipping nodal plane as the true fault plane. In Table 1, we also allowed a slight change in the orientation of fault plane solution between the two subevents. This change is introduced to explain opposite polarities between the two bursts of moment release observed at near nodal stations such as at KEV for P phases and at COL and COR for SH phases.

Thus the overall geometry of the rupture shows two *en echelon* regions of seismic moment release at slightly different depths (Figure 4). Another important aspect of this configuration is the uneven rupture speed. While the apparent rupture speed between subevents is only 1–2 km/s, the speed within each subevent is near 4 km/s, or approximately 75–80% of the shear wave speed in the source region. Assuming a radially propagating rupture front, the high rupture speed and the spatial distribution of two regions of large moment release reported by *Kuo et al.* [1995] in a recent abstract are similar to our results. The only difference is that *Kuo et al.* [1995] assumed that the rupture occurred over a single sub-horizontal plane, precluding any difference in focal depths between subevents.

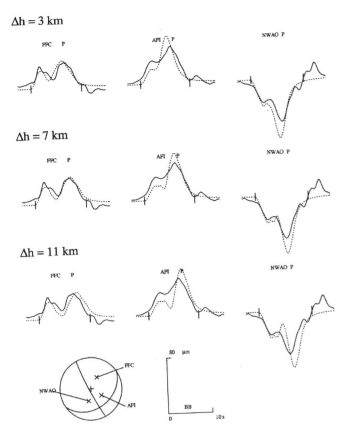
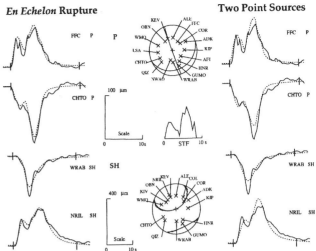

Fig. 3a. Comparison between observed (broad-band ground displacement, solid traces) and synthetic *pP* phases (dashed traces) at selected stations, showing the effect of differential focal depth (Δh) of subevent 2 relative to subevent 1. This comparison constrains Δh to be approximately 7±4 km. After normalization to an epicentral distance of 50°, both observed and synthetic seismograms are plotted with a common absolute amplitude scale indicated in the figure. Orientations of nodal surfaces (first subevent only) and locations of observations on the focal sphere for the direct *P* phases are shown in equal area projection of the lower hemisphere of the focal sphere.

The Great Bolivian Earthquake of June 9, 1994

This great earthquake beneath Bolivia is the largest known deep-focus earthquake [e.g., *Kirby et al.*, 1995a]. Azimuthal variations in the relative timing of distinct arrivals require separation in space and time among the centroids of subevents. Such an analysis indicated that major bursts of seismic moment release occurred to the east of the epicenter (subevent 1). Later subevents (subevents 3 and 4) are displaced to the northwest of the onset of major moment release (subevent 2) over a sub-horizontal plane (Figure 5d).

Fig. 3b. Comparison between observed (solid traces) and synthetic direct *P* and *SH* phases (dashed traces) generated by two different source models at selected stations. Layout is similar to that of (a) but the location of all observed *P* and *SH* phases on the focal sphere are plotted. Vertical bars on seismograms indicate time windows used in the inversion. The shape of the source-time function (STF) is also plotted.

While there is a general consensus on the spatial distribution of subevents (Figure 5), the apparent rupture speed and the exact configuration of faulting have been points of contention (see papers in a Special Issue of *Geophys. Res. Lett.*, August 15, 1995). Taking the ratio between the spatial separation and the time delay of later subevents with respect to the first subevent, several authors estimated a slow apparent rupture speeds of 1–2 km/s [*Kikuchi and Kanamori*, 1994; *Silver et al.*, 1995].

Kikuchi and Kanamori's work is the first published result on this event using waveform analysis (Figure 5e). However, they used no data in the southern hemisphere of the focal sphere. Subsequently, *Chen* [1995] included data from station PMSA due south of the epicenter and found a modified solution involving only distributed point sources (Figure 5d). *Chen* [1995, Figure 3] compared three different types of source models and found that a sequence of discrete subevents (Figure 5d), without additional Doppler effect within each subevent, cannot explain either the pulse-widths at PMSA or the absolute amplitude of high-frequency arrivals at SJG due north of the epicenter (Figure 6). In contrast, he reported that four subevents, each of the latter three as line sources in the azimuth of 0±30°N offers an improved explanation for the observed *P* and *SH* waveforms (Figures 5c and 6c). The rupture speeds of these line sources range from 3 to 3.5 km/s,

TABLE 1. Summary of Source Models[a]

Source Model[b]	Depth, km	Seismic Moment, 10^{18} N m	Strike, deg	Dip, deg	Rake, deg	Duration s	Time Delay s	Horiz. Dist.,[c] km	Azimuth[c] deg	Rupture Speed km/s	Rupture Azimuth deg	Variance of Misfit[d]
2PS	470[e]	35±2	155±7	85 (+5, -10)	55 (+10, -5)	3.5	-	-	-	-	-	1.00[f]
	477[e]	67±3	160 (+8, -4)	88 (+6, -8)	67 (+5, -10)	4.5	4.0±0.5	12±4	160±20	-	-	
EER	470[e]	35±2	155±7	85±10	55 (+10, -5)	4.2	-	-	-	4.0 (+1, -2.5)	0 (+60, -30)	0.93
	477[e]	67±3	160 (+8, -4)	88 (+6, -8)	67±10	4.5	4.0±0.5	6±4	120 (+35, -60)	4.0 (+1, -1.5)	180 (+25, -15)	

[a] For the event on July 21, 1994: Origin time, 1836:31.7 UT; epicenter at 42.34°N and 132.87°E; focal depth 471 km; m_b~6.5 (U.S.G.S. Preliminary Determination of Epicenters).
[b] 2PS: Two discrete point sources (circular ruptures) with spatial separation and time delay; EER: En echelon ruptures. Parameters for each subevent are given in successive rows.
[c] The relative location in space (given by horizontal distance, Horiz. Dist., and azimuth) of each subevent with respect to the epicenter.
[d] Normalized to the largest variance among different models listed.
[e] The second subevent is 7±4 km deeper than the first, whose absolute depth is uncertain by approximately ±5 km.
[f] 16% reduction when compared with the single point source (centroidal) solution.

approximately 60% of the shear wave speed in the source region. Moreover, in the most extreme case, the entire width of the rupture zone could be as narrow as 10–15 km over a length of approximately 70 km.

The discrepancy in estimated rupture speeds reported by different researchers mainly arises from the peculiar geometry of moment release during this event, and to a less extent, from differences in analysis and choice of data. It is important to bear in mind that by definition, as rupture propagates over a finite fault, fault slip must have occurred in the region already swept by the front of the rupture. The spatial distribution of slip, however, is often uneven during a large, complex earthquake (Figure 7).

For the Bolivian event, the first subevent released less than 5% of the total moment and its centroid essentially coincides with the initiation of rupture (Figure 5). Significant amount of slip did not develop until the rupture front reached close to regions marked as subevents 2 and 3 in Figure 7. If one assumes that the rupture propagated radially toward the northeastern quadrant (Figure 7a), the apparent rupture speed would be slow, close to 2 km/s or less [Antolik et al., 1995]. Antolik et al. reported that most of the seismic moment was released over an area of approximately 60x40 km², with its long-axis trending roughly north-south (Figure 5a).

Since the amount of slip between the initiation of rupture and subevents 2 and 3 (Figure 7), if any, is too small to be resolved by observation [Antolik et al., 1995], an alternative view is that subevent 1 is a precursory event whose rupture area did not overlap with those of later subevents (Figure 7b). For instance, such a configuration could arise if the later subevents were triggered by the passing wavefronts generated by the precursory event [Kirby et al., 1995b]. In this alternative view, the rupture during subevents 2 and 3 propagated northward at a high speed, close to 3.5 km/s [Chen, 1995]. This scenario was also investigated by Beck et al. [1995] who reported a bilateral rupture, trending along the azimuth of -10°N at a slow rupture speed of 1–2 km/s (Figure 5b).

To further investigate these issues, we compared observed waveforms with synthetic seismograms generated from these three models at four distinct azimuths (Figure 6). Antolik et al. [1995] assumed a rupture configuration depicted schematically in Figure 7a and used only the long-period (≥ 10 s) portion of the signals whose amplitudes have been normalized (Figure 6a). Even at these long-periods, misfit in timing or pulse-widths of up to 3 or 4 s seems apparent in this case. It is interesting to point out that Antolik et al. [1995] also tested the case in which signals from the small subevent 1 were excluded in their inversion. This procedure resulted in a rupture speed of greater than 2 km/s with no apparent upper bound.

In Figure 6b, we simulate the solution of Beck et al. [1995] who also excluded subevent 1 to isolate the effect of rupture propagation during the main phase of moment release. Their solution consisted of two later subevents, roughly corresponding to subevents 2 and 3 in Figure 7b, whose centroids are marked as E1a and E3 in Figure 5b,

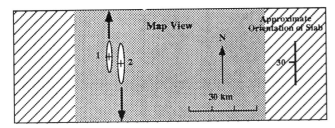

Japan Slab (July 21, 1994; 42.3°N, 132.9°E; 474 km; M_w=7.3)

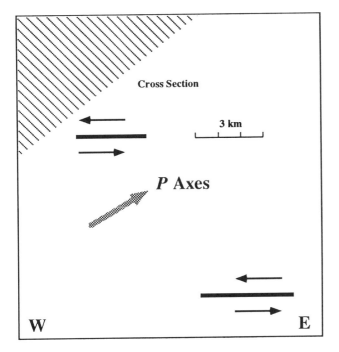

Fig. 4. Schematic diagrams showing the extent of *en echelon* ruptures during the July 21, 1994 earthquake. In a map view of a horizontal plane at the source depth, *en echelon* regions of fault slip are shown as ellipses whose precise width (east-west) is unknown. The centroid of the first subevent is placed at the epicenter because data cannot confidently resolve the separation between the nucleation point and the first centroid. Centroids of subevents are marked by crosses. Arrows show approximate direction of rupture propagation. The entire rupture region is placed within the metastable olivine wedge (shaded region) which is assumed to be approximately enclosed by the 700°C isotherms. Assuming a thickness of approximately 100 km prior to subduction, the hatched regions show portions of subducting lithosphere most of which has already undergone the olivine-spinel transition. In this case, the apparent width of subducted slab is approximately 200 km and the position of the metastable olivine wedge is based on the estimate of *Kirby et al.* [1991]. In cross section view, the geometry of the ruptures and the sense of slip (paired arrows) across the transformational faults (solid lines) are based on results shown in Table 1. The distance between the top of the metastable olivine wedge and the earthquake source is arbitrary.

respectively. Subevent 2 propagated toward azimuth 170°N while subevent 3 propagated toward -10°N, both at a constant speed of 1.5±0.5 km/s. The overall length of their rupture is over 120 km. Beck et al. used a tomographic inversion in which absolute amplitudes are not taken into account. Furthermore, they did not include important data from stations PMSA and SJG, due south and north of the epicenter, respectively.

When those data are included, their solution predicted too narrow a pulse at SJG, largely because the spatial separation between the two centroids is as large as 55 km (Figure 5b). Hence we modified their solution by setting the centroid of subevent No. 2 at position E1 in Figure 5b (where they deemed as the onset of subevent No. 2), reducing the total length of the rupture to be close to 70 km. Synthetic seismograms from this modified model are shown in Figure 6b. When compared with the model proposed by *Chen* [1995, model EER], the variance of misfit for the modified model of Beck et al. is higher by approximately 10%. Their model offered a good match for the first two large peaks at PMSA (Figure 6b). However, Chen's model provided a better match to the absolute amplitudes and pulse-widths of individual peaks in the later portion of the waveform at the same station. At stations to the north of the epicenter such as SJG and CCM, only Chen's solution explains the high frequency content of observed signals (Figure 6c).

It should be emphasized that the Doppler effect manifests itself as variations in absolute amplitudes and frequency content of waveforms over azimuths and take-off angles. If such variations are ignored, Doppler effects associated with rupture propagation can never be detected.

Due to complexity of the waveforms, it is difficult to ascertain whether there is a small difference (<5–10 km) in focal depths of subevents. Arrivals along ray paths that leave the source near the nodal surfaces indicate that there is a slight difference in fault plane solutions between subevents. Such differences, however, are too small to affect the waveforms at most other stations [e.g., *Goes and Ritsema*, 1995]. As such, even though the source seems to involve three sub-parallel, elongate regions of moment release (Figure 5c), and we suspect that the three *en echelon* ruptures may have occurred at different depths, we cannot prove this point.

A similar situation was reported by *Glennon and Chen* [1995] who systematically analyzed the source kinematics of many large deep-focus earthquakes that occurred along the northwestern Pacific between 1987 and 1992. The rupture during the largest event along the Izu-Bonin arc, on September 7, 1988 (M_w=6.7), also showed two sub-parallel regions of moment release with sub-horizontal propagation

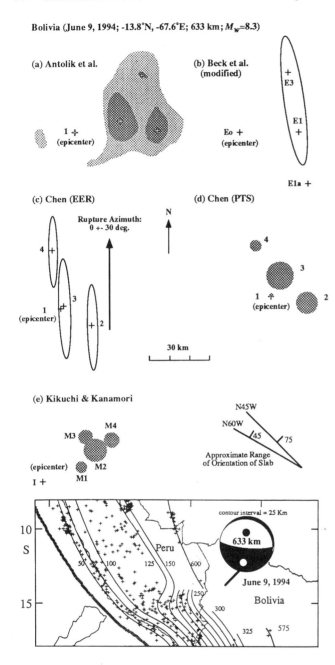

Fig. 5. A comparison of five selected models for the source configuration of the Bolivia event of 1994, shown in map view. The centroid of the first subevent is indistinguishable from the hypocenter. The approximate range in strike and dip of subducted lithosphere near the source is shown next to case (e). Insert in the bottom shows distribution of background seismicity [*Cahill and Isacks*, 1992]. Layout is similar to Figure 4. (a) Distribution of fault slip estimated by *Antolik et al.* [1995]. Regions with slip over 5 and 9 m are shown in light and dark shades, respectively. Crosses in the shaded regions show locations of maximum slip (subevents). Also see Figures 6a and 7a. (b) Distribution of subevents (E0, E1, E1a, and E3) and a bilateral rupture (ellipse) by *Beck et al.* [1995]. Also see Figure 6b. (c) Multiple line sources of *Chen* [model EER, 1995]. Ruptures are taken arbitrarily as sub-parallel to each other. Also see Figure 6c. (d) Distributed point sources of *Chen* [model PTS, 1995]. (e) Distributed point sources of *Kikuchi and Kanamori* [1994]. In the latter two cases, the relative radii of circles are proportional to seismic moments but the absolute scales are arbitrary.

of ruptures. Unfortunately, Glennon and Chen could not confidently resolve any difference between the depths of subevents in that case either.

Even in the interior of subducted lithosphere where space is limited, rupture associated with a large earthquake can be easily accommodated over a single long, narrow fault, oriented either sub-vertically or sub-horizontally along the strike of the slab. Instead, large amount of seismic moment was often released over more than one sub-parallel ruptures whose total length is less than that of a single, long rupture (cf. Figures 5b and 5c). The azimuths of rupture propagation, or equivalently those of the long-axes of the rupture areas, range from sub-parallel to the strike of the Japan slab to nearly normal to the strike of the Izu–Bonin slab. And in the former case, the ruptures have *en echelon* configuration (Figure 4).

IMPLICATIONS ON SEISMOGENESIS

Ideally, to test models of seismogenesis, both the configuration of earthquake sources and that of the subducted lithosphere must be known. Since hydrothermal alteration is expected to concentrate along the top of subducting lithosphere, the depth of hydrous alteration seems too small to explain the spacing between double seismic zones of up to 40 km for deep- and intermediate-focus earthquakes [*Kao and Chen*, 1994; *Wiens et al.*, 1993]. Thus amorphization of serpentine, as proposed by *Meade and Jeanloz* [1991], does not appear to be a suitable model for seismogenesis in many subduction zones.

In another recent model, earthquakes deeper than 350 km are associated with transformational faulting which, in turn, is caused by the transition of metastable α-olivine to β- or γ-spinel [e.g., *Kirby*, 1987; *Green and Burnley*, 1989; *Kirby et al.*, 1991]. Metastable olivine is expected to be the dominant mineral only in the cold interior of subducted lithosphere at depths below approximately 350 km [*Sung and Burns*, 1976; *Goto et al.*, 1985]. Thermal assimilation of subducted slab is likely to result in thinning of the metastable olivine wedge with increasing depth, from approximately 35 km thick at 500 km depth to only about 10 km thick near 600 km [e.g., *Kirby et al.*, 1991]. However, the exact thickness of the wedge is unknown.

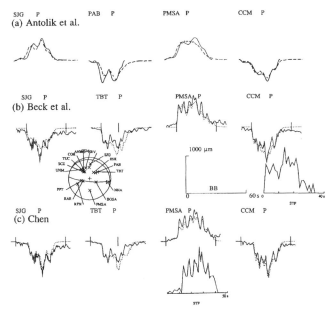

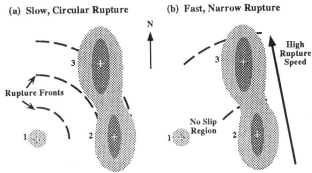

Fig. 6. A comparison of between observed (solid traces) and synthetic direct P phases (dashed traces) generated by three source models for the great Bolivian event of 1994. Layout is similar to that of Figure 3b. Up-going direct phases are plotted as open triangles at azimuths 180° from their true azimuths. (a) The result of *Antolik et al.* [1995], who assumed a rupture configuration depicted schematically in Figure 7a, was reproduced. Instead of station TBT not considered by them, result at nearby station PAB is shown. They low-pass filtered all signals with a cut-off frequency of 0.1 Hz and normalized all signals to have the same area under the synthetic pulses. Notice the misfit in timing or pulse-widths of up to 4 s. In cases (b) and (c), the Nyquist frequency is 2 Hz. In case (b), the onset of inversion excludes signal from the precursory subevent in order to simulate the solution of *Beck et al.* [1995] (Figure 5b). Case (c) is taken from *Chen* [1995]. See text for details.

Fig. 7. Schematic maps showing two different view points of modeling a complex earthquake source over a planar region. Regions of major moment release, or fault slip, are shaded. Crosses mark the centroids of such regions (subevents). (a) A single rupture is assumed to have initiated at location 1 and propagated at a uniform speed [*Antolik et al.*, 1995] (Figure 5a). (b) A small precursory event occurred at location 1, followed by the onset of major moment release at location 2 and propagating toward location 3. This is essentially the approach adopted by *Beck et al.* [1995] (Figure 5b), and *Chen* [1995] (Figure 5c).

Along the Japan slab, the rupture during the earthquake of July 21, 1994 is sub-parallel to the strike of the slab so the overall length of the rupture does not constrain the thickness of the seismogenic zone. However, the apparent *en echelon* geometry of the rupture can be compared with certain aspects of the anticrack model for the nucleation of transformational faults (Figure 4). In this model, lenticular micro-structures, filled with extremely fine-grained spinel, act as anticracks that promote the nucleation and coalescing of transformational faults [*Green and Burnley*, 1989]. Exactly how such microstructures develop rapidly into a major earthquake over an area of over 1,000 km² is a matter of debate [e.g., *Jiao and Wallace*, 1995, *Kirby et al.*, 1995b].

In laboratory specimens, the acute angle between spinel lenses (anticracks) and transformational faults (shear cracks) is close to 60° [*Green and Burnley*, 1990; *Burnley et al.*, 1991]. In other words, since anticracks preferentially develop over planes perpendicular to the axis of maximum compression [*Green and Burnley*, 1989], the acute angle between this axis and the transformational faults is close to 30°. This value is close to the angle between the extremely consistent orientation of the P axes for all events summarized in Figure 2b and the sub-horizontal ruptures depicted in Figure 4. In Figure 4, the exact width of each rupture plane is unknown. The width of each plane, however, does not affect the estimated angle between the P axes and any sub-horizontal plane. Notice that such a geometry between the P axes and rupture planes cannot be reconciled with the anticrack model, if sub-horizontal ruptures occurred in nearly vertical slabs.

The great Bolivia deep-focus earthquake of 1994 occurred in a region of low background seismicity and the configuration of the Wadati-Benioff zone is not particularly well-understood (Figure 5) [e.g., *Kirby et al.*, 1995a]. Based on sparse background seismicity and the distribution of aftershocks of the 1994 sequence [e.g., *Silver et al.*, 1995], the Wadati-Benioff zone appears to have a strike of approximately N60°W, dipping approximately 45° toward north-northeast. Interpretation of travel-time tomography, on the other hand, suggested a strike of approximately N45°W and a steep dip of approximately 75° toward northeast [*Engdahl et al.*, 1995].

Taking into account uncertainties in the orientation of the Wadati-Benioff zone and those in the source parameters, lower bounds on the thickness of the seismogenic zone in the slab range from 10 to 40 km, depending on the exact orientation of rupture propagation and the local configuration of the slab [*Chen*, 1995]. As such, there seems to be no firm conclusions as to whether the source region can be accommodated in a volume rich in metastable olivine or even within the subducted lithosphere (see Special Issue of *Geophys. Res. Lett.*) The issue is further complicated by the possibility that the slab may have thickened considerably near the base of the transition zone [e.g., *Stein*, 1995]. This controversy leads to two additional issues.

First, can large deep-focus earthquakes initiate from a thin wedge in the interior of the slab but propagate beyond the region in which metastable olivine is abundant? Observations from regions where the Wadati-Benioff zone is well defined by background seismicity indicate that ruptures during large deep-focus earthquakes can extend beyond the volume marked by background seismicity. For instance, both the distribution of aftershocks and the slipped region during the main shock seem to indicate that rupture during the deep earthquake sequence of 1994 in northern Tonga extended beyond the double seismic zone as defined by background seismicity [*Wiens et al.*, 1994; *McGuire et al.*, 1995].

Along the Izu–Bonin slab, *Glennon and Chen* [1995] found three events whose direction of rupture propagation is sub-parallel to the dip direction of a steep dipping slab. Consequently, the length of rupture constrained the minimum thickness of the seismogenic zone to be 27±12 km. This value is close to the predicted thickness of the wedge at a depth of 500 km, but is somewhat larger than the thickness of the Wadati-Benioff zone of 15–20 km [*Iidaka and Furukawa*, 1994; *Wiens et al.*, 1993]. One view is that large deep-focus earthquakes are triggered by micro-mechanisms (anticracks or otherwise) within the metastable olivine wedge. The rupture can then propagate beyond the wedge. This view is reminiscent of models for shallow earthquakes in the brittle regime: the rupture must nucleate in the stick-slip region but can propagate downward into the stable-sliding region.

Second, do large deep-focus events have unusually low rupture speeds? If so, does the low speed reflect processes that are responsible for extending the rupture beyond a small volume of nucleation? For instance, it has been proposed that a slow rupture speed of 1–2 km/s may indicate melting induced by shear heating [e.g., *Kikuchi and Kanamori*, 1994], a mechanism that could be responsible for extending an initially small rupture over a large area [e.g., *Jiao and Wallace*, 1995].

In the case of the Bolivian event, it appears that a high rupture speed of 3–3.5 km/s within subevents can explain observed variations in pulse-widths and frequency contents of waveforms better than models invoking low speeds (Figure 6). Because the no- or low-slip region between subevents 1 and 2 (Figure 7) emitted no observable signal [*Antolik et al.*, 1995], it is impossible to resolve the issue illustrated in Figure 7 for this event. On the other hand, using two different approaches of waveform inversion (Figure 3 and *Kuo et al.* [1995]), there is agreement on a high rupture speed of close to 4 km/s for the event in July of 1994 along the Japan slab. *Willemann and Frohlich* [1987] compiled the apparent rupture speeds of subevents relative to the onset of first arrival for approximately 30 deep earthquakes and found that only three events had unusually slow rupture speeds. Approximately ten events, on the other hand, have apparent rupture speeds exceeding half of the shear wave speeds in the source region. We conclude that there is little evidence for unusually slow rupture speeds during deep-focus earthquakes.

CONCLUSIONS

Based on the inversion of broad-band P and SH waveforms, we showed several cases in which the effect of directivity during large deep-focus earthquakes can be well explained by long, narrow regions of moment release over one or more sub-horizontal planes (Figures 4 and 5c). Such ruptures can accommodate a large amount of moment release over a restricted source region, a condition expected to exist in the interior of subducted lithosphere.

Apparent rupture speeds are often non-uniform during a single large event associated with multiple ruptures, varying from only 1–2 km/s between subevents to 3.5–4 km/s within each subevent. Contrary to the results of other researchers who did not allow for Doppler effect within subevents, there is no conclusive evidence for a particularly slow rupture speed during large deep-focus earthquakes. In the case of the July 21, 1994 earthquake in the Japan slab, two left-slip, *en echelon* ruptures, separated by approximately 7 km in depth, are likely to have occurred on two sub-horizontal planes. This configuration can be understood in the context of transformational faulting during deep-focus earthquakes.

In general terms, large fluctuations in rupture speeds and the likelihood that *en echelon* ruptures do not overlap can be construed as indications that each subevent involves the formation of a new fault, lending further credence to the transformational faulting model. At the same time, the possibility exists that the seismogenic zone during large deep-focus earthquakes extended beyond the expected metastable olivine wedge. Whether transformational faults, once nucleated in the metastable olivine wedge, can propagate a large distance beyond the wedge is unknown.

Acknowledgments. We benefited from discussions with M. Antolik, S. Beck, H. Green, H. Houston, S. Kirby, J. Nábelek, and D. Wiens. S. Kirby, S. Stein, and an anonymous reviewer provided helpful reviews. We also thank M. Antolik for a preprint. Seismograms are collected from the IRIS DMC and data centers of the GEOSCOPE and Pre-POSEIDON projects. This research was supported by NSF grant EAR93-16012.

REFERENCES

Antolik, M., D. Greger, and B. Romanowicz, Empirical Green's function source analysis of the large deep focus Fiji and Bolivian earthquake of 1994 (Abstract), *EOS. Trans. Am. Geophys. Union*, 75, 468, 1994.

Antolik, M., D. Greger, and B. Romanowicz, Finite fault source study of the great 1994 deep Bolivia earthquake, *Geophys. Res. Lett.*, in press, 1995.

Beck, S., P. Silver, T. Wallace, and D. James, Directivity analysis of the deep Bolivian earthquake of June 9, 1994, *Geophys. Res. Lett.*, 22, 2257–2260, 1995.

Burnley, P. C., H. W. Green, and D. J. Prior, Faulting associated with the olivine to spinel transformation in Mg_2GeO_4 and its implications for deep-focus earthquakes, *J. Geophys. Res.*, 96, 425–443, 1991.

Cahill, T., and B. L. Isacks, Seismicity and shape of the subducted Nazca plate, *J. Geophys. Res.*, 97, 17,503–17,529, 1992.

Chen, W.-P., *En echelon* ruptures during the great Bolivian earthquake of 1994, *Geophys. Res. Lett.*, 22, 2261–2264, 1995.

Engdahl, E. R., R. D. van der Hilst, and J. Berrocal, Imaging of subducted lithosphere beneath South America, *Geophys. Res. Lett.*, 22, 2317–2320, 1995.

Frohlich, C., The nature of deep-focus earthquakes, *Ann. Rev. Earth Plane. Sci.*, 17, 227–254, 1989.

Giardini, D., and J. H. Woodhouse, Horizontal Shear flow in the mantle beneath the Tonga arc, *Nature*, 319, 551–555, 1986.

Glennon M. A., and W.-P. Chen., Systematics of deep-focus earthquakes along the Kuril-Kamchatka Arc and their implications on mantle dynamics, *J. Geophys. Res.*, 98, 735–769, 1993.

Glennon, M. A., and W.-P. Chen, Ruptures of deep-focus earthquakes in the northwestern Pacific and their implications on seismogenesis, *Geophys. J. Int.*, 120, 706–720, 1995.

Goes, S., and J. Ritsema, A broadband *P* wave analysis of the large deep Fiji Island and Bolivia earthquake of 1994, *Geophys. Res. Lett.*, 22, 2249–2252, 1995.

Goto, K., H. Hamaguchi, and Z. Suzuki, Earthquake generating stresses in a descending slab, *Tectonophys.*, 112, 111–128, 1985.

Green, H. W., and P. C. Burnley, A new self-organizing mechanism for deep-focus earthquakes, *Nature*, 341, 733–737, 1989.

Green, H. W., and P. C. Burnley, The failure mechanism for deep-focus earthquakes, in *Deformation Mechanisms, Rheology and Tectonics*, edited by R. J. Knipe and E. H. Rutter, *Geol. Soc. Spec. Publ.*, London, 54, 133–141, 1990.

Haskell, N. A., Total energy and energy spectra density of elastic waves from propagating faults, *Bull. Seisml. Soc. Am.*, 54, 1811–1841, 1964.

Iidaka, T., and Y. Furukawa, Double Seismic Zone for Deep Earthquakes in the Izu–Bonin subduction zone, *Science*, 263, 1116–1118, 1994.

Isacks, B., and P. Molnar, Distribution of stresses in the descending lithosphere from a global survey of focal-mechanism solutions of mantle earthquakes, *Rev. Geophys.*, 9, 103–174, 1971.

Jackson, J., and D. McKenzie, The relationship between plate motions and seismic moment tensors, and the rates of active deformation in the Mediterranean and Middle East, *Geophys. J. R. Astron. Soc.*, 93, 45–73, 1988.

Jiao, W.-J., and T. C. Wallace, Slip-weakening induced by partial melting: I. A possible dynamic mechanism for deep-focus earthquakes (Abstract), *EOS Trans. Am. Geophys. Union*, 76, F365, 1995.

Kao, H., and W.-P. Chen, The double seismic zone in Kuril–Kamchatka: The tale of two overlapping single zones, *J. Geophys. Res.*, 99, 6913–6930, 1994.

Kikuchi, K., and H. Kanamori, The mechanism of the deep Bolivia earthquake of June 9, 1994, *Geophys. Res. Lett.*, 21, 2341–2344, 1994.

Kirby, S., Localized polymorphic phase transformations in high-pressure faults and applications to the physical mechanism of deep earthquakes, *J. Geophys. Res.*, 92, 13,789–13,800, 1987.

Kirby, S. H., W. B. Durham, and L. A. Stern, Mantle phase changes and deep-earthquake faulting in subducting lithosphere, *Science*, 252, 216–225, 1991.

Kirby, S. H., E. A. Okal, and E. R. Engdahl, The June 9, 1994 Bolivian deep earthquake: An exceptional event in an extraordinary subduction zone, *Geophys. Res. Lett.*, 22, 2233–2236, 1995a.

Kirby, S. H., E. A. Okal, and E. R. Engdahl, Ultra-large, very deep earthquakes: Dynamical triggering of transformational faults in isolated regions of grossly metastable peridotite? (Abstract), *EOS Trans. Am. Geophys. Union*, 76, F606–F607, 1995b.

Kuo, C., M. Antolik, D. Dreger, and B. Romanowicz, Finite fault models of recent large deep earthquakes (Abstract), *EOS Trans. Am. Geophys. Union*, 76, F367, 1995.

Meade, C., and R. Jeanloz, Deep-focus earthquakes and recycling of water into the earth's mantle, *Science*, 252, 68–72, 1991.

McGuire, J. J., D. A. Wiens, P. J. Shore, M. G. Bevis, K. Draunidalo, G. Prasad, and S. P. Helu, Rupture properties of the March 9th, 1994 deep Tonga earthquake and its aftershocks (Abstract), *EOS. Trans. Am. Geophys. Union*, 75, 466, 1994.

McGuire, J. J., D. A. Wiens, and P. J. Shore, Temperature controls on the initiation and propagation of deep earthquake rupture – Implications of the March 9, 1994 Tonga event (Abstract), *EOS. Trans. Am. Geophys. Union*, 76, F367, 1995.

Nábelek, J. L., Determination of earthquake source parameters from inversion of body waves, Ph.D. thesis, Mass. Inst. of Technol., Cambridge, 1984.

Scholz, C. H., *Mechanics of earthquakes and faulting*, 439 pp., Cambridge Univ. Press, Cambridge, 1990.

Silver, P. G., S. L. Beck, T. C. Wallace, C. Meade, S. C. Myers, D. E. James, and R. Kuehnel, Rupture characteristics of the deep Bolivian earthquake of 9 June 1994 and the mechanism of deep-focus earthquakes, *Science, 268,* 69–73, 1995.

Stein, S., Deep earthquakes: A fault too big?, *Science, 268,* 49–50, 1995.

Sung, C. M., and R. G. Burns, Kinetics of high-pressure phase transformation: Implications to the evolution of the olivine-spinel transition in the downgoing lithosphere and its consequences on the dynamics of the mantle, *Tectonophysics, 31,* 1–32, 1976.

Wiens, D. A., J. J. McGuire, and P. J. Shore, Evidence for transformational faulting from a deep double seismic zone in Tonga, *Nature, 364,* 790–793, 1993.

Wiens, D. A., J. J. McGuire, P. J. Shore, M. G. Bevis, K. Draunidalo, G. Prasad, and S. P. Helu, A deep earthquake aftershock sequence and implications for the rupture mechanism of deep earthquakes, *Nature, 372,* 540–543, 1994.

Willemann, R. J., and C. Frohlich, Spatial patterns of aftershocks of deep earthquakes, *J. Geophys. Res., 92,* 13,927–13,943, 1987.

Zhou, H.-W., Observations on earthquake stress axes and seismic morphology of deep slabs, *Geophys. J. Int., 103,* 377–401, 1990.

Wang-Ping Chen and Li-Ru Wu, 1301 W. Green St., 245 NHB, Urbana, IL, 61801; e-mail: w-chen@uiuc.edu

Mary Ann Glennon, 2628 York Court, Woodridge, IL 60517

Imaging Cold Rock at the Base of the Mantle: The Sometimes Fate of Slabs?

Michael E. Wysession

Department of Earth and Planetary Sciences, Washington University, St. Louis, Missouri

We present a review of the arguments in support of the intriguing idea that the eventual resting place of some subducted lithosphere is within the D" layer at the base of the mantle. The results of current models of seismic velocities in the lowermost mantle correlate well with the history of subduction since the break-up of Pangea. Significant support exists to suggest that while slabs do not always penetrate into the lower mantle, in some cases they do. If the lower mantle does not involve a significant increase in density relative to the upper mantle and transition zone, then the cold slab material will reach the base of the mantle, where it will eventually heat up and rise back to the upper mantle in the form of hot spot plumes. The large lateral variations seen in D" seismic velocities are consistent with this hypothetical model, as is the existence of the discontinuous seismic increase in velocity atop D", that could represent a combination of thermal, chemical, and/or phase boundaries.

1. INTRODUCTION

This paper serves as a postscript to the rest of the SUBCON volume, and is very different from the rest for several reasons. First, because it focuses on the base of the mantle as opposed to the top, the tools and techniques required to view the CMB through the nearly 3000 km of intervening rock are necessarily different. Second, while the subducted oceanic lithosphere is initially a very well-defined entity due to its thermal rigidity, any "slabs" at the base of the mantle would have lost their thermal rigidity, having reached temperatures on the order of 2500 K. Third, while we are quite certain that oceanic lithosphere is subducting at the sites of oceanic trenches, what happens next is not at all certain, and the details as well as the very premise of this paper remain speculative. We present one possible scenario for the fate of subducted lithosphere that cannot yet be presented with the same surety and authority as the scenario by which oceanic lithosphere is first brought into the mantle.

The SUBCON conference began with the creation and maturity of oceanic lithosphere, which greatly controls the processes occurring within subduction zones. Mass balance demands, however, a process that takes the material from the bottom of subduction zones and brings it back to mid-ocean ridges to form new lithosphere. This process still remains a topic of discussion and disagreement among geoscientists and is the focal point of the debate between whole-mantle and layered-mantle convection.

The evidence of seismology has been used to support both sides of the slab penetration argument, but the recent results of some regional tomographic studies may have superceded the rest. Putting geodynamic and mineralogical discussion aside, the recent seismic tomographic models of *van der Hilst et al.* [1991], *Fukao et al.* [1992], *Grand* [1994], *van der Hilst* [1995], and *Engdahl et al.* [1995] show in many cases a continuity of fast seismic velocities into the lower mantle that is hard to explain with other mechanisms such as thermal coupling. It is possible that future geologists will look back on these results as the final evidence to verify the occurrence of lower mantle slab penetration. We will assume for the rest of the discussion that some lithosphere can and does penetrate and become part of the lower mantle.

Once in the lower mantle and converted to perovskite, the thermal density anomaly of the slab should allow it to reach the base of the mantle, provided that the bulk-element chemical change across the 660-km discontinuity is small. The correlation between subducted lithosphere and lowermost mantle seismic heterogeneity [*Chase*, 1979] has been well documented at very low spherical harmonics [*Richards and Engebretson*, 1992; *Ricard et al.*, 1993]. Recent seismic studies, at both large and small wavelengths, continue to find fast seismic velocities at the base of the mantle in regions that correlate with paleotrenches. Correlation doesn't imply causality, but the coincidence is striking.

But there is a problem in ascribing the large seismic variations observed at the base of the mantle purely to thermal effects. Even very long wavelength seismic models like the recent degree-12 spherical harmonic model of *Liu et al.* [1994] are showing as much as 8% lateral variation in seismic shear wave velocity in D'', the lowermost mantle layer. This might require temperature variations well in excess of 1000°C, which is much more than the likely temperature difference between remnants of subducted lithosphere and the ambient lower mantle. One solution stems from two effects of post-slab rock on the unusual structure of D'' as a thermal and chemical boundary layer. It is probable that a significant thermal boundary layer (TBL) exists within D'', across which heat from the outer core conducts. The temperature difference across this TBL may well exceed 1000°C [*Boehler*, 1994], and when post-slab rock arrives at the core-mantle boundary (CMB) it would laterally displace the hotter rock. If the post-slab was only slightly colder than the lower mantle adiabat, this process could provide CMB lateral temperature variations of at least 1000°C. In addition, if any denser chemical phases accumulated in D'', such mantle dregs would also be displaced laterally and forced to aggregate in the regions of hotter D'' rock. The higher densities of such dregs, as would occur with iron enrichment, would likely cause slower seismic velocities and augment the effects of the thermal perturbations.

The final chapter of this scenario involves a reheating of the post-slab rock at the CMB until thermal instability occurs, with the return of this rock to the upper mantle as mantle plumes [*Duncan and Richards*, 1991; *Sleep*, 1992], and eventual reentrainment into mid-ocean ridge systems. There is much evidence, both theoretical and observational, to support all stages of this scenario, although the process as a whole cycle remains hypothetical. The main focus of this paper will involve a consideration of the observations of fast seismic velocities at the base of the mantle. Much attention has recently been paid to the structure of D'', which is a very dynamically active layer that sustains lateral variations at a level equal only to the Earth's crust and lithosphere (for discussions, see *Young and Lay* [1987], *Lay* [1989], *Jeanloz* [1990], *Lay et al.* [1990], *Bloxham and Jackson* [1991], *Jeanloz and Lay* [1993], *Wysession* [1995a,b], and *Loper and Lay* [1995]). It is a striking observation that regions of fast seismic velocities within D'' correspond well to those locations of the lower mantle that have underlain active subduction during the past 180 million years. We must be excused if we cannot resist the aesthetically attractive interpretation that the lithosphere and D'' are geodynamically linked by a whole-mantle convective cycle involving sinking slabs and rising plumes.

2. THE FATE OF SLABS

There are many good reviews of the differing arguments for and against slab penetration [*Silver et al.*, 1988; *Lay*, 1994; *Poirier*, 1991; *Davies and Richards*, 1992]. The premise of this paper, that slabs can reach the CMB, is based on the assumption that slabs do penetrate into the lower mantle, and therefore begs the question. There has been increasing evidence that the upper and lower mantles are truly different entities, with different mineralogical phases [*Bina*, 1991], densities [*Dziewonski and Anderson*, 1981], and viscosities [*Peltier and Jarvis*, 1982]. There is even some evidence for small amounts of chemical heterogeneity between the two [*Bina*, 1995]. All this suggests that the amount of communication between the upper and lower mantle cannot be above some maximal limit, and that the two operate largely independently, with the less-viscous upper mantle convecting at a much faster rate. The amount of this maximal limit is debated, however, as it still leaves room for some degree of slab penetration.

The lower mantle is perhaps two orders of magnitude more viscous than the upper mantle and transition zone [*Hager*, 1984; *Hager et al.*, 1985], and this combined with the endothermic phase change in the $(Mg,Fe)_2SiO_4$ system from γ-spinel to $(Mg,Fe)SiO_3$ perovskite and $(Mg,Fe)O$ magnesiowüstite represents a significant barrier to slab penetration. But this barrier is not insurmountable, and convection models incorporating the viscosity increase and mineralogical phase change can still demonstrate slab penetration [i.e., *Zhong and Gurnis*, 1994].

This is also suggested by recent high-resolution seismic tomographic studies for several of the world's major subduction zones. In *van der Hilst et al.* [1991] a study of several subduction zones in the Western Pacific found some cases where the fast seismic velocities associated with the subducting lithosphere took a sharp horizontal bend at the 660-km discontinuity and some continued straight down into the lower mantle. In the cases of the southern Kuril, Japan and Izu-Bonin subduction, where the slab dip was shallower, the slab seemed to be laying down horizontally at the bottom of the transition zone. In the cases of Kuril-

Kamchatka and Mariana subduction, however, where the angle of descent was steeper, the seismic velocities continued down into the lower mantle to at least 1000 km. Figure 1a shows the seismic velocities for the Mariana subduction.

It is possible that the occurrence of trench migration (roll-back) may also play a role. In *van der Hilst* [1995], where no trench rollback was occurring (Kermadec) there was no kink in the fast seismic velocities as they continued into the lower mantle. To the north, where active trench rollback is associated with the opening of the Lau Basin, the fast velocities extended horizontally before they also continued into the lower mantle, shown in Figure 1b. This behavior has also been observed in convection modeling [*Zhong and Gurnis*, 1995]. The observation that the Tonga slab temporarily lies horizontally within the transition zone before sinking into the lower mantle is also observed in the modeling of *Zhong and Gurnis* [1995]. This penetration could occur continually [*Jordan et al.*, 1993; *Bunge et al.*, 1995] or with periodic mantle overturning [*Honda et al.*, 1993; *Tackley et al.*, 1993; *Weinstein*, 1993].

Grand [1994] shows a tomographic model of the seismic velocities beneath North and Central America extending across the entire mantle. A sheet of fast seismic velocities, correlating with the subducted Farallon plate, is seen extending into the lower mantle all the way to the CMB. This and the Pacific studies mentioned above are carefully-done high-resolution studies with detailed efforts taken to demonstrate the resolvability of the data, and they present compelling evidence that slabs can penetrate into the lower mantle. Alternative interpretations of these seismic anomalies are not very convincing. The recurrence of the phenomenon eliminates a coincidence of isolated lower mantle downwelling beneath upper mantle subduction zones. Thermal coupling is also unlikely. Cold slabs resting at the bottom of the transition zone could cause a greater thermal gradient across a mid-mantle thermal boundary layer (if one existed), resulting in a more rapid conductive flux of heat out of the lower mantle. This, in turn, could drive a lower mantle downwelling that would resemble the continuation of the subducted slab. However, because the thermal conductivity of mantle silicates is so low a huge accumulation of slab must rest in one location for a long time for sufficient heat to conduct out of the lower mantle, and this extra piling is not seismically observed. This is especially true for Tonga subduction, which is very recent and has evolved rapidly over roughly the past 20 million years.

There is a possibility that subducted lithosphere might not make it all of the way to the CMB. If the lower mantle is chemically denser than the upper mantle, the slab could have enough thermal negative buoyancy to enter the lower mantle, but would stagnate there and eventually return to the

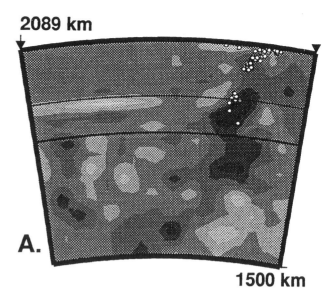

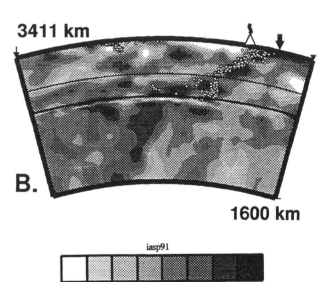

Fig. 1. (a) Seismic *P* velocities in a cross-section perpendicular to the Mariana subduction zone, from *van der Hilst et al.* [1991]. The image is obtained using a regional tomographic inversion of *P* and *pP* arrival times, and shows a continuity of fast seismic velocities across the 660-km discontinuity without a significant amount of piling-up. (b) Seismic *P* velocities in a cross-section perpendicular to the northern Tonga subduction zone, from *van der Hilst* [1995]. Here the regional tomographic image shows the Tonga slab being laid down horizontally along the 660-km discontinuity and then continuing down into the lower mantle.

upper mantle. While the lower mantle may be enriched in iron and silica relative to the upper mantle [*Bina and Silver*, 1990], recent work argues against this [*Bina*, 1995]. *Bina and Liu* [1995] show that a possible silica-enriched lower mantle would involve only a 1% chemical density contrast across the 660-km discontinuity, whereas a 5% density contrast is needed to maintain chemical stratification. Long-lived isolated chemical heterogeneities in the mantle can also occur without chemical stratification [*Gurnis and Davies*, 1986; *Metcalfe et al.*, 1995]. There are possibilities of chemical density contrasts deeper in the lower mantle, perhaps corresponding to seismic discontinuities observed at depths of 900, 1050, and 1200 km [*Wicks and Richards*, 1993]. However, regional tomographic studies like *Grand* [1994] and global studies like *Su et al.* [1994] show a continuity of fast seismic velocities extending continuously from the top of the lower mantle to the CMB.

3. PALEOSUBDUCTION

Assuming that subducted lithosphere actually makes it to the CMB, where would we expect to see it? This question was addressed by *Richards and Engebretson* [1992] and *Ricard et al.* [1993] using plate reconstructions over the past 180 million years to determine the locations of paleosubduction zones in the hot spot reference frame. They projected the accumulating subducting lithosphere down into the lower mantle from the locations of paleo-subduction. Because the rate of subduction during the mid-late Cretaceous and early Cenozoic was much higher than it is now, the lower mantle should show disproportionately more "slab" than if present day rates were projected into the past.

Such an accumulation of fossil slabs had previously been correlated with long wavelength geoid variations by studies like *Chase* [1979], and *Chase and Sprowl* [1983]. *Ricard et al.* [1993] showed that a strong correlation exists at spherical harmonic degrees 2 and 3 between the projected lithosphere graveyard and lower mantle seismic velocities, reproduced in Figure 2. The projected lower mantle slab distribution also correlated well with the global distribution of hot spots and explained 73% of the geoid variance. The suggestion of their study is that we would expect to find fast seismic velocities in regions of the lower mantle where lithosphere has been subducted over the past 200 Ma.

Many factors prevent us from expecting an exact correlation between paleo-subduction and lower mantle seismic velocities. Slabs can travel horizontally through the transition zone before penetrating into the lower mantle. Slabs have different ages when they subduct and different histories of reheating at the base of the mantle. There are large uncertainties in the reconstruction of very old plate motions. But the paleo-subduction studies give seismologists a sense of what they should be looking for: a broad ring of fast velocities along a great circle path that covers the poles and crosses the equator at about 100°E and 80°W, with slow velocities lying in between.

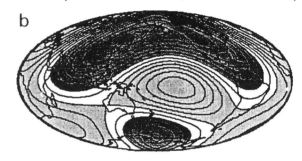

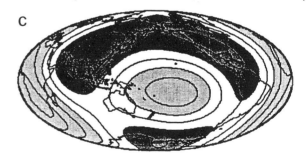

Fig. 2. A map showing the regional accumulation of subducted lithosphere over the past 180 Ma, assuming that all slabs have entered the lower mantle, taken from Figure 7 of *Ricard et al.* [1993]. The top two images show the predicted density variations at a depth of 2000 km inferred from past plate motions (through degrees 15 and 3), and the bottom image shows density variations inferred from the 3-D mantle seismic model SH425.2 of *Su and Dziewonski* [1991]. The contours are 2.0 kg/m^3 apart. The distribution of subducted lithosphere has predominantly occurred along a near-great circle path that extends over the poles and across the Americas and Eastern Asia. In the speculative scenario that this subducted lithosphere has entered the lower mantle and sunk to the core-mantle boundary, studies of the lowermost mantle should see a similar distribution of fast seismic velocities.

4. SEISMIC IMAGES OF D''

4.1. Tomographic Inversions

Seismic images of the lowermost mantle do indeed show very nearly what is predicted by lithospheric paleo-reconstruction assuming that slabs eventually reach the base of the mantle. This significant result is seen in maps of the CMB from both seismic tomographic models and regional teleseismic studies. In general, fast seismic velocities in D'' underlie the rim of the Pacific ocean and slow velocities underlie the central Pacific ocean and western African plate.

Seismic tomography is a powerful tool for mapping the internal structure of the Earth. Its strength comes from the incorporation of large amounts of different types of data, with a best-fit model obtained through linear inversion. A disadvantage is that the resolution and reliability are very variable across the model space, and there is a possibility of artifacts like smearing along the ray paths that result from the inversion. Unfortunately, the resolution also tends to be near its worst at the base of the mantle due to limited earthquake-station geometries [*Pulliam et al.*, 1993]. Good discussions of mantle seismic tomography can be found in *Romanowicz* [1991], *Dziewonski and Woodward* [1992], *Montagner* [1994], and *Ritzwoller and Lavely* [1995].

Recent models of the base of the mantle show great similarities, even when using very different data sets. Figure 3 shows a map (Figure 4h of *Ritzwoller and Lavely* [1995]) of the seismic velocity variations at a depth of 2800 km from four different studies: SH12/WM13 of *Su et al.* [1994], SH.10c.17 of *Masters et al.* [1992], L02.56 of *Dziewonski* [1984], and MDLSH of *Tanimoto* [1990]. Along with other current models, like that of *Inoue et al.* [1990], *Pulliam et al.* [1993] and *Liu et al.* [1994], these images differ slightly in the details, but the main features are similar to all: fast seismic velocities in D'' extending in a ring from the south to north poles, extending across the Americas, and eastern Asia. The regions in between, beneath west and southwest Africa and the Pacific Ocean, display slow velocities.

Previous studies like *Dziewonski et al.* [1993], *Forte et al.* [1994], and *Woodward et al.* [1994] have interpreted the seismic variations in terms of thermal variations, and therefore used them to compute buoyancy forces and convection patterns within the mantle. *Dziewonski et al.* [1993] labelled some of the large D'' scale features as convective "Grand Structures." These long-wavelength seismic structures consist of (1) the Pangea Trench, fast rock extending pole-to-pole beneath the Americas, (2) the Tethys Trough, fast rock extending from beneath India to southeast of Australia, (3) the China High, the seismically fast remnants of subduction beneath the eastern coast of Asia, (4) the North Pacific

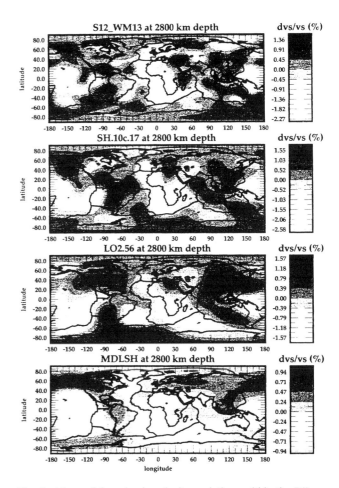

Fig. 3. Maps of the seismic velocity variations within the D'' region at a depth of 2800 km for four tomographic models, taken from *Ritzwoller and Lavely* [1995]. The studies are, from top to bottom: SH12/WM13 of *Su et al.* [1994], SH.10c.17 of *Masters et al.* [1992], L02.56 of *Dziewonski* [1984], and MDLSH of *Tanimoto* [1990]. All models are of *S*-velocity variations except for L02.56, which is for *P* velocities. The main features of all are similar, even though they were done at different times with different data, and correlate well with the distribution of subducted lithosphere shown in Figure 2.

High, fast rock beneath eastern Siberia, Alaska, and western Canada, (5) the Equatorial Pacific Plume Group, slow rock from the eastern Pacific to Indonesia correlating with the many Pacific hotspots (Easter, Galapagos, Hawaii, Marquesas, Pitcairn, Samoa, etc.), and (6) the African Plume, slow rock correlating with hot spots beneath the African plate and surrounding regions (Afar, Ascension, Azores, Canary, Cape Verde, Iceland, Reunion, St. Helena, Tristan, etc.).

These features are labelled on a new map (Figure 4) of D'' variations made using the differential travel times of *Pdiff* and *PKP-DF* phases. Described in *Wysession* [1996], this

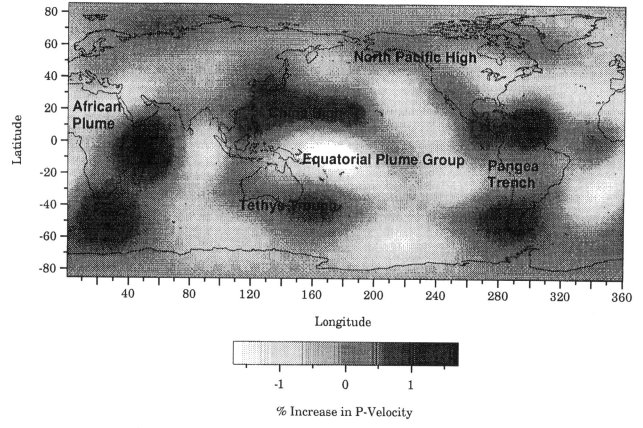

Fig. 4. A map of P velocities at the base of the mantle, taken from 545 PKP-Pdiff differential travel times. The paths of PKP and Pdiff are corrected for mantle path heterogeneity and inner core anisotropy, and the differential times determined through a waveform cross-correlation with reflectivity synthetic counterparts [Wysession, 1996]. The times are converted into velocity anomalies, distributed along the Pdiff 20 s Fresnel zones at the CMB, and then inverted for the seismic P-velocity variations across a grid of 660 spherical nodes. Labelled on the image are the Grand Structures identified by Dziewonski et al. [1993], which correlate well with surface hot spot regions and paleosubduction zones.

new technique isolates the Pdiff path around the core for earthquake-seismometer distances of 120°-165°. The differential times are converted into velocity anomalies and distributed along the 20 s CMB Fresnel zones of the Pdiff waves. These Fresnel zones, the regions that the waves sample coherently, are as large as 2000 × 5000 km, so only very long wavelength signal is retrievable. But the 120°-165° distance range provides coverage of very different parts of the CMB from other seismic studies, and since our velocities are determined by waves travelling horizontally through D″, they provide a complementary study to the tomographic models, which use phases that reflect off or refract across the CMB. The fast velocities in Figure 4 beneath the northwest Pacific correlate well with the distribution of subducted slabs shown in Figure 2a, and is a well-resolved region. This method also provides better coverage of the southern hemisphere than other seismic studies of D″.

4.2. Differential Travel Time Studies

This PKP-Pdiff study is one example of the technique of using differential travel times, which eliminates many uncertainties concerning the hypocenter and its rupture process, and combined with mantle path corrections, can largely isolate the seismic structure of D″ [Wysession et al., 1995b]. This method can incorporate seismic phases that diffract around the core (Sdiff, Pdiff), reflect off of the core (ScS, PcP, sScS), or refract through the core (PKP, SKS, SKKS), as shown in Figure 5. Most of these regional higher-resolution examinations of the lowermost mantle agree very well with the global tomographic images while discerning additional features, and also suggest a correlation with the history of lithospheric subduction.

In Wysession et al. [1992, 1993] and Valenzuela et al. [1993], extended profiles of core-diffracted Sdiff and Pdiff

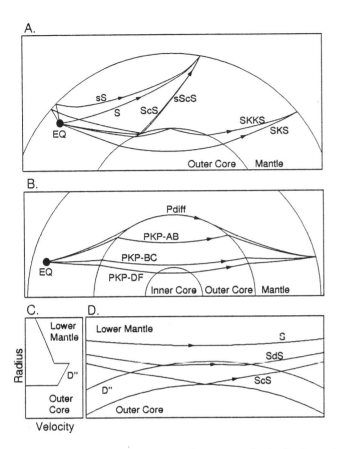

Fig. 5. Cartoons showing many of the ray-traced seismic phases used for examining the seismic velocity structure of the lowermost mantle. (a) Core-reflected phases *ScS* and *sScS* (and their differential counterparts *S* and *sS*), as well as core-refracted *SKS* and *SKKS*. (b) Paths of the three branches of core-refracted *PKP* waves, as well as core-diffracted *Pdiff*. The differential times of these phases were used to produce the map shown in Figure 4. (c) Lowermost mantle velocity structure with a discontinuous increase in velocity at the top of D″, and (d) the resulting triplicated seismic phase *SdS*.

phases isolated regional D″ variations. The fastest D″ *S* velocities were beneath northern North America, correlating with the North Pacific High and Pangea Trough, and beneath Southeastern Asia, corresponding to the Tethys Trough. The slowest D″ *S* velocities were beneath the Indonesian region, correlating with the western extension of the Equatorial Plume Group. The *P*-velocity variations were similar, but with some marked differences (discussed later).

Studies using core-reflected waves also give the expected distribution of fast and slow seismic velocities. The global study of *Woodward and Masters* [1991] found variations in *ScS-S* differential travel times of ±8 s that displayed a coherent regional pattern when plotted at the locations of the *ScS* CMB bounce points. This data provides provides an im-

portant part of the lowermost mantle signal of the tomographic shear wave models of *Su et al.* [1994] and *Liu et al.* [1994]. *Wysession et al.* [1994, 1995a] used *ScS-S* and *sScS-sS* relative arrival times (see Figure 5a) to map out features of the dramatic D″ anomalies beneath the western Pacific. The large number of earthquakes along the western Pacific makes such a map possible, shown in Figure 6. By distributing the inferred seismic velocity perturbations along the *ScS* and *sScS* paths through D″ we were able to resolve a region of the Equatorial Plume Group with velocities as much as 3% slower than PREM [*Dziewonski and Anderson*, 1981] centered beneath Micronesia, which was bordered to the south and west by regions of 3% fast rock. The fast region to the south is the Tethys Trough, and the fast region to the north is the China High.

Similar results are found for *P* velocities using *PcP-P* differential times [*Zhu and Wysession*, 1996]. Mapping 78,793 *PcP-P* differential times from the bulletin of the International Seismic Centre provides sufficient resolution of D″ *P*-velocities in a limited number of regions. These regions include D″ beneath (1) the Hawaiian Pacific (velocities 1.0-1.5% slower than the IASP91 model [*Kennett and Engdahl*, 1991], (2) the northern Pacific rim (velocities fast by 1.0-1.5%), (3) southeast Asia and the westernmost Pacific (with the familiar east-to-west transition from slow-to-fast velocities), and (4) Central America (slightly fast velocities).

Results are consistent for studies of core-refracted phases. The differential times of the *PKP-AB* and *PKP-DF* can map D″ anomalies because *PKP-DF* travels nearly vertically through D″ while *PKP-AB* spends much of its path within it (see Figure 5b). *PKP-AB* paths that sample D″ beneath the Pacific are best fit by a model with a 1.5% reduction in D″ *P* velocity [*Song and Helmberger*, 1993]. *McSweeney and Creager* [1994] used differential *PKP-DF* and *PKP-AB* phases to identify D″ lateral variations, and found fast velocities beneath South and Central America, as well as East Asia. *Garnero and Helmberger* [1993] found that the travel times of *S* and *SKS* were consistent with the tomographic models of D″ beneath the eastern Pacific of *Tanimoto* [1990] and *Su et al.* [1994], while identifying smaller wavelength variations. used The relative times of *SKS*, *SKKS*, and *SPdKS*, the latter diffracting along the CMB, have also revealed fast velocities beneath the Americas and a thin ultra-slow layer beneath the Pacific slow velocities regions [*Garnero et al.*, 1993b; *Garnero and Helmberger*, 1996; *Mori and Helmberger*, 1995].

4.3. Thermochemical Variations in D″

If the seismic variations in D″ are due to slab-induced thermal variations, then very large temperature differences

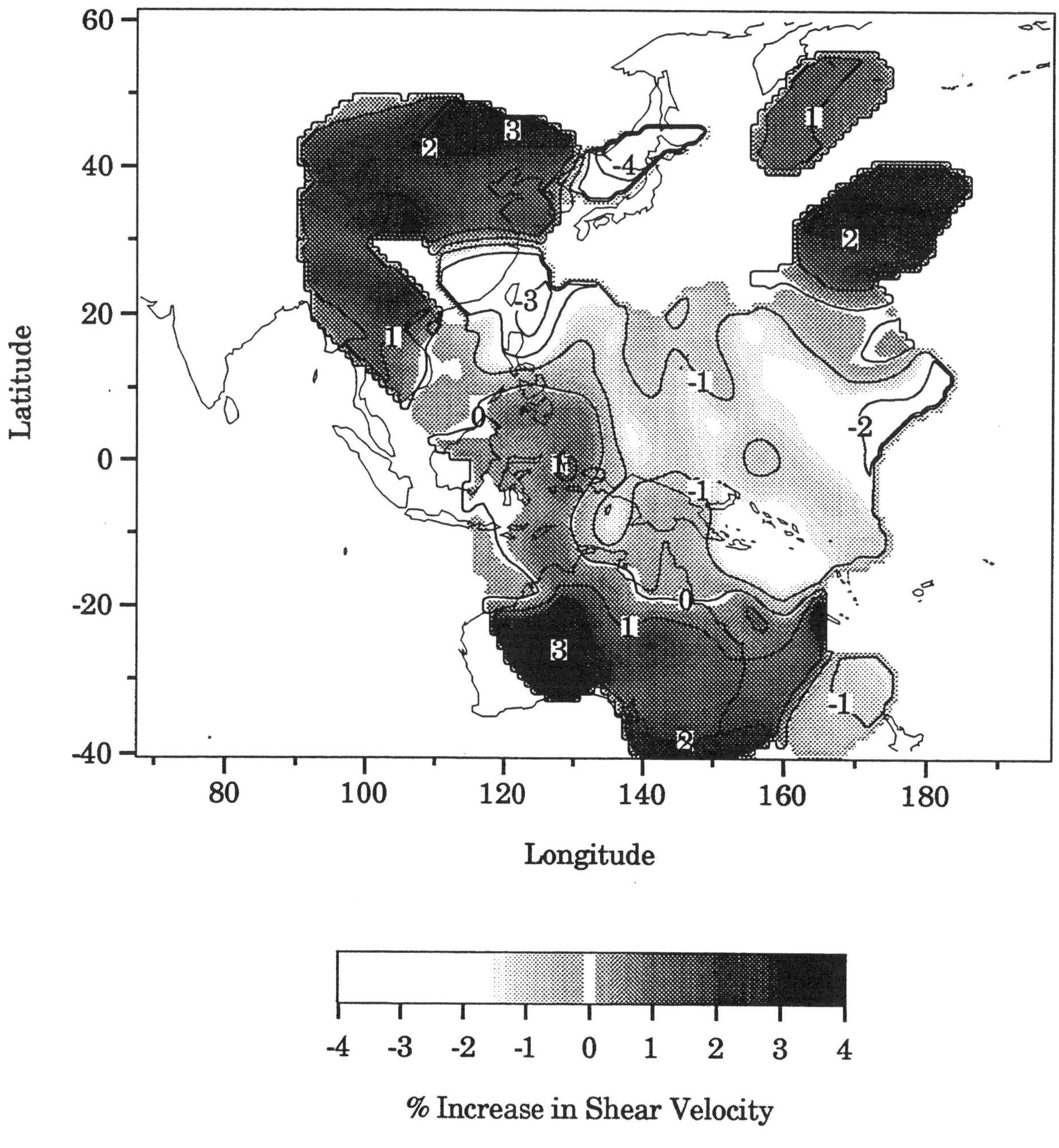

Fig. 6. A map of S-velocity variations in the lowermost mantle beneath the southwestern Pacific Ocean from 747 differential ScS-S and sScS-sS times, taken from *Wysession et al.* [1994, 1995a]. The central low-velocity region corresponds to the western end of the Equatorial Plume Group, the southern high-velocity region to the to the Tethys Trough, and the northwestern high-velocity region to the China High.

are required. While the amount of lateral heterogeneity in D″ varies with different studies, and in general S variations are greater than P variations by roughly a factor of 2.5 [*Robertson and Woodhouse*, 1996], they are on par with Earth's upper mantle. The model of *Liu et al.* [1994], which has very good resolution of D″ through the incorporation of large numbers of computed SKS-S and SKS-Sdiff differential travel times, shows a total range of D″ S-velocity variations of 8%. A computation using a third-order Birch-Murnaghan equation of state with currently available thermoelastic parameters for perovskite and magnesiowüstite showed that a 8% change in S velocity could require more than a 1000°C change in temperatures [*Wysession et al.*, 1992].

Such large temperature variations seem inconsistent with the expected thermal signature of post-slab material. The center of the subducting slab may be more than 1000°C colder than the surrounding mantle when the slab reaches the 660-km discontinuity [*Schubert et al.*, 1975], but this thermal difference will become much smaller by the time the slab crosses to the bottom of the mantle. Using the thermal model of *Stein and Stein* [1995], the core of the slab might be roughly 75% of ambient mantle temperatures, or 600-700 °C colder, when it reaches the base of the mantle. And this calculation does not take into consideration the greater viscosity of the lower mantle, which will increase the time required for slabs to reach the CMB.

Alternatively, there are chemical mechanisms for seismic velocity variations, such as changes in the iron/magnesium or silicate/oxide ratios. Increasing the amount of iron in a perovskite-magnesiowüstite assemblage will reduce seismic velocities, but it must triple before seismic observations are satisfied [*Wysession et al.*, 1992]. Explaining the seismic variations entirely in terms of large changes in the chemical components is actually very difficult to do without invoking enormous bulk chemical variations. No evidence for such bulk chemical heterogeneity is observed in hot spots basalts, though this could also result if hot spots do not originate at the CMB or if the chemical heterogeneities are too dense to be entrained in plumes [*Kellogg and King*, 1993].

One solution may be that either our equations-of-state or our estimates of the thermal derivatives of the incompressibility, rigidity and density are significantly wrong, meaning that large velocity changes are actually compatible with smaller temperature changes. Another solution may be found in the geodynamic models of mantle convection from several studies like *Tackley et al.* [1993, 1994] and *Davies* [1993]. If sinking rock reaches the CMB, it must laterally displace the hot rock that had previously been beneath it. This means that the limit on lateral thermal variations is the radial temperature difference that can be dynamically supported across the thermal boundary layer, and since this temperature difference could be as great or greater than 1000°C [*Boehler*, 1994], such large lateral temperature variations are possible. The sinking of cold plumes would serve to push together the rock that had been conductively heated by the core for long periods, and once consolidated, this hot rock would be the source for rising mantle plumes that would recirculate the rock back to the upper mantle. The lateral flow would also sweep any dense mantle dregs that might exist as part of a D'' chemical boundary layer to the base of the plume [*Davies and Gurnis*, 1986], further accentuating the slow velocities.

There is additional seismological evidence such as CMB topography and Poisson ratio variations that might help identify whether or not the seismic variations in D'' have a thermal or chemical origin. The CMB will be elevated if slow velocities are the result of anomalously hot rock but depressed in the case of anomalously iron-rich rock. Indications from models of CMB topography such as *Morelli and Dziewonski* [1987], *Doornbos and Hilton* [1989] are somewhat conflicting, but show a general correlation between slow seismic velocities and elevated topography, supporting a dominant thermal component. This correlation is not seen, however, in *Rogers and Wahr* [1993]. There is a clear need for more work in the area of CMB topography.

The Poisson ratio is not expected to vary greatly if the P and S velocity variations are the result of thermal anomalies, but this may be different for chemical variations [*Wysession et al.*, 1993]. There is evidence from core-diffracted waves [*Wysession et al.*, 1992], core-reflected waves [*Wysession et al.*, 1994, 1995b], and tomographic models that P and S velocity variations in D'' do not always vary in tandem. *Robertson and Woodhouse* [1996] find that the correlation between P and S velocity variations in the mantle from tomographic studies steadily decreases with depth to a value of 0.3 at the CMB. Geographical variations in the D''Poisson ratio computed from the models of *Su et al.* [1994] and *Pulliam et al.* [1993] range from 0.295 to 0.310, for a total variation of of 5%. The lowest value (0.295) corresponds to D'' beneath Alaska where *Wysession et al.* [1992] also found their lowest Poisson ratios. It is unclear whether these variations represent a real change in material properties or a difference in the wavelengths and therefore sampling resolution of the P and S waves used. Lateral variations in D'' Poisson ratios may eventually be used to identify and classify regional chemical and thermal characteristics, as they are for the crust [*Zandt and Ammon*, 1995].

5. A SEISMIC DISCONTINUITY AT THE TOP OF D'': POSSIBLE MECHANISMS

5.1. Seismic Observations

The most common feature of D'' in seismic studies is the discontinuous increase in velocity at the top of the layer. First detailed for S waves bottoming beneath Central America and Northern Asia [*Lay and Helmberger*, 1983], it was quantified through the identification of a triplicated phase, SdS, which refracts at the top of D'' and appears as an arrival between S and ScS (see Figures 5c and 5d). This phase was later identified for S waves in other parts of the Earth [*Young and Lay*, 1990; *Gaherty and Lay*, 1992], for P waves [*Wright et al.*, 1985; *Houard and Nataf*, 1992; *Vidale and Benz*, 1993; *Weber*, 1993], in long-period ScS reverberations [*Revenaugh and Jordan*, 1991], and in the decay of *Pdiff* amplitudes [*Valenzuela et al.*, 1994]. *Kendall and Shearer* [1994] looked at the appearance of *SdS* in global digital records and found that the feature appeared in all parts of D″ with adequate coverage, even in regions associated with slow seismic velocities, supporting earlier suggestions that the D″ discontinuity is a global and not regional feature [*Nataf and Houard*, 1993]. But this matter is far from being resolved [*Loper and Lay*, 1995]. There is also strong evidence that the depth of the discontinuity varies greatly, ranging from 130 to 450 km above the CMB [*Revenaugh and Jordan*, 1991; *Vidale and Benz*, 1993; *Weber*, 1993; *Kendall and Shearer*, 1994].

5.2. Thermal Mechanisms

While we would like to say that slabs are the simple cause of the D″ discontinuity, there are several possible mechanisms, shown in Figure 7. A simple box-model for convection (Figure 7*a*) is not sufficient, but thermal, chemical, and mineralogical phase variations can be invoked to satisfy the seismological observations. Slabs could be the culprit if cold rock ponding at the the base of the mantle could push hot less-viscous rock up and over it [*Stevenson*, 1993], resulting in abnormally hot (slow) rock overlying slightly cold (fast) rock (Figure 7*b*). While this could give the necessary discontinuous velocity increase, such a structure may not be sufficient to satisfy the required increase in velocity (relative to ambient mantle conditions) observed in some locations of D″. The thermal mechanism is also not sufficient if the D″ discontinuity is a global feature.

5.3. Chemical Mechanisms

It is possible that the D″ discontinuity could be the result of dense chemical heterogeneities that have settled to the bottom of the mantle. Variations in the thickness of this chemical boundary layer would be due to its being swept laterally by horizontal mantle flow [*Christensen*, 1984; *Davies and Gurnis*, 1986; *Hansen and Yuen*, 1989]. The existence of these chemical variations in the form of laminar sheets [*Kendall*, 1995] could even provide a mechanism for explaining observations of seismic anisotropy in D″ [*Lay and Young*, 1991; *Vinnik et al.*, 1995]. The difficulty, however, is to arrive at a dense chemical phase that would be both seismically fast and sufficiently abundant. Silicate phases which are denser because of an increase in iron are seismically slow, so core-mantle reaction products as suggested by *Knittle and Jeanloz* [1989] and *Boehler* [1994] could be swept laterally to form laminar aggregates [*Kellogg and King*, 1993], but their seismic velocities would not be faster than the ambient lower mantle.

A candidate for a seismically fast D″ chemical boundary is the eclogitic ocean crust from subducted slabs (Figure 7*c*), which could separate from the rest of the slab material either on route to or at the D″ thermal boundary layer [*Gurnis*, 1986; *Olson and Kincaid*, 1991; *Christensen and Hofmann*, 1994; *Weber*, 1994]. *Christensen and Hofmann* [1994] computed that the eclogite, assuming a range of mid-ocean ridge basalt compositions, would be 1.5 - 2.3% denser than a lower mantle assemblage based on garnet lherzolite, so the rock would preferentially settle at the base of the mantle. The post-eclogite rock could contain up to 25 wt% SiO_2 stishovite, very seismically fast, that could provide the seismic velocity increase at D″. We calculate the seismic velocities for the garnet lherzolite, MORB1 and MORB2 compositions of *Christensen and Hofmann* [1994], using a third-order Birch-Murnaghan equation of state in the manner of *Wysession et al.* [1992, 1993]. Using the garnet lherzolite as a model for average lower mantle composition, and the MORB1 and MORB2 compositions as examples of D″ mineral assemblages from the eclogite of the ocean crust, we find that the post-eclogite material provided a 2-3% velocity increase, sufficient to explain the seismic discontinuity. For shear waves, the MORB1 and MORB2 compositions were 1.8% and 1.6% faster than the garnet lherzolite at D″ conditions, but were 2.2% and 2.0% faster than a pyrolite composition. For *P* waves, the MORB1 and MORB2 compositions were 2.1% and 2.2% faster than the garnet lherzolite composition, and 2.6% and 2.7% faster than a pyrolite composition. The high-pressure phase of oceanic crust is a viable candidate for the discontinuous increase in velocities at the top of D″.

There are some problems with using post-eclogite rock to explain the D″ discontinuity. Because the *SdS* and *PdP* phases are refracted and not reflected off of the discontinuity, a significant amount of the post-eclogite rock is required, especially if the discontinuity is a global feature. Modeling by *Christensen and Hofmann* [1994], however, suggests that only about 6% of D″ should contain pools of eclogitic material, assuming the pools are 300 km thick. A solution to this discrepancy might exist if the eclogitic material was swept toward but not entrained in plumes that form at the CMB and return to the surface. This would be consistent with the suggestion of *Kendall and Shearer* [1994] that D″, as defined by the discontinuity, is thinner in regions of fast velocities and thicker in regions of slow velocities. *Revenaugh and Jordan* [1991] showed that the seismic D″ would be depressed beneath regions of cold downwelling as a chemical boundary layer was pushed aside, and the seismic D″ would thicken beneath upwelling as the chemical boundary layer was swept towards the plume. This is still under discussion, however, as *Garnero*

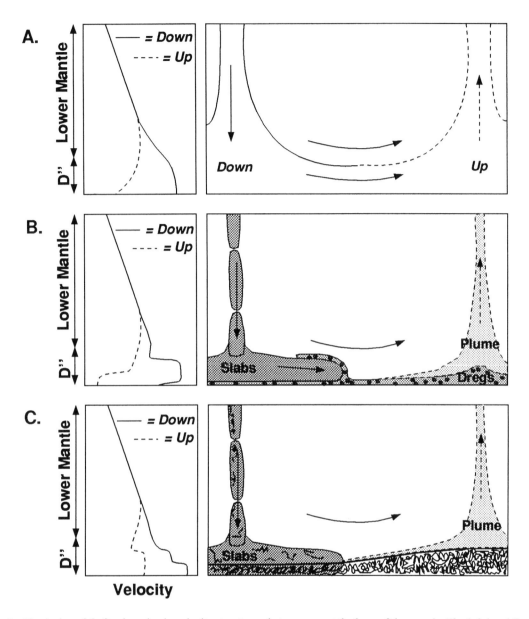

Fig. 7. Physical models for the seismic velocity structures that are seen at the base of the mantle. The left-hand figures are the seismic velocities (either P or S) that would result from the dynamic models shown in the right-hand figures. The dashed lines in the velocity plots show the expected velocity structures near the sites up upgoing plumes, and the solid lines show expected velocity structures near the sites of descending slabs. (a) A box-style convection pattern provides lateral variation in D″ seismic velocities, but not the seismic discontinuity seen atop D″. (b) In the case of a descending plume which ponds at the CMB, hot rock from the base of D″ may be displaced up and over the ponded plume, providing a discontinuous velocity increase. Laterally displaced CMB dregs would accumulate beneath regions of upwelling, explaining seismic observations of a narrow and very slow velocity layer just above the CMB in areas that already show slower than average D″ velocities. (c) A seismic discontinuity atop D″ could be the result of a bulk chemistry change, such as would result from the post-eclogite phase of rock that was once oceanic crust, if it delaminates from the rest of the slab and accumulates at the base of the mantle.

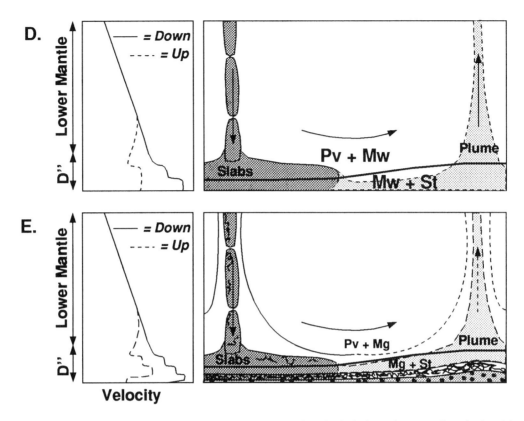

Fig. 7 (continued). (d) Even with no change in bulk chemistry, a mineralogical phase change such as the breakdown of perovskite into magnesiowüstite and stishovite could provide a discontinuous increase in seismic velocities. (e) A hypothetical combination of all of the above scenarios, resulting in several thermal and chemical D'''s.

et al. [1993a] found a D'' layer in a region beneath the central Pacific that displays some of the slowest D'' velocities.

5.4. Mineralogical Phase Transition

A third explanation for the D'' discontinuity (Figure 7d) could be a mineralogical phase change: the breakdown of $(Mg,Fe)SiO_3$ perovskite into $(Mg,Fe)O$ magnesiowüstite and SiO_2 stishovite [*Stixrude and Bukowinski*, 1990]. While the phase relations are still not well constrained, there is a possibility that the top of D'' may represent the bottom of the stability field for lower mantle perovskite. We use recent thermoelastic constants for perovskite and stishovite [*Hemley and Cohen*, 1992; *Sherman*, 1993] in addition to those used by *Wysession et al.* [1992, 993] to computed the increase in seismic velocities that would occur if a pyrolitic composition were to break down into oxides:

$$2 \cdot (Mg_{0.97}, Fe_{0.03})SiO_3 + (Mg_{0.76}O, Fe_{0.24}O) \Rightarrow$$
$$3 \cdot (Mg_{0.9}, Fe_{0.1})O + 2 \cdot SiO_2$$

The velocity of the magnesiowüstite and stishovite assemblage was 3.0% higher for S and 3.2% higher for P, relative to the mantle above it, so this phase transformation is a viable candidate for the D'' discontinuity. The phase transformation model, like the chemical model, could give rise to the seismically observed narrow discontinuity, but has the added advantage that it would be a global feature.

A phase transformation could also explain the variation in the height of the seismic discontinuity atop D'' [*Nataf and Houard*, 1993]. The breakdown of perovskite occurs at lower pressures for higher iron/magnesium ratios, so if the lowermost mantle was richer in iron than the downwelling rock from the upper mantle, the height of the discontinuity would lower, compatible with the results of *Kendall and Shearer* [1994]. Such a depression of the D''discontinuity would also occur if the reaction of perovskite to magnesiowüstite/stishovite were endothermic (like at the 660-km discontinuity). The colder temperatures of the downgoing slab would inhibit the transformation, pushing the discontinuity deeper. While it seems fortuitous to have a phase transition so near the CMB, the high thermal gradient in D" accentuates the likelihood of crossing such an instability.

All three processes are possibly occurring - thermal, chemical and phase - resulting in complex three-dimensional structure within D" [*Vidale and Benz*, 1993; *Weber*, 1993]. A hypothetical mixture is shown in Figure 7e. Discontinuities may not only have varying depths, but there may be more than one. There are many uncertainties that remain in the thermochemical interpretation of the seismic velocity structure of D", but in the current context the important observation is that there are several mechanisms by which the observed structure of D" is compatible with the process of bringing slabs from the upper mantle to the CMB.

6. SUMMARY

Seismology has provided a clear indication that there is a strong correlation between fast seismic velocities at the bottom of the mantle and the geographic distribution of subducted lithosphere. We have described some of the arguments that support the idea that D" may contain the remnants of subducted slabs which penetrate into the lower mantle and sink to the CMB. We stress that this geodynamic process is still highly speculative. The degree of mass transfer between the upper and lower mantles is highly debated, as is whether penetrating slabs can reach the CMB. Nonetheless, the correlation is compelling, and there are several lines of evidence in support.

Over the past 180 Ma, a significant amount of oceanic lithosphere has been subducted as Pangean has broken up and dispersed. *Ricard et al.* [1993] showed that the paleosubduction was concentrated in narrow regions that correlate well with long-wavelength seismic tomography. With the refinement of seismic tomographic models and advances in regional CMB seismic investigations this correlation has improved. We see a consistent pattern of a great-circle path of fast velocities from pole-to-pole, across the longitudes of the Americas and Eastern Asia. While regional tomography shows that some slabs do not penetrate into the lower mantle, it appears that some do, and in the case of subduction beneath the Americas, fast velocities can be followed from the upper mantle all the way to the CMB.

If there is not a larger-than-expected density increase atop or within the lower mantle relative to the upper mantle, the thermal signatures of slabs have the negative buoyancy to reach the base of the mantle. If this ponding rock laterally displaces hot rock of the D" thermal boundary layer, then lateral temperature variations of greater than 1000°C are theoretically possible. This thermal signature can then explain the very large seismic velocity variations of ± 4% seen in recent tomographic models like *Liu et al.* [1994]. A common feature of D" is a 2-3% increase in seismic velocity seen at a radius of about 200-300 km above the CMB. The mechanism for the generation of this seismic discontinuity is still not known, but a reasonable candidate is a combination of any of (1) the displacement of hot thermal boundary-layer rock up and over cold ponded rock from the upper mantle, (2) post-eclogitic rock of the oceanic crust that has delaminated from the rest of the slab rock and accumulated in D" due to its greater density, (3) a mineralogical phase change resulting from the breakdown of $(Mg,Fe)SiO_3$ perovskite to (Mg,Fe) magnesiowüstite and SiO_2 stishovite, and (4) something entirely different and not yet understood.

There is strong evidence to suggest that subducted lithosphere may provide the fast seismic velocities found in D" at the base of the mantle.

Acknowledgments. We thank Emile Okal, Craig Bina, and Tim Clarke for helpful discussions and assistance with seismic and thermodynamic codes. We thank Mark Richards, Michael Ritzwoller, and Rob van der Hilst for contributing figures. We thank the Cooperative Studies of the Earth's Deep Interior (CSEDI) for providing interdisciplinary forums for discussions about the structure and dynamics of the core-mantle boundary. We thank Ed Garnero, Steve Kirby, Mark Richards and an anonymous reviewer for very helpful comments and suggestions. Our work was supported by NSF-EAR9205368, NSF-EAR9417542, and The David and Lucile Packard Foundation.

REFERENCES

Bina, C. R., Mantle Discontinuities, *Rev. Geophys., U.S. National Rep. to International Union of Geodesy and Geophysics 1987-1990*, 783-793, 1991.

Bina, C. R., Confidence limits for silicate perovskite equations of state, *Physics and Chemistry of Minerals*, 22, 375-382, 1995.

Bina, C. R., and M. Liu, A note on the sensitivity of mantle convection models to composition-dependent phase relations, *Geophys. Res. Lett.*, 22, 2565-2568, 1995.

Bina, C. R., and P. G. Silver, Constraints on lower mantle composition and temperature from density and bulk sound velocity profiles, *Geophys. Res. Lett.*, 17, 1153-1156, 1990.

Bloxham, J., and A. Jackson, Fluid flow near the surface of Earth's outer core, *Rev. of Geophys.*, 29, 97-120, 1991.

Boehler, R., Experimental and thermodynamic constraints on ΔT across the CMB, *Eos Trans. AGU Suppl.*, 75, 654, 1994.

Bunge, H.-P., M. A. Richards, and J. R. Baumgartner, A sensitivity study of 3-D spherical mantle convection; Effects from depth-dependent viscosity, phase transitions and heating mode, *Eos Trans. AGU*, 76(46), Fall Meeting Suppl., 605, 1995.

Chase, C. G., Subduction, the geoid, and lower mantle convection, *Nature*, 282, 464-468, 1979.

Chase, C. G., and D. R. Sprowl, The modern geoid and ancient plate boundaries, *Earth Planet. Sci. Lett.*, 62, 314-320, 1983.

Christensen, U. R., Instability of a hot boundary layer and initiation of thermo-chemical plumes, *Ann. Geophys.*, 2, 311-320, 1984.

Christensen, U. R., and A. W. Hofmann, Segregation of subducted oceanic crust in the convecting mantle, *J. Geophys. Res.*, 99, 19,867-19,884, 1994.

Davies, G. F., Cooling the core and mantle by plume and plate flows, *Geophys. J. Int.*, 115, 132-146, 1993.

Davies, G. F., and M. Gurnis, Interaction of mantle dregs with convection: lateral heterogeneity at the core-mantle boundary, *Geophys. Res. Lett.*, 13, 1517-1520, 1986.

Davies, G. F., and M. A. Richards, Mantle convection, *J. Geology*, 100, 151-206, 1992.

Doornbos, D. J., and T. Hilton, Models of the core-mantle boundary and the travel times of internally reflected core phases, *J. Geophys. Res.*, 94, 15,741-15,751, 1989.

Duncan, R. A., and M. A. Richards, Hotspots, mantle plumes, flood basalts, and true polar wander, *Rev. Geophys.*, 29, 31-50, 1991.

Dziewonski, A. M., Mapping the lower mantle: determination of lateral heterogeneity in *P* velocity up to degree and order 6, *J. Geophys. Res.*, 89, 5929-5952, 1984.

Dziewonski, A. M., and D. L. Anderson, Preliminary reference earth model, *Phys. Earth Planet. Inter.*, 25, 297-356, 1981.

Dziewonski, A. M., and R. L. Woodward, Acoustical imaging at the planetary scale, *Acoust. Imaging*, 19, 785-797, 1992.

Dziewonski, A. M., A. M. Forte, W.-J. Su, and R. L. Woodward, Seismic tomography and geodynamics, in *Relating Geophysical Structures and Processes: The Jeffreys Volume*, edited by K. Aki and R. Dmowska, pp. 667-105, AGU, Washington, D. C., 1993.

Engdahl, E. R., R. D. van der Hilst, and J. Berrocal, Imaging of subducted lithosphere beneath South America, *Geophys. Res. Lett.*, 22, 2317-2320, 1995.

Forte, A. M., R. L. Woodward, and A. M. Dziewonski, Joint inversions of seismic and geodynamic data for models of three-dimensional mantle heterogeneity, *J. Geophys. Res.*, 99, 21,857-21,878, 1994.

Fukao, Y., M. Obayashi, H. Inoue, and M. Nenbai, Subducting slabs stagnant in the mantle transition zone, *J. Geophys. Res.*, 97, 4809-4822, 1992.

Gaherty, J. B., and T. Lay, Investigation of laterally heterogeneous shear velocity structure in D'' beneath Eurasia, *J. Geophys. Res.*, 97, 417-435, 1992.

Garnero, E. J., and D. V. Helmberger, Travel times of S and SKS: Implications for three-dimensional lower mantle structure beneath the central Pacific, *J. Geophys. Res.*, 98, 8225-8241, 1993.

Garnero, E. J., and D. V. Helmberger, Seismic detection of a thin laterally varying boundary layer at the base of the mantle beneath the central-Pacific, *Geophys. Res. Lett.*, in press, 1996.

Garnero, E. J., D. V. Helmberger, and S. Grand, Preliminary evidence for a lower mantle shear wave velocity discontinuity beneath the central Pacific, *Phys. Earth Planet. Int.*, 79, 335-347, 1993a.

Garnero, E. J., S. P. Grand, and D. V. Helmberger, Low P-wave velocity at the base of the mantle, *Geophys. Res. Lett.*, 20, 1843-1846, 1993b.

Grand, S. P., Mantle shear structure beneath the Americas and surrounding oceans, *J. Geophys. Res.*, 99, 11,591-11,622, 1994.

Gurnis, M., The effects of chemical density differences on convective mixing in the Earth's mantle, *J. Geophys. Res.*, 91, 11,407-11,419, 1986.

Gurnis, M., and G. F. Davies, The effect of depth-dependent viscosity on convective mixing in the mantle and the possible survival of primitive mantle, *Geophys. Res. Lett.*, 13, 541-544, 1986.

Hager, B. H., Subducted slabs and the geoid: Constraints on mantle rheology and flow, *J. Geophys. Res.*, 89, 6003-6016, 1984.

Hager, B. H., R. W. Clayton, M. A. Richards, R. P. Comer, and A. M. Dziewonski, Lower mantle heterogeneity, dynamic topography, and the geoid, *Nature*, 313, 541-545, 1985.

Hansen, U., and D. A. Yuen, Dynamical influences from thermal-chemical instabilities at the core-mantle boundary, *Geophys. Res. Lett.*, 16, 629-632, 1989.

Hemley, R. J., and R. E. Cohen, Silicate perovskite, *Annu. Rev. Earth Planet. Sci.*, 20, 553-600, 1992.

Honda, R., H. Mizutani, and T. Yamamoto, Numerical simulation of Earth's core formation, *J. Geophys. Res.*, 98, 2075-2089, 1993.

Houard, S., and H.-C. Nataf, Laterally varying reflector at the top of D'' beneath northern Siberia, *Geophys. J. Int.*, 115, 168-182, 1993.

Inoue, H., Y. Fukao, K. Tanabe, and Y. Ogata, Whole mantle P-wave travel time tomography, *Phys. Earth Planet. Int.*, 59, 294-328, 1990.

Jeanloz, R., The nature of the Earth's core, *Annu. Rev. Earth Planet. Sci.*, 18, 357-386, 1990.

Jeanloz, R., and T. Lay, The core-mantle boundary, *Scientific American*, 268, 48-55, 1993.

Jordan, T. H., P. Puster, G. A. Glatzmaier, P. J. Tackley, Comparisons between seismic Earth structures and mantle flow models based on radial correlation functions, *Science*, 261, 1427-1431, 1993.

Kellogg, L. H., and S. D. King, Effect of mantle plumes on the growth of D'' by reaction between the core and mantle, *Geophys. Res. Lett.*, 20, 379-382, 1993.

Kendall, J.-M., Seismic anisotropy in the lowermost mantle (abstract), *Eos Trans. AGU*, 76(46), Fall Meeting Suppl., 403, 1995.

Kendall, J. M., and P. M. Shearer, Lateral variations in D'' thickness from long-period shear-wave data, *J. Geophys. Res.*, 99, 11,575-11,590, 1994.

Kennett, B. L. N., and E. R. Engdahl, Traveltimes for global earthquake location and phase identification, *Geophys. J. Int.*, 105, 429-465, 1991.

Knittle, E., and R. Jeanloz, Simulating the core-mantle boundary: an experimental study of high-pressure reactions between silicates and liquid iron, *Geophys. Res. Lett.*, 16, 609-612, 1989.

Lay, T., Structure of the core-mantle transition zone: a chemical and thermal boundary layer, *Eos Trans. AGU*, 70, 49, 54-55, 58-59, 1989.

Lay, T., The fate of descending slabs, *Ann. Rev. earth Planet. Sci.*, 22, 1994.

Lay, T., and D. V. Helmberger, A lower mantle S-wave triplication and the velocity structure of D″, *Geophys. J. R. astron. Soc., 75*, 799-837, 1983.

Lay, T., and C. J. Young, Analysis of seismic *SV* waves in the core's penumbra, *Geophys. Res. Lett., 18*, 1373-1376, 1991.

Lay, T., T. J. Ahrens, P. Olson, J. Smyth, and D. Loper, Studies of the Earth's deep interior: Goals and trends, *Physics Today, 43*, 44-52, 1990.

Liu, X.-F., W.-J. Su, and A. M. Dziewonski, Improved resolution of the lowermost mantle shear wave velocity structure obtained using *SKS-S* data (abstract), *Eos Trans. AGU, 75*(16), Spring Meeting Suppl., 232, 1994.

Loper, D. E., and T. Lay, The core-mantle boundary region, *J. Geophys. Res., 100*, 6397-6420, 1995.

Masters, G., H. Bolton, and P. Shearer, Large-scale 3-dimensional structure of the mantle (abstract), *Eos Trans. AGU, 73*(43), Fall Meeting Suppl., 201, 1992.

McSweeney, T. J., and K. C. Creager, Global core-mantle boundary structure inferred from *PKP* differential travel times (abstract), *Eos Trans. AGU, 75*, 663, 1994.

Metcalfe, G., C. R. Bina, and J. M. Ottino, Kinematic considerations for mantle mixing, *Geophys. Res. Lett., 22*, 743-746, 1995.

Montagner, J.-P., Can seismology tell us anything about the convection in the mantle?, *Rev. Geophys., 32*, 115-138, 1994.

Morelli, A., and A. M. Dziewonski, Topography of the core-mantle boundary and lateral heterogeneity of the liquid core, *Nature, 325*, 678-683, 1987.

Mori, J., and D. V. Helmberger, Localized boundary layer below the Mid-Pacific velocity anomaly identified from a *PcP* precursor, *J. Geophys. Res., 100*, 20,359-20,366, 1995.

Nataf, H.-C., and S. Houard, Seismic discontinuity at the top of D″: A world-wide feature?, *Geophys. Res. Lett., 20*, 2371-2374, 1993.

Olson, P. L., and C. Kincaid, Experiments on the interaction of thermal convection and compositional layering at the base of the mantle, *J. Geophys. Res., 96*, 4347-4354, 1991.

Peltier, W. R., and G. T. Jarvis, Whole mantle convection and the thermal evolution of the Earth, *Phys. Earth Planet. Int., 29*, 281-304, 1982.

Poirier, J.-P., *Introduction to the Physics of the Earth's Interior*, 264 pp., Cambridge University Press, Cambridge, England, 1991.

Pulliam, R. J., D. W. Vasco, and L. R. Johnson, Tomographic inversions for mantle *P* wave velocity structure based on the minimization of l^2 and l^1 norms of International Seismological Centre travel time residuals, *J. Geophys. Res., 98*, 699-734, 1993.

Revenaugh, J., and T. H. Jordan, Mantle layering from *ScS* Reverberations, 4, The lower mantle and core-mantle boundary, *J. Geophys. Res., 96*, 19811-19824, 1991.

Ricard, Y., M. Richards, C. Lithgow-Bertelloni, and Y. Le Stunff, A geodynamic model of mantle density heterogeneity, *J. Geophys. Res., 98*, 21,895-21,909, 1993.

Richards, M. A., and D. C. Engebretson, The history of subduction, and large-scale mantle convection, *Nature, 355*, 437-440, 1992.

Ritzwoller, M. H., and E. M. Lavely, Three-dimensional seismic models of the Earth's mantle, *Rev. Geophys., 33*, 1-66, 1995.

Robertson, G. S., and J. H. Woodhouse, Ratio of relative *S* to *P* velocity heterogeneity in the lower mantle, *J. Geophys. Res.*, in press, 1996.

Rogers, A., and J. Wahr, Inference of core-mantle boundary topography from ISC *PcP* and *PKP* traveltimes, *Geophys. J. Int., 115*, 991-1011, 1993.

Romanowicz, B. A., Seismic tomography of the Earth's mantle, *Annu. Rev. Earth Planet. Sci., 19*, 77-79, 1991.

Schubert, G., D. A. Yuen, and D. L. Turcotte, Role of phase transitions in a dynamic mantle, *Geophys. J. R. Astr. Soc., 42*, 705-735, 1975.

Sherman, D. M., Equation of state and high-pressure phase transitions of Stishovite (SiO_2): Ab initio (Periodic Hartree-Fock) results, *J. Geophys. Res, 98*, 11,865-11,873, 1993.

Silver, P. G., R. W. Carlson, and P. Olson, Deep slabs, geochemical heterogeneity, and the large-scale structure of mantle convection: investigation of and enduring paradox, *Ann. Rev. Earth Planet. Sci., 16*, 477-541, 1988.

Sleep, N. H., Hotspot volcanism and mantle plumes, *Ann. Rev. Earth Planet. Sci., 20*, 19-43, 1992.

Song, X., and D. V. Helmberger, Effect of velocity structure in D″ on *PKP* phases, *Geophys. Res. Lett., 20*, 285-288, 1993.

Stein, C., and S. Stein, Thermal state of oceanic lithosphere and implications for the subduction process, in *Dynamics of Subduction*, edited by G. E. Bebout, D. Scholl, S. Kirby, and J. P. Platt, this volume, AGU, Washington, D. C., 1996.

Stevenson, D. J., Why D″ is unlikely to be caused by core-mantle interactions (abstract), *Eos Trans. AGU, 74*(43), Fall Meeting Suppl., 51, 1993.

Stixrude, L., and M. S. T. Bukowinski, Fundamental thermodynamic relations and silicate melting with implications for the constitution of D″, *J. Geophys. Res., 95*, 19,311-19,325, 1990.

Su, W., and A. M. Dziewonski, Predominance of long-wavelength heterogeneity in the mantle, *Nature, 352*, 121-126, 1991.

Su, W., R. L. Woodward, and A. M. Dziewonski, Degree 12 model of shear velocity heterogeneity in the mantle, *J. Geophys. Res., 99*, 6945-6981, 1994.

Tackley, P. J., D. J. Stevenson, G. A. Glatzmaier, and G. Schubert, Effects of an endothermic phase transition at 670 km depth in a spherical model of convection in the Earth's mantle, *Nature, 361*, 699-704, 1993.

Tackley, P. J., D. J. Stevenson, G. A. Glatzmaier, and G. Schubert, Effects of multiple phase transitions in a three-dimensional spherical model of convection in Earth's mantle, *J. Geophys. Res., 99*, 15,877-15,901, 1994.

Tanimoto, T., Long-wavelength *S*-wave velocity structure throughout the mantle, *Geophys. J. Int., 100*, 327-336, 1990.

Valenzuela, R. W., J. L. Butler, and M. E. Wysession, Velocity structure at the core-mantle boundary from digital diffracted shear waves (abstract), *Eos Trans. AGU, 74*(43), Fall Meeting Suppl., 406, 1993.

Valenzuela, R. W., M. E. Wysession, and T. J. Owens, The velocity gradient at the base of the mantle from broadband diffracted shear waves (abstract), *Eos Trans. AGU, 75*(44), Fall Meeting Suppl., 664, 1994.

van der Hilst, R., R. Engdahl, W. Spakman, and G. Nolet, Tomographic imaging of subducted lithosphere below northwest Pacific island arcs, *Nature*, 353, 1991.

van der Hilst, R., Complex morphology of subducted lithosphere in the mantle beneath the Tonga trench, *Nature, 374*, 154-157, 1995.

Vidale, J. E., and H. M. Benz, Seismological mapping of fine structure near the base of the Earth's mantle, *Nature, 361*, 529-532, 1993.

Vinnik, L., B. Romanowicz, Y. Le Stunff, and L. Makeyeva, Seismic anisotropy in the D″ layer, *Geophys. Res. Lett., 22*, 1657-1660, 1995.

Weber, M., P and S wave reflections from anomalies in the lowermost mantle, *Geophys. J. Int., 115*, 183-210, 1993.

Weber, M., Lamellae in D″? An alternative model for lower mantle anomalies, *Geophys. Res. Lett., 21*, 2531-2534, 1994.

Weinstein, S. A., Catastrophic overturn of the Earth's mantle driven by multiple phase changes and internal heat generation, *Geophys. Res. Lett., 20*, 101-104, 1993.

Wicks, C. W., and M. A. Richards, Seismic evidence for the 1200 km discontinuity (abstract), *Eos Trans. AGU, 74(43)*, Spring Meeting Suppl., 550, 1993.

Woodward, R. L., and G. Masters, Lower-mantle structure from *ScS-S* differential travel times, *Nature, 352*, 231-233, 1991.

Woodward, R. L., A. M. Dziewonski, and W. R. Peltier, Comparisons of seismic heterogeneity models and convective flow calculations, *Geophys. Res. Lett., 21*, 325-328, 1994.

Wright, C., Muirhead, K. J., and A. E. Dixon, The P wave velocity structure near the base of the mantle, *J. Geophys. Res., 90*, 623-634, 1985.

Wysession, M. E., The inner workings of the Earth, *American Scientist, 83*, 134-147, 1995a.

Wysession, M. E., Seismic images of the core-mantle boundary, *GSA Today, 5*, 237, 239-240, 256-257, 1995b.

Wysession, M. E., Large-scale structure at the core-mantle boundary from diffracted waves, *Nature*, submitted, 1996.

Wysession, M. E., E. A. Okal and C. R. Bina, The structure of the core-mantle boundary from diffracted waves, *J. Geophys. Res., 97*, 8749-8764, 1992.

Wysession, M. E., C. R. Bina and E. A. Okal, Constraints on the temperature and composition of the base of the mantle, in *Dynamics of the Earth's Deep Interior and Earth Rotation, Geophys. Monogr. Ser.*, edited by J.-L. LeMoüel et al., pp. 181-190, AGU, Washington, D.C., 1993.

Wysession, M. E., L. Bartkó, J. Wilson, Mapping the lowermost mantle using core-reflected shear waves, *J. Geophys. Res., 99*, 13,667-13,684, 1994.

Wysession, M. E., L. Bartkó, and J. Wilson, Correction to "Mapping the lowermost mantle using core-reflected shear waves", *J. Geophys. Res., 100*, 8351, 1995a.

Wysession, M. E., R. W. Valenzuela, L. Bartkó, A.-N. Zhu, Investigating the base of the mantle using differential travel times, *Phys. Earth Planet. Int., 92*, 67-84, 1995b.

Young, C. J., and T. Lay, The core mantle boundary, *Ann. Rev. Earth Planet. Sci., 15*, 25-46, 1987.

Young, C. J., and T. Lay, Multiple phase analysis of the shear velocity structure in the D″ region beneath Alaska, *J. Geophys. Res., 95*, 17,385-17,402, 1990.

Zandt, G., and C. J. Ammon, Continental crust constrained by measurements of crustal Poisson's ratio, *Nature, 374*, 152-154, 1995.

Zhong, S., and M. Gurnis, Role of plates and temperature-dependent viscosity in phase change dynamics, *J. Geophys. Res., 99*, 15,903-15,917, 1994.

Zhong, S., and M. Gurnis, Mantle convection with plates and mobile, faulted plate margins, *Science, 267*, 838-843, 1995.

Zhu, A.-N., and M. E. Wysession, Mapping global D″ P-velocities from ISC *PcP-P* differential travel times, *Phys. Earth. Planet. Int.*, in press, 1996.

M. E. Wysession, Department of Earth and Planetary Sciences, Washington University, St. Louis, MO 63130